Biology for Engineers

Second Edition

Biology for Engineers

Second Edition

Arthur T. Johnson

CRC Press
Taylor & Francis Group
Boca Raton London New York

CRC Press is an imprint of the
Taylor & Francis Group, an **informa** business

CRC Press
Taylor & Francis Group
6000 Broken Sound Parkway NW, Suite 300
Boca Raton, FL 33487-2742

First issued in hardback 2019

ISBN-13: 978-1-1380-6789-9 (hbk)
ISBN-13: 978-1-3511-6564-8 (eBook)

Visit the Taylor & Francis Web site at
http://www.taylorandfrancis.com

and the CRC Press Web site at
http://www.crcpress.com

Piglet noticed that even though he had a Very Small Heart, it could hold a rather large amount of Gratitude.

A. A. Milne

Angels come in many forms, and I would not have been able to come this far without being the beneficiary of the goodness of many, many angels. They started with my parents, grandparents, and family, who began me on the path my life has taken. Then, there were my teachers and fellow students in grade school and high school who helped shape my life in innumerable ways. On to college, and I had a great education that lifted me from the impecunious existence that I had experienced until then (although short of material things, my upbringing was never devoid of hope and confidence). My professors and advisors were great. I was extremely fortunate to come back from Vietnam unscathed. My professional and personal life, as chaotic (in the mathematical sense) as it was, was filled with wonderful people who helped me as much or more than I helped them. There are so many good folks out there who have inspired and motivated me to be the person that I have become that I cannot even name them all. Thanks to everyone. If you read this book, I hope you can see a little of yourselves in it. It could not have been done without you.

Dedication to First Edition

This book is dedicated to a handful of cherished true friends:

Jim Caldwell

Bernie Mullin

Adel Shirmohammadi

And, especially, to

Cathy Johnson

Contents

Section IV Scaling Factors

Section V Utilizing Living Systems

List of Examples

List of Tables

List of Boxes

Chapter 8

Preface for First Edition

> Technology made large populations possible; large populations now make technology indispensable.
>
> —**Joseph Wood Krutch**

Each of my books has been different from any other book on its topic, but more so with *Biology for Engineers* than the others. There just was not a book out there that looked at the engineering of biology. I had been involved in the earliest stages of the biological engineering movement in the American Society of Engineering Education and the American Society of Agricultural Engineers, and had written extensively about the field in general and about its philosophical foundations. Others had as well. But filling in the details and specifying exactly what should constitute education in the field was another matter. That was coming extremely slowly, and it almost seemed as if the great minds involved in the establishment of biological engineering as a distinct discipline could not divorce themselves from their own engineering education paradigms and define what should be the educational requirements for a discipline of engineering broadly based upon the science of biology.

One thing I knew for sure: I had received a very broad-based agricultural engineering education at Cornell University, augmented by elective courses that supported my own interest in the emerging field of biomedical engineering. I wanted to be sure that the breadth of my education was transposed into educational breadth for budding biological engineering students. So, when I realized that no one that I knew was going to write a biology book in a way that emphasized the engineering view of science and how science is utilized in a creative way, I began to write what I thought should be included in such a text.

The way that I have found to write a text that I can be satisfied with is to teach a course on the topic. I start with a framework for material to be included in the course, and let the students assist me by letting me know what is good about the approach and, most importantly, what is missing. They are usually not shy about giving feedback.

So, this text was pieced together like ornaments on a Christmas tree, one piece at a time and placed strategically on the proper branches. When the framework was not inclusive enough to accommodate new facts, then the framework was expanded. This process continued for nine years.

During this time, I have learned a huge amount by reading a lot of books and articles. Every time I came across something that I thought should be in a book on *Biology for Engineers*, I wrote it down and placed it appropriately in the next draft. Whenever I read about a new advance in biology or engineering related to biology, I gave myself a test, and this test took the form of a question: Is this topic covered in the book? At least the fundamentals of the topic had to be introduced so that students, when they came upon more details in later readings, would have seen the basic terms and introductory concepts. I knew that I could stop writing further drafts when I could continually say to myself, after reading newer articles, that all the new topics were already included in the text. I have reached that point.

One real challenge to writing a text of this kind, in a science that is progressing as rapidly as is biology, is to keep it current. It seemed as if new biological information was being discovered faster than I could include it in my class. How could I possibly presume to write a text that wouldn't be out of date in one or two years?

I now feel more confident about that. Although there will continue to be new biological information generated, many of the fundamentals are known. The science of biology is entering a phase of technological adaptation or utilization, and many of the fundamental processes, the real surprises, have been discovered and are covered in this text. At least I hope they are. We'll see in a couple of years.

Preface for Second Edition

In an atmosphere of liberty, artists and patrons are free to think the unthinkable and create the audacious; they are free to make both horrendous mistakes and glorious celebrations.

—**Ronald Reagan**

A few years ago, I realized that if I ever had a chance to update this book that I would have to get going, because advances in biology and technology related to biology were happening so fast that I would never be able to catch up if I didn't start soon. So, I read, and as I read, I wrote. This edition is how those notes came together.

In particular are the many new forms of shorthand for various biotechnological terms. I have tried to include many of them in this book, because their definitions may be hard to come by once they become generally used.

I have also filled in some voids that I realized, belatedly, were present in the first edition. I believe that to be a really good engineer dealing with biological subjects one must know more about the subjects than just the topics in one's specialty. One might never know how other advances can help solve a bio-based engineering problem if one is unaware of progress made elsewhere. One problem with a book as general as this one is the fact that keeping track of all the progress being made in fields as broad as genetic editing all the way to environmentally green roofs is not an easy task, and there are, undoubtedly, important facts that have been missed herein. I also realize that there are not as many generalists in the bioengineering field as I would hope, so reading or teaching from this book may be difficult for some. That's OK; this book is meant to be an introduction or a reference: it has a little of a lot of things in it. But, further details can be found elsewhere. If this book has provided enough background so that the reader or the student knows enough about a topic to know where to look and what for, then it will have served its purpose.

In some of my essays printed in other publications, I have commented on STEM, short for Science, Technology, Engineering, and Mathematics. STEM is a popular term lately when considering studying for careers. However, as an engineer, I have a problem with the term STEM. I prefer STEAM, STEM with the addition of Arts. Engineering is part STM with the addition of subjectivity and creativity. The practice of engineering combines elements of STM with creative talent that can be fostered by studying the softer subjects. You will see some of that philosophy in the contents of this book. One cannot know everything, but having a good knowledge of where bits of information fit into the overall scheme of things is not a bad trait to have when making engineering judgments, or any judgments, for that matter.

I stand by my prediction that the basics of biology had largely been discovered by the time of the first edition, but that applications of those basics to products and processes would continue to grow. And grow they have, more and more, faster and faster, it seems. This is an exciting time to be involved with biological technology. The fun has just begun...

Acknowledgments for First Edition

I would like to thank Ms. Erika Lopresti and Mr. Pete Mazzocchi for contributing to the figures in this book, and the many students in my classes who offered constructive criticisms for its improvement. Most of all, I thank Cathy Johnson for spending untold hours typing and retyping various drafts of this book. It certainly wasn't easy.

Acknowledgments for Second Edition

> No one who achieves success does so without acknowledging the help of others. The wise and confident acknowledge this help with gratitude.
>
> —**Alfred North Whitehead**

Almost all of the figures in the second edition of this book were redrawn by the talented hand of Ms. Catherine Montgomery, and she did a great job, as you can see for yourself.

Author

Arthur T. Johnson attended Cornell University for his undergraduate and graduate degrees. His PhD was awarded in 1969, and he immediately began serving as an officer in the US Army, eventually serving in Vietnam at the rank of captain. He was awarded the Army Commendation Medal and Bronze Star Medal. He joined the faculty of the University of Maryland in 1975, and was professor from 1986 until 2009, when he became professor emeritus. He was cochairman of the committee to found the American Institute for Medical and Biological Engineering (AIMBE) from 1988 to 1992, and served as the executive director of AIMBE in 2004. He has been president of the Alliance for Engineering in Medicine and Biology (1984–1988), Institute for Biological Engineering (1998), and International Society for Respiratory Protection (2004–2006). He was the secretary of the Biomedical Engineering Society from 2004 to 2009. He has been on the board of directors of the American Society for Agricultural and Biological Engineers (1995–1997). He is a founding fellow of the American Institute for Medical and Biological Engineering (1992), life fellow of the American Society for Engineering Education (1996), life fellow of the American Society for Agricultural and Biological Engineers (2002), fellow of the American Industrial Hygiene Association (2005), fellow of the Biomedical Engineering Society (2005), fellow of the Institute for Biological Engineering (2009), and the life fellow of the Institute for Electrical and Electronics Engineers (2010). He is a member of Phi Kappa Phi, Sigma Xi, Tau Beta Pi, and Alpha Epsilon. He has written three books: *Biomechanics and Exercise Physiology: Quantitative Modeling; Biological Process Engineering: An Analogical Systems Approach to Fluid Flow; Heat Transfer, and Mass Transfer, and Biology for Engineers*, and coauthored a fourth, *Design of Biomedical Devices and Systems*, with Paul King and Richard Fries. He is the author or coauthor of more than 150 peer-reviewed publications. He holds four patents related to respiratory mechanics and measurement. His research interests are human performance wearing respiratory protective masks, respiratory mechanics and measurement, and transport processes. He has taught many course topics, including electronic design, transport process design, and engineering in biology. He is currently working to continue development of the Airflow Perturbation Device (APD) as a noninvasive measurement of respiratory resistance. He is also operating a 49-acre farm producing organically raised fruit for sale at the local farmers' market.

Section I

Introduction

1

Introduction

1.1 Introduction

> I hear, I forget…I see, I remember…I do, I understand.
>
> **—Confucius**

Engineers and scientists can each study biology. Yet, the ultimate purpose for this study is different for the two groups. Understanding the characteristics and purposes of engineers and bioscientists can explain the approach taken toward the field of biology in this text. Thus, we begin by contrasting these two fields and distinguishing between them.

1.2 Science and Engineering

> Science cannot answer all questions.… It can, however, give some good indications, exclude certain hypotheses. Engaging in the pursuit of science may help us make fewer mistakes. It's a sort of gamble.
>
> **—Francois Jacob**

As the links between the fields of engineering and biology, biological engineers must appreciate the identities and personalities of both groups. Differences between science and engineering that should be appreciated by both sides fall into three different perspectives: (1) phylogeny, (2) motivation, and (3) methods (Johnson and Phillips, 1995).

1.2.1 Phylogeny

> Fate makes our relatives, choice makes our friends.
>
> **—Jacques Delille**

The evolution (phylogeny) of technology usually occurs within at least four distinct phases: (1) A random phase where events occur by chance and observation occurs haphazardly. The major outcome of this phase is to make the observers aware of the phenomenon being observed. (2) A descriptive phase where cause and effect relationships are established. The result of this phase is that the observed phenomenon no longer remains random, but can be expected whenever a series of foretold events happens. The phenomenon is still not

TABLE 1.2.1

The Four Phases of Technology

Phase	Description	Physical Example	Biological Example
Random	Phenomena are encountered haphazardly	Heavenly bodies are observed to move	Differences and similarities are noted in animals and plants
Descriptive	Cause-and-effect relationships are established	Apparent heavenly movement appears to be related to seasonal changes	Genetic material is discovered, and transgenic organisms are developed
Quantitative	Measurements are refined, and dependencies are given numerical values	Kepler's laws describe planetary motion	Optimal microbial growth environments are determined
Control	Modeling and predictive equations lead to knowledge of useful substance amounts, design of systems, and applications to achieve desired ends	Satellites orbit around the Earth, moon, and other planets	Transgenic microbial production of biochemicals becomes reality

Source: Johnson, A.T. and Davis, D.C., *Eng. Educ.*, 80, 15–18, 1990.

able to be brought about at will, but its appearance is at least expected. (3) A quantitative phase wherein measurements are refined, and dependencies are given numerical values. These values may be deterministic or probabilistic, but, during this phase, there is a growing knowledge about the intensity of the phenomenon as related to the strength of the precursor variables. (4) A control phase where modeling and predictive equations lead to knowledge of useful substance amounts, design of systems, and applications to achieve desired ends. The results of this stage are products and processes using the phenomenon. Examples are given in Table 1.2.1.

For some sciences, the early phases began long ago. The science of mechanics, for example, entered its descriptive phase before the time of Aristotle; however, the science of electricity was still partially random in the time of Benjamin Franklin and the science of genetics entered a long descriptive phase in the time of Gregor Mendel.

The first two of these four phases clearly belong to the field of science. Engineering contributes primarily in the control phase by using quantitative information to design useful products. The overlap between science and engineering generally occurs during the quantitative phase. Early attempts at quantification are largely made by scientists, but engineering researchers, usually motivated by the need for design information, can accelerate the quantitative process. Engineering is involved more with the latter stages of technology than with the earlier stages where science dominates.

1.2.2 Motivation

> Children are born engineers. Everything they see, they want to change.... Grown-up engineering, which is as old as civilization, maintains the youth, vigor, and imagination of a child.
>
> **—Henry Petroski**

Scientists and engineers can both be highly motivated, but the sources of work-related interests are often different for each group. Neglecting the recent trend toward entrepreneurship in both groups, the major source of motivation and satisfaction for engineers comes from the final products or processes as a result of their efforts. Engineering is largely creative, forming things that never were, and engineers, like artistic painters, become highly motivated by the tangible realization of their ideas and concepts. Additionally, if there are visible groups that can benefit from these realizations, a strong drive and sense of urgency can develop within the engineer.

Biologists, generally more removed from the ultimate applications of their work than are engineers, are often motivated by the subjects of their study. They may feel empathy toward these subjects, and study them because they are interested. This study, of course, leads to more interest, and a strong bond can develop between the observer and the observed. Biologists are thus motivated more by their involvement with their subjects, and engineers by their involvement with the things they produce.

Of course, the relatively recent trend toward studying cells and subcellular components has taken away some of the attachment from the biologist and the object of study. It is hard to feel close to a cell. The idea behind this kind of study may also be to launch a commercial success. Nonetheless, biologists do not usually approach their studies in the same way that engineers do. Biologists may take the results of their research and use them to produce products essentially unchanged from their natural state, while engineers often use the same results to produce products modified more or less from their original forms. Thus, one would expect a biologist who identifies a key enzyme to offer the enzyme for sale. The engineer might take the same enzyme and use it to produce fuel or food in a more efficient or effective way.

1.2.3 Methods

> If I have ever made any valuable discoveries, it has been owing more to patient attention than to any other talent.
>
> **—Isaac Newton**

There is a fundamental difference in methods used by scientists and by engineers. Biological scientists often perform experiments to ascertain new facts. Since many of their observations are related to phenomenon description, the pattern of scientific experimental episodes may be determined more by the observed phenomenon than by any regular scientific plan. Such is often the case while observing various life-forms in their natural habitat, for example, observations about eating only occur when the object of the attention decides to eat. Any attempt to tamper with the behavior of the being would result in criticisms of methods and observations, rendering them practically invalid.

Engineers rarely, if ever, become involved with their experimental objects at the descriptive phase and, hence, are often remote from these types of experiments. The impatience of most engineers would not allow them to observe phenomena without trying to tinker with the experiment to see what happens. Engineers are not educated to be distanced, impartial observers; they are educated to become involved, to attempt to predict or control an outcome, and to synthesize fragments that may not naturally fit together.

The advent of biotechnology and genetic manipulation (see Section 8.2.3, Biotechnology) has seen a fundamental change in the way scientists approach their objects of study.

Scientists working with *functional genomics* (linking functions to specific genes) are much more likely to tamper with their study subjects. They may attempt to turn off certain genes (see Section 5.3, Genetics), change genetic sequences (see Section 8.2.3, Biotechnology), or manipulate genetic material to observe the results of their trials. In this respect, scientific methods may be considered to be coming closer to engineering methods. The essential difference between the two; however, remains that scientists alter their subjects to discover new scientific knowledge, whereas engineers make their changes to enhance the performance of their products.

There is a difference between typical scientific literature and typical engineering literature. Scientific experiments beyond the completely descriptive phase are conducted for specific sets of conditions, with as many variables controlled as possible. To cover an entire scientific field with scientific observations requires a very large number of specific experiments, wherein control over the multitude of variables may be either tightened or relaxed, but many, if not most, combinations of the imposed conditions must be tested before a phenomenon is considered to be well understood.

There are very few, if any, surprises appearing in scientific papers of this sort, but these papers have scientific value by extending the realm of the known by additional increments. The differences between scientific papers, cited and uncited, related to a particular field are often few, and they all form a congealing mass that establishes scientific truth by the weight of consistency of experimental results.

Science, therefore, is inductive. Scientific facts accumulate until an overall unifying concept emerges as irrefutable. In science, the conceptual framework is induced from the many facts that precede it (Figure 1.2.1).

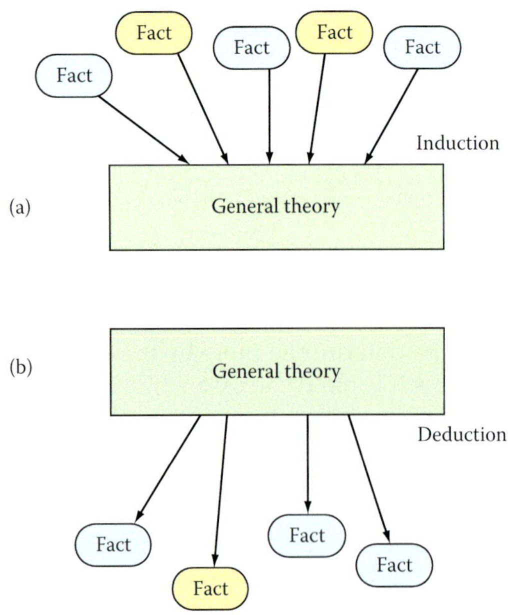

FIGURE 1.2.1
Science (a) is largely inductive, with many accumulated experimental facts contributing to an overall general theory. Engineering (b) is usually deductive, with theory presented first, and predicted facts derived from the theory.

SUSHI SCIENCE AND HAMBURGER SCIENCE

(Excerpted from Motokawa, 1989)

I had always regarded science as universal and believed there are no differences in science at all between countries. But I was wrong. People with different cultures think in different ways, and therefore their science also may well be different. Let me explain:

A visitor to the United States from Japan tried several seafoods. Most of them were deep, deep-fried denatured protein once called fish; a blackened red fish: it was nothing but charcoal. The conclusion he drew was that the cuisine of the West is overcooked. Japanese dishes seem to have no art of cooking at all. Although sashimi and sushi use uncooked fish meat, they are one of the most difficult dishes to prepare among Japanese cuisines. A lot of skills are hidden behind the no-cook. This is really an art, and definitely a different kind of art than that found in Western cooking.

Similar differences are found also in science. Western science is hypothesis oriented. A hypothesis is a personal interpretation using words about how a universal rule works in a particular matter of interest. The hypothesis should be big: the final rule should be one, and therefore the biggest and most general hypothesis is the best one. This drives the hypothesis to become abstract.

Eastern science is fact oriented. It tries to communicate with the truth not through generality and abstraction as Western science does, but through specificity and objectivity. A specific fact represents the absolute truth. Interpretations and hypotheses should be avoided because human discursive intellects conceal the reality.

When Western people read papers written by Japanese scientists they will often have difficulties in understanding what the authors wanted to say, even if the article is written in English. One obvious cause is poor English; another cause is the difference in "logic." Western logic is quite clear: it has a structure in which each statement is tightly connected and linearly arranged to reach a conclusion. Japanese logic is not so clear. Westerners may well find no logic at all. Japanese people talk about something and, without stating a conclusion, move the discussion to another topic. These two topics often have no logical connection, although they are related in the mind of Japanese people. What Japanese are trying to do is to describe one fact from various points of view. Each view is connected by imagery to others, not by strict logic such as syllogism.

Every scientist in the West tries to establish his ego. Each scientist has to put forward his own hypothesis to establish his raison d'être, even if he knows his hypothesis is not the absolute truth. Scientists have to advertise their hypotheses and their re-created world in a loud voice to be visible and to persuade other scientists. In the East, scientists have no big hypothesis to advertise. To advertise "I" and "my something" is quite bad manners in the East. Scientists seek a highly specified fact. Once they have obtained such a fact, they do not have to speak because the fact speaks. In natural sciences, it is more natural and therefore better to let nature speak for itself than to let man speak; scientists should keep silent. This is the aesthetics of the East. Similar aesthetics are found in various activities of Eastern people, such as the cooking I have referred to.

The engineering approach is different. Engineers generally try to conceptualize first then fit facts within this established framework. Engineering is thus deductive.

This method suits engineers well because it tends to reduce all knowledge to a small set of fundamental principles, among these the conservation of matter and energy, Newton's laws of motion, the laws of thermodynamics, and Maxwell's equations. Thus, engineering designs are based upon a rather limited set of simple principles or concepts. Given the choice between one of these fundamental principles and a conflicting fact, engineers nearly always chose the principle.

Such a fundamental methodological difference between scientists and engineers inevitably leads to misunderstandings. Scientists are often bothered by the tendency of engineers to simplify, while engineers wonder why scientists can't see readily apparent connections.

1.2.4 Synthesis

> As simple ideas are observed to exist in several combinations united together, so the mind has a power to consider several of them united together as one idea.
>
> **—John Locke**

Although science and engineering are separated by dominant domain, methodology, and approach, engineering is complementary to science and science is supplementary to engineering (Table 1.2.2). Engineering represents the ultimate application of the facts generated by science. And, engineering approaches are having their effects on scientific methods (Figure 1.2.2). Science, on the other hand, not only discovers the basic phenomena that are the subjects of later engineering models, but also discovers pertinent variables for inclusion in those models.

Relative merits of experimental and conceptual (or model) approaches to a scientific phenomenon are well known. Each approach is so compelling that the ideal means to study the phenomenon is to incorporate both approaches. It is the willingness of scientists over the last 30–40 years to include modeling and conceptualization in their work that has enabled the rapid application of scientific knowledge by (usually) engineers (Shneiderman and Hendler, 2017).

TABLE 1.2.2

Summary of Contrasts between Science and Engineering

	Science	Engineering
Phylogeny	Random phase through quantitative phase	Quantitative phase and control phase
Motivation	Objects of study	Objects of creativity
Methods	Inductive: large numbers of facts suggest a unifying concept	Deductive: a small set of basic principles leads to specifics
Literature	Incremental	Conceptual
Synthesis	Scientists need engineers to show eventual applications	Engineers need scientists to identify basic facts

Source: Johnson, A.T. and Phillips, W.M., *J. Eng. Educ.*, 84, 311–318, 1995.

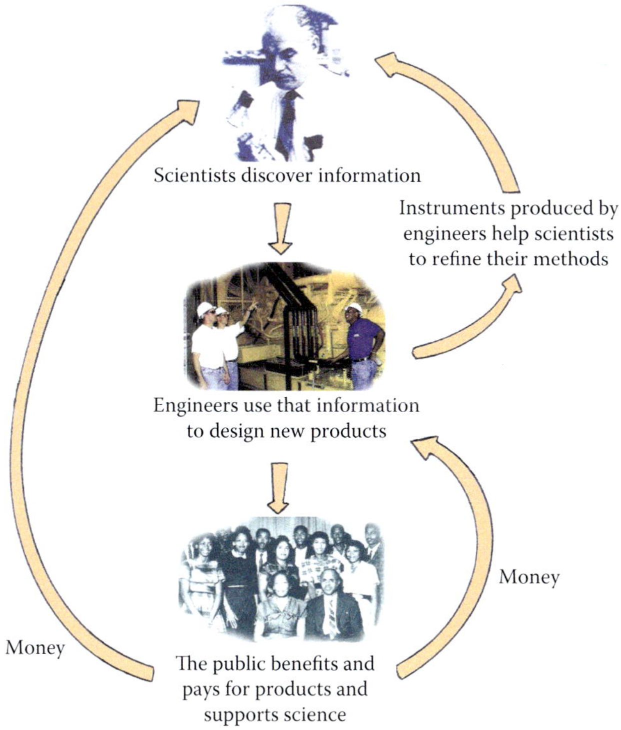

FIGURE 1.2.2
Simplified diagram of the interactions occurring in the technology loop.

Although biological scientists are often capable of generating the information necessary for the design of a new product or process involving a biological system, they don't often deliver the information in a form suitable to make design trade-off decisions. Biological engineers are in positions to function as key participants in the synthesis of biological science and engineering to produce results useful to humankind.

1.3 Scientific Method

> It takes more than a village to raise a scientist. It takes a village full of scientists.
>
> —**Brian Hayes**

The scientific method is fundamental to the inductive process used by scientists in their work. It is the basis for all scientific knowledge understood today. Although the act of discovery (the random phase described earlier in Section 1.2.1, Phylogeny) does not depend on the scientific method, the establishment of the truth and reproducibility of modern scientific observations depends almost entirely on this process. Science made progress (mostly before the early 1800s) without the scientific method, but the rigor of scientific truths has been enhanced through the use of this method.

Understanding how science develops information allows the student to appreciate the rigor of this process. The scientific method is a relatively simple set of steps that uses established knowledge as the basis to achieve new knowledge, and that in turn can be used to acquire even newer knowledge, and so on. The scientific method is important to a careful and methodical approach to progress in science.

The basis for the scientific method is this: from previous observations, a hypothesis is formed. An experiment is then planned to test this hypothesis. The test may either be one to substantiate the hypothesis or it may be a test to refute the hypothesis. The latter usually forms the more compelling evidence. Sometimes, for a particularly strong proof, especially if the subject of the experiment is controversial, both experiments are conducted (see Hellman (2001) or de Kruif (1926) for examples). In the conduct of experiments, new observations are made that lead to additional hypotheses, so the pattern

Hypothesis
Observe
Revise hypothesis
Observe
etc.

continues (Figure 1.3.1). The scientific method enhances scientific knowledge and refines understanding. In some sense, the final truth about a subject may never be learned, but once the state of knowledge reaches a certain point, a technology of useful things can develop from it.

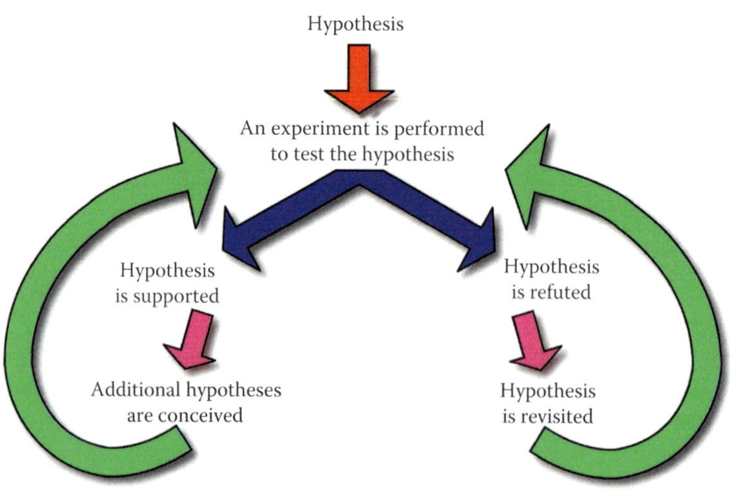

FIGURE 1.3.1
The scientific method diagrammed.

SCIENTIFIC FACTS IN BIOLOGY

Scientific hypotheses, when dealing with a scale large enough, are also called theories. And theories are tentative explanations for events that await further substantiating evidence. As evidence accumulates to support a theory, it becomes transformed into a "law," or "principle." However, just as there are no absolutely complete works, there are no laws that are completely irrefutable. All laws and scientific principles are derived from empirical data that depend on some arbitrary and relative system of measurements. Likewise, a theory may be as true or may become as well-established as a law, but only if enough evidence can be collected to support it.

The dependence on empiricism is the ultimate weakness of all of modern science because evidential facts can change as the technology of measurement advances. As an example of this, consider the measurement of an environmental toxin. Using present methods, we might not be able to detect any toxin, but to call the amount "zero" would be incorrect. Advances in measurement techniques may someday reveal extremely small but, nonetheless, measurable amounts of toxin present. The amount present is presumed to be unchanged (although we cannot know this for sure), but the amount may be measurable someday whereas it is not measurable now.

Likewise, all other measurements depend on definition for their correctness. There are international standards for time, length, mass, light intensity, and many other basic units. These are definitions, and if they somehow change, then all dependent measurements change as well. As long as the definitions remain unchanged, there is always something somewhere to compare other measurements to. All measurements, and theories and laws dependent upon these measurements, are indeed relative as well.

It was because he wanted to create a system of logic and philosophy on an absolute basis that Descartes uttered his famous, "I think, therefore I am." From this statement and other inferences, Descartes was able to construct a non-empirical philosophy. However, the world of real objects and events remains empirical and relativistic.

The formulation of theories as a means to explain related scientific facts contains both elements of absolutism and relativism. The facts themselves are empirical, and thus relative. However, the theories are often based upon abstract notions about how we can tie those facts together. The theories, then, have an element of absolutism based upon ideals.

It is tempting to believe that a system behaves the way it does because of the way in which we think it should work. We interpret behaviors (in other words, measurements and data, or facts) to mean that the supposed theory is true because there is agreement between the theory and the facts. But theories often go beyond facts by postulating mechanisms of action. Thus, we can believe that a biological system is optimized because the known data seems to indicate that it is optimized. However, the same results may have been able to be observed from a set of random responses conforming to a set of environmental constraints. The optimization theory in this case is useful because it explains the known data, but it may be incorrect because it attributes the explanation to the wrong underlying mechanism.

(*Continued*)

SCIENTIFIC FACTS IN BIOLOGY (Continued)

It is interesting to speculate on the reason that the nightingale sings in the night. Is it because the nightingale exhibits a passion for the moon, loves the dark, has deep sexual urges that cannot be satiated, or boasts his splendorous voice? None of these, it seems. The nightingale sings because he is hungry, and must eat every four to five hours (Landau, 1984). Each of these theories attributes to the nightingale different human characteristics. Who is to say that any of these theories is correct or incorrect? It is not easy to tell as long as they cannot be distinguished based upon the measurable facts.

Laws, theories, hypotheses...they differ only in their degrees of acceptance stemming from the amount of supporting evidence. When explaining the behavior of some system, especially a biological system, there is a tendency to project ourselves onto the system. That doesn't necessarily make it so.

Example 1.3.1: Development of the Scientific Method

Robert Koch (1843–1910) was a German bacteriologist who was the first to actually *prove* that a disease was caused by a specific microbe. He worked on anthrax, bubonic plague, and sleeping sickness, and established causative agents based on his set of postulates (de Kruif, 1926):

1. The suspected agent microorganism is isolated from the victim of the disease and must be present in every case of the disease.
2. The microbe must be isolated from the victim and cultured in the laboratory.
3. The cultured agent is inoculated into other healthy hosts, where it produces the original disease.
4. The agent is isolated from these animals, cultured in the laboratory, and identified as the same as the suspected microbe.

Koch repeated these procedures, attempting to disprove them, to prove beyond a doubt that this was, indeed, the causative agent.

Remark

This is a powerful procedure. Yoon and his coworkers used it to show that at least some cases of diabetes are caused by a Coxsackie virus that damages the Islets of Langerhans (Maurer, 1979).

Example 1.3.2: Statistical Inference

An experimental relationship suggesting a cause-and-effect relationship should be viewed with utmost suspicion, especially if experimental variables have not or cannot be controlled (Lave and Seskin, 1979). Many epidemiological studies of human disease draw conclusions based on statistical inference. Yet, there are often many covariables that are not always easily recognized. People who are health conscious in some matters are often health conscious in others, so these two effects are difficult to separate. Although epidemiological studies are important to determine long-term effects of various environmental factors on human, animal, and plant health, they never should be confused with controlled experiments constituting the scientific method. One of the most convincing methods in the use of the scientific method is repeatable results and, especially, results that are repeatable despite attempts to disprove them.

Example 1.3.3: Stem Cell Donations

Stem cells are harvested from neonatal umbilical cord blood, multiplied in a special brew, which contain disabled skin cells from the potential recipient (these cells supply the needed biochemical mix), and are then injected into the blood stream of the recipient. The cells seem to know where to go to correct existing problems.

Children with metabolic diseases lack critical enzymes to utilize complex sugars in various cells. As the sugars accumulate in vital organs, cells become damaged and die. Children receiving stem cell therapy seem to recover liver, heart, and brain function rapidly. How do you prove that the stem cells are taking the place of native cells?

Solution

Both positive and negative aspects must be shown. Children must improve after receiving the stem cells. This demonstrates that the stem cells are a necessary component of the cure, but it doesn't prove that these cells are directly responsible. For that, a DNA analysis of the functioning cells is necessary. If the DNA of these cells differs from the DNA of the child and is the same as the DNA of the donor, then these cells are the ones that have migrated to the site and began to assume their intended function.

The steps that Koch outlined for microbial causes of diseases cannot be followed completely because the life of the recipient child must be protected. It would be unthinkable to somehow remove the stem cells from the child once they were functioning correctly. Nonetheless, by proving that the stem cells are necessary for a cure, and are the cure, the case is logically and scientifically proven.

Example 1.3.4: Flu Vaccine Effectiveness

Quite a few studies in the scientific literature have demonstrated that seniors who were vaccinated against influenza had nearly one-half the death rate of seniors who were not vaccinated. The implication of this is that influenza vaccination is highly effective in preventing influenza-related deaths. But is the vaccination really so effective?

Upon further investigation, it can be seen that these studies looked at seniors, either vaccinated or not, and how many died during flu season. Because it is difficult to know which senior died of the flu, they looked at deaths from any cause. Included in the death rate had to be causes such as stroke, car accidents, heart attacks, and other such causes not affected by a flu vaccination (Saulnier, 2009).

In order to see this big difference between seniors vaccinated and not vaccinated, there had to be other differences between these two groups besides the fact that one received vaccination and the other did not. Those who chose to be vaccinated were generally healthier. They were mobile, they took care of themselves, they were younger, and they were less likely to suffer from chronic diseases. Frail and less-functional seniors were less likely to seek vaccination.

So, the classical double-blind studies (those in which neither investigators nor the subjects know who is receiving which treatments) were not able to distinguish adequately between groups because the groups were not matched for all factors except the one of interest, in this case, vaccination or no vaccination.

Correct interpretation of test results cannot always be made without looking carefully at the methods used to obtain those results. In this case, flu vaccination was made to look to be much more effective than it really was because confounding factors also contributed to final results.

1.4 Mathematical Modeling

1.4.1 The Value of Models

> Building a model is like eating an elephant: it's hard to know where to begin. As with almost all problems, it is helpful to break a big problem into smaller, more manageable pieces. We do this with model formulation by first creating a qualitative model and then converting this to a quantitative model. Qualitative model formulation, then, is the conversion of an objective statement and a set of hypotheses and assumptions into an informal, conceptual model. This form does not contain explicit equations, but its purpose is to provide enough detail and structure so that a consistent set of equations can be written. The qualitative model does not uniquely determine the equations, but does indicate the minimal mathematical components needed. The purpose of a qualitative model is to provide a conceptual framework for the attainment of the objectives. The framework summarizes the modeler's current thinking concerning the number and identity of necessary system components (objects) and the relationships among them.
>
> **—J. W. Haefner**

Just as the scientific method is fundamental to the work of scientific research, mathematical modeling is fundamental to engineering research. A mathematical model may be simple or complex: it may consist of no more than one mathematical equation or it may involve hundreds of equations. Its subject may be a comprehensive overview of a total system or it may be the tiniest piece of a microminiature subsystem.

As Grodins (1981) states:

> [Models]…clarify our thinking about a problem by explicitly identifying and clearly stating every assumption and limitation and…set the stage for a rigorous analysis usually expressed in mathematical language.… They provide a compact, clear, rigorously integrated summary of current conventional wisdom about how some natural system works.… Textbooks in the biological sciences are often swollen with detailed verbal descriptions, which do not depart very far from raw experimental observations. Textbooks of physics, on the contrary, are compact because they contain descriptions of models almost exclusively.…
>
> The archival function of models implies that they should also serve a valuable teaching function, as indeed they do in the physical sciences. Dynamic respiratory models, especially in their computerized interactive format, should be very valuable in teaching physiologists, medical students, and physicians the essence of normal and pathological pulmonary physiology.…
>
> Finally, models provide a mechanism for rigorously exploring the observable implications of physiological hypotheses and, thus, can help to design experiments to test them. Investigators must know what a particular hypothesis commits them to in terms of experimental observations before they can test it. In a complex system with many interacting variables, which cannot be experimentally isolated, rigorous modeling may be the only way to obtain them. Such predictions may sometimes turn out to be unexpected and counterintuitive. If they survive an exhausting recheck of model formulation and computation, this surprising behavior of models is one of their most valuable attributes in testing the hypothesis.

Starfield et al. (1990) state that mathematical models are like caricatures: they overly emphasize some aspects at the expense of others to make conspicuous those results due to the emphasized aspects. Thus, models are not always general descriptions of a phenomenon.

Indeed, a thorough mathematical description of some scientific phenomenon would be as complicated as the original phenomenon itself, and serve very little purpose. It is often difficult for a scientist to truly believe what value is contained in a model that does not predict all scientific observations related to a particular phenomenon.

AN ENGINEERING APPROACH TO TRANSLATIONAL MEDICINE

(Excerpted from Liebman, 2005)

In the years since the completion of the Human Genome Project, physician-scientists have applied new energy to translating findings from the laboratory into better treatments for patients. Yet this accelerated, unidirectional transfer of knowledge from the bench to the bedside, a practice that goes by the name of translational medicine, is hitting an obstacle: The generation of data is far outstripping scientists' ability to convert it into usable knowledge. For example, scientists can now correlate a disease with a specific pattern of gene expression. Such experiments are straightforward and fairly quick when the tools are available, and they provide a massive quantity of data. However, by diverting limited resources of time, money and personnel, mining this wealth of data may actually lead investigators away from grasping the governing laws from which they could build predictive models of the disease.

As clinical investigators, we stand to reap significant benefits on behalf of society by expanding our focus and viewing translational medicine not through the eyes of a scientist, but as an engineer might. Why an engineer? Because an engineer uses the fruits of science to feed the appetite of technology. Unlike scientists, who tend to approach problems from a "bottom-up" perspective by collecting data and seeking patterns, engineers take a "top-down" approach, probing a specific system for clues, taking it apart and considering how each component can be handled in a tailored solution. An engineer is a problem solver rather than a hypothesis generator.

The two perspectives are neatly symbiotic in physics and chemistry, for which fundamental laws yield predictive models. But in the life sciences, biologists, including physicians, must be more aware of the gap between science and technology—we still know too little about the complexity of living systems to make many generalizations from first principles.

An engineering approach, what might be called "real systems analysis," may be a better way for scientists to identify and develop solutions for biomedical problems. This kind of problem solving requires that translational-medicine research place more emphasis on going from the bedside to the bench, rather than the other way around.

Disease is a process, not a state. For the purposes of diagnosis, analysis and experimentation, academic physicians tend to focus on disease at a single point in time. But disease needs to be treated as a process that evolves over time through the interaction of genetic, environmental and lifestyle factors. This view puts a premium on understanding the complex history of a patient, and it acknowledges that most disease cannot be tied to a single cause.

When physicians make a diagnosis, it's natural to focus on the patient and symptoms at the time of presentation. The doctor's knowledge of a patient's past is typically limited to major illnesses, allergies and family history. Yet clinical assessments could be much more meaningful if we understood the way that genes and environment interact to produce disease.

1.4.2 Types of Models

Imagination is more important than knowledge.

—**Albert Einstein**

There are several ways of classifying mathematical models. One way is to split them into theoretical or empirical models. The theoretical model is based on well-established basic principles, such as Ohm's Law ($I = E/R$) or Newton's Second Law ($F = ma$). Constructing a model from one of these would then involve relating the parameters, (I, E, and R) or (F, m, and a), to the target object, and then providing further mathematical descriptions of these parameters in terms specifically related to the object. For instance, electrical resistance (R) can be related to the length, diameter, and resistivity of a cylindrical object. Theoretical models tend to be idealistic, linear, and relatively simple. They also tend to be relatively easy to solve mathematically for parameters of interest. They often do not reproduce good details of the object's responses to input modifications. They are, however, conceptually satisfying and relatively easy to defend.

Empirical models are mathematical descriptions of observations and often involve a good deal of fitting curves to data. A large number of mathematical models of biological systems are empirical because the subject matter is often very complex and far removed from the simple application of basic principles. To construct an empirical model, one would begin with a set of numerical observations and attempt to fit the data with a mathematical expression that preserved the essence of the variation of the data without reproducing the unessential (or *noise*) aspects of data variation. The form of the mathematical expression used to fit the data is usually left up to the person forming the model, and there are many shortcomings of this approach. One particular disadvantage is the lack of confidence that the fitted curve will adequately describe data outside its original range. One should always graph the data and the curve to judge the adequacy of the fit (see Section 4.2).

Because basic principles are ultimately based upon experimental observations, the distinctions between theoretical and empirical models ultimately disappear. It is most common for mathematical models of biological systems to include both theoretical elements (many in the form of Balances, see Section 2.2) and empirical elements mixed together.

Another distinction among mathematical models relates to the centrality of the computer to the formation and solution of numerical values. Nearly all modern mathematical models require computers to solve for numerical values of various parameters. In the equation-based models described earlier, the essence of the model is contained in the equations, and the computer is used only as a convenient means to obtain numerical results. However, some models require the computer at the formative stage since the model is written specifically for computer solution. This type of model, including finite element examples, can be difficult to understand from the basic equations included. Numerical results in graphical form are required in order to understand essential model information.

An important class of models is physical models based upon mathematical calculations and design criteria to produce useful aids for learning, testing, and substitution. These models take the forms of tangible items that may or may not resemble the prototype that they are meant to mimic. For instance, a physical model can be used for learning and training for injection procedures in the eye (Mukherjee and Sivaprakasam, 2015). Other physical models are used for artificial heart testing and lung replacement. Prosthetic devices are actually physical models that perform the same functions as the limbs or organs they are meant to replace.

There are other model classifications, including stochastic versus deterministic models, compartmental models, and others. Whole courses are devoted to the means of constructing, obtaining results, and properly managing models (Starfield et al., 1990).

1.4.3 Steps in the Modeling Process

> There are no scientific authorities. There are scientific experts, but there should be no authority figures whose statements are not subject to question by anyone...One of the greatest experiences scientists, indeed anyone, can have is to have some truly and deeply cherished idea proved wrong by the evidence of reality, for only in this way can we learn to look beyond our *a priori* prejudices and be willing to judge the world for the way it is, not the way one would like it to be.
>
> **—Lawrence Krauss**

Just as there is a prescribed set of steps required for the scientific method, so, too, there is a set of steps recommended for modeling. These are

1. Conceptualize
2. Separate model elements
3. Capture the essence of each element
4. Maintain interface ability
5. Mathematically describe each element
6. Solve numerically for parameter values (calibration)
7. Compare model results against experimental results (validation)
8. Revise model

In the conceptualization phase, the model creator decides the objectives of the model (different objectives require different kinds of models), the important properties of the model, and what things are to be included. This is a very important stage, one in which an overview of the model is worked out. The conceptualization phase should not be underemphasized, or else it is likely that later work will have to be repeated. Material in this text is especially relevant to the conceptualization phase.

Thereafter, the grand scheme of the model is dissected into elements or, at least, into multi-element modules. It is here that the mathematical details of the model are formulated. Again, only those details important to the objectives of the model will be included. Unnecessary details only add complexity and computational hazards. For instance, a model intended to reproduce the action of the heart during exercise would not include details of the endothelial cells lining the blood vessels. A model concentrating on the development of atherosclerosis would contain these details, but the heart would be omitted entirely. A model designed to look at the interactions among endothelial intracellular constituents would even ignore the blood flowing outside.

The essence of each element means that only the most important and essential means of describing that element should be included, no more and no less (Figure 1.4.1). For instance, the respiratory systems of amphibians, reptiles, birds, and mammals are composed of airways, lung tissue, and the respiratory muscles. Mechanical descriptions of the respiratory system include mass (which has inertia), the elastic tissues (which have nonlinear pressure–volume characteristics), and the geometry of the airways (which determines

FIGURE 1.4.1
The essence of an element is obtained by throwing out all descriptive qualities not necessary to the purpose of the model. Too often, students cannot tell for sure the difference between essential and nonessential qualities. Nonessential qualities may be essential in other contexts.

resistance to flow). These qualities are distributed throughout the respiratory system in some non-uniform fashion. Distributed parameters add a great deal of computational complexity to the model and are usually unnecessary for most purposes. The essence of the model can usually be captured by describing one inertance value, one resistance value, and one elastance value, and, in addition, these parameters can usually be linearized with no loss of model utility. Some respiratory system models include inertance, resistance, and elastance values for the three elements of airways, lung tissue, and chest wall for each lung. For other models, even this complexity is too great, and one inertance, resistance, and elastance parameter is assigned to the entire respiratory system inclusive of the airways, lung tissue, and chest wall. At normal breathing rates, inertia may usually be neglected, so the essence of the entire respiratory system can be reduced to one resistance value and one elastance value. Depending on the objectives of this model, this simplification makes a lot of sense and can adequately describe respiratory system function.

It is important, when adding mathematical details, to remember that model elements must be put back together to constitute the entire model for visualization. Thus, the input and output variables of each element must be satisfied by the surrounding elements (Figure 1.4.2). Thus, if predator population is a required computed input parameter for one element of a model, it must be computed as an output parameter of another element.

Calibration and validation are important steps in model development. *Calibration* is the process of fitting a model to a certain set of data by adjusting numerical values of parameters to obtain the best fit. The experimental data against which the model is calibrated are substituted for model input and output data, and the model is run

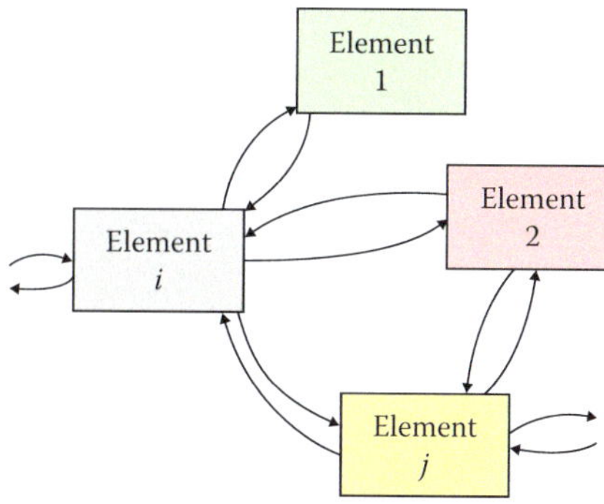

FIGURE 1.4.2
Each element of a model must interface with other elements and pass information between them. The input parameters of each element must be satisfied by the output parameters of interfaced elements. All elements need not interface with all the others.

backward or forward in order to obtain the best numerical values for model variables. That means two things:

1. Model calibration is limited by the quality of the experimental data.
2. The model will probably fit calibration data the best of any data that it can be tested against.

Thus, model results plotted for the calibration data set usually are very impressive, but mean little.

That's where validation comes in. *Validation* is the process of trying the model on an independent set of input data and seeing how well the model output data match the actual set. If the match is good, then the model is validated, at least somewhat. The more independent experimental data sets are that result in good matches between model output and experimental data, the more valid the model becomes. Thus, there are degrees of validation.

There are other steps that may be taken in modeling. Sensitivity analyses show how much model outputs change for incremental changes of model parameters. A model is said to be *robust* if it can accept many changes in values for a wide range of input conditions. All these steps are further explained in other courses and texts on modeling.

1.4.4 Models and Empirical Observations

> Although we humans often judge ideas on their plausibility, plausibility is not a rigorous test of the validity of scientific ideas.
>
> **—Pat Shipman**

Although modeling is a fundamental tool of engineers, empirical observations are still necessary to assure that the models are based in reality. Both models and empirical observations are necessary for modern technological advancements (Table 1.4.1). Models by themselves only allow deductions and are only compact descriptions of what is already known. Empirical observations allow for the inductive discovery of new knowledge, but do not organize known facts into a useful form. Thus, as long as there is new knowledge

TABLE 1.4.1
Functions of Models and Empirical Observations

Models	Experiments
Lead to predictions of experimental events	Guide development of models
Show what experiments need to be conducted and what parameters require measurement	Calibrate models
Form a framework into which to form experimental results	Validate models
Make some experiments unnecessary, either because	
1. All information is known to predict outcomes	
2. Some things can be shown to be impossible	

to be discovered, we will need real-world observations; as long as we require engineers to design new products and processes, we will need mathematical models.

Example 1.4.1: Mathematical Model of an Infectious Disease (Marchuk, 1983)

Disease is caused when certain types of antigens (for example, bacteria, viruses, and so on) overcome immune responses in an organism (see Section 6.20.3, Redundancy). The interaction between antigens and immune responses can be modeled at different levels, from macroscopic to intracellular genetic. The type of result the modeler seeks will often determine the type of model that is constructed, and this dictates which properties are included and which are ignored. For example, these models have been formulated:

1. Equations to describe the change in the number of circulating antibodies as a function of the number of specific plasma cells (Hege and Cole, 1966).
2. Probabilistic models of the interaction of antigens with immunocompetent B cells (Jilek, 1971).
3. Model describing antigen–antibody relations as predators and prey (Bell, 1971).
4. Bilinear system theory interpretation of immune response as a heterogeneous immunocyte population varying with time and affinity (Bruni et al., 1975).

In each of these cases, the models developed would have different appearances resulting from different starting points. Various computer-based models might also extend the range of models for disease development.

Example 1.4.2: Localized Estrogen Delivery Affects Neural Plasticity

Male songbirds have better-developed neural song centers in their brains than do females. The more imbalance there is between song delivery by males and females, the more difference there is in their brains. This difference begins to appear during adolescence, when males begin to sing and, apparently, influenced by local estradiol levels. The estradiol (one of the estrogen hormones) is apparently formed in the brain from circulating testosterone produced in the testes outside the brain, but the estradiol is not produced in the song centers where it is needed to influence neuronal structures. Instead, the estradiol is stored in presynaptic boutons that connect to neurons in the song centers.

The amount of stored estradiol varies by a certain amount, the amount released during an action potential can vary, and the fate of the estradiol in the postsynaptic cleft is not completely certain. In addition, the effect of estradiol that manages to reach the target neuron is also variable. To model this problem requires four probabilities: (1) the probable amount stored (p_1), (2) the probable percentage of stored estradiol that is actually released (p_2), (3) the probability that released estradiol reaches the target neuron (p_3), and (4) the probable effect that a certain concentration of estradiol will have on the target neuron (p_4). The overall probability of a certain neural outcome is the product of the individual probabilities:

$$p_{\text{tot}} = p_1 \cdot p_2 \cdot p_3 \cdot p_4$$

as long as each probability is independent of the others. If there is a dependence of one or more of the probabilities on prior probabilistic events, or if nonlinearities occur so that probabilities are not constant, but vary in some definable way, then the calculation of the overall probability of an outcome is much more complicated.

Each of these probabilities p_1 through p_4 has some variability. Just as with flipping a coin, we would not expect the first few trials to yield the same fraction of specific outcomes as would a large number of trials. So, to model the effects of estradiol on target cells requires looking at the outcomes of a given number of trials and noting the results.

In order to do this, we must use a random number generator with the same probability of an occurrence as we expect would happen with the estradiol. In addition, the random number generator should have the same variability (called variance) as the biological process. And, we need to have four random number generators acting simultaneously.

These random number generators could be coin flips or die tosses, but it would be hard to adjust the probabilities and variances to the necessary exact values. Instead, computers usually run random number generators.

The outcome of the model is not known ahead of time. Instead, the model is run as many times as we are interested in, and the number of successes (perhaps a certain level of target neuronal response) is noted. Then, the model is run again, and again, and again, each time for the number of trials of interest. Over time, an overall average success rate may emerge, and that will be considered the model result. Notice that the model does not give a definite answer (nondeterminate), and the results must be obtained numerically. These are the characteristics of *stochastic* (probabilistic) models.

Example 1.4.3: A Grass-Deer Ecosystem

Model the carbon flow in an ecosystem defined as grass and some deer that eat the grass.

Solution

The qualitative solution is based upon this conceptual framework (Haefner, 1996):

1. Grass will be assumed to have a constant rate of growth. Growth is defined as the mass of carbon newly fixed in grass plants per unit mass of carbon already present in the plants. Total amount of carbon fixed is therefore related to the total amount of carbon present.
2. The only loss to the quantity of carbon present in the grass population is by deer consumption.
3. Deer compete with one another for grass, so that each deer receives less carbon as the number of deer increases.
4. Deer excrete or respire a fixed proportion of their existing carbon as either atmospheric carbon or solid/liquid waste.

None of these stipulations is detailed enough to allow a complete set of mathematical equations, but this is the framework that provides structure to the model to be formulated. From these four conditions, equations can be developed either based on first principles or on empirical evidence. Stating these conditions in this way breaks the overall concept of the model into smaller pieces that can then be worked on one at a time, being sure that the developed equations fit together as a package and that they exclude unnecessary detail.

Example 1.4.4: Animal Models for Drug Testing

Animal models are often used to test predictions of the effects of drugs intended for human use. However, animals meant for this purpose have many disadvantages (Dove, 2010):

1. There are ethical and moralistic concerns about animal use and care.
2. Animals often give poor predictions of human responses.
3. Animals are expensive and inconvenient.
4. Animal populations used for drug testing usually have uniform genotypic and phenotypic characteristics, and so do not display idiosyncratic reactions.

For these reasons, alternatives have been sought. Human cell cultures hold promise; they can be fast and inexpensive, but cultured human cells rarely act the same as identical cells in the body. Mathematical models of drug reactions are another possibility; but, until model results are validated by animal experiments, there is little confidence in model predictions. Indeed, validation of mathematical models could actually increase animal usage rather than decrease it.

1.5 Biological Engineering

The emerging discipline-based biological engineering has the potential of using biological materials and living processes in designing systems that are more in harmony with nature.

—Brahm Verma

Engineering is a profession. In other words, it is an occupation that involves a liberal education and mental rather than manual labor. Biological engineering is a discipline within the engineering profession. A discipline *is characterized by a distinct body of knowledge and commonly accepted methods*. The body of knowledge for biological engineering includes fundamentals of engineering practice, including

Analysis
Computation
Design

skills, along with a working knowledge of the science of biology, including

Methods
Principles
Properties

applicable to utilization. Biological engineering methods include

Systems approach
Modeling techniques
Black-box viewpoint

A systems approach means a broad, conceptual consideration of all the possible influences and characterizations affecting a biological system, whether that system is the interior of a cell, a group of tissues in an organism, or the entire earthly biome. Modeling techniques capture the essentials of the biological system pertinent to the goals of the model and deal with biological principles and simplifications. Models usually reduce the biological system to mathematical form, although mechanical, electrical, chemical, or thermal models are also possible. The black-box viewpoint is used to replace complex biological elements with output–input relationships, thus avoiding unnecessary complications in forming the model.

In addition to this knowledge and these approaches, there is an enthusiastic passion that biological engineers exhibit toward biology, in general, and a wonderment at the interconnectedness and sense of apparent order present in the biological world. It is both the complexity and the apparent simplicity of life that inspires biological engineers to work with biological systems on an intimate basis. Passion and creativity are essential attributes of successful engineers, and living systems elicit these attributes in biological engineers.

Engineering disciplines can be classified into two categories: applications-based and science-based. Applications-based engineering disciplines serve particular economic sectors such as petroleum, mining, military, medicine, and agriculture. Science-based engineering disciplines are much broader, more fundamental, and are based upon particular sciences. The foundational engineering disciplines were all based on some portion of physics: electricity, mechanics, and heat. Chemical engineering added the science of chemistry to engineering. Biological engineering adds biology, although it also retains interests in physics and chemistry.

The earlier definitions and descriptions can be very abstruse for many people. Certain descriptors have been used to explain biological engineering, and one or more of these can help to understand what biological engineering is. The following are not definitions, but are included here for elucidation:

Familiar with both engineering and biology
Not identified with any particular application
Act as bridges between engineering and biology
Doesn't just work with biology, but has a substantial knowledge of, and continuing interest in, the field of biology
Should be to the science of biology as chemical engineers are to chemistry, electrical engineers are to electricity, and mechanical engineers are to mechanics
Broad, fundamental, integrative, and unspecialized
A specialist in technical diversity (my particular favorite).

Table 1.5.1 lists desired attributes of an engineer from several important sources (McMasters and Cummings, 2004). There are no attributes that can be specifically identified as being associated with biological engineering, nor are they identifiable as related to any other

TABLE 1.5.1

Desired Attributes of an Engineer by Boeing, Lockheed Martin, and the National Academy of Engineering

- A good understanding of engineering science fundamentals
  - Mathematics
  - Physical and life sciences
  - Information technology (far more than "computer literacy")
- A good understanding of design and manufacturing processes (i.e., understands engineering)
- A multidisciplinary, systems perspective
- A basic understanding of the context in which engineering is practiced
  - Economics (including business practice)
  - History
  - The environment
  - Customer and societal needs
- Good communication skills
  - Written
  - Oral
  - Graphic
  - Listening
- High ethical standards
- An ability to think both critically and creatively—independently and cooperatively
- Flexible, agile, and resilient—the ability and self-confidence to adapt to rapid or major change
- Curiosity and a desire to learn for life
- A profound understanding of the importance of teamwork
  - Management skills
  - Leadership abilities
  - Organizational skills
  - Ability to mentor
- DIVERSITY—wanted and needed!

engineering discipline. Nevertheless, Table 1.5.1 provides a good checklist for the qualities that any engineer, including those primarily dealing with biological systems, should, ideally, possess. Qualities in Table 1.5.1 also remind us that engineering is part science and part art.

The engineering attributes in Table 1.5.1 illustrates that engineering integrates many skills with knowledge from many different sources. Biological engineering is even more multidisciplinary than most engineering disciplines because of the broad range of potential applications. Thus, you will find in this book a range of topics not normally found in any engineering or biology text. Nonetheless, the biological engineer should appreciate the many internal and external influences shaping any engineering design involving living things.

1.6 Expectations for Biological Engineers

> ...one definition of engineering might be that it is the avoidance of failure....The engineer ensures that...failures do not occur by analyzing the design on paper, and the objective of the analysis is to calculate the intensity of forces in the structure and compare them with limiting values that define failure....The nature of engineering

> design is such that emerging fields such as bioengineering...can be expected to follow similar paths as have the older and more traditional fields, in that design errors will be made, failures will occur, and designs will evolve in response to real and perceived failures. We can only hope that when those failures occur, loss of human life will not be the result.
>
> **—Henry Petroski**

We have seen how engineering involvement with biology is largely at the technological stage where products and processes are produced to provide useful products to fill some predetermined need (Petroski, 2001). Others besides engineers may also operate at this technological stage. For them, the study of biology must involve the consideration of the many complexities that living systems have to offer. However, many scientific details of living systems are not necessary to create required products and processes. Instead, those who use living systems as part of their creative domain must rely upon principles and generalizations, which reduce the field of choices from an infinite whole to a limited set. For instance, a certain pollutant could be removed from the environment using any living entity, or a chemical approach, or a physical approach. Knowing that bacteria may use the pollutant as an energy source would open the possibility that bacteria could form part of the solution to the problem. Knowing that bacteria often need moisture, oxygen, and other nutrients as well as an energy source can give a quick idea about how a pollutant-removal solution can be constructed. Although these generalizations are rather simple, they are powerful enough to limit the range of possibilities rather quickly.

Thus, we might expect three things from engineers who deal with biological systems:

1. The knowledge of biological principles and generalizations that can lead to useful products and processes
2. The ability to transfer information known about familiar living systems to those unfamiliar
3. The ability to avoid or mitigate unintended consequences of dealing with any living system

To the third expectation, we add that living systems are not passive: they move, they change, and they influence their surroundings. Thus, they cannot be used blindly without expecting other changes to happen. Anticipating these other changes can distinguish those who are experts in biological engineering from all others. Whether the process involves installing an artificial heart into a sick human patient or introducing a new law to limit harvesting of a wild food species, there will be other unrelated and perhaps unseen consequences.

Technological advances have made unintended consequences almost inevitable. Like a phantom in a bag that pops out in every direction that isn't held, secondary effects that are masked by primary effects assume much more importance when the primary effects are conquered (Tenner, 1996). Chronic illnesses such as cancer, silicosis, and cumulative trauma disorder probably were not recognized centuries ago as important because acute illnesses such as typhoid, plague, and pneumonia killed so many. After anesthetics allowed painless surgery, the number of surgical procedures skyrocketed, and the total amount of pain experienced by the total human population is higher because of it.

This message is discouraging, because it implies that unintended consequences can never be avoided. As more acute problems affecting humans and their environment are tackled and cured, problems that had seemed inconsequential can suddenly become limiting. Nevertheless, the experienced biological engineer should be more able than others to anticipate the likely consequences of his or her technological fixes and to be prepared to deal with them.

It is the intention, then, that the approach toward the life sciences taken in this text should support the three expectations given earlier.

Example 1.6.1: Environmental Conditions and Human Disease

There is a direct link between environmental conditions and human diseases. Diseases such as hantavirus in the southwest United States, cholera in South Asia, dengue fever in Vietnam, and malaria in Peru all seem to be related to El Niño, the periodic warming of the tropical waters of the southern Pacific Ocean. During El Niño years, Peru's mountainous valleys are warmer and rainier; this promotes the reproduction and growth of sand flies called *lutzomyia,* which prefer meals of human blood. While sucking, they transmit *Bartonellosis* bacteria that enter the red blood cells of its victims and destroy them. Between 40% and 60% of its victims die; the rest develop a rash and bleeding warts. Knowing this environmental connection to human disease can make prevention and control easier by concentrating resources for the times of most peril (Roylance, 2002).

This is just one example, among many, of the link between the environmental portion of the biosphere and individual organisms. This is one reason why the connections among biological components need to be known: understanding these connections can point to key steps where cycles can be disrupted, and disease outbreaks controlled.

Example 1.6.2: Sickle Cell Anemia

Sickle cell anemia is a disease caused by a gene that makes defective hemoglobin. The defective hemoglobin molecules form long, sticky polymers that cause the red blood cells to be sickle-shaped rather than round. These abnormal cells clog the blood passageways and starve vital organs of oxygen. Gene therapy has been suggested to cure this disease; placing the desired gene in the shell of a neutralized virus vector could introduce the gene into the patient's cells. However, this technology is mostly hit-or-miss, and may be dangerous to the people receiving the cure. Suggest alternative means to treat the disease.

Solution

Several possible treatments present themselves. On an organismal scale, altering the blood or breathing hyperbaric oxygen so that the blood can carry more oxygen than normal might be a possibility. Or, perhaps a substance could be added to the blood to make the sickle cells less likely to clump. Blood transfusions with healthy blood certainly could help.

Another possibility exists. There is a gene present in all fetuses that makes a protein to draw oxygen from the mother's blood into its own. This is fetal hemoglobin, and it has a higher affinity for oxygen than does adult hemoglobin. This gene never causes sickling, but the gene is turned off at birth. If the environmental switch that turns off the gene can be reversed, then the sickle cell patient may be cured of the disease.

Remark

When attempting to solve a problem involving living things, there are usually many possibilities for a solution. The biological engineer must know enough about all aspects of biology to be able to enumerate as many possibilities as she or he can. Then the best choice can be made.

Example 1.6.3: Unintended Consequence of GMO Squash

Cultivated squash plants are susceptible to a variety of viral diseases that cause infected plants to grow more slowly and produce misshapen fruits. In the mid-1990s, genetically modified (GMO) squash that were resistant to three of the most important viral diseases were approved for release to the agriculture industry.

Plants without the virus-resistant transgene contract the viral disease. Consequently, they do not grow as well as the GMO plants. Cucumber beetles feed on both types, but prefer the healthier GMO plants. The cucumber beetle feeding spreads bacterial wilt disease. Hence, plants resistant to viral diseases are those most likely to suffer from bacterial wilt disease. This is an example where solving problems one at a time may not be the most successful approach (Lancaster Farming, 2009).

Example 1.6.4: Unintended Consequence of Roundup-Ready Crops

Glyphosate (commercially sold with the name "Roundup") is a very popular herbicide because it kills both grasses and broadleaf weeds at any stage in their development. It works as a chelating agent that ties up minerals like iron, manganese, zinc, calcium, nickel, and copper. Some of these are essential for critical enzyme formation. Glyphosate kills plants by blocking the essential shikimic acid enzyme pathway required for plant respiration. It also interferes with enzymes necessary for disease resistance (Lehnert, 2010).

A plant gene conferring resistance to glyphosate had been inserted into popular genetically engineered crops in the mid-1990s. This allowed these crops, called "Roundup Ready," to be grown with minimal herbicide use. One spray of glyphosate replaced four or five sprays of other, more environmentally damaging herbicides. Glyphosate-resistant cotton and soybean crops became dominant due to their economic and management advantages. More than 143 million acres were planted to Roundup Ready crops worldwide, and 92% of the US soybean crop was planted to glyphosate-resistant varieties in 2010 (Lancaster Farming, 2010).

Such reproductive pressure has rapidly led to the emergence of glyphosate-resistant weeds. There were no resistant weed species in 1995; there were 19 resistant species in 2010 (Lancaster Farming, 2010). Resistant weeds were a problem on 2.4 million acres of crops in 2007, 5.4 million acres in 2009, and 11.4 million acres in 2010.

The biological realm depends on redundancy for effectiveness. Scientists who developed Roundup Ready crops inserted just one gene into their plant genomes. The probability that effectiveness of one gene can be overcome is small, but finite. If there were two or three genes to overcome, the probability of developing resistance would be nearly zero. Biotechnologists, engineers, and scientists who wish to avoid the unintended consequence of making their creations widely ineffective need to depend on more than one mechanism to safeguard their works.

1.7 About Predictions

You may wish to become a physician and save people one at a time or to become a biomedical engineer and save them a thousand at a time.

—Raj Tonnosh

Modern biology is headed more and more toward being predictable. That is, as new knowledge is gained and integrated into those things already known, patterns emerge that make

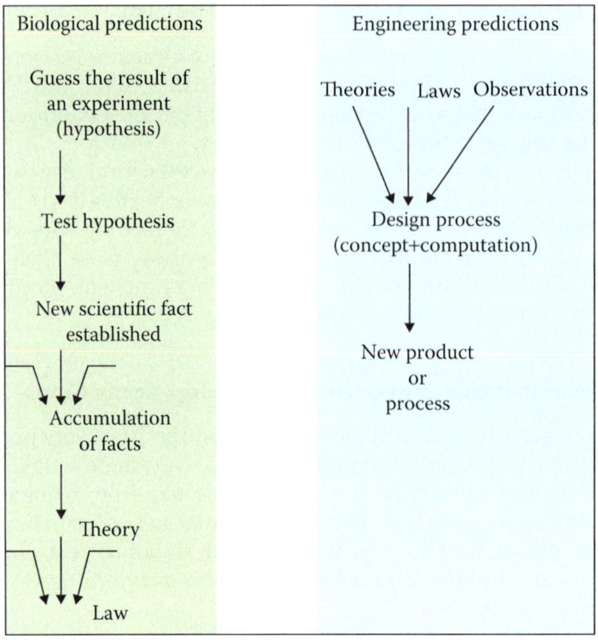

FIGURE 1.7.1
Biological predictions lead to the establishment of new knowledge, whereas engineering predictions lead to a successful application of existing knowledge. Designs requiring new knowledge are never attempted.

possible the foretelling of future facts. When the degree of sophistication of this process reaches a sufficient level, then vaticinatory models can be formulated, and further experimental observation becomes redundant rather than exploratory.

Hypotheses have always been important in biology, as in the rest of science. The scientific method is based upon the cycle of hypothesis–test–hypothesis–test, and hypotheses made, whether proved true or false, indicate a state of knowledge somewhat above complete ignorance.

It is against this background that we wish to distinguish between predictions made in the exercise of the scientific method and those made by technologists in the applications of or with biological systems (Figure 1.7.1).

Successful predictions depend very strongly on the amount of knowledge we already possess. Take the case of predicting the location of the next data point in a time series. If we have no knowledge whatsoever, there is nothing we can base our prediction on. With one known data point, the next is probably located somewhere in the same vicinity, but we have no idea about the direction locating the second data point from the first (Figure 1.7.2). With two data points, the number of likely directions is reduced somewhat. With many data points, we can often make a pretty good prediction of the location of the next data point (even if the data points are randomly scattered). Thus, we can conclude that the more knowledge we have of biology, the better able we can be to utilize biological systems for some useful purpose. A better prediction means a more confident design is possible.

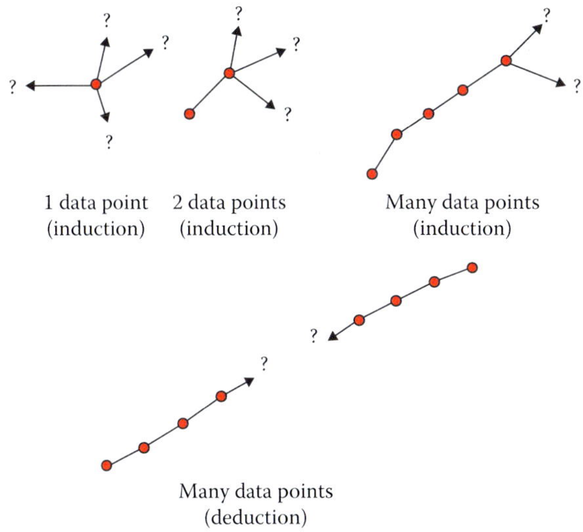

FIGURE 1.7.2
The more data points that are known, the better the prediction of the location of the next data point. If the prediction involves deduction rather than induction, then the prediction can be made very precisely.

Better yet, if the next data point is located within the confines of a data set, then the prediction of the next data point can be made with very high assurance of correctness. This illustrates the difference between induction (or extrapolation) and deduction (interpolation). Engineers who use already-existing knowledge to predict the behavior of biological systems can do so with a high degree of certainty, especially if they know how biological systems normally react and how they react in the extreme.

Biologists would say that "scientific theories are built by testing their predictions of *new* findings, not simply by explaining existing knowledge" (Beckwith, 2001). Technologists including biological engineers might use the word "prediction" a little differently; prediction to them would be most useful if it could lead to a successful biological application. In other words, prediction is a term that describes the summarization of all that is known about a subject in order that it be usefully applied. Predict, therefore, is what one does when one designs a new product or process using *old* knowledge in a new way.

This book has the intentional purpose of interpreting biology in a manner most useful to biological engineers and others who engage in activities at the control stage of technology (see Section 1.2.1, Phylogeny). Thus, the predictions made in this text may seem obvious and mundane. They are not intended to expand the state of knowledge, but to consolidate what is already known. If this approach disappoints the biological scientist, at least it will guide the engineer toward a successful design career.

Engineers need to know about typical responses, some idea about the range of responses, and important exceptions. These pieces of information are not normally referred to as "predictions." However, in order to demonstrate the utility of the information presented in each subsequent section, predictions, both sublime and ridiculous, have been included at the ends.

Example 1.7.1: Predictions about Water Temperature Control Downstream from Dams across Spawning Rivers

The Rogue River in southwest Oregon has been widely known for its runs of Pacific salmon and steelhead trout. Following severe flooding, however, three large dams were constructed to control the river's outflow. These dams cut off the upstream waters from migrating salmon. To mitigate the loss of spawning waters, thousands of adult migrants are collected each year below the dam and brought to a hatchery to reproduce. Juvenile fish raised in the hatchery are released downstream from the dam later in the year.

In the lake behind the dam is a $20 million free-standing, 256-foot tower with water intake ports at four widely spaced elevations. It had been predicted that water temperature downstream from the dam could be controlled by drawing water from different levels, and that, by controlling water temperature, growth of the fish would be enhanced.

What actually happened was that water temperature could not be controlled as well as was expected. Summertime release of thermally suboptimal water slowed the growth of juvenile salmon; release of warm water in the summer accelerated upstream migration of adult spring Chinook salmon and overwhelmed the hatchery; sudden changes in the rate of water release activated downstream migration of juvenile fish. Peak releases in the spring and summer caused downstream migrating juveniles to crowd downstream areas where temperatures were too high, and juveniles emigrated prematurely to the sea where their survival chances were greatly reduced.

The predictions that technological remedies could compensate for the presence of the dams were far off target. In this case, these predictions served as hypotheses in costly time-consuming experiments and were found to be wrong (Larson, 2002).

1.8 About This Book

One of the gravest charges ever made against science is that biology has now put it into our power to corrupt both the body and the mind of man.

—**Peter Medawar**

The engineering design process begins with a concept and continues to completion using various tools of engineering, such as mathematics, physics, engineering sciences, computers, and models. Some have suggested that a book on biology for engineers ought to be chock full of equations, quantitative models, and numbers. That approach has not been taken here because the very beginning of an engineering solution, the art of engineering, is the concept, or the vision of what the solution should do and how it should do it. In order to produce an engineering design involving living things, one must be familiar first with how living things work. Then, and only then, should an engineer investigate the question of "how much?"

No one reading this book will become an expert in biology. Additional texts and courses will be necessary for that. And, even the experts must read constantly in order to stay abreast of the many new developments that are happening these days. This book is intended to give perspective—to give an appreciation for the entire interface between technology and the life sciences.

Biological systems are not exempt from the laws of physics, chemistry, and other sciences. Therefore, the (Section II, Principles from the Sciences) deals with scientific principles relevant to biology. It would be extremely presumptive to expect that all relevant principles are included, and no such claim is made. However, an attempt was made to identify and explain those principles that form the basis for biological responses appearing later in the text.

Following the scientific synopsis is a section dealing with biological responses (Section III, Responses of Living Systems). A "black-box" approach has been used; impose a set of conditions on the black-box and a certain response is expected. When the opportunity presents itself, some additional mechanistic explanation has sometimes been given, but it is the input–output responses that are of primary concern here. The reader will find that the range of topics covered in Section III is very broad. Some will question why it isn't more focused. Biological engineers these days must be able to integrate vast amounts of knowledge into a final solution (design). That is why there is material included in this book from such diverse fields as psychology, human factors, genetics, plant and animal physiology, imaging, control systems, actuary, medicine, and others not normally found in a biology book. All of these can be relevant to biological engineering work.

Scaling of biological responses and attributes is important to be able to make quantitative predictions, especially when few data are available. The section on scaling (Section IV, Scaling Factors) is the most mathematical of all in this book, but although somewhat inconsistent with the general nature of this text, is included because a thorough presentation of scaling relationships is hard to find elsewhere.

Finally, the last section of the book (Section V, Utilizing Living Systems), short as it is, provides a classification of different types of applications related to biological systems. It is in this section that the engineer or technologist can see the broad relevance of this predictive approach to biology.

The appendix has tables of useful information that is difficult to find elsewhere. They are included to add to the value of this book as a reference. The reader may need some of this information for future research. Look over these tables just to see the range of information that appears there, and remember that they are accessible for later use.

This material may be either a review or a preview for students. As a review, it can help to tie together many loose ends from many other courses; it can add perspective. As a preview, it can help to show why later courses are necessary; it can add perspective. There is no ideal placement of this text in a curriculum. It can be helpful both at the beginning and at the end, and the student may wish to use it that way; read it first for motivation for studying all those other topics, and read it last to see why you had studied them.

Given the earlier, there is no ideal background for this text. To understand everything included, a host of other courses would have been required. However, if this text is to be used as an introduction to biology, then the student should not be expected to learn all details included herein. The student should try to read for perspective and to form general concepts, realizing that further study is necessary to fill in the details.

Lastly, the amount of mathematics included in this text has been minimized in order that the conceptual nature of the material not be lost. This is not to say that math isn't important to the application of biology, because it is. However, in these times that emphasize multimedia learning, there are few opportunities to exercise the conceptual mind. It is at the concept level that this material is targeted.

Questions

1.2.1 Give at least two biological examples of the stages of technology.

1.2.2 What do we mean by the "control phase" of technology? What does this phase mean for biological engineers?

1.2.3 Describe what you think engineering is.

1.2.4 Are biologists who manipulate genes really "genetic engineers"?

1.2.5 What are the differences between inductive and deductive fields of study?

1.2.6 How would you go about explaining an idea to a scientist? What if the listener were an engineer?

1.2.7 Make a list of engineering contributions that have enabled scientific progress.

1.2.8 Give an example where an understanding gap is likely to exist between a scientist working in a particular area and an engineer who also wishes to work in that same area. How could you, as a biological engineer, help each to understand the other?

1.3.1 Why is the scientific method so powerful? Can any kind of science proceed without using the scientific method?

1.3.2 Why is Koch's method of proving a connection between a microbe and a disease so powerful? Are there other possible causes of the disease if microbes identified by his procedure pass all tests?

1.3.3 Name some diseases whose cause would not be able to be identified through Koch's method.

1.3.4 If you are told that a certain drug is effective against a disease, what kind of evidence should you look for in order to be convinced?

1.3.5 State a principle of science. Trace its evolution from isolated facts through theory to its present form.

1.3.6 Ohms law ($I = E/R$) was originally published with an additive constant term ($I = E/R + C$). Ohm spent the rest of his life trying to amend his original mistake. If you were George Simon Ohm, how would you go about doing this?

1.3.7 Give an example where some physical or biological phenomenon is explained with a human motivation. How could you go about proving or disproving the explanation of the event?

1.4.1 Give an example of a model that includes both theoretical and empirical elements. How can one be distinguished from the other? When is it necessary to use empirical models? When is it desirable to base models on theory?

1.4.2 Someone tells you that they have an equation that almost perfectly fits the data, and so that equation is the best description of the phenomenon. You inquire and find out that parameter values for the equation were obtained from the data set for which the fit is nearly perfect. Why should you be suspicious about the value of the equation? What could you do to determine the validity of the equation?

1.4.3 How can a mathematical model help an experimentalist? How can experimental facts be used to guide model development?

1.4.4 Why might animal testing for medical purposes be banned? What might replace it?

1.5.1 Give your own additional descriptions for biological engineering.

1.5.2 What part of engineering is science and what part is art?

1.5.3 What is meant by a "specialist in technical diversity?" Why a "specialist?"

1.6.1 Give an example where unintended consequences resulted from some attempt to fix a problem.

1.6.2 Should there be other expectations of biological engineers? If so, what and why?

1.6.3 If the gene that controls the protein drawing oxygen from the mother's blood were able to be turned on to cure sickle cell anemia, what unintended consequences would you suspect would happen?

1.7.1 What number am I thinking of? If the number is somewhere between 1 and 10, what is the number? If the number is odd, and between 1 and 10, what is the number? If the number is evenly divisible by the integers 3 and 9, is odd, and is between 1 and 10, what is the number? What does this game have to do with engineering predictions?

1.7.2 Why can it be said that an engineer must be able to predict the future? How does prediction relate to design?

1.7.3 List five biological attributes that can be predicted and five that cannot. Justify each.

1.7.4 Give an example of an inductive argument and an example of a deductive argument.

1.8.1 Give an example of a product of engineering design and the original concept that led to the final product.

1.8.2 What expectations do you have for this book? What do you hope to learn?

1.8.3 Look through the table of contents. Are there topics that particularly interest you?

1.8.4 Contrast the knowledge that you already possess about biology with the contents of this book. Does this look to you like a biology book? What, do you suppose, is the difference, if any?

Section II

Principles from the Sciences

2

Principles of Physics

> None of us knows—at the level of consciousness—how to walk, or breathe, or throw a baseball. If we had to take charge of these movements, issuing commands to all the hundreds of muscles in just the right sequence, who would not collapse in a quivering mass?
>
> **—Brian Hayes**

Physics is a science that deals with matter and energy, and their interactions. It is a quantitative science of measurement, experiment, and systematization of experimental results. The scientific method is used in its most classic sense to formulate principles or laws that can be widely applied in predicting results to be achieved in engineering applications.

The study of physics is the foundation of the manipulation and control of our physical surroundings. The movements we make, the processes we use to extract energy, and the strengths of body members are all physical topics. When a biological unit (BU) (be it a cell, organ, organism, or population) is considered, it is subject to interactions with its physical, chemical, and biological environments (Figure 2.0.1). These environments determine the responses of the BU and are also affected, more or less, by the presence of the living entity.

Physics is the basic science upon which all engineering rests. The quantitative descriptions of physical events are models for engineering analysis and design. Physics is the paradigm for a science that uses empirical observations and transforms them into fundamental principles. From these come many of the basic equations used in engineering. Indeed, physics can be likened very much to engineering, except that classical physics does not emphasize the creativity and design inherent in engineering.

Some of the physical environment is changeable, but some basic physical principles are unchanging. Living things obey the laws of physics and adapt to them as well as they can.

The understanding of physical laws, then, is important in order to understand the context in which biology has developed.

This chapter will present some of the principles that are most applicable to predict the behavior of biological systems. For this, a descriptive approach will be used that minimizes quantitative considerations. However, the science of physics can also help to determine how much the biological system is influenced by its physical environment.

There are many important principles in physics. Then from these several can be selected as being of paramount importance in the understanding of BU. Among these are

1. *There are places with higher potential and places with lower potential. Things move from higher to lower potential.* "Potential" is a word used to describe the tendency of something to move, so potential, which can be either positive or negative, can either attract or repel moveable things. These things can be particles, energy, or BU themselves, as explained in Section 2.1 (Effort and Flow Variables).
2. *The maintenance of order requires energy.* Living things are highly ordered instead of random. As a consequence, living things require constant expenditure of energy. See Section 2.6 (Entropy).

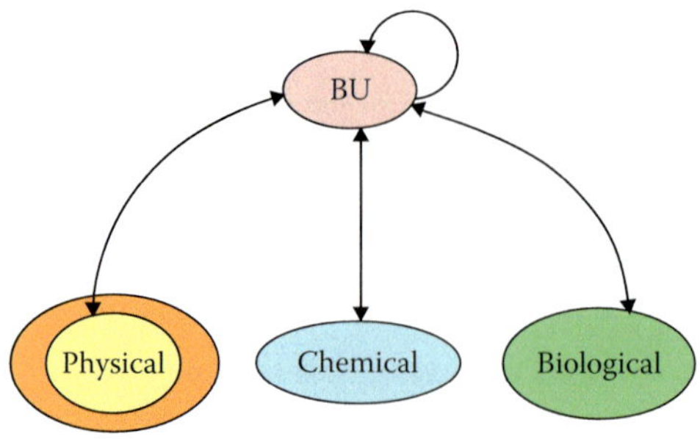

FIGURE 2.0.1
The biological unit (BU) is affected by interactions with its physical, chemical, and biological environment. Likewise, environmental elements are affected by the BU. Self-adjustment is also a possibility.

3. *What goes in but doesn't come out is stored inside.* This principle covers the idea of balance, and the conservation of matter or energy, and is covered in Section 2.2 (Balances).
4. *Different forms of energy can be used to perform mechanical work.* Mechanical work is extremely important for BU, and this work is used to define the meaning of energy. See Sections 2.4 (Work and Energy) and 2.8 (Mass Movement).
5. *The transfer of something from one place to another depends directly on the surface area and is inversely proportional to the distance between the two places.* No matter whether the "something" is heat, mass, electrical current, or something else, BU must deal with the constraints imposed by geometry. See Sections 2.4 (Work and Energy), 2.8 (Mass Movement), and 2.11 (Electricity).
6. *Mechanical strength depends on geometrical configuration, the amount of material present, and the properties of the material.* Various BU maintain their physical integrities by balancing these three considerations, some by their shapes, others by deposits of strong materials, and others by using materials with unusual properties. See Sections 2.9 (Fluids) and 2.10 (Solids).
7. *Unbalanced mechanical forces cause acceleration.* If there are no net forces, then a body in motion continues to remain in motion and a body at rest stays at rest. See Section 2.10 (Solids).
8. *Heat is the ultimate nonspecific form of energy.* Low grade waste heat is useless to perform any kind of work. See Sections 2.4 (Work and Energy) and 2.12 (Temperature).
9. *Hydrostatic pressures are equal in all directions.* Strong walls are unnecessary if hydrostatic pressures are equal on both sides. Pressure can distort objects or may support them. See Section 2.9 (Fluids).
10. *Flowing fluids require energy to overcome resistance.* Fluid movement is important in biological systems to supply oxygen, nutrients, and control chemicals, and to remove wastes. The capacity of the heart to move these fluids depends mostly on the ability to overcome resistance, and the configuration of the flow path achieves a balance between high and low resistances. See Sections 2.4 (Work and Energy) and 2.9 (Fluids).

In the following sections, further explanations will be found related to these principles. The understanding of physical constraints is important to understand the nature of biological systems.

2.1 Effort and Flow Variables

Knowledge is more than equivalent to force.

—**Samuel Johnson**

There are two basic kinds of variables that describe the action of a physical system. Effort variables are those things that *cause* an action to occur. Flow variables are the *responses* to effort variables, usually involving movement but not always (Table 2.1.1). For the simple case of a running animal, the effort variable is the force required to propel the animal; the flow variable is the velocity of movement. Heat loss from that same animal, which is the flow variable, occurs in response to a temperature difference, an effort variable. Sexual attraction to an animal of the opposite sex (effort variable) can result in a wide range of activities, including copulation (a flow variable). Hunger (an effort variable) can result in feeding (a flow variable). Thus, there are a wide variety of causes and effects related to biological activity, and these can be thought about in terms of effort and flow variables, which tend to simplify the concepts of biological activities. For any activity of a biological organism or system, searching for the effort variable, the flow variable, and relationships between these two can make it easier to comprehend not only why and how the activity occurs, but also the intensity of the activity.

It is important to note that flow only occurs from a higher level of effort variable to a lower level, never the other way around without some additional energy input. Thus, heat will spontaneously flow from a higher temperature to a lower temperature. If it is desired that heat flows from a lower temperature to a higher temperature, as in an air conditioner, then extra energy must be added to make that happen. As another example, molecules of sugar will always flow spontaneously from a region of higher concentration to a region of lower concentration, never the other way around. Cellular processes may concentrate the sugar, but energy will be expended in the process. Systems where the action cannot occur spontaneously in both directions are called *irreversible*; when flow can occur either way without an additional energy penalty, we call that process *reversible*. From the earlier discussion, the reader would surmise that all processes are irreversible, and, to some degree, the reader would be correct.

TABLE 2.1.1

Various Effort and Flow Variables

System	Effort Variable	Flow Variable
Electricity	Voltage	Current
Fluid	Pressure	Fluid flow
Heat	Temperature	Heat flow
Mass	Concentration	Mass flow
Mechanics	Force	Velocity
Optics	Photon pressure	Photon flow
Magnetics	Magnetomotive force	Flux
Nutrition	Hunger	Eating
Reproduction	Sexual attraction	Seek mate

2.1.1 Resistance

> Understanding is joyous.
>
> —**Carl Sagan**

There are several basic relationships between effort and flow variables that should be introduced. The first is resistance.

Resistance is the ratio of effort to flow variable amounts. Mathematically, resistance is given as

$$\text{Resistance} = \frac{\text{Effort}}{\text{Flow}} \tag{2.1.1}$$

Resistance describes the limit of the flow variable for any given effort variable amount. Without resistance, the amount of flow that resulted from even a small amount of effort would be limitless. Resistance is what limits the spread of disease, the speed of a nerve action potential, and the speed of a bicyclist, just to name a few (Figure 2.1.1).

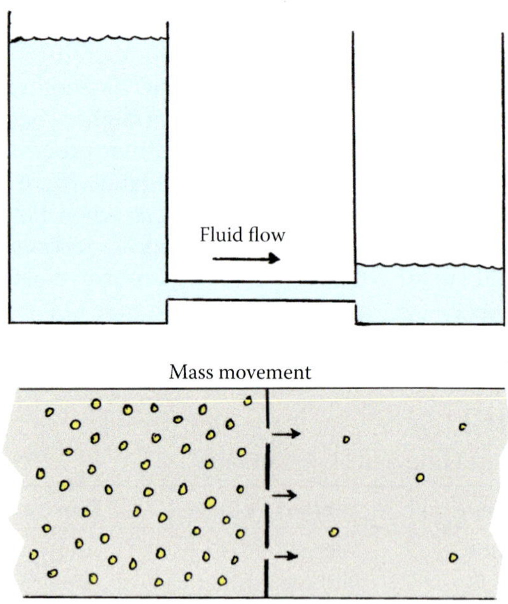

FIGURE 2.1.1
Two cases of effort and flow variables. In the top case, the differences of pressures between the two water tanks cause fluid to flow between them. In the bottom case, molecular concentrations are different between the two chambers, and thus molecules move left to right through a membrane. Flow is limited by the resistances of the pipe and membrane, respectively. The amount of water stored in the tanks and the number of molecules in the chambers represents capacity. When the heights of the liquids in the two tanks are the same, and when molecular concentrations in the two chambers are the same, capacities in the two will be equal and there will not be any effort variable differences. Net flow will then cease, although movement from one chamber to another can still continue as long as it is equal in both directions.

2.1.2 Capacity

> Symmetry is fascinating to the human mind, and everyone likes objects or patterns that are in some way symmetrical. Even the animal and vegetable worlds show some degree of symmetry, although the symmetry of a flower or of a bee is not as perfect or as fundamental as is that of a crystal.
>
> **—Richard Feynman**

Capacity is the ability to accumulate, or store, flow over time. Capacity is given mathematically as the accumulated flow divided by the effort variable value:

$$\text{Capacity} = \frac{\text{accumulated flow}}{\text{effort}} \tag{2.1.2a}$$

In terms of calculus (see Section 4.3, Calculus),

$$\text{Capacity} = \frac{\int (\text{flow})\,dt}{\text{effort}} \tag{2.1.2b}$$

As capacity becomes greater, more stuff can accumulate in a given time for any amount of effort. Thus, more heat can be stored in a body with higher heat capacity, more information can be stored in something with higher information capacity, and a higher population of plants and animals can be found in a region with higher carrying capacity.

2.1.3 Inertia

> One curious result of this inertia, which deserves to rank among the fundamental laws of nature, is that when a discovery has finally won tardy recognition it is usually found to have been anticipated, often with cogent reasons and in great detail.
>
> **—Ferdinand Schiller**

Inertia is the term that describes resistance to a change in flow over time. Mathematically, inertia is given as the amount of effort required to produce a rate of change of flow variable with time:

$$\text{Inertia} = \frac{\text{effort}}{\text{rate of change of flow}} \tag{2.1.3a}$$

In terms of the calculus (see Section 4.3),

$$\text{Inertia} = \frac{\text{effort}}{d(\text{flow})/dt} \tag{2.1.3b}$$

Inertia is what keeps changes from happening instantaneously. Inertia is the reason a person doesn't accelerate immediately from rest to her steady state running speed; inertia is the reason biological fluids don't change speed immediately. Inertia is related to Newton's first law of motion to be described in a subsequent section.

2.1.4 Efforts Required

> For a successful technology, reality must take precedence over public relations, for nature cannot be fooled.
>
> **—Richard Feynman**

We can determine the effort required to accommodate resistance:

$$\text{effort} = (\text{flow})(\text{Resistance}) \tag{2.1.4}$$

Or to fill capacity:

$$\text{effort} = \frac{\text{accumulated flow}}{\text{capacity}} = \frac{\int (\text{flow})\, dt}{\text{capacity}} \tag{2.1.5}$$

Or to overcome inertia:

$$\begin{aligned} \text{effort} &= (\text{rate of change of flow})\,(\text{Inertia}) \\ &= \frac{d(\text{flow})}{dt}(\text{Inertia}) \end{aligned} \tag{2.1.6}$$

We can determine the flow through resistance:

$$\text{flow} = \frac{\text{effort}}{\text{Resistance}} \tag{2.1.7}$$

Or accumulate in capacity:

$$\begin{aligned} \text{flow} &= (\text{rate of change of effort})\,(\text{Capacity}) \\ &= \frac{d(\text{effort})}{dt}(\text{Capacity}) \end{aligned} \tag{2.1.8}$$

Or be affected by inertia:

$$\begin{aligned} \text{flow} &= \frac{\text{accumulated effort}}{\text{Inertia}} \\ &= \frac{\int (\text{effort})\, dt}{\text{Inertia}} \end{aligned} \tag{2.1.9}$$

However, we will see in the next section that effort variables cannot accumulate. Therefore, Equation 2.1.9 is meaningless.

Systems are often made of multiple elements of different kinds. They can be any combination of resistances, capacities, and inertias (Figure 2.1.2). Thus, the effort and flow variable magnitudes are dependent on the exact nature of the system of interest.

- Wanting to drive from one place to another is the *effort variable*. The movement of the automobile is the *flow variable*.
- A narrow winding road offers more *resistance* to traffic than does a super highway. Thus, speed is slower on the narrow road.
- The *capacity* of a parking lot limits the number of cars that can be parked there.
- Moving again after stopping at a traffic light requires extra gasoline to accelerate and does not happen immediately (*inertia*).

FIGURE 2.1.2
A traffic example.

APPLICATIONS AND PREDICTIONS

1. Warm-blooded animals will lose more heat than cold-blooded animals.
2. Fur will act as resistance to limit heat loss or gain.
3. Living things will fill the capacity of the environment to sustain them.
4. Breeding females are a powerful attractant for the males of the same species. The males of other species will be unaffected.
5. Fish must use energy to accelerate themselves (inertia) and overcome the friction (resistance) of the water.
6. No blood will flow unless there is a heart to produce pressure.
7. As long as there is blood flow and resistance, there will be blood pressure.
8. Sodium and potassium ion concentration differences across a cell membrane can lead to movements of these ions during a neural action potential.
9. Animal growth is related to the concentration of growth hormone in the blood.
10. A desire to graduate (effort variable) acts on college students, producing academic work (flow variable), while they must overcome the difficulty of the work (the resistance).
11. Asthma attacks increase airways resistance and decrease flow of air to dangerously low levels.

2.2 Balances

A man must have a certain amount of intelligent ignorance to get anywhere.

—Charles F. Kettering

When there is a defined space with identifiable boundaries, energy or material that flows across those boundaries is related by the balance:

$$\text{rate of stuff in} - \text{rate of stuff out} + \text{rate of stuff generated} = \text{rate of stuff stored} \tag{2.2.1}$$

This balance applies to all flow variables that cross the boundaries. It applies to tigers in the jungle, blood flow through a kidney, water through a fish's gill, or sodium through a cellular membrane. A flow balance may be written for a heterogeneous substance such as waste water, or for a homogeneous substance such as chloride ions. For each of these, if the rate of stuff out is smaller than the rate of stuff in plus the rate of stuff generated, then the extra stuff must be stored inside the boundaries.

If the substance of interest is a chemical compound, for example the enzyme amylase, then an amylase balance does not completely describe the entire story. Other elements and compounds must pass across the boundaries of the producing cells in order that the amylase can be generated inside the cells. Thus, other balances are needed to describe the entire process. In this case, the generation of amylase inside the cell is positive, but the generation of substrates that are used to form amylase is negative.

In calculus notation, the balance in Equation 2.2.1 becomes

$$\frac{d}{dt}(\text{stuff in}) - \frac{d}{dt}(\text{stuff out}) + \frac{d}{dt}(\text{stuff generated}) = \frac{d}{dt}(\text{stuff stored}) \tag{2.2.2}$$

Often, when substance generation can be considered to be uniformly distributed, the rate of stuff generated is given on a unit volume basis. Then the rate of stuff generated is the uniformly distributed rate $q.'''$ times the volume (V).

For a steady state, there can be no change in stored stuff, so the right-hand side of the Equation 2.2.1 must be zero. When that happens, the net efflux of stuff from the volume of interest must be equal to the rate of stuff in plus stuff generated. When, in addition, there is no generation, the rate of stuff in must equal the rate of stuff out.

The balance in Equation 2.2.1 is true at all times as long as the "rate of" is included. This is not the same as saying "stuff in – stuff out + stuff generated = stuff stored," because there may already be stuff stored in the volume of interest. An absolute balance, without the "rate of," is only true if it can be applied from the time where there is no stuff stored in the volume of interest.

This is especially true of an energy balance. Energy is a flow variable and can be treated similarly to other flows such as fluid flow and mass flow. Consider a heat balance:

$$\text{rate of heat in} - \text{rate of heat out} + \text{rate of heat generated} = \text{rate of heat stored} \tag{2.2.3}$$

Without the "rate of" words in this balance, the balance

$$\text{heat in} - \text{heat out} + \text{heat generated} = \text{heat stored} \tag{2.2.4}$$

is not correct unless the heat account began when temperature of the body was absolute zero. Nowhere in the realm of biological science does this ever happen.

Balances written for effort variables are somewhat different from balances written for flow variables. Flow variables can usually accumulate, or become stored, in some volume of interest. Thus, heat can accumulate, and stored heat makes sense; electrical current can accumulate as charge; sodium ions can accumulate, and so can cholesterol.

Effort variables, on the other hand, cannot accumulate. Pressure, voltage, and temperature, all effort variables, are not stored; their corresponding flow variables, fluid, current, and heat are the entities stored. Thus, an effort variable balance lacks the "rate of stuff stored" term in Equation 2.2.1. Since this term represents past activity of the system, flow variable balances exhibit memory whereas effort variable balances are always immediate.

Also, effort variable values are not usually considered to be generated like flow variable values are. Heat, a flow variable, can be generated from frictional work or from metabolism of carbohydrate; temperature is not generated from either, in the same way, but instead appears as a manifestation of the presence of heat. Although we may talk of force generation, force is usually considered to be applied rather than generated, and the "generation of…" term in the effort variable balance equation is usually omitted.

Without the storage term in effort variable balances, one may use either "rate of…" or actual amount for effort balances. Thus, a force balance on an object can be either

$$\text{rate of force in} - \text{rate of force out} = 0 \tag{2.2.5a}$$

or

$$\text{force in} - \text{force out} = 0 \tag{2.2.5b}$$

Sometimes balances given for effort variables and flow variables look somewhat similar, but have an inverted appearance. For instance, using the definitions for resistance, capacity, and inertia already given, an effort balance would be

$$\begin{aligned}\sum \text{applied efforts} - \sum (\text{resistance})(\text{flow}) - \sum \frac{\text{accumulated flow}}{\text{capacity}} \\ - \sum (\text{Inertia})\,(\text{rate of change of flow}) = 0\end{aligned} \tag{2.2.6a}$$

where Σ denotes a summation of all similar terms in field of interest. Given in more mathematical form,

$$\sum \phi - \sum R\omega - \sum \int \omega \frac{dt}{C} - \sum I \frac{d\omega}{dt} = 0 \tag{2.2.6b}$$

where:

ϕ is the general effort variable
R is resistance
ω is the general flow variable
t is time
C is capacity
I is inertia, all in appropriate units

A flow balance would give

$$\sum \text{flows} - \sum \frac{\text{effort}}{\text{Resistance}} - \sum (\text{Capacity})\,(\text{rate of change of effort}) = 0 \qquad (2.2.7a)$$

Because we have recognized that effort cannot accumulate, there is no accumulated effort term in Equation 2.2.7a. If there had been one, it would have been (accumulated effort)/Inertia. In mathematical symbolic form, Equation 2.2.7a becomes

$$\sum \omega - \sum \frac{\phi}{R} - \sum C \frac{d\phi}{dt} = 0 \qquad (2.2.7b)$$

where
- ω is the general flow variable
- R is resistance
- C is capacity
- t is time, all in appropriate units

Notice, in these two balances, the terms are similar, except that "accumulation of…" and "rate of change of…" switch places. Also note that, although efforts cannot be stored or accumulated, there is a storage term in the effort balance. The flow variable is accumulating, but not the effort variable. Also, there is no apparent storage term in the flow balance. The Σ (Capacity)(rate of change of effort) term is actually a storage term. And lastly, note that all terms in the effort balance are given in terms of efforts and all terms in the flow balance assume the units of flow. All terms in each of these equations must have the same units in order to be added together.

Example 2.2.1: Butterfly Balance

The very rare Saint Francis' satyr butterfly (Figure 2.2.1) has an extremely limited range in eastern North Carolina where it is still found. Scientists have worked to understand the habits of this butterfly in order to save it from extinction. They found that caterpillars of

FIGURE 2.2.1
The Saint Francis' satyr butterfly in eastern North Carolina.

the butterfly fed on a particular sedge plant, but that the population of that plant species was shrinking as wetlands where it grew were drained for other purposes. They set out to restore the wetlands where these sedges can grow. When they did so, they found that they had also created habitat for predators of the butterfly. Birds, dragonflies, and other insects thrived in the new wetlands. The result is that, although the butterfly reproduction rate could increase with greater food availability, so could the attrition of butterfly numbers. On balance, the wetland restoration to encourage Saint Francis' satyr butterfly population growth was met with only limited success (Burke, 2015; Aschehoug et al., 2015).

Example 2.2.2: Growing Trees Store Carbon

Carbon dioxide (CO_2) levels in the atmosphere have reached historic levels, and, due to the ability of that gas to reflect long-wave radiation that otherwise would have escaped into space, contributes to warming of the Earth. Carbon dioxide is a product of combustion, so human activities have contributed a large proportion of the carbon dioxide increase.

Removal of CO_2 from the atmosphere has become a goal in order to reduce global warming. One means to do this is to plant more trees, because trees convert CO_2 into carbon-based compounds such as cellulose and lignin. This is called *carbon sequestration*.

Young, growing trees can remove significant amounts of CO_2 from the atmosphere and lock it up in woody tissues. Older trees, however, die and rot, releasing their carbon back into the atmosphere. Old-growth forests that are populated with both young and old trees are in equilibrium, with younger trees sequestering carbon and dead trees releasing it. Except for perhaps seasonal fluctuations in carbon sequestration as leaves either grow or fall, there is a CO_2 balance in an old-growth forest (Figure 2.2.2).

The means to locking up carbon is through selective tree harvesting and use. Removing an older tree from a forest and using it, possibly for wood so that it does not decompose, allows a new tree to take its place and begin to sequester its own carbon.

This is what a carbon balance tells us: trees remove CO_2 from the atmosphere and it can only be permanently removed if the carbon is retained in its chemically locked form.

FIGURE 2.2.2
Forests such as these can store carbon from the atmosphere as long as trees are actively growing faster than trees in the forest are dying.

APPLICATIONS AND PREDICTIONS

1. The amount of water stored in a wetland can be determined by accounting for the inflows and outflows. Water is not generated in a wetland, although water does evaporate (negative generation).
2. Cardiac output in a human can be determined using a mass balance on a tracer substance in the blood.
3. If the pressure difference between mouth and respiratory alveoli remains constant, then respiratory airflow will decrease to zero as the lung fills.
4. The time required to remove heavy metals from the soil can be determined using a balance on heavy metal ions.
5. The cellular transmembrane potential can be determined from a charge balance on ion concentrations inside and outside the cell.
6. The rate of oxygen perfusing the tissue can be obtained from the rate of oxygen entering the capillary bed minus the rate of oxygen leaving the capillary bed.
7. Weight gain is related to food intake minus the indigestible part.
8. Energy is conserved.

2.3 States of Matter

> We are an intelligent species and the use of our intelligence quite properly gives us pleasure. In this respect the brain is like a muscle. When it is in use we feel very good.
>
> **—Carl Sagan**

Matter important to biological systems can exist as solids, liquids, or gases. Each of these plays important roles in living systems.

2.3.1 Gases

> All I need is the air that I breathe and to love you.
>
> **—The Hollies**

Gases are the most energetic of the states of matter (Figure 2.3.1). Their molecules travel relatively long distances between collisions with other molecules, and, as a consequence, are relatively unconfined. Gases have low densities, and so must be used in relatively large volumes in order to result in modest amounts of mass. For example, air is about 700 times less dense than water. Thus, to obtain enough mass of oxygen to support metabolism requires a volume of about 700 times more than if the air had the density of water. In addition, the density of gases is temperature dependent, and the higher the temperature, the less dense is the gas.

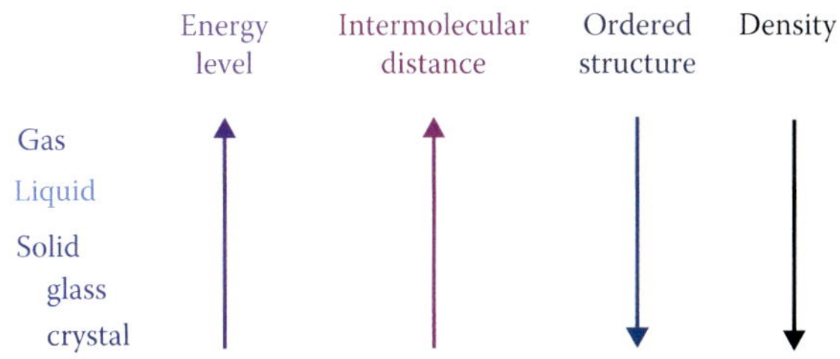

FIGURE 2.3.1
General properties of various states of matter.

The ideal gas law expresses the relationships among variables important to gases:

$$\text{(pressure)(volume)} = \frac{\text{(gas constant) (mass) (absolute temperature)}}{\text{(molecular mass)}} \tag{2.3.1}$$

Absolute temperature in Equation 2.3.1 is referenced at absolute zero, where gases would exert no pressure on their environment. Instead of degrees Celsius or Fahrenheit, degrees Kelvin or Rankine must be used. The molecular mass is the number of kilograms, grams, or pounds-mass that equals the equivalent of the atomic weight of the gas. The actual mass of the gas divided by the molecular mass gives the number of moles of the gas (see Section 3.1, Periodic Table). The gas constant is a measured quantity that completes the equation. It has different values depending on the system of units used.

Because density is mass per unit volume, gas density is

$$\text{density} = \frac{\text{mass}}{\text{volume}} = \frac{\text{(pressure) (molecular mass)}}{\text{(gas constant) (absolute temperature)}} \tag{2.3.2}$$

From this we can see that gas density depends on pressure (in other words, the gas is compressible) and on absolute temperature. At very high pressures or very low temperatures, the density of a gas can approach or even exceed that of a liquid, and it is then difficult to distinguish the two states based on physical properties. However, such high pressures or low temperatures would be lethal to life, so we will not be concerned about these conditions here.

2.3.2 Liquids

> Great works are performed not by strength but by perseverance.
>
> —**Samuel Johnson**

The second state of matter is a *liquid*. Liquids are fluid, just like gases, but, unlike gases, they are nearly incompressible and maintain the same volume throughout a wide range of pressures and temperatures. There is no equation of state for liquids like Equation 2.3.1 for gases, but if there were, it would probably be something like

$$\text{density} = \text{constant} \tag{2.3.3}$$

Liquids are less energetic than gases, meaning that intermolecular distances are significantly less in a liquid. When energy, especially in the form of heat, is added to a liquid, the liquid molecules acquire some of this heat and change into a gas. This phase change is called evaporation or vaporization, and the added heat is called the heat of vaporization. The temperature of the liquid does not change during the transition from liquid to gas, as it does when heat is added to the liquid below the transition. The heat contributes solely to the change of state.

Vaporization can occur spontaneously at any temperature for some of the liquid molecules. This is because in any given population of liquid molecules, there is a distribution of energy levels. Molecules with higher levels of energy can attain enough energy to escape the liquid surface if they are involved in collisions with other energetic molecules. The presence of gaseous molecules above the liquid surface exerts a pressure known as the *vapor pressure* for that substance. Vapor pressure increases with temperature and only very slightly with total pressure. When the vapor pressure reaches the value of atmospheric pressure, the liquid is said to *boil*, which means that the average energy level of all liquid molecules is enough for them to pass freely into the gaseous state. Thus, boiling temperature depends on local atmospheric pressure, and will generally be higher at lower elevations where atmospheric pressure is higher.

2.3.3 Solids

> The proper role of the engineer is in problem simplification, evaluation, and design.
>
> **—Edwin N. Lightfoot, Jr.**

The least energetic state of matter is the *solid*. There is so little space between molecules that solids are mostly rigid and not easily penetrated. The solid-state density is therefore usually the highest of the three states of matter. A notable exception, however, occurs in the case of water, where the crystal structure of ice is less dense than non-crystalline liquid water. This is fortunate, because a layer of ice floating on liquid water allows aquatic life to be sustained in the liquid, and the ice acts as an insulating barrier to trap heat in the liquid below.

The solid state is really not just one state, but many. Certain solids form *crystal* structures in regularly repeating arrangements of atoms. Crystals are often very hard, somewhat brittle, tend to cleave along certain planes, and are either transparent or translucent. This latter property is due to the fact that, despite the dense packing of atoms, there is still enough space between atoms to allow light to pass unimpeded.

Other solids form *glasses*. These are the irregular arrangements of atoms that have no particular structure, but are densely packed, often more densely packed than are crystals of the same material. Glasses are very slightly fluidic in nature; they respond to external forces and, over long periods of time, can assume different shapes in response to these forces.

Metals are closely packed arrangements of similar atoms that share electrons rather loosely. Metals are often lustrous or reflective. They are ductile and may be bent without breaking. They are often excellent conductors of heat and electricity, in contrast to crystals or glasses that usually are poor conductors.

Polymers are chains of (usually) organic building blocks that exhibit plastic properties. Polymers have particular strength in one direction, along the axis of the chain. When these chains intertwine, strength is imparted in all directions.

Composites are mixtures of several materials that can be used to exploit the advantages of each. Many modern materials are made of fibers (for instance, glass or carbon fibers) embedded in a glassy epoxy matrix. The fibers add tensile strength and the matrix adds compressive strength. Steel reinforced concrete acts similarly. There are many composite materials in living systems, because not only are there strength and weight advantages to composite materials, but there are also certain materials that trap ions, repel water, pass lipid-soluble compounds, insulate against heat, allow light to pass, and have other useful properties. In fact, the entire human body itself can be considered a composite material, with its internal skeleton adding strength and rigid structure, and other tissues adding other properties.

2.3.4 Gas Plasma

> The word "scientist" did not even exist prior to 1840; it was coined by the English scholar William Whewell.
>
> **—Hal Hellum**

A gas plasma is sometimes called the fourth state of matter. Gas plasma consists of gas ions surrounded by a cloud of highly energized electrons, and is produced when a gas under a partial vacuum is subjected to an electrostatic or electromagnetic field. Gas plasmas have many uses, including imparting particular properties to surfaces (Winter, 2006).

2.3.5 Phase Changes

> Facts do not cease to exist because they are ignored.
>
> **—Aldous Huxley**

Changing from the solid state to the liquid state is called *fusion* or *melting*, and from liquid to solid is called *freezing*. Because both liquids and solids are less energetic than liquids and gases, the amount of heat required to melt a solid is normally less than the amount of heat required to evaporate a liquid. This state transition is similar to the liquid-to-gas transition in that it occurs at a constant temperature for a pure substance.

Liquid solutions usually freeze at lower temperatures than the pure liquids, with the *freezing point depression* (for aqueous solutions) being directly related to the amount of solute dissolved. Certain proportions of two metals often freeze at a temperature much lower than for other proportions. The proportions of these two metals that freeze at the minimum temperature is called the *eutectic* mixture. One useful eutectic mixture is tin/lead in a 60/40 mixture. Pure lead melts at 327°C; pure tin melts at 232°C. The 60/40 mixture, called solder, melts at 190°C, lower than for either metal alone. For this reason, 60/40 solder is often used for constructing electric circuits.

Some solid molecules can attain enough energy to change directly from solid to gas. Such a transformation is called *sublimation*. Because of sublimation, there is a vapor pressure of the material that is present over a solid. The water vapor pressure of ice is normally considered to be dependent only on temperature. The heat of sublimation for ice to steam is slightly greater than the sum of the heat of fusion for ice to water plus the heat of vaporization from water to steam.

State change is not the only possibility with solids. There are also phase changes that occur under different conditions of temperature and pressure. Phase changes may be crystal to glass, or one crystalline structure to another crystalline structure. When these occur, there are definite changes of the physical properties of the material, and heat may either be evolved or absorbed.

APPLICATIONS AND PREDICTIONS

1. The air handling respiratory system of the body must occupy a larger volume than the liquid handling circulatory system.
2. If the respiratory and circulatory systems were organized similarly (both either in a closed circuit or reciprocating), the circulatory system would require more energy.
3. Cold air will expand as it enters the lungs.
4. Crystalline ice formed when water freezes will disrupt cellular structure.
5. Sugar that forms glass will have properties different from sugar that forms crystals.
6. The evaporation of sweat from skin can be used to cool the body.
7. The composite structure of a cell membrane allows it to be selectively permeable.
8. To cool something without producing a liquid, use dry ice.
9. Blood will boil if the atmospheric pressure is low enough.
10. Cooking food by boiling at high altitude takes longer than at low altitude.

2.4 Equivalence of Work and Energy

> An experiment is a question which science poses to Nature, and a measurement is the recording of Nature's answer.
>
> **—Max Planck**

2.4.1 Work

> At the simplest level, we can consider the body as a thermodynamic system absorbing energy from its environment and in turn releasing heat and doing mechanical work.
>
> **—Edwin N. Lightfoot, Jr.**

When speaking of work, we often mean *mechanical* work, where something is moved from one place to another. The mechanical work to move this thing is calculated as

$$\text{work} = (\text{force})\,(\text{distance moved}) \tag{2.4.1}$$

The force may be that required to overcome friction, including air resistance, or to accelerate the body, or to move it against the pull of gravity, for instance. The distance moved is determined between any two points while the force was being applied. Because the force may not be constant during the movement, and the movement itself may be sporadic, mechanical work depends on the way in which the force and distance happen together. When force is not constant, the total amount of work is the sum of small increments during

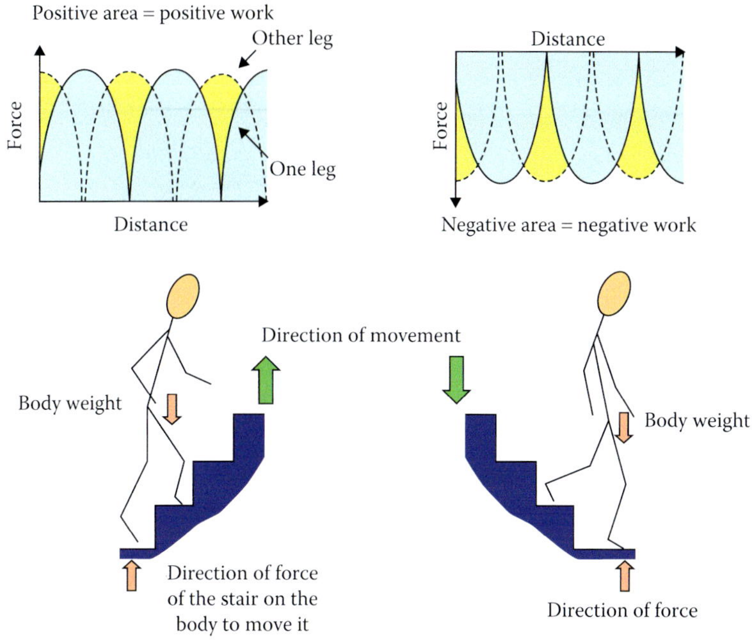

FIGURE 2.4.1
Diagram of positive and negative work. The leg muscles produce a force acting on the stairs, but it is the difference between the stair reaction force and body weight that determines the direction of movement. When the resultant force and direction of movement are the same, there is positive work done by the body. When the resultant force and the direction of movement are in the opposite direction, the body does negative work.

which force can be considered constant times the distance moved during that increment. Thus, work is considered to be a *path function*, the value of which depends on the particular set of increments from start to finish.

There are several polarities of physical work (Figure 2.4.1). If the system does work on its surroundings, for example, then force is headed in the same direction as the movement, and the work is said to be positive. A muscle lifting an object, and the object being moved in the direction of the force being applied, produces positive work. Climbing upstairs produces positive work because the weight of the body is being lifted by the leg muscles in the same direction as the movement.

If the force is not in the direction of the movement, then the surroundings are said to do work on the system, and the physical work is considered to be negative. Walking down stairs produces negative work because the leg muscles support the body weight with an upward force, but the direction of movement is downward. Negative work is usually encountered because the organism wishes to maintain control. If the leg muscles did not exert an upward force while walking down stairs, then the body would accelerate uncontrollably, probably ending in a catastrophe at the bottom of the stairs.

If positive work results from a force in the direction of the movement, and negative work results from a force opposite to the direction of movement, what of a force applied at right angles to the movement? In this case, no work is done: no positive work and no negative work.

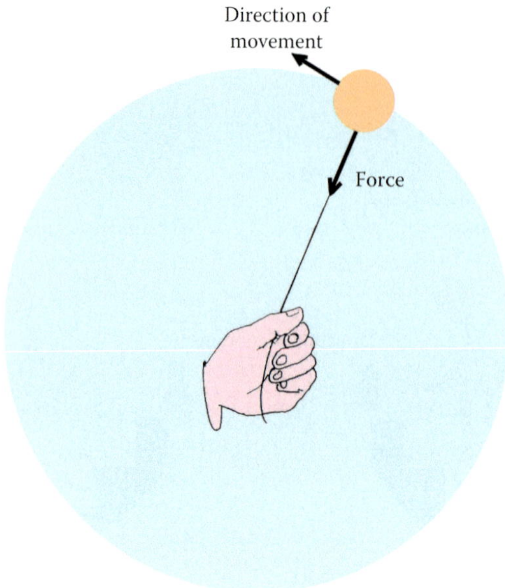

FIGURE 2.4.2
Weight on a string. There is theoretically no work done if the directions of force and movement are at right angles.

If a weight attached to a string is swung around in a circle, there is a force exerted on the weight that is at right angles to the direction of the movement of the weight (Figure 2.4.2). No work is done by the string on the weight. So, if there is no work done by the string on the weight, why does the weight accelerate to its final speed? In order to accelerate the weight, the string cannot be entirely straight, and there is a component of force in the direction of the movement of the weight. This component of force in the direction of the movement of the weight causes work to be done on the weight. When the weight reaches its final speed, no additional work is done by the string on the weight, except, perhaps, a small amount to overcome air resistance. Again, the string cannot be perfectly at right angles to the direction of the weight.

2.4.2 Energy

> I was seldom able to see an opportunity until it had ceased to be one.
>
> —**Mark Twain**

Energy is defined as the ability to perform useful work. Thus, energy can be converted into work, and vice versa. There is thus an equivalence between work and energy.

Energy can be classified as either *kinetic* energy or *potential* energy. Kinetic energy is the energy of motion. It is present as either linear motion or rotating motion or some combination of the two. When a wheel rotates on its axis, it contains rotational kinetic energy. If it rotates in a frictionless environment, then it does no work on the environment, and the environment does no work on the wheel. But, if the wheel is caused to slow because it comes in contact with another member, the wheel can do work. Thus, a rotating wheel can be used to sharpen a knife, or to heat a brake pad, or to generate electric current that drives an electric motor. Doing these things, however, slows the wheel as it converts energy into work.

The same wheel rolling down a hill has both rotational and translational (linear) kinetic energy. If that wheel hits another object, it can deform the other object by translating its energy into work. It may not be useful work, but it is work nonetheless.

Translational kinetic energy is given as

$$\text{kinetic energy} = \frac{1}{2}(\text{mass})\,(\text{velocity})^2 \tag{2.4.2}$$

and, therefore, the energy of a moving object quadruples as its velocity doubles. A collision between a moving object and a stationary object is thus much worse as the speed of the moving object increases.

Rotational kinetic energy is calculated similarly, but the velocity of a moving body increases in proportion to the distance between the body and its center of rotation. Thus,

$$\text{kinetic energy} = \frac{1}{2}(\text{mass})\,(\text{radius})^2(\text{angular velocity})^2 \tag{2.4.3}$$

where the product of the radius (or distance from the center of motion) and the angular velocity is analogous to the translational velocity in Equation 2.4.3.

The total energy of a composite mass is the sum of the translational and rotational kinetic energies of its components. Thus, the kinetic energy of a sprinting person is the translational kinetic energy of the body plus the translational and rotational kinetic energies of the arms and legs.

At the atomic level, the random movement of atomic particles is another form of kinetic energy called *heat*. This kinetic energy can be transferred to another object as long as the temperature of the donor object is higher than the temperature of the acceptor object (that is, from higher potential to lower).

Heat is thus a form of kinetic energy that is convertible into work. Consider this: heat can be used to boil water and turn it into steam. This steam can push a piston and force it to move in the direction of the push. The piston can be connected to a rod which then performs useful mechanical work. Heat energy is equivalent to work.

Another type of energy is potential, or stored, energy. Potential energy can take the form of a weight on a high place. If the weight were to fall, its potential energy would be converted into kinetic energy, and it could do work on its environment. Thus, the weight used to power a grandfather clock falls throughout the day, and as it falls, it loses potential energy. The energy is transferred to the gears, pendulum, and hands of the clock.

Potential energy can take many forms. In the case of the weight poised on some high place, the potential energy is mechanical. If the potential energy is in the form of a hot object, which could transfer its heat to a cooler object if contact is made with one, then the potential energy is thermal. Likewise, we could have potential energy as pressure, electrical charge, or a magnetic field.

One important form of potential energy important in biological systems is chemical potential energy. Energy in the form of chemical compounds is used to power all biological systems. We call these compounds *food*.

FOOD TRANSFORMATIONS

The process of digestion is used to decompose very complex organic compounds into simpler ones. This is necessary for the food to pass into the interior of the organism where it can be used. Otherwise, the intestines would need to be permeable to a very large number of digested ingredients.

Catabolism is the reduction of complex molecules into simpler ones. *Anabolism* is the formation of complex substances from much simpler chemical building blocks. Catabolism takes place during digestion and before absorption; anabolism takes place in the body after digestion and absorption.

Both of these processes generally require energy to occur, and so usually proceed against an effort variable gradient. The energy necessary to process food comes from the food itself, and is generally not available as energy useful to the organism.

2.4.3 Efficiency

> Important principles may and must be flexible.
>
> —**Abraham Lincoln**

The ratio of useful energy to required energy defines *efficiency*:

$$\text{efficiency} = \frac{\text{useful energy}}{\text{required energy}} \tag{2.4.4}$$

The difference between required energy and useful energy is the energy that is not able to be used effectively. This energy usually ends up as waste heat. Muscle tissue, for instance, produces physical work, but it also produces a great deal of heat, a fact that was used to demonstrate the conservation of energy (Fick, 1881). Thus, very inefficient processes produce more waste heat than do efficient processes.

Heat is the lowest common repository for energy (Figure 2.4.3). At each step in a process, there is some inefficiency. Thus, some heat is produced. This heat can be used to produce work as long as it is at a high enough temperature, but, in order to produce work, there must be a lower temperature reservoir to accept the heat resulting from the conversion of heat to work. Hence, heat moves continually from a higher temperature to a lower temperature at each step in the process. At some point, there is not enough energy remaining in the heat reservoir, and so additional work of a meaningful amount cannot be accomplished.

In this entire discussion, we have implied that energy could be converted from one form to another with some degree of equivalence. Thus, kinetic energy can be converted into potential energy and vice versa. Electrical energy can be converted into mechanical energy that can be converted into heat energy that can be converted into chemical energy, etc., and each of these can be converted into work either done on the environment or by the environment. The conversion is exactly equivalent for a reversible process that is 100% efficient and less exact for irreversible processes less than 100% efficient.

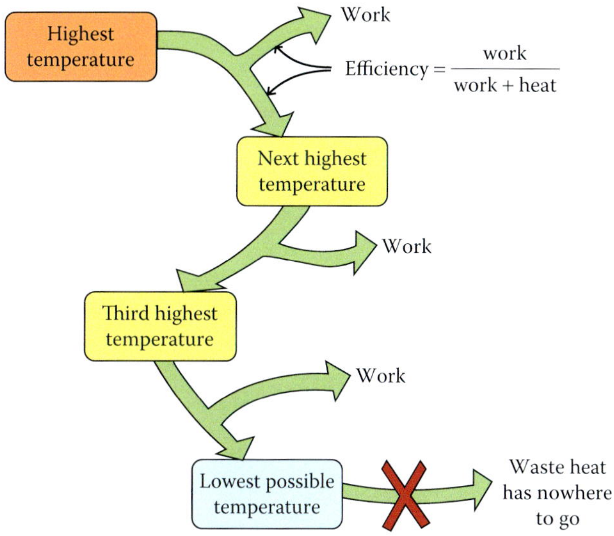

FIGURE 2.4.3
Diagram of heat from high temperature to waste. When heat moves from a high to a low temperature, it can generate work; more for a larger difference between the source and the temperature sink. When there is no lower temperature sink, the heat can no longer flow, and work is no longer produced.

Another way of expressing this is the conservation of energy, which states that "The total amount of energy in a closed system remains constant."

The closed system we are talking about is one in which energy and mass do not pass across its boundaries. As long as the system is closed, the sum of all energies, including mechanical, kinetic, and potential energy; chemical potential energy; heat energy; and others, must be constant. Energy can be neither created nor destroyed.

Thus, when a muscle produces useful work, it is usually no more than 25%–30% efficient (see Section 7.4.9, Walking and Running), and this efficiency applies to large muscles producing large forces. Smaller muscle movements are usually less efficient because they are usually required to be well controlled. This control is achieved by *antagonistic muscles* that produce forces against the forces produced by the primary muscles. In this system, with some muscles producing positive work, and other muscles producing negative work, efficiencies in the range of 0%–5% are common.

Metabolism is the sum of conversion processes whereby useful energy is produced, chemical compounds are changed, and the organism is maintained in a healthy state. Thus, metabolism includes anabolism, catabolism, and the conversion of chemical energy to muscular work, among other things. The metabolic cost of useful work is often called *physiological work*, which is equivalent to the required energy term in Equation 2.4.4.

Chemists have traditionally used calories as the unit of chemical energy. Nutritionists have used kilocalories, or 1000 calories, to express food energy. Ergonomists have used Joules as the metric unit of work. Because energy forms are interchangeable, these units can each be converted into the others, allowing us to calculate the amount of muscular work able to be theoretically obtained from a hamburger.

Summarizing this section

1. Work equals force times distance.
2. Work and energy are equivalent.
3. Different forms of energy can be converted from one to another.
4. Kinetic energy is the energy of movement and potential energy is stored energy.

BIOMASS CONVERSION EFFICIENCY

Except for a few extremophilic ecosystems deriving their initial energy supplies from chemicals, sunlight is the ultimate source of energy. Plants convert solar energy into biomass with a photosynthetic efficiency of about 1%–2%. Primary consumers (herbivores) eat the plant biomass and digest about 50% of the energy contained in their food. The remainder is passed as feces. Of the 50% incorporated into the body, two-thirds is utilized for cellular respiration (maintenance of cellular function) and one-third is used for growth (Campbell et al., 1999). Thus, only one-sixth of the ingested biomass is converted into biomass of the primary consumer. Of course, the feces remain in the ecosystem and can be utilized by other organisms. Contrarily, the formation of new biomass becomes less as the consumer reaches maturity.

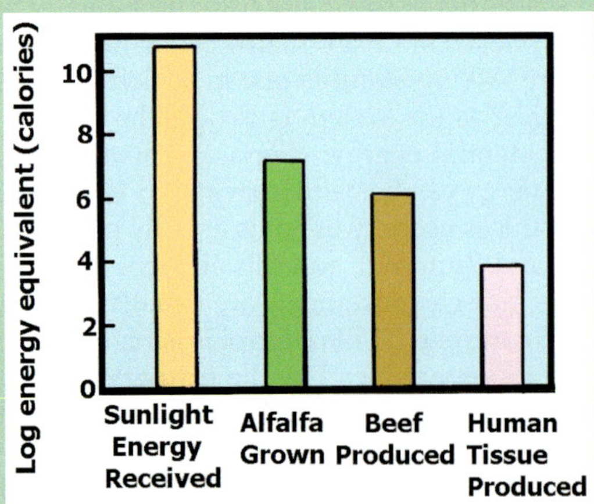

Total energy is reduced at each step in a process. (Redrawn from Haynie, D.T., *Biological Thermodynamics*, Cambridge University Press, Cambridge, UK, 2001.)

Secondary consumers eat the primary consumers and produce feces waste, biomass, and cellular respiratory products. Respiration totally degrades the biomass, usually producing water and carbon dioxide. These compounds cannot be further exploited for their energy values and so are the ultimate energy sink for an ecosystem.

Considering other losses and wastes, an average of about 10% biomass-to-biomass conversion efficiency is usually assumed for each level of consumption.

APPLICATIONS AND PREDICTIONS

1. All living systems will require energy sources.
2. All living things will generate heat.
3. Energy stored in one species can be used by another.
4. Energy used for growth cannot be used for reproduction.
5. Energy used for defense cannot be used for reproduction.
6. Death and waste products are useful to provide energy for other living things.
7. Athletes who train hard will need more food energy than sedentary people.
8. Animals will usually be more active during breeding season. Thus, they will either need to eat more during that time or they will need to have enough stored energy to use.
9. The same energy cannot be used twice.
10. Cold-blooded animals will absorb heat energy from their surroundings.
11. Living things have different modes of energy absorption, generation, and transformation depending on their evolutionary environment.
12. Hibernation saves energy.

2.5 Free Energy

I would rather be attacked than unnoticed.

—Samuel Johnson

Biological systems do not consume energy; they transform it. Energy available as food for animals, sunlight for plants, and energy-rich compounds for microbes is transformed from whatever form it is at the source into a form more useful to the living system. There is always waste (up to 90% or more) in these transformations.

However, many energy transformations never happen spontaneously. Cold objects don't heat themselves, blood does not rush throughout the veins and arteries without a heart to pump it, and dust on the floor does not self-organize into a beautiful plant. Some things just don't happen by themselves.

Why is this so? It was explained in a previous section on effort and flow variables (Section 2.1) that flow always occurs from points of higher effort to lower, never the other way around. Perhaps that explanation is sufficient. We usually have enough experience with real effort variables (pressure, gravity, temperature, and so on) to expect that the impossible just won't happen. And it doesn't.

Thermodynamics is the science of energy transformations, and thermodynamicists have a different method they use to explain the spontaneity of energy transformations. They call it *free energy*, and the sign of free energy determines whether or not the transformation will proceed; the amount of free energy determines whether or not the transformation will take place with gusto.

Because engineers and technologists will encounter the concept of free energy in their classes, readings, and conversations, free energy will be explained briefly here. Let it be noted that free energy is an explanation for the spontaneity of energy transformations, and, as such, it stands alongside the explanation that (for example) flow occurs from points of higher potential to lower. In concept, neither of these explanations is superior to the other. In fact, free energy is often thought of as an effort variable. Free energy, however, is more highly developed mathematically, and can be used to give quantitative estimates of energy transformations.

There are several definitions to be presented, and these can be quite confusing the first time they are encountered. This is natural, and should not be taken as an excuse to forget this whole approach. With time and acceptance, you will probably find yourself at ease with the free energy approach.

2.5.1 Internal Energy

> Time: You have written more than 45 books. What has driven you to be so prolific?
> William F. Buckley: The fear that the enemy will write more than I do.

The first definition is that of *internal energy*. This is the energy within the boundaries of a system. Because the boundaries can usually be drawn anywhere we want them to be—although drawing them some places may make the enclosed system easier to deal with than drawing them differently—the amount of internal energy depends upon the locations of the boundaries.

Energy is the capacity of the system to perform mechanical work. Heat can be transformed into work. So, in a physical sense, the amount of internal energy is changed by adding either heat or work. By convention, the internal energy of a system will increase either by delivering heat to it or by doing work on it (Haynie, 2001). The First Law of Thermodynamics expresses this by

$$(\text{change in internal energy}) = (\text{heat added}) + (\text{work done on system}) \qquad (2.5.1)$$

Internal energy is a *state function* of a system, which means that internal energy differences can be calculated by subtracting values at any two points without knowing about events happening between the two points. If such knowledge were necessary, it would be called a *path function* (as we have seen in Section 2.4, Work and Energy).

2.5.2 Enthalpy

> There is a mask of theory over the whole face of nature.
>
> —**William Whewell**

The next definition we need is that for *enthalpy*, which is the heat absorbed by a system at constant pressure. Mathematically, enthalpy is given as

$$(\text{enthalpy}) = (\text{internal energy}) + (\text{pressure})\,(\text{volume}) \qquad (2.5.2)$$

You may recognize (pressure) (volume) as the mechanical work performed in a system where pressure and volume can change. By the convention stated previously, the work is negative if it is done by the system. Thus, Equation 2.5.1 becomes

$$(\text{internal energy change}) = (\text{heat added}) - (\text{work done by system}) \qquad (2.5.3)$$

and Equation 2.5.2, for constant pressure, becomes

$$\begin{aligned}(\text{enthalpy change}) &= (\text{internal energy change}) + (\text{pressure})\,(\text{volume change}) \\ &= (\text{internal energy change}) + (\text{work done by system})\end{aligned} \tag{2.5.4}$$

The enthalpy of a system at constant pressure changes by an amount equal to the added heat.

2.5.3 Entropy

> There is nothing more difficult to take in hand, more perilous to conduct, or more uncertain in its success, than to take the lead in the introduction of a new order of things.
>
> **—Niccolò Machiavelli**

Entropy is a mathematical measure of disorder. Living systems are highly ordered, if nothing else. Hence, living systems will have a negative entropy. When the living thing dies, and its elements scatter to the environment, its entropy increases.

Because entropy was defined first within the science of thermodynamics, its calculation is traditionally given in terms of heat. We know that, due to the kinetic energy imparted to the particles of an object, the state of disorder of that object increases as heat is added. We also know that the state of disorder is smaller at lower temperatures. Adding heat to an object at lower temperature can increase its disorder more than adding the same amount of heat to a hotter object. This is the basis for the calculation of entropy as (Haynie, 2001)

$$(\text{entropy change}) \geq \frac{(\text{heat added})}{(\text{absolute temperature})} \tag{2.5.5}$$

Notice that there is an inequality in Equation 2.5.5. This means that the minimum amount of entropy change that occurs in a process is given by the heat added divided by temperature. This only happens for a *reversible* process.

2.5.4 Gibbs Free Energy

> Disorder increases with time because we measure time in the direction in which disorder increases.
>
> **—Stephen Hawking**

Gibbs free energy is defined mathematically as

$$(\text{Gibbs free energy}) = (\text{enthalpy}) - (\text{absolute temperature})\,(\text{entropy}) \tag{2.5.6}$$

This equation can be interpreted to mean that the free energy of a system is equal to the capacity of the system to do work or to add heat (given by enthalpy), minus the waste energy that cannot be used (given by (temperature)(entropy)). If the Gibbs free energy is negative, the process will proceed spontaneously; the environment can supply the energy required to drive the process to completion. If the Gibbs free energy is positive, the process will not proceed spontaneously; additional energy must be supplied from another source for the process to proceed (Figure 2.5.1).

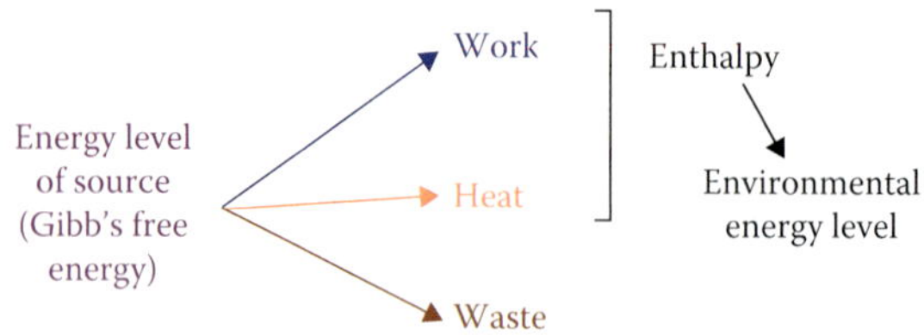

FIGURE 2.5.1
Gibbs free energy consists of the ability of a system to do useful work on the environment or to add heat to the environment. If the Gibbs free energy is negative, the environment will be able to affect the system and the process will thus happen spontaneously.

Physical parameters involved are heat, mechanical work, and the state of disorder (or structure). Any of these can be used to drive a process to completion. Many experimental measurements have been made to quantify the components of Gibbs free energy. Thus, calculations can indicate to the engineer if the process can be expected to yield energy or require energy.

Example 2.5.1: Controlling Hydraulic Energy in Streams and Near Shorelines

Moving water carries with it a lot of energy that can damage stream banks and lake or ocean shorelines through erosion. Natural settings can accommodate the actions of waves or flowing water by dissipating the extra energy that would normally dislodge and carry away soil or sand particles. Biological engineers with special knowledge of environmental techniques attempt to control erosion by returning a stream bank or ocean shoreline to as close to a natural state as possible. In most cases, this means reestablishing gentle slopes and growing native plant species as close to the shoreline as realizable. Root systems of native plants are rugged enough to resist wave action and retain soil particles, while their foliage and branches reduce erosion caused by rainfall and winds.

Besides the advantages of erosion prevention, there are also the ecological benefits of allowing native vegetation to stabilize river or stream banks and shorelines. Fish and frogs often spawn in the silt at the bottom and near the shore; vegetation provides nesting spots for birds and food for insects, waterfowl and aquatic mammals; fallen logs and branches provide shelter and hunting areas for fish and mammals, while turtles use them to sunbathe; turtles and amphibians have free access to land; the shoreline's natural vegetation acts as a filter, preventing sediment, fertilizers, and pesticides from entering the water body.

APPLICATIONS AND PREDICTIONS

1. Water will freeze spontaneously at absolute temperatures less than 273 K. Therefore, the Gibbs free energy of freezing water will be negative.
2. Enzymes will help to transform other biomolecules. Thus, enzymes will cause the Gibbs free energy to be negative.
3. The Gibbs free energy for me to run a 4-minute mile will be positive.
4. Hydrolyzing ATP to ADP + P occurs spontaneously in the power stroke of muscular contraction.

2.6 Disorder and Entropy

> Description of "vuja dé": I've never been here before and it does not look at all familiar.
>
> —**George Carlin**

The consequence of the fact that a flow can never spontaneously occur against an effort variable gradient is very profound for biological systems. Any biological system, from the subcellular to the biomic levels, represents an ordered system. That is, particular structures are maintained, particular activities are maintained, and particular relationships are maintained. Without these, living systems are unsustainable.

However, we know the energy level is degraded at every step of a process because energy can never spontaneously flow from a lower potential to a higher potential. That is, total available energy always becomes less at each stage (see Section 2.4). Without sufficient energy, biological structures, activities, and relationships cannot be maintained, and the living system dies.

The fact that a flow can only occur spontaneously from a higher to lower effort variable value means that

1. Electric current never flows from a lower voltage to a higher voltage,
2. Heat never flows from a lower temperature to a higher temperature (second law of thermodynamics),
3. An object never moves against a net applied force,
4. Energy is never created spontaneously,
5. A solution never becomes more concentrated,
6. A fluid never flows against a pressure gradient,
7. Information never spontaneously arranges itself into books (this one took energy!),
8. Dark objects never transmit light to brighter objects,
9. A cell never reproduces without energy input,
10. Life is not spontaneously created,

and others. However, despite these statements, each of them can be made to reverse if sufficient energy of the correct form is added to the system under consideration.

Living biological systems represent order. Proteins are synthesized in cells, and stable biomes depend upon a stationary balance of species. Dead biological systems quickly degrade to simple elements and molecules, and after a time, we are incapable of recognizing these as coming from any structure, engaging in any biological activities, or sustaining any meaningful relationships.

Biological systems seem to contradict the notion that disorder is the most natural state of things. Biological systems, after all, are stable and sustaining. They concentrate material species against density gradients, they separate random external environments from ordered and complex interiors, and they act in complex, but mostly predictable, manners. And these patterns have been sustained for eons.

The only way that living systems can maintain their states of order is if there is a continuing flow of energy into each and every biological component (Figure 2.6.1). Biological systems, then, are net absorbers of energy. Thus, photosynthetic plants need sunlight energy, herbivores need plant energy, and carnivores need herbivorous energy.

FIGURE 2.6.1
As the state of order increases, entropy decreases. Entropy of the entire universe increases constantly as free energy is consumed, but biological systems maintain a low entropy, and high state of order, by consuming energy from the environment. (Redrawn from Harris, S., *Am. Sci.*, 89, 2001. Copyright credit to: ScienceCartoonsPlus.com, with permission.)

Viewing this from a grand scale, therefore, each energy transformation that occurs within or between biological systems must result in less available energy overall, with the net loss occurring in the environment. In the simple example given previously, the sun is a net energy donor, and loses energy at a higher rate than biological systems gain. If a living system were to be frozen to absolute zero, the entropy contained in the system would still not be zero, because the orderliness of its structure would remain. The structures of its compounds, and the way they are arranged is an important form of information contained in the system (Haynie, 2001). When the biological unit is allowed to thaw, no new structural information is added; in fact, some information is lost because the heating eventually leads to total disorder.

In thermodynamic terms, living things are heat engines that absorb energy from heat sources at high temperatures and expel a lesser amount of heat to heat sinks at lower temperatures in order to maintain a local region of negative entropy.

APPLICATIONS AND PREDICTIONS

1. Blood pressure will be highest at the aorta.
2. Biological complexity will be greater than that of the environment.
3. Cells will require a continuous energy supply to maintain their configurations and activities.
4. There will be subcellular structures for specialized functions. The largest of these will be less mobile inside the cell than smaller subcellular structures.

(*Continued*)

APPLICATIONS AND PREDICTIONS (Continued)

5. Energy is required to bend the material forming the cell membrane so that it will enclose the entire cell.
6. Dormitory rooms spontaneously increase entropy.
7. The kidney regulation of healthy blood concentrations requires ATP usage.
8. Electromagnetic energy is required to arrange hydrogen atoms in a unique direction during an MRI scan.
9. Energy is required to concentrate a solute.
10. Forming proteins requires a large amount of ATP.

2.7 Heat Transfer

Power is the great aphrodisiac.

—Henry Kissinger

The final form of energy is heat. All other forms of energy flow from higher potential to lower, and eventually degrade into heat. Heat may be transformed into other forms of energy, but the transmogrification is never 100% efficient, and so some heat is left over. This heat can do work as it moves to a lower temperature heat sinks, but once it reaches the lowest temperature possible, there are no lower temperature heat sinks available, the heat cannot move anymore; it is incapable of producing any other kind of energy.

This unique position for heat is the reason why heat, of all the forms of energy, needs to be studied in further detail.

Heat transfer is thus important to biological organisms. Excess heat must be removed, and heat deficiencies must be filled. Otherwise, organisms will not survive.

Heat may be transferred through four common mechanisms: (1) conduction, (2) convection, (3) radiation, and (4) change of state. Conduction requires contact for heat to move from one object or fluid to another. Convection requires that there be a moving fluid (liquid or gas) that can heat or cool and move the heat along with the fluid. Radiation requires only that two objects be in line-of-sight contact; heat is moved through electromagnetic means. Change of state heat transfer is important when a substance, such as water, evaporates, condenses, melts, or freezes.

Conduction, convection, and radiation heat transfer happen according to

$$\begin{pmatrix}\text{rate of heat}\\ \text{transfer}\end{pmatrix} = \frac{(\text{surface area perpendicular to flow})\,(\text{temperature difference})}{(\text{distance})\,(\text{insulation value})} \tag{2.7.1}$$

This relation is somewhat different for the three different heat transfer modes, but dependence on these parameters is important for each.

All heat is transferred due to a difference in temperature. As the temperature difference increases, more heat can be transferred from the stuff at higher temperature to the stuff at the lower temperature.

All heat is transferred from a surface area. The larger the surface area, the more heat that can be transferred. When an animal needs to conserve heat, it curls into a ball with low surface area. When an animal needs to lose heat, it spreads out to expose as much surface area as possible to the environment.

Lastly, all heat is transferred through a resistive medium that acts as insulation. If there were no resistance, the very smallest temperature difference would cause an enormous amount of heat to flow. Biological organisms may control their heat exchanges through the active manipulation of the resistance. Thus, birds either fluff their feathers to insulate themselves or position their feathers to allow air to flow through to lose heat. Humans either vasodilate or vasoconstrict the blood vessels of their skin to insulate themselves or promote heat transfer. Changing surface color from dark to light would either promote or inhibit radiant heat transfer.

Metabolic heat is usually considered to be generated throughout the volume occupied by an organism. There may be localized regions of higher or lower metabolic rate (for instance, the liver and the human brain have particularly high metabolic rates), but, in general, the amount of heat generated by an organism depends upon its volume.

The surface area to volume ratio thus expresses the ease of heat transfer compared to heat generation. With a high surface area to volume ratio, the arms and legs are good candidates to lose heat to the environment. Arctic animals generally have shorter legs than do tropical animals to minimize heat loss from the legs. Similarly, ears are particularly good appendages for heat loss. Animals from colder areas usually have much smaller ears than do animals from warmer climes.

Heat removal can be critical to an organism. Overheating can cause severe damage to critical enzymes and the loss of vital functions. Because heat loss depends upon the amount of surface area available in relation to the volume of tissue generating the heat, larger animals have developed circulatory systems to increase the surface area internally and to bring the heat to the exterior body surface by the convection of the blood. During times when heat must be conserved, blood circulation is cut off to the parts of the body, and then the overlying layers of tissue act as insulation for internal tissues.

Sometimes these mechanisms are not sufficient to remove all excess heat. Biological organisms then turn to the evaporation of water to aid in cooling. Evaporation removes a large amount of heat for the mass of water involved.

Those animals with developed circulatory systems use the blood to bring the cooling power of evaporation to the interior of the body. Plants, without the necessary complex circulatory systems, use the large surface areas of the leaves from which to evaporate water and cool the plant. Desert plants that must conserve moisture but also survive high temperatures have usually adapted to having low surface areas and a high tolerance for heat.

Example 2.7.1: Conduction Heat Transfer Can Cool Koalas

Koala fur insulates the Australian mammal from its thermal environment. At times, however, temperatures rise so high that koalas must find a means to remove excess body heat. They can use panting or fur-licking to enhance evaporative heat loss, but these two methods use a lot of precious water. Instead, they lose heat by pressing their bodies against the trunks of trees.

Tree trunks can be more than 12°F (7°C) cooler than the air. Koala belly fur is relatively thin, so, when they hug trees (Figure 2.7.1), they cool their body temperatures by heat conduction more efficiently than by panting or fur-licking (Briscoe et al., 2014).

Example 2.7.2: Countercurrent Heat Exchange Saves Body Heat in Warm-Blooded Animals

When an artery carrying warm blood from the warm core of the body lies next to a vein returning cool blood from the periphery, temperature differences between the two

FIGURE 2.7.1
A koala bear hugging a cool tree trunk can lose excess body heat to the tree by conduction.

vessels drives heat from the artery to the vein. This action conserves body heat because arterial blood is cooled by the venous blood before it is delivered to the cool body surface and venous blood is warmed by the arterial blood before it returns to the warm core. This type of heat transfer is called *countercurrent heat exchange* because of the opposite directions of the fluid flowing in the adjacent vessels (Johnson, 1999).

Humans have simple countercurrent heat exchange in their arms and legs to save body heat in cool environments (Figure 2.7.2). There are more efficient vascular structures, called *retes*, in some warm-blooded arctic animals that spend much of their time in cold air or water.

There is a deep-water predatory fish called the *opah* that has special blood vessels in its gills that operate as a counter current heat exchange system (Wegner et al., 2015).

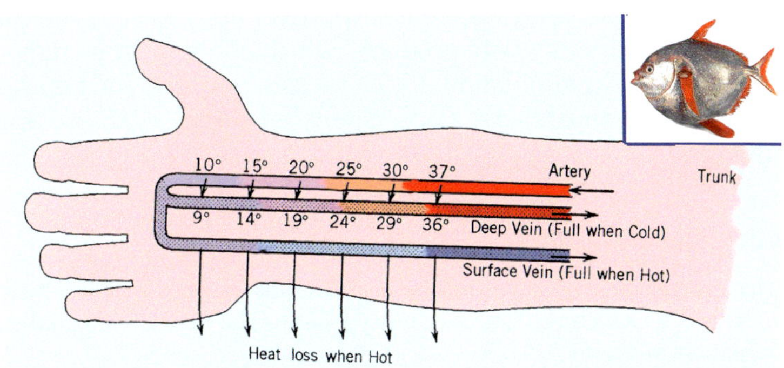

FIGURE 2.7.2
Countercurrent heat exchange in the human arm saves body heat when warm arterial blood transfers its heat to the cool venous blood returning to the trunk of the body. If the environment is warm enough to require body cooling, then blood returning to the body core is shunted to veins closer to the skin surface where heat can be lost to the ambient environment. The Opah fish (insert) uses countercurrent heat exchange in its gills to save body heat and sustain a very active lifestyle.

Oxygen-poor warm blood from its working muscles flowing toward its gills normally would lose its heat as it picked up oxygen in the gills, but, instead, the heat is transferred to the oxygenated blood returning from the gills. As such, unlike most other fish, the opah is warm-blooded. Its eyesight and muscle performance are much sharper than its sluggish, cold-blooded prey. This gives the opah a strong competitive advantage over other predators in the region.

APPLICATIONS AND PREDICTIONS

1. The thermal comfort state of an animal can be inferred from its posture.
2. The skin coloring of tropical animals will be more intense than that of arctic animals.
3. Arctic animals and plants will tend to be shorter and more stout than tropical animals and plants.
4. Fingers and ears will be the body parts most prone to frostbite.
5. Immune system activity contributes to body temperature increase.
6. Warm-blooded animals need a more complex circulatory system than cold-blooded animals in order to conserve heat.

2.8 Movement of Materials

It has become quite a common proverb that in wine there is truth.

—**Pliny the Elder**

The movement of matter is critical to living systems at all levels. The sources of energy often take the form of energy-rich compounds, and these must move from wherever they are located into cells. There are other compounds necessary for the well-being of the living system. These include enzymes, coenzymes, minerals, and chemical building blocks (such as amino acids). These must also be transported into the cells. The consequence of metabolism is the production of waste products, which either must be removed or stored in immobilized form. The accumulation of these waste products (including carbon dioxide, lactate, urea, alcohol, and others) can severely harm the organism, and so must be transported away.

2.8.1 Convection and Diffusion

The first recognized heart–lung machine was built by Jacobi in 1895. This machine appears to have been a technical success in oxygenating the blood, but a medical failure in terms of patient survival.

—**Edwin N. Lightfoot, Jr.**

Mass movement occurs by several mechanisms. Macroscopically, mass can move through convection, which is what happens when fluid surrounding a living entity moves and carries included materials with it. Convection is the process that carries smoke particles from a chimney, or disperses cream in your morning coffee. Mass movement by convection is important not only to whole animals and plants, where gases or nutrients are moved in air or water, but also to cells and tissues where stirring surrounding fluid helps to maintain healthy local environments (as in bioreactors used to grow certain cells).

On a smaller scale, diffusion is the chief mechanism for mass movement. Diffusion causes materials to move when a concentration difference exists between any two points. In terms of effort and flow variables, concentration difference is the effort variable, and the mass rate of flow is the flow variable.

Fick's first law states that

$$\begin{pmatrix}\text{mass rate}\\ \text{of flow}\end{pmatrix} = \frac{\begin{array}{l}\text{(diffusion coefficient) (area perpendicular to the flow)}\\ \text{(concentration difference)}\end{array}}{\text{(distance)}} \tag{2.8.1}$$

In other words, the mass rate of flow between any two locations is directly proportional to the concentration difference between the locations, inversely dependent on the distance between the locations, and proportional to the area through which the flow occurs. The diffusion coefficient is a constant of proportionality used to make the calculation correct. It is empirically determined. Values of the diffusion coefficient are highest in gases, considerably lower in liquids, and very small in solids. Thus, diffusion mass transfer is fastest in gases, slower in liquids, and often impossibly slow in solids.

The dependence of diffusion on area perpendicular to the mass flow has led to some adjustments in biological entities. If a particular material is necessary for the cell or other entity and must be supplied at a high rate, then the surface area of the cell is often found to be large. This can happen with surface foldings and invaginations (Figure 2.8.1).

Because mass transfer rates are inversely proportional to distance, smaller distances are advantageous for the cell. Thus, there is a practical limit to the sizes of most cells, as long as they depend on diffusion to carry materials from one place to another.

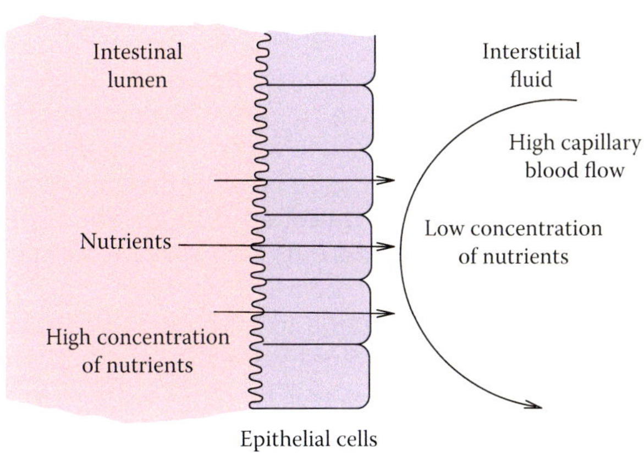

FIGURE 2.8.1
The large surface area of the intestine and the high concentration difference between the lumen and intestine wall makes a very efficient mass transfer system. (Redrawn from Johnson, A.T., *Biological Process Engineering: An Analogical Approach to Fluid Flow, Heat Transfer, and Mass Transfer Applied to Biological Systems*, John Wiley & Sons, New York, 1999.)

Considering that resistance is the ratio of effort variable to flow variable, diffusion resistance is

$$\text{resistance} = \frac{\text{distance}}{(\text{area})\,(\text{diffusion coefficient})} \tag{2.8.2}$$

Thus, long distances, small areas, and small values for the diffusion coefficient lead to high resistances to mass transfer. The higher the resistance, the harder it is to move materials either toward or away from the cell.

Concentration difference is usually not an easily controlled variable. For any given location, the concentration in the environment is often not directly influenced by the cell; the concentration at the cell surface can be controlled by absorbing or emitting material, but that has limited effect on the concentration difference. If too much material is absorbed or emitted, then the distance between surface concentration and environmental concentration can increase considerably. Sometimes an organism can influence the environmental concentration of one type of material by selectively manipulating another related material. For example, the availability of nitrogenous materials is affected by pH (acidity), and some plants seem to increase the availability of nitrate over ammonium ions by changing the pH of the surrounding soil. Mobile organisms, of course, can change their locations to ones where necessary materials are more abundant and, thus, overcome some of the material limitations that they encounter in other locations. However, as we shall see, this strategy leads to more competition for available resources and may not result in a net gain. We must conclude that numerous strategies are necessary for living systems to obtain the materials they need. Influencing concentrations is one possible means, but reducing resistance may be more economical.

The diffusion of oxygen, nutrients, and metabolic wastes is one of the most definitive determinants of organismal configuration (LaBarbera and Vogel, 1982). The maximum distance that oxygen can diffuse into oxygen-consuming tissue is typically 20–100 μm (Secomb et al., 2004). Because the rate of diffusion, varying according to Fick's Law (Equation 2.8.1), depends on the concentration difference over a given distance, the concentration gradient is crucial. If the movement of material is based on diffusion alone, an organism faces severe geometrical constraints. The diffusion distance from the environmental source of material must be comfortably small. Different organisms have met this problem in different ways. The coelenterates (jellyfish, sea anemones, corals, and hydroids) have no tissue layer more than two cells thick. Other multicellular organisms either have one flattened dimension, are threadlike, or are branched. Some have channels or invaginations that serve to bring the environment to the cellular material.

Another course of action is to use convection to overcome the limitations of diffusion (Figure 2.8.2). The convective process moves fluid in bulk so that fluid depleted in essential materials is replaced by new fluid. This is the basis for the blood flow in the cardiovascular system of larger animals. This system takes a number of forms in different animals, but in mammals, oxygen passes into the blood in the lung alveoli and is pumped by the heart to the remainder of the body tissues. In order to be effective, there must be many blood vessels in intimate contact with the cells. Thus, no matter how far away from the lungs these cells are physically located, from an oxygen standpoint, they are right next to the lung. The principle here is that the fluid transport system must effectively reduce diffusion distances between points within an organism or between a point within an organism and the external environment.

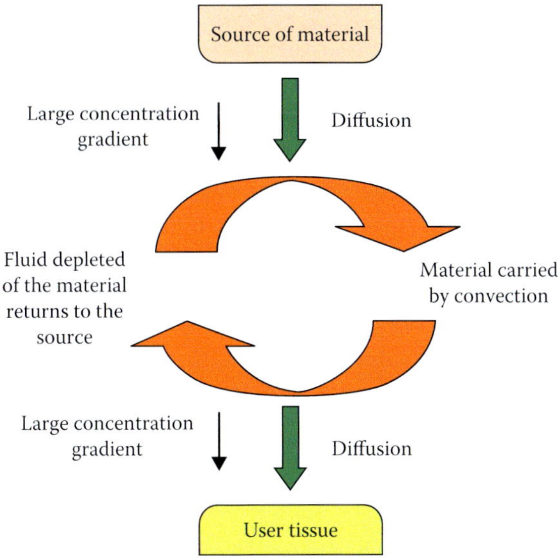

FIGURE 2.8.2
To overcome the limitations of diffusion, convection is used to bring the required material close to the tissues where it is used.

Another effect of the fluid transport system is that it acts to homogenize nutrient concentration within an organism. That is, within narrow limits, the concentration of the essential compound is the same within the entire organism. Because of this, the concentration of oxygen in the lung is nearly the same as the concentration of oxygen in the working muscles. Thus, the concentration difference over the small alveolar-to-capillary distance is as large as possible, and this promotes the maximum rate of oxygen transfer in the lung.

Of course, a third possibility is that as the organism grows large, it can tolerate a smaller diffusion rate from the environment to the most remote tissues. For the case of oxygen, the only way to accommodate this is to lower the mean metabolic rate. In Section 7.4.4 (Metabolism), it can be seen that, indeed, metabolic rate per unit body mass does decrease as organisms grow larger, but the effect is not as drastic as it would be without the internal bulk movement of fluid.

Within the cell, vesicles (storage vessels) are sometimes utilized to store a certain excretory or secretory chemical in relatively pure form. Rather than release this chemical within the cell and rely on diffusion to carry it to the membrane and beyond, the vesicles are actively carried to the periphery of the cell, where their membranes fuse with the cell membrane in a process called *exocytosis*. The vesicle contents are then released to the outside. The same principle relating to exocytosis, carrying the material in concentrated form in a storage vessel, and moving the entire vessel, is used for the movement of oxygen throughout the body in red blood cells. In the opposite process of *endocytosis*, the cell can incorporate material by surrounding it and forming a vesicle to either store or transport it to the interior of the cell. Endocytosis is important for lymphocytes to ingest invading microbes, and for other cells to transport materials too big to pass through membrane pores from the environment into the cell. Endocytosis is more likely to be developed in adherent cells than in suspended cells.

West et al. (2000) have given three basic general principles for the design of biological network transport systems:

1. In order for the network to supply the whole volume of the organism, a hierarchical branching pattern to fill the entire space is required.
2. The final branch of the network, where nutrients are exchanged (e.g., the capillary of the circulatory system or the petiole of a plant), is a unit of fixed size.
3. Organisms have evolved so that the energy required to transport materials through the network is minimized.

From these principles, with the interplay between physical and geometric constraints, come scaling laws that relate properties and functions from one organism to another (see Chapter 7, Allometry).

2.8.2 Osmosis

> Natural ability without education has more often raised man to glory than education without natural ability.
>
> **—Cicero**

Related to the diffusion of materials is the process of *osmosis*, which occurs mainly across semipermeable membranes found in living things. Membranes are used to separate different molecules and to control local environmental conditions. There are membranes surrounding the cell and many of its inclusions.

Most biological membranes are porous to some extent. They have holes that are often large enough to allow the free passage of water (18 Da in size; a Dalton (Da) is a unit of molecular size corresponding to its molecular weight) and other small molecules, but are not large enough to allow many of the more complex biochemical molecules (often 50,000 Da or larger) to pass.

If there is a higher concentration of a solute on one side of the membrane, there is consequently a lower concentration of water (as long as total pressures on both sides of the membrane are approximately equal). In other words, if the mass fraction of solute is higher, the mass fraction of water must be lower.

Because water is the freely moving material, the solute being blocked from moving through small membrane pores, water moves from its higher concentration to its lower concentration. It continues moving until the hydrostatic pressure on the other side exactly balances the tendency for water to move from higher to lower concentration. The amount of hydrostatic pressure needed to exactly balance flows in both directions through the membrane is called the *osmotic pressure*, and the combination of solute, water, and semipermeable membrane results in *osmosis* (Figure 2.8.3).

From our effort and flow variable perspective, there are two effort variables that act in concert, either adding or subtracting. These are concentration (or osmotic) pressure and hydrostatic pressure (see Section 2.9, Fluids). Either one can cause the flow of water through the semipermeable membrane (Figure 2.8.4).

Osmosis is important in living things because it helps to maintain water balance, and, sometimes when solute concentrations are too high either inside the cell or outside, can cause cellular death.

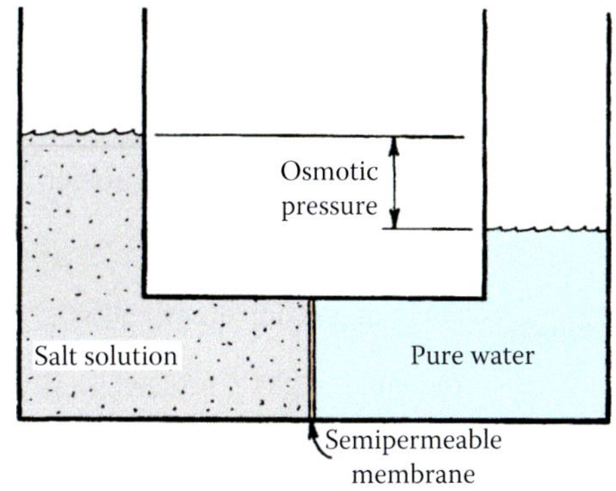

FIGURE 2.8.3
Placing a semipermeable membrane between an aqueous solution and pure water results in a movement of water from the pure water side to the salt solution side that is halted when the difference in heights of the two liquids counterbalances the osmotic pressure.

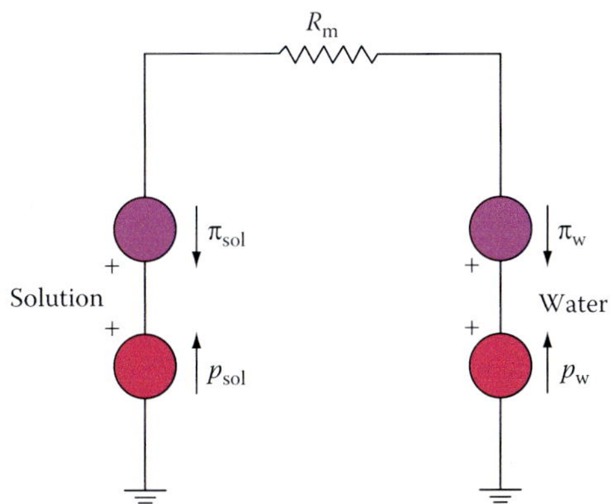

FIGURE 2.8.4
Systems diagram for the apparatus in Figure 2.8.3. The flow variable is the flow of water, and the effort variable is pressure. Two pressure sources appear in series on both sides of the membrane. One pressure source of each pair is osmotic pressure (π) and the other is mechanically applied hydrostatic pressure (p). In all cases, $\pi_w > \pi_{sol}$, and, therefore, to stop the flow requires that $p_{sol} > p_w$. The membrane resistance R_m limits the flow rate when pressures on the two sides are unbalanced.

Example 2.8.1: Neonatal Tidal Liquid Ventilation

Very premature newborns are often lacking in respiratory surfactant. It is the surfactant layer on the inside of lung tissues that keeps the lungs from collapsing during breathing. Infants suffering from scarce or absent surfactant are normally placed on mechanical ventilation until they mature enough to produce surfactant and can breathe on their own.

There is benefit to ventilation with perfluorocarbon (PFC) instead of air. Eliminating the liquid–air interface in the alveoli (or, in the case of very young neonates, saccules that are the precursors of alveoli) increases lung compliance and improves alveolar recruitment (alveoli remain open).

Corno et al. (2004) have developed a model for the liquid ventilation of neonates. The objective of this model was to properly specify ventilation parameters (breathing rate, tidal volume, inspiratory-to-expiratory time ratio, flow waveforms, PFC oxygen content, and so on) to add adequate oxygen to the blood and remove carbon dioxide.

As part of the model, Corno et al. (2004) modeled gas transfer in the lung. Their basic gas transport equation was

$$\begin{pmatrix}\text{rate of change of}\\ \text{gas concentrations}\end{pmatrix} = \text{change of gas flux along the axial dimension } (z)$$
$$+ \text{ the rated of gas produced locally}$$

Mathematically, their equation was written as

$$\frac{\partial c}{\partial t} = -\frac{\partial J}{\partial z} + R$$

They coupled gas exchange with mechanical and fluid properties of the lungs and simulated ventilator conditions on a premature infant. Such a model can become an effective tool to support technicians deciding ventilation strategies in newborns.

Example 2.8.2: Osmoregulation in Asian Clams

The Asian clam was introduced into San Francisco Bay from ship ballast water in the mid-1980s. It has since spread prolifically throughout the Bay area, reaching a peak population density of 10,000 per square meter (Werner et al., 2003).

The Asian clam is an osmo-conformer. That is, clams can rapidly adapt to changes in salinity by increasing or decreasing the intracellular concentrations of the amino acids alanine and glycine betaine. In this way, the clam remains in osmotic equilibrium with its surroundings.

Although osmo-conformation allows the clam to survive in highly saline environments, the processes of protein synthesis and repair are energy-intensive, and are estimated to cost 20%–25% of total energy expenditure under non-stressful conditions. The energy used to accumulate amino acids in response to environmental elevated salinities is not available for other survival and reproductive purposes, and probably makes the clam susceptible to succumb to other stresses.

Example 2.8.3: Solute Transport through the Endothelial Intercellular Cleft

Endothelial cells lining the inside of blood vessels play an important role in the transport of materials to the intercellular fluid outside the blood vessels. *Hydrophobic* (water insoluble) materials can pass directly through the lipoprotein double layer of the endothelial cell membranes, so the entire cell surface is available for transport. *Lipophilic* (soluble in fats or lipids) O_2 is one of these substances.

Albumin and low-density lipoprotein are among *hydrophilic* (water soluble) materials that cannot pass directly through the endothelial cells. Instead, these materials pass between cells through the intercellular cleft, typically 20 nm wide (Hodgson and Tarbell, 2002). The small area of these clefts significantly throttles mass transport (Hodgson and Tarbell, 2002). Also, because endothelial cells lining the blood vessels are typically aligned to the direction of flow, most of the resistance to transport of hydrophilic materials through the vessel wall is in the wall itself and not in the fluid inside the vessel supplying the materials.

Example 2.8.4: Passive Diffusion through a Cell Membrane

Many ions, water, and small molecules move across the cell membrane by passive diffusion. The rate at which they accumulate inside the cell is thus limited by

1. Surface area of the membrane. Smaller cells have lower diffusion rates.
2. Concentration difference between the inside of the cell and its external environment. As materials accumulate inside the cell, this concentration difference decreases. As more accumulates inside the cell, there is less material outside the cell unless the outside amount is replenished. If the rate of replenishment is slow compared to the rate at which material enters the cell, the concentration difference decreases. It will eventually decrease to equal the rate of replacement outside the cell.

Other considerations for passive diffusion have to do with the charge density on the ion, size of the ion, and the degree of lipophilicity of the molecule. If the molecule is highly lipophilic, then it can move through the cell membrane in addition to through membrane pores.

Example 2.8.5: Water Bears Need No Circulatory or Respiratory Systems

Water bears (also called *Tardigrades*) are extremely small aquatic animals found almost everywhere on Earth (Miller, 2011). Terrestrial water bears live in damp places in moss, lichens, leaf litter, and soil. They have amazing survival strategies, including *cryptobiosis*, or the ability to cease all metabolic activity in the absence of water. They are so small (Figure 2.8.5) that they contain only a little over 1000 cells, with enough access for water, oxygen, and nutrients that they lack circulatory or respiratory systems.

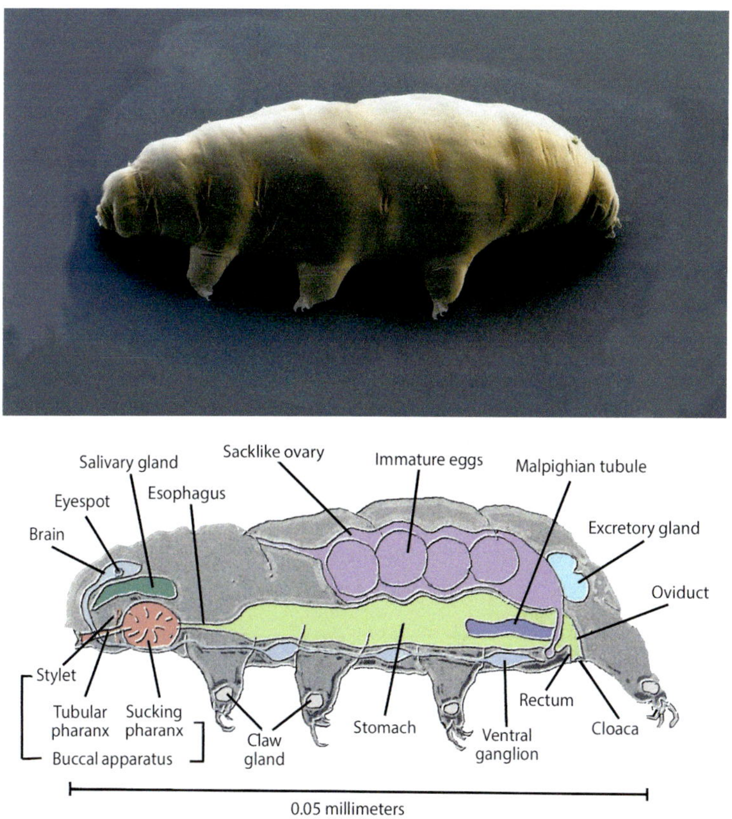

FIGURE 2.8.5
Water bears are microscopic animals with extreme survivability. They are small enough that they need no respiratory or circulatory systems.

Example 2.8.6: Desert Crust Conserves Moisture

Desert soils are covered by free-living cyanobacteria and lichens. These soil crusts may be very thin over hard, clay soils or up to a centimeter thick in sandy, fluffy soil. Microbes in the crust photosynthesize and fix ambient nitrogen. They produce sticky, gooey polysaccharide starch that holds on to nutrients and moisture, and that holds soil particles together to resist blowing away in the wind. This crust acts as an effective impermeable layer to prevent water movement from the soil into the air. Human activity, vehicles, or wandering livestock disrupt this crust. It may take from 15 years to hundreds of years for damaged crust to recover (Borrell, 2012).

APPLICATIONS AND PREDICTIONS

1. Epithelial cells in the walls of the intestines will have highly folded surfaces on the side facing the lumen of the gut.
2. Oxygen supplied to a tissue may be more than adequate, but the oxygen at the cell location may not be adequate because of the distance involved.
3. There will be many small blood capillaries to supply nutrients close to the cells that need the nutrients.
4. Single cell and small multicellular organisms will not require circulatory systems to supply oxygen and nutrients.
5. The walls of capillaries will be thinner than the walls of arteries and veins through which materials are not expected to diffuse.
6. Respiratory alveolar walls will be very thin.
7. Cells that pump sodium outside the cell produce concentration gradients that aid the transport of glucose into the cells.
8. The flattened shape of a red blood cell, with its high surface-to-volume ratio, allows for the faster diffusion of oxygen than if it were a sphere.
9. Transdermal medications rely on convection and diffusion in the body for distribution.
10. Many animals will have filaments, hairs, or other means to promote the flow of fresh fluid over absorptive surfaces.
11. Plant root hairs will be small for the same reason that capillaries are small.
12. Larger organisms will be composed of larger numbers of cells rather than fewer numbers of larger cells in order to satisfy the surface area to volume necessary for oxygen, nutrient, and metabolite transfer to and from the cells.
13. Plants placed in salt water will dehydrate.

2.9 Fluid Mechanics

What can a river teach us about thinking? ...In general, when we are in a problem-solving mode our thinking is like the flow of water through the fast stretches – convergent and highly linear.... After a period of fast, convergent, sometimes turbulent progress,

> it slows, diverges, and eases along between the wider shores, considering, perhaps, how to proceed when it enters the next constriction.... We need to do the same in our thinking.
>
> **—Lyle Feisel**

Fluids are important to biological organisms because they fill interior spaces and completely surround every organism. Water and air are the two most important fluids for biological systems; air because it supplies oxygen, acts as a sink for excess carbon dioxide, and can dry the organism; and water because it acts as a solvent, attaches to dissolved molecules (see Section 3.2, Chemical Bonds), and is relatively thick.

2.9.1 Viscosity

> A man who does not benefit the world by his life does so by his death.
>
> **—Anonymous**

The physical properties of the fluids are very important. Thin fluids are easy to move or move within. Thick fluids are just the opposite. The physical property that measures fluid thickness is called *viscosity*, and the viscosity of water is about 50–100 times as large as the viscosity of air. Dissolving or suspending materials in the water can enhance its viscosity very much. The viscosity of whole human blood is about 10 times that of water.

Suspended particles composed of long-chain molecules are very important in biology. Proteins and polysaccharides are two types of molecules that fit this description. The viscosities of suspensions of these particles are not constant, as it would be for water at constant temperature, for instance. Instead, viscosity decreases as relative flow rate increases, a fact put to good use in the circulation of liquids. A great deal less energy is expended to circulate the blood than would be the case if viscosity did not decrease at high relative flow rates.

2.9.2 Fluid Movement

> Everything does not happen continuously at any one moment in the universe. Neither does everything happen everywhere in it.
>
> **—Père Teilhard**

Fluids move (the flow variable) due to a difference in pressure (the effort variable). The rate of flow is limited by the resistance located in the flow path. The amount of this resistance depends on whether the flow is laminar or turbulent. Laminar flow is smooth, and flow occurs in streamlines. There is little mixing between layers of fluid motion in laminar flow (Figure 2.9.1).

Turbulent flow is agitated, tumultuous, and wild. There is much eddying, with no semblance of the layered flow found in laminar flow (Figure 2.9.2). There is much mixing between regions of higher velocity flow and lower velocity flow in turbulent flow, so the differences in velocity tend to disappear, and all places within the conduit tend to have the same net velocity.

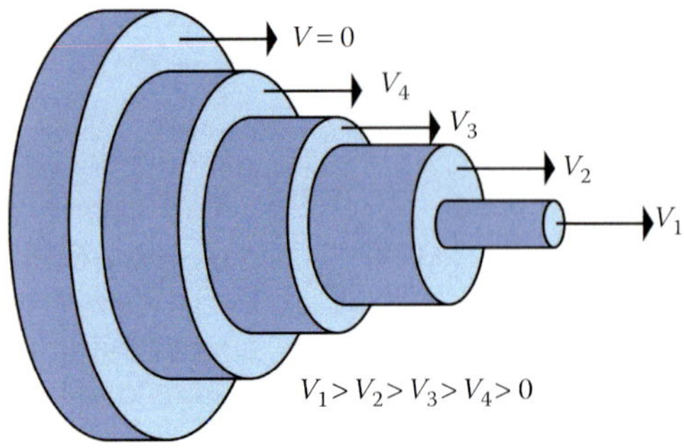

FIGURE 2.9.1
Laminar flow occurs in layers, diagrammed here schematically as five layers. The center layer is moving the fastest, but the outside layer is not moving at all.

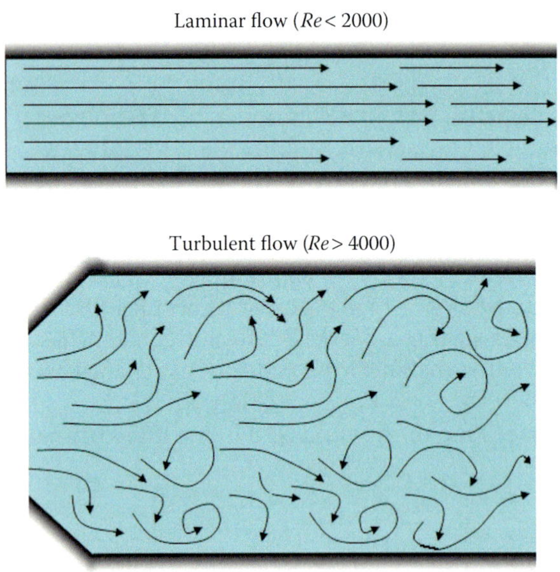

FIGURE 2.9.2
Turbulent flow is characterized by eddying and mixing. Velocities of the fluid in the pipe are in all directions, but the net flow is from a region of high pressure to a region of lower pressure downstream.

The Reynolds number is the ratio of inertial forces (related to the mass of the fluid and its momentum) to the viscous forces (related to fluid thickness and the resistance it causes):

$$\text{Reynolds number} = \frac{\text{inertial forces}}{\text{viscous forces}} = \frac{(\text{fluid density})\,(\text{pipe diameter})\,(\text{fluid velocity})}{\text{fluid viscosity}} \tag{2.9.1}$$

The Reynolds number is a predictor of the transition from laminar to turbulent flow. Flow tends to be laminar at low Reynolds numbers and tends to be turbulent at high Reynolds numbers.

For steady laminar flow in straight pipes, resistance is given by the Hagen–Poiseuille relationship:

$$\text{Resistance} = \frac{128(\text{fluid viscosity})\,(\text{pipe length})}{\pi(\text{pipe diameter})^4} \tag{2.9.2}$$

The dependence of resistance on the inverse of the fourth power of the pipe diameter makes resistance very sensitive to the size of the pipe. This property is important to biological organisms. For instance, a reduction in arteriolar diameter by 16% is sufficient to double resistance and thus halve blood flow. This mechanism is used in the body to direct blood from one pathway to another to bring oxygen to muscles, nutrients from the gut, or to conserve heat loss from the skin (Figure 2.9.3). On the other hand, a similar reduction in bronchiolar diameter during an asthma attack can cause extremely difficult breathing.

Only very rarely is blood flow or respiratory air flow steady and laminar, so the resistance determination by Equation 2.9.2 is only approximate. Turbulence causes higher energy losses and thus higher resistances. Changes in pipe diameters, branching, or pipe directions can also add resistance. Thus, animals that depend on speed are usually seen to be streamlined.

We are familiar with locomotion on a large scale because that is the kind of locomotion that we use. We, and other animals like us, move through a relatively thin fluid (air or water) where the viscous forces (those tending to slow us down) are relatively small and our body momentum (tending to keep us moving) is relatively large. The Reynolds number of humans moving in air or fishes moving in water can be as high as 10,000 or more. For microbes moving in water, the Reynolds number calculates to 0.1 or smaller. Thus, inertial forces (momentum) completely dominate viscous forces for humans, whereas friction forces (viscous) completely dominate inertial forces for microbes. This would be akin, for us, to try to swim in a pool filled with thick molasses. We would soon exhaust ourselves under these circumstances.

When we walk, or when fish swim, we push on a solid surface or against a fluid, and the reaction force propels us forward. As long as the fluid through which we move is not

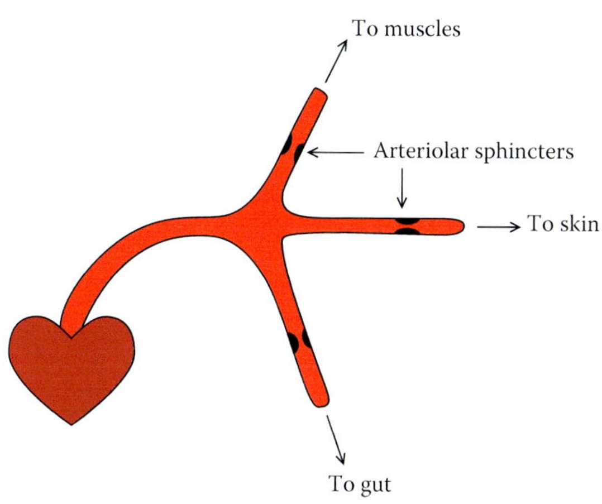

FIGURE 2.9.3
Contraction of arteriolar sphincter muscles can provide exquisite control of blood flow to different regions of the body. As the sphincters contract, they decrease diameters and increase resistances of their arterioles. Increased resistance decreases blood flow.

too viscous (or "thick"), our inertia keeps us moving until the next step or flick of the tail. This system works well for large bodies with large masses. It takes a lot of friction to slow them quickly.

Microbes, on the other hand, have small bodies and small masses. Thus, the viscous drag present in water, although small for us, is enough to slow a microbe almost immediately. Microbes must be in continuous motion to move.

Most single-cell organisms have fine hair-like organelles that extend from their surfaces. Short ones are called *cilia* and long ones are called *flagella*. Large ciliate protozoa have many short motile cilia that function as oars: they have a rigid forward thrust and a flexible bending return similar to a bird's wing in flight. The more oars (cilia), the faster these protozoa can swim. Numerous cilia work in concert to form a coordinated wave over the entire microbial surface (Yates, 1986). Bacteria, however, are much smaller, and must deal with the viscosity of water in a different way. They could not use the oar principle because the oars would not slip by the water on the return stroke. Not only would rowing be harder, but it would also not go anywhere. Thus, bacteria have devised a flagellum that rotates at its base like a miniature propeller (Bonner and Horn, 2000). Viscous fluid resistive force theory shows how this rotational motion can be translated into the forward motion of the microbe.

2.9.3 Fluid Energy

> A pessimist is one who makes difficulties of his opportunities and an optimist is one who makes opportunities of his difficulties.
>
> —**Harry Truman**

Energy relationships within fluids are composed of both potential and kinetic energy terms (Section 2.4, Work and Energy). The potential energy of a fluid is related to the height of a fluid relative to some datum, and also to the pressure exerted either by or on the fluid. As examples of these, consider the height of sap within the tallest Sequoia tree. In order to raise the sap to this height above the roots, work must be done to the fluid. The taller the tree, the more work that must be done. The potential energy of the sap at the top of the tree would be available to perform mechanical work if it could be harnessed.

Fluid pressure is likewise able to be used. Pressure can move the fluid through a conduit or can distort the walls of a container.

Kinetic energy depends on fluid velocity (see Section 2.4, Work and Energy), and is nonlinearly higher for higher velocities. Friction (or viscous losses related to resistance) tends to slow the fluid. In order to accelerate fluid from rest to some higher velocity, energy must be used, and this is often in the form of pressure potential energy. Thus, an energy balance, which must be satisfied within the fluid, would state that kinetic energy increases must be accompanied by potential energy decreases:

$$\text{(kinetic energy gained)} = \text{(potential energy lost)} \tag{2.9.3}$$

Fluid pressures vary around an immersed body moving relative to the fluid. At the blunt front end, relative fluid velocity slows to nearly zero. The kinetic energy possessed by the moving fluid (here, it doesn't matter whether the actual movement is by the body or by the fluid; it's the combination that matters) is converted into potential energy, and the pressure rises. As the fluid flows around sides of the body, potential energy converts into kinetic energy, and the pressure falls. Thus, fluid pressure around the moving body may be higher than, equal to, or less than static pressure in the fluid surrounding the unmoving body.

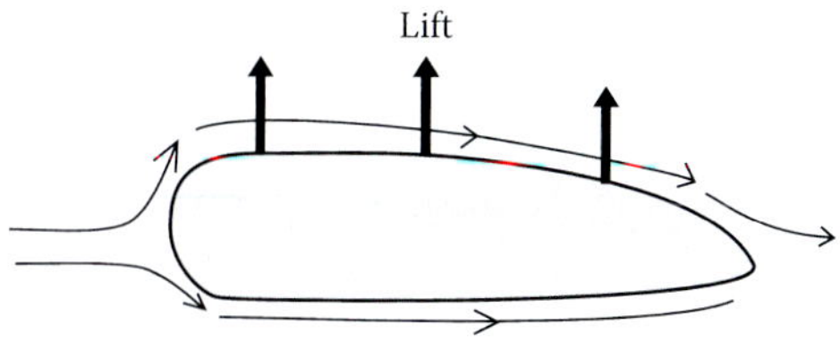

FIGURE 2.9.4
The cross-section of an airplane wing shows that the upper surface is longer than the bottom surface. Flow splits at the front and some goes over and some under the wing. Velocity is higher at the top, so pressure is lower.

This is the physical mechanism behind the lift of an airplane wing (Figure 2.9.4). The upper surface of the wing is curved more than the lower surface, and the distance for fluid to travel is longer from front to the rear of the wing. With a longer distance to travel in the same amount of time, fluid flows faster over the top of the wing than over the bottom. Faster fluid velocities mean lower pressures, so there is lower pressure on the top of the wing than on the bottom. The wing is lifted by a net force determined by the area of the wing times the difference in pressure.

Many organisms can use flow-induced pressure differences to their advantage (Vogel, 1988). Swimming fish, for instance, have a range of pressures over the lengths of their bodies (Figure 2.9.5). High pressures at the mouth and low pressures just behind the head help to

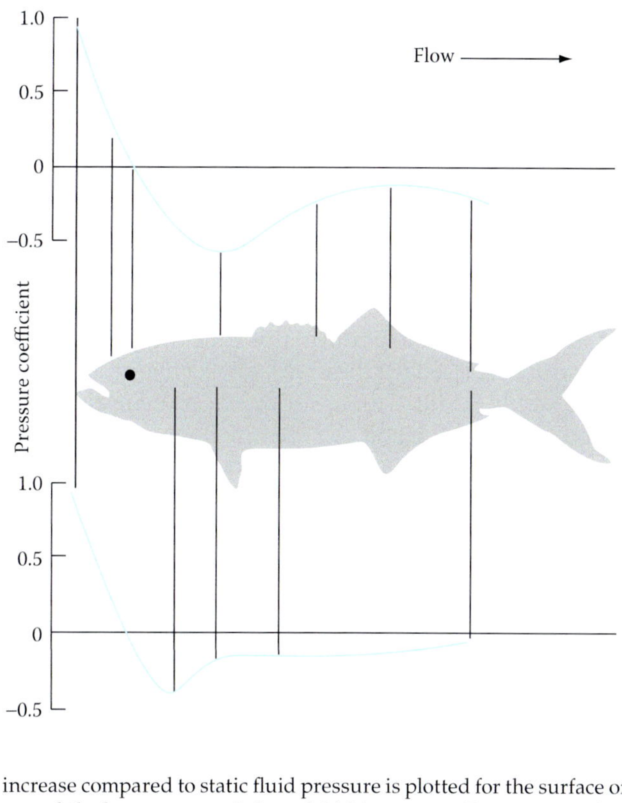

FIGURE 2.9.5
The relative pressure increase compared to static fluid pressure is plotted for the surface of a fish. The top curve is dorsal (top) pressure, and the bottom curve is lateral (side) pressure. Pressures are highest in the front and lowest near the outlet of the gills. (Redrawn from Vogel, S., *Am. Sci.*, 76, 28–34, 1988.)

force water through the gills that they need in order to breathe. Some fish, like adult tuna, mackerel, and sharks, are obligate ram ventilators who must swim continuously to bathe their gills.

Actual pressures produced during swimming depend on speed. Fish eyes are located at a point where pressures do not change, so visual distortion doesn't occur.

Squids, whales, and other water animals use pressures to function correctly. Birds, bats, and insects also fly in the air because of flow-induced pressures. This is a phenomenon important to many creatures.

2.9.4 Circulatory System

> The heart, consequently, is the beginning of life; the sun of the microcosm, even as the sun in his turn might be designated the heart of the world; for it is the heart by whose virtue and pulse the blood is moved, perfected, made apt to nourish, and is perceived from corruption and coagulation; it is the household divinity which, discharging its function, nourishes, cherishes, quickens the whole body, and is indeed the foundation of life, the source of all action.
>
> **—William Harvey**

The circulatory systems in plants and animals conform to physical principles. Each system has a means to propel the fluid, because, from Newton's laws of motion, fluid will not move by itself. Power is required to move the fluid. In plants, this power is supplied by a combination of water evaporation in the leaves, an osmotic gradient from root to leaf, and capillary action. In animals, this power is usually supplied by one or more hearts through which the blood flows.

The power expended by the heart has no particular advantage for an organism; this power does not directly aid reproduction, nor does it directly improve survival. As long as there is adequate blood flow, the organism can both survive and reproduce. Any extra power expended by the heart actually reduces the capacity of an individual to survive a physical challenge, so there is both a survival and reproduction advantage to reduce expended power as much as possible and still provide adequate circulation.

The main objective of circulating blood is to provide oxygen and essential nutrients while removing metabolic wastes. In order to perform this function adequately, blood vessels must be in intimate contact with tissue cells. If these vessels were large and still in intimate contact with tissue cells, there would be little room for the cells themselves, and the body would consist mainly of blood vessels and blood. All the functions of the body, including muscular movement, digestion, chemical functions, etc., would be crowded out by the blood circulation. Thus, the vessels that serve the immediate needs of the tissues must be small.

Small vessels have some interesting characteristics. First of all, a flow balance shows that, for the same volumetric flow rate, the velocity of flow through a vessel of smaller diameter must be much greater than the velocity of flow through a larger vessel. Because flow velocity is related to volumetric flow rate through the cross-sectional area, velocity in a blood vessel depends on the square of the vessel diameter. Therefore, a vessel twice as small will have a flow velocity four times as large as a larger vessel with the same volume rate of flow.

Fluid that moves faster requires more power to propel it. Thus, there is a penalty to be paid for the small vessels in the circulatory system.

Power is calculated as the product of pressure and volumetric flow rate:

$$\text{Power} = (\text{volume flow rate})\,(\text{pressure}) \tag{2.9.4}$$

and pressure in a tube with laminar flow is related to volumetric flow rate through the inverse of the fourth power of the vessel diameter (Equation 2.9.2):

$$(\text{pressure}) = \frac{128(\text{vessel length})(\text{fluid viscosity})(\text{volume flow rate})}{\pi(\text{vessel diameter})^4} \tag{2.9.5}$$

Fluid power is thus related to the vessel length, the square of the volumetric flow rate, and vessel diameter to the fourth power. To minimize this power, we could make the vessel shorter, reduce the volumetric flow rate, or make the vessel larger. We already know that we cannot make the vessel larger and still maintain adequate body function. However, we can make the vessels larger in places where materials exchange is not necessary; larger vessels can be located to move blood from one region of the body to another before the blood enters the smaller exchange vessels.

We can also make the exchange vessels short. Because they are small, the capacity of the blood in these vessels to carry oxygen, nutrients, and metabolic wastes is very limited, so there is no advantage to long exchange vessels. Keep them short to reduce pumping power required.

Reducing volumetric flow rate can be accomplished if the configuration of the smaller vessels allows many of them in parallel. Thus, the flow will split, each smaller vessel carrying only part of the volume flow in the larger feeder vessels.

What we end up with is a circulatory system consisting of large arteries moving blood from the heart to other parts of the body, many small, short capillaries to facilitate materials exchange with the tissues, and large veins to collect the blood and move it back to the heart. This is the configuration that performs the required function of materials exchange while still minimizing power to pump the blood (LaBarbera and Vogel, 1982).

Regions of little or no flow are prone to accumulate particulate deposits. In the case of the circulation of blood, clots may form in these areas of reduced flow. If these clots break loose, they can be transported to places in the circulatory system where they obstruct flow, and this may be life-threatening. It is important, therefore, for circulatory prostheses to avoid stagnant flow regions where clots can form.

2.9.5 Static Pressure

Ability is nothing without opportunity.

—Napoleon

Pressure in a fluid is exerted equally in all directions. Pressure outside of a cell must be resisted by an equal pressure inside a cell in order that the cell doesn't collapse. If pressure inside a cell increases beyond the amount that just balances outside pressure, the cell either expands until it bursts or until it reaches rigid boundaries that support the cell by providing externally applied pressure (Figure 2.9.6).

Because hydrostatic pressure is exerted in all directions simultaneously, it can be used to add rigidity to otherwise limp structures. For instance, the blue crab has an external skeleton that we know as a shell. When the crab grows larger than the shell can accommodate, the crab must molt, removing the outmoded shell and growing a new one. If human skeletons turned soft, we would all collapse in a heap of quivering protoplasm until our bones hardened again. Newly molted blue crabs could be in a similar predicament. However, soft-shell crabs remain capable of vigorous motion despite the lack of a hardened shell because they

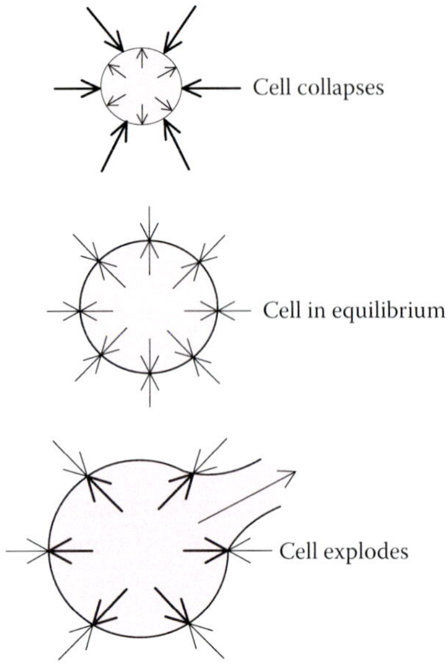

FIGURE 2.9.6
A cell surrounded entirely by fluid must resist pressure exerted by the fluid. If the cell cannot resist the external pressure, it collapses; if internal pressure is higher than external pressure, the cell can explode.

use hydrostatic pressure inside the soft shell to maintain its stiffness (Figure 2.9.7). Internal water pressure of freshly molted crabs surges to as much as 13 times normal when they move. This keeps their limbs rigid and functioning. As the shells harden, peak hydrostatic pressure returns to normal. Hydrostatic pressure also substitutes for a rigid mechanical skeleton in the squid during propulsion and during male penile erection. Substituting internal fluid pressure for mechanical rigidity is the basis for movements of soft robots made with flexible materials (Mengüc, 2017).

The Law of Laplace expresses the relationship between pressure and geometry at equilibrium:

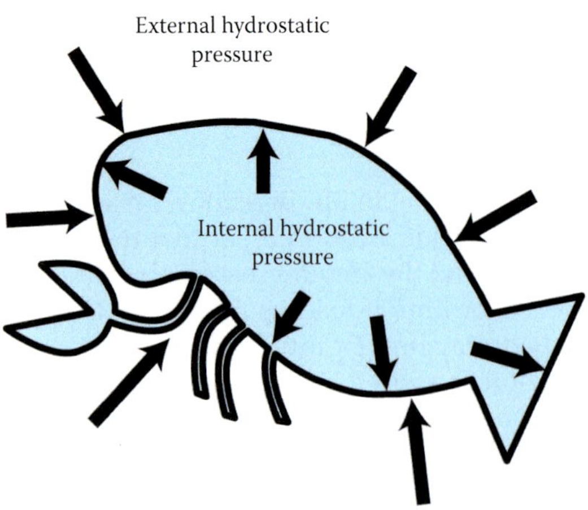

FIGURE 2.9.7
A soft-shell crab retains the ability to move despite the lack of a rigid shell because internal hydrostatic pressure elevates well above normal.

$$\text{(internal pressure)} = \frac{2\text{(wall tensile stress)}\,\text{(wall thickness)}}{\text{(radius of wall)}} \tag{2.9.6}$$

Thus, the pressure that can be developed inside a cell, or resisted by a cell, will depend on the strength of the wall material, the wall thickness, and inversely with the wall radius. Those cells required to resist higher pressures can be expected to have thicker walls of stronger materials and they will be smaller in size. However, if the pressure inside the cell is the same as that outside the cell, then thicker, stronger walls and smaller sizes would not be required.

Counteracting high internal pressures requires strong surrounding structures. Bacteria, fungi, plants, and some archea have tough, sometimes flexible outer cell walls that offer structural support, protection, and limited filtering. The cell wall prevents the cell from bursting when too much water enters the cell in a hypotonic (low solute concentration) environment. Different bacteria have generally two different kinds of cell walls that either stain or not with a standard blue dye. These are called gram-positive or gram-negative. Gram-positive bacteria have stronger cell walls than do gram-negative bacteria.

The Law of Laplace can be used for cylindrical shapes without the factor 2. Thus, smaller conduits such as the capillaries would be expected to resist high blood pressures with thinner walls than would the arteries just because the capillaries are smaller in radius.

Example 2.9.1: Outsmarting Beavers

Beavers build dams to trap water, but in conflict with human activities in the process (Figure 2.9.8). Beaver damage to cropland and woodland has been estimated at several billion dollars. Eliminating the beavers by trapping and hunting is not always an option, but controlling the extent of their ponds takes ingenuity. When beavers sense moving water, they work to plug the leak. Thus, any means to control the level of water in the pond must minimize water velocity and not be able to be repaired by a beaver.

FIGURE 2.9.8
The culprit. Beavers like to flood low-lying areas to produce an environment to their liking.

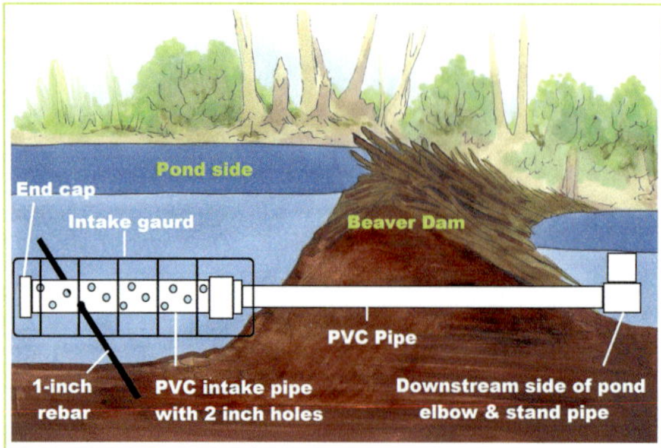

FIGURE 2.9.9
The Clemson Pond Leveler allows beavers to build and maintain their dams but limits the economic costs to woodlands and croplands. (Redrawn from Miller, D., *Prog. Farmer,* 118, 70–71, 2003.)

Solution

The Clemson pond leveler (Figure 2.9.9) suppresses the flooding of timber or cropland but still preserves the pond for waterfowl, plants, and other wildlife. It lets water move through the dam quietly, and doesn't attract the attention of beavers (Miller, 2003).

Example 2.9.2: Enhanced External Counterpulsation

Pressure pulses propel the blood through the cardiovascular system. Pressure pulses provided by the healthy heart are enough to maintain blood circulation at necessary levels, but when the heart weakens, as in congestive heart failure, not enough pressure is produced to keep blood circulating. With a weakened heart, blood tends to pool in the peripheral veins and doesn't return to the right atrium in sufficient amounts. The result is fluid build-up in the body and insufficient circulation to the myocardium. Angina, a painful sign of cardiac circulatory insufficiency, can result.

External pressure applied from outside the skin can be used to help propel pooled blood back to the heart. Oversized blood pressure cuffs applied to the legs and buttocks are pulsed in synchrony with the heartbeat. Insufficient cardiac pressure pulses are thus assisted with this external squeezing action, and blood is returned to the heart. The heart often responds to this extra flow of blood by producing tiny blood vessels to nourish the heart better. An Enhanced External Counterpulsation (EECP) procedure takes an hour a day, 5 times a week, for 7 weeks. The benefits of EECP often last several years.

With EECP, the heart and cuffs act as pressure sources in series to move blood around the circulatory system. In a similar manner, cardiopulmonary resuscitation (CPR) is used to apply external pressure to the chest and heart to restore some circulation when the heart has failed. Pressure, from whatever source, is needed to overcome the resistance of the vasculature to make blood circulate.

Example 2.9.3: Instant Hot Water in Your Hotel Room

Did you ever wonder how you can have hot water as soon as you open the tap in your hotel room? Hot water in pipes cools with time, and with a heater located some distance from the faucet, the water would have to run a while before it turned hot. A hot water heater located in the room would solve the delay problem, but would be expensive to implement.

What is done instead is that there is a hot water circulation loop that constantly replenishes hot water available to the room. As the water cools in the pipe, it is pumped to a remote heater, where it is reheated and recirculated.

The blood circulatory system of the body acts the same way. If not for the looping circulation of the blood, all the nutrients and oxygen could be used by the first tissues contacted by the blood, and downstream tissues would be exposed to metabolic wastes produced by the upstream tissues. A circulation system is a good technique for supplying constantly fresh fluid to a range of remote locations.

APPLICATIONS AND PREDICTIONS

1. Microbes living in environments where pressures vary from low to high will have thicker, stronger walls and smaller sizes.
2. Plant cells, with strong cell walls of cellulose, will be able to resist relatively low outside pressures.
3. Microbes will use different modes of locomotion than larger animals.
4. The blood flow regions of high velocity will have low pressures.
5. Very humid air that is less dense than dry air produces less lift on a wing. Flying birds must expend more energy to fly on such days.

2.10 Solid Mechanics

> The reasonable man adapts himself to the world; the unreasonable one persists in trying to adapt the world to himself. Therefore, all progress depends on the unreasonable.
>
> —**George Bernard Shaw**

Force is the effort variable and velocity is the flow variable in the field of mechanics. Within that context, force balances are very important to determine the mechanical state of an object.

2.10.1 Inertia

> According to you [Voltaire], morality is a very slight thing and ought to be subjected to physics. I say that physics ought to be subjected to morality.
>
> —**John Needham**

Isaac Newton stated three basic principles that all bodies must obey. His first principle states that

> A moving object will continue to move in a straight line at a constant speed, and a stationary object will remain at rest, unless acted on by an external unbalanced force.

This is called Newton's law of inertia because the rate of change of velocity must be zero unless there is a net force acting on the object. The object does not change either the magnitude or direction of velocity without a net force. There is an idea here for which the importance may not be apparent. That idea is that it takes a force, not only to accelerate or slow an object, but also to change its direction. Before Newton, it was thought that a heavenly body traveling in a circular orbit at a constant speed was not being affected by external forces. However, if you have ever twirled a weight on the end of a rope, you know that you must hold the rope and pull on it to keep the twirling object from flying off in a tangential direction. Let go of the rope, and the object flies away. Thus, only straight-line motion is maintained if there are no unbalanced external forces.

The reason that *unbalanced* external forces are mentioned is that as long as they are balanced, forces may be brought to bear on an object without any visible effect. Balanced forces mean that they are applied in opposite directions (see Section 6.20.4, Antagonistic Action). Thus, the force balance results in equal magnitude forces opposite in sign that sum to zero. It should never be assumed that because the sum is zero there are no applied forces. There may be applied forces that sum to zero for antagonistic muscular contractions (to maintain exquisite control) or for isometric exercise (where the balancing force is supplied by an object external to the body).

Forces must be balanced in all orthogonal directions to result in no velocity changes. Orthogonal directions are ones with no components in other primary directions. The x–y–z (length–width–height) Cartesian directions are orthogonal because they are each at right angles to both other directions. The r–θ–z (radius–angle–height) directions are orthogonal in the radial coordinate system. There are other orthogonal systems; forces must be balanced in each of the orthogonal directions for no velocity change to take place.

2.10.2 Acceleration

> One had to be a Newton to notice that the moon is falling, when everyone sees that it doesn't fall.
>
> **—Paul Valéry**

Newton's second law states that

> The acceleration produced on a body by an unbalanced external force is proportional to the magnitude of the force and inversely proportional to the mass of the object.

In equation form, this law becomes

$$\text{Force} = (\text{mass})(\text{acceleration}) \tag{2.10.1}$$

Because force is the effort variable and acceleration is the time rate of change of the flow variable, mass can be seen to be the inertia (or inertance) of the object.

From Newton's first and second laws, we can see that if there are no unbalanced forces, the object continues moving at a constant velocity; if there is an unbalanced force, then the body either speeds up, or slows down, or changes direction.

One force that is always present is friction, and friction is always in a direction that opposes movement or the tendency to move. Friction force is proportional to the normal force between the surfaces, and the coefficient of proportionality is the friction coefficient:

$$\text{Friction force} = (\text{friction coefficient})(\text{normal force}) \tag{2.10.2}$$

where the normal force is the force tending to clamp the two surfaces together.

Biological systems use a variety of schemes to either enhance or decrease friction forces. Flies walking on a wall use a sticky substance on their feet to maintain a high friction coefficient so that they don't slip and fall; ridges in the surface of the skin of the fingers (fingerprints) help to grip surfaces with the hands; fluids in knee and hip joints help to reduce friction to reduce energy consumption and joint wear.

Friction is usually smaller in fluids than between dry surfaces. Thus, oil is used as a lubricating medium, and most of the sliding motion takes place within the thin oil layer.

Sometimes it is desired to enhance friction differentially in one direction as opposed to another. A sawtooth surface where the incline is more gradual in one direction than another allows material to slide more easily in the direction of the less steep incline.

2.10.3 Reaction Forces

> Be careful reading health books; you may die of a misprint.
>
> —**Mark Twain**

Newton's third law states that

> For every action there is an equal and opposite reaction.

In other words, any externally applied force is met with an equal internal force in the opposite direction. An accelerating bird exerts a force on the air; the reaction force from the air on the bird causes the bird to accelerate forward. The cilia (small surface hairs) on a paramecium (a one-celled microbe) are moved when the surrounding fluid flows by; the cilia exert forces in return on the fluid, slowing the fluid. A runner pushes off against the floor; the floor pushes back and supplies the unbalanced external force needed to accelerate the runner. Should the runner start on a slippery surface, the reaction force cannot be made large enough to accelerate the runner (the famous Bambi scene on the frozen pond comes to mind).

UNIDIRECTIONAL MOTION

Movement in random directions is not usually beneficial to organisms. Instead, the organisms usually head toward food or favorable environments, or away from dangerous or depleted environments. To move in the desired direction, then, requires a means, and, usually, as fast or efficiently as necessary. There are a number of adaptations that make this possible. Engineers often use similar methods to accomplish their design goals of movement of machines or materials in the desired directions.

Awns are the stiff bristles attached to the seeds of wheat, rye, and some other grain seeds. As the seeds dry, the awns coil; when the humidity increases, or the seeds become wetter, the awns straighten out. This coiling and uncoiling motion helps to dig the seed into the ground. Backward facing hairs on the awns keep theseed from moving back toward the surface of the soil.

(*Continued*)

UNIDIRECTIONAL MOTION (Continued)

Sawtooth-shaped mechanisms have higher friction in one direction than the other, allowing movement in one direction only. Engineers use sawtooth-shaped mechanisms to accomplish the same purpose. They move straw from one end to the other in combine harvesters (these are called *straw walkers*) or unload heavy materials from truck beds (these are called *bed walkers*).

Cilia are small hairs in human lung airways and middle ears that beat rhythmically and push mucous with deposited particles toward the tube openings, where they are then removed from the protected organ. Cilia extend and stiffly push on the power stroke and fold and bend on the return or recovery stroke. Female mammals have beating cilia in the Fallopian tubes to move the ovum from the ovary to the uterus. Many bacteria have cilia attached to their outer membranes; these cilia beat to propel the bacterium in the forward direction. Similarly, humans use oars on boats to propel them along the water.

Peristaltic motion in the gut moves the contents in one direction from entrance to exit by pulsing a circumferentially squeezing motion that moves as a wave down the gut, moving a bolus of material ahead of the squeezed section. Peristaltic motion motivates mammalian male semen toward ejaculation. Similarly, peristaltic pumps move fluid inside a tube from the inlet to the outlet. One advantage of this type of pump is that the fluid never contacts the pumping mechanism, and so can remain in sterile condition.

Some bacteria have corkscrew tails, called *flagella*, which they use to propel themselves forward as the flagella rotate. A rotating flagellum pushes on the fluid surrounding the bacterium and sends it backward. Reaction forces then drive the bacterium forward. The same mechanism is used by engineers in worm drives and screw conveyors to move materials or mated gears in one desired direction.

Flapping wings can push a bird forward in flight, but only if the flapping motion is nonsymmetrical in forward and reverse directions. The flight stroke of bird wings traces a complex ellipsoidal path, with the down-stroke providing the majority of the thrust, and the up-stroke with the wing slightly folded inwards to reduce drag and reposition the wing for the next down-stroke. Birds change the angle of attack of their wings continuously within a flap. Although attempts have been made by many inventors, including Leonardo da Vinci, to reproduce wing action for mechanical flight, called *Ornithopters*, these attempts have largely failed.

Fluids, such as blood, are propelled by pulses generated by a beating heart. Without valves between heart chambers and between the heart and output vessels, the relaxation (or *diastolic*) phase of the heart beat would result in at least some flow back into the heart chamber that pushed the blood forward during contraction (or *systole*). This would be a very inefficient means to pump blood, and would be a severe evolutionary penalty.

Valves are also present in human veins, especially in the legs. These valves prevent pooling of the blood in the lower legs. Contraction of muscles surrounding the veins squeeze the veins and send the blood moving toward the heart. Malfunctioning valves cause a condition called *varicose veins*. Engineers use valves in similar situations when backflow must be eliminated.

(*Continued*)

UNIDIRECTIONAL MOTION (Continued)

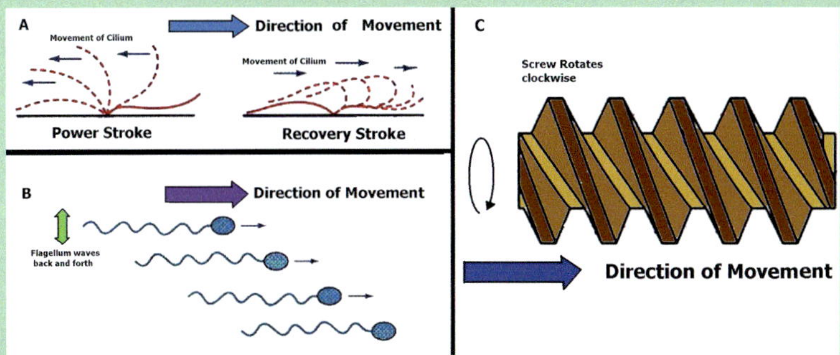

(A) Cilia propel a mass forward by pushing in one direction and folding on the return stroke, similar to the use of oars in a rowboat that are lifted out of the water during return, or to the path of a swimmer's arms or legs in the water. (B) Flagella act as propellers that push against a fluid as they rotate and reaction forces move the bodies to which the flagella are attached. (C) A screw acts similarly to a flagellum except that it is fixed and moves a material forward as it rotates.

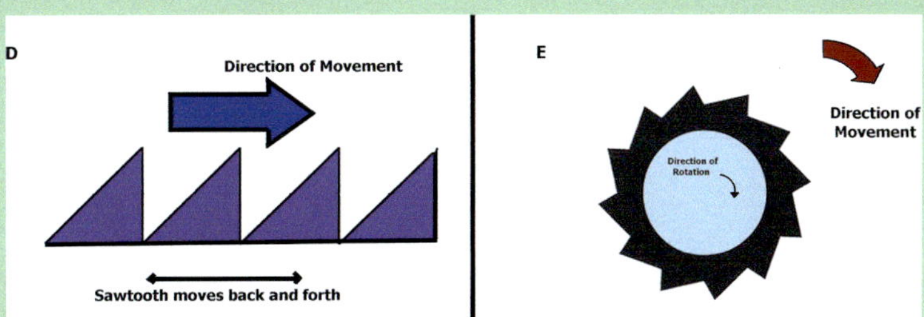

(D) A sawtooth-shaped device acts similarly to barbs on living things by offering a higher coefficient of friction in one direction than in the reverse direction; materials are thus moved from one end of the sawtooth to the other as the sawtooth shakes back and forth. (E) The sawtooth shape can be wrapped around a wheel.

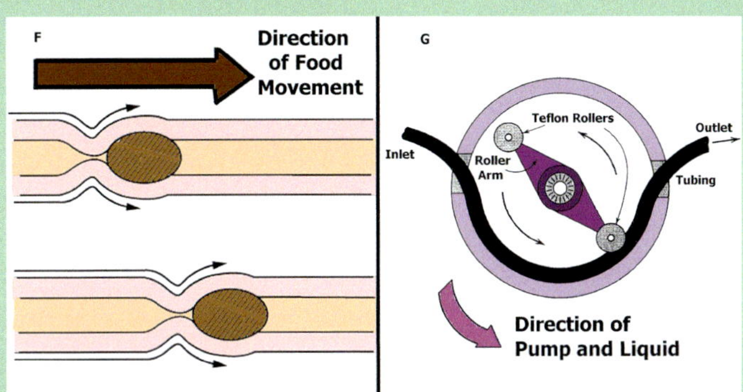

(F) Peristaltic motion moves mass by squeezing behind it; the squeezing proceeds from one end to the other. (G) The same peristaltic principle is used to pump liquids.

(Continued)

UNIDIRECTIONAL MOTION (Continued)

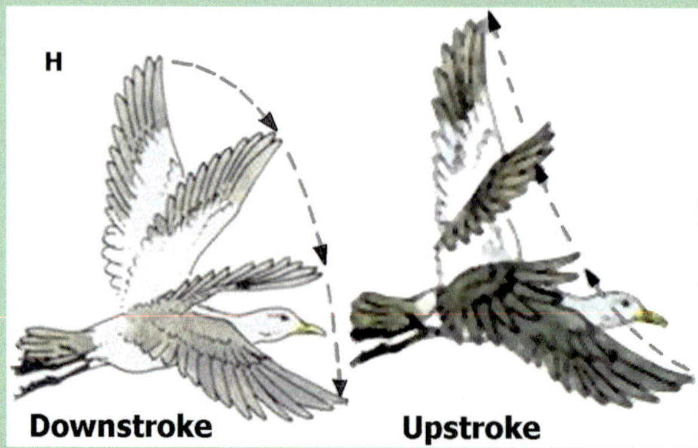

(H) A bird in flight moves its wings in a complex motion, unfolding them to provide lift and thrust during the upstroke, and folding them to reduce drag during the upstroke. The path of the wings is similar to that of the cilia in (A).

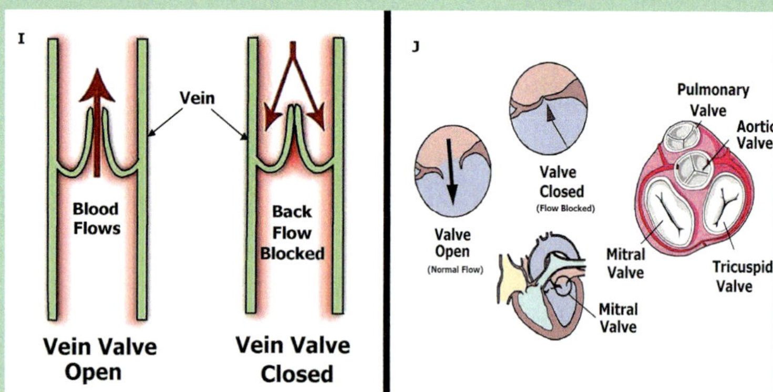

(I) Correctly operating vein valves assure that blood is pushed back to the heart rather than pooling in the legs due to gravitational action. (J) Valves between heart chambers keep blood from flowing backward during return (diastole) of the heart muscle.

2.10.4 Stress

The mark of true theories is their fruitfulness.

—Louis Pasteur

Force divided by area is pressure, and various materials are limited in the amount of pressure that they can resist (Figure 2.10.1). If the force tends to elongate the piece of material, then it is called a tensile force. Dividing this force by the cross-sectional area (that area perpendicular to the direction of the force) gives the tensile stress. Some materials, such as metals, have relatively high tensile stress values before they stretch to the breaking point. Others, such as concrete or a piece of bread, have low allowable tensile stresses.

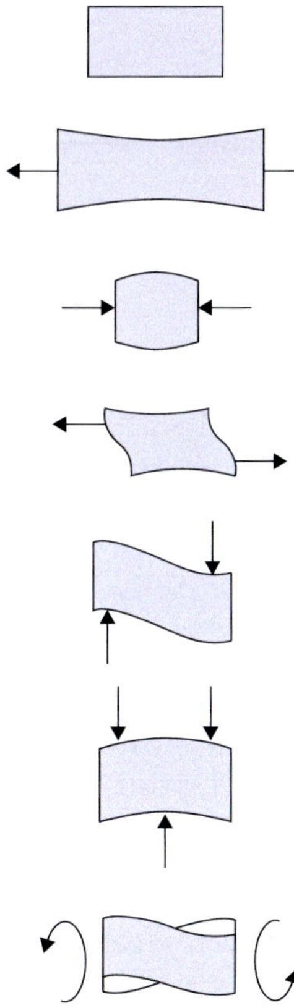

FIGURE 2.10.1
Different modes of mechanically loading a member. Force divided by area is pressure, or stress. The proper area to use depends on the type and direction of loading.

If the force is exerted in the compressive direction, then it is a compressive force. Dividing by the cross-sectional area gives the compressive stress. Materials such as concrete and diamond have relatively high allowable compressive stress values. Others, such as mud or flesh, have low allowable compressive stress values.

If the force is tangential to the object, and tends to rip the object apart, then it is a shear force. Dividing by the area in the direction of the force gives the shear stress. Shear stresses tend to deform the shape of an object, or make the object fail along the direction of the force. A piece of paper or peanut butter have very low shear stress values before they fail. Polymers have low allowable shear stresses in the direction of the polymer chains but higher allowable shear stresses perpendicular to the direction of the polymer chains.

There are other stresses that are important to some living systems. Tortional stress is one of these where the material is subjected to a twisting motion. Bending stress incorporates some tensile stress and some compressive stress in the material.

There is ample evidence that biological systems react to various stresses by strength enhancement, often by thickening of tissues where stress is concentrated.

A very typical biomaterials response to mechanical distortion is shown in Figure 2.10.2. The S-shaped curve indicates that the force needed to deform the material is extremely small at low deformations and extremely high at high deformations. The resting state of the material is usually near the middle of the curve where the slope is greatest, although

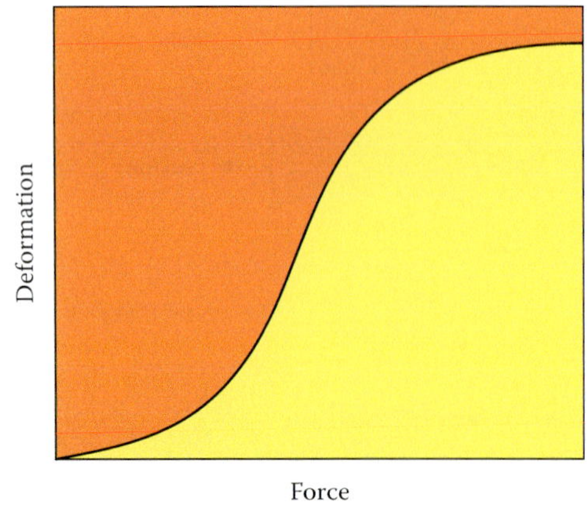

FIGURE 2.10.2
Force–deformation curve typical of biological materials. Deformation of the resting state of the material is usually somewhere in the center of the curve where the slope is greatest.

there may be some offset from the exact center (see Johnson, 1995). The S-shaped curve is also typical for other biological processes, such as bacterial growth and plant growth.

The curve represents a material that acts as a spring. As it deforms from its resting state, it stores energy, and this energy is recovered when the material returns to its resting state. The slope of the curve is related to the spring constant.

There is a limit to how much a biological material will deform without breaking. The material becomes stiffer, meaning that a much greater force is required to deform the material by even a little bit compared to the force required in the center of the curve.

Many biological materials exhibit this kind of characteristic curve. Many fresh plant materials show this behavior. So, does passive muscle tissue.

The inflations of the lungs and capillaries are similar. Indeed, this curve is so typical that it should be expected for all biological tissues. The S-shaped curve also describes microbial growth in a bioreactor and grass growth in a field. It fits enzyme activity on a substrate and many other biological phenomena.

BONE IS A PIEZOELECTRIC MATERIAL

Piezoelectric materials combine electrical and mechanical properties. Mechanical distortion causes these materials to generate electrical currents, and electrical current causes them to change shape. Quartz is a good example of a piezoelectric material used to maintain timing in an electrical device by vibrating at a frequency determined by the size and shape of the quartz crystal whenever an electrical current is applied. The same material can be used in pressure and force transducers in that they generate small amounts of current when deformed.

Bone is also piezoelectric, generating small currents whenever force is applied to them, which means for every step taken while walking or running, for instance. These small currents, in turn, seem to be necessary to maintain bone strength and health. Reversing the piezoelectric process, small electrical currents externally generated have been found to assist with bone healing. This interplay between electrical and mechanical properties helps maintain *homeostasis* of the human skeletal system.

Example 2.10.1: Insertion of Needles into Liver Tissue

The insertion of needles into soft tissue is needed for the medical procedures of drug delivery, minimally invasive surgery, and biopsy. Knowledge of needle insertion forces is necessary when contemplating robot-assisted medical procedures. When needles puncture soft tissue, the total force is the sum of stiffness force (mainly of the membranous capsule that surrounds the tissue), friction force (along the side of the needle as it slides along the punctured tissue), and cutting force (as the needle slices its way through). Stiffness of bovine liver tissue has been measured as the tissue deforms (Okamura et al., 2004).

Figure 2.10.3 shows these results. It can be recognized that the lines are curved, similar to the initial section of the general force–deformation curve in Figure 2.10.2. Because the curves end at the deflection where puncture occurred, the reverse curve at higher forces and deflections is not shown.

Example 2.10.2: Measuring Physical Properties of Biological Samples

Often the biggest challenge to make physical measurements on biological samples is the need to hold them, attach to them, or immobilize them without damaging them. Imagine trying to attach to a skin tissue sample in order to make a tensile strength measurement, or trying to measure the surface area of a banana, or tethering a honeybee in order to measure its metabolism while it flies. Vivek Jayaraman and his team at the Janelia Farm Reach Campus at the Howard Hughes Medical Institute found a way to immobilize a fly's head in a two-photon microscope while its wings and legs moved freely (Kresge, 2014; Seelig and Jayaraman, 2013). This research was an important milestone toward understanding how the fly brain extracts relevant information about a visual scene to guide its behavior.

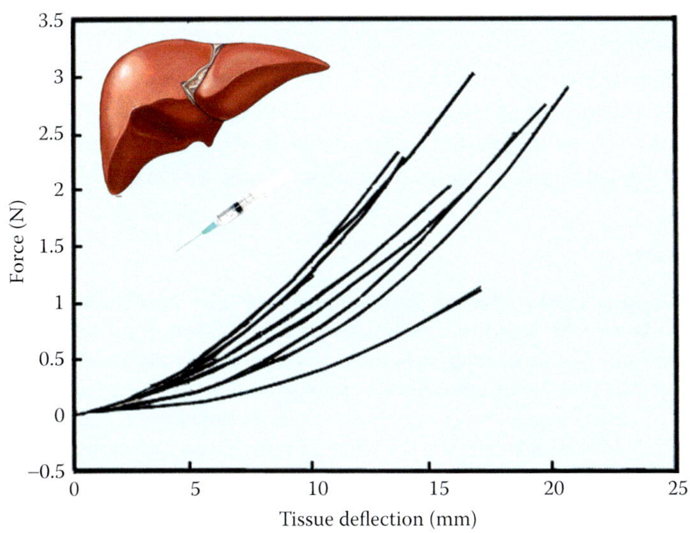

FIGURE 2.10.3
Force–deflection curves for multiple probes of a single bovine liver. The curves are truncated at the deflection where puncture occurs. Different curves indicate differences in blood vessel locations as well as other tissue inhomogeneities. (Redrawn from Okamura, A.M. et al., *IEEE Trans. Biomed. Eng.*, 51, 1707–1716, 2004.)

APPLICATIONS AND PREDICTIONS

1. Swimming in circles will require force.
2. Larger trees will sway less in the wind.
3. The friction coefficient when walking on ice may be too small to permit enough force to propel a body forward.
4. The bones of the joints will wear excessively if not lubricated.
5. Hair, especially wet hair, will reduce friction.
6. Bones must be unusually strong in compression.
7. The trunks of trees blown by high winds will be thicker than are those of trees not blown by winds. Most coconut palm trees must grow in places where the wind doesn't blow hard.
8. Erosion wears away hard surfaces.
9. The backbone is constructed to withstand compression, shear, and torsion.
10. Saliva reduces friction between teeth.
11. The cross-linking of biomolecules improves dimensional stability.

2.11 Electricity

I do not wish to give the impression that I think there is no mystery about consciousness. There is, for instance, something of a paradox connected with any attempt to localize it.

—Alan Turing

Responses to charged particles and electricity are sometimes important at the organismal level, but are very important at the subcellular level of biological systems. Charged particles arise when electrons are added or removed from atoms or molecules to form ions. *Anions* are negatively charged particles, whereas *cations* are positively charged.

2.11.1 Electrostatics

In this decade-long feud [between Jean Voltaire and John Needham], Voltaire took it upon himself to expose Needham, a well-known English naturalist, as a "dangerous biological thinker." Voltaire's methods included one that is not unknown in our own day: He suggested that Needham was a homosexual. Needham, in answer, scornfully referred to "so-called sages" who rigorously profess, but do not practice, celibacy – a shot at Voltaire's several love affairs, the latest of which was with Voltaire's own niece.

—Hal Hellum

The simple law of electrostatics is as follows:

Like charges repel
Unlike charges attract

Therefore, it is difficult to concentrate any given ionic species for two reasons: first, the concentration of any species must occur against a potential gradient that tends toward dispersion and, second, similar ions repel each other, which also tends toward dispersion.

Certain types of reactive cells are extremely sensitive to ionic charge. They act as electrostatic sensors to determine the presence of electrostatic fields. Other cells use electrostatic charges to transmit information (such as neurons). These cells concentrate certain types of ions, usually sodium, potassium, and, in the case of plants, calcium cations, and release them to the surrounding fluid whenever events are favorable. The mutual repulsion of similarly charged ions aids in this release.

PROTEIN SEPARATION BY GEL ELECTROPHORESIS

Charge and mass differences among proteins are used to advantage in gel electrophoresis. A mixture of proteins is introduced into a gel (typically agarose or polyacrylamide) and the mixture is then introduced into an electrostatic field. Negative charges on the proteins are attracted to the positive electrode. The gel not only acts as a lubricating medium in which the proteins can move, but it also supplies enough resistance to movement that the proteins are slowed. That allows the proteins with the largest charge to mass ratio to move faster than others. After a certain amount of time, proteins have separated sufficiently that they occupy different positions in the field. These are called *blots*. Identifying specific proteins requires calibration of the blots by performing

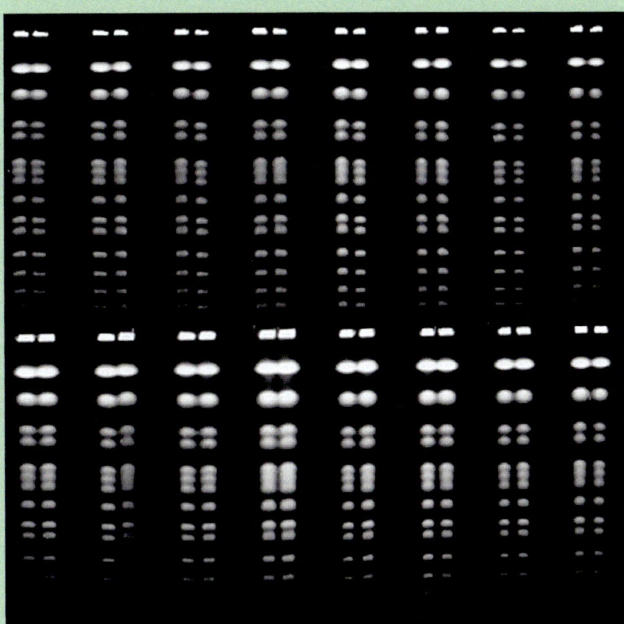

Gel electrophoresis is used to determine DNA snippets present. The snippets are produced when DNA strands are cut by restriction enzymes. The mixture is then mixed with an agarose or other gel and kept in an electrostatic field for a certain length of time. Identification of exact DNA components can be made by comparing DNA lines with known standards.

(Continued)

PROTEIN SEPARATION BY GEL ELECTROPHORESIS (Continued)

the electrophoretic process on proteins that are known. Times and distances may have to be adjusted in order to distinguish some protein lots from others. The calibration process is easier if the proteins to be separated differ by large net charges or masses.

Blots are usually made visible by the way light is used, or with nonspecific (polyclonal) antibodies to which chemoluminescent agents are attached. Specific (monoclonal) antibodies may be used to visualize individual protein blots.

Dielectrophoresis is a phenomenon whereby particles suspended in liquid or air can be made to locate in particular regions of a variable electric field (Hughes, 2003). It can be used to count and sort single cells for desired properties. It has the advantage of being able to separate individual cells without mechanical contact (Müller et al., 2003). Cell sorting by dielectrophoresis can be used as part of the cloning process. Individual cells can be manipulated, held in a stable position, oriented, spun, and observed from arbitrary sides by rotating in all three axial directions. High frequency electric fields are required because when the transmembrane potential, normally about 100 mV, exceeds about 1 V, the cell membrane breaks down (Müller et al., 2003). Higher frequencies have lower responses.

2.11.2 Electrical Current

A mule has neither pride of ancestry nor hope of posterity.

—**Robert G. Ingersoll**

An electrostatic field is usually measured as a difference in voltage. The electrical potential gradient, then, is a difference in voltage at any two or more points. When an electrical potential gradient exists, there is a tendency for current in the form of charged particles to flow from the regions of higher potential to lower. The rate of flow will depend on the electrical resistance between the points and the total amount of flow will depend on the number of charges residing in the system. We have identified this earlier as capacity.

Because there is a tendency for electrical current to flow, like charges away from each other, and unlike charges toward each other, the natural state is electroneutrality, where there are equal numbers of like and unlike charges. Any deviation from electroneutrality is either a transient situation or requires energy to maintain.

Ohm's law is the equation used to predict the flow of electric current:

$$\text{electric current} = \frac{\text{voltage difference}}{\text{resistance}} \tag{2.11.1}$$

If heat and temperature difference are substituted for electric current and voltage difference, then this equation would describe the flow of heat. If mass rate of flow and concentration difference were used instead, the equation would be equivalent to Fick's law for mass movement. Thus, Ohm's law is universal in form, and demonstrates the analogies among effort and flow variables introduced in Section 2.1 (Effort and flow Variables).

Electrical current in the body is usually the result of the movement of charged ions. Electrical currents in the body are important for the development of configuration of organs, appendages, and general structure. The DNA of all somatic cells is identical, yet, for most people, certain organs are located on the left and some on the right side of the body. There are electrical gradients in the developing embryo from one side to the other. This may either be the determinant of the position of organ growth, or a side effect, but both seem to be related.

In addition, for those species capable of appendage regeneration (such as various amphibians), electrical gradients and associated cellular ion channels, also seem to be involved. Manipulation of bioelectric signals in these animals can cause the growth of extra limbs or different placement of other organs. The hope is that someday, regeneration of human lost appendages may be possible (Piore, 2017).

DNA electrical conductivity can be used as the basis for biosensors giving information about the state of the organism they are associated with. DNA has an electrical conductivity equivalent to that of other conductive polymers, but the actual conductivity value depends on DNA configuration and composition, which is why DNA configuration can be detected electrically. Genetic mapping can be achieved by stretching the DNA molecule and then pulling through a micropore. Electrical resistance of each nucleotide can be measured as each pass through the pore. Alternatively, specific dyes can be attached to the DNA molecule and the individual regions identified by fluorescence microscopy (Bogas et al., 2017).

Neural signals and muscular contraction both involve the flow of electric current, so can be used to monitor the activities of these tissues inside the body. The *electrocardiogram* (ECG) measures heart muscle activity, the *electromyogram* (EMG) measures skeletal muscle activity, and the *electroencephalograph* (EEG) measures brain activity. These measures are usually taken with electrodes placed somewhere near, and on either side of, the source of the signal. They are usually voltage measurements made without significant current flow.

When current does flow through the body, there can be trouble. This can occur when people or animals are connected to a recording instrument, or even through an inadvertent electric pathway from the source of electricity to ground. The biological engineer who designs or uses electrical devices (both machinery and instruments) around humans and animals must be especially careful to isolate the patient from electric shock. The easiest way to do this is to be sure that there is not a complete path from source to ground with the patient in between and completing the circuit. Without a complete electrical pathway, there is no current flow.

Current introduced to the body is often held to low values by the electrical resistance of dry skin, between 5,000 and 10,000 Ω (Kantrowitz, 1972). Wet skin will have electrical resistance levels as low as one-tenth of this. Where electrical leads are introduced subcutaneously, resistance falls to 10–50 Ω, and this situation is potentially dangerous. Subcutaneous leads are used for heart pacemakers, artificial hearts, transcutaneous electrical neural stimulation (TENS) to block pain, and subdural leads for the control of Parkinson's disease.

Current density is also important, because an electrical current spread throughout a large area will have less effect than a current concentrated in a particular spot. For the very worst condition, concentrated current can burn the affected area by overheating. In Table 2.11.1 are listed various current limits and their effects. Remember that some people are more sensitive than others to electric shock.

Implanted electrodes are sometimes used to sense neural signals from the brain. These signals need to be amplified to be detected in a useful form. The amplification system should consume very little power to keep heat generation to a minimum.

The safety limit for cranial use is 30 milliwatt (mW) of power, and the surrounding tissue temperature should be raised by no more than 1°C; otherwise cells will be killed (Patel, 2009).

As frequency increases higher than 50–60 cycles/s, the danger of electric shock decreases dramatically (Figure 2.11.1). Currents that could cause death at 60 cps are

TABLE 2.11.1

Physical Effects of Electrical Shock at 60 Hz

Current Value	Effects
Macroshock (external electrodes)	
0–500 μA	No sensation
0.5–2 mA	Sensation threshold; faint tingle
2–10 mA	Pain becomes objectionable; muscle contraction
5–25 mA	Current becomes more painful; mechanical injury, due to powerful muscle contractions; heart and respiratory functions unaffected (let-go current limit, where it becomes impossible to release contact, occurs at about 16 mA, with men's thresholds 20% higher than women's thresholds)
25–100 mA	Respiratory paralysis; death possible
100–3000 mA	Ventricular fibrillation and subsequent death unless cardiac resuscitation follows promptly; nerve damage
Above 3000 mA	Cardiac standstill resulting from depolarization of entire heart muscle; severe tissue destruction by heating; death probable
Microshock (internal electrodes)	
0–10 μA	Safe for a normal heart
10–800 μA	Ventricular fibrillation threshold

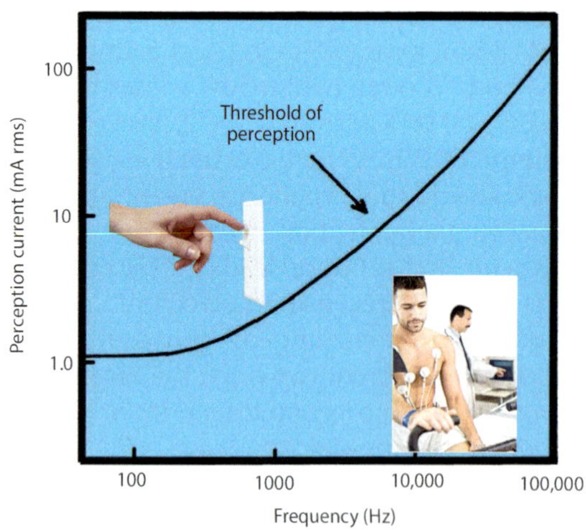

FIGURE 2.11.1
Sensation of electrical current at different frequencies. The most sensitive region is located at normal power line frequencies. (Redrawn from Dalziel, C.F., *IEEE Spectr.*, 9, 41–50, 1972.)

imperceptible at very high frequencies. This is because there is a short time after neural or muscle depolarization during which the nerve or muscle is insensitive to further stimuli (see Section 4.4.3).

2.11.3 Electrical Power

> If we define biomedical engineering as the systematic application of existing science and technology to medical problems, our field has its real beginning in the renaissance period. Its early development was inspired by and dependent upon the simultaneous development of physics; it was based on the belief that living organisms obeyed the same laws as inanimate objects and could be described in terms of these laws.
>
> **—Edwin N. Lightfoot, Jr.**

Electrical power is the product of current and voltage. Power is useful because it can be transformed into work rate, or, more biologically, into heat. Electrical power delivered to a tissue causes heating for either tissue destruction or healing. Focused electrical power, usually at very high frequencies, can be used to ablate small tumors and scar tissue, and has been used to correct vision through the process of *conductive keratoplasty* (Panescu, 2004). In this procedure, the heating of the periphery of the cornea denatures tissue and shrinks collagen fibers. The peripheral tightening results in a flatter cornea and improved optical focus.

In order for focused electrical power to have the desired effect, heat must be generated faster than local heat transfer mechanisms can carry it away. For the cornea, this is relatively easy because there is no blood supply to the cornea (in order to maintain transparency), but for tumors that typically have well-developed vascularity to supply the great metabolic needs of tumor tissue, heat can be removed rapidly by blood convection. It is difficult, under such circumstances, to kill all tumor cells without damaging surrounding tissue.

CONTROLLING ELECTROSTATIC DISCHARGE

Charges can easily be generated on the hands of personnel who slide, walk, or otherwise move across nonconductive materials. When these people contact sensitive electronics, they can easily damage such devices. The problem becomes much worse when the humidity of the air is very low because accumulated charge cannot leak off into the air (see the following table). With the inclusion of microprocessors and other sensitive electronic components in modern medical devices, electrostatic discharges can not only damage the devices, but also compromise the health of patients. It is estimated that electrostatic discharge causes many billions of dollars of damage per year worldwide (Vermillion, 2008). Procedures need to be established to minimize the risk of damage. Such things as washing hands, touching grounded objects, or isolating sensitive components from direct contact with the hands are often used.

(*Continued*)

CONTROLLING ELECTROSTATIC DISCHARGE (Continued)

Voltages from Charging at Low and High Relative Humidity

Action	20% RH (kV)	80% RH (kV)
Walking across vinyl floor	12	0.25
Walking across synthetic carpet	35	1.5
Arising from foam cushion	18	1.5
Picking up poly bag	20	0.60
Sliding styrene box on carpet	18	1.5
Removing mylar tape from PC boards	12	1.5
Shrinking film on PC boards	16	3
Triggering vacuum solder remover	8	1
Aerosol circuit freezer spray	15	5

Example 2.11.1: Electroporation Allows Genetic Material to Penetrate the Cell

Genetic manipulation, in order to introduce new genetic material into target cells, requires that the large nucleic acids making up the DNA must first move past the cell membrane. Cellular membranes have evolved to protect cells by excluding foreign materials. Under normal circumstances, the cell membrane is an impenetrable obstacle to large molecules.

Various means have been attempted to breach this barrier. Viruses can be used as vectors to carry genetic material into the cell through natural means. Mechanically disrupting the membranes with acoustic energy or by shooting inert tungsten metal pellets through the membrane (called *biolistics*) can sometimes be successful. Another means is electroporation, in which the cells are subjected to a relatively large electrical impulse leading to the temporary formation of pores in the cell membrane. Molecules, such as DNA, RNA, ATP, proteins, peptides, and other small molecules can pass into the cell before the pores are repaired. Electroporation pores can close relatively rapidly after an electrical pulse ends, thus limiting the types and quantities of materials that can transfect the cell. When the pores close, materials are locked inside the cell. Certain cell-specific transfection solutions help stabilize the pores formed by electroporation and permit more materials to enter the cell. Once inside the cell, genetic material finds its way into the nucleus and may then combine with native genetic material. *Non-thermal irreversible electroporation* (NTIRE) uses brief pulses of electricity to ablate tumor cells.

Bulk electroporation used to introduce genetic material or proteins into a host of target cells can stress or kill many of the cells, and only a small proportion of the cells successfully incorporate the introduced material. *Tissue nanotransfection* involves a chip containing an array of tiny channels to create one or more tiny holes in the cell membrane so that DNA can be injected in a controlled manner (Gallego-Perez et al., 2017). Reprogrammed cells can then secrete the introduced materials into extracellular vesicles, which affect other cells that have not been originally targeted. These extracellular vesicles have been found to be as effective as the chip.

Example 2.11.2: Use of Electroporation to Increase DNA Vaccine Effectiveness

Vaccines typically use an inactivated pathogen injected into a host to elicit an immune response. When subsequent exposure to a live pathogen happens, the immune system is primed to respond and fight the infection (see Section 6.20.3, Redundancy). New approaches to vaccine development use snippets of DNA called plasmids (see

Section 5.3.13, Plasmids) instead of the entire pathogen. These plasmids enter the cells of the host, produce proteins identified as belonging to the pathogen, and elicit an immune response without any danger that could come from the entire pathogen. The host's own cells do the work of eliciting the immune response (Morrow and Weiner, 2010).

It is the protein produced from the plasmid that provokes the immune system, not the DNA plasmid itself. Thus, a critical step in the process is moving plasmids into enough of the host's cells so that the cells can begin producing the protein in large enough quantities.

DNA material does not easily translocate across the cell membrane. For this, a vector is needed. Vectors can include adenoviruses (common cold viruses) for humans or agrobacteria for plants. However, adenoviruses by themselves provoke an immune response that may not allow the plasmids to be incorporated into host cells.

Electroporation momentarily opens pores in the cell membranes to allow injected plasmids to translocate into host cells. This is thus a preferred method to move DNA past the cell membrane barrier.

DNA vaccines have been developed to treat or protect against human immunodeficiency viruses (for humans), West Nile virus (horses), infectious hematopoietic virus (salmon), melanoma (humans and dogs), fetal loss (pigs), hepatitis C (humans), human papillomavirus (humans), and liver cancer (humans).

Example 2.11.3: Deep Brain Stimulation

Deep brain stimulation via an electrode array implanted at some location in the brain is a standard treatment to calm the tremors of Parkinson's disease. It can also be used as an intervention for depression, obsessive-compulsive disorder, or other psychological disorders. Magnetic coils are placed against the scalp and induce electrical currents in the brain in response to magnetic pulses. These can either increase or decrease brain activity. This technique has been used successfully to block migraine headaches when they first appear.

Example 2.11.4: Functional Electrical Stimulation Helps Overcome Spinal Cord Injuries

Functional Electrical Stimulation (FES) can deploy pulses of electricity to activate nerves controlling limb muscles rendered dormant when spinal cord connections are severed. Simple FES systems use surface electrodes that stick to the user's skin and faintly stimulate nerves by transmitting pulses in the general direction of the nerves. More precise control can be achieved with implanted wires contacting specific nerve fibers to be stimulated. Stimulation patterns can be computer coded to realize the type of muscle action desired. The result is that spinal cord patients can regain much of their pre-injury functionality.

The Cybathlon is a competition combining FES stimulations with athletic events. Paralytic and amputee participants, called "pilots" rather than "athletes," take part in competitive events that showcase human-machine collaborations. Cybathlon events include routine tasks of daily life incorporated into the overall competition in order that gear developed for the competition is useful beyond the competition itself. For instance, the race for people with powered leg prostheses includes stair climbing and walking across stepping stones, and the obstacle course for amputees with arm prostheses includes slicing of bread loaves and opening jars of jam (Strickland, 2016).

Example 2.11.5: Reversed Electroporation Kills Cells

Electroporation can be used in reverse to kill unwanted cells that would otherwise form scars over burn wounds. A pulsed electric field (PEF) creates holes in cell membranes that allow cell contents to leak out, thus destroying the unwanted cells. PEF techniques are also used to ablate tumors and kill pathogenic bacteria in food (Allen, 2016).

APPLICATIONS AND PREDICTIONS

1. Large proteins with many local surface charges will attract ions of the opposite charge.
2. The presence of a charge will repel ions with a like charge.
3. ATP, the universal energy-rich chemical compound, can have both chemical and physical effects. The physical effect comes from local surface charge.
4. Electrophoresis can be used to separate proteins.
5. The maintenance of a cell potential of about −50 to −70 mV (inside negative relative to outside) will require energy.
6. Electrostatic effects will usually be felt at a small scale rather than a large scale.
7. Electric charge will cause a person's hair to rise.
8. Electric currents through the body will cause pain or damage at low frequency; nothing at high frequency.
9. Conductive fluids present in hospitals pose risks of electric shock. Dry skin is a good insulator.

2.12 Temperature Effects

> Almost all our perceptions have corresponding sensations which constantly accompany them, and, on that account, are very apt to be confounded with them. ...When I smell a rose, there is in this operation both sensation and perception.
>
> **—Thomas Reid**

Temperature is a measure of the heat content of a substance and has profound effects on the physical world. Many physical properties are highly temperature dependent. This influences the rates at which certain processes proceed and the amounts of energy needed to complete them. For example, the viscosity (or thickness) of gases increases with temperature but the viscosity of liquids decreases, sometimes dramatically, as temperature increases. While the viscosities of gases are usually small enough that the resistance to movement of gases, or through gases, is usually small, resistance to movement of, or through, liquids is much larger, with consequently requiring large amounts of energy. Living systems that move liquids or move through liquids therefore can do this more efficiently at higher temperatures.

The ease at which substances dissolve in solvents is temperature dependent. Gases dissolve in water much more readily at lower temperatures. Contrarily, liquids and solids usually dissolve more readily at higher temperatures. The water solubility of a material may be profoundly affected by temperature.

Other physical properties can be extremely temperature dependent. The density of a gas, for instance, is inversely proportional to absolute temperature (Equation 2.3.2).

Mass diffusion is greater at higher temperatures because the mass diffusivity increases with higher temperature. This result is more profound for gases than liquids. Thus, mass transfer is facilitated at higher temperatures.

Heat is transferred more rapidly from higher temperatures. Because at higher temperatures, the removal of heat is often more important than the acquisition of heat, enhanced heat transfer is beneficial. Because higher temperatures imply higher heat contents, systems at higher temperatures are usually more energetic than at lower temperatures.

Temperature also affects the state of matter. Higher temperatures favor more energetic states of matter, such as liquid over solid and gas over liquid. The change from one state to another often occurs at a very abrupt temperature for a pure substance. Heat is liberated if the transition is from gas to liquid or from liquid to solid. The temperature of the mixture of the two phases does not change during the transition until there is a clear predominant phase in the mixture. Thus, liquid water turning to ice will linger at 0°C until nearly all the liquid is gone.

The more complex substances found in biological systems have more complex phase transitions as well. Even a mixture of one solute dissolved in water will freeze over a range of temperatures because, as the liquid water freezes, the remaining solution becomes more concentrated. Indeed, this method has been profitably used to concentrate solutes, such as table salt and apple juice. For the solutions or suspensions of many solutes, the phase transition from liquid to solid can be very gradual.

There are also temperature effects on different solid or liquid phases. Transitions from one type of solid or liquid phase to another can occur at different temperatures. For example, at certain temperatures, the material might exist as a crystal and, at other temperatures, it might exist as a glass. Both of these are solids.

APPLICATIONS AND PREDICTIONS

1. Molasses will be thicker in January.
2. Oil or grease can be cleaned more easily with hot water than cold water.
3. The skin will usually be warmer when the body is overheated.
4. Residual sweat after exercise can sometimes cool the skin enough to seriously impair the loss of excess body heat.
5. Gases or liquids will mix more rapidly at high temperatures. To keep them separate longer, lower the temperature.
6. To discourage mold or bacterial growth, heat the air to prevent water condensation.
7. Body metabolism tends to increase as body temperature increases.
8. Skin blood flow is controlled by temperature. Reduce swelling by cooling the area.

Questions

2.0.1 Explain in your own words how knowledge of physics aids the understanding of biological systems.

2.0.2 How is biological engineering design related to the study of physics?

2.0.3 Physical, chemical, and biological components of the environment interact with living things. Give examples of the physical environment of organisms.

2.0.4 Can you think of additional physical principles that relate to biology? If so, list them.

2.1.1 Describe the behavior of a biological organism. In this description identify the effort and flow variables. Remember that effort variables don't describe things that move; that is what flow variables do.

2.1.2 Describe the effects of resistance. If resistance did not exist in a plant or an animal, what would be different?

2.1.3 Give examples of ways living things have to overcome resistance in some of their functions.

2.1.4 Give examples of ways living things have to reduce resistance in some of their functions.

2.1.5 Give specific examples of the appearance of resistance in an ecological community.

2.1.6 Describe the meaning of capacity. Give examples of the appearances of capacity in living things.

2.1.7 Give examples where capacities in biological systems should be minimized. Give examples where it should be maximized.

2.1.8 Describe the effects of inertia on living things. Give examples of inertia.

2.1.9 Why is inertia important in biological systems? What effect does inertia have on the amount of energy needed by a biological system?

2.1.10 Why are the absolute values of physical quantities nearly (if not absolutely) impossible to make?

2.1.11 Add to the list of Applications and Predictions.

2.2.1 Describe the number of cells in the human body in terms of a balance.

2.2.2 If the concentration of cells in a bioreactor is increased, how would this affect an oxygen balance written for the bioreactor? How would it affect a heat balance?

2.2.3 If resistance, capacity, and inertia of the circulatory system are known, what additional information must be obtained in order to determine blood pressure?

2.2.4 Describe an experiment to measure the rate of a chemical movement into a cell, given that the rate cannot be measured directly. What assumptions must be made?

2.2.5 Under what conditions can the rate of energy storage in a microbe be zero if the organism is incapable of generating its own energy?

2.2.6 Explain how the rate of oxygen perfusing a tissue is related to a balance on the capillaries.

2.2.7 Add to the list of Applications and Predictions.

2.3.1 What is the concentration of oxygen in air? How does concentration relate to density?

2.3.2 What is the vapor pressure of boiling alcohol?

2.3.3 Look up atmospheric pressure variation with altitude. At what height would human blood boil?

2.3.4 Classify the type of solid represented by

- Cell membrane
- Bone
- Tendon
- Muscle
- Skin

2.3.5 Phase change in a biological material does not usually occur at one specific temperature but, rather, occurs over a range of temperatures. What do you think is the explanation for this?

2.3.6 Describe a gas plasma and state where gas plasmas are important in biological engineering.

2.3.7 Add to the list of Applications and Predictions.

2.4.1 What distinguishes between a path function and a function not dependent on path? Why can the value of a path function not be totally determined from its values at the start and finish?

2.4.2 Give the difference between positive and negative work. What happens to potential energy during each of these types of work?

2.4.3 Diagram the process of converting chemical potential energy in the form of food into mechanical energy in the form of running.

2.4.4 Does a growing plant perform mechanical work as it lengthens? Why or why not?

2.4.5 Diagram the relationship among chemical energy in food, physiological work, and mechanical work.

2.4.6 What do we mean by efficiency? How would we determine the energy efficiency of a clam?

2.4.7 Add to the list of Applications and Predictions.

2.5.1 There is a spontaneously occurring process (let's say, for example, the oxidation of biological wastes). What does that say about the relative magnitudes of enthalpy and waste heat? Does the process result in work being done on the environment or by the environment?

2.5.2 Consider Figure 2.5.1. What constitutes waste, and how does it differ from heat and work?

2.5.3 Add to the list of Applications and Predictions.

2.6.1 If living things are ordered, and order requires energy, what living things are the most likely to survive the longest without additional energy?

2.6.2 What types of biological processes require energy expenditure? How do these lead to maintenance of order?

2.6.3 Add to the list of Applications and Predictions.

2.7.1 Explain why human hands and feet feel cooler than the trunk of the body on a cold day.

2.7.2 How do animals manipulate their surface areas to increase or decrease heat exchange? Contrast the expected heat transfer responses by a dog and a lizard on a sunny, cold day.

2.7.3 How does circulating blood contribute to heat exchange with the environment?

2.7.4 Animals living in hot climates are usually leaner than animals living in cold climates. Considering heat exchange issues, would you expect microbes adapted to hot climates to be shaped differently from those adapted to cold climates?

2.7.5 How does circulating blood help to maintain the correct heat balance of the body?

2.7.6 Explain what counter-current heat transfer is. Why is it important for animals? Would you expect it to be present in plants?

2.7.7 Add to the list of Applications and Predictions.

2.8.1 Speculate on what would happen to body size if atmospheric oxygen percentage increased to 50%.

2.8.2 Would you expect a faster diffusion rate to the environment from a larger plant or a smaller plant?

2.8.3 Estimate the relative magnitudes of resistance to diffusion in a plant leaf, stem, and roots.

2.8.4 State the ways to enhance material movement from the environment into an organism.

2.8.5 Give a physical justification for circulating fluid in a body.

2.8.6 Given the principles stated by West et al. (2000), compare circulatory systems of small and large animals.

2.8.7 Describe osmosis. If water is the only molecule that can pass freely from one side of the membrane to the other, how is equilibrium achieved?

2.8.8 What types of materials are expected to move easily (without large energy expenditure) through the cell membrane?

2.8.9 Add to the list of Applications and Predictions.

2.9.1 If viscosity of water were higher, what would happen to the required size and strength of the heart?

2.9.2 Does laminar or turbulent flow occur at low fluid velocities? If it was desired to mix a hormone in blood so that it would have a reasonably constant concentration, would laminar or turbulent flow be required?

2.9.3 Given that laminar flow occurs at low Reynolds numbers, if it was desired to maintain laminar flow in a channel, suggest means to promote laminar flow.

2.9.4 Describe what it would be like for us to move around if we had the same Reynolds number as a microbe.

2.9.5 A bison faces into the wind of a storm. On what part of the body would you expect the fluid pressure to be the highest? Why?

2.9.6 Describe the engineering trade-offs involved in the design of a natural heart. Compare these with the design trade-offs for an artificial heart.

2.9.7 Why are smaller blood vessels short? Why are larger blood vessels short?

2.9.8 If internal hydrostatic pressure can substitute for shell rigidity in a soft-shell crab, why does the shell need to harden?

2.9.9 Of what importance is the Law of Laplace to biological organisms? How does the Law of Laplace explain shapes common in biology?

2.9.10 If moving bacteria had the same Reynolds numbers as fish, how fast would they swim?

2.9.11 How would you outsmart the beaver?

2.9.12 Give other examples of places where a circulatory system can help deliver something uniformly to various locations.

2.9.13 Add to the list of Applications and Predictions.

2.10.1 Will a car traveling straight use more, less, or the same amount of fuel as the same car going the same speed along a winding track?

2.10.2 Isometric muscular exercise results in no visible movement. What is the mechanical efficiency of this exercise? What happens to the energy used? Because there is no movement, can we assume that there is no force generated by the muscles?

2.10.3 Is Newton's second law theoretical or empirical? What is the difference?

2.10.4 List advantages and disadvantages of living in a world without friction. Make your list from the standpoint of a human, a worm, and a fungus.

2.10.5 Speculate on the relative magnitudes of the reaction forces for a small bird and a large bird. What effects do these birds have on the surrounding air?

2.10.6 Living organisms have unique ways to deal with large mechanical stresses. List some of them.

2.10.7 Why is springiness important in biology? What would be the consequences if biological materials were not elastic?

2.10.8 Why do you suppose most biological materials have the shape of the force–deformation curve given in Figure 2.10.2?

2.10.9 If you were to design a device to insert a needle automatically into the liver, what mechanical considerations would you have to account for?

2.10.10 Add to the list of Applications and Predictions.

2.11.1 Give instances where ionic charges can help or hinder an engineering design.

2.11.2 How is gel electrophoresis used to identify genetic matches or mismatches?

2.11.3 How are electrophoresis and dielectrophoresis different?

2.11.4 List effects of electrical current flow in a living body. What kinds of charged particles carry the flow?

2.11.5 If electrical currents are much less dangerous at high frequencies, why is electrical power transmitted at 50 or 60 cps?

2.11.6 At what intensity does electrical current become lethal? Justify your answer.

2.11.7 What things limit the use of focused electrical power for the removal of unwanted bodily tissue?

2.11.8 Propose a means to deliver sufficient electrical power to a small volume in the pancreas to kill cancer cells.

2.11.9 How does electroporation work?

2.11.10 Explain how DNA vaccines work. How is that different from normal vaccines?

2.11.11 Add to the list of Applications and Predictions.

2.12.1 If the viscosity of blood decreases with temperature, would you expect the work of the heart to increase or decrease as environmental temperature increased? Be sure to examine thoroughly the changes that occur at high temperatures.

2.12.2 Would flagellated microbes find it easier or harder to move within cooler liquids? Why?

2.12.3 What kinds of stresses on aquatic life forms are changed as the water around them is heated? Are the changes greater or less?

2.12.4 Your neighbor calls to warn you that the temperature during the night is supposed to drop to –1°C. He urges you to pick the remaining apples from your apple tree before they freeze. You say that you are not concerned that they will freeze. Why do you say that?

2.12.5 Distinguish between the effects of low temperature and negative heat flow.

2.12.6 Add to the list of Applications and Predictions.

3

Principles of Chemistry

> ...every day is bringing us to look more and more to chemistry to explain the physiology of our own bodies.
>
> **—Peter Mere Latham**

Chemistry is the branch of knowledge that deals with the composition of substances, the relation of properties to composition, and interactions among substances. There is a certain overlap of chemistry with physics because each of these substances is present in the physical world and subject to the world's physical laws. However, much of what distinguishes chemistry from physics is the attention given to electrons and their energies in the field of chemistry. It is the interactions among electrons from different elements that cause compound materials to form.

The reason to study chemistry is to understand many of the basic functions of life. General chemistry is the study of the interactions among different substances, especially those involving the electrons whizzing around the nuclei of those substances. Biochemistry relates more to configurations of molecules, and how they interact because of their shapes. Physical chemistry is the field that considers physical environmental effects, especially temperature, pressure, and interactions among mixtures of compounds on chemical activities. All of these are important in the understanding of intracellular processes and of the responses of the living entity (the biological unit) to its physical, chemical, and biological environs (Figure 3.0.1). In this chapter, we consider some of the details of chemical processes important for life and important to understand how living systems can be manipulated. Such is the fundamental interest of a biological engineer.

Relevant chemical principles are given as follows:

1. *There is a periodicity of properties of elements, and these properties are related to numbers of electrons and their energy states in elemental atoms.* A classification of elements is possible, based on similarities and differences. Biological substitution of one element for another is thus possible when the preferred element is scarce, as explained in Sections 3.1 (Periodic Table) and 3.2 (Chemical Bonds).
2. *Elements can combine to form compounds with different properties.* Compounds, especially those based on carbon, are the basis for most of the biochemical complexity of living things. See Sections 3.2 (Chemical Bonds) and 3.6 (Carbon).
3. *Molecular configurations determine usages.* Highly complex molecular structures may be formed when many atoms representing various elements are joined together. Particular configurations are created when locally unbalanced electrical charges either attract or repel other charges, either in the same compound or in neighboring compounds. Functions of many biochemicals depend on the ways in which they physically fit together with other compounds. See Section 3.7 (Water Chemistry).

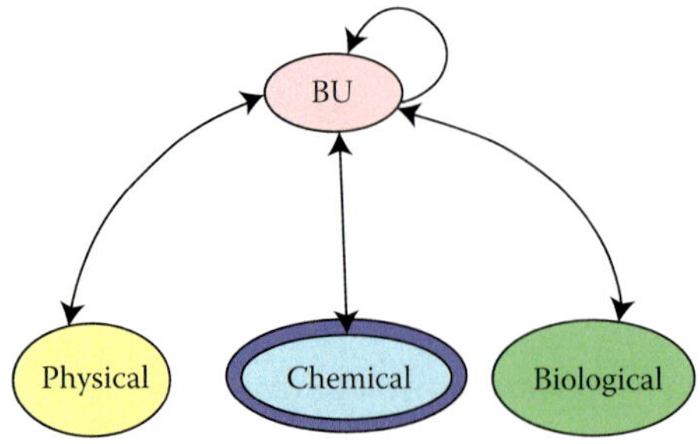

FIGURE 3.0.1
The biological unit (BU) is affected by interactions with its physical, chemical, and biological environment. Likewise, environmental elements are affected by the BU. Self-adjustment is also a possibility. The chemical environment determines many basic responses.

4. *Chemical reactions occur spontaneously when they yield energy to the environment.* These reactions are called exothermic, or heat generating, reactions. They are important because they are the sources for energy in biological systems. Certain compounds can act as chemical energy stores, and the energy can be made available for other uses. Especially important are those chemical reactions that require energy from the environment to proceed, as explained in Sections 3.4 (Acids and Bases), 3.9 (Enzymes), and 3.11 (Temperature and Pressure).
5. *Reaction rates depend on reactant concentrations, temperatures, and pressures.* In order for a chemical reaction to occur, there must be contact between the reactants, and the contact must be energetic enough to overcome atomic repulsion. Biochemical reactions are notable because compounds, such as enzymes, bring reactants in close proximity and increase local concentrations, thus promoting energetic contact. Temperature increases molecular kinetic energy, and pressure increases concentration. See Sections 3.5 (Reaction Rates) and 3.10 (Energy).
6. *Intermediate reactions are most important to living things.* This includes intermediate energy and intermediate rates. Fast reactions, or ones that deliver too much energy, or those with products that are too stable, cannot be readily controlled. Slow reactions use enzymes to speed them. A common means has not been developed to slow fast reactions. Biochemicals cannot be too basic or too acidic. Weak acids and bases can be buffered readily; strong acids and bases cannot, and so tend to extremes. See Sections 3.4 (Acids and Bases), 3.7 (Water Chemistry), 3.9 (Enzymes), and 3.10 (Energy).

3.1 Periodic Nature of Elements

> To say...that a man is made up of certain chemical elements is a satisfactory description only for those who intend to use him as a fertilizer.
>
> —**Herbert J. Muller**

When the elements are arranged by increasing mass, it soon becomes apparent that there is a periodic nature to elements with similar properties. This periodicity is even more readily apparent if the elements are arranged according to increasing number of electrons they possess (Figure 3.1.1). For instance, most elements react with other elements, but there are some elements like helium, neon, argon, krypton, xenon, and radon, with 2, 10, 18, 36, 54, and 86 electrons, respectively, which do not react. Each of these exhibits similar physical properties as well; each of these is a gas, for instance. The elements that directly follow the inert gases—lithium, sodium, potassium, rubidium, cesium, and francium, with 3, 11, 19, 37, 55, and 87 electrons, respectively—are all metals with shiny surfaces and are good conductors of heat and electricity. They are called the alkali metals because they all form solutions in water that counteract acids.

The periodic table of the elements

Period \ Group	1	2	3	4	5	6	7	8	9
1	1 1.008 H Hydrogen								
2	3 6.941 Li Lithium	4 9.012 Be Beryllium							
3	11 22.99 Na Sodium	12 24.30 Mg Magnesium							
4	19 39.10 K Potassium	20 40.08 Ca Calcium	21 44.96 Sc Scandium	22 47.87 Ti Titanium	23 50.94 V Vanadium	24 52.00 Cr Chromium	25 54.94 Mn Manganese	26 55.85 Fe Iron	27 58.93 Co Cobalt
5	37 85.47 Rb Rubidium	38 87.62 Sr Strontium	39 88.91 Y Yttrium	40 91.22 Zr Zirconium	41 92.91 Nb Niobium	42 95.94 Mo Molybdenum	43 98.91 Tc Technetium	44 101.1 Ru Ruthenium	45 102.9 Rh Rhodium
6	55 123.9 Cs Cesium	56 137.3 Ba Barium	*	72 178.5 Hf Hafnium	73 180.9 Ta Tantalum	74 18 .8 W Tungsten	75 186.2 Re Rhenium	76 190.2 Os Osmium	77 192.2 Ir Iridium
7	87 223.0 Fr Francium	88 226.0 Ra Radium	~	104 261.1 Rf Rutherfordium	105 262.0 Db Dubnium	106 263.0 Sg Seaborgium	107 262.0 bh Bohrium	108 265.0 Hs Hassium	109 266.0 Mt Meitnerium

Alkali metals; Alkali earth metals; Transition metals

The lanthanide series *	57 138.9 La Lanthanum	58 140.1 Ce Cerium	59 140.9 Pr Praseodymium	60 144.2 Nd Neodymium	61 146.9 Pm Promethium	62 150.4 Sm Samarium
The actinide series ~	89 227.0 Ac Actinium	90 232.0 Th Thorium	91 231.0 Pa Protactinium	92 238.0 U Uranium	93 237.0 Np Neptunium	94 239.1 Pu Plutonium

Rare earth metals

(a)

FIGURE 3.1.1
(a, b) Periodic chart of the elements.

(Continued)

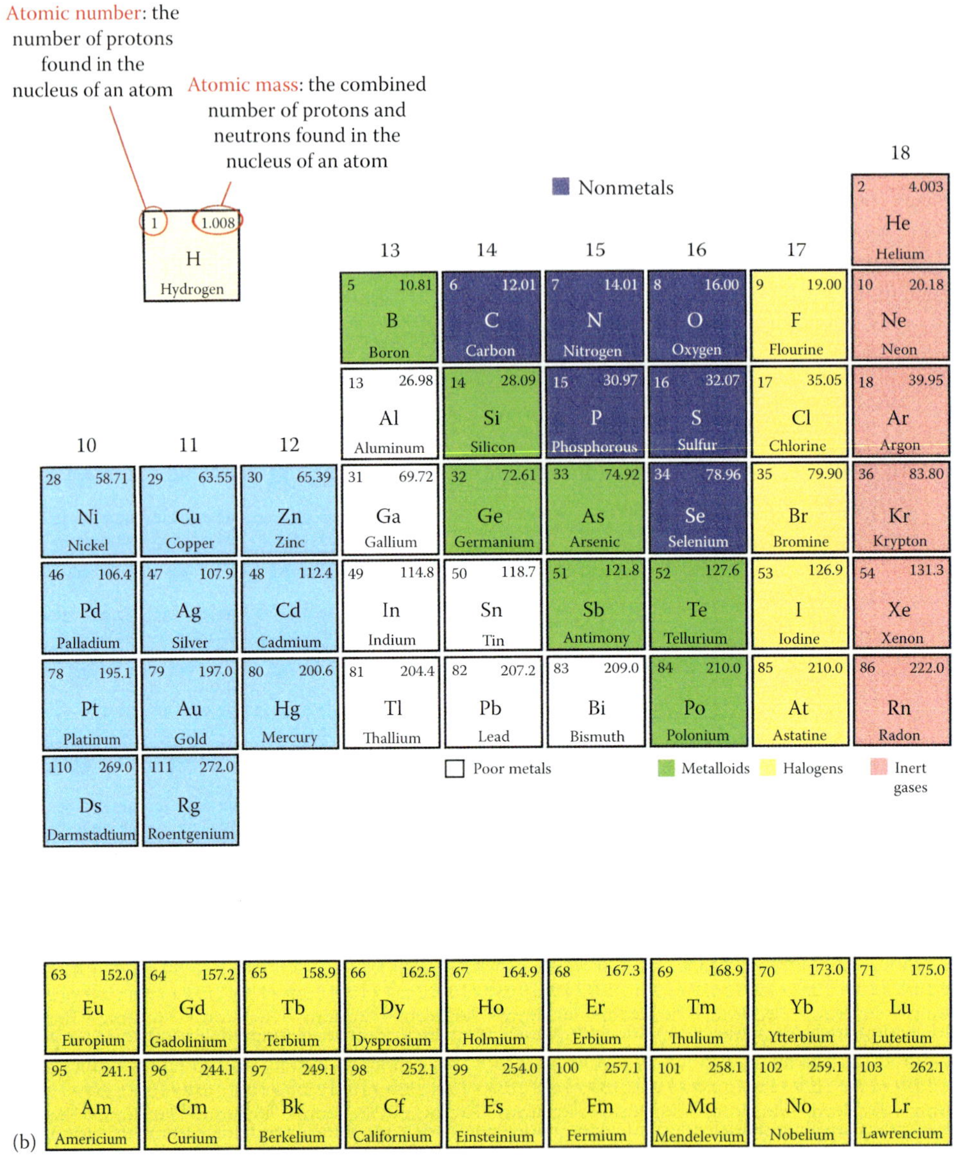

FIGURE 3.1.1 (Continued)
(a, b) Periodic chart of the elements.

The elements that directly precede the inert gases—fluorine, chlorine, bromine, iodine, and astatine, with 9, 17, 35, 53, and 85 electrons, respectively—also resemble each other. They are nonmetals, poor heat and electric conductors, and form salts when reacted with the alkali metals. Therefore, these elements are called *halogens*, or salt-formers, which form acidic solutions in water.

The elements that fall between the alkali metals and the halogens show a progressive gradation of properties between the two extremes. For instance, the elements such as magnesium (12 electrons), aluminum (13), silicon (14), phosphorus (15), and sulfur (16), which fall between sodium (11) and chlorine (17), show a decrease in metallic character and an increasing tendency to form less alkaline and more acidic solutions in water.

Those elements that lie midway between the alkali metals and the halogens are special to the life sciences. These elements—carbon (6), silicon (14), germanium (32), tin (50), and lead (82), include elements that have neither strong metallic nor strong alkali properties. The elements silicon and germanium are especially important in the electronics industry because, as semiconductors, they have controllable electrical conduction properties. Carbon is the basic element for all known life.

Before English physicist Henry Moseley, the periodic table of elements was ordered according to atomic mass. Moseley, using the frequencies of X-rays reflected from material samples, was able to detect the nuclear charges of different elements. From these, he ordered the periodic table based on atomic number (Scerri, 2014).

The periodic table of the elements is an arrangement of elements that accents the regular recurrence of similar physical and chemical properties of the elements. It shows elements that are in the same family, or group, with similar properties, and it shows the different periods, with gradations of physical properties from the alkali metals to the halogens. While the periodic table does not explain all property variations and is more complicated than expected in certain transition regions, it is a good means to classify general chemical properties in an organized way.

Families (or groups) of elements are arranged vertically in the periodic chart. These are the elements with similar chemical and physical properties, although these properties weaken from top to bottom. Elements at the top of the chart are lighter and less dense; elements at the bottom have greater masses and are denser. Elements arranged horizontally form a recurring pattern of transition from highly alkaline to highly acidic natures. Elements at the right and left edges of the chart are most chemically active, whereas those in the middle do not react with other elements very vigorously. The last group at the far right of the chart has elements that are completely inert chemically. They do not take part in chemical reactions, so they are neither necessary nor harmful for life (although xenon has been found to form bonds with halogens, oxygen, carbon, and gold, under certain conditions). Looking at the periodic table, flammability decreases from left to right, and toxicity increases from top to bottom. Elements heavier than atomic number 53 (iodine) are apparently not used in living things (see Section 6.3.1, Essential Elements), and those heavier than 92 are not found in nature.

It sometimes happens in living systems that, when the primary element is not immediately available, a substitution can be made with an element with similar properties. Thus, an element from the same group can sometimes be substituted for an element not readily available. For instance, strontium (38) can sometimes be found to substitute for calcium (20) in the bones when strontium is more available than calcium. Radioactive strontium isotopes are sometimes released in nuclear accidents, so the incorporation of strontium into the bones is a cause for concern. The relative stability of lead compounds in the body is the result of its appearance in the same group as carbon and silicon. However, lead has toxic neurological effects that make it a dangerous element.

Lead incorporated into the body can have severe effects on the kidneys and the nervous system. Effects of lead poisoning are:

1. Reduced fetal growth with elevated lead levels in the mother's blood
2. Increased blood pressure
3. Hypertension
4. Essential tremor

Over time, an excess amount of lead leads to a decline of health and to neurological disruptions. It is thought that one of the contributing causes to the decline of the Roman Empire was the widespread use of lead water pipes among the upper classes. Over time, the ruling classes became less able to rule, less vigorous, and the civilization declined.

Arsenic (33) is dangerous because it can substitute for phosphorus (15), and phosphorus is nearly everywhere in living things. Although toxic to almost all life, arsenic can apparently substitute for phosphorous without ill effect in a bacterium called GFAJ-1 (Wolfe-Simon et al., 2010). The arsenic was incorporated in the nucleic acids (as the DNA backbone), proteins, and cell membranes of healthy bacteria.

ARSENIC AND OLD LACE

In Joseph Kesselring's 1939 play *Arsenic and Old Lace*, two misguided spinster sisters decide to help lonely old men relieve their loneliness by inviting them to their house for dinner and poisoning them with homemade elderberry wine laced with arsenic, strychnine, and just a pinch of cyanide.

Arsenic is acutely poisonous. It has been known as the "poison of kings" because it is nearly odorless and tasteless and could be discretely slipped into food or drink to assassinate persons of wealth or power. It was not able to be detected as a poison until sensitive tests were developed in the nineteenth century. The Borgias of fifteenth and sixteenth century Italy used arsenic-tainted wines to rid themselves of influential popes and cardinals in order to further their own ambitions.

Arsenic competes with phosphorus to interfere with cell metabolism. Acute arsenic exposure causes necrotic cell death. Long-term arsenic exposure can lead to cancer. Arsenic poisoning is now considered to be an environmental toxin of particular harm to fetuses and very young children, in whom it can cause neuropathological conditions and death to develop (Yosim et al., 2015).

Selenium is a micronutrient required by grazing animals. Sheep require 0.1 ppm (parts per million) selenium in their total diet, with 0.3 ppm being better, and levels more than 3 ppm being toxic. Forage plants do not require selenium at all, but because selenium is chemically similar to sulfur (being in the same chemical family), plant roots will absorb it from the soil and incorporate it into amino acids in place of sulfur. When selenium is present in the soil in large enough quantities, forage plants supply the selenium needs of the sheep grazing on them.

The periodic nature of chemical properties is explained by the distribution of energy levels of the electrons found outside the nucleus of each atom. Quantum mechanics successfully predicts the discrete energy levels of these electrons and the number of electrons that possess these levels of energy (Figure 3.1.2). For instance, only 2 electrons can have the lowest level of energy, 8 electrons can have the next lowest level, and 18 electrons can have the third lowest energy level. The maximum number of electrons for these atoms is 2 (lowest energy level filled), 10 (2 at the lowest level plus 8 at the next level), and 28 (2 at the lowest level plus 8 at the second lowest level plus 18 at the third lowest level). If the number of electrons at each of these levels is equal to the maximum, then the elements exhibit inert properties; the inert gas helium has 2 electrons and neon has 10.

Thereafter, things become more complicated because there are many higher electron energy levels and each of these has sublevels. But the chemical properties of an element are mainly determined by the number of electrons at the highest energy level. There is a

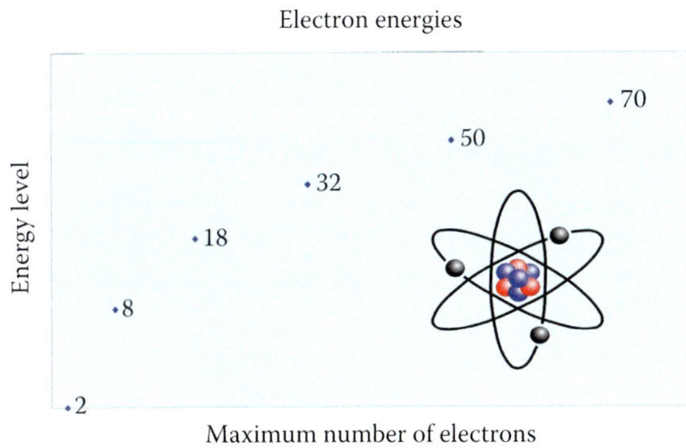

FIGURE 3.1.2
The maximum number of electrons at different energy levels is fixed. There are no allowable intermediate energy levels.

tendency for atoms to complete the maximum number of electrons at the highest energy level for that element. Thus, if an atom has 2 electrons at its lowest energy level and 7 electrons at the next energy level, which is the highest energy level for that atom, then the atom tends to want to combine with another atom that has 1 electron at the highest energy level. The addition of that 1 electron completes the maximum number (8) of electrons at the second energy level for the first element (Figure 3.1.3). In this combination, both atoms are able to achieve the desired number (8) of electrons at the highest energy level for each atom. This combination involves fluorine (2 + 7 electrons) and lithium (2 + 1 electrons).

In a similar fashion, an atom of oxygen, with 6 highest energy level electrons combines readily with beryllium with 2 highest level electrons. Each atom then has 8 electrons in its highest energy level. The same result could be achieved among oxygen (2 + 6 electrons) and two lithium (2 + 1 electrons) atoms.

Similar chemical properties are found whenever an element requires the same number of electrons to complete its highest energy level. Thus, fluorine, chlorine, bromine, iodine, and astatine all lack 1 electron from completeness. Lithium, sodium, potassium, and rubidium each have 1 extra electron. Each of these families of elements exhibits many similar properties among the individual elements. They are not all alike, however, but they do share many properties.

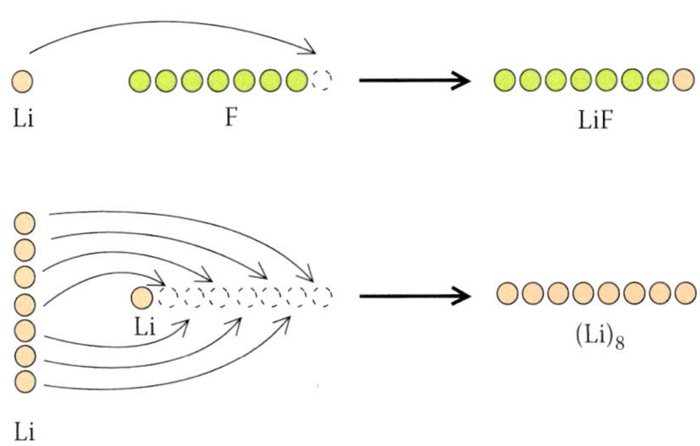

FIGURE 3.1.3
The addition of 1 electron at the highest energy level to 7 electrons at the same level results in 8 electrons, a very stable configuration. Eight electrons can be satisfied by lithium (Li = +1) and fluorine (F = +7, or −1) together, or by eight Li atoms.

THE MOLE

Atoms and molecules are extremely small and difficult to deal with on a one-to-one basis. Then, too, there are attractions, repulsions, or other interactions among these units that make it difficult to treat them as independent. Thus, there is a clear advantage to grouping atomic or molecular units into larger parcels.

Chemists have long agreed that the sizes of these parcels should have a mass value that scales from the molecular mass (number of protons and neutrons in the nuclei of the atoms making up the molecule). The mass scale used is left to the discretion of the individual chemist. Thus, mass units commonly used are pounds, grams, and kilograms; others may also be used.

A mole is the mass of a substance in grams that corresponds to its molecular mass, expressed in grams. Thus, a mole of water (H_2O) has a mass of 18 grams (2H = 2, 1O = 16, total molecular mass = 16 + 2 = 18). This mole may also be called a gram-mole for clarity.

A kilogram-mole is the mass of a substance in kilograms that equals the molecular mass (still expressed in grams) of the substance. A kilogram-mole (kg-mole) is 1000 times as large as a gram-mole. A pound-mole is 454 times as large as a gram-mole.

There is a definite number of atoms or molecules in a mole. This is called the Avogadro number and is equal to 6.0235 × 1023. This many molecules or atoms are in a sample containing 1 mole of a substance. A sample with one kg-mole would contain 6.0235 × 1026 atoms or molecules.

Valence is the term given to the excess number of electrons that an element needs to complete its highest energy level. Thus, sodium has a valence of +1, whereas fluorine has a valence of −1. Fluorine could also be assigned the valence of +7, but this is a very unlikely state for fluorine. Sodium could also be given a valence of −7, but, again, it is unlikely that sodium would contribute 7 electrons to complete the highest energy level of another element. It is much more likely that it would complete its own highest energy level by donating its excess electron to another element with valence of −1.

For elements that appear in the middle of the periodic table, valence levels are +3, +4, or +5. These elements may appear to be able either to donate or to accept electrons. Thus, phosphorus may have a valence of either −3 (most likely) or +5 (less likely). Carbon with a valence of either +4 or −4 can act as either an electron donor or acceptor. This property makes carbon a very versatile element and we shall see later how this makes carbon special to life on Earth.

Example 3.1.1: Phosphates in Detergents Cause Algal Bloom

Phosphorous as phosphate (phosphorous with the addition of four oxygen atoms) is an important constituent of many detergents. Cleaning agents remove dirt from dishes or clothes, and phosphorous binds to the dirt and keeps it suspended in water. The problem that results, however, is that phosphorous added to waste water acts as a fertilizer that stimulates algal growth. When the algae die, they decompose and absorb oxygen from the water, suffocating other forms of aquatic life.

Example 3.1.2: Geoengineering to Reduce Atmospheric Carbon Dioxide

In the Southern Ocean, phytoplankton growth is limited by iron availability. There have been proposals to seed the South Pacific Ocean with iron to promote phytoplankton growth and enhance photosynthesis, using carbon from the atmosphere as a result. Schemes such as this that are meant to solve problems on a global scale, especially regarding climate change, are called *geoengineering*. There is a good deal of skepticism about these schemes, especially whether or not unintended consequences could result from their implementation.

APPLICATIONS AND PREDICTIONS

1. There will be many biochemical variations that differ in one element substituting for another similar element.
2. Elements with higher numbers of electrons will usually have larger physical sizes than elements with smaller numbers of electrons. Larger sizes will usually mean that they are not as mobile.
3. Ionic charge of elements can be predicted by the location of an element on the table of elements. Elements on the left will tend to form positively charged ions and elements on the right will form negatively charged ions.
4. Elements with higher numbers of electrons, and protons in their nuclei, will be larger and less able to penetrate cell membranes.
5. Properties of a certain element can be inferred from its position in the periodic table of elements.

3.2 Chemical Bonding

Slowly, slowly, slowly we oxidize!
Become old and rusty,
Fungoid and musty,
Diminish in size;
Reputation decreases and self-conceit ceases –
Cares fret and wear out facial lines incessantly,
Yet doctors grow old rather pleasantly!

—Isaac N. Himes

There are several ways that atomic elements bond together to form molecular compounds. The end result of most of these chemical reactions is that the atoms that participate end up with 8 electrons (an octet) in their uppermost energy levels. This electron octet appears to be a very stable atomic configuration.

3.2.1 Ionic Bonds

My name is Bond, Ionic Bond; Taken, not shared!

—**Caren Thomas**

The most stable compounds formed are those where electrons are transferred from an electron donor atom to an electron acceptor atom. Energy is required to remove an electron from a neutral atom to form a positive ion (called a *cation*). Energy is released from a neutral atom when it forms a negative ion (called an *anion*). If the electron affinity energy of one atom is greater than the energy required to strip electrons from another, then the reaction proceeds spontaneously, producing excess energy in the form of heat. If the energy required to strip one or more electrons is greater than the electron affinity energy, then the reaction will not proceed spontaneously.

The alkali metals have low energy requirements for electron removal. This is because removal of the one excess electron leaves a stable 8 electrons at the highest energy level. It is thus relatively easy to form cations from this group with a positive charge. An example of this is sodium, which forms a cation with charge of (+1) according to the following chemical equation:

$$\mathrm{Na} + 5.1\,\mathrm{ev} = \mathrm{Na}^{+} + \mathrm{e}^{-} \tag{3.2.1}$$

where:

Na is the symbol for sodium (from the Latin name Natrium)
Na^+ is the symbol for a sodium ion with (+1) charge
e^- is the symbol for an electron with a negative charge

The 5.1 eV term is an energy unit of 5.1 electron-volts (each eV is equivalent to 3.8×10^{-20} cal). Without the energy term on the left-hand side of the equation, both sides would not be equivalent.

Metals with two electrons at the highest energy level (valence of 2) also have relatively low-energy requirements to remove the two excess electrons to form cations with (+2) charge. The halogens, with 7 excess electrons, have high-energy requirements to remove all 7 electrons, but release relatively large amounts of energy when an electron is added to produce the stable octet. This forms ions with (–1) charge, as illustrated by chlorine:

$$\mathrm{Cl} + \mathrm{e}^{-} = \mathrm{Cl}^{-} + 3.75\,\mathrm{ev} \tag{3.2.2}$$

where:

the symbol Cl denotes a chlorine atom
Cl^- denotes a chloride ion with a (–1) charge
e^- denotes an electron with (–1) charge

Note that 3.75 eV of energy is liberated by this action.

Elements with 6 excess electrons (such as oxygen and sulfur) are similar to the halogens, but form anions with (–2) charge.

When an Na^+ and a Cl^- contact each other, they form NaCl, or common table salt, which releases enough energy that the entire process of stripping an electron from the sodium atom, adding an electron to the chlorine atom, and combining the two resulting ions into an NaCl molecule (a process that liberates energy) results in the liberation of considerable amounts of heat. This type of chemical bond is called an *ionic bond*.

In terms of effort and flow variables, electron affinity represents one effort variable, electron removal energy represents a second effort variable, and combination energy represents a third effort variable. The flow variable is liberated heat. If the sum of the three effort variables is positive, heat is liberated; if the sum is negative, then heat is absorbed. Resistance, or the ratio of effort to flow variables, is proportional to the spontaneity of the chemical reaction.

Ions are very important in biological processes because they represent the charge carriers in biological organisms. In the chemical soup inside biological entities, electrical current is carried by ions rather than by solitary electrons.

REDOX REACTIONS AND ELECTRON TRANSFER

A biologically very important class of chemical reactions is called reduction-oxidation, or *redox* for short. In these reactions, electrons are transferred from one reactant to another:

$$A^+ + B \rightarrow A + B^+$$

Here an electron is transferred from an atom of B to a cation A^+. In the process, A^+ is said to be *reduced* and B is said to be *oxidized*.

Oxygen is one of the most common (although not the only) oxidizing agents. When oxygen is added to another atom, it forms an oxide, and the other atom is said to be oxidized:

$$4Fe + 3O_2 \rightarrow 2Fe_2O_3$$

Here, iron is oxidized to form iron oxide. Each iron cation in the molecule carries a charge of +3 and each oxygen anion in the molecule has a charge of −2.

Removing the oxygen means reduction. If the earlier chemical reaction were to be driven backwards, for instance by the addition of a lot of heat energy to the iron oxide, elemental iron would be formed, and the iron would be said to be reduced.

Although oxygen was recognized as the first oxidizing agent, it is not the only such substance. Other common oxidizing agents include sulfur, phosphorus, and chlorine. The common property of each of these is that they lose electrons in the process. So, generally speaking, when a substance (such as the aforementioned iron) loses electrons, it is said to be oxidized. When the substance gains electrons it is said to be reduced:

$$\text{electrons} \xrightarrow{\text{lost}} \text{oxidation, positive charge increases;}$$

$$\text{electrons} \xrightarrow{\text{gained}} \text{reduction, positive charge decreases}$$

An *oxidizing agent* is a substance, such as oxygen, which can oxidize something else. A *reducing agent* is the substance that can reduce something else. In every redox

(*Continued*)

REDOX REACTIONS AND ELECTRON TRANSFER (Continued)

reaction, there is always at least one oxidizing agent and one reducing agent. The oxidizing agent is reduced, and the reducing agent is oxidized. In the iron example,

$$4Fe + 3O_2 \rightarrow 2Fe_2O_3,$$

the oxygen is the oxidizing agent for iron, and the iron is the reducing agent for the oxygen.

Hydrogen is a common reducing agent because it is an electron donor. When hydrogen is added to a molecule, an electron is transferred from the hydrogen atom to the molecule: the molecule is reduced, and the hydrogen is oxidized (even without the presence of oxygen).

NAD (nicotinamide adenine dinucleotide) is a coenzyme common in respiratory chemical reactions. The oxidized form is NAD, and the reduced form is NADH. This illustrates that removal of a hydrogen atom is in itself an oxidative process.

Summarizing, then,

A substance that loses electrons is oxidized.

A substance that gains electrons is reduced.

3.2.2 Covalent Bonds

New functions come from new molecules.

—Bruce Yu

The second type of chemical bonding is the *covalent bond*, which occurs whenever one atom is not strong enough to cause an electron to shift completely from one atom to another. In that case, the two (or more) atoms share the electrons needed to complete their octets. Such covalent electron sharing occurs when two atoms of the same element combine, as with O_2, N_2, or H_2. Covalent bonding is especially important for the elements in the middle of the periodic table, those with 3, 4, or 5 electrons at their highest energy level.

Covalent bonds vary between the extremes of equal sharing of electrons to nearly complete donor–acceptor ionic bonding. Depending on the equality of sharing and the geometrical configuration of the molecule, a molecule may exhibit an unequal charge distribution from one side to the other. Such a molecule with a positive end and a negative end is called a *polar* molecule.

3.2.3 Electronegativity

To the ass, or the sow, their own offspring appear the fairest in creation.

—Latin proverb

Electronegativity is the term used to describe the relative affinity of an atom to attract shared electrons. The higher the electronegativity of an element, the more tightly will electrons be bound to that type of atom. Many different energies contribute to the electronegativity,

<1.0 | 1.0–1.4 | 1.5–1.9 | 2.0–2.4 | 2.5–2.9 | 3.0–4.0

1 H 2.1																
3 Li 1.0	4 Be 1.5											5 B 2.0	6 C 2.5	7 N 3.0	8 O 3.5	9 F 4.0
11 Na 1.0	12 Ma 1.2											13 Al 1.5	14 Si 1.8	15 P 2.1	16 S 2.5	17 Cl 3.0
19 K 0.9	20 Ca 1.0	21 Sc 1.3	22 Ti 1.4	23 V 1.5	24 Cr 1.6	25 Mn 1.6	26 Fe 1.7	27 Co 1.7	28 Ni 1.8	29 Cu 1.8	30 Zn 1.6	31 Ga 1.7	32 Ge 1.9	33 As 2.1	34 Se 2.4	35 Br 2.8
37 Rb 0.9	38 Sr 1.0	39 Y 1.2	40 Zr 1.3	41 Nb 1.5	42 Mo 1.6	43 Tc 1.7	44 Ru 1.8	45 Rh 1.8	46 Pd 1.8	47 Ag 1.6	48 Cd 1.6	49 In 1.6	50 Sn 1.8	51 Sb 1.9	52 Te 2.1	53 I 2.5
55 Cs 0.8	56 Ba 1.0	57 La 1.1	72 Hf 1.3	73 Ta 1.4	74 W 1.5	75 Re 1.7	76 Os 1.9	77 Ir 1.9	78 Pt 1.8	79 Au 1.9	80 Hg 1.7	81 Tl 1.6	82 Pb 1.7	83 Bi 1.8	84 Po 1.9	85 At 2.1
87 Fr 0.8	88 Ra 1.0	89 Ac 1.1														

FIGURE 3.2.1
Table of electronegativities. Elements with much different electronegativities will tend to have high ionic affinity for one another. Elements with similar electronegativities will tend to share electrons.

including energies associated with the nuclear charge, the size of the nucleus, and partial shielding of nuclear charge by other electrons.

In general, electronegativities increase from left to right in the periodic table of the elements (see Section 3.1, Periodic Table) and decrease from top to bottom (Figure 3.2.1). Inert elements placed on the far right of the table have electronegativities of zero.

Elements with low electronegativities (but not zero) have relatively low electron affinities, which means that they can lose electrons readily. Elements with high electronegativities hold tightly to their electrons and can readily attract electrons from other elements. Thus, two elements with very different electronegativities would be expected to form ionic bonds where electrons are donated by the low electronegative element to the element with high electronegativity. Furthermore, this combination would be expected to be very polar, with excess electrons located at the position occupied by the more electronegative atom.

Elements with nearly equal electronegativities would be expected to form covalent bonds and share their electrons nearly equally. A molecule with this combination would be expected to be nonpolar.

3.2.4 Water as a Polar Molecule

> Money is the best cosmetic.
>
> **—Gregory Nunn**

Water is a polar molecule (Figure 3.2.2). There is an angle of 104.5° between the two hydrogen atoms with the oxygen atom at the vertex. The covalent bonds in the water molecule result in the oxygen atom being negative and the hydrogen atoms being relatively positive. If there were 180° between the hydrogen atoms with the oxygen atom in between, then there would be a positive charge on each end of the water molecule, and the molecule would be *nonpolar*. Such is the structure of carbon dioxide, which is nonpolar. Water, however, does not have a 180° angle between the hydrogen atoms and is polar.

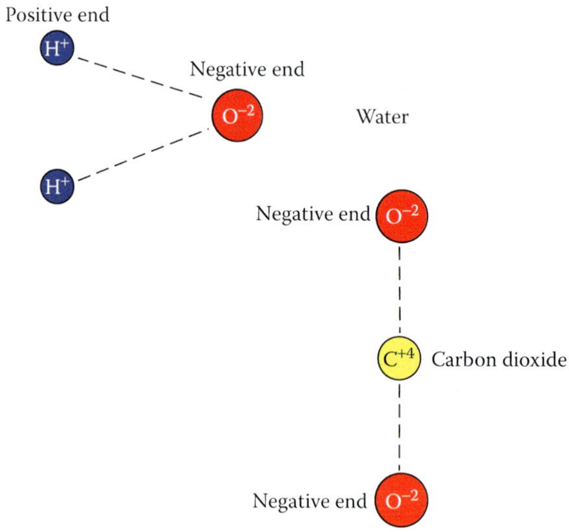

FIGURE 3.2.2
Water as a polar molecule. Carbon dioxide as a nonpolar molecule.

Polarity is important in biological systems because polar molecules are sometimes attracted to other polar molecules on the surface of cells, and helps to bind these together. When this happens, we call the surface *hydrophilic*, meaning that it attracts water and other polar substances.

When several water molecules are brought together, there is an attraction between the negative oxygen end of one molecule with the positive hydrogen end of another. Unlike charges attract each other. The result is that water tends to be structured even as a liquid. The hydrogen atoms arrange themselves as a tetrahedron around each oxygen atom. In a container with vast numbers of water molecules, this structure continues in three dimensions.

Ice is a crystal with this solidified structure (Figure 3.2.3). The arrangement of ice is such that it is honeycombed with hexagonal channels that gives ice its relatively low density. When ice melts, the structure becomes less orderly but is not completely destroyed. The arrangement of tetrahedra is more random and constantly changing. Nonetheless, some of the structure still remains.

It is fortunate for life that solid ice has a lower density than liquid water. Life can exist in warmer water beneath a frozen layer of colder ice that floats above because of its lower density. If ice were denser, the water would be exposed to colder environments and would certainly all freeze, killing many of the organisms suspended in it.

3.2.5 Hydrogen Bonds

> A world that includes the Human Element, along with hydrogen, oxygen and the other elements, is a very different world indeed. Suddenly, chemistry is put to work solving human problems. Bonds are formed between aspirations and commitments. And the energy released from reactions fuels a boundless spirit that will make the planet a safer, cleaner, more comfortable place for generations to come.
>
> **—Dow**

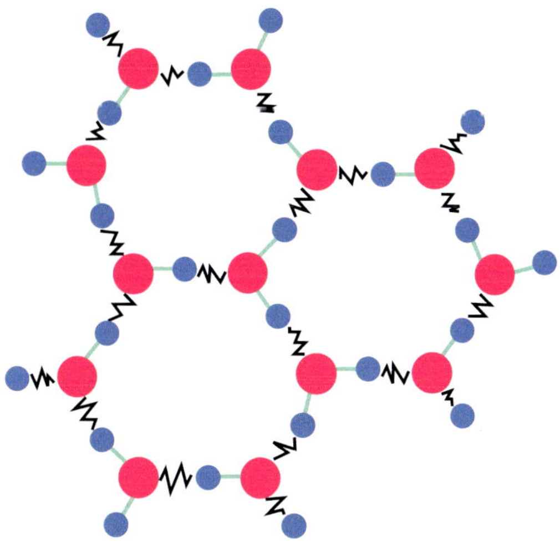

FIGURE 3.2.3
Two-dimensional representation of the structure of ice. Each oxygen atom is covalently bound to two hydrogen atoms and hydrogen bonded to two additional hydrogen atoms. The resulting structure is a tetrahedron not able to be shown completely in two dimensions. Ice has a hexagonal cellular structure that has relatively low density compared with water.

Positive hydrogen ions can act as bridges between two negative atoms, as they do between two oxygen atoms in water. This bridging is called a *hydrogen bond*, and is the result of polarities in molecules. Hydrogen bonding occurs in other molecules besides water, but the hydrogen bond is only found between two electronegative atoms. The result of hydrogen bonding, besides the structure already discussed for water, is a higher boiling point than would otherwise be expected. The structures of proteins, too, depend on hydrogen bonds that bridge between nitrogen and oxygen atoms.

Hydrogen bonding is what gives liquid water its structure. When a solute is introduced into liquid water, considerable energy is required to tear water molecules apart from their neighbors in order to make room for solute particles. Hydrogen bonds can be broken between water molecules and established between a water oxygen atom and a portion of a solute molecule. In the case of ammonia (NH_3), a hydrogen bond is established between the N of NH_3 and the O of H_2O. A hydrogen bond between the O of C_2H_5OH (ethanol) and the O of H_2O, and a hydrogen bond between the O of $C_{12}H_{22}O_{11}$ (sucrose) and the O of H_2O allow ethanol and sucrose to dissolve readily in water.

Many solutes, such as the ethanol and sucrose mentioned previously, exist in solution as whole molecules. They are assisted in becoming dissolved solutes by the hydrogen bonds that form between them and water molecules. Otherwise, they would not liberate enough energy to break apart the structure of liquid water; they would not dissolve; and they would remain separate from the water.

Other solutes, called *electrolytes*, consist of molecules that dissociate to form charged ions in aqueous solution (they do not dissociate unless water is present). Thus, HCl is found in water as H^+ and Cl^- ions; NaCl is found in water as Na^+ and Cl^- ions. The polar water molecule becomes attracted to these ions, the negative oxygen end to the H^+ ions (to form a H_3O^+ complex) and the positive hydrogen end is attracted to the Cl^- ions.

Water is, thus, attached to each ion because of the electropolar nature of the water molecule and the charged ion in solution. Water is always associated with any dissolved species and will affect its properties. There may be an amelioration of the effects of ionic charge on its surroundings because of the shielding effect of the water molecules surrounding the ion. There is always a size enhancement to any dissolved particle because the surrounding water clings to each dissolved particle. For example, a bare hydrogen ion is nothing but a proton of essentially zero size. The H_3O^+ ion, which is the form that H^+ ions actually take, has a volume about 10^{15} times bigger than H^+, and is comparable in size to other ions.

Because solutes disrupt the normal pattern of a water structure, solutions have freezing temperatures lower than that of pure water (Feeney, 1974) and solutions have boiling temperatures higher than that of pure water. Differences in freezing and boiling points are determined almost solely by the number of solute particles present and not by their properties. Total dissociated compounds, such as table salt (NaCl), will have twice as much of a freezing point depression and boiling point elevation as will nondissociated compounds, such as sucrose. Incomplete dissociated compounds will have intermediate effects.

BAD HAIR DAY

Did you ever have a bad hair day when your hair just won't lie right or stay curled or some other bothersome thing? One thing you might then do is to wet your hair and let it dry in the fashion desired, and, when dry, your hair retains its shape. Did you ever wonder why?

Hair strand shape is largely determined by hydrogen bonds between locations on the proteins that constitute hair. Water disrupts these hydrogen bonds, and they are reestablished as the hair dries. The newly established hydrogen bonds retain the shape of the drying hair.

Likewise, wrinkled clothes can be sprinkled with water and pressed with an iron to look neat again. The water disrupts the hydrogen bonds in the cloth and the iron dries the cloth and reestablishes the hydrogen bonds so that the correct shape is retained. This works best with cotton, consisting of mostly cellulose, and wool, made of proteins. Polyesters do not have the same hydrogen bonds as do the other types of cloth, and are usually wrinkle-free.

3.2.6 van der Waals Forces

Every artist was first an amateur.

—Ralph Waldo Emerson

Even nonpolar molecules form attractive forces among themselves. Due to the nonstationary nature of electrons moving about the nuclei of the atoms making up the molecule, polarities may be temporarily formed in what over a long time may be considered to be a nonpolar molecule. When this happens, it influences electron positioning in neighboring molecules. Electrons repel electrons; protons attract electrons. Attraction between molecules occurs when polarities among neighboring molecules become more or less synchronized. These forces are called *van der Waals forces*.

van der Waals forces are present in all molecules. They are relatively weak attractive forces, certainly weaker than covalent bonds, which, in turn, are weaker than ionic bonds. Higher temperatures, indicating more molecular kinetic energy, weakens the relative strength of van der Waals forces, but higher pressures, pushing molecules closer together, increases their relative strength.

GECKOS' FEET

Geckos are lizard-like creatures that can scamper up walls and hang from ceilings. Whether the surface is made from wood, metal, or glass, the gecko can climb it. What enables the gecko to do this? Is it glue, suction, or a thin film of water? None of these, as it turns out.

The gecko has feet with millions of microscopic hairs sprouting from the bottom. The hairs, called setae, split into as many as 1000 hairlets, each capped with a triangular bacterium-sized pad called a spatula (the shape is reminiscent of a hamburger flipper). These spatulae contact the climbing surface and induce van der Waals forces. Each spatula is capable of supporting the weight of an ant. Together, the spatulae could theoretically hoist a 130-kg man.

Although van der Waals forces are very weak, enough of them together serve the gecko very well. And now that the secret of the gecko's climbing is known, new glues can be made with artificial spatulae (Autumn et al., 2002).

Example 3.2.1: Wastewater Bioreactor Troubleshooting

Municipal wastewater can be treated biologically in a bioreactor, consisting of a basin containing wastewater and activated sludge (solids with active microbes). In a batch mode, (1) the reactor is filled with wastewater, (2) the mixture of wastewater and sludge is aerated, (3) the contents are allowed to settle, (4) effluent is decanted and some sludge is removed, and (5) the system idles until the next cycle. Sometimes there are problems with anoxic conditions—odor, low capacity, and poor effluent quality. What are the possible causes and solutions for these problems?

Solution

All of these problems are related, and solving them can be counterintuitive. One possible solution would be to clean the bioreactor and begin again, but this may be very expensive and not entirely satisfactory.

Anoxic conditions are caused by a large demand for oxygen in the wastewater. Demand can be caused either by biological substances (BOD, or biological oxygen demand), or by chemicals (COD, or chemical oxygen demand). Adding more wastewater will not dilute BOD or COD, and may make matters worse. The solution to this problem is to lengthen step 2 of the cycle.

Anoxic conditions favor sulfur instead of oxygen as an oxidizing agent. Hydrogen sulfide (H_2S) is formed that has a rotten egg odor. Raising pH reduces the concentration of H_2S compared with other sulfide species, thereby reducing odor.

Temperature has a profound effect on bioreactor performance. The rate at which gases, such as oxygen, can be dissolved decreases rapidly as temperature increases; chemical reaction rates increase exponentially with temperature; thermal agitation increases water turbidity as temperature increases. Because of these, anoxic conditions are more likely to occur at higher temperatures.

The solutions to these problems are, consequently, to improve aeration, monitor pH, and cool the bioreactor (Marx et al., 2003).

Example 3.2.2: Natural Halocarbons as Bioactive Compounds

In the search for useful biochemicals naturally produced by plants and animals, the class of halocarbons is still in the process of being explored for exploitation. These compounds, combining one or more halogens with carbon, are usually quite stable and have interesting uses. Some of these compounds are already in use and are synthesized industrially. Others are still being found, and uses are being determined. Some of these compounds are shown in the following table (Gribble, 2004):

Compound	Produced by	Use
Brominated dioxin	Sponges, corals	Prevent barnacle growth seaweeds
Panacene	Sea hare	Taste discourages being eaten
Methyl bromide	Cabbage, broccoli, turnips, canola, evergreen trees, and potatoes	Natural pesticide
Fluoroacetic acid	Australian and African plants	Poisonous to grazers
Thyroxine	Many vertebrates	Metabolic regulator
Chloroform	Termites	Metabolic waste product
Epibatidine	Small frog	Painkiller
Vancomycin	Soil bacteria	Antibiotic

Other compounds are finding uses as hormones, pheromones, toxins, bleaches, anticancer agents, anesthetics, and others. As more of these compounds become known, who knows what benefits might be derived from the effort?

Example 3.2.3: Detection of Skin Cancer by Classification of Raman Spectra

Malignant melanoma skin cancer is often lethal. Unfortunately, even experienced dermatologists misdiagnose this type of cancer 25%–35% of the time.

Raman spectra are obtained by pointing a laser beam at a sample. Molecules in the sample become excited and scatter energy back to the probe. Frequency shifts in the spectra depend on the types of molecules present, and frequency shifts can be used as signatures identifying specific molecules.

Raman spectra can be used as inputs to neural networks for the purpose of diagnosing skin cancer types (Sigurdsson et al., 2004). Signals must be preprocessed to reduce variability and background from skin fluorescence. Although complicated, these steps improve the reliability of skin lesion classification.

Example 3.2.4: Nitric Oxide Messenger

Nitric oxide is a small gaseous molecule that is synthesized in many animals from barnacles to humans. It plays a role as a biological messenger with a broad range of physiological processes, including neurotransmission, blood clotting, blood pressure control, and immune system response. *Hydrogen sulfide* plays a similar messenger role (Wang, 2010).

APPLICATIONS AND PREDICTIONS

1. Compounds formed from ionic bonds will usually be extremely stable, and are rarely formed by living systems.
2. Polar molecules will be hydrophilic. Nonpolar molecules will be hydrophobic.
3. Water inside the cell will be structured, not free.
4. The expansion of freezing water will disrupt cellular structure and will kill the cells.
5. Compounds with hydrogen bonds will be firmer, slower to melt, and harder to boil than compounds without hydrogen bonds. The fats and oils deposited in the hands, where it is relatively cool, will have fewer hydrogen bonds (are said to be less saturated) than do fats and oils deposited in the center of the body.
6. Hydrated Na^+ will be much larger than hydrated K^+.
7. Organisms adapted to cold environments will use dissolved solutes as antifreeze compounds.
8. Water will be an excellent solvent for polar molecules or ions.
9. Polar compounds (like water) will not mix well with nonpolar compounds (like oil).
10. Chemical bonding controls the movement of energy from one compound to the next.
11. Energy can be stored by forming chemical bonds; energy can be released by breaking chemical bonds.
12. The strongest bonds in protein formation will be covalent bonds.
13. The most durable compounds found in living things will have the strongest chemical bonds; the compounds most likely to be manipulated will have the weakest bonds.
14. Elements with similar electronegativities will tend to form covalent bonds.

3.3 Chemical Equilibrium

> The ultimate aim of the modern movement in biology is in fact to explain all biology in terms of physics and chemistry.
>
> —**Francis Crick**

A general chemical equation can be given by the form

$$mA + nB = pC + qD + \text{energy} \tag{3.3.1}$$

where *m*, *n*, *p*, and *q* are coefficients needed to balance elements in the compounds A, B, C, and D, respectively. The amount of energy may be either positive (*exothermic*, or heat producing) or negative (*endothermic*, or heat using). If all compounds are gases, then the *equilibrium constant* (*K*) for this equation equals the product of concentrations of the chemical products raised to appropriate powers divided by the product of concentrations of reactants raised to appropriate powers:

$$K = \frac{(\text{concentration of C})^p(\text{concentration of D})^q}{(\text{concentration of A})^m(\text{concentration of B})^n} \tag{3.3.2}$$

where concentrations are normally given in moles per liter, and the exponent values are the same as the coefficients in the general Equation (3.3.1). For instance, the oxidation of hydrogen to form water vapor,

$$2H_2 + O_2 = 2H_2O + 2.42 \times 10^5 \text{ N} \cdot \text{m} \tag{3.3.3}$$

has a chemical equilibrium given by

$$\begin{aligned} K &= \frac{(\text{Concentration of water})^2}{(\text{Concentration of hydrogen})^2(\text{Concentration of oxygen})} \\ &= 1.11 \times 10^{40} \end{aligned} \tag{3.3.4}$$

that defines the ratios of concentrations of products (H_2O) to reactants (H_2 and O_2). The large value of *K* means that almost no excess hydrogen and oxygen remain in the system; all are converted to water vapor as long as the original number of moles of hydrogen was twice that of oxygen.

If all reactants and products are in solution, then the equilibrium constant can assume several meanings. If the substance in solution is an acid or base that dissociates in solution, then the equilibrium constant becomes the *dissociation constant*. For example, some acetic acid molecules dissociate into hydrogen (H^+) and acetate ($C_2H_3O_2^-$) ions:

$$C_2H_4O_2 \leftrightarrow H^+ + C_2H_3O_2^- \tag{3.3.5}$$

with a dissociation constant of 1.8×10^{-5} at 25°C:

$$K = \frac{(\text{Concentration of } H^+)(\text{Concentration of } C_2H_3O_2^-)}{(\text{Concentration of } C_2H_4O_2)} \tag{3.3.6}$$
$$= 1.8 \times 10^{-5}$$

Similar equilibrium constants have been found and tabulated for solubilities and precipitations.

CONCENTRATION MEASURES

Chemical concentrations are given in different units depending on the field of technology they come from. They can be converted from one to another, if need be, through suitable conversion equations given in Johnson (1999). Numerical values of equilibrium constants depend on the concentration units used, so the conventional concentrations are given in moles per liter. Other commonly used concentration units are:

M	Molarity	moles solute/liter solution
m	molality	moles solute/kilogram solvent
μ	mole fraction	moles solute/moles solution
N	normality	gram equivalents/liter solution
ψ	grams/liter	grams solute/liter solution
ppm	parts per million	grams solute/10^6 grams solvent or liters solute/10^6 liters solvent
v%	volume percent	liters solute/100 liters solution
w%	weight percent	grams solute/100 grams solution
X	mass fraction	grams solute/grams solution
ρ	density	kilograms solution/cubic meter solution

With such a large number of concentration terms used, one must be quite sure that units are specified and understood.

Example 3.3.1: Systemic Pesticides in Fruits and Vegetables

Pesticides are sprayed on fruits and vegetables to protect against damage by insects and diseases. Many modern nonorganic pesticides are systemic in nature; that is, they penetrate the plant tissues and reside throughout the plant, protecting from the inside out. It is impossible to remove all traces of these pesticides by washing or removing the outside skin. Concentrations of these chemicals are considered below toxic levels because of the dilution that occurs as the pesticides spread throughout the tissues. However, the entire chemical dose for the eater is the amount of chemical that resides inside the fruit or vegetable.

APPLICATIONS AND PREDICTIONS

1. Acetic acid will be a weak acid.
2. The heat energy that accompanies the formation of water from hydrogen and oxygen will be so large that it can damage living tissue unless dissipated.
3. A chemical reaction at equilibrium will remain at equilibrium until energy is added.
4. If not in chemical equilibrium, a reaction will either create energy or need energy to proceed.
5. Chemical reactions will be important in bioreactors; the stomach is one form of bioreactor.
6. The physical environment (temperature and pressure) will change the outcome of a reaction.

3.4 Acids and Bases

The greater the difficulty, the more glory in surmounting it.

—Epicurus

Acids and bases are very important in the biological world. An *acid* is a chemical that increases the concentration of hydrogen ions (H^+) when dissolved in pure water. In more general terms, acids are proton (equivalent to H^+) donors. A *base* is a substance that increases the concentration of hydroxide ions (OH^-) when dissolved in pure water. In more general terms, a base is a proton acceptor.

3.4.1 Strong and Weak Acids and Bases

Life is nothing but an electron looking for a place to rest.

—Albert Szent-Györgyi

Acids and bases may be either strong or weak. In a dilute water solution, strong acids and bases dissociate almost completely into anions and cations:

$$HX + H_2O \rightarrow H^+ + X^- + H_2O \rightarrow H_3O^+ + X^- \tag{3.4.1}$$

where:

HX is any typical acid containing hydrogen (H) and another element (X)
H_3O^+ is the hydronium ion composed of an H^+ cation and an associated water molecule

The hydronium ion is the common form of H^+ in water. For bases, the general reaction is

$$YOH \rightarrow Y^+ + OH^- \tag{3.4.2}$$

where YOH is any typical base containing hydroxide (OH) with another element (Y). Hydroxide ions do not form complexes in association with water analogous to the hydronium ion. Weak acids and bases dissociate only partly in water, and range from those with almost no dissociation to those with nearly complete dissociation:

$$HX + H_2O \leftrightarrow H^+ + X^- + H_2O \rightarrow H_3O^+ + X^- \quad (3.4.3)$$

Here, the double-ended arrow indicates that the dissociation proceeds both ways. Under certain conditions, the dissociation into ions proceeds more completely, and under other conditions, the undissociated form predominates. Most acids and bases are weak. Such weak acids or bases can be used as *buffers* against large changes in acidity of a solution.

One such buffer substance commonly employed in living organisms is carbonic acid (H_2CO_3), which is weakly dissociated into hydrogen ions and bicarbonate ions (neglecting the H_3O^+ complex actually formed):

$$H_2CO_3 \leftrightarrow H^+ + HCO_3^- \quad (3.4.4)$$

Depending on the acidity of the blood (which means the presence of H^+ ions from other sources), more H^+ ions or less H^+ ions may be released from carbonic acid. The more acidic the blood, the more H^+ ions that are present, and the more the reaction will be driven back to the undissociated H_2CO_3 form. Thus, this chemical acts to oppose large changes in acidity levels, and internal optimal conditions are maintained.

3.4.2 Salts

> Remember, if you're not part of the solution, you're part of the precipitate.
>
> —**Unknown**

Acids and bases react together to form salts:

$$HX + YOH \leftrightarrow H^+ + X^- + Y^+ + OH^- \leftrightarrow XY + H_2O \quad (3.4.5)$$

where the chemical XY is a salt.

Some salts are readily soluble in water, and some are not (Kotz and Treichel, 1999). If they are insoluble, then they precipitate and have no further influence on the solution. If they are partly or completely soluble in water, then they often dissociate into ionic form, a process called *hydrolysis*. For instance,

$$NaCl \rightarrow Na^+ + Cl^- \quad (3.4.6)$$

The product of anion concentration and cation concentration in a *saturated solution* is a form of equilibrium constant (Equation 3.3.2), except that the solid form of the salt has no meaningful concentration term. The *solubility product constant*, K_{sp}, expresses the ability of a salt to dissolve and dissociate in water:

$$K_{sp} = (\text{Anion concentration})(\text{Cation concentration}) \quad (3.4.7)$$

If NaCl is the only salt in solution, then the concentrations of Na^+ and Cl^- would be identical. However, in the complex solutions typical of biological systems, both Na^+ and Cl^-, as well as other ions, are contributed by other chemicals, and it is unlikely that the two concentrations would be the same.

If a salt is the product of a weak acid and a strong base, then, when the salt dissolves and dissociates in water, it will react with the water to form a strongly basic solution.

$$XY + H_2O \rightarrow X^+ + Y^- + H_2O \rightarrow HY + X^+ + OH^- \leftrightarrow H^+ + Y^- + X^+ + OH^- \tag{3.4.8}$$

The acid HY weakly dissociates, and more OH^- ions are produced in the water than are H^+ ions. Thus, the solution becomes basic.

The opposite is true for a salt formed from a strong acid and a weak base. The salt in solution forms an acidic solution.

Sodium chloride, shown previously, is the salt product of a strong acid (HCl, hydrochloric acid) reacting with a strong base (NaOH, sodium hydroxide). Thus, NaCl in water solution will not significantly affect the acidity of the solution.

3.4.3 pH

Chemistry will liberate the laborer.

—**Johann Most**

Water by itself tends to dissociate somewhat into ions:

$$2H_2O \leftrightarrow H_3O^+ + OH \tag{3.4.9}$$

The equilibrium constant for this reaction is

$$K = \frac{(\text{Concentration of } H_3O^+)(\text{Concentration of } OH^-)}{(\text{Concentration of } H_2O)^2} \tag{3.4.10}$$

Because the concentration of water is nearly constant at 55.5 M (molar, see Section 3.3), the equilibrium constant for water (K_w) is designated to be

$$\begin{aligned} K_w &= (\text{Concentration of } H_3O^+)(\text{Concentration of } OH^-) \\ &= 1.0 \times 10^{-14} \end{aligned} \tag{3.4.11}$$

Pure water is neither acidic nor basic, being called *neutral*. In this case, the concentrations of H_3O^+ and OH^- are equal. In other words, the concentration of H_3O^+ is 1.0×10^{-7} M at 25°C.

The most common measurement of acidity is *pH*, given as the negative logarithm of the hydrogen ion concentration:

$$\text{pH} = -\log_{10}(\text{concentration of } H_3O^+) \tag{3.4.12}$$

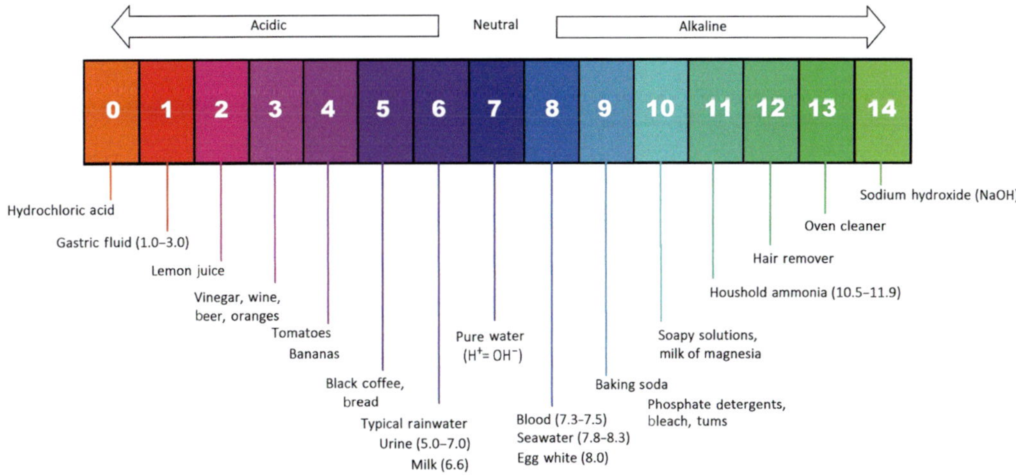

FIGURE 3.4.1
pH values of common substances. (Redrawn from Starr, C., *Biology: Concepts and Applications*, Brooks/Cole, Stamford, CT, 2000.)

A pH of 7.0 is neutral; pH values less than this are acidic, and solutions with pH values greater than this are basic (Figure 3.4.1). Because the product of the H_3O^+ and OH^- concentrations is constant, when pH decreases, the concentration of H_3O^+ increases and the concentration of OH^- must decrease.

pH has a profound effect on biological systems. Solubility of most biochemicals is affected by pH, so the availability of certain ions in the environment and inside organisms is very sensitive to pH. The concentrations of some metal ions might be too low to sustain life at some pH values, or so high as to be toxic at other pH values. Activities of many enzymes are pH sensitive, and thus metabolic reactions can change dramatically with slight shifts in pH. H^+ and OH^- can combine with different molecules and directly alter their functions. At both extremes, chemical burning can result.

Because of these effects, living things contain elaborate buffer systems to maintain constant pH levels. These buffer systems depend on the dissociations of weak acids and bases to perform their functions. The pH of normal human blood is usually taken to be 7.40 (slightly basic). Even a change in blood pH of 0.1, easily occurring in working muscles, causes a large shift in the availability of oxygen.

Immediate buffering of the blood uses bicarbonate/carbonic acid, as mentioned earlier. Removal of carbon dioxide through the respiratory system can help restore pH balance in the short term. Over a longer time, kidney-selective excretion of different ions is meant to restore blood pH levels to normal.

A common means to inhibit microbial growth is to preserve foods with low-pH liquids. Many canned fruits have pH low enough that microbial spoilage is extremely unusual. Additional organic acids, such as ascorbic acid, are routinely added to assist preservation and prevent enzymatic oxidation.

APPLICATIONS AND PREDICTIONS

1. One means to remove a toxic chemical from the environment is to precipitate it.
2. There will be a high concentration of bicarbonate in the blood.
3. Strong acids and bases can be used as disinfectants.
4. Most organisms will have pH values close to neutral.
5. Some substances will rise to toxic levels at abnormal pH values.
6. Because pH is a logarithmic scale, the acidity of the stomach (as low as pH 1.0) is millions of times more acidic than cellular fluids (typically near pH 7.0).
7. Strong acids will be neutralized by strong bases.

3.5 Reaction Rates

A drug is a substance which when injected into a guinea pig produces a scientific paper.

—Unknown

3.5.1 Collision Theory

Man fools himself. He prays for a long life, and he fears an old age.

—Chinese proverb

The *chemical collision theory* has been able to explain many of the observations related to chemical kinetics. The assumption made for this theory is that particles must physically collide for a chemical reaction to occur. In addition, these collisions must be effective; without sufficient force, the electrons surrounding the nuclei of the atoms involved would just repel each other and the atoms would not combine. Thus, the rate of any step in a reaction is directly proportional to (1) the number of collisions per unit time and (2) the fraction of these collisions that are effective.

The number of collisions for gases at standard temperature and pressure (STP: 0°C and 1 atm) has been calculated to be more than $10^{30}\ s^{-1}$. If all these collisions were effective, then reaction rates would be extremely fast. However, this is not true because only a small fraction of collisions are effective. The extra amount of energy required in a collision to overcome interatomic repulsive forces and produce a chemical reaction is known as the *energy of activation*. Its magnitude depends on the properties of the reactants.

3.5.2 Intermediate Reactions

Our biomedical engineering world is like that. We have the mundane, the ordinary, and the commonplace. We have hard work, unsolvable problems, and so many distractions. But we also have sublime moments when we realize that we have made progress, that what we are doing is important, and what we are doing is good.

—Raj Tonnash

It is common to express a chemical reaction in its simplest possible terms. Thus,

$$A + B \rightarrow AB \tag{3.5.1}$$

is the net chemical balance normally cited. However, there may be several intermediate steps:

$$A + B \rightarrow C \tag{3.5.2a}$$

$$C + B \rightarrow D \tag{3.5.2b}$$

$$D + A \rightarrow AB \tag{3.5.2c}$$

where C and D are intermediate products that may be so short-lived as to be unobservable under normal circumstances. Each step (a), (b), and (c) has its own reaction rate and its own activation energy. Thus, the reaction rate for Equation 3.5.1 depends on the slowest reaction rate (a), (b), or (c), and the activation energy for Equation 3.5.1 depends on the activation energies for (a), (b), and (c). If there are several slow intermediate steps and several high activation energies, then Equation 3.5.1 could depend on a combination of (a), (b), and (c). Enzymatically mediated reactions important in living things usually reduce activation energies and greatly increase rates.

When reaction rates are determined, they can only be determined experimentally because of the inability to observe all intermediate steps. It is known that reaction rates depend on some combination of the concentrations of the reactants:

$$\text{rate} = (k)(\text{Concentration of A})^m(\text{Concentration of B})^n \tag{3.5.3}$$

where:

k is a measured *constant specific rate*
m and n are exponents whose values must be experimentally determined

Sometimes the exponents m and n equal the coefficients in front of A and B in Equation 3.5.1, sometimes they do not. For the case of a *reversible* reaction,

$$A + B \leftrightarrow C + D, \tag{3.5.4}$$

the compounds C and D are the products of the forward reaction with A and B as reactants. However, the compounds A and B are the products of the reverse reaction with C and D as reactants. Each direction of the chemical equation has its own rate. At equilibrium, these two rates are equal:

$$\text{forward rate} = k_f\,(\text{concentration of A})^m\,(\text{concentration of B})^n \tag{3.5.5a}$$

$$\text{reverse rate} = k_r(\text{concentration of C})^p\,(\text{concentration of D})^q \tag{3.5.5b}$$

Thus,

$$K = \frac{k_f}{k_r} = \frac{(\text{Concentration of C})^p(\text{Concentration of D})^q}{(\text{Concentration of A})^m(\text{Concentration of B})^n} \tag{3.5.6}$$

where K is the *equilibrium constant*, a measured quantity as given in Section 3.3 (Chemical Equilibrium). Because K is constant, changing the concentration of one or more of the constituent compounds will change the others in the system.

3.5.3 First-Order Reactions

The way we name things inevitably affects how we perceive those things.

—Robert Dorit

Reaction rates depend on concentrations of reactants that change during the reaction. Thus, the rate of a reaction will usually be greatest at the beginning of a reaction and will decrease with time until equilibrium is established. One common type of chemical reaction is the *first-order reaction*, in which the rate of disappearance of a reactant is proportional to the reactant concentration. This leads to an exponentially decreasing reaction rate as time proceeds (see Section 4.3, Calculus). Other types of kinetics are also possible.

3.5.4 Enzyme–Substrate Reactions

Science is by no means incompatible with finding joy, meaning, and purpose in the world.

—Robert T. Pennock

A substrate is any substance on which an enzyme can act to form a product. Rates of simple enzyme–substrate reactions are often described by the Michaelis–Menten construction. It has been found that the rate of product formation depends directly on the substrate concentration, but that the dependence is small for low substrate concentrations and for high substrate concentrations, and the dependence is highest somewhere between. The curve of rate of product formation plotted against substrate concentration forms an "S" shape (Figure 3.5.1). The steeper the "S," the higher is the affinity of the enzyme for the substrate. A typical enzyme-substrate system has a rate of product formation equal to half the maximum (saturation) rate at about 5 mM at room temperature. Changing the concentration of an enzyme will affect the position of the curve.

Example 3.5.1: Faster Enzyme-Linked Immunosorbent Assay

One quick and popular immunoassay is the enzyme-linked immunosorbent assay, or ELISA (Kemeny, 1991). In ELISA, a surface is coated with antigen or antibody and dried. An antibody is then added to the test (see Section 6.20, Control), usually suspended in water. Chemically linked to the antibody is an enzyme, usually one of three: horseradish peroxidase, alkaline phosphatase, or β-D-galactosidase. When the sample to be tested is added, an antibody–antigen complex is formed and becomes fixed to the surface. Excess

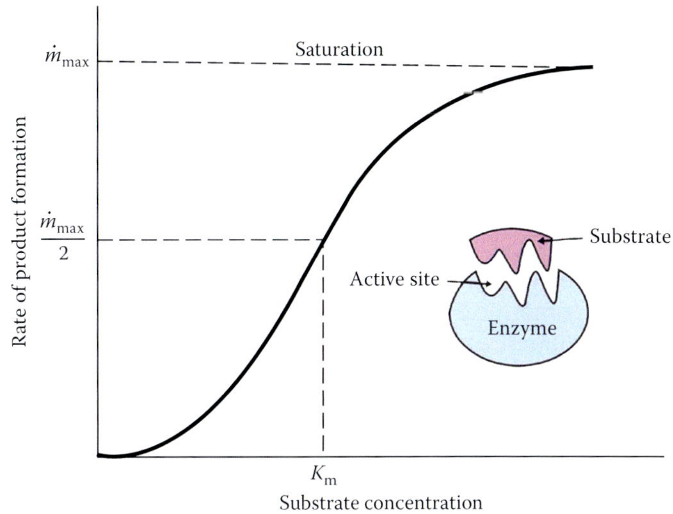

FIGURE 3.5.1
Michaelis–Menten reactions. (Redrawn from Johnson, A.T., *Biological Process Engineering: An Analogical Approach to Fluid Flow, Heat Transfer, and Mass Transfer Applied to Biological Systems*, John Wiley & Sons, New York, 1999.)

enzyme-labeled antibody is then washed from the surface. A chromogen substrate is added that combines with the enzyme linked to the antibody. When it does, it forms a colored compound that can be quantified by optical density. The optical density is thus dependent on the amount of enzyme remaining in the assay, and that amount, in turn, depends on the concentration of antibody in the sample tested.

This assay usually takes several hours to complete, but there is interest in speeding it up. The time taken by the assay is composed of diffusion time, antigen–antibody binding (very quick), and enzyme–substrate kinetics of the Michaelis–Menten type (except that the substrate concentration does not limit the rate). Thus, a faster assay will come only as a result of minimizing each component of time (Johnson, 1999).

APPLICATIONS AND PREDICTIONS

1. To drive a reversible chemical reaction to one side and produce the largest number of products, greatly increases the concentration of reactants.
2. To increase the rate of a chemical reaction in aqueous solution without increasing the amounts of reactants, decrease the amount of water.
3. Enzymes will increase reaction rates by increasing local concentrations of reactants.
4. To ameliorate environmental effects of reactive chemicals, dilute them.
5. Removing the product of a chemical reaction will increase the rate of reaction.

3.6 Carbon Chemistry

> Some types of molecular chains outcompeted other molecular chains for the planet's resources, and gradually they led to the kind of molecules that life depends upon—all this before the first living thing oozed forth.
>
> **—Joel Achenbach**

Carbon occupies a unique place and is essential to life as we know it. More than 90% of all known compounds contain carbon, and the number of hydrocarbons and derivatives number close to one million.

Carbon has a valence of 4, and, therefore, can either gain 4 electrons or lose 4. Carbon usually forms covalent bonds where it shares electrons with another atom. What makes carbon so unique is that it readily forms covalent bonds with other carbon atoms, making long chains possible.

Diagrammatically, the sharing of a pair of electrons is usually denoted by a dash between letters standing for particular atoms. Thus, –C–C– denotes a single pair of electrons shared between two carbon atoms. The lines extending radially outward from each carbon atom indicate that pairs of electrons are shared with four other atomic species. The number of shared electrons for each carbon atom is 8.

3.6.1 Many Possible Configurations

> Organic chemistry is the chemistry of carbon compounds. Biochemistry is the study of carbon compounds that crawl.
>
> **—Mike Adam**

Carbon can also form a double bond with another carbon, C == C, and a triple bond with another carbon, —C ≡ C—. The total number of shared electrons is still 8.

Carbon compounds can form chains, –C–C–C–C–; they can form branches, –C–C–, and rings:

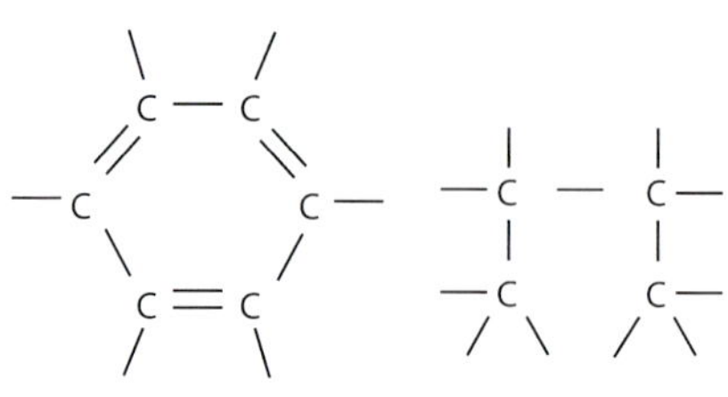

In all of these configurations, the number of shared electrons is 8.

WHAT'S SO SPECIAL ABOUT CARBON?

(Reprinted from Berger, 2002)

Why should there be a separate field ("organic chemistry") for compounds of one element, carbon, while the other ninety-five or so chemically important elements are lumped together as "inorganic chemistry?" Worse yet, more than a thousand known compounds include carbon for each one that does not—even after more than fifty years of determined effort by synthetic inorganic chemists. And it is carbon, always and only carbon, which forms the basic molecular framework of living systems. Although there has been and continues to be speculation about inorganic biochemistries, most professional exobiologists (who search for signs of extra-terrestrial life) do not give them a moment's thought.

How can one element so dominate the provincial imaginations of chemists that exobiologists rarely bother to speculate about life based on any other element? Let's do a systematic search for *other* elements on which life might be based. They only need to fulfill a few characteristics:

1. They must be common in the universe.
2. They must be chemically reactive.
3. They must be able to form complex, branched compounds similar to those we know in organic biochemistry.
4. The two most common, chemically reactive elements in the universe are hydrogen and oxygen. Our candidates must have compounds that are reasonably stable in the presence of hydrogen and oxygen.

Let's see how many elements we can find that fill the bill.

If an element is rare, there will not be enough of it to form the chemical basis of large- or even small-scale ecosystems. As a reasonable cutoff, we will require life-elements to have at least one atom per billion hydrogen atoms. Of the 85 or so elements found in nature, 22 are common: hydrogen, helium, nitrogen, carbon, oxygen, fluorine, neon, sodium, magnesium, aluminum, silicon, phosphorus, sulfur, chlorine, argon, potassium, titanium, chromium, manganese, cobalt, nickel, and iron.

Three of our 22 candidates are completely inert chemically, so much so that they form no known stable compounds: Helium, neon, and argon. This pares our list to 19: hydrogen, nitrogen, carbon, oxygen, fluorine, sodium, magnesium, aluminum, silicon, phosphorus, sulfur, chlorine, potassium, titanium, chromium, manganese, cobalt, nickel, and iron.

To form the chemical basis of life, an element ought to be able to form large and complex molecular structures, including branched rings and chains. It would be simplest if the element were able to do this by bonding to itself.

Most elements are metals, bonding to themselves only in infinite, three-dimensional arrays, and the "metallic bonds" in these arrays are not localized between pairs of atoms. This means that they cannot be selectively broken and re-formed during biochemical processes, and small molecular structures with metal–metal bonds are almost unknown. We have 10 metals left in our list: sodium, magnesium, aluminum,

(*Continued*)

WHAT'S SO SPECIAL ABOUT CARBON? (Continued)

potassium, titanium, chromium, manganese, cobalt, nickel, and iron; almost all of the 63 elements that we eliminated earlier as being too rare are also metals.

This leaves us nine elements: hydrogen, nitrogen, carbon, oxygen, fluorine, silicon, phosphorus, sulfur, and chlorine. All of these can bond to themselves, by localized "covalent bonds" that could be precisely manipulated by biochemical processes.

But to form branched rings and chains, an element has to be able to form more than one or two bonds. Hydrogen, fluorine, and chlorine cannot form more than one bond, and so if they bond to another atom they are done. Oxygen and sulfur can form only two bonds; they can make chains and rings such as –O–O–O–O– or –S–S–S–S–, but that is all. So, from our original list of a quarter of the elements found in nature, we are left with only four candidates for life: nitrogen, carbon, silicon, and phosphorus.

And we have to throw out nitrogen, too. When nitrogen bonds to itself, it overwhelmingly prefers to make all three bonds to one other nitrogen atom: N ≡ N. Nitrogen–nitrogen single and double bonds are unstable enough to make such compounds explosive.

What if nitrogen formed single bonds in which it alternated with another element? For example, single-bonded compounds with alternating boron and nitrogen atoms, or alternating nitrogen and phosphorus atoms, are stable; nitrogen forms strong bonds to carbon, silicon, and phosphorus. Unfortunately, this is essentially the same as making boron (a rare element), carbon, silicon, or phosphorus the basis of life. Besides, there is small likelihood of finding compounds lying around with the regular alternation required. So, we can leave combinations of other elements with nitrogen out of consideration, at least for the moment.

We are left with only three candidates for the element of life: carbon, silicon, and phosphorus. All of our remaining candidates form bonds to themselves, and all prefer to form single bonds so that bonds are left over for rings and branched chains.

Hydrogen and oxygen are respectively the most common and third-most common elements in the universe, and are quite reactive besides. To have a fighting chance of forming any sort of life, compounds of other elements must be reasonably stable in the presence of these two elements. This stability can be approximated by bond strength: how strong are bonds to hydrogen or oxygen compared with bonds of the element to itself? If a bond is reasonably strong, there will be little if any advantage in forming a different kind of bond; and the bond will be more difficult to break in order to form a different kind of bond.

Our remaining candidates for "element of life" are carbon, silicon, and phosphorus. All of them form stronger bonds to oxygen and hydrogen than they do to themselves. But how much stronger?

Silicon's bond to oxygen is two-and-a-half times the strength of a silicon–silicon bond: so strong that in nature, silicon is *exclusively* bound to oxygen. We could conceive of a biochemistry based on silicon–oxygen polymers of the form –Si–O–Si–O–Si–O–; artificial polymers of this type (silicones) are common. Unfortunately,

(*Continued*)

WHAT'S SO SPECIAL ABOUT CARBON? (Continued)

the silicon-oxygen bond is too strong. It is *hard* to break—as one would want to in order to work with biomolecules.

Phosphorus, like silicon, is almost always found in nature combined with oxygen. It can form stable, polymeric compounds with nitrogen (phosphazenes) that are similar to silicones. But the formation of such compounds is unlikely without strenuous help, given nitrogen's overwhelming preference for itself. (Besides, silicone and phosphazene chains and rings are oxygen-stable only if they have carbon-based side groups!)

Carbon, too, is more stable when bonded to oxygen, *but only by about 10%–20%.* Carbon–oxygen bonds are common, but they are relatively easy to break in favor of carbon–carbon or carbon–hydrogen bonds.

Furthermore, under "reducing conditions" (an excess of hydrogen), phosphorus and especially silicon are not able to maintain bonds to themselves in favor of bonds to hydrogen: silicon–hydrogen bonds are 60% stronger than silicon–silicon bonds. Carbon–carbon bonds, on the other hand, are about 90% as strong as carbon–hydrogen bonds; hydrocarbons (compounds with both carbon–carbon and carbon–hydrogen bonds) are no more reactive than pure carbon.

To avoid infinite chains, a "chain cap" is needed, an element that forms only one bond. The capping element should form strong bonds to our life-element, but not too strong: it should be stable in the presence of oxygen, but not be too much stronger than the bonds of the element to itself. Because it is so common, hydrogen is the most likely candidate.

Only carbon is able to work well with hydrogen. Carbon–carbon bonds are strong in comparison with carbon–hydrogen bonds, so that hydrogen will not break up carbon chains. And carbon–hydrogen bonds are nearly as strong as carbon–oxygen bonds, so that hydrocarbons need a considerable energy "boost" to start reacting with oxygen. Silicon and phosphorus, on the other hand, form bonds to hydrogen that are too strong—SiH_4 and PH_3 are much more stable than the same number of Si–Si or P–P bonds—*and* not strong enough—SiH_4 and PH_3 burn spontaneously in the presence of oxygen.

The other common elements suitable as chain caps are fluorine and chlorine, but they are worse than hydrogen. Carbon, silicon, and phosphorus form extremely strong bonds to fluorine—so strong that silicon–silicon and phosphorus–phosphorus bonds cannot compete. And, except for carbon, bonds to chlorine cannot compete with bonds to oxygen.

Not only does carbon fit our requirements best of all the possible elements, it often forms stronger bonds to other elements than those elements do to themselves. This allows these elements to be incorporated into carbon-based biochemical structures, and makes the organic structure of life far richer. In fact, three-quarters of the chemically active elements we began with are involved in biochemistry!

3.6.2 Functional Groups

> The biological properties of a molecule are determined by the molecule itself - the atoms in it and the way they are bonded to one another, its flexibility and shape.
>
> —**Roald Hoffmann**

Carbon can share its electrons with any number of other elements or functional groups. Each of these must be able to share 2 electrons. For instance, carbon readily bonds with hydrogen to form hydrocarbons:

$$\begin{array}{c} \mathrm{H} \\ | \\ \mathrm{H}-\mathrm{C}-\mathrm{H} \\ | \\ \mathrm{H} \end{array}$$

, or CH_4, is methane, the simplest hydrocarbon. Hydrocarbons may be either *saturated* or *unsaturated*, depending on whether there are additional carbon–carbon bonds that can be replaced by other elements or functional groups. Thus, $\overset{|}{\underset{|}{\mathrm{C}}}=\overset{|}{\underset{|}{\mathrm{C}}}$ is unsaturated; $-\overset{|}{\underset{|}{\mathrm{C}}}-\overset{|}{\underset{|}{\mathrm{C}}}-$ is saturated. Hydrogen is often the additional element added to the unsaturated compound to produce the saturated compound.

Instead of a simple hydrogen atom, the functional group linked to carbon may be an –OH hydroxyl group (Table 3.6.1). Addition of the –OH to carbon produces an alcohol. The alcohols act as weak acids and have polar molecules. They can dissolve both ionic and nonionic solutes.

The –CHO functional group produces an aldehyde, which is odoriferous and has various uses, including disinfection (formaldehyde). The –CO functional group produces a ketone, the most well-known of which is acetone. The –COOH carboxyl functional group produces weak organic acids (CH_3COOH is acetic acid). Organic acids with few carbons are soluble in water, but the acids with more carbons are less soluble because their polarities are diluted more by the hydrocarbon remainder of their molecules. The carboxyl functional group –COO– produces esters, the product of a reaction between an organic acid and an alcohol. Esters are pleasant tasting and smelling and are the principal flavor constituents of many fruits. The –O– functional group produces an ether, which can be formed when two alcohol molecules come together so that an H from one and an OH from the other split off to form water as a by-product. The extraction of water in this process is called a *condensation reaction*, and it does not occur spontaneously in water solution. Ethers are good organic solvents that dissolve a large variety of nonpolar substances that are only sparingly soluble in water. The amino functional group $-NH_2$ produces amines as building blocks for proteins. Protein contains approximately one part by weight of nitrogen for each 6.25 parts of protein (Lane, 2010). Additional complexity is possible considering that sulfur, phosphorus, iron, and other elements can also locate in organic compounds.

Of particular importance are the carbonyl and amino groups (Table 3.6.1). Each of these has two forms depending on pH: an ionized and a nonionized form. This will be discussed further in Section 3.7.4 (Isoelectric Point).

3.6.3 Amino Acids

> The best and most efficient pharmacy is within your own system.
>
> —**Robert Peale**

Amino acids are extremely important to living systems because they form the bases of the many proteins necessary for life. There are 20 common amino acids that function as protein building blocks (Table 3.6.2). There are at least 31 additional amino acids not

TABLE 3.6.1

Organic Functional Groups

Functional Group	Formula	Name of Compounds
Hydroxyl	—OH	Alcohols
Carbonyl	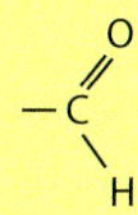	Aldehydes
	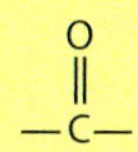	Ketones
Carboxyl	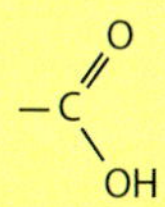 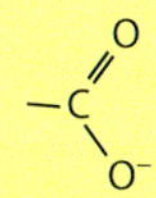 (Nonionized) (Ionized)	Carboxylic acids
Amino	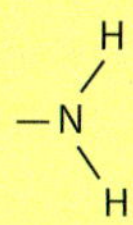 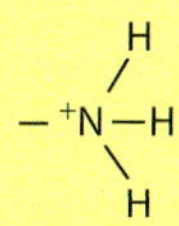 (Nonionized) (Ionized)	Amines
Sulfhydryl	—SH	Thiols
Phosphate	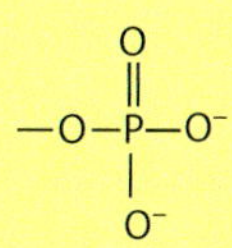	Organic phosphates

commonly part of natural proteins; among these are some hormones and neurotransmitters (Garrett and Grisham, 1999).

Amino acids polymerize to form peptides and proteins by joining amino and carboxyl groups. The structure of carbon-based organic compounds is extremely important because functional properties depend on structure. Compounds with the same chemical composition but different structures are called *isomers*. Amino acids exist in mirror-image forms (*stereoisomers*) that do not function identically. Only the L-isomers of amino acids commonly occur in nature (the two *chirality* designations are L, for *levorotatory*, and D, for *dextrorotatory*; levorotatory molecules rotate plane-polarized light in a counter-clockwise direction). Animals are made of proteins from L-amino acids, coded for by DNA built from D-sugars. L-amino acids and D-sugars are slightly more stable than their *enantiomers* (mirror image molecules) due to weak chemical interactions (Everts, 2013; MacDermott and Tranter, 1990).

3.6.4 Macromolecule Types

> Assembling a review panel is analogous to finding the ingredients for a batch of minestrone soup that takes a pinch of this and a spoonful of that to taste just right.
>
> **—Thomas Cech**

TABLE 3.6.2

Twenty Common Amino Acids

	Name	Acronym
Nonpolar		
	Leucine	Leu, L
	Proline	Pro, P
	Alanine	Ala, A
	Valine	Val, V
Polar, uncharged		
	Glycine	Gly, G
	Serine	Ser, S
	Asparagine	Asn, N
	Glutamine	Gln, Q
Acidic		
	Aspartic acid	Asp, D
	Glutamic acid	Glu, E
	Methionine	Met, M
	Tryptophan	Trp, W
	Phenylalanine	Phe, F
	Isoleucine	Ile, I
	Threonine	Thr, T
	Cystine	Cys, C
	Tyrosine	Tyr, Y
	Histidine	His, H
Basic		
	Lysine	Lys, K
	Arginine	Arg, R

The four basic types of macromolecules in living systems are given as follows:

1. *Carbohydrates*: These serve as fuels (glucose, for example) and building materials (cellulose). Carbohydrates can exist as simple sugars, monosaccharides, and polysaccharides.
2. *Lipids*: These are hydrophobic molecules used for dense energy storage, structural components of cell membranes, and steroidal hormones.
3. *Proteins*: These complex molecules give structural support to cells, store energy, assist in transport of other substances, serve as signal markers, result in mechanical movement and force generation, and act as enzymes in biochemical reactions. Proteins are usually structured with a long carbon chain backbone with various organic functional groups hung like ornaments on a tree. The particular sequence of amino acids in a protein determines its shape and function (see Section 3.7).
4. *Nucleic acids*: These molecules act as information storage units for the cell and are the basis for genetic encoding. Nucleic acids are polymers of nucleotides (complex organic compounds composed of pentose sugars, organic bases, and phosphate groups), and have the property that a single strand of nucleic acid can only form its conjugate image when duplicating (see Section 5.3).

MICROBIAL STOICHIOMETRY

Stoichiometry is the branch of chemistry concerned with proportions of chemical elements or compounds involved in reactions. Stoichiometry has also been associated with the use of microbes for *bioremediation* of environmental pollutants.

The composition of simple organisms such as bacteria, yeast, and algae are very similar. Microbial protoplasm has a relatively constant stoichiometric composition of (Chamberlain, 2003):

$$C_{106}H_{263}O_{110}N_{16}P_{1}.$$

If algal protoplasm is viewed in terms of carbon dioxide (CO_2), nitrate (NO_3^-), phosphate (HPO_4^{-2}), and water (H_2O), then the following stoichiometric balance applies (Strumm and Morgan, 1981):

$$106\ CO_2 + 16\ NO_3^- + HPO_4^{-2} + 122\ H_2O + 18\ H^+ \leftrightarrow \left[C_{106}H_{263}O_{110}N_{16}P_1\right] + 138\ O_2.$$

Notice that the masses of all elements and electrical charges are balanced on both sides of the equation. Energy in the form of light, and some trace elements are required on the reactant (left) side of the reaction for it to proceed. This balance simply expresses the means for algal protoplasm and oxygen to be formed from photosynthesis.

Sometimes a bioremediation design using microbes requires the addition of nutrients to promote microbial growth. The objective is to add scarce nutrients that limit microbial growth and force the microbes to use available nutrients found in pollutants. Hence, as the microbes grow and metabolize, they deplete pollutant levels. Macronutrients are often expressed in terms of carbohydrates (CH_2O, a carbon source), ammonia (NH_3, a nitrogen source), and orthophosphate (H_3PO_4, a phosphorous source). The only additional macronutrient commonly added to plant fertilizer is a potassium source, which doesn't appear in the stoichiometric formula for protoplasm. However, potassium, sodium, calcium, chloride, and other ions are necessary to maintain normal activity of living cells (see Section 4.4.3, Communication).

In terms of these macronutrients, microbial protoplasm can be expressed as:

$$\left[(CH_2O)_{106}\ (NH_3)_{16}\ (H_3PO_4)\right],$$

and this formulation can be used to determine how much of which limiting nutrient to add to either an in-situ remediation project or to a bioreactor to promote cellular growth.

Anaerobic bacteria are sometimes used in anaerobic digesters to change human, animal, or plant wastes into more useful or less toxic products. Anaerobic bacteria have a stoichiometric form of (Chamberlain, 2003):

$$C_{60}H_{87}O_{23}N_{12}P_1.$$

3.6.5 Polymers

> If any student comes to me and says he wants to be useful to mankind and go into research to alleviate human suffering, I advise him to go into charity instead. Research wants real egotists who seek their own pleasure and satisfaction, but find it in solving the puzzles of nature.
>
> **—Albert Szent-Györgi**

Polymers are giant molecules composed of many small molecules, called *monomers*, arranged in a repetitive chain. Polymer molecular weights range from thousands to millions (Kotz and Treichel, 1999). Many polymers are formed from identical monomers; the molecules thus formed are long and thin. These compounds act as strands and can be highly elastic. Rubber, in fact, is an *elastomeric polymer*. Some polymers have *cross-linking* chemical bonds between strands or between portions of the same strand, and so have greater rigidity or dimensional stability. Hydrogen bonds or sulfur bonds are often the major stabilizing bonds present.

Polysaccharides, proteins, and nucleic acids are all polymers. Polymers such as collagen are biologically important, and the elastic properties of collagen are often used to advantage for energy storage in tendons.

3.6.6 Melting and Boiling Points

> Glamour and rock-steady dependability do not easily walk hand-in-hand.
>
> **—Raj Tonnash**

An important physical property of organic compounds is their melting and boiling points. The smaller the organic molecule, the lower is its melting and boiling temperatures. Thus, larger molecules are less volatile and more likely to exist in solid form at room temperature. This has important implications for biological systems: those that exist at lower temperatures are more likely to contain low molecular weight fats that are more pliable (less stiff) at those temperatures than are higher molecular weight fats. Fats deposited in the appendages of the human body, with normal temperatures lower than body temperature, are usually of lower molecular weight than those deposited in the body core.

3.6.7 Organic Reactions

> Those who have an excessive faith in their theories or in their ideas are not only poorly disposed to make discoveries, but they also make very poor observations.
>
> **—Claude Bernard**

Two important features of reactions between organic compounds are as follows:

1. They are extremely slow, taking hours or days to complete.
2. Only part of the molecules take part in the reactions.

The result of the first feature is that heat and enzymes can be used to great advantage to speed the reaction, so it occurs at a more useful pace. The result of the second is that the organic compounds are often constructed a piece at a time in a step-by-step fashion; working on one part of the organic molecule does not usually alter the remainder. Both of these features are true whether the organic reaction takes place *in vivo* (in the living body) or *in vitro* (in glass, or outside the body).

PRIMITIVE FORMS OF LIFE

There is mounting evidence that comets could have provided the young Earth with its water, atmosphere, and carbon compounds that seeded prebiotic life (Delsemme, 2001; de Duve, 1995). But prebiotic Earth would have been much different from present-day Earth. It would have been rocked by volcanic activity, and the land would have been barren. The oceans were probably acidic. The atmosphere would have been devoid of oxygen and rich in hydrogen sulfide, carbon dioxide, nitrogen, and water vapor. Atmospheric pressure would have been high enough to suppress complete loss of oceanic water through boiling, and the atmosphere would have been electrically charged. Sunlight was less strong in those days, but unfiltered ultraviolet light would have bathed the surface of the planet. It was not a place conducive to the formation of life except by the bombardment by rocks from the sky that contained primitive organic compounds.

Life, however, could have evolved under these conditions. That life is resilient has been seen in the recovery of living communities around hydrothermal vents following volcanic eruptions in the mid-Pacific. Like terrestrial biomes after a disturbance, hydrothermal vent communities develop in a succession of stages (see also Section 6.21.1, Regeneration). At first there was an abundant growth of chemosynthetic microbes that feed on H_2S, H_2, Fe^{++}, and other reduced species. Then followed the tubeworm (*Tevnia jerichonana*) that lived symbiotically with the microbes (the tubeworm ingested the microbes, but did not digest them). This tubeworm was supplanted by a larger tubeworm (*Riftia pachyptila*), and then the mussels (*Bathymodiolus thermophilus*). It had taken eight years to reestablish a fully functioning symbiotically based modern ecological community of life (Lutz et al., 2001).

Evolutionary Highlights of the Earth's Biosphere. (a) a protosolar disk at 0 years, (b) proto-Earth accretion at 40 million years, (c) formation of the moon at 50 million years, (d) heavy cometary bombardment from 50 million years to 600 million years, (e) ancient sedimentary rocks form and first life appears at 800 million years, and (f) more advanced life exists at 4540 million years.

Chemical reactions involve the breaking of old bonds and the formation of new ones. When organic compounds, with their covalent bonds are reacting, the compound A–B may break to form two ions, A^+ and B^-, or they may break to form two neutral fragments, A and B, called *radicals*, each with an odd number of electrons. A radical is a molecule with an unpaired electron in its outer shell (highest energy level). Without the requisite number of 8 electrons, each of these radicals is highly reactive. Usually one or other of these radicals is of immediate use; the other, the *free radical*, is a by-product that must be dealt with in some way. Free radicals can be destructive to carbohydrates and other essential biochemicals inside the cell (Capra, 1996), but do assist in defending against microbial invaders of the human body.

ANTIOXIDANTS

Free radicals seek to combine with electrons from stable compounds and thus produce more free radicals in the process. The cell membrane is one of the most vulnerable structures to free radical damage. Also affected are low density lipoproteins, other proteins, and DNA. Free radicals alter functions of these molecules or cause mutations in DNA. There are mechanisms in place to repair free radical damage, but the repair is not completely effective and becomes less so with age. The result is heart disease, cancer, arthritis, cataracts, and aged skin (Elson, 2009).

Antioxidants derived from foods help to protect against free radical damage. Carotenoids are a very important class of antioxidants, as are vitamins A, C, and E. Carotenoids, including alpha-carotene, beta-carotene, lycopene, lutein, and zeaxanthin, have been demonstrated to protect against skin cancer, prostate cancer, and atherosclerosis, among others. Higher blood antioxidant levels have been correlated with lower blood levels of C-reactive protein (CRP), and thus lower inflammation of the blood vessels (and less atherosclerotic plaque).

But, like a lot of biology, the story of free radicals and antioxidants may be much more complicated than commonly believed. Some level of free radicals in certain locations may be necessary to stimulate cellular repair and defensive mechanisms. Free radicals at higher levels and at other places in the body may lead to cellular damage (see Figure). Just as with other substances, the concentration of antioxidants may be important, with a *hormetic* effect (beneficial at low concentrations but toxic at higher concentrations). Antioxidants in foods seem to be healthy; antioxidants as vitamin supplements are not always so (Yang and Hekimi, 2010; Moyer, 2013). A low level of free radicals in the central nervous system (CNS) causes a mild neural stress that induces the neurons to react and become fortified against more threatening stresses from other sources. Thus, free radicals appear to have a hormetic effect on CNS neurons quite different from their ability to damage cellular DNA and proteins (Mattson, 2015).

Whereas antioxidants are generally thought to be good for health, they also appear to be good for metastasizing cancer cells (Piskounova et al., 2015). Metastasizing melanoma cells experience very high levels of oxidative stress, which leads to the death

(*Continued*)

ANTIOXIDANTS (Continued)

of most of the metastasizing cells, but ingested antioxidants relieve oxidative stress and let more cells survive. Cancer patients getting the antioxidants were found to die faster than those not treated with antioxidants.

The antioxidant status of a patient can be determined quickly and easily by Raman spectroscopy administered to the palm of the hand. This test is noninvasive and painless, yet accurately measures serum carotenoid concentrations.

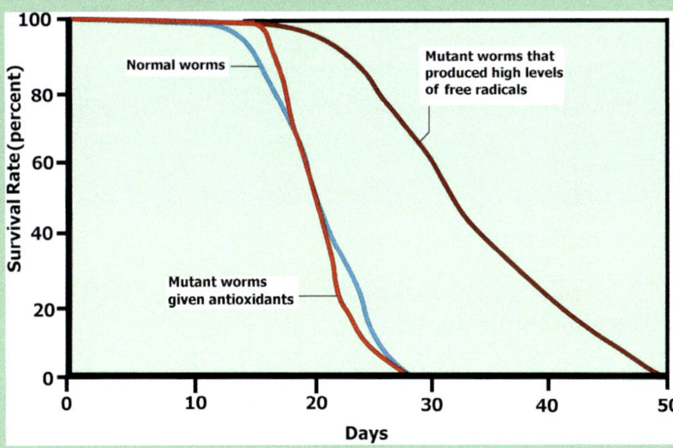

Round worms genetically modified to produce high levels of certain free radicals lived 32% longer than unmodified worms. Feeding Vitamin C to the worms shortened their life spans. (Redrawn from Moyer, M.W., *Sci. Amer.*, 308, 62–67, 2013.)

THE IMPORTANT PORPHYRIN RING

(Reprinted from Dayan and Dayan, 2011)

The porphyrin pathway is ubiquitous in the biochemical realm, serving throughout the plant and animal kingdom as the assembly line for the most abundant pigments in nature. The ring-shaped porphyrin molecules bind an array of metal ions, with each combination associated with different biological functions. Chlorophylls bind magnesium to play a pivotal role in photosynthesis. Heme binds iron to coordinate molecular oxygen and carbon-dioxide transport chains necessary for cellular respiration and contributes to the catalytic activities of many enzymes. Porphyrins bind nickel to form coenzyme F430, which plays critically important roles in bacteria that metabolize methane. Vitamin B12 is formed from the binding of cobalt to a derivative of porphyrin; lack of the vitamin can result in pernicious anemia and impair the function of the brain and nervous system. Taken together, these porphyrin-derived pigments can be called the "colors of life," in the sense that these rings are necessary to sustain key activities in nearly all organisms (Figure).

(Continued)

THE IMPORTANT PORPHYRIN RING (Continued)

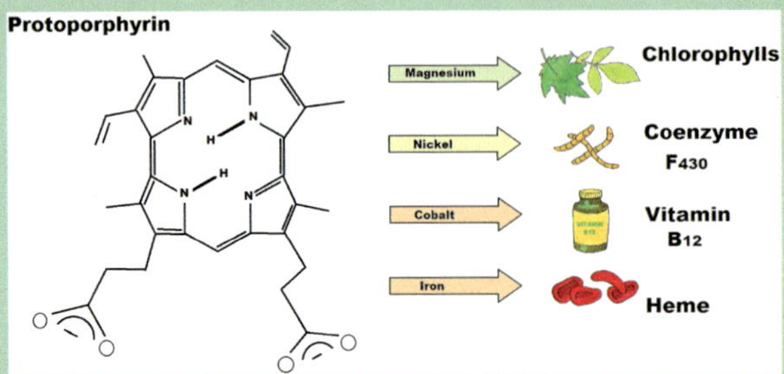

The porphyrin ring-shaped molecule can accept different metals to produce pigments essential to metabolism in various organisms. Heme and vitamin B_{12} are vital to animals, chlorophyll to plants, and coenzyme F_{430} to methane-processing bacteria.

Example 3.6.1: Macromolecule Sieves

Some molecules, namely, H_2O, O_2, CO_2, and even insulin, important to living things are relatively small in size. Other molecules, notably proteins, poly carbohydrates, and some fatty acids, are large in size. Substances that identify a particular individual, called antibodies or immunoglobulins (abbreviated as Ig, see Section 6.20.3, Redundancy), are among the larger macromolecules.

Molecular sizes are measured in terms of Daltons, which correspond to molecular weight. Antibodies are larger than 100 kilodaltons (kDa), whereas proteins such as insulin are smaller.

Transplanting foreign cells into a body is a method proposed to compensate for the failure of native cells to produce essential biochemicals. Type I diabetics, for instance, lose the ability to produce sufficient amounts of insulin.

Cells taken from other people, or from other species, could be used to supply necessary insulin. These cells, however, carry markers identifying them as nonnative. The body's immune system would attack them and kill them, leaving the body again without a source of insulin.

Artificial membranes can be made containing holes of a certain size. If these membranes allow water, oxygen, carbon dioxide, glucose, and insulin to pass through, but block marker proteins and antibodies from contacting each other, then nonnative cells can be implanted into the body without subsequent rejection. The only concern here is to make the membrane out of a material that is itself not identified as a foreign body.

Membranes with specifically designed pore sizes can be used to separate larger from smaller molecules for many purposes. They are used in the food industry, in wastewater treatment, in bioreactors, and in biomedicine.

They act as molecular sieves and provide relatively inexpensive solutions to many practical problems.

Example 3.6.2: Carbon:Nitrogen Ratio for Composting

Composting is used to turn discarded or infected organic waste into a pathogen-free soil amendment. Composting is used to remediate pathogenic threats from animal carcasses as well as from infected plant matter. Composting uses mesophillic (medium warm temperature loving) and thermophilic (hot temperature loving) bacteria to metabolize organic matter, producing heat in the process. If the temperature rises high enough, pathogenic bacteria are killed and weed seeds are inactivated. The end product is a nitrogen and micronutrient rich fibrous material.

Nitrogen is required for the composting bacteria to form microbial tissue. Nitrogen and other essential nutrients are supplied by the organic matter to be composted. Carbon, however, must be added to supply energy for aerobic metabolism. Carbon may come from any number of sources: sawdust, straw, chopped newsprint, or other cellulosic material. Oxygen must also be available to achieve aerobic digestion. Anaerobic digestion, due to insufficient oxygen, results in an odorous product without killing pathogens.

The ideal ratio of carbon to nitrogen is in the range of 20 or 30 to 1. Table 3.6.3 gives a list of carbon to nitrogen ratios of common compost materials. It is apparent from this table that the required C:N ratio must be achieved with a mix of carbon-rich and nitrogen-rich sources.

TABLE 3.6.3

Common Compost Materials

Compost Material	C:N
Animal carcass	5:1
Broiler litter	14:1
Cattle manure	19:1
Corn stalks	60–73:1
Cottonseed meal	1–7:1
Finished compost	30–50:1
Grass clippings	17:1
Hay-general	15–32:1
Horse manure-general	30:1
Rice hulls	121:1
Sawdust	442:1
Sheep manure	16:1
Soybean meal	4–6:1
Straw-general	80:1
Straw-wheat	127:1
Swine manure	14:1
Turkey litter	16:1

Source: Payne, J. and Pugh, B., *The Shepherd*, 26–29, 2010.

APPLICATIONS AND PREDICTIONS

1. A carbon source will be necessary for life as we know it.
2. There are many organic compounds still undiscovered.
3. Fats and oils found in trout (that thrive in cool water) will be of smaller molecular weight than those found in *tilapia* (a high-temperature fish).
4. Enzymes will greatly enhance the speed of organic chemical reactions.
5. Organic compounds in living things will usually be built piece-by-piece, and the order in which they are built will be important.
6. Amino acids are necessary for life.
7. Polymers and isomers give variety and diversity to organic molecules.
8. In organic reactions, there usually will be a by-product that affects the environment.

3.7 Physical Chemistry in Water

> I believe that any bio-based engineer who doesn't appreciate how any living being (or system) interacts with, reacts to, and is affected by its total chemical, physical, and biological environment is not well prepared.
>
> **—Raj Tonnash**

Water is such an active substance that it profoundly affects other chemicals it associates with. Water can tear molecules apart or greatly affect their physical properties. In this section, we will learn of some of these effects.

3.7.1 Solutions

> Avogadro's Number is the fundamental physical constant that links the macroscopic world of objects we can see and feel with the submicroscopic, invisible world of atoms.
>
> **—Ronald Fox and Theodore Hill**

A great deal of the chemical activity of the cell takes place in solution. When a solid goes into solution:

1. The physical structure of the solid must be disrupted.
2. The physical structure of the liquid is disrupted.
3. The *solute* (or solid going into solution) must be attracted to the *solvent* (or liquid dissolving the solid).

When the process of a solution is completed, solute particles are surrounded by solvent particles; the solute is hydrated (if water is the solvent) by the process called *solvation*.

Water as a polar molecule is attracted to other molecules with locally charged regions (Figure 3.7.1). Thus, when sugar dissolves in water, hydrogen bonds are established between water and sugar molecules. Sugar molecules thus become surrounded by water molecules bound with intermediate-strength hydrogen bonds. Substances such

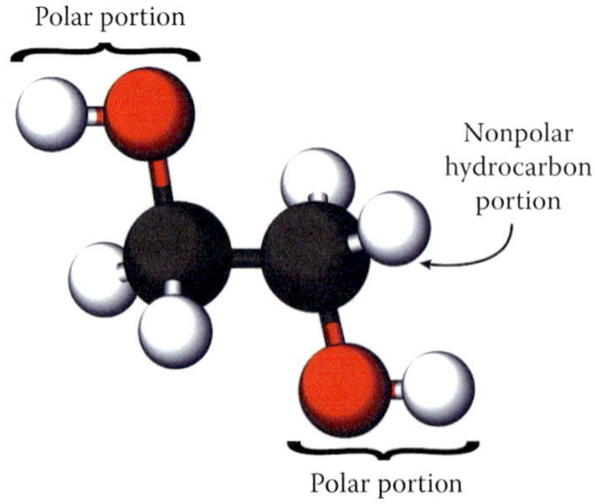

FIGURE 3.7.1
A molecule of ethylene glycol is a two-carbon alcohol with two hydroxyl groups that make it soluble in water.

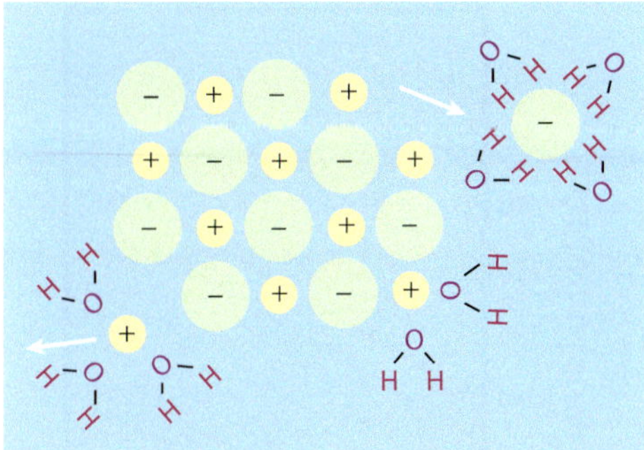

FIGURE 3.7.2
As NaCl is dissolved, the NaCl crystal is torn apart by water molecules.

as lipids are not polar and do not display surface charges the way polar molecules do. Nonpolar molecules dissolve very poorly, if at all, in water or other polar solvents such as ethanol. Solutes that dissolve in polar solvents are called *hydrophilic;* solutes that do not dissolve in polar solvents are called *hydrophobic.* In general, hydrophilic solutes dissolve readily in polar solvents, whereas hydrophobic solutes dissolve readily in nonpolar solvents.

Some solutes *dissociate* when they dissolve; other don't. Salt, for example, dissociates, but sugar doesn't. Salt (NaCl) forms the ions Na^+ and Cl^-, and so forms twice as many particles in solution as does sugar (Figure 3.7.2).

Liquids can also dissolve in other liquids. These are termed *miscible.* When liquids do not dissolve in other liquids, they are called *immiscible.* These liquids will remain separate. Mixtures of polar and nonpolar liquids are generally immiscible.

Alcohol is a versatile solvent because part of the molecule is polar whereas the rest is not. Thus, some nonpolar solutes and some polar solutes can dissolve in alcohol. Alcohol relates both to water and oil.

Each of the three aforementioned steps to the formation of a solution has associated energy requirements. Breaking the structure of the solute requires energy. Attraction of the solvent for the solute liberates energy. In addition, if the solute dissociates, energy is required for that process also. The sum of energy requirements and surpluses must be positive in order for the process of solution to succeed. Whenever more energy is liberated than is required, heat is released, and the solution heats; the solution is *exothermic.* Whenever the required energy exceeds the amount liberated, the solution cools; the solution is *endothermic.*

The maximum amount of solute that will dissolve in a standard quantity of solvent is termed the *solubility* of the solute. Exceeding the solubility of the solute will not result in any additional amount of solute dissolved. The solubility of various solutes is affected both by temperature and pressure. Most solid solutes increase their solubilities as temperature increases (Figure 3.7.3), whereas the solubilities of gases decrease as temperature increases There is less oxygen in warm water than in cool water, so aquatic species may sometimes asphyxiate in warm water.

The solubility of gases increases as pressure increases according to Henry's law:

$$\text{(Gas solubility)} = \text{(Henry's law constant) (Partial pressure of the gas)} \quad (3.7.1)$$

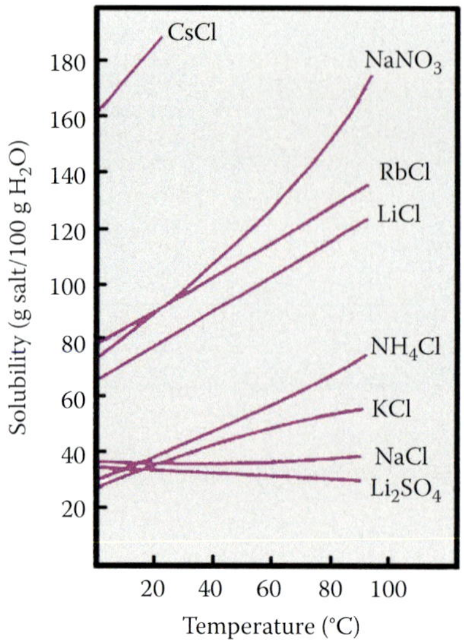

FIGURE 3.7.3
The temperature dependence of the solubility of some ionic compounds in water. Most compounds increase in solubility with increasing temperature.

Thus, large amounts of carbon dioxide can be made to dissolve in water at high pressure. If the pressure is suddenly released, the water fizzes. This is commonly utilized for carbonated beverages.

The vapor pressure of either the solvent or the solute is the pressure of that substance existing in equilibrium with the liquid phase. Thus, there is water vapor in the atmosphere above a container of pure water. The pressure exerted by the water vapor depends on the ability of the liquid water molecules on the liquid surface to evaporate and escape into the air. At the same time, as liquid is evaporating, some vapor is condensing, so an equilibrium is established where evaporation and condensation are exactly equal (Figure 3.7.4). The vapor pressure at balance is the *saturated vapor pressure* of the liquid.

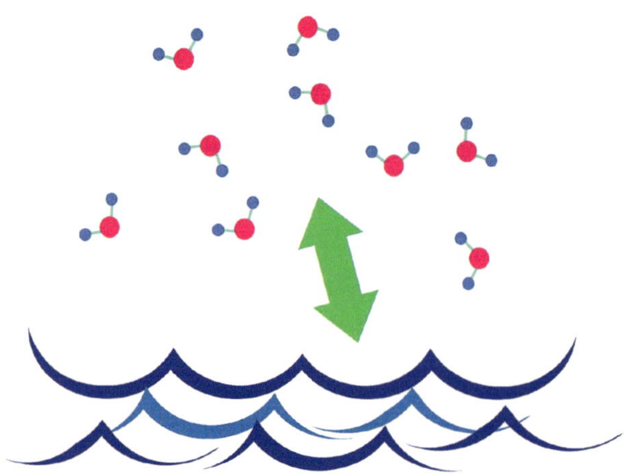

FIGURE 3.7.4
Net liquid evaporation or condensation depends on which physical process is greater. Both are occurring simultaneously at all times.

It takes energy to evaporate liquid. This energy comes from the kinetic energy of the liquid molecules. The higher is their temperature, the greater is the amount of kinetic energy that they possess. Hence, the saturated vapor pressure of a liquid depends on temperature (and only extremely slightly on pressure). At some temperature, the saturated vapor pressure reaches atmospheric pressure, and the liquid boils.

The surface of a solution is randomly populated by solvent and solute particles. The opportunity for a solvent molecule to escape by evaporation is thus lowered by the fraction of solute molecules present in the solution. It could then be expected that the vapor pressure of the solvent above the solution is less at any given temperature than it would be for a pure solvent. This leads to *Raoult's law*:

$$\begin{gathered}(\text{Solvent vapor pressure in solution})\\ =(\text{Mole fraction of solvent})(\text{Vapor pressure of pure solvent})\end{gathered} \tag{3.7.2}$$

With a lower vapor pressure for any given temperature, a higher temperature is thus required for the solution to boil than for the pure solvent. Thus, boiling food will happen at temperatures higher than 100°C, making it more likely that microbes will be killed, and also making it more likely that some of the nutritional biochemicals in the food will be altered.

At the other end of the temperature range, freezing of a solution occurs at a temperature lower than the freezing point of the pure solvent. This is because freezing requires the establishment of a solid structure for the solvent. Solute particles among the solvent particles make it more difficult to establish this structure. The more concentrated is the solute solution, the more the freezing point is depressed. Thus, most biological materials will not begin to freeze until they reach a temperature of about −3°C. Some cold-water species of plants and animals contain high concentrations of glycoproteins that lower their freezing points a good deal further.

Freezing of a mixture of substances does not occur at one temperature, but rather over a temperature range (Figure 3.7.5). In many cases, it is possible that pure water will freeze

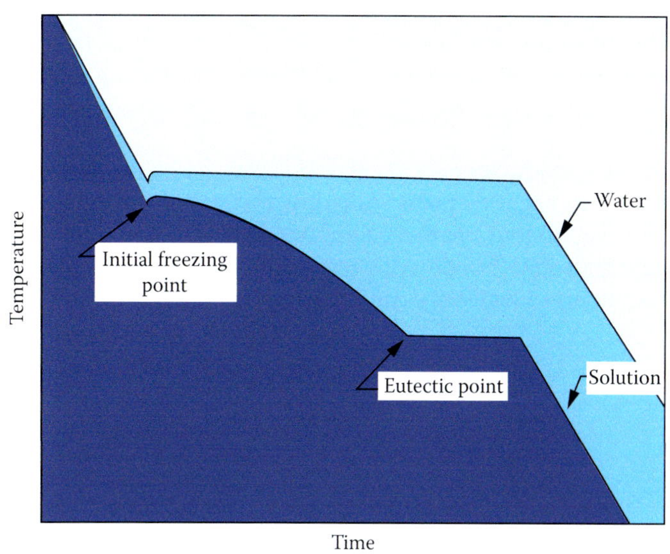

FIGURE 3.7.5
Schematic freezing curves for pure water and for an aqueous solution with one solute. Solutions with many solutes, found in real foods, have much more complex freezing curves.

first, and then the remaining unfrozen portion would contain more concentrated solutes, which depress the freezing point even further. Thus, the temperature at which the solution begins to freeze is perhaps much higher than the temperature at which freezing is complete. The more kinds of solutes present, the more complex is the freezing process. For certain mixtures of liquids, where both liquids dissolve in each other, there may be formed a *eutectic* mixture that has a freezing temperature lower than either liquid by itself. Eutectic mixtures can be useful to protect against freezing or to promote changes of state at useful temperatures. For instance, heat can be stored in the liquid form and liberated when the substance solidifies. In order to make this heat available at the desired temperature, a eutectic mixture could be used.

3.7.2 Gels

> The hypotheses we accept ought to explain phenomena which we have observed. But they ought to do more than this: our hypotheses ought to foretell phenomena which have not yet been observed.
>
> **—William Whewell**

Gels are soft, solid, or solid-like materials that consist of two or more components, one of which is a liquid of some abundance (Pollack, 2001). In the gels important for life, the liquid is water and the most important other component is protein.

Proteins are large complex molecules with molecular weights from 10^4 to more than 10^6. They are built of amino acids joined together by peptide bonds (Figure 3.7.6); peptide bonds are carbon–nitrogen bonds that join the carboxyl group (–C–OH) of one amino acid to the amino group ($-NH_2$) of another amino acid. In this way, they form a long, unbranched, polypeptide chain as the primary structure. Proteins form complex folded structures that expose carbonyl groups with negative surface charges and amino groups with positive surface charges. Side chains may have either charge. As formed, the overall protein charge is neutral, but the large separation of positive and negative charges allows local regions to be anything but neutral. Protein in water dissociates somewhat, losing positive ions (see Section 3.7.4, Isoelectric Point). The surrounding water spatially buffers the protein molecules and prevents cations from reassociation. Thus, proteins in the cell carry a net negative charge.

Water, as we have already seen, is itself a dipolar molecule, with charge separation between its positive and negative poles. Adjacent water molecules tend to align, forming hydrogen bonds with up to four nearest neighbors. Liquid water is composed of molecules that form transient hydrogen bonds with one another on a timescale of 10^{-11} s (Pollack, 2001). Clusters of structured water are continually forming and dissolving.

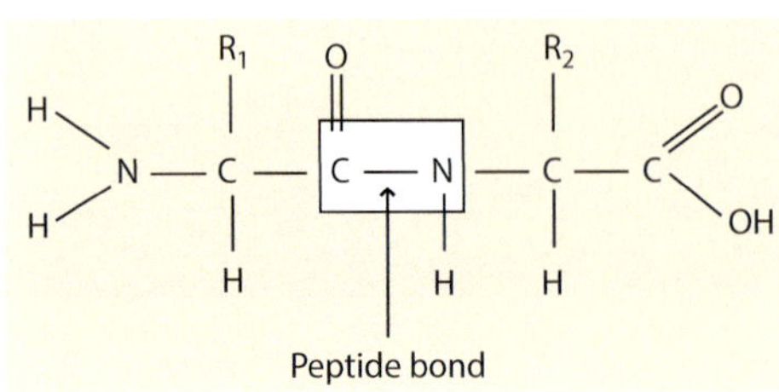

FIGURE 3.7.6
The peptide bond.

Surface charges on the protein attract water molecules, which adsorb on the protein surface. Whenever one water molecule aligns with protein surface charge, adjacent water molecules align with the first. This induces the layer of water molecules to form a loose structure with positive ends of water molecules attracting negative ends of other water molecules (Figure 3.7.7).

Because water–water hydrogen bonds are relatively weak, the influence of a single protein surface charge diminishes with distance from the charge. Proteins, however, pack many surface charges into a dense region.

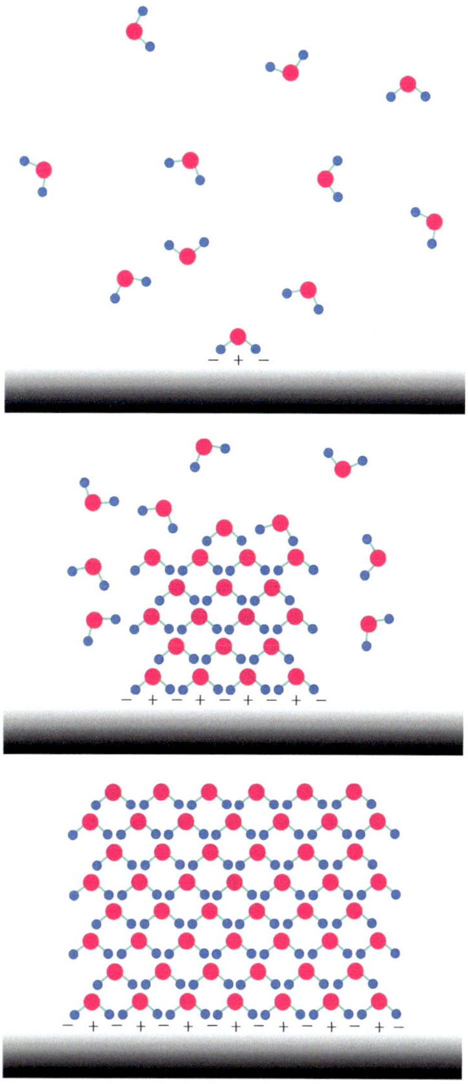

FIGURE 3.7.7
Surface charge on a protein strand attracts a water dipolar molecule (top). The adsorbed dipole attracts additional dipoles to form a structured organization within the liquid (middle). Additional surface charges on the protein reinforce the external dipole network (bottom). (Redrawn from Pollack, G.H., *Cells, Gels, and the Engines of Life: A New, Unifying Approach to Cell Function*, Ebner and Sons, Seattle, WA, 2001.)

Each charge, therefore, reinforces the influence of all other charges, and water becomes very tightly bound. The regularly repeating nature of protein surface charge tends to capture water molecules and hold them tightly to the protein surface (Pollack, 2001).

Not only does protein surface charge capture one layer of water, but, due to the dipolar nature of the water molecule, its influence extends outward through many water layers. Evidence has accumulated that indicates that water may be stratified to distances of 400–500 layers of single water molecules (Pollack, 2001). Thus, cellular water is different from bulk water; cellular water is highly stratified and immobile. The *cytoplasm* (that part of the cell not included in the nucleus) is packed with proteins (Figure 3.7.8)—all of which tend to keep water structured and fixed in place.

This is likely the source of the gel-like nature of cellular contents (Pollack, 2001). Because of this gel, cellular contents do not spill out when damage to the cellular membrane occurs, nor is there loss of integrity due to the many holes that are naturally present in the membrane.

Structured water is a very poor solvent. Non-water molecules included within this structure tend to disrupt the structure. The larger the molecule or ion, the more disruption it will cause, and the less likely it will be to gain entry. The structured water state is at a minimum energy level, and any molecule forced into the structure will require a considerable

FIGURE 3.7.8
Cytoplasm packed with proteins and other molecules. Molecules must perform their tasks under the cell's very crowded conditions. This cross section through the cytoplasm of a typical human cell depicts macromolecules. Macromolecules typically occupy 20%–40% of cytoplasm volume (Clabby, 2011). The large reddish molecules are ribosomes that are reading genetic information from the small white snaky messenger RNA molecules. At the same time, brown, L-shaped transfer RNA molecules are aligned on the ribosome in order to build a new protein. Actin filaments and intermediate filaments crisscross the space, providing support to the cell and a scaffold for hundreds of different enzymes, which are working on their metabolic tasks. The spaces between the macromolecules are also filled with small molecules and water, not depicted here for clarity, forming a very busy environment. This picture is a static snapshot of the cell, but, in reality, all of these components are in rapid motion. (Redrawn from Goodsell, D.S., *Am. Sci.*, 88, 230–237, 2000.)

amount of energy to be inserted. Thus, we would expect ions to be pretty much excluded from the cytoplasm, but, because diffusion pressure depends to a great extent on concentration, cytoplasmic ionic concentration should be related to ionic concentration outside the cell. That is, if ionic concentration of potassium (K^+) outside the cell doubles, then it is expected that the ionic concentration of K^+ inside the cell will double, although the inside concentration of K^+ will remain at a small fraction of the K^+ concentration outside the cell. Hydrated sodium ions (Na^+), because they are larger than hydrated potassium (K^+) ions, will have a smaller concentration inside the cell than will K^+.

Because protein surface charge is the basis for the gel that forms inside the cell, any change in protein structure can disrupt water structuring. Pollack (2001) has provided a hypothesis for cell activity that involves a phase transition of protein structure from long and straight to short and coiled. The long, straight protein configuration allows water to stratify regularly to maintain the gel and ion exclusion. A change of protein structure to short and coiled masks much of the protein surface charge and frees much of the surrounding water (Figure 3.7.9). The result is that the water acts much more like bulk water. It will tend to dissolve molecules and ions, and will not exclude relatively small substances from the interior of the cell.

Controlling the phase transitions allows the cell to perform its functions. Divalent ions such as Ca^{2+} and the adenosine triphosphate (ATP) molecule that contains a high charge density tend to draw adjacent protein strands together and restore structure from the disordered state. Once proteins are restored to their long configurations, water inside the cell will once again form the gel structure that apparently is required to maintain cellular integrity.

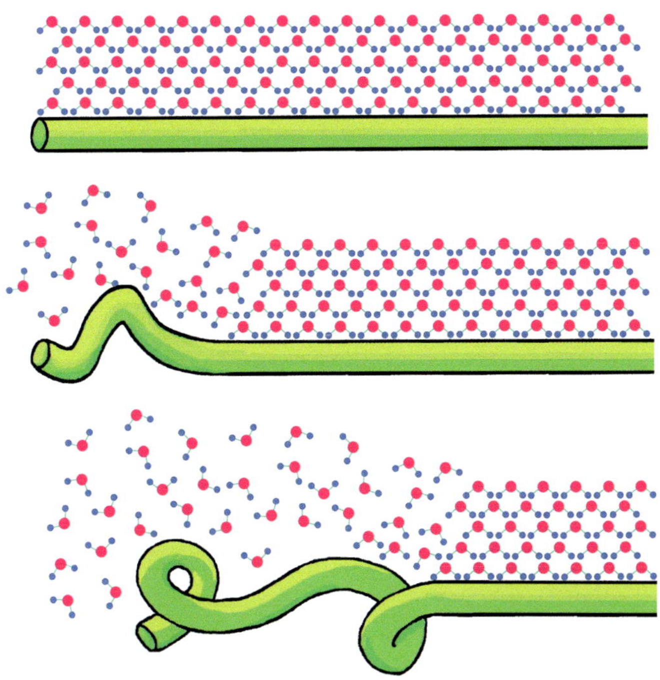

FIGURE 3.7.9
Protein phase transitions that cause structural changes also disrupt water layers surrounding the protein. Water structure disappears, and this allows further condensation of protein strands by coiling. The unstructured water is also free to dissolve molecules and ions. (Redrawn from Pollack, G.H., *Cells, Gels, and the Engines of Life: A New, Unifying Approach to Cell Function*, Ebner and Sons, Seattle, WA, 2001.)

CELLULAR RESTING POTENTIAL

All cells exhibit a negative internal voltage of –50 to –200 mV relative to the outside. This voltage, called the *resting potential*, has been explained since about 1950 by means of a cellular membrane that selectively pumps sodium ions (Na^+) out of the cell and potassium ions (K^+) into the cell (see Figure). The pump, it is thought, consists of a transport protein embedded in the cell membrane that translocates both Na^+ and K^+ against their concentration gradients. Because the concentration of Na^+ is higher outside the cell than inside the cell, Na^+ flows passively back into the cell. K^+ flows passively, too, but from inside to outside. This so-called *ion-pump* requires energy in the form of ATP to keep operating.

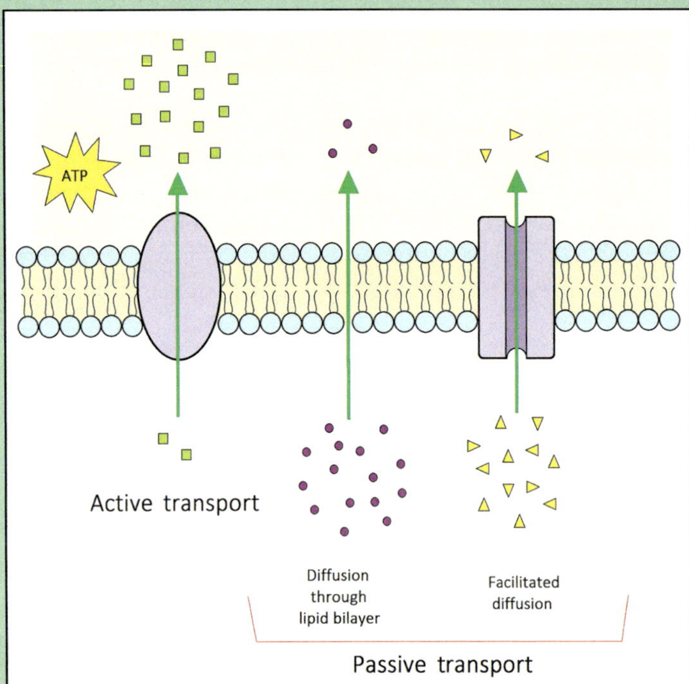

Ions move through the cell membrane by means of passive diffusion from high concentration to low. The traditional view is that hydrophobic molecules and very small uncharged polar molecules diffuse directly across the membrane. Hydrophilic substances diffuse actively, facilitated by transport proteins. Active transport occurs against a concentration gradient and requires energy. (Redrawn from Campbell, N.A. et al., *Biology*, 5th ed., Addison Wesley Longman, Menlo Park, CA, 1999.)

According to this explanation, the resting potential, sometimes called a membrane potential, is the net result of all the charges inside and outside the cell. Immobile proteins have a net negative charge (anions), but cannot move outside the cell, chloride ions (Cl^-) are nearly in equilibrium across the membrane, K^+ is at a higher concentration inside than out, Na^+ is at a higher concentration outside, and calcium ions (Ca^{++}) are present in concentrations that have largely been ignored.

(*Continued*)

CELLULAR RESTING POTENTIAL (Continued)

The problem with this scenario is that the integrity of the membrane is not necessary to maintain the resting potential (Pollack, 2001). Large holes can and do appear in the membrane, and if the Na^+–K^+ ion pump was necessary to maintain the resting potential, it surely would be overwhelmed by passive diffusion of K^+ and Na^+. Nonetheless, the resting potential does not disappear. We can no longer call it a membrane potential if it doesn't depend on the membrane.

According to Pollack (2001), much of the negative charge inside the cell is due to the immobile proteins, with some small additional negative charge from glycoproteins, nucleic acids, and other constituents. In the gel that constitutes the cytoplasm of the cell, there is some room for relatively small hydrated K^+ ions, but hardly any room for relatively large hydrated Na^+ ions. Some Cl^- ions are present (they are relatively small), but together with the K^+ ions they do not offset the overall negative charge of the internal proteins. Additionally, the internal concentration of Na^+ is much smaller than the internal concentration of the K^+ ions, all without need to resort to an ion-pump explanation.

3.7.3 Suspensions

> Jan Baptista van Helmont believed that anyone could make mice by mixing dirty rags with wheat. How could anyone doubt this? All that was needed was to put the two items together in an open container, wait an appropriate time, and sure enough, the mice would appear.
>
> —**Hal Hellman**

Suspensions consist of relatively large particles in a liquid. The particles eventually settle unless disturbed by shaking or turbulent flow. Suspensions can be important in biology because the large particles sometimes scatter or reflect light necessary for aquatic species, or they may form a sediment when they settle that can either entrap smaller individuals or choke filtration mechanisms. Suspensions can sometimes be stabilized by detergents that attach the particles to water or by the formation of a gel structure.

3.7.4 Isoelectric Point

> Never doubt the power of a doggedly curious mind.
>
> —**Camille Rey**

Amino acids are made of an amino group (NH_2) and a carboxyl group (COOH). These are both diagrammed in Table 3.6.1. There are, in addition, other attached organic groups that give each amino acid its unique character. When there is an abundance of H^+ ions (protons) in the surrounding water (pH < 7, or acidic solution), the amino group acts as a proton acceptor and forms ionized NH_3^+. When the surrounding water contains an abundance of OH^- ions, the carboxyl group acts as a proton donor and becomes ionized COO^-. Both of these ionized forms are present in Table 3.6.1.

Amino acids can thus act as either weak acids (proton donors) or weak bases (proton acceptors), depending on the pH. Such molecules are termed *amphoteric*. Weak acids or bases, as we have seen in Section 3.4.1 (Acids and Bases), dissociate only partly and can act as buffers. They can also carry a net positive or negative charge.

There is a pH value at which the positive charges on the NH_3^+ and the negative charges on the COO^- exactly balance. It is at this point where the amino acid molecule is electrically neutral. At all other pH values, the molecule carries either a net positive or a net negative charge. The pH value for electrical neutrality is called the *isoelectric point*.

Proteins are composed of combinations of amino acids as well as other functional groups. Due to a preponderance of weak acidic groups, proteins are nearly all negatively charged at neutral pH. Thus, proteins trapped inside living cells will usually carry a net negative charge.

The greater the pH deviates from the isoelectric point, the greater will be the charge on protein molecules. If proteins are suspended in water with varying pH, and an electrically negative cathode and positive anode are in place, the protein will move toward the cathode at low pH, will not move at all at the isoelectric point, and will move toward the anode at high pH values. Proteinaceous materials such as wool, silk, and gelatin will exhibit this behavior.

The solubility of the protein is minimal at the isoelectric point. Also, because protein shape is determined very much by intramolecular electrical attraction (see Section 3.8), the shapes of proteins can be dramatically altered by pH. Such abnormally shaped proteins are called *denatured* and no longer function for their intended purposes.

Example 3.7.1: Clarifying Wine

Fermentation affects the sugars in grape juice, but the proteins remain in suspension with the result that the wine appears cloudy. To clear the wine, a fine clay-like substance, called Bentonite, can be added. Bentonite attracts positively charged particles and precipitates them. This process is called *fining*.

Wine is usually very acidic and below the isoelectric point of the suspended proteins. At a pH of 3.2, 100% of the proteins are positively charged and are capable of being removed with a Bentonite fining agent.

Isoelectric points for proteins in wine are between 3.6 and 7.1 (Zoecklin, 2000). In some years, wine pH is at or above 3.6. For those proteins previously their isoelectric points, they become negatively charged and Bentonite has no effect. In that case, it becomes much more difficult to clarify the wine.

APPLICATIONS AND PREDICTIONS

1. Increased pressure will cause more salt to dissolve in water.
2. To dissolve a lipid, a detergent must be added to bridge between the hydrophobic lipid and polar water molecules.
3. Increased temperature will usually cause more salt or sugar to dissolve in water.
4. Suspensions will eventually settle. Using a centrifuge will speed the settling.
5. pH can have a profound effect on the net charge of a protein.
6. Constant motion of intracellular components maintains the unstructured state of water molecules, allowing water to be a better solvent.
7. The gel structure inside a cell protects cell integrity when injecting materials into a cell.
8. The resting potential of a cell will be changed based on the pH of the surrounding solution.
9. Cells with membranes partially destroyed can still function.
10. Shooting genetic material or drugs through cell membranes will not destroy the cells unless it also destroys the gel-like structure of the cytoplasm.

3.8 Protein Folding

> Life, mind, and self are all constituted not by biochemistry, but by the higher-level patterns that biochemistry makes possible.
>
> **—Douglas Hofstadter**

Proteins are very diverse and complex. The intestinal bacterium *Escherichia coli* contains about 2400 different proteins, with an average length of 320 amino acids (King et al., 2002). The simple nematode *Caenorhabditis elegans* has 14,261 different proteins, ranging from 40 to 2,000 amino acids long. Human beings are estimated to have 30,000 genes, each coding for a different protein. The muscle protein *titin* is one of the longest of these proteins, containing 10,000 amino acids.

The challenge in *proteomics* is to learn how final protein structures are formed, and the steps involved in their formations. With that knowledge, important proteins can be synthesized in the lab and in industry, improvements in protein enzyme activity can be attempted, and the proteins derived from genetic deficiencies may be able to be corrected. Protein study challenges are listed in Chen and Sivachenko (2005).

Proteins are formed by the intracellular ribosomes (see Section 5.3.2, RNA) with instructions from the DNA, where amino acids are strung together to form long chains. Depending on the organism, amino acids may be strung together at a pace of 20 s^{-1} in bacteria or 5–10 s^{-1} in humans (King et al., 2002). The resulting proteins begin to fold even before they are completely pieced together, but the exact nature of the processes or intermediate products are as yet unknown.

Intricate three-dimensional shapes are crucial to protein functioning. These shapes come about by cross-linking between various biochemical functional groups (mostly amino acids) on the long-chain protein molecule (King et al., 2002). It is the cross-linking among these elements that provides the motivation and the stability of protein structure.

Protein folding is the result of interactions among very weak chemical attractions. Proteins have evolved to fold in conditions present in the cells. In some cases, *chaperone proteins* are required to help other proteins to fold properly. The structural proteins actin and tubulin cannot fold without their specific chaperonins present (King et al., 2002).

This problem provides a particular challenge to biotechnologists attempting to produce functional proteins *in vitro*. Although the sequence of amino acids and side chains on the protein backbone structure may be known and duplicated, proper folding can only be achieved under conditions determined empirically. Natural conditions within the cell are extremely complex, with water, salts, other proteins, complex molecule building blocks, and other biochemicals, all at the proper temperature. Tens of thousands of interactions are probably involved in folding a single protein (King et al., 2002). As might be expected, protein folding and other natural processes proceed more slowly in the crowded environment within the cell than they do in a test tube (Ebbinghaus et al., 2010).

There are four separate levels of protein structure that constitute protein shape (Figure 3.8.1). Not all proteins have all four levels.

The primary protein structure is its unique sequence of a chain of amino acids (Campbell et al., 1999), called a *polypeptide backbone*. This sequence is determined by the genetic code of DNA.

Even a small change of one amino acid substituted for another can drastically alter the configuration of the protein. On the other hand, some amino acid substitutions have relatively minor effect and the protein may still be able to function nearly normally.

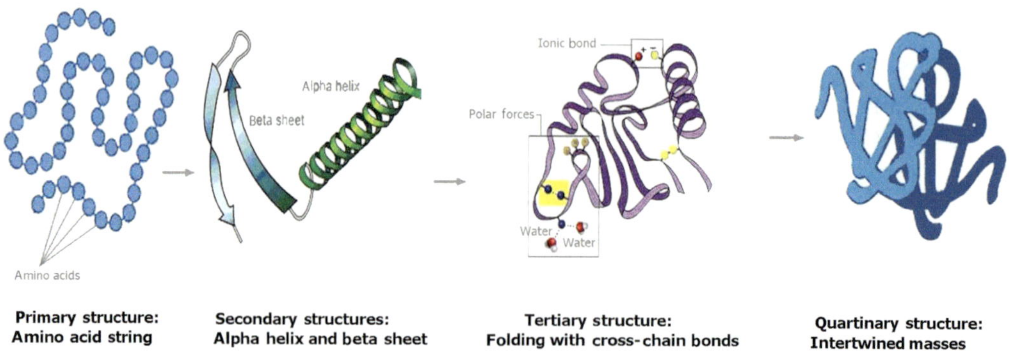

FIGURE 3.8.1
Diagram of four levels of protein structure.

The difference between normal red blood cells and sickle cell red blood cells is one strategically placed amino acid in hemoglobin.

The secondary protein structure, formed once the primary structure is in place, results from hydrogen bonding between weakly positive hydrogen atoms and weakly negative oxygen and nitrogen atoms appearing along the polypeptide backbone of the protein. Although each hydrogen bond is very weak, a multitude of bonds strengthens the entire structure.

Secondary structures can take two forms: an α-helix is a spiral shape, whereas the pleated sheet looks like a piece of paper folded many times. The parallel polypeptide chains cross-linked by hydrogen bonds form an extremely tough structure. Silk is an example (Hale et al., 1995).

Tertiary protein structure is superimposed on the others, and appears as irregular contortions in protein shape. Various chemical bonds form the basis for tertiary structure. Hydrophobic reactions cause nonpolar parts of the protein to congregate in clusters inside the core of the protein away from contact with polar water molecules. van der Waals forces between nonpolar groups strengthen the shape. Hydrophobic reactions occur almost immediately after the proteins are formed, and may be one of the primary instigators of protein folding. Hydrogen bonds between polar side chains and ionic bonds between charged elements appearing at various sites help to consolidate the structure. Very strong disulfide bridges form where two sulfur atoms are located adjacent to one another. The three-dimensional tertiary globular structure is represented by enzymes, antibodies, most blood proteins, and myoglobin.

Quaternary structure occurs with proteins formed from two or more polypeptide chains acting together. These chains may be coiled around one another or tangled together. An example is hemoglobin (Hale et al., 1995).

In Figure 3.8.1 are shown diagrams of these four structures. The diagram is drawn in popular ribbon form that emphasizes the polypeptide backbone of the protein. The protein itself looks more like the diagram in Figure 3.8.2, where all atoms are bunched together. It is the entire configuration of the protein that makes it functional, and defects in any part of the shape can make the entire protein unable to be useful in cellular processes.

Functional proteins that are folded the wrong ways are not useful. Such misfolding can result from genetic mutations or from faulty processing of the multi-amino protein backbones from which all proteins are formed. Abnormal folding can occur because of protein overabundance, temperature changes, oxidative stress, and some cellular signals (Conn and Janovick, 2005). It is believed that proteins do not spontaneously fold into their final, active formations.

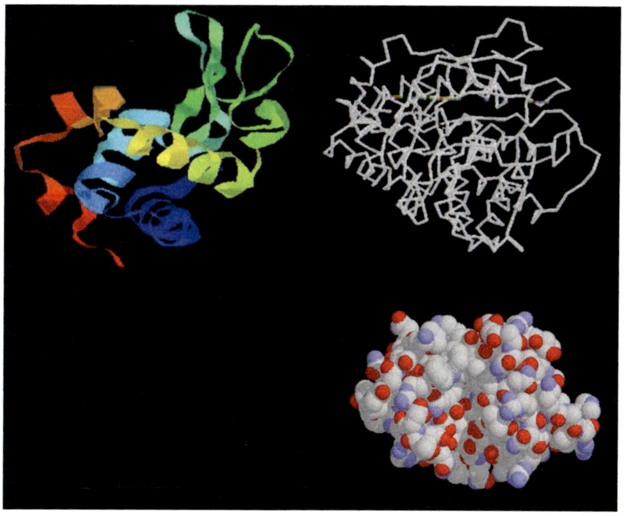

FIGURE 3.8.2
Three schematic models of protein structure. The ribbon model (upper left) reveals some of the underlying protein shape, the stick model (upper right) shows the connections among atoms, and the solid sphere (lower right) gives an idea of its total shape.

Proteins are originally synthesized inside ribosomes in the cytoplasm or inside the *endoplasmic reticulum* (*ER*). The latter synthesize proteins intended to be embedded in the cell membrane (Conn and Jamovick, 2005). When released from the ribosomes, the protein backbone first folds in groups of four or more amino acids to limit interference from bulky or charge-bearing portions of the molecule. Chaperone molecules assist the proteins to fold into their correct three-dimensional shapes. This is especially difficult among the molecules crowded inside the cell. Chaperone molecules work by encasing an unfolded protein in an isolation compartment. Safely inside, the proteins fold correctly by themselves. Without the isolation, the proteins clump together and kill the cell (Goforth, 2012).

All cells, from yeasts to humans, and at least some bacterial cells, attach complex sugars to their proteins in a process called *glycosylation*. These sugars help the proteins fold and function properly.

Mutant proteins, those incorrectly folded, are flagged by the cell's quality control mechanism and destined either for destruction or reprocessing (Figure 3.8.3). Terminally misfolded proteins are marked by the addition of many copies of a small protein called *ubiquitin*, and are then digested into fragments by *proteasomes*. Cellular mechanisms for destroying misfolded proteins weaken with age, allowing proteins to accumulate and cause diseases.

There is an intracellular feedback loop that regulates protein folding. The ER inside eukaryotic cells serves as a cellular factory where newly synthesized proteins are folded into their proper structures. If the ER cannot keep up with demands, then unfolded or misfolded proteins pile up awaiting processing. This glut of unfolded proteins is detected by a series of enzymes that then activates genes to

1. Expand the ER
2. Reduce new protein synthesis in the ribosomes
3. Speed up protein degradation

and the system of correct protein formulation returns to its balance (Chen, 2008).

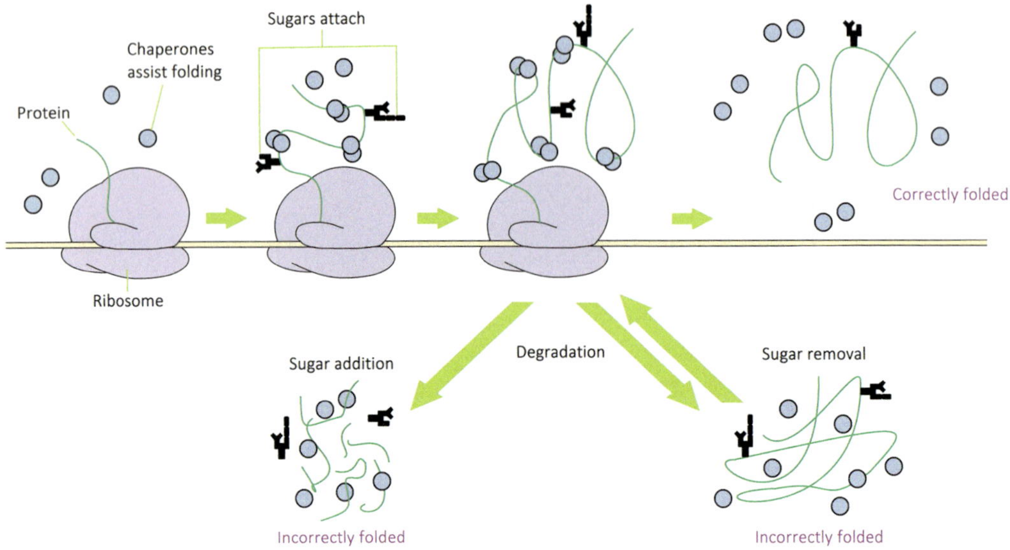

FIGURE 3.8.3
As proteins are formed, chaperone molecules bind to them and assist them to fold in the proper way. These same chaperone molecules can recognize features of improperly folded proteins. Addition or removal of sugar molecules aids in the diagnosis. Incorrectly folded proteins may be either decomposed or fixed. (Redrawn from Conn, P.M. and Janovick, J.A., *Am. Sci.*, 93, 314–321, 2005.)

In the event that equilibrium cannot be reestablished, the unfolded protein response prompts the cell to commit suicide (called *apoptosis*).

One of many characteristics of misfolded proteins is the presence of hydrophobic portions on their surfaces; normally only hydrophilic portions appear on the outside. The problem is that the cell's quality control mechanism relies on chemical, not functional, tests. Thus, a protein labeled for destruction or reprocessing may actually be functionally equivalent or superior to a normal protein.

If the protein fails to fold properly, its shape is incorrect, and it cannot perform its intended function. Aberrations in protein folding appear to contribute to human and animal diseases. Among these are Alzheimer's disease, prion diseases, emphysema and cirrhosis, amyotrophic lateral sclerosis (Lou Gehrig's disease), cystic fibrosis, some tumors, osteogenesis imperfecta (King et al., 2002), and transmissible spongiform encephalopathy (TSE). Unclustered prion molecules are harmless and found in all mammals. When bunched together, however, these same prions become infectious (Wilham et al., 2010).

Prions are misfolded proteins. When a protein converts to a prion, it polymerizes from coiled alpha helices that dominate normal protein structure into an aggregate of tightly packed beta sheets (Saltus, 2010a). The prion that is implicated in the cause of bovine BSE (mad cow disease), for instance, appears as a pleated sheet rather than a smooth helix. Prions have the unusual property that they can replicate without DNA or RNA (Saltus, 2010a). This gives the possibility that prions can act as an evolutionary route parallel to the genetic code for the organism. Whereas some prions have been identified as causing debilitating diseases, others in yeast, have been shown to change the pattern of gene expression in cells, at times enabling the cell to cope with radically different environmental pressures.

The ability to self-replicate may also be important in the retention of memories through prion alteration of frequently used neural synapses.

Prions are instrumental in stabilizing neural synapses involved with long-term memory formation. When an electrical impulse is passed from one neuron to another in the brain, a chemical change in an associated prion allows the synaptic connection to be retained even after the initial stimulus has been removed. Thus, the memory of the stimulus remains (White-Grindley et al., 2014).

Treating some of these diseases at the protein level may be easier and more ethical than using gene therapy. The strategy would be to restore missing or nonfunctional proteins with pharmacological chaperone molecules inserted into the cells.

HEAT SHOCK PROTEINS

Exposure to high temperatures poses a danger to living organisms because essential proteins and enzymes unfold, a process called *denaturation*. When this happens, the compounds no longer mate with complementary shape receptors or chemicals, and normal biological activities cease.

Organisms have developed a common heat stress response. There is a rapid production of *heat shock proteins* that act as molecular chaperones to ensure that newly formed proteins are folded properly, and damaged proteins are either repaired or removed from the cell (Taiz and Zeiger, 1998). In addition to stabilizing unfolded protein precursors and either repairing or degrading damaged proteins, heat shock proteins translocate proteins across cellular membranes and dissolve protein aggregates (Werner et al., 2003). This is an energy-intensive process: repair of each damaged protein molecule requires as much as 100 ATP molecules.

Heat shock proteins associate with a wide range of target proteins and perform a wide range of different functions. They help newly formed amino acid chains fold to their proper protein shapes, they dismantle damaged proteins, and they escort proteins to combine with the correct other molecules. Heat shock proteins also assist the immune system to recognize foreign invaders. The appearance of heat shock proteins outside cells may be a mechanism to alert the immune system of danger (Srivastava, 2008).

Although the heat stress response appears to be universal across both animals and plants, it is in plants, which cannot move to escape the heat, where heat shock proteins are especially important for survival. Plants thus have more complex heat stress response components than animals. Different heat shock proteins are localized to the nucleus, mitochondria, chloroplasts, endoplasmic reticulum, and cytosol.

Heat shock proteins are not unique to high temperature stress; some are induced by other environmental stresses such as water deficit, chemical exposure, wounding, low temperature and salinity (Taiz and Zeiger, 1998). Exposure to one type of stress can cause protection against exposure to others. Therefore, production and isolation of heat shock proteins in bioreactors has commercial value when they are subsequently applied to stressed organisms.

Genes for heat shock protein production are encoded in the cell, but are not activated until other protein intermediaries sense damaging stresses and switch them on (ASAE, 2002). Thus, this is a wonderful example of the complex feedback that occurs between the environment and the basic genetic regulation of cell activity.

CARRIER PROTEINS

Within the crowded cell interior, there are molecules, usually proteins, which are critical to cell functioning and must be located in the correct place, be it the cell nucleus, the outer cytoplasm, or the cell membrane. These necessary molecules are transported along actin microtubules inside the cell by other carrier proteins. It is the function of these carriers to pick up the necessary cargo molecules and facilitate their transport within the cell. One carrier protein, *kinesin*, similar in structure to myosin, carries molecules into the interior of the cell; another, *dynein*, carries molecules in the other direction from the interior to the outer cytoplasm. Still other molecules freeload on the actin as it streams within the cell. In these ways, necessary proteins are delivered to the places where they are needed, and in a timely fashion. If these transport processes are disrupted, cells cease functioning correctly and physical disorders result (Williams, 2013B).

Example 3.8.1: Biomarkers as Indicators of Environmental Stresses

Maintaining natural populations of aquatic organisms in lakes, rivers, and streams is not simple. Human activities have so modified temperature, pH, salinity, and toxin concentrations that plants, insects, fish, and amphibians living in those bodies of water face a multitude of difficulties. The problem is especially acute for sessile organisms that cannot escape to less stressful environs.

Proper management requires that stresses be known. If a human is under stress, she or he can tell the investigator. But asking a fish is out of the question. How can environmentalists determine the level of stress in a fish?

Solution

An indirect measure of stress can be made by determining levels of all physical and chemical stressors in the environment. However, not only is this an expensive proposition, but it is not easy to translate these physical measurements into a stress index. It would be much better to directly measure the level of stress in the living fish.

Stress proteins (also called heat shock proteins) are meant to protect organisms against damaging environmental effects. They do this by repairing damaged proteins. Stress proteins can act as biomarkers of integrated environmental stress effects.

Stress proteins are nearly the same across a wide range of species. Thus, it is not necessary to search for specific stress proteins for each type of organism. The hsp70 family of stress proteins can be monitored in the target species as an input for proper wildlife management (Werner et al., 2003).

Remark

Werner et al. (2003) found that the sensitivity of fish embryos to toxic chemicals in the San Francisco Bay area is age-dependent. Very young embryos do not manufacture sufficient stress proteins to correct environmentally damaged proteins. Thus, susceptibility to environmental stress in early development is critical to survival of these species.

Example 3.8.2: Biomarkers as Indicators of Disease

A biomarker has been defined by the National Institutes of Health as "a characteristic that is objectively measured and evaluated as an indicator of normal biologic processes, pathogenic processes, or pharmacologic responses to therapeutic interventions."

Biomarkers can be antecedent (indicates the proclivity to develop a disease), diagnostic (assists in the identification of a disease), or prognostic (predicts the progression or cure of the disease). Biomarkers can be as simple as the appearance of tumors in cancer or as complex as a protein profile in lupus. Advances in molecular imaging and compound detection have advanced the state of biomarker identification. When biomarker technology is sufficiently advanced, biomarkers can be used as rapid indicators of drug efficacy in the treatment of disease.

Liquid biopsies are made when looking for biomarker molecules shed by specific gene mutations and circulating in the blood. Liquid biopsies can be used to detect cancer, among other conditions, without resorting to invasive tissue biopsies.

APPLICATIONS AND PREDICTIONS

1. Substituting a similar amino acid (polar, charged, or uncharged) in a polypeptide chain will have a much less drastic effect on the protein function than will the substitution of a much different amino acid
2. Improperly formed proteins can cause disease
3. Easily measured biomarkers should be sought for any biological condition of interest
4. Understanding of chaperone molecules can help cure disease

3.9 Shape Effects and Enzymes

> The chemical reaction should go, but refuses to do so? Bang on it—with heat, with light, with pressure. Or, so much better, find a catalyst—an ingenious partner in a reaction that cuts down an energy hill that is in the way, or gets involved with the reactant molecules, takes them in hand, so to speak, and guides them in a path around that hill.
>
> **—Roald Hoffmann**

Reactions among organic compounds are extremely slow if left to themselves. They are too slow, in fact, to sustain life. Because living things exist, they have obviously found a way to circumvent the problem.

Before two compounds can react chemically, they must contact each other. Collision theory has been formulated to explain various factors on chemical reaction rates. Part of this theory is the Law of Mass Action, which states that a chemical reaction will occur at a rate proportional to the concentration of its reactants. Thus, anything that can be done to concentrate reaction precursors in the presence of each other will favor the desired chemical combination to occur.

One means to accomplish concentration is by exploiting surface effects of chemical compounds. Surface atoms do not have their electron needs totally fulfilled. They tend to be very reactive and attract other molecules with unusually strong forces. Charcoal or carbon is a substance with especially strong surface forces, although there are a large number of other substances with strong surface attractiveness as well. Many of these are used as glues or cements. What makes carbon so unique is the many channels and pores that can be formed in it to increase its surface area by orders of magnitude. Carbon tends to attract many ionic or polar molecules to its surface, and is largely nondiscriminatory. Carbon can thus be used to concentrate many different chemical compounds. At temperatures low enough to reduce the energies of gaseous particles, nonpolar atoms such as O_2 can be attracted in large numbers to carbon, probably as a result of van der Waals forces.

A more important strategy for living beings is the use of enzymes. Enzymes are proteins that are shaped by twisting and bending such that they fit extremely well with the shapes of certain chemical reactants. One part of an enzyme molecule may fit well with one reactant and another adjacent part may fit with a second reactant. With the two reactants in such close proximity, they readily combine. Once combined, the product of the reaction is removed from the enzyme and the enzyme is ready to repeat the process. The enzyme is not changed by the reaction; enzyme action is largely physical.

Enzymes are characterized by their actions. There are enzymes that link two compounds together (called *ligases*), enzymes that transfer a chemical group from one compound to another (*transferases*), enzymes that catalyze hydrolysis (*hydrolases*), enzymes that remove a chemical group from substrates (*lyases*), enzymes that oxidize some compounds while reducing others (*oxidoreductases*), and enzymes that convert one isomer to another (*isomerases*). In general, there is an enzyme for every chemical reaction important to living systems. Some enzymes are very specific, yet others may have several functions. For instance, the enzyme maltase is specific to the sugar maltose and produces glucose as a result of its action. Chymotrypsin, on the other hand, is a general enzyme for the digestion of proteins, and can operate on many proteins.

Because enzymes are classified by their actions rather than by their actual chemical composition, many different enzymes can be identified by the same name. There are four different hexokinases that occur in tissues. Each operates on a glucose substrate, but each has a different sensitivity to glucose concentrations.

Most enzymes catalyze reactions of their substrates only in the presence of particular organic compounds called *coenzymes*. The B vitamins are important coenzymes, for instance. Coenzymes can be considered to be cosubstrates because they frequently take part in the chemical reactions, and often change form; if the substrate is oxidized, the coenzyme may be reduced, for example, or the coenzyme may act as the receptor, donor, or carrier for some functional group that is modified on the substrate.

Many enzymes also show a specific requirement for a particular metal ion. These include various ions of iron, copper, cobalt, molybdenum, zinc, manganese, magnesium, calcium, and potassium. The metal ion is thought to have a binding function for the substrate, coenzyme, and enzyme. Thus, these metals are needed in the diet in trace amounts.

Enzyme action may be decreased by various competitive and noncompetitive inhibitors. An inhibitor is a compound that combines with an enzyme, which cannot be converted into a product. The result is an enzyme-inhibitor combination that changes the configuration of the enzyme enough that the enzyme is no longer effective. Competitive inhibitors are affected by the relative amounts of substrate and inhibitor; the more substrate available, the less inhibition that occurs. Noncompetitive inhibitors bear no structural resemblance to the substrates and so are not affected by substrate concentration.

Competitive inhibitors that block enzyme reactions in a parasite are potent chemotherapeutic agents. For example, sulfanilamide will block folic acid (a B vitamin) synthesis, and the resulting deficiency of this essential vitamin is fatal to the organism. Because humans cannot synthesize folic acid, but require it as a nutritional supplement, humans are not harmed by sulfanilamide ingestion. Vitamin B, however, is often in short supply in nature.

There are also enzyme inhibitors used as defenses by an organism. Pepsin and trypsin inhibitors are found in the pancreas and keep the pancreas from being digested by its own enzymes. Some enzyme inhibitors also act as antibodies against microbial infections (see Section 6.20.3).

Not all enzyme interactions are competitive. One cooperative enzyme reaction takes place when a substrate primes an enzyme with multiple reactive sites to accept additional substrate molecules. Thus, the presence of substrate enhances the enzyme and makes it more effective.

With enzyme performance dependent on molecule configuration, anything that disrupts that configuration will interfere with enzymatic activity. The helical structure of the proteins comprising the enzyme is maintained largely through hydrogen bonding and sometimes with cross-link bonds involving elements other than hydrogen. These bonds are relatively weak compared with the covalent carbon bonds present in the carbon chain, and are therefore affected by temperature. At higher temperatures, these bonds begin to break and reduce enzyme effectiveness. The temperature at which this begins to happen depends to some extent on the normal operating temperature of the organism. An organism acclimated to a higher environmental temperature will possess enzymes that maintain their effectiveness at higher temperatures than those organisms operating normally at lower temperature environments. The actual enzyme proteins used by these two organisms would be different, although their functions (and thus their names) would be the same. At very high temperatures, these cross-link bonds cannot be reestablished, and the compounds lose all enzymatic functionality. This is called *denaturation.*

One characteristic of enzyme action is that a large chemical change is usually brought about in small steps. Thus, one chemical reactant is not changed all at once into a final product. Instead, a series of specific enzymes change the initial reactant into intermediate products, each closer to the final product. The *law of constant heat sums* states that the total amount of heat produced or consumed when a chemical system changes from an initial state to a final state is independent of the way in which the change is brought about. For example, glucose is oxidized to carbon dioxide and water by the following equation:

$$C_6H_{12}O_6 + 6O_2 = 6H_2O + 6CO_2 + 2.82 \times 10^6\ N \cdot m, \tag{3.9.1}$$

which is the process by direct burning. With the enzyme zymase, glucose is oxidized in two steps:

$$C_6H_{12}O_6 \rightarrow 2C_2H_5OH + 2CO_2 + 0.08 \times 10^6\ N \cdot m, \tag{3.9.2}$$

$$2C_2H_5OH + 6O_2 \rightarrow 6H_2O + 4CO_2 + 2.74 \times 10^6\ N \cdot m \tag{3.9.3}$$

The resulting products and energy sum is the same for both pathways.

Enzymes are the regulators of biological metabolism. Chemical reaction rates in the absence of enzymes are too slow to maintain life. Enzymes allow intracellular processes to adjust to new environmental factors and to reach the point where growth can occur. This system can be regulated in two ways: (1) the stimulation or inhibition of enzyme activity and (2) the induction or repression of enzyme synthesis (Figure 3.9.1). The product of enzyme activity, or compounds derived from those products, can act as inhibitors of enzyme activity. There are two possible mechanisms for this (Figure 3.9.2): (1) the presence

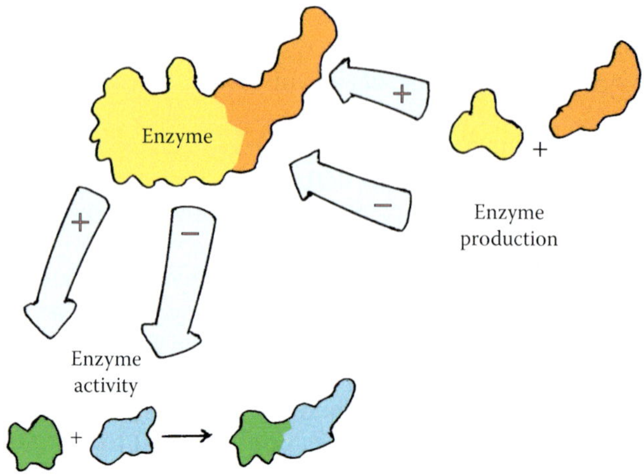

FIGURE 3.9.1
The two methods to regulate enzyme effectiveness are through the production or activity of enzymes.

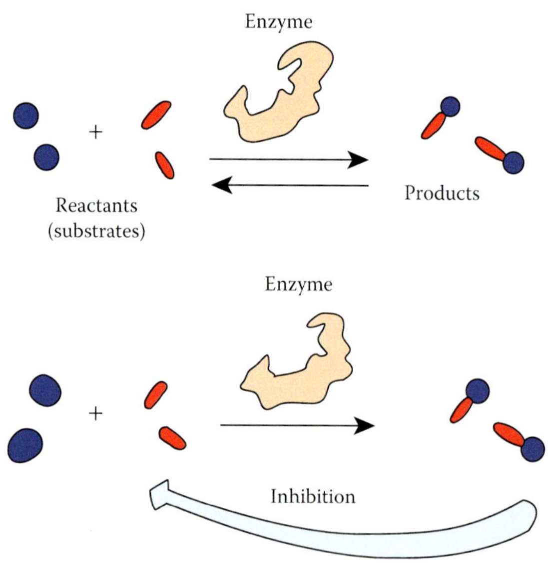

FIGURE 3.9.2
Inhibition of enzyme activity can be due to a high concentration of the product, or through direct inhibition of the product on the substrates, or through product repression of enzyme synthesis.

of large amounts of product can inhibit the conversion of substrate through the law of mass action and a reversible biochemical reaction, or (2) the product may be a potent competitive inhibitor for the substrate (called *feedback inhibition*).

The product may also inhibit the formation of the enzyme, in a process called *repression*. In this way, a relatively small molecule (the product) has a direct effect on the production of a large molecule (the enzyme), and thereby, an indirect effect on the formation of more product. Such exquisite control allows an organism to respond to a variety of substrates in a wide range of conditions to display adaptability and survival.

UNFOLDED, UNSTRUCTURED, AND MEANT TO BE THAT WAY

Most cellular proteins have a particular structure that they assume in order to serve their specific purpose. Enzymatic proteins are this way, and are ineffective if misfolded. Elaborate mechanisms inside cells detect misfolded proteins and either destroy or repair them.

However, about 35% of human proteins have long unstructured regions with no particular shape (Dunker and Kriwacki, 2011). Nearly all eukaryotic cells, but few, if any prokaryotes, depend upon these unstructured proteins for signaling and regulation functions. Enzymes must still form highly regulated structures in order to be effective, but unstructured proteins are functionally flexible because they can wrap around many different types of molecules. When they wrap around enzymes, for instance, the enzymes are rendered ineffective. Hence, their functions are regulated by proteins that can assume any necessary shape. The ability to fold or unfold as the need arises presents many different possibilities.

Long, thin proteins can thread through small openings in the cell and carry other molecules with them. They can thus help regulate the availability of necessary chemicals, or help signal the cell of the chemical nature of the surrounding environment.

Not all proteins must have rigid structures to be effective; some work best if they can assume a multitude of shapes.

APPLICATIONS AND PREDICTIONS

1. Activated carbon can be used to enhance reaction rates.
2. The rate of a chemical reaction involving enzymes will be limited by both amounts of reactants and enzymes. Without enzymes, some reactions would practically not occur.
3. The name of an enzyme will indicate its function, but will not designate a particular compound.
4. An unwanted enzymatic reaction can be inactivated by introducing an inhibitor.
5. Bacteria will make their own folic acid. Inhibiting this synthesis can be used as an antibacterial strategy.
6. High temperatures will interfere with proper functioning of an enzyme.
7. Without inhibitors, organisms will digest themselves.
8. Enzyme regulation will be a complex process that can be disrupted for the control of unwanted organisms in the environment, or production of preferred biochemicals in a reactor.
9. Enzymes will be denatured by harsh chemicals as well as by heat.
10. Vitamins will be necessary to form certain enzymes.

(*Continued*)

APPLICATIONS AND PREDICTIONS (Continued)

11. Enzymes will be able to use similar related compounds when the original compounds are not available.
12. Any substrate that bonds tightly with an enzyme will inactivate the enzyme.
13. Controlling reaction rates within limits requires the presence of inhibitors as well as enzymes.
14. Some useful drugs act as enzyme inhibitors.

3.10 Energy-Rich Compounds

> If we gather huge numbers of unconscious elements together in the right way and give them the right tasks to perform, then at some point, something happens, and consciousness emerges.
>
> **—David Gelernter**

The reason that dynamite can be used to blast through rocks and remove tree stumps is that dynamite contains an energy-rich compound. Oxidation of nitroglycerin, the active ingredient of dynamite, yields a surplus of energy that can be used to produce mechanical work. Such a chemical reaction is called *exergonic*, or *exothermic*. If the net result of a reaction is that energy is absorbed, as in photosynthesis, the reaction is said to be *endergonic*, or *endothermic*.

Nearly all Earthly life depends ultimately on energy from the sun. However, only certain photosynthetic organisms can capture the sun's energy and store it for future use. Most other life must utilize the energy that these primary energy-transforming organisms store as chemical compounds. Whereas the utilization of energy in as dramatic a fashion as with a dynamite explosion would be quite damaging to living beings, the release of energy in smaller quantities could be used to drive the many processes needed to sustain life. Thus, chemical energy of the right form and amount would be quite useful.

Long-term chemical energy storage resides mainly in carbohydrates, fats, and proteins. Liberation of energy accompanies oxidation of these compounds.

There is a small family of molecules that is utilized by nearly all living things to shuttle energy between the oxidative reactions of primary storage compounds and the chemical processes that maintain life (Figure 3.10.1). Most of these contain high-energy phosphate bonds. Of these, ATP is the most important. ATP is formed from adenosine, which is a nucleotide formed from the double ring purine, adenine, with the 5-carbon sugar ribose and three phosphate groups. The three phosphates are attached to one another (de Duve, 1995).

The two bonds linking the three phosphates are high-energy bonds, called *pyrophosphate bonds*. The formation of these bonds is accompanied by the removal of a water molecule, called a *condensation reaction*. When the pyrophosphate bond is split, water is added, and this is called *hydrolysis*. Hydrolysis liberates energy (the new bond has lowered energy

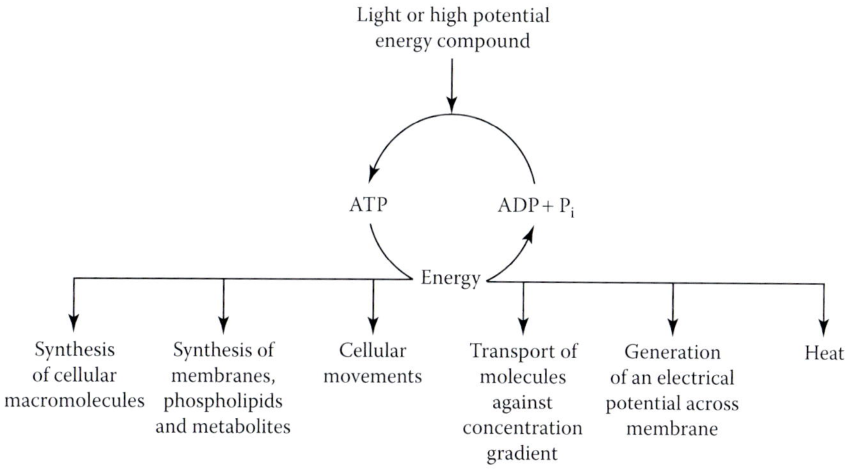

FIGURE 3.10.1
ATP is central to all life processes. (Redrawn from Haynie, D.T., *Biological Thermodynamics*, Cambridge University Press, Cambridge, UK, 2001.)

level) and condensation absorbs energy (the new bond has a higher amount of stored energy) in a water-filled environment. The pyrophosphate bond thus acts as a chemical storage battery, with energy available to drive other chemical reactions.

The other chemical reaction (whatever it is) will occur as long as it requires less energy than the energy stored in ATP. If, on the other hand, the other reaction takes more energy than the formation of the pyrophosphate bond, then the pyrophosphate bond will be formed. In terms of effort and flow variables, the bond energy is the effort variable, and the chemical reaction is the flow. The reaction will only occur if the effort variable goes from higher to lower.

THE PHYSICAL ACTION OF ADENOSINE TRIPHOSPHATE

Adenosine triphosphate (ATP) is the universal energy carrier for life. It is ATP that stores energy to be used on demand, and this energy can actually take both chemical and physical forms (Pollack, 2001). The close proximity of its three phosphate complexes results in a high concentration of negative charge. Because of this, it has a high affinity for protein surfaces; ATP association for the protein myosin, for instance, is practically irreversible.

Such high charge concentration can induce structure by linking to protein strands that form the gel inside living cells. ATP negative charge can then repel negative charges on adjacent protein strands and cause them to extend to their high-energy conformation (see Section 3.8, Protein Folding). These extended strands are the ones with structured water between them. Thus, ATP confers energy to the cell by causing order and structure.

Without ATP, muscle is in a constant state of tension. Its actin and myosin proteins are cross-linked to one another and contracted (Pollack, 2001). ATP breaks actin-myosin cross-links, and restores the proteins to their extended state. ATP does this despite the fact that its binding sites are located remotely from the sites of cross-linking and contraction. ATP drives the system to its high energy state.

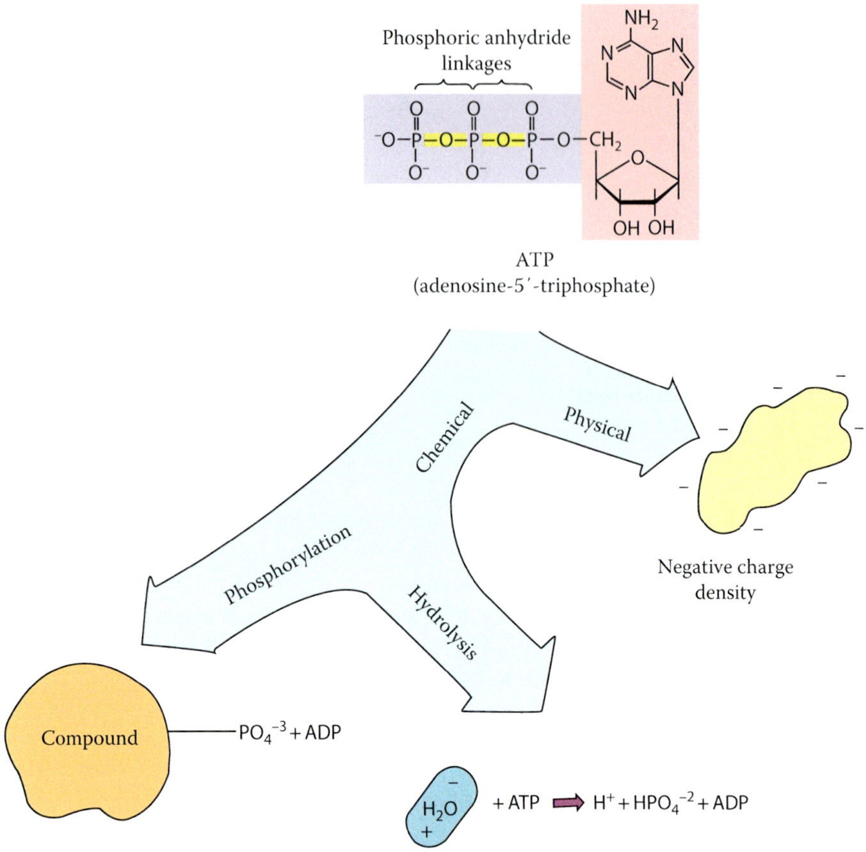

FIGURE 3.10.2
Energetics of ATP. There are multiple mechanisms for ATP activity, including both physical (charge density) and chemical (hydrolysis and phosphorylation).

Most of the bonds in biological substances require less energy for their formation than do the pyrophosphate bonds in ATP. The bonds of proteins and other natural substances are low energy bonds in comparison. This is why ATP is so valuable to life (de Duve, 1995).

ATP can transfer energy to another molecule in a variety of ways (Figure 3.10.2). The most common way is the transfer of a phosphate group to the molecule that requires energy (Fogiel, 1999). This reaction liberates adenosine diphosphate (ADP), and is known as a *phosphorylation reaction*. It is usually catalyzed by enzymes called *kinases*. If two phosphate groups are transferred, adenosine monophosphate (AMP) remains.

The *hydrolysis* of ATP, wherein the phosphate group is transferred to water, can also be used to transfer energy to another molecule (Fogiel, 1999). In addition, the adenosine group can be transferred with the liberation of pyrophosphate and inorganic phosphate.

The source of muscular energy is in the form of ATP. Each mole of ATP represents 46,024 joule (J) of energy (the net energy available from each mole of ATP depends on the chemical pathway utilized), but there is not enough ATP to supply more than 0.5 seconds of muscular energy use. Creatine phosphate provides additional energy to produce more ATP as needed, but for no more than 2-seconds worth of energy. Glucose can be oxidized to produce longer-term energy, and glycogen (a storage form of glucose) provides even more energy stores. If the energy in these chemicals was not equivalent, then there could be no smooth transition from one energy source to another, but the energy is equivalent, and the muscle can produce both short-term and sustained work from these various energy sources.

ATP is a molecule present in very primitive life-forms, such as amoebas and worms. ATP receptors discovered in these organisms suggest that, in addition to its energy storage role, ATP also has had a communications-signaling role since early evolutionary time (Khakh and Burnstock, 2009).

Phosphate groups are not the only energy-rich chemical compounds used by living systems. Sulfur can also form high-energy bonds as part of thioesters. Sulfur was probably prevalent in the prebiotic Earth, and was probably used by the most ancient of organisms to provide energy for life (de Duve, 1995). Even today, there are primitive life forms that surround sources of sulfur-containing gases from deep within the Earth and use these gases as their primary energy sources (see Sections 6 [BRIC], 6.2 [Oxygen], and 6.5.1 [Heat Sources]).

RESPIRATION OF GLUCOSE

(Starr, 2000)

Glucose is the starting point for respiration, and energy for biochemical reactions is the end point. In between, adenosine triphosphate (ATP) is the substance that can store energy for immediate use; it is the energy currency that makes possible all the processes contributing to life.

Respiration may be either aerobic (requiring oxygen) or anaerobic (proceeding in the absence of oxygen). The first stage in energy-releasing reactions is anaerobic, and takes place in the cellular cytoplasm. This is a very primitive respiratory stage, probably evolving in the first protists (microbes), and is still used today by organisms where oxygen is not abundant.

This first stage involves *glycolysis*, where enzymes split each six-carbon chain into two three-carbon molecules of pyruvate. Some energy is necessary to start this reaction, and the net result is the production of two ATP molecules.

Aerobic respiration starts with the anaerobic stage described previously and adds two additional stages. The second stage in aerobic respiration proceeds according to the Krebs, or citric acid, cycle. This reaction takes place in the mitochondrion, a subcellular inclusion not present in the most primitive microbes (see Section 5.5.1, The Cell). Pyruvate from glycolysis enters the mitochondrion and is shepherded through the cycle by coenzymes NADH and $FADH_2$. Carbon dioxide is formed from the two pyruvate molecules, and two more ATP molecules are formed by phosphorylation from two ADP molecules. Electrons and hydrogen are placed on various coenzyme molecules to set up the third stage.

(Continued)

RESPIRATION OF GLUCOSE (Continued)

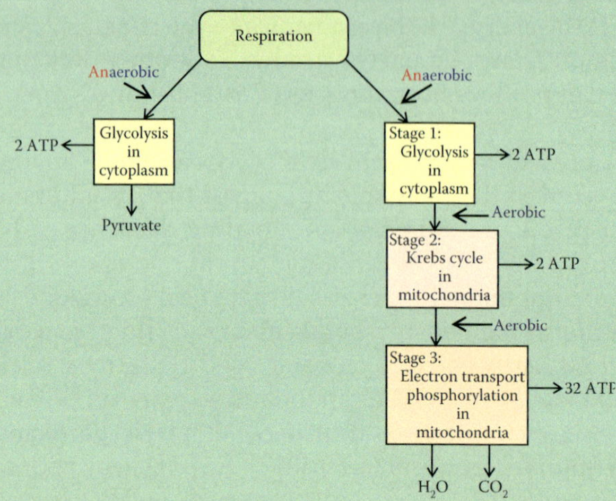

The Krebs cycle is central to nearly all cellular metabolism, and its history may date prior to the first cells or even before the first enzymes (Trefil et al., 2009). These metabolic reactions result in products with overall energy levels less than those of the reactants. In cells that derive their energy from oxidation of carbon-based molecules, the Krebs cycle results in high-energy electrons that power the cells through ATP formation and use. For certain primitive cells that derive their energy from chemical sources, the Krebs cycle can run backwards to synthesize biochemical products necessary for the life and health of the cells. Energy is thus used to form products.

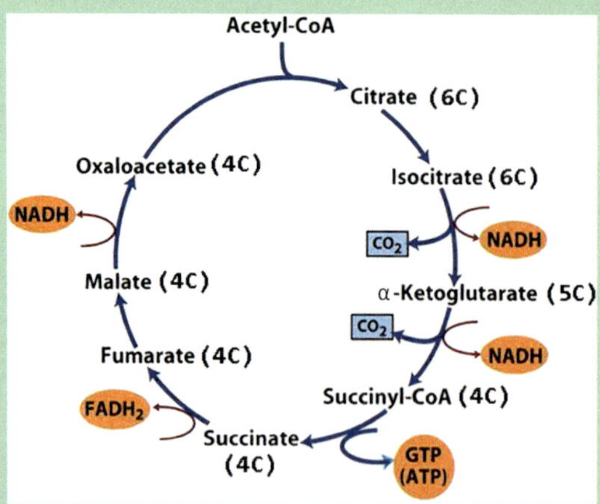

The Krebs cycle. Pyruvate from the anaerobic metabolism stage is converted into acetyl and coenzyme A (coA) and enters the cycle as a 6-carbon compound (citrate). Two of the carbons are removed by oxidation as the cycle proceeds, resulting in the evolution of two molecules of carbon dioxide. NAD (nicotinamide adenine dinucleotide) is a coenzyme that acts as an energy carrier in the cell. NADH is the reduced form of NAD. FAD (flavin adenine dinucleotide) is an electron carrier in the cell, and transports electrons to the electron transport stage of glucose metabolism.

(Continued)

RESPIRATION OF GLUCOSE (Continued)

The third aerobic respiration stage takes place at the mitochondrial membrane, and is called electron transport phosphorylation. A hydrogen ion gradient is established and H^+ flow drives formation of ATP from ADP. Oxygen serves as the ultimate electron sink and produces water as a result. This stage yields 32 ATP molecules.

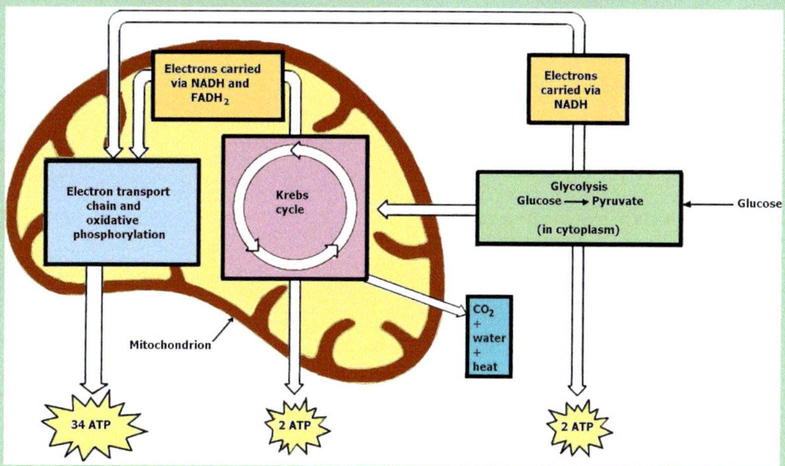

The entire glucose metabolic scheme is shown here. The net result is 36 molecules of ATP, two molecules of carbon dioxide, heat, and water.

The typical yield of ATP molecules is 36, but fewer ATP molecules may be formed if some of the intermediate substances are in short supply for one reason or another. Each mole of glucose represents 686 kcal of energy. Each mole of ATP represents 7.5 kcal. Thus, the 36 moles of ATP coming from complete respiration of a mole of glucose contain (36) (7.5) = 270 kcal. The energy efficiency of this process is 270/686 = 39%.

Anaerobic organisms derive their energy only from the initial glycolysis step. Animals may also use this as a temporary energy-producing pathway when oxygen is limited. However, only two ATP molecules are produced, and this is not enough to sustain large, active, multicellular organisms.

NATURAL VARIATIONS ON A GLUCOSE THEME

Excerpted from Lane (2007)

Let's start with the basics: the simple sugar molecule called *glucose*. Glucose contains six carbon atoms with five of these atoms arranged in a ring and the #6C branching off from the #5C. Five of these carbons are also linked to a hydroxyl group (–OH). Glucose is a critical molecule for animal and plant metabolism—it's the primary source of energy for most metabolic processes as well as the starting molecule for synthesizing thousands of other biological compounds. Our interest here, however, is how it combines with other glucose molecules.

If we connect two glucose molecules by linking the #1C of one glucose with the #4C of the other glucose, we'll create a two-glucose molecule called *maltose*, which is a disaccharide (*di* = two, *saccharide* = sugar). Two other disaccharides are quite familiar to us: the milk sugar *lactose*, which contains one glucose and one galactose, and *sucrose*, which contains one glucose and one fructose, which we also sweetly call *table sugar*.

If we connect thousands of glucose molecules in a straight-line chain like a string of pearls, each connected with a C1–C4 linkage, we'll have a polymer of glucose called a polysaccharide (*poly* = many). But here's the rub: the C1–C4 bond can exist in two possible arrangements: (1) when the C1 bond points downwards, the bond is called an *alpha* linkage, and (2) when the C1 bond points upwards, the bond is called a *beta* linkage.

If the thousands of glucose units are linked by alpha 1–4 bonds, then that molecule is called *starch*. Plants use starch as their primary storage compound, and all of our common food seeds—corn, wheat, barley, oats, and others—are really just packages of starch. A smaller version of this glucose polymer also occurs in animals as the energy-storage molecule called *glycogen*.

On the other hand, if the thousands of glucose units are linked by beta 1–4 bonds, then that molecule is called *cellulose*. Plants use this polymer as part of their structural material for building stems and leaves. The geometry of these beta linkages allows the long cellulose chains to align themselves side-by-side and thus create strong fibrous strands.

In the broadest sense, cellulose is sunlight captured in a fibrous form—a compound that is universal, nutritious, and completely renewable.

Do all those quadrillion tons of cellulose in the world represent a good source of nutrition? Well, yes and no. Sure, cellulose contains energy that is potentially available to animals. However, cellulose is also a very large molecule. Since large molecules can't be absorbed directly across the gut wall, the digestive tract must first enzymatically break them into smaller units that can be absorbed and transported into the blood. But there is a slight problem: no mammal (or bird, reptile, insect, etc.) has the digestive enzymes that can break that 1–4 linkage, which means that mammals are unable to digest cellulose directly.

(*Continued*)

NATURAL VARIATIONS ON A GLUCOSE THEME (Continued)

Although mammals don't have enzymes to digest cellulose, certain bacteria do have these enzymes. In fact, these cellulolytic bacteria thrive by using cellulose as their primary food source. Capitalizing on this feature, evolution has come to the rescue.

Over millions of years, some animals evolved a specialized pouch in their digestive tracts to house these bacteria. This pouch, a fermentation sac, can either be located at the front end of the digestive tract, where it is called the *rumen*, or at the far end of the digestive tract, where it is called the *large intestine* or *cecum*. In these fermentation sacs, the cellulolytic bacteria happily do their thing and in turn, produce nutritional products that can be absorbed and used by the host animal for its own metabolism. Thus, the ruminant, such as a horse, alpaca, or any animal housing these bacteria (including, to a small extent, humans), can obtain nutritional value from cellulose.

Actually, mammals aren't the only animals which evolved fermentation sacs for cellulolytic bacteria. Termites have done it too, for exactly the same reason.

In nature, plant fibers are actually very complicated structures that contain a few types of fibrous substances. Cellulose is only one component of this fiber matrix, although it is generally the largest one. Other types of common fiber molecules include hemicellulose, which is a complex cellulose-like molecule with lots of side-chains, and lignin, which is a strong indigestible compound that acts to reinforce the entire fiber structure. So, in nature, cellulose occurs as an integrated part of a complex package, rather than a stand-alone compound.

With one notable exception—cotton. The cotton plant produces a fruit called a *boll*, which contains seeds in a mesh-work of cotton fibers. These cotton fibers are essentially pure cellulose.

Every glucose molecule has five carbons linked to hydroxyl groups (–OH). For glucose molecules bound into cellulose, two of these carbons are used in the glucose-to-glucose linkage (#1C and #4C), so they don't have hydroxyl groups. But that still leaves three carbons that do have hydroxyl groups.

If we take the #2 carbon in every glucose unit and replace its hydroxyl group (–OH) with something called an *acetylamine group* ($-NHCOCH_3$), the resulting polymer is called *chitin*. Like cellulose, chitin molecules can align themselves side-by-side and form tightly condensed flexible sheets. But chitin can be harder and stronger than cellulose because the acetylamine side-chains allow the parallel strands to cross-link better. In its flexible form, chitin forms the outer layers of certain insects and fungi. But some organisms combine chitin with various proteins to make it extremely tough and impenetrable. Chitin is the material that forms the hard-exterior shells of insects and also crustacea such as lobsters, shrimp, and crabs.

Cellulose, starch, glycogen, and chitin are all-natural variations on a glucose theme.

FATS AND FATTY ACIDS

Excerpted from Lane (2008)

The term *fat* can refer to a wide array of substances—oil, tallow, lard, fatty acids, all the *glycerides* (tri-, di-, and mono-), and so on. All of these are nutritionally available, and all contain the same high level of energy, of about 9 calories/gram, while carbohydrates and protein only contain 4 calories/gram.

Fat molecules contain *fatty acids*, which are primarily strings of carbons connected by single bonds. Although details vary among different fatty acids, all fatty acids share the same basic structure: they are all composed of a single string of carbon atoms like individual links in a chain, with a methyl group ($-CH_4$) at one end of the chain and a carboxyl group (–COOH) at the other end of the chain. Also, because most of these molecules contain a chain of 16–18 or more carbons, they are called *long-chain fatty acids*. In nature, these long-chain fatty acids generally don't exist as free-standing molecules. They are usually linked together with a glycerol backbone (glycerol is a simple 3-carbon alcohol) to form a molecule with three fatty acids that is called a *triglyceride*.

If two of those carbons are linked with a double bond, the fat is called *unsaturated*. If the fatty acid contains two or more double bonds, the fat is *polyunsaturated*. A fatty acid with no double bonds is called *saturated*.

We also use a shorthand code to describe fats. For example, an 18-carbon fatty acid with two double bonds would be identified as C18:2. A common name for such a fat is *linoleic acid*.

When fats are fed to an animal with a rumen, they are exposed to rumen microbes in an environment that contains lots of excess hydrogen (low pH) and no free oxygen. The rumen bacteria and protozoa don't metabolize fat directly—they are unable to use the fats for their own energy—but they can alter the structure of some fats.

The nutritional fats in grains, vegetable oils, and forages are a combination of saturated and unsaturated fatty acids. Once these fats enter the rumen, the rumen bacteria use the excess hydrogen atoms to reduce some double bonds to single bonds, and by the time these fat molecules exit from the rumen, they have become more saturated.

This phenomenon does not occur in animals without rumens, which explains why the fats of sheep and cattle are more saturated than the fat of pigs. Ruminant fat reflects the saturation of fats in the rumen; pig fat mirrors the relatively unaltered vegetable fats consumed by that animal.

Saturated fats tend to have higher melting points than unsaturated fats, which means that at room temperature, ruminant fat tends to be harder than pig fat. Because of the difference, rendered fat from cattle and sheep is called tallow, and the rendered fat from hogs is called lard.

MICROORGANISMS CAN GENERATE ELECTRICAL CURRENT

(Bennetto, 1990)

Energy-rich compounds, such as carbohydrates, carry electrons that can be liberated to form electric currents. For instance, the respiration of glucose that proceeds chemically as:

$$C_6H_{12}O_6 + 6\,O_2 \rightarrow 6\,CO_2 + 6\,H_2O$$

can be described electrochemically as:

$$C_6H_{12}O_6 + 6\,H_2O \rightarrow 6\,CO_2 + 24\,H^+ + 24\,e^-$$

$$6\,O_2 + 24\,H^+ + 24\,e^- \rightarrow 12\,H_2O.$$

This second set of chemical equations can be seen to be equivalent to the first.

In order to realize the electrochemical equation set, it is only necessary to oxidize glucose without the presence of molecular oxygen (O_2). Various steps in the oxidation process thus proceed through a series of intermediate steps, with the electrons handed off from one intermediary to the next. This is how glucose respiration naturally occurs by enzymatic reactions.

Electrons may be diverted from the respiratory chain by a redox mediator, which enters the microbe cell and becomes reduced (acquires extra electrons) before it leaves the cell. It then shuttles electrons to an electrode, where they are liberated as the mediator becomes oxidized. This electrode is the cathode, delivering electrons to an external circuit (Figure).

Connected electrically but not chemically, another electrode (the anode) delivers electrons to an oxidizing material that can be oxygen or some other oxidizing reagent. As the electrons move from the cathode to the anode (remember, electrical current is assumed to be composed of positively charged particles, not electrons), they can perform electrical work, such as operate a motor or light a bulb.

As long as the oxidizing agent flows to the distil side of the membrane and an energy-rich substrate flows to the microbes, they can maintain the external electrical current. This is a form of microbial *fuel cell*.

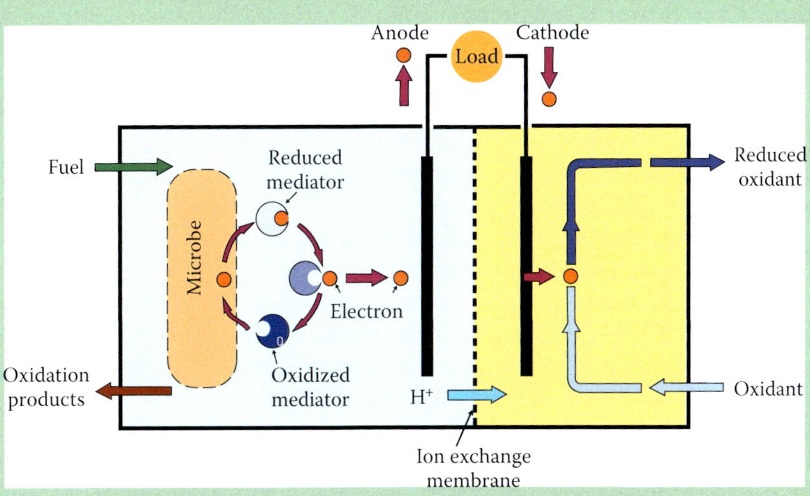

APPLICATIONS AND PREDICTIONS

1. To search for evidence of biological activity, assay for ATP.
2. Microbes surrounding thermal vents will metabolize sulfur-containing compounds for their energy source.
3. Excess glucose storage capacity must be present to supply energy continuously.
4. To kill organisms, interfere with ATP production or utilization.
5. Cells deprived of oxygen will produce energy less efficiently than cells with oxygen.

3.11 Temperature and Pressure Effects

> I believe that for most diseases, prevention by control of their origins is cheaper, more humane, and more effective than intervention by treatment after they occur.
>
> **—Thomas McKeown**

Temperature and pressure sometimes have profound effects on reaction rates and chemical equilibrium. One reason that biological organisms are more active and grow faster at higher temperatures is that, for every 10°C rise in temperature, a doubling or tripling of reaction rates can generally be expected. Animals that maintain constant body temperatures (*homeotherms*) have a metabolic advantage (as long as food is not scarce) over animals whose body temperatures vary with environmental conditions (*poikilotherms*) because they can maintain higher metabolic rates. There is a limit to the body temperature that can be maintained, however, because of enzymatic inactivation or denaturation at temperatures approaching 45°C.

Some chemical reactions are *exothermic*; that is, they produce heat, as in the oxidation of glucose to form water and carbon dioxide. Other chemical reactions require heat to proceed; these are termed *endothermic*. An increase in temperature increases the reaction rate for every reaction, whether it be endothermic or exothermic. However, the rates of endothermic reactions are always increased more than the rates of exothermic reactions for any given rise in temperature.

Chemical equilibria are also affected by temperature, and the specific changes depend on the type of reaction. Raising the temperature of an exothermic reaction decreases the equilibrium constant, whereas the same temperature increase affects an endothermic reaction by increasing the equilibrium constant.

Increasing temperature or pressure increases the likelihood of the number and force of collision between atoms and thus favors reactions moving more rapidly to completion. Le Châtelier's principle states that if a stress is applied to a system at equilibrium, then the system readjusts, if possible, to reduce the stress. Thus, an endothermic reaction will be enhanced by an increase of temperature because the reaction absorbs heat. Increasing the concentration of one of the reactants drives a reaction toward completion because this action reduces stress on the system.

Increasing pressure on a system can have several different effects, depending on whichever action tends to reduce stress. For instance, for the chemical system that forms gaseous hydrogen iodide from hydrogen gas and gaseous iodine,

$$H_2 + I_2 \leftrightarrow 2HI \tag{3.11.1}$$

there is no net effect of pressure on the equilibrium of this reaction because there are two molecules of gaseous product that replace two molecules of gaseous reactants. The rate may increase, but there is no change in equilibrium.

For the reaction forming gaseous water vapor from hydrogen and oxygen,

$$2H_2 + O_2 \leftrightarrow 2H_2O \tag{3.11.2}$$

increasing pressure tends to form more water vapor because two molecules of gaseous product are formed from three molecules of gaseous reactants. Thus, the overall pressure stress placed on the reaction is reduced by forming more water vapor. If liquid water is formed, the forward reaction is favored even more because the volume occupied by liquid water is even less than for water vapor, thus relieving pressure stress even further.

CRICKET THERMOMETER

Crickets make a chirping sound, called *stridulation*, when they rub their tiny wings together. The higher the temperature, the faster crickets chirp. This is an example illustrating how metabolism, and actions metabolically based, depend on temperature. Dolbear's law states that temperature in degrees Fahrenheit can be determined accurately from the number of chirps counted in 15 seconds:

$$\text{degrees Fahrenheit} = (\text{chirps in 15 sec}) + 40.$$

Temperature in degrees Celsius can be obtained from:

$$\text{degrees Celsius} = (\text{degrees Fahrenheit} - 32)\ 5/9.$$

Other insects also use chirps as a means for communicating. Katydids, for instance, can be used to determine temperature from:

$$\text{degrees Fahrenheit} = 60 + (\text{chirps in 1 min} - 19)/3.$$

APPLICATIONS AND PREDICTIONS

1. Poikilotherms should be able to survive food scarcity better than homeotherms.
2. Snakes and fish will be poikilothermic.
3. Animals with higher body temperature will produce more body heat.
4. Optimum temperatures will exist for growth, reproduction, conversion of feed to flesh, and other input–output relationships.
5. Enzymes present in poikilotherms are less sensitive to temperature changes than are those in homeotherms.

3.12 Free Energy

Facts are not science—as the dictionary is not literature.

—**Howard Fabing and Ray Marr**

The concept of free energy was discussed in a physical context in Section 2.5. Living systems, however, exist in an environment with contributions from both physics and chemistry (actually, there is also a biological component that has yet to be incorporated fully into the free energy concept). Chemical contributions to free energy determinations can give an overall view of the spontaneity of physicochemical processes. Living systems can exploit those processes that would occur spontaneously; they could obtain energy from them. On the other hand, there are certain physicochemical processes that are necessary, but energetically expensive. An example of these is the formation of many of the biochemical regulators of life. Living systems have developed ways to make these processes less expensive (e.g., by using specialized enzymes). Enzymes may not reduce the overall energy of formation of these compounds, but they can help to overcome energy obstacles that stand in the way of moving from one step to the next (Figure 3.12.1).

Molecules that should react with one another often do not because they lack sufficient energy. For every reaction, there is a specific activation energy, which is the minimum amount of energy that two molecules must have before a collision between them will be successful in leading to a reaction. Were it not for activation energy, all reactions would proceed quickly to equilibrium, and life as we know it would be impossible.

Enzymes enhance the rate of a reaction by lowering the energy of activation, thereby ensuring that a greater proportion of molecules will have sufficient energy to collide successfully and undergo reaction. A primary feature of an enzyme is that it is not permanently changed or consumed as the reaction proceeds.

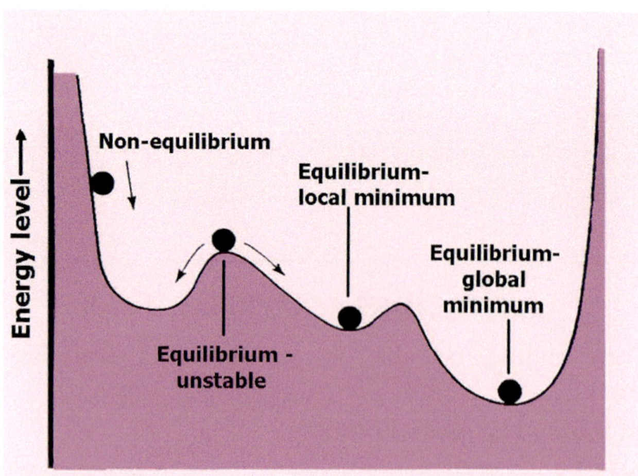

FIGURE 3.12.1
Before chemical reactions can take place, activation energies must usually be overcome. These appear in this diagram as energy hills between two valleys. If the reaction is exothermic, the valley on the other side of the hill is at a lower elevation, releasing energy as heat. If the reaction is endothermic, the far valley is at a higher elevation than the starting point. (Redrawn from Haynie, D.T., *Biological Thermodynamics*, Cambridge University Press, Cambridge, UK, 2001.)

Free energy is mainly a quantitative approach. Therefore, to calculate free energy values and determine whether a certain chemical reaction is favored, we need to have measurements of chemical reactions, and these must be made on some bases that allows comparison. Thus, it is that measured *chemical activity* of compounds that has been made at standard conditions of 298 K, 1 atm of pressure, 1 mol of substance, and 1 molar concentration if the reaction occurs in aqueous solution. These values can be mathematically corrected for temperature by (Haynie, 2001):

$$(\text{Chemical activity at some temperature}) \propto (\text{Absolute temperature})$$

Any biochemical reaction that occurs spontaneously or that emits a net amount of energy will have a negative free energy value. In this form, chemical potential and Gibbs free energy can be used interchangeably for isothermal (constant T) and isobaric (constant p) chemical processes.

It is likely that pressure and temperature can change during a chemical reaction (e.g., for products that form smaller volumes of gas than the reactants, or for exothermic reactions, respectively). These physical changes need to be taken into account, and, in some cases, can eliminate the spontaneity of the reaction.

BIOCOMPATIBILITY OF MATERIAL SURFACES

Xenobiotic materials (materials not biological in nature) for implants must satisfy several requirements (Baier, 2002):

1. Must be able to adhere
2. Not diminished by biological actions
3. Cannot diminish host functions
4. Forms an infection-free seal
5. Able to remain for a long time

Implants must be able to function in a wet, salty, biochemically active environment.

Surfaces of these materials are critical to their successful use. For some implants, such as artificial bones or joints, the surface should promote healing through growth of host tissue. In other cases, materials to be used in a biological environment should repel growth. Examples are surfaces of food-handling equipment, and the bottom surfaces of ships. Dental implants (artificial teeth) have both requirements: the root must promote growth to anchor it into the gum and jaw, while the top must repel plaque formation by bacteria.

The surface free energy is the energy of binding of adherent molecules to the surface of the material. Materials that promote adhesion have high surface free energies. Those that do not promote growth have low surface energies.

Materials in contact with biological environments become coated with glycoproteinaceous macromolecules in highly hydrated solution states. These bound compounds form films of considerable complexity, and can support complex bacterial colonies in the open environment (called biofilms, see Section 6.16.4, Communal

(*Continued*)

BIOCOMPATIBILITY OF MATERIAL SURFACES (Continued)

Benefit). In the body, early stage attachment to the glycoproteins is by macrophages, and at later states by fibroblasts. Capsules of tissue are eventually formed around the inclusions.

An operational surface free energy of about 40 dyne-cm/cm^2 appears to divide surfaces that promote growth from those that don't. Surface energies in the neighborhood of 20–30 dyne-cm/cm^2 prevent macromolecule adhesion (Baier, 2002). These low energies are difficult to achieve. Surfaces must be cleaned either by flaming or corona discharge, held in boiled (out-gassed) triply distilled water, and coated with stearates or silicones. Coating is not necessary if the implant is to be healed into the body.

Substrate	Typical Surface Energy (dyne-cm/cm^2)
Aluminum	45
Polyamide	33–46
Polycarbonate	46
Polyetherimide	40–45
Polyimide	40–50
Polymethylmethacrylate	38
Polysulfones	41
Polyurethane	43
Silicone Elastomer	24
Stainless Steel	700–1100
Titanium	>250

Source: Values from Riegler, B. and Rhodes, K., *Med. Dev. Diagn. Ind.*, 24, 80–90, 2006.

APPLICATIONS AND PREDICTIONS

1. Most biochemical reactions with a negative free energy value will not occur spontaneously because activation energies will have to be supplied before the reactions can occur.
2. Enzymes can provide surface free energy and promote binding sites for molecules.
3. Surface free energies can be selected depending whether surface growth is wanted or unwanted.
4. Enzymes lower activation energies.

Questions

3.0.1 Distinguish among the different subdisciplines of chemistry.

3.0.2 In which branch of chemistry would you expect to study the following:

a. Proteins
b. Arsenic
c. Ionic bonds
d. Osmosis
e. Properties of cell membranes

3.0.3 Physical, chemical, and biological components of the environment interact with living things. Give examples of the chemical environment of organisms.

3.0.4 Why is the study of chemistry important for biological engineers?

3.0.5 Explain why the study of chemistry is necessary to understand the workings of biological systems.

3.0.6 Someone has said that the study of biology is the same as the study of chemistry. Why is this statement both true and false?

3.0.7 Describe in your own words differences among different branches of chemistry and how they relate to biology.

3.1.1 Discuss the advantages to knowledge of families of elements with similar features.

3.1.2 Estimate the number of electrons in a wart hog.

3.1.3 Will an atom of barium (element number 56) combine most readily with bismuth (Bi, no. 83), polonium (Po, no. 84), astatine (At, no. 85), or radon (Rn, no. 86)?

3.1.4 Is carbon an alkaline or acidic element? Why?

3.1.5 What is the heaviest element with a clear function in living things? Does this mean that heavier elements are not found in living things?

3.1.6 Add to the list of Applications and Predictions.

3.2.1 Rank the following bonding types in terms of stability of bonds:

Covalent bonds
Hydrogen bonds
Ionic bonds
van der Waals bonds

3.2.2 Compare chemical ionic bonding to covalent bonding. Is there a clear distinction between the two?

3.2.3 The element nitrogen forms combinations with many other elements. The charges on nitrogen range from −3 to +5, inclusive. Give examples of as many compounds as you can by illustrating different charges on the nitrogen atom.

3.2.4 What biological functions would you expect to be performed by compounds composed of elements with high electronegativity? Low electronegativity?

3.2.5 What are the characteristics of polar and nonpolar molecules that are exploited in living things?

3.2.6 The relatively low-density structure of ice is fortunate for the survival of aquatic species. How does the same structure harm single cells?

3.2.7 Speculate on ways in which the void spaces in hexagonal-structured ice can be exploited.

3.2.8 Where are hydrogen bonds of importance in living things?

3.2.9 If the freezing point depression of a single solute solution is known, how can the degree of dissociation of the solute in solution be determined? What other information is necessary?

3.2.10 Give an example from everyday experience that illustrates by analogy how van der Waals forces operate.

3.2.11 For what purposes could the mechanism utilized by the Gecko be used by biological engineers?

3.2.12 Suggest other uses for Raman spectra.

3.2.13 Add to the list of Applications and Predictions.

3.3.1 Under what conditions would exothermic or endothermic reactions be advantageous?

3.3.2 If the equilibrium constant of a chemical reaction is high, what can you say about the completeness of the reaction? If the equilibrium constant is low, what can be done to drive the reaction toward higher production of products?

3.3.3 What information does the dissociation constant give about the presence of ionic forms in solution?

3.3.4 Add to the list of Applications and Predictions.

3.4.1 If the dissociation constant of an acid is small, is the acid strong or weak?

3.4.2 Why is carbonic acid such an important buffer compound in living things? What does this have to do with metabolism?

3.4.3 Why would you expect solutions in living systems to be almost devoid of strong acids and bases?

3.4.4 What information is contained in a measurement of pH? What can a measurement of pH tell about the biological state of: blood, wastewater, food, groundwater, rain, bioreactor medium, air conditioner condensation, and gel for growth of microbes?

3.4.5 Add to the list of Applications and Predictions.

3.5.1 Would you think fast or slow chemical reactions are important biologically? Why?

3.5.2 How is the equilibrium constant for a reversible chemical reaction related to the forward and backward rates?

3.5.3 How can the reaction rate of a first-order reaction be maintained at a high level? How can the rate be controlled at any level?

3.5.4 How do reactions described by Michaelis–Menten kinetics relate to biological processes?

3.5.5 List some specific uses of ELISA.

3.5.6 Add to the list of Applications and Predictions.

3.6.1 Suggest other elements besides carbon on which an organic chemistry could be based.

3.6.2 For what special purposes would the substitution of another element for carbon be appropriate in certain biochemicals?

3.6.3 Make a list of biochemical functional groups and give a short description of their characteristics.

3.6.4 How does an amino acid differ from a hydrocarbon?

3.6.5 To what valuable use can a D-amino acid be put?

3.6.6 List the four basic types of macromolecules, and their purposes.

3.6.7 For what purposes can the equivalent chemical formulas for living tissues be used?

3.6.8 Why does it make sense that lower molecular weight fats are located in the peripheral areas of the body?

3.6.9 If you wanted to find a natural source of high melting point fats, would you look in a plant growing in warm or cold climates?

3.6.10 List the two important features of reactions between organic compounds. Why is each important?

3.6.11 What is a free radical? What harm can free radicals do? What good can they do?

3.6.12 Describe a process to form a membrane that can let insulin molecules through but exclude antibodies.

3.6.13 Add to the list of Applications and Predictions.

3.7.1 Why is physical chemistry in water so important for the study of biology?

3.7.2 If the solubility of a substance, say oxygen, is too limited to be useful, explore other means to carry adequate amounts of the substance to distant tissues.

3.7.3 When the vapor pressure of a liquid has reached an equilibrium, what changes of state of that liquid are still occurring?

3.7.4 How is the gel structure of water and proteins related to the integrity of a cell?

3.7.5 If a protein molecule is electroneutral, why does water adhere so firmly to it?

3.7.6 When the gel structure of water is disrupted by a protein phase change, what effect would you expect there to be on the ionic composition inside the cell?

3.7.7 How is the ordered state of water restored after a phase transition?

3.7.8 How can the resting potential of a cell be measured?

3.7.9 Suggest applications that take advantage of the isoelectric point of proteins.

3.7.10 Considering the pH effects on proteins, why would it be normally disadvantageous for the interior of a living organism to have an uncontrolled interior pH level? Would there be any unusual circumstances that would benefit from large pH changes?

3.7.11 Add to the list of Applications and Predictions.

3.8.1 Of what commercial value are heat shock proteins?

3.8.2 Describe the four types of protein folding.

3.8.3 Describe a process to deliver pharmacological chaperone proteins to treat cystic fibrosis.

3.8.4 Describe a hypothetical process that intervenes in the body's natural protein quality-control mechanism to improve health.

3.8.5 Add to the list of Applications and Predictions.

3.9.1 Describe an experiment you could perform to determine the amount of energy required to liberate a chemical from the surface of carbon.

3.9.2 Hexokinase is an enzyme involved in glucose metabolism. Hexokinase in the liver has different kinetics than hexokinase in the brain. How can that be?

3.9.3 Explain how enzymes promote chemical reactions.

3.9.4 Explain how vitamin B can be used to identify chemicals that act as competitive inhibitors.

3.9.5 What is the difference between a natural protein and a denatured protein?

3.9.6 Why does the law of constant heat sums make sense for living things?

3.9.7 The mechanisms of many modern therapeutic drugs affect enzyme actions. Some drugs work by inactivating essential enzyme action, thus depriving microbes of necessary metabolic steps. Other drugs use the enzymes to produce products toxic to unwanted cells. Discuss the possibilities to influence enzymatic actions to cure diseases.

3.9.8 Add to the list of Applications and Predictions.

3.10.1 Would you call ATP a high, low, or intermediate energy compound? Why must it be so?

3.10.2 Compare the chemical energy action of ATP to the electrical charging and discharging of a storage battery.

3.10.3 At what stage are respiration of glucose requiring oxygen and respiration not requiring oxygen the same?

3.10.4 How would you expect the activity level of anaerobic microbes to compare with that of aerobes? Why?

3.10.5 Add to the list of Applications and Predictions.

3.11.1 Use Le Châtelier's principle to predict how the following would be affected by an increase in (1) temperature and (2) pressure (separately changed):

Respiration of glucose

Condensation of water

Oxygen dissolved in water

Photosynthesis

Production of proteins

3.11.2 The other night I heard no crickets chirping. What was the temperature?

3.11.3 Add to the list of Applications and Predictions.

3.12.1 Besides using enzymes, what are other possible means to overcome activation energy requirements for a chemical reaction?

3.12.2 What is meant by surface free energy of biomaterials, and how can it be minimized?

3.12.3 Explain why a catalyst should have high surface energy.

3.12.4 Compare the surface energies of stainless steel and polyimide. Which would you use if you wished to discourage biochemical attachment to the surface? Which would you use to encourage attachment?

3.12.5 Add to the list of Applications and Predictions.

4

Principles of Mathematics and Engineering Sciences

> The idea of "proof" is the guiding light of mathematics. No matter how many examples you can give for the reality of your theorem, if you cannot offer a valid proof, then your theorem is merely a conjecture.
>
> **—Devlin M. Gualtieri**

The engineering sciences, such as statics, dynamics, strength of materials, transport processes, and electricity, can be described as refinements or adaptations of physical fundamentals. However, some others, such as information theory and control systems, are conceptual in nature and lend themselves well to mathematical manipulation. Hence relevant mathematical attributes have been included here along with engineering sciences not directly related to physics.

Intellectual insight can be obtained from mathematical manipulation, and this insight is directly related to the understanding of interactions of biological units with their physical, chemical, and biological environments (Figure 4.0.1). Furthermore, mathematical constructs, such as statistics, can help to describe biological events and predict outcomes.

Although mathematical models are, at best, idealizations, and suffer from a certain degree of unreality, they can still be used to gain understanding and to yield predictions. Engineering sciences are highly dependent on these mathematical models; their idealizations are necessary for simplified analyses of biological characteristics and responses.

Mathematics is not by itself a science, and so its contribution to the biological sciences is different from physics and chemistry. It is through mathematics that concepts and organization of principles can be applied; mathematics provides the tools for this process. As some required concepts are not entirely mathematical, we must also draw upon the engineering sciences such as control theory to help understand biological function. In this section, material relevant to biological systems is presented.

1. *There is an element of randomness in biology.* Whether we are looking at some response (such as movement) of a biological system or making a measurement of a physical attribute (such as weight), biological systems seem never to be completely predictable. There are always differences with time, or space, or between individual organisms. If enough was known about the determinants of the measurement, perhaps much of the apparent randomness could be explained, but there is ultimately, at a small enough scale, an inevitable random element always present. Randomness is manageable if the probabilities of occurrences are known for entire populations. See Section 4.2 (Probability).
2. *Appropriate responses require control systems, and each of these needs sensors, actuators, processing, and information pathways.* Survival requires that adequate responses are made to environmental stimuli. Information important to the biological unit must be sensed, sent on to some structure that determines the appropriate response, and

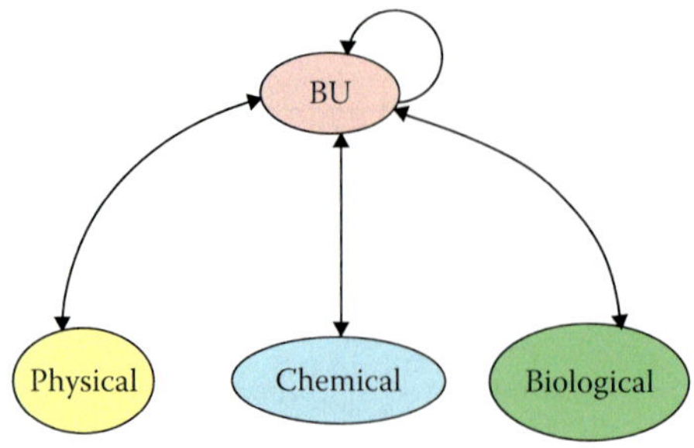

FIGURE 4.0.1
The biological unit (BU) is affected by interactions with its physical, chemical, and biological environment. Likewise, environmental elements are affected by the BU. Self-adjustment is also a possibility. Mathematics and engineering sciences help to describe and analyze these interactions.

then action taken. These controls may be simple or elaborate, but the fact that they occur at all in biology is a sign of life. See Sections 4.3 (Calculus) and 4.4 (Controls).

3. *Optimization conserves resources.* Most resources necessary for life are limited to some extent, and cannot be wasted if life is to continue to thrive. Optimization is a term that has come to be overused and misused because many things we call optimized are not optimized at all. To be optimum, a process must have some criterion that is to be minimized or maximized, and then the process must minimize or maximize the criterion. Biological systems often have many of these criteria that should be optimized simultaneously. Priorities must then be established. See Section 4.5 (Optimization).
4. *Information implies order.* Shannon's information theorem that relates information content to the probability of occurrence has been applied mainly to electronic communications systems. Information, however, is extremely important to biological units because information allows the unit to survive, grow, and reproduce. Thus, Shannon's theorem is relevant to biology and to maintaining the integrity of biological systems. See Section 4.6 (Information).

4.1 Equality

It depends on what the meaning of "is" is.

—William J. Clinton

Basic to mathematical manipulations is the concept of equality. This leads to the algebraic notion of an equation, wherein an equal sign is placed between mathematical expressions. There are actually three types of equations, and it is important to understand the differences between them:

1. *Unequivocal equality.* The equal sign in this case means that, under all circumstances, the mathematical expressions on both sides must always be equal. This is a powerful concept. It states that not only must these mathematical expressions be equal in magnitude at sometimes, or under certain conditions, but they must

be equal under all conditions. If there are limits to the range of validity of the equation, then these are stated explicitly outside the equation, and the equation is considered to be true for any set of circumstances within these limits.

An example of this type of equation is the well-known $F = ma$, which expresses a fundamental relationship between the force on an object, the mass of that object, and its acceleration. Another example is $y = 2x + 1$, where the value of y is related to the value of x for all values of x.

Not only does an equal sign denote equal magnitudes, but it also indicates equal dimensions and identical units for each of the mathematical expressions appearing in the equation. Thus, if one expression is eight oranges and another is eight apples, these two cannot be set equal to each other despite the fact that the numbers are the same (Figure 4.1.1).

Energy and power values for biological systems are often expressed in several different sets of units. Among these are watts, calories, kilocalories, British Thermal Units (BTU), ergs, horsepower, electron-volts, and others. Unless the same units are used on both sides of the equation, and unless the same units are used for each expression on one side or other of the equation, then the equation is not valid.

2. *Conditional equality.* The equal sign in this case indicates that equality is achieved only for certain values of the variables, and is often used as the means to solve for roots. An example of this type of equation is $x \cot x = 1$, which is only true for certain discrete values of x (x = 0, 4.4934, 7.7253, 10.9041, 14.0662, 17.2208…) (Carslaw and Jaeger, 1959).
3. *Replacement.* The equal sign in this case means to replace the current value of the left-hand variable with the calculation of the term on the right. This kind of

FIGURE 4.1.1
Terms involving apples and oranges cannot be part of the same equation.

equation is used in computer programming, and an example is $y = x + 1$. The value of the variable y is given as the value 1 added to the value of the variable x. The value of the variable x was previously calculated.

Equations may express theoretical concepts, or they may express empirical information. The division between these two is not very clear when closely scrutinized because most theoretical concepts and so-called fundamental principles were based in their formative stages on direct observation. As time went on, and as it became clear that there were means of expressing these observations so that they could generally predict further observations, they were then elevated to the status of principles. These principles may still be subject to modification as further experimental observations cause the principles to be re-evaluated. Such was the case for Newton's laws of mechanics when it became clear that they did not predict actions occurring at the subatomic level.

Some information must remain empirical and will never become a generally accepted first principle. After all, there can only be a small number of fundamental principles (according to the presiding conceptual framework of science). Empirical information is useful nonetheless, especially in the designs of utilitarian devices to be used with living systems. Designs of artificial kidney machines, bioreactors, automobiles, hospital ventilators, and even light bulbs are based on empirical information that serves to produce better products.

Equations are important in the study of living systems because they are often used to express input-output-storage relationships, such as the balance equation (Equation 2.2.1). There are numerous inputs and outputs at the interface between a living entity and its environment. Each of these may be the object of considerable study, and there may be no other means to calculate its value than to employ equations expressing all other inputs, all other outputs, and quantities stored in the organism. Indirectly, then, the term of interest may be found.

DIMENSIONAL ANALYSIS

Dimensional analysis has been used in engineering to formulate relationships between various dependent variables and the parameters they depend upon. The results have been useful for analysis and design of products and processes. The basis for dimensional analysis rests upon the fact, as stated earlier, that each term in an equation must have the same dimensions. Mazumdar (1989) has given succinct rules for dimensional manipulations:

1. Quantities added or subtracted must have the same dimensions.
2. Quantities equal to each other must have the same dimensions.
3. Any quantity may be multiplied or divided by any other quantity without regard to dimensions. However, the resulting product or quotient must have appropriate dimensions so that the earlier rules are not violated.
4. The dimensions of an entity are entirely independent of its magnitude.
5. Pure numbers, such as Avogadro's number, exponents, and the base of natural logarithms (e), have no dimensions.

(Continued)

DIMENSIONAL ANALYSIS (Continued)

If all the parameters relevant to a problem are known, then dimensional analysis can be used to form a complete set of dimensionless numbers to formulate a general relationship among output and input parameters. This can be done by expressing the parameters in terms of their basic dimensions and then determining the required form of dimensionless groups.

For example, Mass (*M*), Length (*L*), and time (*T*) are the three basic mechanical dimensions. If we wished to consider the case of a simple pendulum, the relevant variables and their dimensions are:

r = pendulum length (L)
m = mass (M)
θ = initial angle of displacement from the vertical (dimensionless)
t = period of oscillation (T)
g = acceleration due to gravity (L/T^2)

Any product of these variables must be of the form:

$$m^a r^b \theta^c t^d g^e$$

where a through e are exponents. Dimensions of this group are

$$[M]^a[L]^b[\,]^c[T]^d[L/T^2]^e$$

In order for the product of the earlier form to be dimensionless,

$$a = 0 \quad \text{from Mass terms}$$

$$b + e = 0 \quad \text{from Length terms}$$

$$d - 2e = 0 \quad \text{from Time terms}$$

We have three equations and five unknowns, so we can make two arbitrary choices. If we set $e = 0$ and $c = 1$, then

$$a = b = d = 0$$

giving one dimensionless group equal to θ.

A second independent dimensionless product is obtained when $e = 1$ and $c = 0$, yielding

$$a = 0$$

$$b = -1$$

$$d = 2$$

the second dimensionless group is gt^2/r.

(Continued)

DIMENSIONAL ANALYSIS (Continued)

These dimensionless groups are called pi terms, and the number of independent pi terms is equal to the number of variables involved minus the number of dimensions in which those variables may be measured (Murphy, 1950). In this example there were five variables and three dimensions, giving two pi terms. Because the formulation of pi terms in the aforementioned example was based on arbitrary choices, the pi terms are not unique. However, they are complete.

The relationship between input and output parameters thus takes the form:

$$F(\theta, gt^2/r) = 0$$

where F ( ) indicates some functional relationship, usually determined by experiment.

The advantage of this technique is that it can be used to form a general relationship that is more efficient than one obtained from considering each variable by itself. Varying, in this case, any part of gt^2/r will give the same relationship, whether the variation comes about by changing period of oscillation or pendulum length (it is not likely that we can vary the acceleration of gravity). This technique has been found to be very useful in problems involving fluid flow, heat transfer, and mechanics. It has not yet been used extensively for problems involving biology, due mainly to the lack of identifiable independent variables, but there have been instances where functional relationships have been based on dimensionless terms in biology (DiMilla et al., 1991).

Example 4.1.1: Flow in the Pulmonary Vein (Johnson, 1999)

The pulmonary vein is a very distensible blood vessel. The vein has a zero-pressure diameter of about 0.5 cm and a length of about 15 cm. The mean blood pressure is about 1200 N/m^2 where blood enters the pulmonary vein, and is about 0 N/m^2 where the vein empties into the left cardiac atrium. Assume that the distensible tube diameter is given by a simple relationship: $D = D_0 + D_1 p$ (Figure 4.1.2). Determine the tube compliance constant value (D_1) when the vein carries a volume flow rate of 83 mL/s.

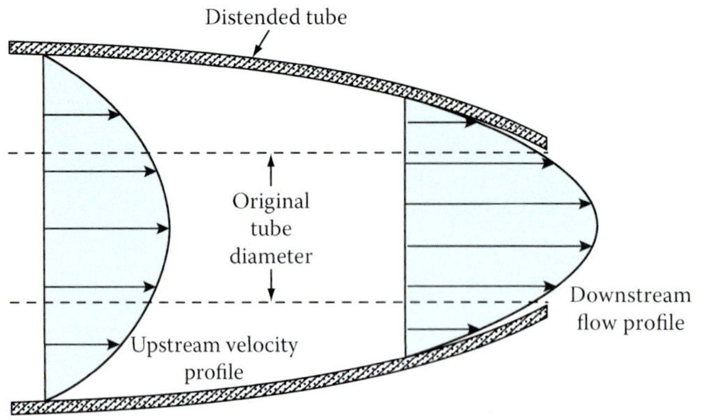

FIGURE 4.1.2
An elastic tube distends nonuniformly due to pressure differences within a flowing fluid. (Redrawn from Johnson, A.T., *Biological Process Engineering: An Analogical Approach to Fluid Flow, Heat Transfer, and Mass Transfer Applied to Biological Systems*, John Wiley & Sons, New York, 1999.)

Solution

The equation for volume flow rate through a distensible tube is (Johnson, 1999)

$$\dot{V} = \frac{\pi}{640\,\mu L D_1}\,[D^5(L) - D^5(0)]$$
$$= \frac{\pi}{640\,\mu L D_1}\,\{[D_0 + D_1 p(0)]^5 - [D_0 + D_1 p(L)]^5\}$$

The process of *iteration* is one that uses an equality to mean replacement of the value on the left-hand side of the equation with the value calculated on the right. Thus, one would make an initial guess for D_1 and calculate $\dot{V}$. When the value for $\dot{V}$ does not equal the known value, another guess for D_1 is made. Eventually, the known $\dot{V}$ value is obtained. At that point, the value used for D_1 is correct. Because the pressure at the distal end of the pulmonary vein is about 0 N/m², and the term D_0^5 is small, we can obtain an initial estimate for D_1 by approximating the earlier equation.

$$\dot{V} \approx \frac{\pi}{640\,\mu L D_1}[D_1 p(0)]^5 = \frac{\pi D_1^4 p^5(0)}{640\,\mu L}$$

Thus

$$D_1^4 = \frac{(83\times 10^{-6}\ \text{m}^3/\text{s})\,(640)\,(4.5\times 10^{-3}\ \text{N s/m}^2)\,(0.15\ \text{m})}{\pi(1200\ \text{N/m}^2)^5}$$
$$= 8.2\times 10^{-6}\ \text{m}^3/\text{N}$$

We check this result by calculating $\dot{V}$ from the equation and comparing against the known value of 8.3×10^{-5} m³/s.

$$\dot{V} = \frac{\pi}{640(4.5\times 10^{-3}\ \text{N s/m}^2)\,(0.15\ \text{m})\,(8.2\times 10^{-6}\ \text{m}^3/\text{N})}$$
$$\times\{[0.005\ \text{m} + (8.2\times 10^{-6}\ \text{m}^3/\text{N})(1200\ \text{N/m}^2)]^5 - (0.005\ \text{m})^5\}$$
$$= 6.4\times 10^{-4}\ \text{m}^3/\text{s}$$

By trial and error, we generate the following values:

D_1 (10^{-6} m³/N)	$\dot{V}$ (10^{-4} m³/s)
8.2	6.4
5	2.3
3	1.1
2	0.69
2.5	0.86
2.4	0.83

The value for D_1 is thus about 2.4×10^{-6} m³/N.

APPLICATIONS AND PREDICTIONS

1. If two things are said to be equal, they must have the same numerical values and units.
2. Biological entities will sometimes be said to be equivalent only within certain ranges.
3. Convection heat exchange in the circulatory system can be found by summing all other components of a heat balance equation.
4. Natural frequencies of biological oscillations can be found by finding the roots of certain equations.
5. Pressure balance equations can be used to model the respiratory system.
6. Mass movements into and out of the cell can be found to conform to a mass balance equation.
7. Total energy in a fluid system can be obtained by summing potential and kinetic energy terms.
8. Not all equations are true all the time.
9. When programming a computer, the variable whose value is to be updated must always appear in simple form on the left side of the equal sign.
10. Relationships among several variables can be derived by equating their dimensions.

PETO'S PARADOX

If every living cell in an organism theoretically has an equal probability of developing cancer, then large animals with more cells should have higher cancer rates compared to smaller animals with fewer cells. However, it has been found that all mammals, regardless of size, have similar rates of cancer. This effect is called *Peto's Paradox* (Sneed, 2014).

4.2 Randomness and Probability

Chance favors the prepared mind.

—Louis Pasteur

There are few characteristics more basic to living systems than the element of random variation. There may be two or more measurements of some biological activity that appear to be identical in value, but there are many more that differ. There appears to be a variation that cannot be directly predicted in almost any measurement made on a biological system, and this is the random variation of which we speak. There may come a time when models of biological systems are so complex that they can account for effects of genetic variations, present and past environments, and spontaneous acts of free will. From these models may come the understanding that all biological activity is deterministic and not stochastic, predictable and not random; but that time is far off. For now, biological variation must be accepted as inevitable.

The scale of measurement can influence the amount of variation seen. There is, for instance, one and only one known planetary ecological system. Within that there are many biomes, the number of which may change with time, definition, or physical scale. The numbers of individuals of a certain species within similar biomes will vary either over time or even at some constant point in time. Responses of individuals of a species to environmental factors such as temperature begin to show much more variation, and so on. As the scale becomes finer, the perception of variation becomes greater.

4.2.1 Probability Distributions

> I'm tired of all this nonsense about beauty being only skin-deep. That's deep enough. What do you want, an adorable pancreas?
>
> —**Jean Kerr**

The more variation in the results of an experiment, the harder it is to conclude that there are any meaningful differences between treatments and untreated control. Mathematicians deal with variation by talking about the probability of occurrence. The probability of occurrence when plotted over the range of possible measurement values forms a probability distribution (Figure 4.2.1). Dealing with variability can be made easier by accepting that variation will occur and then characterizing the probability distribution.

So, what is probability? It is the fraction of all the measurements that will occur between two measurement values. It is defined in this way, rather than to say that it is the fraction of all the measurements that assume a particular value, because of the scaling problem illustrated earlier. As the measurement scale becomes finer and finer, the probability of occurrence of any one measurement value falls toward zero. However, the range of measurement values can be made small, and we can then determine the probability that any one measurement will be found within that range.

There are independent and conditional (or dependent) probabilities of occurrence. Independent probabilities are ones where there is no linkage among several events, and conditional probabilities are just the opposite. Almost all probabilities in living systems are conditional probabilities. Growth rate, for instance, is usually dependent on the abundance of nutrition.

There is, however, a range of exhibited growth rates, depending on genetic character and local environment. The probability of the occurrences of different growth rates will depend on many factors, both internal and external to the living system.

There is a tendency for most biological measurements to have a higher probability of occurrence around the average, or *mean* value. This gives some validity to using mean values of data because most individuals will have values somewhere in the neighborhood of the mean. However, if we wish to be sure that we include most of the population within a range of measurements, then we must consider the spread, or range, of the data. Spread is characterized by the term *variance*. Data spread over a wider range will have higher variance values.

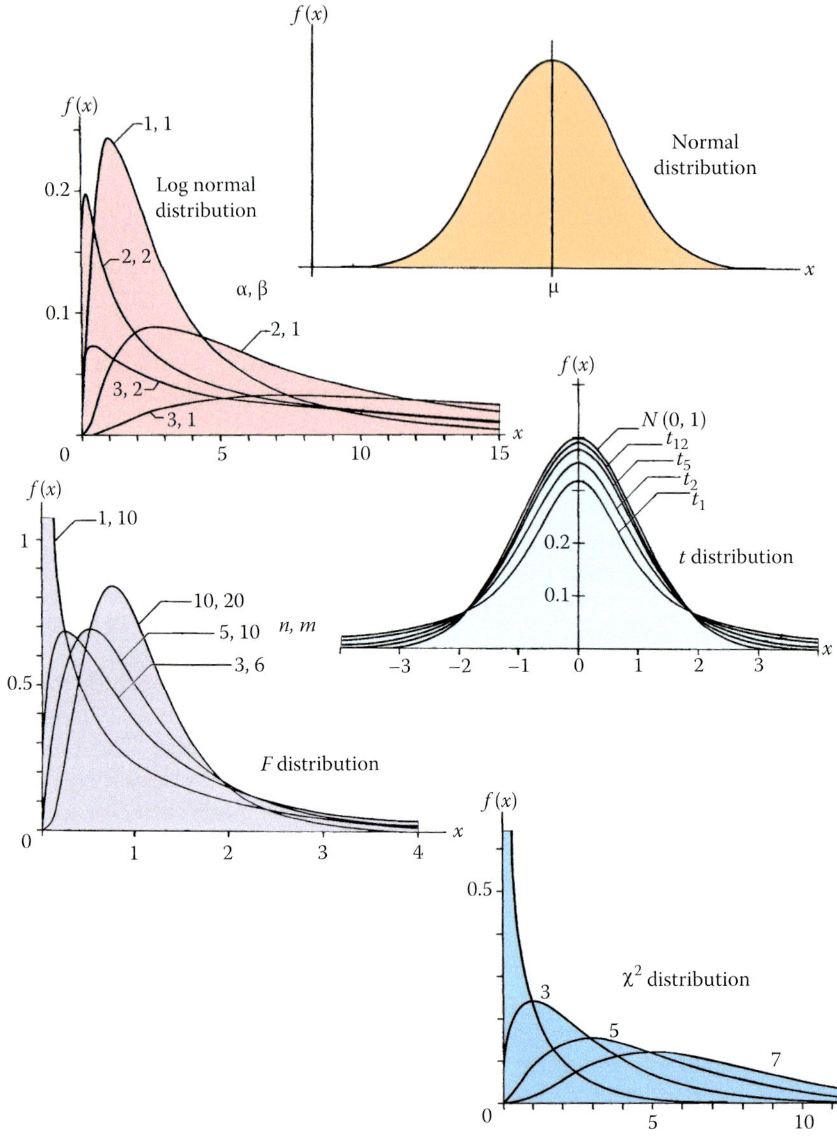

FIGURE 4.2.1
Various probability distributions important in biology. The normal distribution is used for most applications. The t-distribution is used for small sample sizes from a normal distribution. The log normal distribution fits some data better than a normal distribution. The F distribution is used to check equality of variances, and the χ^2 (chi square) distribution is used to check expected values of data. The curves shown here are for various values of distribution parameters. (Redrawn from Barnes, J.W., *Statistical Analysis for Engineers and Scientists: A Computer-Based Approach*, McGraw-Hill, New York, 1994.)

Rates of diseases, drug use, crime, and other measures have this in common: extremes, whether high or low, are more likely to be found in units with low populations (Figure 4.2.2). This reflects the tendency for larger populations to cluster around the mean. It is important to remember when testing engineering designs that the smaller the sample size tested the more likely it is that the results will not reflect average responses of the entire population (Wainer, 2007).

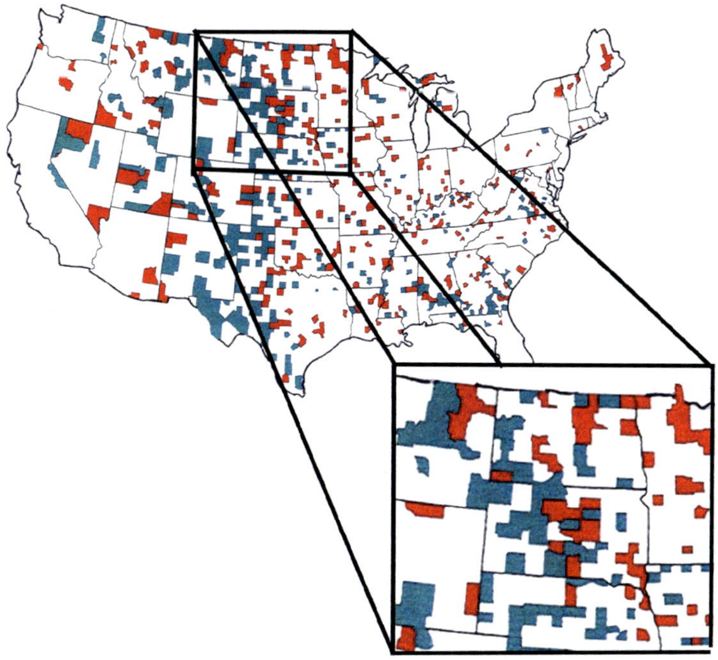

FIGURE 4.2.2
A map of the counties in the United States with the lowest kidney cancer death rates (green) and the highest kidney cancer death rates (red) for the years 1980–1989. Both extremes are in rural areas with low populations and are often adjacent to one another. This illustrates the fact that there is more rate variation for smaller sample sizes than there is for larger sample sizes. Larger sample sizes tend more toward the mean. (Redrawn from Wainer, H., *Am. Sci.*, 95, 249, 2007. With permission.)

There are a number of probability distributions that have been found to be important by mathematicians when describing real-world data. These have been idealized by assuming extremely large sample data sets, so that the plotted distributions are smooth curves and the mathematical equations describing them are continuous. For small sample data sets, the probability distributions are far from ideal, do not plot as smooth curves, and cannot be described by continuous mathematical expressions.

By far the most widely assumed probability distribution applicable to biological data is the Normal Probability Distribution, or Gaussian distribution. When plotted, this distribution forms the familiar bell-shaped curve that is symmetrical about the mean. The mathematical expression describing this distribution is

$$p(x) = \frac{1}{\sigma\sqrt{2\pi}} e^{-(x-\mu)^2/2\sigma^2} \tag{4.2.1}$$

where:

- x is any particular value of data
- $p(x)$ is the probability of occurrence of this data
- μ is the mean of the data
- σ is the variance

The term $e^{-(x-\mu)^2/2\sigma^2}$ stands for the base of Napierian logarithms (e) raised to an exponent of $-(x-\mu)^2/2\sigma^2$.

The values for the mean (μ) and variance (σ) are not known until they are calculated for a particular data set. Once these two parameters have known values, the probability distribution is determined, and the mathematical expression (Equation 4.2.1) can be manipulated to give useful predictive results, at least in an abstract sense.

There are cases where the data cannot be evenly distributed about the mean. Two instances of this are where negative data values are impossible or where there are limited numbers of choices for ordinal values. Other probability distributions have been developed to describe these situations, and they may have more descriptive parameters than just the mean and variance to completely determine the shape and size of the plotted curve.

WHAT IS BEAUTY?

> The absence of flaw in beauty is itself a flaw.
>
> **—Havelock Ellis**

When we see, hear, smell, feel, or taste beauty, we seem to know it. We find some landscapes beautiful, or some sunsets, or the faces of some people, or certain paintings or sculptures. We listen to music that is beautiful, and certain pieces of music evoke deep emotion within ourselves. Foods are not always distinguished as either beautiful or not, but they can be either attractive or unattractive. Combinations of tastes and textures and aromas mark very attractive foods, and some national haute cuisines emphasize contrasts to enhance attractiveness.

We know what it is when we experience it, but just exactly what is beauty?

Beauty has not been completely defined, and it is often personal. We do know this, however: beauty is formed from some intermediate states between completely predictable and completely random extremes.

Completely predictable landscapes, for instance, are boring, very uninteresting. Completely predictable music is tedious, and, for that reason, many people cannot tolerate music from minimalist composers (some compositions by Phillip Glass come to mind as examples). Completely predictable food (tofu, grits, or milk perhaps) has nothing in it to make it interesting, and so it is best used in combination with other foods.

Completely random sights, sounds, tastes, smells, or feels are too chaotic for patterns to be discerned. They cannot be figured out, and so are relegated as noise. There is no information for us in a completely random input, so it is dismissed as of no interest.

In between these extremes, however, are experiences that are mostly predictable, but with enough surprises to keep our attention. Think of some of the best jokes you have heard. You probably had tried to predict the punch line (which would have ruined the joke!), but the ending contained surprise—it was not predictable after all. Think also of some of the worst jokes you have heard: could it be that they didn't make sense to you, and you dismissed them as too random?

(*Continued*)

WHAT IS BEAUTY? (Continued)

Interpersonal relationships seem to follow the same trend. We seek people who are largely predictable in character but varied enough to be interesting. They are predictable, yet with a random, or surprise, element. The ratio of randomness to predictability has yet to be modeled and is probably itself a random variable.

The need for balance between order and randomness extends also to scientific studies. If the experimenter controls the experiment completely, then the outcome is known before the experiment is conducted. Conversely, if nothing is controlled, then there is no expected outcome (Shapin, 2004). Scientific discovery depends very much on study controlled enough to be useful yet with enough randomness to allow for unexpected results. This is called serendipity.

Music, food, landscapes, faces, jokes, aromas, textures—the really interesting and attractive ones contain elements of predictability coupled with random surprises. Hence, it seems that we are programmed to deal with some amount of randomness in our experiences. The same is true for other animals (see Section 6.22), and indicates that beauty is, indeed, in the eye of the beholder.

4.2.2 Self-Similar Data

> There's luck in odd numbers.
>
> **—Samuel Lover**

> There is divinity in odd numbers.
>
> **—William Shakespeare**

Some biological data do not fit any of these descriptions. These are data that are self-similar at different levels of magnification. Branching patterns of retinal nerve cells, blood vessels in the retina, and airways in the lungs have spatial self-similarity that seem to form repeating patterns from a larger scale to a smaller scale (see also Sections 7.4, Fractals, and 7.5, Tissues and Organs). The electrical voltage across a cell membrane of a T-lymphocyte and ionic current through cell membrane channels in pancreatic β cells are self-similar in time. These are called *fractals*, and have the same features over a broad range of sizes or times (Figure 4.2.3). Fractal data have no means, because the mean value changes as the scale of measurement changes; fractal data may have no variances because the variance values do not converge to any particular value as the sensitivity of the measurement changes.

4.2.3 Pseudorandom Data

> The fundamental difference between [engineering and science] is that science acknowledges uncertainty, while engineering avoids it.
>
> **—Michael A. Russell**

Some biological data (and perhaps much more than we think) is not truly random, but is determined in such a way that it appears to be random. An example of this is the time between beats of chick heart cells, where it can be determined that the period between

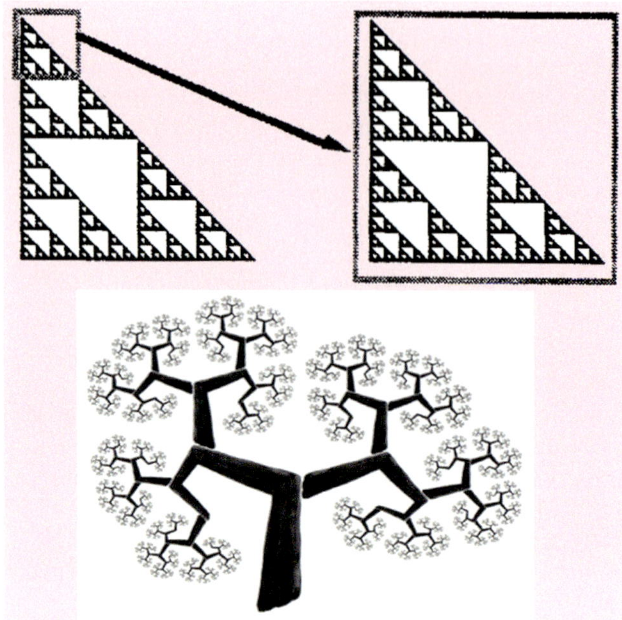

FIGURE 4.2.3
Fractals are those patterns repeated at different scales.

beats depends on the previous interbeat interval. Although the pattern of interbeat intervals looks random, it is actually deterministic. Such a system is called *chaotic*.

Scientists and engineers usually consider data from biological systems to be distributed randomly, unless shown otherwise. The use of the normal probability distribution pervades studies of all biological systems at all levels.

CHAOS

Chaos is an unfortunately chosen name for something that is not chaotic, anarchistic, or confused. Chaos is, instead, a term used to mean a response that appears to be random but which, in reality, is deterministic. As an example of this (Liebovitch, 1998), consider the case $x_{n+1} = 3.95\, x_n\, (1 - x_n)$. In this example, the value of x at any sample time $(n + 1)$ depends on the value of x at the previous sample time (n).

Comparing the time series of sampled data from $x_{n+1} = 3.95\, x_n\, (1 - x_n)$ to the time series of data from a truly random event shows that they both appear to be random. Both sets of data bounce around from higher to lower values, and no pattern is apparent. When, however, the values of x_{n+1} are plotted against the values of x_n, the random data show no relationship between the two, but the example data show a parabolic relationship. This plot is called a *phase space*, and suitably choosing the variables to be plotted reveals the underlying relationship.

Because the value of x_{n+1} depends on x_n, the phase space is said to be *one-dimensional*. A relationship between x_{n+2} and x_n would be called two-dimensional. Chaotic phase space is always low dimensional.

(*Continued*)

CHAOS (Continued)

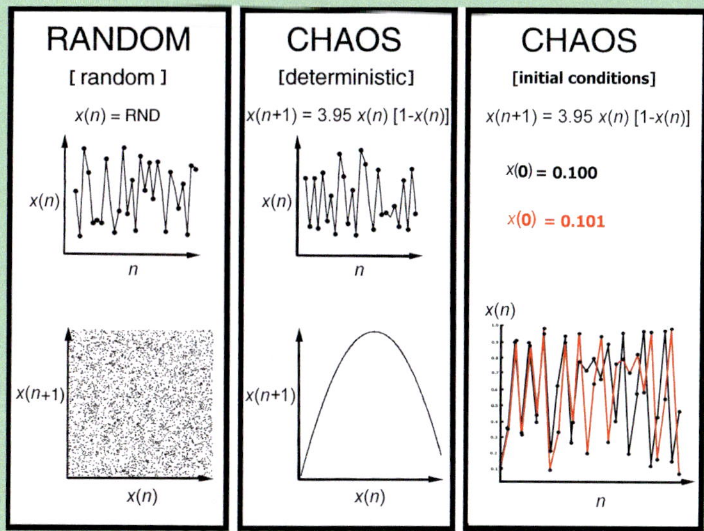

Characteristics of Chaos. The leftmost panel shows a completely random function, as demonstrated by the plot of $x(n + 1)$ with its neighbor $x(n)$. In the central panel is plotted a chaotic function, and the plot of $x(n)$ against $x(n + 1)$ shows that the two are related despite the appearance of the function with index, n, plotted earlier. The rightmost panel demonstrates the extreme sensitivity of a chaotic function upon initial conditions. Just a small difference of 1% in starting value resulted in entirely different values that differed by almost 700% at times.

Chaotic systems do not give random outputs, but they have a random appearance. The output is predictable (deterministic) if the underlying relationship is known.

Chaotic outputs are very sensitive to initial conditions. Nearly identical initial values can result in very different final values because chaotic system outputs depend on previous outputs and inputs.

Biological systems may act in a chaotic manner. The outcome of a human life, for example, depends not only on genetic predisposition, but also on environmental factors such as nutrition, education, and opportunity. So, the outcomes (accomplishments, life styles, number of offspring, etc.) may be dependent not only on initial conditions but also on conditions along the way.

It would not be in an organism's best interest to be totally at the mercy of environmental conditions, so biological systems are not totally chaotic. Whenever possible, biological units attempt to regulate their responses through active control mechanisms. It can thus be said that biological systems bring order out of chaos.

4.2.4 Statistics

What is, is, and it is impossible for the same thing to be and not to be.

—John Locke

Statistics is a mathematical tool used to separate random from nonrandom (usually intentional) effects. We have seen in Section 1.3 (Scientific Method) that the scientific method

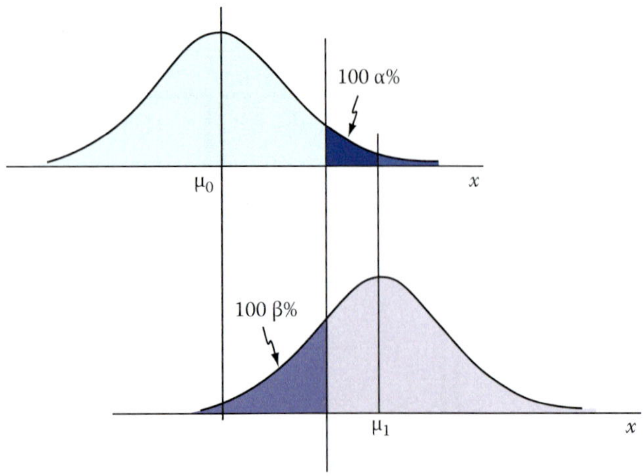

FIGURE 4.2.4
An illustration relative to Type I and Type II errors. The top frequency distribution is of the original untreated population. The bottom frequency distribution is of the population after application of the treatment. The treatment changed the mean, but not the variance, of the population. If there were no differences between means of untreated and treated populations, and data were obtained from the shaded area in the upper distribution, one would conclude that the treatment was effective when it really was not (Type I). Alternatively, if there were a real difference between treatment means, but the data were taken from the shaded portion of the lower distribution, then the conclusion that the treatment was ineffective would be false (Type II).

involves the repeated steps of hypothesis and testing. After the experimental data have been gathered, how can we be sure if the hypothesis has or has not been supported?

Experimental data drawn as a sample from an underlying population is used with the process of *induction* [see Sections 1.3 (Scientific Method) and 1.7 (Predictions)] to determine if the hypothesis is true for the entire population. That is, induction involves some elements of guesswork; statistical tests are used as a guide to make the best possible guess about the hypothesis. There are four possibilities (Figure 4.2.4):

1. The hypothesis is really true, and it is supported by the data. In this case, the hypothesis is accepted correctly.
2. The hypothesis is really false, and the data supports its falsity. In this case, the hypothesis is rejected correctly.
3. The hypothesis is really false, but the data indicates that it is true. In this case, the hypothesis is accepted in error. This is called a Type I error.
4. The hypothesis is really true, but the data does not support it. In this case, the hypothesis is rejected in error. This is called a Type II error.

Most statistical procedures are concerned with minimizing Type I errors, and the probability of the occurrence of a Type I error is determined prior to conducting the experiment (within the limitations of certain assumptions made about probability distributions of errors). The level of one Type I error being made for every 20 experiments ($p = 0.05$) has generally been accepted by the scientific community. This error rate has usually been designated by the symbol α.

Type II errors are more difficult to predict. The concept of a Type II error is important to determine the sample size of an experiment to detect a difference of stated magnitude.

Statisticians talk about unbiasedness and robustness of their methods. An *unbiased* procedure is supposed to be a fair and honest estimate of the parameter, whereas a *robust* procedure can be applied in nearly all cases without difficulty. Often, both unbiasedness and robustness depend on the assumptions underlying the development of the procedure.

Statistical tests depend upon mathematical models of expected results. Models related to the effect on the mean (average) of the population by the treatment are given in general by

$$\text{sample mean} = \text{population mean} + \text{treatment effect} + \text{error} \tag{4.2.2}$$

The statistical test to accept or reject the hypothesis (treatment has an effect) is based on a test of the sample mean.

There is error incorporated in every measurement. The error term in statistical tests is usually considered to be random, and the underlying population error is usually assumed to be a Normal distribution. If either of these conditions is violated, then modifications must be made in standard procedures. Statistics has matured enough to allow the development of many specialized tests.

The Student's t test is a widely used test of experimental means. For equal numbers of unpaired observations of two treatments, one of which may be no treatment at all, the sample means ($\bar{x}_1$ and $\bar{x}_2$) for the two samples (x_{1i} and x_{2i}) are calculated, where i stands for each individual measurement:

$$\begin{aligned}\bar{x}_j &= \frac{\text{sum of observations}}{\text{number of observations}}\\ &= \sum \frac{\bar{x}_{ji}}{n_j}.\end{aligned} \tag{4.2.3}$$

An estimate of the sample variance is also calculated, and the square root of the variance estimate, called the *standard deviation* (s), is used to calculate a t statistic

$$t_{\text{calc}} = \frac{|\bar{x}_1 - \bar{x}_2|}{s}. \tag{4.2.4}$$

This value is then compared to a table of t values. If the calculated t is greater than the tabled value, the treatment effect is considered to be statistically significant: the hypothesis is accepted with a $p = 0.05$ chance of being wrong.

A normal distribution is also characterized by its variance, and other statistical tests (the F test is one of these) may be used to check whether the treatment has changed the population variance from what it was before the treatment was applied. If the F test is used, the calculated value is compared to the tabled value in a manner similar to the t test mentioned previously. Based on this comparison, a decision can be made on the acceptability of the hypothesis that the treatment is effective.

The study of statistics is, for the most part, a study of various tests and their calculations. However, the application of statistics depends very strongly on a philosophical foundation. In order to avoid bias in the results, the appropriate statistical test must be decided upon before the data are seen. Once the data are known, even partly, it is tempting to apply a particular statistical procedure that seems to fit patterns in the data. Although there is no difference in the outcome of the calculation when done this way, the interpretation of the result can be erroneous. Studying the data before a specific statistical test is applied violates the assumption of random error.

Can what you don't know hurt you? It can, in statistics. Imagine the following scenario: there was a test of the effectiveness of two drugs, A and B. There were 2000 people who were scheduled to take the experimental drugs, but test subjects had to be identified as the opportunity arose; therefore, not all 2000 started at the same time. In the first 300 pairs of subjects, drug A proved much more effective than drug B in 275 of the pairs. Drug B even proved somewhat harmful in 100 of the test subjects. The results were so overwhelmingly in favor of drug A, the test was canceled, drug A was declared to be superior, and drug B was labeled as harmful.

However, if the test were continued, it might have been found that drug B would have been found to be more effective than drug A in 600 of the next 700 subject pairs. Not only that, but drug A may have been harmful for 200 subjects. Incorrect conclusions were drawn, for example there is a small, but finite probability that random events can form patterns some of the time. The philosophical assumptions of basic statistics were violated.

Sometimes statistical methods are used to determine overall treatment effects without accounting for individual or group variations. This is often the case with clinical trials of new drugs, where the results of interest are the benefits and risks on the average population. However, drugs are prescribed not for populations, but for individuals who may or may not react as the population does. Groups of individuals can be analyzed for different expected risk and benefit outcomes in the absence of the drug, and these different groups may give entirely different outcomes when given the drug. Most, if not all drugs (or surgical procedures, or medical devices, or treatments, etc.) convey both risks and benefits. If the overall result from a clinical drug trial shows no benefit from use of the drug, this may be because all participants in the trial showed no effects, but the no-benefit average results could also be derived from one subgroup benefiting very positively and another subgroup being harmed significantly. How the drug is administered as part of a treatment program can differ greatly as a result of this difference (Figure 4.2.5). Analysis of clinical trial results

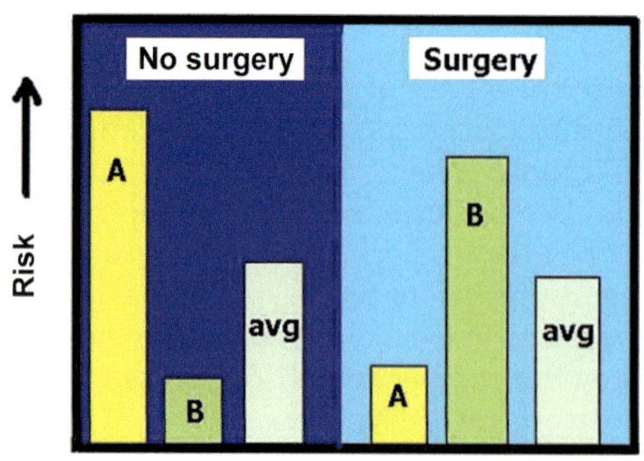

FIGURE 4.2.5
Here are expected outcomes for two patients at risk for a stroke. If no surgery is performed, patient A has a much higher risk than patient B of developing a stroke. If surgery is performed, the risk of a stroke as a result of the surgery is much higher in patient B than in patient A. Patient A benefits from the surgery, but patient B is exposed to higher risk. If these two patients are not distinguished, (average), then there is no indication that surgery should be performed on either.

by risk stratification is very rarely, if ever, done (Kent and Haywood, 2007). However, there is a need to be aware of this procedure in many experiments in biology.

Statistical procedures can also be used to design experiments to be most useful with the least cost. There are classical statistical designs to be used for field experiments, for multiple hypotheses to be tested simultaneously, and for control of *covariates*.

Covariates occur quite frequently, when levels of a second or third variable depend on the magnitudes of a first variable. For instance, growth rate varies directly, although not exclusively, with nutrition level. Disease and genetics are among the other variables that also affect growth rate. Statistical procedures can help separate various effects.

It is often also true that there is a training or a time-dependent effect that cannot be avoided. Just exposing a human, animal, or plant to repetitive experimental conditions results in adjustments that improve performance. This would be called a training effect, and training effects are common in athletic procedures, in learning experiments, and in skill acquisition. Time-dependent effects are always present with living things because they are constantly adjusting, growing, and maturing. Often, time-dependent effects can be ignored if the duration of the experiment is relatively short or if it is long enough to cover the entire life span of the object of the experiment. In between, however, time-dependent effects can appear to reinforce or to counteract effects caused by experimental treatments. One popular experimental design, the Latin Square, attempts to apportion time or training effects equally to all treatment levels (Figure 4.2.6).

Statistical ideas are often used to derive empirical equations from sets of data, and these equations form the basis for mathematical models used by biological engineers (see Section 1.4, Models). The method of least squares is a very popular procedure wherein a curve can be drawn through a set of data in an attempt to extract the essential nonrandom variation of data responding to various levels of an input variant (Figure 4.2.7). Although a very popular technique, the method of least squares is only valid for data that is linear and has a constant variance. For nonlinear relationships or nonconstant variances, modifications of the technique are available, but they, too, have limitations.

Subject number	Treatment level 1	2	3	4	5
1	1	2	3	4	5
2	5	1	2	3	4
3	4	5	1	2	3
4	3	4	5	1	2
5	2	3	4	5	1

FIGURE 4.2.6
One possible Latin Square arrangement. Entries in the interior cells of the table are experimental session numbers, given so that each treatment appears the same number of times in each row and column. This arrangement balances training and time-dependent effects as long as all subjects can be considered the same.

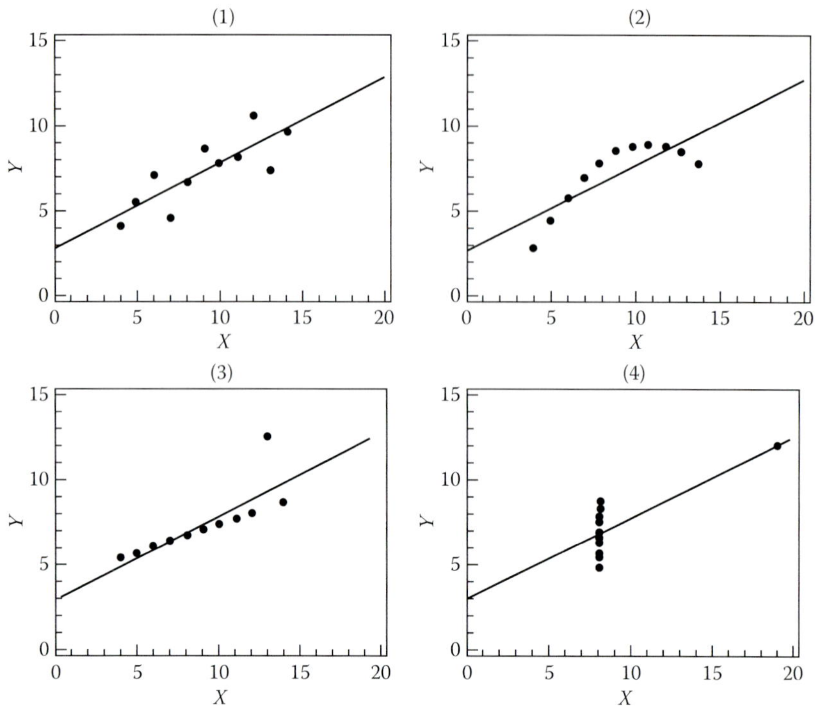

FIGURE 4.2.7
These are four graphs of four data sets with identical statistics, including the best-fit linear line through the data (Johnson, 1991). Only for the first (upper-left) data set is the line appropriate. For the second data set, a parabolic line would be correct, for the third set the outlier should be ignored, and for the fourth set there is no correct line. These graphs illustrate the point that data should always be seen before statistical procedures are blindly applied.

LINEAR LEAST SQUARES METHOD

(Johnson, 1991)

If we were to locate a line by eye through a data set y_i, we would probably try to balance the distance of the line from the data points. We would take distances on the upper side of the line and balance them with distances on the lower side. If we were really good, we would take all data points into consideration, and maybe even decide that a longer distance from the line to that data point way up there is balanced by the many shorter distances to the cluster of data points on the lower side of the line.

The least squares method is a mathematical way of doing what was just described. Vertical distances of the line to the data point $(y_i - \hat{y}_i)$ are used. (Here, y_i represents the ith data value, and $\hat{y}_i$ represents the estimated value of y that lies on the best fit line.) But because distances may be either positive or negative, and so may cancel,

(Continued)

LINEAR LEAST SQUARES METHOD (Continued)

the distances are squared, $(y_i - \hat{y}_i)^2$, to make the numbers all positive. Next, all data points are used: $\left(\sum(y_i - \hat{y}_i)^2\right)$, where Σ denotes the sum of the squared differences inside the parentheses. To find the parameter value to minimize this quantity, its derivative (see Section 4.3.1) is set to zero:

$$\frac{d}{da_j}\sum(y_i - \hat{y}_i)^2 = 0.$$

For instance, if $\hat{y} = a_0 + a_1x + a_2x^2$, and we need to find values of a_0, a_1, and a_2 to produce the best-fit line, then

$$\frac{d}{da_0}\sum(y_i - \hat{y}_i)^2 = \frac{d}{da_1}\sum(y_i - \hat{y}_i)^2 = \sum\frac{d}{da_2}(y_i - \hat{y}_i)^2 = 0$$

$$0 = \sum\frac{d}{da_0}(y_i - a_0 - a_1x_i - a_2x_i^2)^2$$

$$0 = -2\sum(y_i - a_0 - a_1x_i - a_2x_i^2)$$

$$0 = \sum y_i - \sum a_0 - \sum a_1x_i - \sum a_2x_i^2$$

$$0 = \sum y_i - Na_0 - a_1\sum x_i - a_2\sum x_i^2$$

where N = number of data points. Also

$$\frac{d}{da_1}\sum(y_i - \hat{y}_i)^2 = 0$$

$$0 = -2\sum x_i\,(y_i - a_0 - a_1x_1 - a_2x_i^2)$$

$$0 = \sum x_iy_i - a_0\sum x_i - a_1\sum x_i^2 - a_2\sum x_i^3$$

and

$$\frac{d}{da_2}\sum(y_i^2 - \hat{y}_i^2) = 0$$

$$0 = \sum x_i^2y_i - a_0\sum x_i^2 - a_1\sum x_i^3 - a_2\sum x_i^4$$

(Continued)

LINEAR LEAST SQUARES METHOD (Continued)

Thus, the set of least squares equations to determine a_0, a_1, and a_2 values is:

$$\sum y_i = Na_0 + a_1 \sum x_1 + a_2 \sum x_i^2$$

$$\sum y_i x_i = a_0 \sum x_i + a_1 \sum x_i^2 + a_2 \sum x_i^3$$

$$\sum y_i x_i^2 = a_0 \sum x_i^2 + a_1 \sum x_i^3 + a_2 \sum x_i^4$$

These can either be solved simultaneously, or more likely, be solved using matrix methods. By noting the patterns that occur in these equations, they can be written without resorting to differentiation.

Cramer's rule states that the value of a variable in a set of simultaneous equations can be determined by evaluating the value of the coefficient matrix with the column corresponding to the variable of interest, replaced by the last column of the augmented matrix (Sokolnikoff and Redheffer, 1958). For three or fewer variables, this process is particularly easy. The three variables in the set of equations earlier are a_0, a_1, and a_2. In explicit form, the determinant of the coefficient matrix becomes:

$$D = N\left[\sum x_i^2 \sum x_i^4 - \left(\sum x_i^3\right)^2\right] - \sum x_i \left[\sum x_i \sum x_i^4 - \sum x_i^3 \sum x_i^2\right]$$
$$+ \sum x_i^2 \left[\sum x_i \sum x_i^3 - \left(\sum x_i^2\right)^2\right]$$

For a_0,

$$N_0 = \sum y_i \left[\sum x_i^2 \sum x_i^4 - \left(\sum x_i^3\right)^2\right] - \sum x_i \left[\sum y_i x_i \sum x_i^4 - \sum x_i^3 \sum y_i x_i^2\right]$$
$$+ \sum x_i^2 \left[\sum y_i x_i \sum x_i^3 - \sum x_i^2 \sum y_i x_i^2\right]$$

and $a_0 = N_0/D$.
For a_1,

$$N_1 = N\left[\sum y_i x_i \sum x_i^4 - \sum x_i^3 \sum y_i x_i^2\right] - \sum y_i \left[\sum x_i \sum x_i^4 - \sum x_i^3 \sum x_i^2\right]$$
$$+ \sum x_i^2 \left[\sum x_i \sum y_i x_i^2 - \sum y_i x_i \sum x_i^2\right]$$

and $a_1 = N_1/D$.

LINEAR LEAST SQUARES METHOD (Continued)

For a_2,

$$N_2 = N\left[\sum x_i^2 \sum y_i x_i^2 - \sum x_i^3 \sum y_i x_i\right] - \sum x_i \left[\sum x_i \sum y_i x_i^2 - \sum y_i x_i \sum x_i^2\right] + \sum y_i \left[\sum x_i \sum x_i^3 - \left(\sum x_i^2\right)^2\right]$$

and $a_2 = N_2/D$.

Notice that each term in these expressions has the same x exponent and the same y exponent. Notice further that the exponents on x and y correspond to the units for the variables a_0, a_1, and a_2.

Example 4.2.1: Meaning of the Mean

When looking at the normal, or Gaussian, distribution, it is the central hump that beckons for the most attention. It is there that the largest number of individuals is found, and these individuals comprise the group that is considered average, or normal, or usual. Humans that have physical or mental characteristics clustered around the mean are usually easily accepted by others and share many of these characteristics with their friends and acquaintances. This is the situation of the *common man*, of whom Abraham Lincoln said that God must have loved them, for he made so many of them.

It is the tails of the distribution that contain interesting and influential individuals. People with disabilities are at one end of the distribution, and only recently, physical access to public facilities has been guaranteed by the US legal system through the Americans with Disabilities Act (ADA). Medicines have been developed at dosages that must be effective for a vast majority of the population, and that means for people including those at the least-sensitive tail of the population. Many medicines, therefore, are often stronger than needed for the vast majority of people.

Example 4.2.2: Protecting Against Hyperthermia

Core body temperatures of 40° have been considered to be where there are 50% heat casualties in armed combat. In civilian occupations, this high rate of heat casualties is not even thinkable, much less tolerable. Thus, government regulations meant to protect workers must use a more stringent standard, one based on protecting a higher percentage of the population. In the United States, excessive heat strain has been defined as core body temperature greater than 38.5°C for medically selected and acclimatized personnel or greater than 38°C in unselected, unacclimatized workers.

Example 4.2.3: Digestibility of Corn Silage in Sheep and Steers (Steele and Torrie, 1960)

Data for digestibility of corn silage (chopped and preserved corn plants) are given for seven sheep and six steers, as follows. Determine if there is a statistically significant difference between the two types of animals.

Digestibility of Dry Matter, Feed Corn Silage, in Percent

x_{1j} **(Sheep)**	x_{2j} **(Steers)**
57.8	64.2
56.2	58.7
61.9	63.1
54.4	62.5
53.6	59.8
56.4	59.2
53.2	

Basic sums

$$\sum x_{1j} = 393.5 \qquad \sum x_{2j} = 367.5$$

$$\sum x_{1j}^2 = 22,174.41 \qquad \sum x_{2j}^2 = 22,535.87$$

$$\bar{x}_1 = 56.21\% \qquad \bar{x}_2 = 61.25\%$$

Calculation of sample variances

$$\begin{aligned}\sum (x_{1j} - \bar{x}_1)^2 &= \sum x_{1j}^2 - \frac{\left(\sum x_{1j}\right)^2}{n_1} \\ &= 22,174.41 - 22,120.32 \\ &= 54.09 \\ &= (n_1 - 1)s_1^2\end{aligned}$$

$$\begin{aligned}\sum (x_{2j} - \bar{x}_2)^2 &= \sum x_{2j}^2 - \frac{\left(\sum x_{2j}\right)^2}{n_2} \\ &= 22,535.87 - 22,509.37 \\ &= 26.50 \\ &= (n_2 - 1)s_2^2\end{aligned}$$

Estimating the common variance

$$\begin{aligned}s^2 &= \frac{\sum (x_{1j} - \bar{x}_1)^2 + \sum (x_{2j} - \bar{x}_2)^2}{(n_1 - 1) + (n_2 - 1)} \\ &= \frac{54.09 + 26.50}{6 + 5} \\ &= 7.33\end{aligned}$$

The common standard deviation for the difference between the two means is

$$s_d = \sqrt{s^2 \frac{(n_1 + n_2)}{n_1 n_2}}$$
$$= \sqrt{7.33 \frac{(7+6)}{42}}$$
$$= \sqrt{2.27}$$
$$= 1.51\%$$

The t statistic is

$$t = \frac{(\bar{x}_1 - \bar{x}_2)}{s_d}$$
$$= \frac{56.21 - 61.25}{1.51}$$
$$= \frac{-5.04}{1.51}$$
$$= -3.33$$

The number of degrees of freedom (*df*) for this example is the total number of data points less the two means estimated:

$$df = (n_1 - 1) + (n_2 - 1) = 13 - 2 = 11$$

Next, the calculated value of t is compared to tabled values, using $\alpha = 0.05$, and considering that either value could have been larger than the other, is called a two-tailed test.

An excerpt from a table of t values (Steele and Torrie, 1960) is given here:

	Probability of a Larger Value of t		
df	**0.1**	**0.05**	**0.02**
1	6.314	12.706	31.821
3	2.353	3.182	4.541
5	2.015	2.571	3.365
8	1.860	2.306	2.896
10	1.812	2.228	2.764
11	1.796	2.201	2.718
12	1.782	2.179	2.681

Because the magnitude of the calculated t value (−3.33) is greater than the tabled value (2.201) with $\alpha = 0.05$ and $df = 11$, the two means are considered to be statistically different.

Example 4.2.4: Elastic Properties of Heart Muscle

In Figure 4.2.8 is shown a diagram of the isometric elastic properties of heart muscle for a 10 kg dog during systole (Johnson, 1991). Develop an equation to describe the pressure–volume relationship of the muscle.

Solution

A linear least-squares quadratic polynomial equation will be developed. Data were obtained from the curve. For purposes that will be explained later, volume data will be limited to the range of 0–29.2 mL (milliliters). For consistency of notation, pressure will be designated as y and volume as x.

	x	y	x^2	x^3	x^4	yx	yx^2
	0.0	0	0.00	0.00	0.00	0.00	0.00
	2.0	10	4.00	8.00	16.00	20.00	40.00
	5.3	20	28.09	148.88	789.05	106.00	561.80
	9.0	30	81.00	729.00	6561.00	270.00	2430.00
	14.2	40	201.64	2863.29	4065.69	568.00	8065.60
	29.2	40	852.64	24897.09	726994.97	1168.00	34105.60
sums	59.7	140	1167.37	28646.26	775019.71	2132.00	45203.00

The curve up to the point where the last data point was taken appears to be an inverted parabola; therefore, a quadratic polynomial will be chosen to represent the curve. From equations in the box on linear least squares,

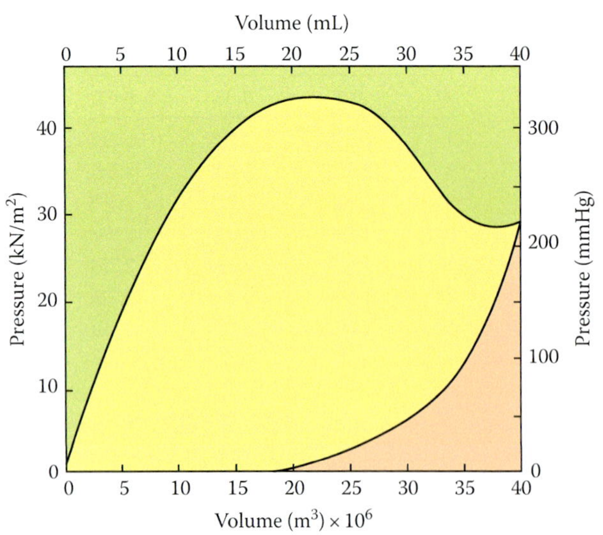

FIGURE 4.2.8
Isometric elastic properties of heart muscle for the 10 kg dog during systole. (Redrawn from Johnson, A.T., *Biomechanics and Exercise Physiology: Quantitative Modeling*, CRC Press/Taylor & Francis Group, Boca Raton, FL, 2007.)

$$\begin{aligned} D &= 6[(1167.37)(775019.71)-(28646.26)^2] \\ &-59.7[(59.7)(775019.71)-(28646.26)(1167.37)] \\ &+1167.37[(59.7)(28646.26)-(1167.37)^2] \\ &= 144512317.53 \end{aligned}$$

$$\begin{aligned} N_0 &= 140[(1167.37)(775019.71)-(28646.26)^2] \\ &-59.7[(2132.00)(775019.71)-(28646.26)(45203.00)] \\ &+1167.37[(2132.00)(28646.26)-(1167.37)(45203.00)] \\ &= 133483814.55 \end{aligned}$$

$$\begin{aligned} N_1 &= 6[(2132.00)(775019.71)-(28646.26)(45203.00)] \\ &-140[(59.7)(775019.71)-(28646.26)(1167.37)] \\ &+1167.37[(59.7)(45203.00)-(2132.00)(1167.37)] \\ &= 593664070.87 \end{aligned}$$

$$\begin{aligned} N_2 &= 6[(1167.37)(45203.00)-(28646.26)(2132.00)] \\ &-59.7[(59.7)(45203.00)-(2132.00)(1167.37)] \\ &+140[(59.7)(28646.26)-(1167.37)^2] \\ &= -13715380.27 \end{aligned}$$

$$a_0 = \frac{N_0}{D} = 0.92$$

$$a_1 = \frac{N_1}{D} = 4.11$$

$$a_2 = \frac{N_2}{D} = -0.095$$

The least squares equation is thus,

$$y = 0.92 + 4.11x - 0.095x^2$$

To check this equation, pressure data from the curve were compared with pressure data calculated by means of the equation:

x_{curve}	y_{curve}	$y_{predicted}$
0	0	0.92
2.0	10	8.7
5.3	20	20.0
9.0	30	30.1
14.2	40	40.0
29.2	40	39.6
36.0	30	25.4
40.0	30	12.9

The last two data points are outside the range of the original data, and the predicted pressure is very inaccurate. This illustrates one problem with polynomial fits of data: one must be very careful to avoid extrapolation outside the original range.

APPLICATIONS AND PREDICTIONS

1. As the precision of the measurement becomes greater and greater the probability of two measurements being exactly equal will decline.
2. 68% of a normal population will fall within ±1 standard deviation from the mean.
3. As the number of measurements increases, the probability data will look more and more like a Gaussian distribution.
4. Average measurements will apply only to a small number of individuals.
5. No matter how a system is designed to work with living organisms, it will occasionally be inadequate because of random variation.
6. Interactions between macromolecules without the presence of enzymes will depend on random collisions.
7. Increasing the number of observations will decrease the uncertainty of the measurement.
8. Many biological activities will be conditional on the presence or absence of other factors, and therefore will not be independent.
9. Fetal development is an example of a chaotic system.
10. The larger is the random component of a measurement, the more measurements must be taken.

4.3 Calculus

A mathematician is a machine for turning coffee into theorems.

—Paul Erdos

One of the greatest advances in mathematics came about with the formulation of methods of calculus. Two men working independently, Isaac Newton and Gottfried Wilhelm Leibniz, discovered calculus nearly simultaneously and the resulting feud over which of the two would be known as the inventor of calculus lasted for many years (Hellman, 1998). Calculus is the branch of mathematics that deals with infinitesimally small changes and the infinite sum of such changes. Thus, calculus deals with things in motion, not only mechanical motion, but also anything that changes with time or space. Because biological systems vary with time and space, calculus has important biological applications.

4.3.1 Derivatives and Differential Equations

Mathematics possesses not only truth, but supreme beauty.

—Bertrand Russell

Derivatives are rates of change of a dependent variable with respect to one or more independent variables. We have already seen derivatives in Section 2.1.2, where capacity was related to the rate of change of voltage with time. Inertia, in Section 2.1.3, was seen to be related to the time rate of change of current.

An equation that contains at least one term with a derivative is called a *differential equation*. If there is but one independent variable, and the derivative appears only with an exponent of 1.0, then the differential equation is called a *linear ordinary differential equation*. If there is more than one independent variable, then the differential equation is called a *partial differential equation*.

4.3.2 First-Order Equations

> It was ultimately recognized, as Poincaré pointed out, that *a complete conspiracy is itself a law of nature*, that it is not possible to discover an ether wind by *any* experiment; that is, there is no way to determine an absolute velocity.
>
> **—Richard Feynman**

Differential equations may be either *first order*, or *second order*, (third order and others are also possible, but less likely), depending on the derivative levels that appear in the equation. For instance, a differential equation commonly appearing in biological systems is the first order constant coefficient linear ordinary differential equation

$$x + \tau \frac{dx}{dt} = 0 \tag{4.3.1}$$

This equation has two terms, a term depending directly on the variable x, and a term involving the rate of change of x with respect to time (dx/dt). Many biological receptors, for instance, have an output that depends on the level of the input and also on the rate of change of the input with time. The variable x may stand for temperature, glucose level, or light, for example (see Section 6.20.1).

If we try as a solution, $x = e^{-t/\tau}$, where e is the base of Napierian logarithms ($e = 2.718\ldots$), then $dx/dt = -(1/\tau)e^{-t/\tau}$, and

$$x + \tau \frac{dx}{dt} = e^{-t/\tau} + \tau(-\frac{1}{\tau})e^{-t/\tau} = 0 \tag{4.3.2}$$

Thus, the equation is satisfied by the trial solution. Situations where the response $x = e^{-t/\tau}$ are called *exponential* responses, and are very common in biology at all levels.

4.3.3 Exponential Responses

> Thomas Malthus relied on an exponential-growth model to make his famous prediction about human population growth.
>
> **—Santiago Schnell**

The rate of unrestricted reproduction of cells depends on the number of cells present. When only a few cells are present, the rate of appearance of new cells is small. As more and more cells are produced, the rate of appearance of new cells increases. This has the effect of producing new cells at a faster rate until such time as reproduction is no longer unrestricted.

Before that happens, however, the growth in the number of cells is considered to be exponential, and can mathematically be expressed as

$$\text{number of cells} = (\text{number of cells at time } 0)e^{t/\tau} \quad (4.3.3)$$

where:

t stands for time

τ is said to be the time constant

When the value of time is equal to the value of the time constant, the ratio t/τ equals 1.0, and $e^{1.0} = 2.718$. As the value of t increases, $e^{t/\tau}$ becomes progressively greater (Figure 4.3.1).

Most biological responses are not as unbounded as this. Indeed, such unconstrained growth represents the lack of control or balance that characterizes most biological systems. Exponential processes are still important, but they act to diminish rather than increase the exponential term as time goes on. This can be accomplished mathematically by inverting the term $e^{t/\tau}$, which is equivalent to making the exponent negative, or $e^{-t/\tau}$. As time increases, the magnitude of $e^{-t/\tau}$ becomes smaller (Figure 4.3.1).

We may consider cell death rather than growth. If the rate of cell death is proportional to the number of cells present, then

$$\text{number of cells} = (\text{number of cells at time } 0)e^{-t/\tau} \quad (4.3.4)$$

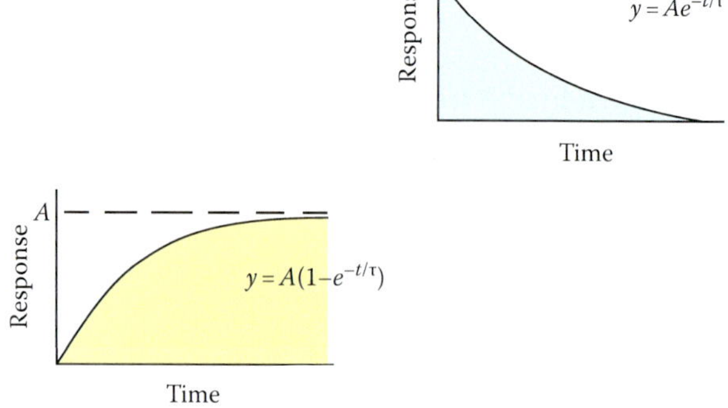

FIGURE 4.3.1
Exponential curves. The upper left curve is an unbounded exponential curve where t/τ is positive. This curve can be used to represent the unrestricted growth or death of cells. The middle curve is exponentially decreasing, and represents some kind of biological decay. The lowest curve is an exponential response to a step input, and is very commonly seen in biology when conditions change suddenly.

The term $e^{-t/\tau}$ never really assumes a value of 0, except abstractly when time is infinite. Because there cannot be fractional numbers of cells present, when Equation 4.3.4 predicts less than one cell, there will be a small probability that a cell survives, but this probability decreases as time goes on. This fact is important in processes that sterilize food or medical products (see Section 6.23, Death).

Biological systems are sometimes asked to respond to sudden changes from one level to another. Such is the case for a sudden change in environmental temperature or a sudden change in peripheral vascular resistance or even a sudden change in interstitial potassium ion concentration. The biological response appears to be exponential in nature, but rather than vary between extremes, as given by Equations 4.3.3 and 4.3.4, the response varies exponentially between one level and another. The equation that expresses this response is

$$\text{response} = \text{level 1 response} + (\text{level 2 response} - \text{level 1 response})\ (1 - e^{-t/\tau}) \qquad (4.3.5)$$

When time is 0, the term $(1 - e^{-t/\tau})$ is $(1 - e^{-0/\tau})$, or 0, and the response equals the level 1 response. When time is infinity, the term $(1 - e^{-t/\tau})$ is 1.0, and the response equals the level 2 response.

4.3.4 Second-Order Equations

> Mathematics possesses not only truth, but supreme beauty—a beauty cold and austere, like that of sculpture, without appeal to any part of our weaker nature, sublimely pure, and capable of a stern perfection such as only the greatest art can show.
>
> **—Bertrand Russell**

Some biologically important differential equations contain second derivatives, d^2y/dx^2, and are called second order differential equations. A simple example of these is

$$x + \frac{1}{A\omega^2}\frac{d^2x}{dt^2} = 0 \qquad (4.3.6)$$

The second derivative of a sine or cosine also contains a sine or cosine term:

$$x = A\sin\omega t \qquad (4.3.7a)$$

$$\frac{dx}{dt} = A\omega\cos\omega t \qquad (4.3.7b)$$

$$\frac{d^2x}{dt^2} = -A\omega^2\sin\omega t \qquad (4.3.7c)$$

Therefore, $x = A \sin \omega t$ can be seen to satisfy Equation 4.3.6. Biological responses displaying this type of behavior are *periodic* in nature.

4.3.5 Periodicity

> We often forget that the burden of the Biological Engineer is to have some level of understanding of all of biology, just as we must be familiar with general engineering approaches to problem-solving.
>
> **—Raj Tonnash**

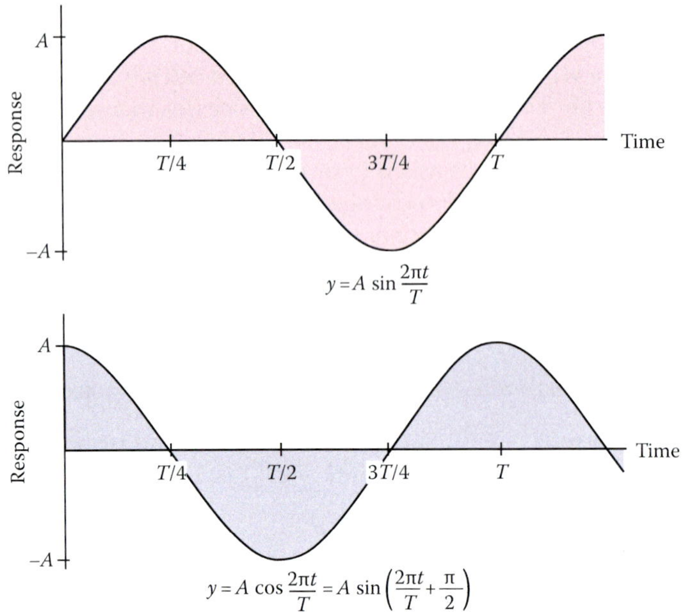

FIGURE 4.3.2
Sine and cosine waves.

Some biological responses are periodic, varying in a predictable manner between two limits. The firing of certain nerve cells and the levels of circulating hormones are two examples of periodic responses. Another example is predator–prey population dynamics (see also Section 6.20.3).

Oscillatory behavior (Figure 4.3.2) is described mathematically by combinations of sines and cosines:

$$\text{response} = (\text{magnitude})\left(\sin\frac{2\pi t}{T}\right) \tag{4.3.8a}$$

$$\text{response} = (\text{magnitude})\left(\cos\frac{2\pi t}{T}\right) \tag{4.3.8b}$$

where:

t is time

T is the period of the response

The difference between the response in Equation 4.3.8a and that in Equation 4.3.8b is that a sine wave varies from zero to a positive magnitude to zero to a negative magnitude, and returns to zero. The cosine wave appears to have the same shape except that it begins and ends at the positive magnitude rather than zero. Inverting the period T gives the *frequency* of the signal. The frequency of a simple sine or cosine wave is called the *fundamental frequency*. Frequencies related to the fundamental frequency by integer multiples are called *harmonic frequencies* (Figure 4.3.3). If the system is nonlinear, as many biological systems are, they may generate *subharmonics*.

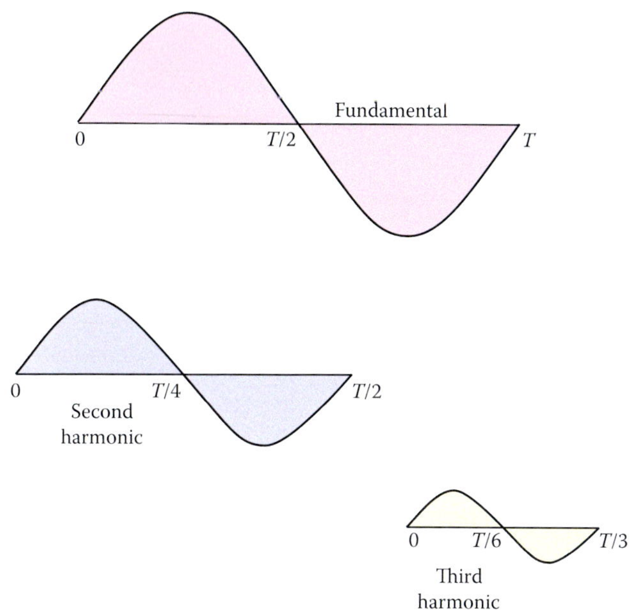

FIGURE 4.3.3
Fundamental and harmonics.

More complex periodic responses require the sums or differences of sines and cosines with different periods. *Fourier series* is a means to express any periodic response in terms of the sum of sine and cosine waves with fundamental and harmonic frequencies (Figure 4.3.4).

4.3.6 Nonlinear and Nonconstant Equations

> Round numbers are always false.
>
> **—Samuel Johnson**

The previous first and second order differential equations were extremely simple. They had derivatives that were raised only to the first power, they had constant coefficients, and they had only one independent variable, time. Not all equations used to describe physical or biological phenomena are so simple.

Many biological happenings are nonlinear. They may oscillate, but not with any set frequency. They may form exponential-like responses, but cannot be characterized by one time constant. Input–output relationships may not follow idealized forms. In these cases, the biological engineer must either resort to nonlinear equations or to numerical solutions to describe these phenomena.

A case in point is adaptive control systems, to be described in the next section. Most control systems are based upon first and second order differential equations, but adaptive systems can change their responses to satisfy special requirements. Biological systems are particularly adept at this: they can often change the type or magnitude of response when simple predetermined responses are no longer adequate. Systems of this sort can be

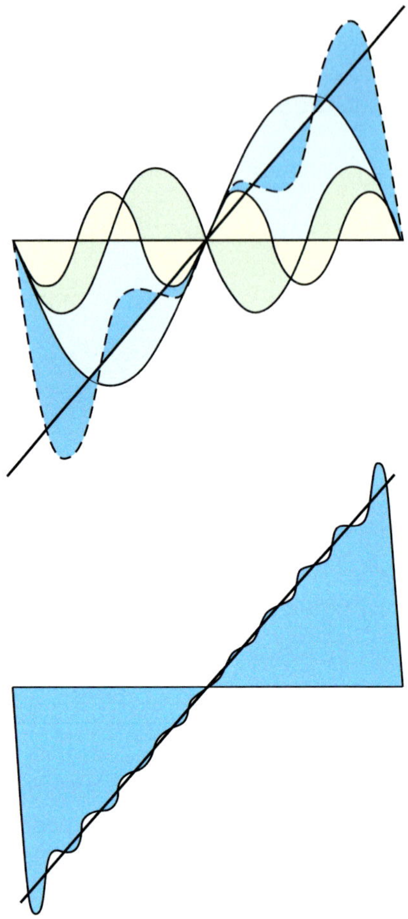

FIGURE 4.3.4
A complex waveform expressed as the sum of sines and cosines. At the top is a ramp function expressed as the sum (shown as a dotted line) of the fundamental, second, and third harmonics (shown individually as solid lines). At the bottom is the same function composed of ten harmonics. The more harmonics that are used, the closer will be the representation to the actual waveform.

described by differential equations the coefficients of which are themselves dependent on magnitudes of the variables. The general approach is this:

1. Determine by experiment the expected form of the response
2. Choose an equation that adequately matches the essence of the response
3. Incorporate the equation in a model that can be used to predict future responses

See Section 1.4, Modeling.

4.3.7 Integration

Nothing someone had ever measured was now or ever could be the same as before.

—Daniel Kehlmann

An *integral* is an infinite sum of infinitesimal elements. Integration is the process of determining the value of an integral.

Integrals are often used to determine areas under curves when the curve is determined by a known mathematical function. One example of where this is important is in the determination of the work required to move a fluid (air in the lungs, blood in the vasculature, or cytoplasm in the cell). The volume flow rate dV/dt is often known over time. The pressure in the system is many times given as the pressure developed in a compliance element, C, across a resistance element, R, and across an inertance, I (see Section 2.1, Effort and Flow):

$$p = \frac{dV}{dt}R + \frac{1}{C}\int \frac{dV}{dt}dt + I\frac{d^2V}{dt^2} \tag{4.3.9}$$

The rate of work, dW/dt, is $p(dV/dt)$; therefore, the amount of work is

$$W = \int \frac{dW}{dt}dt = \int p\frac{dV}{dt}dt = \int \left[R\left(\frac{dV}{dt}\right)^2 + \frac{1}{C}\frac{dV}{dt}\int \frac{dV}{dt}dt + I\frac{dV}{dt}\frac{d^2V}{dt^2} \right] dt \tag{4.3.10}$$

where the symbol $\int$ denotes integration.

Example 4.3.1: Human Population of the World

The world population in 1980 was about 4.432 billion and growing at a rate of 1.7% (Alocilja, 2002). Modeling world population growth as an exponential process,

$$N(t) = (4.432 \times 10^9)e^{0.017t}$$

where $N(t)$ = world population at any time (t, years after 1980). Calculating the world population in 1986, gives

$$N(1986) = (4.432 \times 10^9)e^{0.017 \cdot 6} = 4.908 \text{ billion people}$$

A census of people gives the world population in 1986 to be 4.9 billion people. Thus, the exponential model of world population appears to be a good fit. Exponential growth is also seen to be valid for microbes, plants, birds, insects, cancer cells, or any biological entity not limited by environmental resources.

Example 4.3.2: Classroom Ventilation

Studies of indoor air quality necessarily include ventilation of occupied spaces. Inadequate ventilation is often the cause of sick building syndrome, wherein occupants complain of various ailments associated with their presence inside the building.

Ventilation efficiency can be characterized by measured levels of metabolically generated CO_2 or by the concentration decay of a passive tracer gas. Sulfur hexafluoride (SF_6) is often used for this purpose, and it is introduced into the space under consideration. The faster the decline of SF_6, the greater the ventilation of the space.

Bartlett et al. (2004) have modeled CO_2 in naturally ventilated classrooms occupied by children. The only air exchange present in naturally ventilated classrooms is provided by infiltration and exfiltration, mostly through open doors and windows. Bartlett et al. (2004) gave the basic mass balance equation for CO_2 as

$$a(t)[c(t)-c_{out}]+b(n(t))=\frac{dc(t)}{dt}$$

where:

$c(t)$ is the classroom concentration of CO_2
c_{out} is the concentration of CO_2 in the atmosphere surrounding the classroom
$a(t)$ is the air exchange rate
b is the CO_2 generation rate, dependent upon $n(t)$, the number of people in the room

All parameters with "(t)" appended can vary with time.

Considering this first-order differential equation in light of the mass balance (see Section 2.2, Balances), we see that the terms in the earlier equation correspond to

$$(\text{rate of } CO_2 \text{ in} - \text{rate of } CO_2 \text{ out}) + \text{rate of } CO_2 \text{ generated} = \text{rate of } CO_2 \text{ stored}$$

And, by introducing the term

$$\Theta(t)=\frac{(c(t)-c_{out})}{c_{out}}$$

they transformed the aforementioned mass-balance equation into

$$\frac{d\Theta}{dt}+\alpha(t)\Theta=\beta(t)$$

which is formed from dimensionless terms and has the advantage that the variable Θ does not carry a specific set of units.

Note that this last equation would be of the standard form of a first-order differential equation, with an exponential function solution, except for the time dependencies of the terms $\alpha(t)$ and $\beta(t)$. These terms make this a nonlinear differential equation that most likely must be solved numerically.

Example 4.3.3: Respiratory Work Rate

The work of breathing represents an energy drain on the exercising human or animal. About 8%–10% of the body's oxygen consumption is spent for respiration during heavy exercise (Johnson, 2007). Many adjustments are made in the respiratory system to make respiration more efficient. Among them are the breathing airflow waveshape, which changes from sinusoidal at rest to trapezoidal during exercise.

The work of breathing can be calculated with a suitable model of the respiratory system. Resistance, compliance, and inertance are all present. Thus, the pressure that must be developed by the respiratory muscles is (Johnson, 1993)

$$p=R\dot{V}+\frac{V}{C}+I\ddot{V}$$

where:

p = pressure
R = resistance

C = compliance
I = inertance
V = lung volume
$\dot{V}$ = volume flow rate
$\ddot{V}$ = volume acceleration

Assuming a sinusoidal inhalation flow waveshape,

$$\dot{V} = \dot{V}_{max} \sin \frac{\pi t}{T} \quad 0 \le t \le T$$

where:
t = time
T = time for inhalation to occur
$\dot{V}_{max}$ = peak flow rate

Lung volume can be obtained by integrating flow rate over time:

$$V = \int_0^t \dot{V} dt = \int_0^t \dot{V}_{max} \sin \frac{\pi t}{T} dt = V_0 + \frac{\dot{V}_{max} T}{\pi}\left(1 - \cos \frac{\pi t}{T}\right)$$

where V_0 = initial lung volume.

Volume acceleration can be determined by differentiating flow rate:

$$\ddot{V} = \frac{d}{dt}\left(\dot{V}_{max} \sin \frac{\pi t}{T}\right) = \frac{\dot{V}_{max}\pi}{T} \cos \frac{\pi t}{T}$$

Work can be found by integrating the product of pressure and flow rate over time. Average work rate is the total work divided by inhalation time:

$$\dot{W} = \frac{1}{T}\int_0^T p\dot{V} dt = \frac{1}{T}\int_0^T \left(R\dot{V} + \frac{V}{C} + I\ddot{V}\right)\dot{V}\, dt$$

$$= \frac{1}{T}\int_0^T \left(R\dot{V}^2 + \frac{V\dot{V}}{C} + I\ddot{V}\dot{V}\right) dt$$

$$= \frac{1}{T}\left\{ \begin{array}{l} R\int_0^T \dot{V}_{max}^2 \sin^2 \frac{\pi t}{T} dt + \frac{1}{C}\int_0^T \left(\dot{V}_{max} \sin \frac{\pi t}{T}\right)\left(V_0 + \frac{\dot{V}_{max} T}{\pi}\right)\left(1 - \cos \frac{\pi t}{T}\right) dt \\ + I\int_0^T \left(\dot{V}_{max} \frac{\pi}{T} \cos \frac{\pi t}{T}\right)\left(\dot{V}_{max} \sin \frac{\pi t}{T}\right) dt \end{array} \right\}$$

$$= \frac{R\dot{V}_{max}^2}{2} + \frac{2\dot{V}_{max}^2 T}{\pi^2 C} + \frac{2\dot{V}_{max} V_0}{\pi C}$$

Work rates for other waveshapes can be determined in a similar manner. The sinusoidal waveshape has been found to be 6% more costly than a trapezoidal waveshape during light exercise and 9% more costly during heavy exercise (Johnson, 1993).

APPLICATIONS AND PREDICTIONS

1. The amount of drug remaining in a biological system will follow a decreasing exponential relationship.
2. The rate of growth of an individual can be described as an increasing exponential curve followed by a decreasing exponential curve.
3. Temperature receptors will respond exponentially to a sudden change in temperature.
4. Movement of the eyes will be oscillatory.
5. Most biological responses will be exponential, not oscillatory.
6. Sleeping patterns will be periodic.
7. Disease patterns are often periodic, based upon a periodic environmental fluctuation.
8. Harmonic analysis can be used for diagnosis of heart problems.
9. Membrane activities can be modeled with exponential equations.

4.4 Control Systems

> Because of the great complexity in a biological system, extended chains of mathematical reasoning are less relevant in biology than in physics and engineering....
>
> —**E. Körner and G. Matsumoto**

Biological stability (called *homeostasis*) is achieved through active control. Whether we consider the intracellular production of enzymes or the maintenance of whole-body posture, it is important for a living system to be able to sense the level of the controlled variable and then respond in a manner to correct discrepancies between desired and actual levels.

All control systems require

Sensors
Actuators
Controller
Means to communicate among these elements

4.4.1 Sensors

> When we have no control over our sense organs, we have no control over the world. We become a slave to it. At the beck and call of the world, we let our energies run dry, dissipating all vitality from our personality. What remains is but a carcass of the physical body: a mere biological unit moving about, with its physiological activities intact but with no personality to assert, plan, or achieve. If a society is made up of such exhausted and empty human beings, no scientist can help improve it, no politician can save it, no economist can develop it.
>
> —**Swami Chinmayananda**

Any control system that purports to respond to environmental stimuli must sense those stimuli. It may seem overly simple to realize that an environmental attribute does not really exist for that biological unit if the attribute cannot be sensed. Thus, humans cannot see ultraviolet radiation the way honey bees can, they cannot hear high frequency sound the way bats can, they cannot sense magnetic fields the way migrating birds can (birds have magnetite in their beaks to sense magnetic fields), and they are not aware of electrostatic fields as sharks are. To humans, lack of sensation indicates no information available, and we are not even aware of our unawareness. Signals outside our range of sensation do not exist for us.

Sensors take several forms (see Sections 6.19, Communication, and 6.20, Controls). Sensors, called *receptors* in biology, are *transducers* that change one type of signal into another more easily manipulated or communicated. Animal nervous systems are specialized for communication; therefore, many animal receptors are those that convert chemical, mechanical, or electromagnetic radiation into a form that can be transmitted by neurons.

Often the most critical element in control systems designed to be used by humans is at the sensor level. Sensors must be reproducible and stable over very long periods of time, and these two criteria are difficult to meet. Indeed, one of the biggest impediments of controlled drug delivery systems, as for insulin in diabetics, has been the inability to produce a reliable and reproducible glucose sensor. Biological receptors are often nonlinear; their output signals are not often linear representations of their environmental input signals. In many of the feedback control systems to be subsequently described, this nonlinearity is not a shortcoming. Rather, it is hardly noticed.

Biological receptors are also often sensitive to the rate of change of the stimulus as well as to the stimulus level. Receptor outputs, therefore, often have two thoroughly mixed components representing the level of the stimulus and the rate of change of that level. As we will see, rate of change information can convey an advantage to the biological system.

Surfaces of cell membranes abound with biomolecules called ligands that function normally as outer membrane receptors (Figure 4.4.1) or other structural features (Kim et al., 2004). Each receptor is configured with a pocket into which a specific external molecule can fit. Ligands act as links between the cell and external entities such as viruses, cells, and other bioactive molecules. These ligands are chemically tethered to surfaces

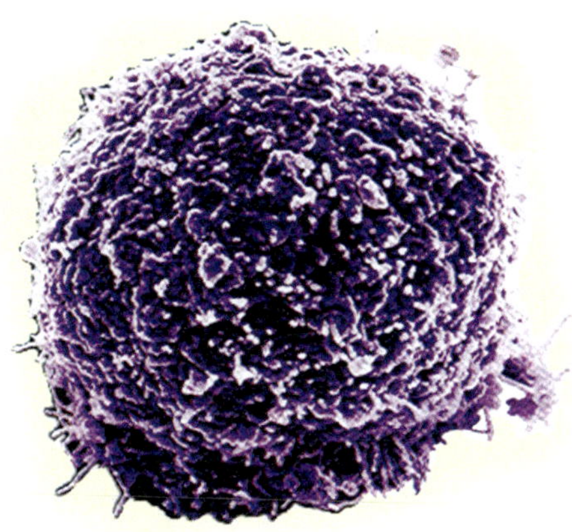

FIGURE 4.4.1
The surface of a cell is rough, and includes many molecules that act as receptors for other substances in the surrounding fluid.

by long (approximately 100 Å) organic linking molecules. Specific surface proteins mark some cells as targets of destruction by apoptosis, whereas other proteins protect those cells from targeting by the immune system. Ligands are important for pathogens to be recognized by the cell, and ligands are important for the cell to be recognized by pathogens. It is this property that can be useful in the design of biosensors. Capture events can be detected almost immediately by intrinsic fluorescence of microbes, toxins, or DNA. There is no waiting for microbial growth necessary for culture plate counts, or for enzyme reactions used in Enzyme-Linked Immunosorbent Assay (ELISA) (see Section 3.5, Reaction Rates), or for replication of DNA used in protein coupled reactors (PCR) (see Section 5.3.4, Genetic Expression). As long as the fluorescence can be detected, the result is real-time pathogen or contaminant detection. Modern drug therapies are now targeting these receptors or the chemicals (often hormones) that bind to them in order to produce desired results (Amábile-Cuevas, 2010).

HOW RECEPTORS WORK

There are many kinds of receptors in the body, and several different kinds of mechanisms that transform the adequate stimuli into a series of action potentials. For those receptors that receive energy in some form (for example, light, mechanical energy, or heat), the energy can be used directly to change the resting potential of the cell. It would seem likely that the energy would somehow upset the gel structure of the cell (Pollack, 2001) by agitation of the straight protein strands that form its foundation. Structured water between the strands would be disrupted and allow Na^+ ions to flood into the cell, at least locally. This will change the cell resting potential (this change is called the *generator potential*), moving the cell potential closer to the threshold potential of the neuron, and making it easier to fire an action potential. In the case of vision sensors, the addition of light actually decreases the frequency of action potentials (Zimmerman, 1995).

Some chemicals can pass into a receptor cell and either bind to specific receptors inside the cell or modify the formation of mRNA from DNA. However, most chemical receptors use a complex cascade known as a *second messenger* system. Most signal molecules are water soluble and too large to pass freely through the plasma membrane of the cell. In the surface of the membrane are located fixed receptor proteins, and these are shaped to be able to accommodate smaller molecules, called *ligands* or *agonists*, like a peg in a hole of complementary shape (see figure). Depending on the shape of the receptor protein, the receptor response may be more or less specific to a type of chemical. When the mating of receptor protein and ligand takes place, there is a conformal change in the protein.

Inside the membrane is one of a number of purine (a double-ringed organic structure) molecules based upon guanine (the same nucleotide base that helps form DNA and RNA). These are called G-proteins.

G-proteins can form complexes with phosphate compounds similar to ATP. The two complexes of importance here are GDP (two phosphates) and GTP (three phosphates). When GDP is bound to the G-protein, it is inactive; nothing happens; this is the resting state. When GTP is bound to the G-protein, then a series of actions follows.

(*Continued*)

HOW RECEPTORS WORK (Continued)

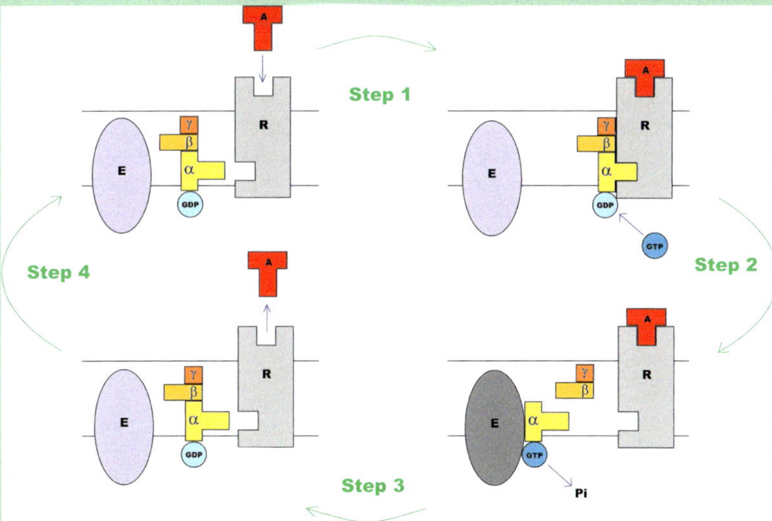

Second messenger receptors. When the agonist (A) binds to the receptor site (R), the receptor loses its affinity for GDP and instead binds GTP. When the GTP binds to the G protein (α, β, γ), the G protein subsequently dissociates and fixes to the target enzyme (E). The target enzyme is either activated or inhibited. Hydrolysis of GTP to GDP causes the Gα-GDP complex to lose its affinity for the target enzyme and it returns to the receptor site. The agonist leaves, returning the receptor to its resting state (Sleight and Lieberman, 1995).

Before the ligand mates with the receptor protein, there is no association between the receptor protein and the G-protein with its bound GDP. When the ligand mates with the receptor protein, and the protein changes shape, the G-protein associates with the receptor protein, and this causes the GDP to be replaced by GTP. Once GTP is attached, the G-protein dissociates from the receptor and binds to a target enzyme present in the cell. The enzyme, in turn, produces many more second messenger molecules that amplify the signal. There is enzymatic action that hydrolyzes the bound GTP to GDP to return the system to its resting state.

G-protein receptor systems are widespread in living systems. They are found in many animals and microbes. Both vision and smell in humans depend on G-proteins (Campbell et al., 1999), and glucagon receptors use the system to produce a second messenger called cyclic AMP (cAMP). Receptors for epinephrine, angiotensin, endorphins, and acetylcholine also use this G-protein second messenger mechanism (Sleight and Lieberman, 1995).

Structural analogs of the natural ligands that bind to receptors are called either *agonists* or *antagonists*, depending on their consequences. Agonists mimic the effects of the ligands; antagonists bind to the receptors but have no biological response (Sleight and Lieberman, 1995). Bacteria that cause cholera, pertussis, and botulism make their victims ill by producing toxins that interfere with G-protein function (Campbell et al., 1999). G-protein coupled receptors (GPCR) are prime targets for therapeutic drugs.

Example 4.4.1: Making Bitter Food Taste Better

The tongue is home to more than 10,000 taste buds. Each of these detects specific chemicals in food and drink. There are taste receptors for sweet, salty, bitter, savory, and sour tasting chemicals.

Not all people taste food in the same way. For some people, very sensitive bitter receptors make certain foods unpleasant to taste.

Aside from avoiding bitter foods, the only thing that has been done to make these foods more palatable is to add salt and sugar to mask the bitter flavors. Now, new chemicals can be added to the food to block the bitter taste receptors. With knowledge about how these receptors work, specific biochemicals with correct configurations can keep bitter flavors from triggering responses from their receptors (Lashinsky, 2007). Blocking receptors from responding to stimuli is one way to manage unpleasant or painful perceptions.

4.4.2 Actuators

Information is physical.

—**Rolf Landauer**

Actuators are the means for a control system to have an effect. Actuators may be local and mechanical, as are muscles, or they may be diffused and chemical, as are glucose respiration processes in cells.

Many of the more visible control actions are concerned with movement. Many microbes and most animals are capable of locomotion (see Sections 2.9, Fluids, 6.9, Movement, and 7.4.9, Walking and Running). The actuators in these cases are the two protein fibrils, actin and myosin, which slide past each other and shorten to produce movement. On a much longer time and space scale, plant populations move in response to climate changes, and, in this case, the actuators are the seeds that disperse in various ways.

MUSCLE TYPES

The contractions of all muscle cells depend on two proteins that appeared in very early life: *actin* and *myosin*. Actin and myosin filaments, in units called *sarcomeres*, slide past each other when a propagating phase change occurs.

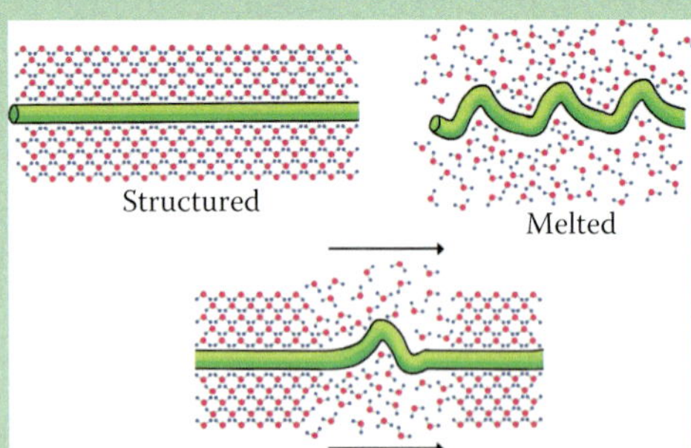

Actin can undergo a phase change from an extended state to a folded state. The surrounding water forms a gel structure in the extended state. It is thought that this or a similar phase change moves the actin filaments along the sarcomere. When actin moves, it displaces in multiples of the 2.7 nm cross-link spacing. Three different types of muscle filaments are capable of shortening. (Redrawn from Pollack, G. H., *Cells, Gels, and the Engines of Life: A New, Unifying Approach to Cell Function*, Ebner and Sons, Seattle, WA, 2001.)

(*Continued*)

MUSCLE TYPES (Continued)

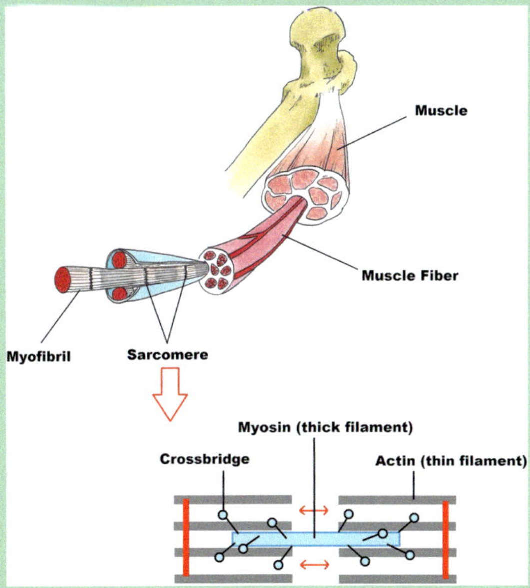

Sarcomeres are the basic muscle units.

Skeletal muscles contract when depolarized by neural signals transmitted at the neuromuscular junction. Each muscle fiber has one synaptic connection, but each motor neuron typically branches and controls several muscle fibers. The larger and more powerful muscles are controlled by neurons branched many times. The contractile apparatus composed of motor neuron and all the fibers it controls is called a *motor unit*.

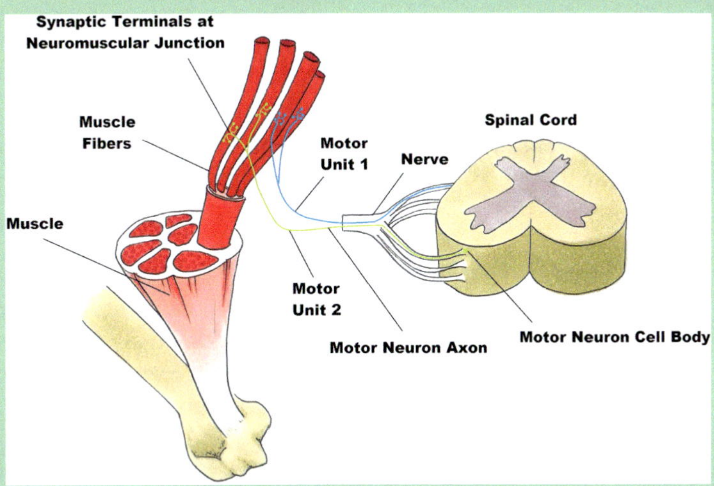

Organization of muscular motor units.

(Continued)

MUSCLE TYPES (Continued)

There are *fast twitch* and *slow twitch* skeletal muscle fibers specialized for different uses. Fast fibers are used for rapid, powerful contractions. They do not fatigue quickly in the absence of oxygen; they utilize anaerobic metabolism. Slow fibers are meant to sustain long periods of repeated contractions without fatiguing; however, they do require a steady supply of oxygen in order to maintain this action. They have many *mitochondria*, a rich blood supply, and an oxygen-storing protein called *myoglobin*. Fast twitch fibers are used for strong voluntary muscular movements such as running or flying; slow twitch fibers are used to maintain posture or to carry heavy weights.

Cardiac muscle (myocardium) is only found in the heart. It has the ability to generate action potentials and to pass depolarization (and subsequent contraction) from cell to cell without neural intervention. Cardiac muscle otherwise has some of the same internal cellular organization as skeletal muscle.

The actin and myosin of smooth muscle is organized differently from the striated (containing sarcomeres) skeletal and cardiac muscle cells. Smooth muscle cannot generate as much tension as striated muscle, but it can contract over a wider range of lengths. Smooth muscle is mostly found in the walls of hollow vessels of the digestive system, respiratory system, and blood vessels. There it can control flow into and through the vessels as it contracts. Smooth muscle in the urinary bladder can produce a nearly constant pressure on the fluid for a wide range of bladder volumes.

Some invertebrates contain muscles similar in many ways, but different in others. The flight muscles of insects, for instance, can spontaneously depolarize and contract very rapidly (Campbell et al., 1999). Clam shell muscles can clamp the shells closed with a low energy consumption for as long as a month (Campbell et al., 1999).

4.4.3 Communications

> When asked what distinguishes a functioning animal from an inviable heap of cells, most biologists would draw attention to the hormonal and nervous systems.
>
> **—Jonathan Cooke**

Without some means to transfer information between control system elements, there can be no precise control. Intact eyes without an optic nerve to convey visual information to the brain are no better than no eyes at all. Cellular insulin receptors without access to blood flow cannot function as intended. Likewise, communication among individuals in an ecological community is just as necessary as scouting reports to an army; together, the group can respond to the information obtained from the outside.

The importance of direct communication connections between sensors and neural perception or control centers can be illustrated with the human sense of taste. Taste receptors are found in the entire gastrointestinal tract as well as in the tongue (Davidson and

Riley, 2015). There are taste receptors also in the pancreas, lungs, and testes (Owen, 2015). These receptors are sensitive to the same chemicals as are the receptors in the tongue, but they are not connected to the taste centers of the brain, and so do not lead to taste perception.

There are two broad classes of communication in a complex animal organism: (1) Neural and (2) Humoral (chemical). Distinguishing between these in the traditional ways, humoral communication generally uses circulating fluids as the medium of intercourse. The circulating blood in animals, cytoplasmic streaming in single cells, or phloem fluids in plants are examples of these. Neural communication uses the nervous system, and can deliver messages to specific locations. The nervous system has been thought to deliver its messages by membrane depolarization of specific target cells. Actually, the nervous system is a chemical delivery system that is very specifically targeted. Only in nerves that end on muscle cells is membrane depolarization of most importance. For other cells, it is the neurotransmitters that are released at the ends of the nerves that initiate action. Thus, both neural and humoral communications mechanisms turn out to be chemical. Humoral communication differs from neural communication in that it is more general throughout an organism, and it is often slower to respond.

Plants and single-celled organisms, of course, do not have nervous systems; therefore, they rely on circulating chemicals, often hormones, to convey messages. This limits the speed at which they can respond to stimuli.

Ecological systems, also, do not communicate by nervous systems. However, there is an analogous system of individual-to-individual communication that enables targeted information transfer. A system of this kind can be as simple as a microbe on the skin of a human or as complex as a nation.

Communications mechanisms must be fast enough to enable the control system to respond in a timely manner. Humoral communications cannot be used for fast responses, and many unmyelinated nerve fibers are also too slow for many purposes. The organism that can transmit information and process it the fastest often has a survival and reproductive advantage; therefore, some specialized communication means (such as nerve fibers in myelin sheathes) have been developed.

Although they are often diagrammed as straight lines or wires, communications systems are much more complex and often display characteristics of a transmission line (Figure 4.4.2). Transmission lines have resistance, capacity, and inertia elements that slow and degrade a signal. This is true for both neural and humoral signals.

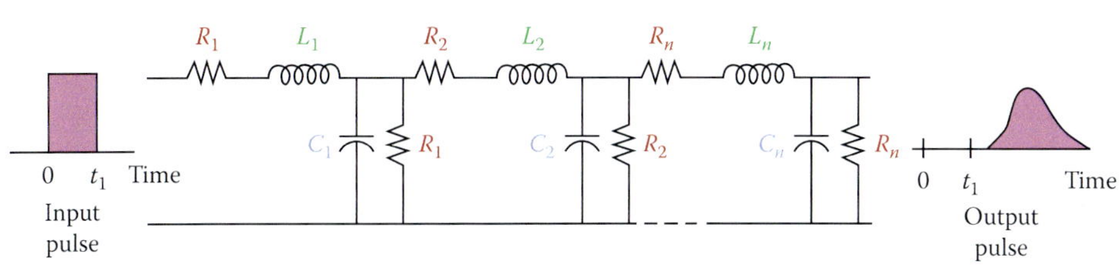

FIGURE 4.4.2
Representation of a transmission line with resistance, capacity, and inertia elements. If a sudden change is made on the left-hand side of the transmission line, the resistances degrade the signal and slow the flow variable, the capacity elements must be filled, and the inertia elements slow the rate of change. Thus, the transmission line delays the output from appearing, and degrades its sharpness.

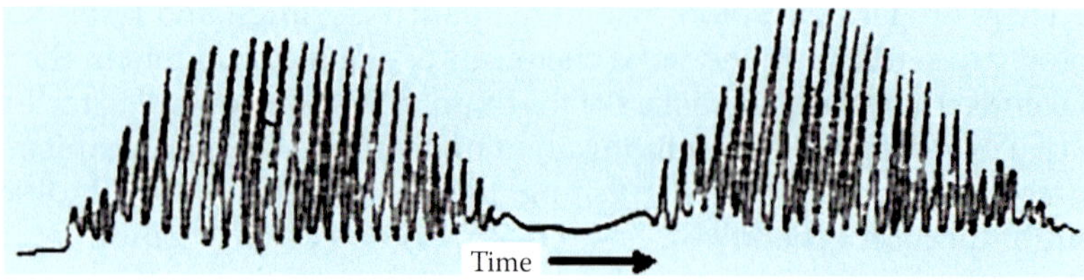

FIGURE 4.4.3
Cheyne-Stokes breathing, shown here as respiratory flow rate with time, is characterized by periods of intense breathing followed by no breathing at all. This condition is caused by a delay in the control system.

Communications delays can sometimes cause instability in a control system. Cheyne-Stokes breathing, characterized by periods of apnea interspersed with maximal respiration (Figure 4.4.3), has been found to be attributed to an abnormal delay in the respiratory controller (Hornbein, 1981). Delays, for whatever reason, cause instability because, by the time corrective action can be taken, the stimulus has already changed significantly.

AUTONOMIC NERVOUS SYSTEM

The peripheral nervous system (those nerves outside the central nervous system) of vertebrates is made up of both *sensory* (afferent) and *motor* (efferent) nerves (Figure). Each of these includes a bundle of nerve cells that connect to different sensors and actuators in the body. Motor nerves can be either *somatic* (mostly voluntary) or *autonomic* (mostly involuntary).

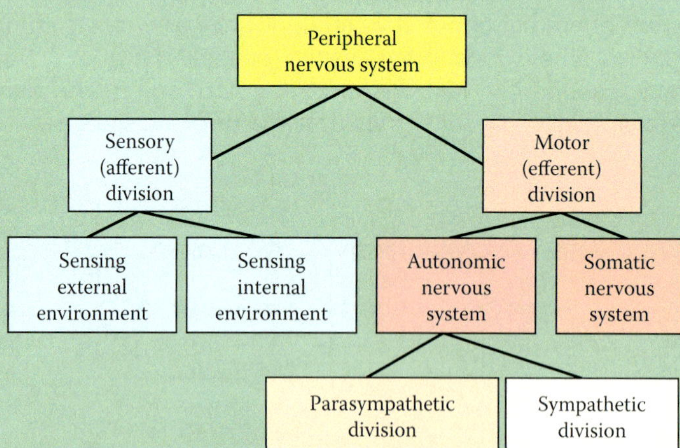

Diagram of the organization of the peripheral nervous system.

(*Continued*)

AUTONOMIC NERVOUS SYSTEM (Continued)

The autonomic nervous system is composed of two groups of nerves that have mostly (although not always) antagonistic effects.

The *sympathetic* system generally produces reactions to respond to emergencies or exigencies, the so-called "fight or flight reactions." The *parasympathetic* system promotes homeostasis, calm, and routine activities. Parasympathetic nerves originate in the brain or the sacral region of the spinal cord. They supply visceral structures in the face and head through the *oculomotor, facial,* and *glossopharyngeal* nerves.

The thorax and upper abdomen is supplied by the *vagus* nerve. The pelvic viscera are supplied by the pelvic branches of the second to fourth sacral spinal nerves. Parasympathetic neurotransmission uses *acetylcholine,* and are thus called *cholinergic* nerves.

Sympathetic nerves form *ganglia* (structures containing neuronal cell bodies) outside the spinal cord. Sympathetic neurotransmission uses *norepinephrine,* and are called *adrenergic.*

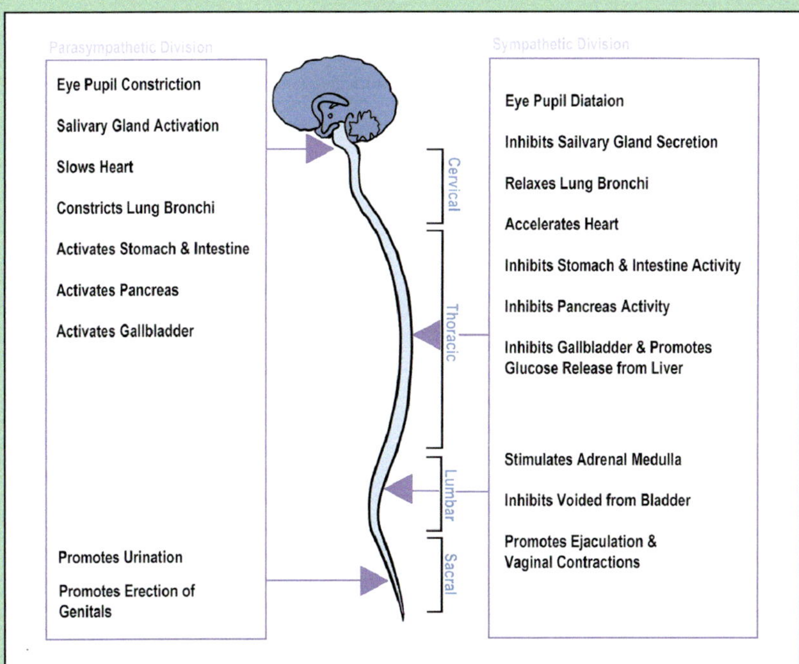

The autonomic nervous system.

ACTION POTENTIALS

Unlike epithelial and other types of cells that maintain a steady cell resting potential (see box, Section 3.8, Water Chemistry), the internal voltages of neurons and muscle cells can be altered by external events. In so doing, information can be transmitted along the cell. Neurons are particularly suited for this type of information transmission. They have relatively large cell bodies with branching, fiber-like extensions called *dendrites* to receive signals, and a long extension, called an *axon*, to transmit signals (Figure). The dendrites are located close to the axon terminals of one or oftentimes many other neurons; they function by receiving signals from the other neurons and integrating them in summation fashion. Some of these inputs may be excitatory, and some inhibitory; the dendrites receive these signals and their cellular potentials vary depending upon how many of each they receive within a short time.

The axon functions as a transmission line, bypassing the slow or unresponsive tissue between the cell body and the terminus of the axon. It is along the axon that the signal is transmitted. Some axons are short (in the central nervous system, or CNS), while some are very long. Some axons of the sciatic nerve extend a meter or more from the lower part of the spinal cord to the lower leg and foot. Here, the word "nerve" refers to a bundle of neurons, or nerve fibers, in a common sheath of connective tissue.

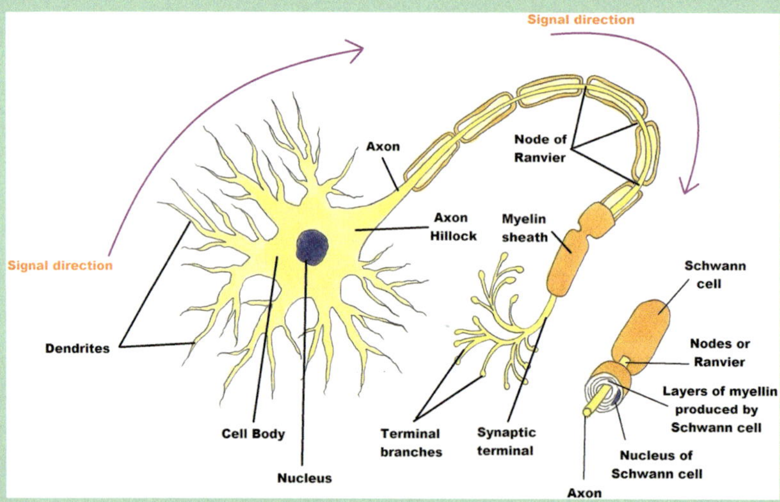

Schematic of a neuron.

The signal itself is a cell potential depolarization that travels in wavelike fashion along the axon. The cell resting potential of about –70 mV actually reverses polarity to become +30 to +35 mV for a very short time. This is called the *action potential*, and it moves from one end of the axon to the other (Figure). The action potential is an all-or-nothing event, similar to flushing a toilet; once a threshold voltage is reached, the action potential starts and can't be stopped until completion. Movement of the action potential is like the crowd doing the wave in a stadium, or like the peristaltic movement along the bowel.

(*Continued*)

ACTION POTENTIALS (Continued)

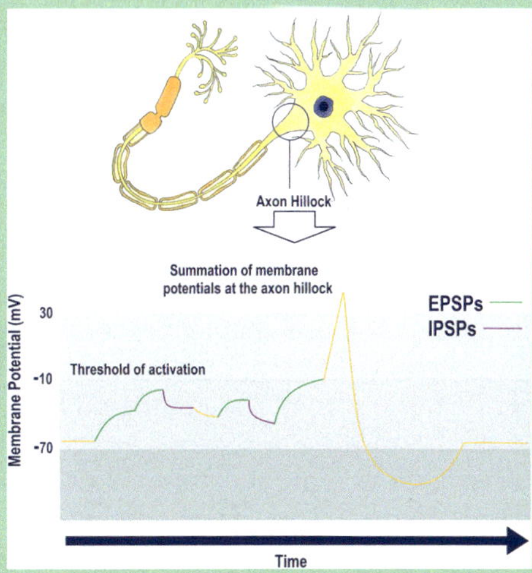

The resting potential of a neuron can either be increased or decreased by external events. If the resting potential decreases to a threshold value, an action potential is generated.

Behind the depolarization is a repolarization that actually hyperpolarizes the cell potential for a while. The action potential normally moves from cell body toward the axon terminus, but it could be propagated in the other direction. The hyperpolarization following the action potential, however, eliminates the possibility of back-propagation because it moves the cell potential farther from the threshold voltage. This results in a latent (refractory) period during which the cell cannot easily form an action potential. The strength-duration curve shows the transmembrane current required to initiate an action potential following a previous action potential (Figure). *Rheobase* is the minimum cell current that would initiate the action potential. *Chronaxie* is the time following an action potential for the current required to initiate the next action potential to be twice the rheobase value (Cuervo, 1976).

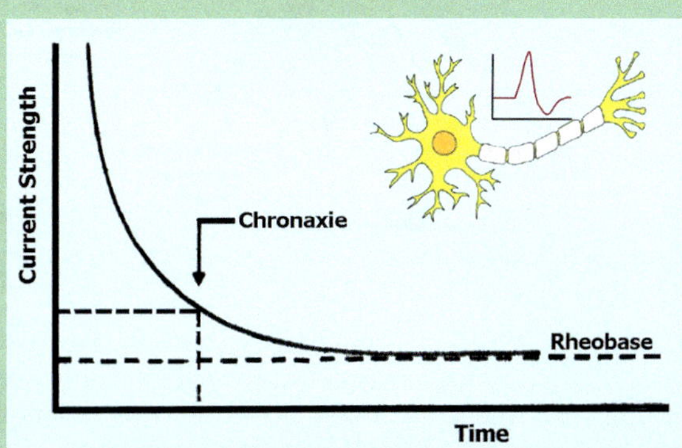

The current strength required to initiate an action potential decreases with time following the previous action potential (Cuervo, 1976).

(Continued)

ACTION POTENTIALS (Continued)

So how does the cell potential reach the threshold voltage? The junction between two neurons is called a *synapse*, and this junction is usually located at the terminus of a presynaptic neuron and the *axon hillock* of the postsynaptic neuron. The axon hillock is the location where the axon joins the neuronal cell body, and it is here that the action potential begins. A signal coming from the presynaptic neuron can trigger an action potential in the postsynaptic neuron by causing the cell potential of the postsynaptic neuron to reach threshold voltage.

There may be connections with several presynaptic neurons, and not all of these lead to action potential generation in the postsynaptic neuron (Figure). Those connections that favor depolarization (moving toward threshold voltage) are called *excitatory postsynaptic potentials* (EPSP); those connections that tend to hyperpolarize (move the cell potential farther from the threshold voltage) are called *inhibitory postsynaptic potentials* (IPSP). There may be both temporal summation and spatial summation of these postsynaptic potentials (Figure). If incoming signals occur from one presynaptic neuron fast enough, the signals can sum to have a larger effect than just one signal would have. If in this instance we consider just EPSPs, then this example of temporal summation will make easier the formation of an action potential in the postsynaptic neuron. Similarly, if signals from two or more presynaptic neurons arrive nearly simultaneously, they can sum to produce a larger effect than if they arrived separately.

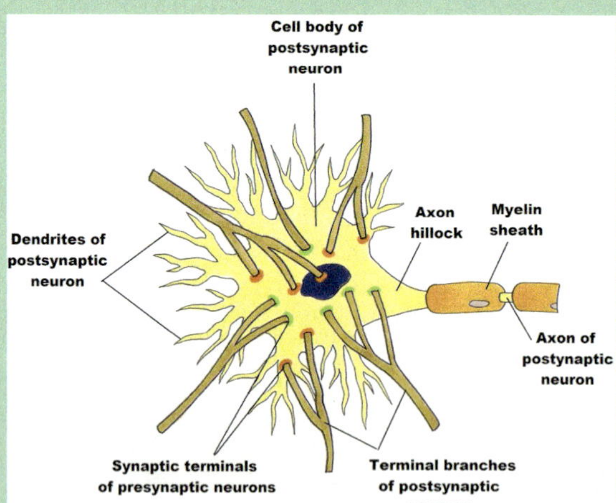

There are many possible dendritic connections to a neuron cell body (Campbell et al., 1999).

Although they cannot form an action potential, the dendrites can influence the cell potential at the axon hillock, and thus raise or lower the neuronal sensitivity to action potential formation. They do this by summing other inputs from other neurons that, again, may be either excitatory or inhibitory.

(Continued)

ACTION POTENTIALS (Continued)

Thus, the action potential is formed from spatial and temporal summation of EPSP and IPSP signals with graded input from the dendrites. This makes it possible that neurons can form complex responses from relatively simple input signals.

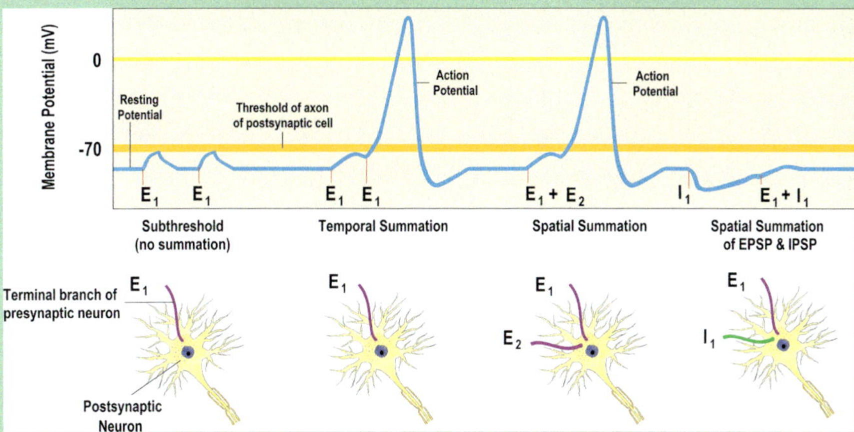

Both temporal and spatial summation of subthreshold potentials are possible (Campbell et al., 1999).

Axons exhibit all the electrical characteristics of a transmission line. In particular, they offer resistance to current flow during the action potential, and, because there is a separation between unlike charges inside the cell and outside the cell, they exhibit capacitance as well. You will recall that in the resting state there are more negative charges inside the cell than outside, and during the action potential there are more positive charges inside the cell than outside. Supplying the current through a resistor to reverse the charge on a capacitor takes some time, and follows an exponential time course (see Section Section 4.3.2, First Order Equations). The time constant of the exponential curve is given by the product of resistance and capacitance. The rate at which action potentials can be propagated along an axon is limited by the time constant of the axon.

For very long axons, especially those located outside the CNS, this time is too slow for adequate information transfer, so reducing the time constant would be of great survival benefit. Resistance can be reduced by increasing the size of the axon, because resistance is inversely related to the cross-sectional area through which current flows. Transmission speed can be increased from several centimeters per second in very thin axons to about 100 m/sec in the giant axons of invertebrates such as the squid and lobster (Campbell et al., 1999).

There are, also, means to reduce capacitance of the axon, and the means used in many vertebrate axons is to increase the distance between charges. Peripheral axons are enclosed in *myelin* sheaths formed by supporting cells called *Schwann cells*; some larger CNS axons are enclosed in myelin sheaths formed by *oligodendrocyte* cells. Myelin is a fatty substance that acts as an electrical insulator preventing current flow.

(*Continued*)

ACTION POTENTIALS (Continued)

There are breaks in the sheath, between supporting cells, called *Nodes of Ranvier*. These allow for currents to flow to support propagation of the action potential. The action potential, and associated currents, proceed by jumping from one node to the next (called *saltatory conduction*) at a rate of up to 150 m/sec.

Until this point, we have avoided details about the nature of the current flow and mechanisms causing the action potential. This is because certain mechanisms are generally accepted as explanations for the action potential, but, related to establishment of the cell resting potential (see box, Section 3.8), there are alternative theories.

Sodium, potassium, chloride, and calcium ions carry the currents associated with the action potential. The two that have seemed to be most important are Na^+ and K^+. According to the traditional explanation, the cell resting potential was maintained by a Na^+- K^+ ion pump (Figure).

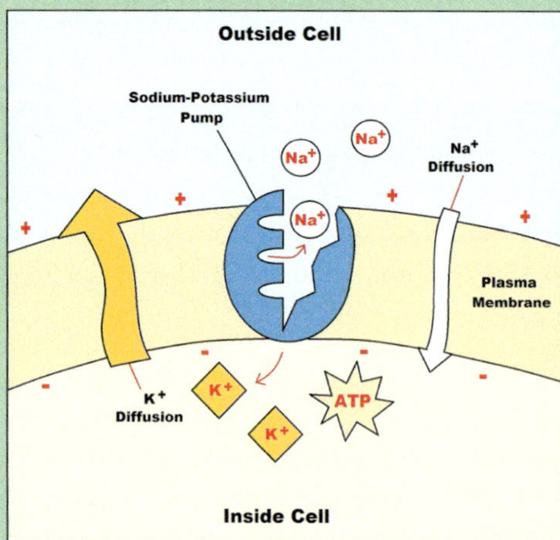

Traditional view of sodium and potassium movement across a cell membrane. Passive diffusion occurs because of concentration gradients across the membrane. A sodium-potassium pump, powered by ATP, moves sodium and potassium against their concentration gradients (Campbell et al., 1999).

When the threshold voltage was reached, or the membrane was disturbed in some way, Na^+ conductance of the membrane increased, which allowed Na^+ to flood into the cell. This reversed the cell potential, and formed the positive-going portion of the action potential (Campbell et al., 1999). A slower increase in K^+ conductance allowed K^+ to flow out of the membrane. This, and the flow of Cl^- into the cell, brought the cell potential back down, forming the trailing negative-going part of the action potential. The refractory period following the action potential was caused when the ion pump reestablished inside and outside Na^+ and K^+ concentrations.

(*Continued*)

ACTION POTENTIALS (Continued)

Pollack (2001) has a different explanation, one that is based more on the physicochemical basis of the cell, and one that explains newer research observations, but one that does not have an explanation for as many details as the traditional hypothesis. You may recall that the cell membrane is very porous, and its integrity is not essential to the presence of the cell resting potential (see box, Section 3.8, Water Chemistry). The resting potential comes about because of the net negative charge on immobile proteins inside the cell, and the structured water (gel) that surrounds these proteins. There is hardly any room for included ions, especially large hydrated ions such as Na^+. Pollack (2001) presents evidence that the action potential requires an intact cellular cytoskeleton. The cytoskeleton is a dense polymer-gel matrix composed mainly of cross-linked actin (a contractile protein present in muscles of all animals from protozoa to vertebrates, and in the microfilaments of all cells) and microtubules (hollow filaments 20–25 nm in diameter found in eukaryotic cells and composed of an actin-like protein called tubulin). The cytoskeleton in nerve cells is present just below the membrane and carries a high negative surface charge. According to Pollack, a phase change in the cytoskeleton during the action potential, accompanied by water absorption and heat liberation, disrupts the structured layers of water in the gel. The liberation of heat would be expected if a highly ordered structure, which takes energy to form, were to become less ordered.

Na^+ would be expected to rush into the unstructured water because of a concentration difference between Na^+ outside the cell and Na^+ inside the cell. However, it has been observed that action potentials can be generated even when Na^+ and K^+ are absent (Pollack, 2001). Ca^{++}, however, must be present.

Ca^{++} is a small divalent ion that can cross-link with different protein strands and hold them together. If Ca^{++} is displaced by Na^+, a monovalent ion that cannot cross-link protein strands, then the protein structure that holds water in place cannot be maintained. Thus, as Na^+ rushes into the cell, it may displace Ca^{++}, and add to the tendency of the gel to become less ordered. Water also rushes in, and the system of protein strands expands. Only when the strands are pushed apart far enough will there be enough recoil to reverse the process. Thus, the action potential is a physicochemical process involving the expansion and contraction of the cytoskeleton matrix.

What can initiate this action? Catecholamines are chemicals formed from a benzene ring, adjacent hydroxyl groups, and an amine group (tyrosine). Examples are epinephrine (adrenalin), dopamine, and norepinephrine. They have specific effects on the nervous system.

Catecholamines are necessary for the repetitive action potentials generated by heart pacemaker cells. Binding of catecholamines in the cell could be the mechanism by which the action potential is initiated.

Indeed, the transmission of a neural signal at the synapse is usually by chemical means (some direct electrical connections are present in the giant axons of crustaceans

(*Continued*)

ACTION POTENTIALS (Continued)

and fishes, but these are rare). At the end of the axon are numerous sacs called *synaptic vesicles*, each of which contains small amounts (but thousands of molecules) of neurotransmitter substances. When the depolarization of the action potential reaches the synapse, neurotransmitter is released into the small gap between the neurons. When the depolarization of the action potential reaches the synapse, a neurotransmitter is released by exocytosis into the small gap between the neurons.

Neurotransmitter Substances

Neurotransmitter Type	Functional Class	Secretion Sites
Acetylcholine	Excitatory to vertebrate skeletal muscles; excitatory or inhibitory at other sites.	CNS, PNS, vertebrate neuromuscular junctions
Biogenic Amines		
Norepinephrine	Excitatory or inhibitory	CNS, PNS
Dopamine	Generally excitatory; may be inhibitory at some sites.	CNS, PNS
Serotonin	Generally inhibitory	CNS
Amino Acids		
GABA (gamma aminobutyric acid)	Inhibitory	CNS, invertebrate neuromuscular junctions
Glutamate	Excitatory	CNS, invertebrate neuromuscular junctions
Glycine	Inhibitory	CNS
Aspartate	Excitatory	CNS
Neuropeptides		
Substance P	Excitatory	CNS, PNS
Met-enkephalin (an endorphin)	Generally inhibitory	CNS

CNS = central nervous system; PNS = peripheral nervous system.

There are several known neurotransmitters (Table). The most common of these is acetylcholine, which, as with many other neurotransmitters, may be either excitatory or inhibitory on the postsynaptic cell, depending on the type of receptors present on different cells. Some of these substances may be released into the blood stream as hormones by endocrine glands. When acting as hormones they have much more general effects on the body.

Psychoactive drugs, including LSD and mescaline, produce their hallucinatory effects by binding to serotonin and dopamine receptors in the brain. Endorphins function as natural analgesics and decrease the perception of pain as well as produce euphoria.

(*Continued*)

ACTION POTENTIALS (Continued)

It is important that neurotransmitters are destroyed soon after release. If they weren't, their effects would be felt too long after they were intended. There are enzymes present in the synapses to hydrolyze neurotransmitters soon after they are released by the presynaptic neuron. Cholinesterase is the enzyme that acts on acetylcholine. A lack of cholinesterase would cause permanent depolarization of the postsynaptic neuron, and this neuron would then become ineffective for transmitting signals. Certain phosphate insecticides (and similar chemical agents) work in this way.

Brain neurotransmission (passing a signal from one neuron to the next) occurs within 0.3–100 milliseconds (synaptic time delay) and over a distance of 30–50 nanometer (synaptic cleft) (Anonymous, 2007). These small scales of time and distance pose challenges for those intending to measure neural activities. New analytical methods are constantly being sought for real-time, small-scale measurements in biological systems.

Light can be used as a neuroengineering tool to detect the firing of individual neurons or to initiate neuron firing. In the first case, a fluorescent dye is inserted into a neuron and the neuron emits light during an action potential. An optical detector is used to identify individual neuronal circuits performing different functions. In the second case, a light-sensitive protein, called an *opsin*, is linked to DNA and inserted into a neuron, usually with a virus vector. The opsin forms an ion channel that can be opened with a pulse of light (Palmer, 2014). This initiates an action potential, causing certain types of behavior to occur. Using this method, specific neural networks in the brain can be linked to responses that they control.

4.4.4 Closed-Loop Feedback Systems

> Much of human behavior is hard-wired. But, unlike the heart, liver, or even our genes, the brain can respond in a dynamic way not only to internal physiological cues but also to unpredictable external ones, and it can embody that response in future behavior.
>
> **—Judy Illes**

We can conceptualize this control system as a loop (Figure 4.4.4). The input level is sensed, the control system acts on the input level, and this, in turn, changes the input level. This type of control system is called *feedback* control, Ie information about the output of the system is fed back to the input in order to maintain stability.

Feedback usually gives very good control. The input level is sensed and compared to a set-point level. The difference between these two is then used to form a correction at the output, and this changes the input level. This is called *proportional control*.

There are times when this feedback can lead to oscillations, especially if there are delays, as discussed earlier. Oscillations in feedback control systems represent loss of control Ie

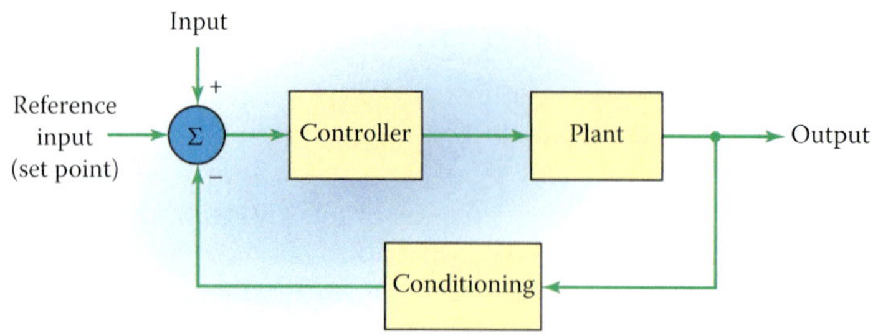

FIGURE 4.4.4
Generalized feedback control loop. A portion of the output signal is fed back to the controller input, where it is subtracted from the reference level input. The controller then acts upon this difference, feeding it to the process to be controlled, known as the plant. The feedback signal may be conditioned before being compared with the reference input.

they are difficult to stop once started. To improve stability, we could add rate of change information somewhere in the control system. A system that responds not only to the input level but also to the rate of change of the input anticipates where the input level will be and corrects ahead of time. Perhaps that is why so many biological receptors are sensitive to the rate of change in addition to the actual level of input.

We have just described proportional (P) control and proportional plus derivative (PD) control. Integration can be added to a controller. It not only gives it reset action, but also can exacerbate instability. There are proportional plus integral (PI) and proportional-integral-derivative (PID) controllers. These classical types are used where the system dynamics (the Plant) are well defined.

It is very difficult to tell the differences in the actions of the different kinds of controllers as long as the feedback loop is unbroken. Differences are only apparent when the loop is opened for some reason.

The box labeled "Plant" in Figure 4.4.3 is the process to be controlled. An example of what we mean might be the insulin control system (see Section 6.20, Controls), where the input signal relates to the rising glucose level of the blood and the beta cells of the pancreas are the sensors. Controller action takes place within the cells, and insulin release is stimulated. The Plant in this system is represented by the glucose uptake in the body cells and liver due to increased circulating insulin.

BIOFEEDBACK

There are many bodily functions unconsciously controlled, such as breathing rate, heart rate, and skin temperature. Conscious control can be established over these functions if the person can be made aware of their intensities. Biofeedback is the process of developing methods to bring awareness of these functions to the individual so that the person can learn to control these normally autonomous activities. Awareness is often made through electronic instruments with appropriate sensors ending with a dial, alarm, or other readout means giving information about the level

(*Continued*)

BIOFEEDBACK (Continued)

of intensity of the activity being measured. Once the person can perceive of the level of activity, then it is possible to learn how to control its level. This learning is not always conscious, but is effective nonetheless.

There are some requirements for the feedback device. It must be specific to the type of activity to be monitored, and it must respond fast enough so that changes in activity level can be detected in time to make the association between control attempts and results of those attempts. For some of the more rapid bodily functions, such as heart rate, the response time of the feedback mechanism must be of the order of a half second or less. For slower bodily functions, such as nighttime incontinence, it is sufficient to sound an alarm within a minute or so of the event.

Biofeedback has been successfully used to control heart rates, skin temperatures, anxiety levels, muscle tensing, chronic pain, attention deficits, and vision problems, as well as other conditions.

4.4.5 Open-Loop Systems

> Make an estimate before every calculation, try a simple physical argument before every derivation, guess the answer to every puzzle.
>
> —**John Wheeler and Edwin Taylor**

An alternative to feedback control is open-loop control, where there is no loop. System output is determined by dead reckoning and there is no means to correct the output should there be an error. Open-loop control is faster, and requires less effort, than feedback control. Open-loop control works well for repetitive situations where there is small likelihood that an error will occur. An example of this is the signals sent to the leg muscles for walking on flat terrain. Only when the ground becomes rough or sloped is feedback normally required.

ARTIFICIAL NEURAL NETWORKS

A human brain continually receives input signals from many sources and processes them in parallel to create appropriate output responses. There are billions of neurons in the brain that interconnect in a myriad of ways to form elaborate neural networks. Artificial Neural Networks (ANN) are an attempt to process information efficiently and quickly using brain neural networks as a model. Like brain neural networks, ANN have many neuron-like nodes that interact with one another; like brain neural networks, ANN must undergo a learning process before they are ready to process information automatically; like brain neural networks, ANN take information from a number of primary inputs and form useful outputs.

(*Continued*)

ARTIFICIAL NEURAL NETWORKS (Continued)

ANN architecture consists of an input layer of nodes, an output layer, and, sometimes, one or more hidden layers between the two (Figure). The input layer receives (analog, not digital) information from a set of primary transducers. For example, different temperature or pressure measurements, or size information, may represent the input data. That data is fed to the ANN input layer.

The output layer typically consists of one node. The data value at the output node is the parameter that the ANN is meant to estimate.

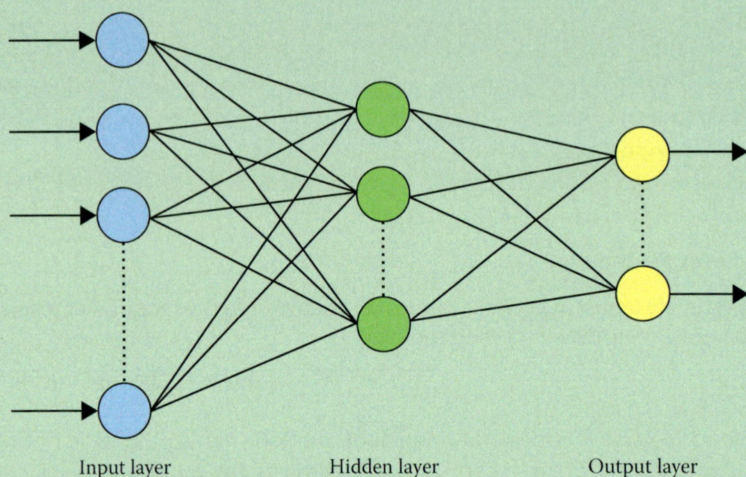

Configuration of a three-layer artificial neural network. Each layer consists of a number of neuron-like nodes, here indicated by open circles.

There can be one or more hidden layers, and each hidden layer can be composed of many different nodes. The number of hidden layers and the number of nodes per layer is empirically determined based upon performance. There is no theoretical basis for choosing the number of layers or the number of nodes. Frequently, one hidden layer is all that is required. The number of nodes for the hidden layer is often about twice the number of input nodes.

Although the ANN is given conceptually in the Figure, the actual ANN is usually a computer program without a physical basis. There are many ANN programs available, and differences among them are related to sizes and computational procedures.

Connections among the nodes are shown in the Figure. Each connection is given a weight and a threshold (or bias) value. These are estimated at the beginning.

The ANN is calibrated through a learning procedure where known input values are applied at the input layer and the output value at the output layer is computed.

(*Continued*)

ARTIFICIAL NEURAL NETWORKS (Continued)

The computed output is compared to the ideal known value, and the weights and biases for each connection are adjusted to reduce the error. There are various computer algorithms available to make these adjustments efficiently. After a process of iteration, with connection weights and biases undergoing successive adjustment, the ANN computed output is nearly the same as the ideal output. At this point, again determined empirically, the learning process is stopped in order to avoid over-learning, where the ANN estimates the output for test conditions almost perfectly, but errors for other conditions become large.

After the learning step, the ANN is used to estimate output values for actual input data. ANN weights and biases are fixed during this process, and the ANN acts as an open-loop feedforward estimator.

4.4.6 Closed-Loop Feedforward Systems

There are no creeds in mathematics.

—**Peter Drucker**

There are times when a living system can anticipate that a corrective change will be required. Such a case might arise for a person living in a hot climate. Sweating can begin before a rise in body temperature because the body has acclimatized to the heat, and the thermoregulatory system knows that sweating will be required to maintain correct body temperature. When the response begins before a change in input level can be sensed, we call this *feedforward* (Figure 4.4.5). Feedforward does not form the same kind of loop as feedback and does not necessarily result in a very stable system. Indeed, feedforward can cause responses to be too extreme at times, but it also can be used to avoid catastrophic loss of control under extreme environmental circumstances, and, because it is anticipatory, the responses are very fast.

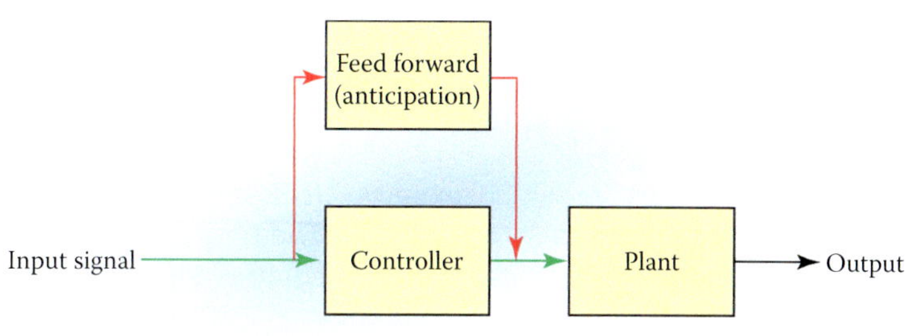

FIGURE 4.4.5
Diagram of feedforward control. The feedforward signal is predicted from the input signal without waiting for slow feedback.

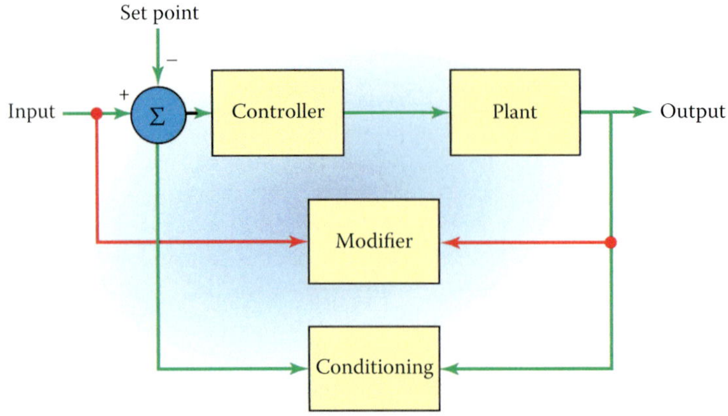

FIGURE 4.4.6
Diagram of an adaptive control system with feedback. The difference between this system and the one in Figure 4.4.3 is the presence of the modifier that changes the controller performance.

4.4.7 Adaptive Control Systems

> The methods of computational biology are now so advanced that it's conceivable to make a computer atlas of the nervous system and map the expression of hundreds of genes.
>
> **—Chris Doe**

An adaptive controller continually and automatically readjusts itself for proper operation in the presence of changing system dynamics or noise characteristics. It combines a parameter estimator and a control scheme that changes the control algorithm as needed. A block diagram of an adaptive system can be seen in Figure 4.4.6. Adaptive control is based on a linear differential equation with nonconstant coefficients, and is often used in drug-delivery systems (Woodruff, 1995) or where patient-to-patient variation is particularly wide.

4.4.8 Fuzzy Control Systems

> Whatever a man prays for, he prays for a miracle. Every prayer reduces itself to this: "Great God, grant that twice two be not four."
>
> **—Ivan Turgenev**

Fuzzy control systems are ones in which experts' decision-making rules are used to produce a control output. The rules may either be mathematically based or not, and the controller output is usually sufficiently correct to perform the intended function. However, control with a Fuzzy system is not usually very precise.

Example 4.4.2: The Potted Rose (Alocilja, 2002)

A healthy rose plant will bloom with profuse, beautiful flowers. The plant requires water, and therefore must be watered. If water is added without regard to the condition of the plant, this is an open-loop system. If water is added only when the plant needs it, or if the amount varies to meet plant moisture requirements, then the system is closed-loop with feedback. If extra water is added because you will be unable to care for the plant for a few days, then this is a feedforward control system.

Example 4.4.3: Time Delays in Closed-Loop Diabetes Control

The insulin used to treat diabetes reaches its peak effect in 60 to 90 minutes, which is too slow for closed-loop sensor and insulin pump feedback loop control. This lag can cause control problems with overshoots and undershoots in diabetic blood sugar levels (Ross, 2015).

APPLICATIONS AND PREDICTIONS

1. Learning will require feedback control; habit will allow open-loop control.
2. Cell growth will be regulated with feedback from its neighbors.
3. Removal of metabolic carbon dioxide during exercise can be considered to be feedforward control.
4. Movement of the limbs and joints will cause a direct and immediate rise in heart rate. This is feedforward control.
5. Hormone levels will be regulated by feedback control.
6. Enzyme-substrate inhibition will act as a feedback loop.
7. Going to the store to buy milk before it is all used up will be feedforward.
8. The stomach will produce gastric juice when hungry because of feedforward action.
9. Different control principles can be used to design artificial intelligence systems.
10. Myelinated axons will propagate signals faster than nonmyelinated axons.
11. Growth receptors in cells provide feedback to halt unchecked growth.
12. Feedback will regulate social interactions.

4.5 Optimization

True science thrives best in glass houses, where everyone can look in.

—**Max Perutz**

Maintenance of life and its many activities is a costly process. Energy is required to move, to maintain health, to compete for food, to reproduce, to grow, and to produce the chemicals of life. During normal times of homeostasis, the energy requirements of these processes are not particularly burdensome, but when additional demands are placed on the system, the organism may find that it cannot readily supply energy for all demands. Thus, adjustments may be made to minimize energy expenditure or to maximize output.

It is not exactly clear how optimization processes in living systems come about. However, mathematics may be used to find the optimum of a process if the mathematical description of that process is known over the range of interest. An optimum point can usually be described as a maximum or the minimum. We find that the rate of change of the variable of interest is zero at a maximum or minimum point. If the mathematical description is in equation form, then all we need do is to determine the rate of change of the variable of

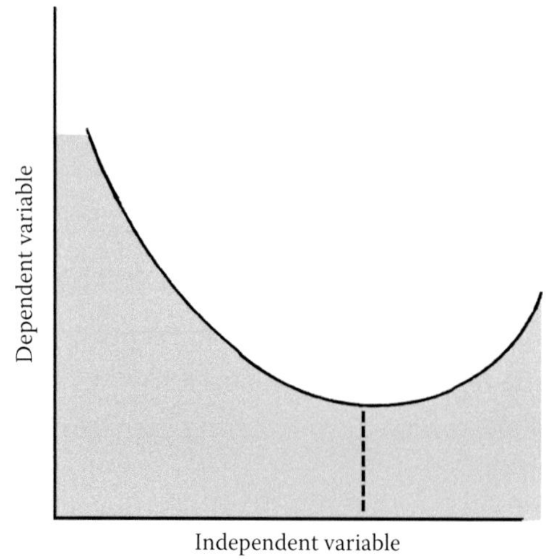

FIGURE 4.5.1
The optimum point is usually obtained where some dependent variable is minimized for some value of the independent variable. For instance, the rate of respiratory work is minimum at a particular breathing rate. A broad minimum, as seen here, means very little penalty is paid for operating away from the minimum.

interest and set the rate of change to zero. Solving for the value of the variable of interest defines either the maximum or minimum point.

There are other tests we may use to determine if this is a maximum or a minimum. One of these is to know ahead of time that the point must be a maximum or must be a minimum, and that the place where the rate of change is zero has to be one or has to be the other, but cannot be either one or the other.

In graphical form, the place where the rate of change is zero is the point where the slope of the curve is zero (Figure 4.5.1). From this we can see that an optimum point may be found graphically where the slope is zero, mathematically where the rate of change is zero (using differential calculus), or numerically, where the optimum is estimated from known data values.

There may be local maxima or minima and global maxima or minima. The difference is that a local maximum or minimum is only the largest or smallest value over a very limited range. The global maximum or minimum is the most extreme value over the whole range. Biological data is usually simple and has only one maximum or minimum over the range of interest.

The maximum or minimum may only exist at one of the extremes of the range of interest. Graphically, this indicates that the highest or lowest value is at one end or other of the graph. If this situation happens with data from biological systems then it indicates that the process is not optimized by the biological system. In order for a process to be optimized, it must be able to be controlled, both from above the optimum or from below, and that measures must be able to be taken to move the process toward the optimum. If the optimum is at an extreme value, it cannot be controlled from both sides.

An optimum point may be either shallow or deep. With shallow optima, the process may operate without significant penalty at a point away from the exact optimum. Thus, the system may be found to deviate one way or the other from the optimum and a relatively large variation in the process operation is possible. With a deep optimum, there is a large penalty paid for deviations from the optimum and the amount of variability of response is small.

APPLICATIONS AND PREDICTIONS

1. Many exercise parameters, for instance the rate of cycling, will be optimized to reduce energy expenditure.
2. When nutritious food becomes scarce, digestion will be optimized to extract as much nutrition as possible to minimize waste.
3. Kidneys save water when water is limited; they will excrete a lot of water when water is plentiful.
4. Animals and plants rest at night and are active during daylight.
5. Some animals hibernate when it is too cold, or food is not available.
6. Whales reduce heart rate and metabolism when underwater.
7. The distribution of branches and leaves in a tree will optimize the use of sunlight.
8. Human hemoglobin is optimized to deliver the maximum amount of oxygen to the working muscles.
9. A neural cell network in the CNS becomes optimized to perform specific tasks.
10. Students eat cheap food to maximize nutrient content for the least cost.
11. Sizes of animal parts will be optimized to perform needed functions without expending too much energy.
12. A plant in the window will bend toward the light to maximize light reception.
13. Bacterial enzymes will adjust to food substrates available.

4.6 Information

If you understand everything, you must be misinformed.

—Japanese Proverb

In order to be able to act on and react to environmental conditions, biological systems must be able to obtain information from the environment, pass information internally to integrative centers where it can be processed, and then send information to actuation structures. Biological systems have developed special structures to obtain information, called sensors, and have developed special mechanisms to transmit this information internally.

There are many modes of information acquisition in biological systems. Information can take the form of chemicals either sensed or produced. In this case, environmental chemicals are sensed, and they may trigger production of other chemicals, and these other chemicals may be circulated throughout the organism, and other cells may use the circulating chemicals to produce other chemicals to react to the presence of the original chemicals in the environment.

The environment we mention here does not need to be external to an organism. It just needs to be external to the sensor.

Information can take the form of *neural* action potentials that travel from one end of a neuron to another and from there across a synapse by chemical means to another neuron. These action potentials can be triggered originally by chemicals or by energy sources such as heat, light, or radiation.

In all information systems there is the problem of recognition of meaningful signals. *Noise*, or meaningless information, is present at all levels, but is especially important where the strength of the meaningful signals is the smallest. Biological sensors must be able to separate the signals from the noise. If not, then the system would expend much wasted energy reacting to meaningless inputs.

By the time a meaningful signal is recognized in some way, it is usually amplified by the system so that the noise becomes relatively smaller. At this point, any additional noise in the system may actually help the organism to maintain its awareness of the environment.

There is a thermodynamic limit to the smallest signal that can be sensed and distinguished from background noise. In general, the more sensitive the sensor, the more energy it takes to maintain that sensitivity. Thus, the most sensitive sensors are those related to environmental conditions most critical to the organism.

The movement of information can be imagined as a flow variable, running from a region of higher concentration to lower (Schneck, 1990). Thus, information, like heat or light, can have a capacity and resistance that store information and limit its flow.

The form in which information is stored in living systems is in the state of order of the various levels of biological organization, from subcellular to ecological. Beginning with the genetic code, moving on to the structures of proteins, the specialization of the many types of cells in the body, and the intricate natural balances of species, there is order in living systems (May, 2006; Schneider, 2006). This order represents information about which we are still learning; it is this knowledge that the biological engineer must know in order to produce a successful design.

As we have previously seen (see Sections 2.5, Energy, 2.6, Entropy, and 3.10, Energy), the maintenance of an ordered state requires energy expenditure, energy extracted from the environment. Entropy has previously been introduced (Section 2.5, Energy) as a concept of disorder. Thus, information storage (as an ordered state) and thermodynamic entropy (as a measure of disorder) are somehow related inversely. Shannon's definition of information is (Shannon and Weaver, 1949; Gatlin, 1972; Loewenstein, 1999):

$$I = \sum_i p_i \log_n p_i \tag{4.6.1}$$

where

I is information content

p_i is probability of a particular occurrence

$\log_n$ is the logarithm using a meaningful base n

The logarithmic base is usually chosen to be 2; in this way the unit of information is in *bits* (standing for *binary digits*), common in digital computers. As Igor Aleksander has said:

> The amount of information I depends on the surprise that the message holds. This is because the mathematical way of expressing surprise is as a probability p; the less probable an event is, the more surprising it is and the more information it conveys.

Thermodynamic entropy of this ordered system is given as (Loewenstein, 1999):

$$\text{entropy} = -(\text{Boltzmann's constant})(\ln 2)\,I \tag{4.6.2}$$

where Boltzmann's constant has a value of 1.3802×10^{-23} N · m/K, and ln 2 = 0.693. By this means, it is possible to equate information storage (as structural order) in a biological

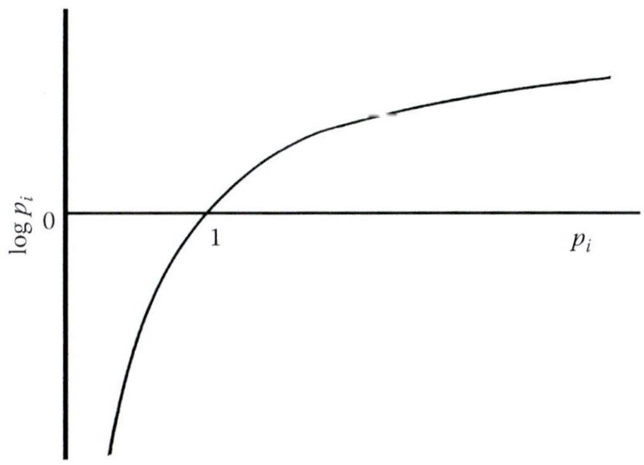

FIGURE 4.6.1
The $\log(p_i)$ races toward $-\infty$ as p_i decreases below 1.0. The magnitude of p_i does not decrease nearly as fast as $\log_2(p_i)$ becomes larger but more negative. Thus, $p_i \log_2(p_i)$ becomes very negative very fast as p_i approaches zero.

system with energy storage in the same system. This also reminds us that, in order to maintain the order represented by a living system, energy must be extracted from the surrounding environment, and overall, the universe becomes less ordered. Energy, information, and entropy (or order) are all related.

The probability of a surprise is very small. The entropy of the surprise is thus very greatly negative because all probabilities are less than 1.0, and the $\log_2(p_i)$ quickly approaches $-\infty$ as p_i approaches zero (Figure 4.6.1). Thus, $p_i \log_2(p_i)$ approaches $-\infty$ as p_i becomes less expected. Remember that negative entropies are directly related to highly ordered states, and that these states contain a lot of information to maintain that order. Thus, the smaller is p_i, the more information that is represented by the occurrence.

JELLYFISH DO WELL WITHOUT BRAINS

(Excerpted from Liubicich, 2010)

Jellyfish are mysterious and amazing creatures. Many people are entranced as they watch these beautiful animals dance gracefully through the water. Although they appear simplistic at first, jellyfish participate in many complex behaviors. The explanation behind this feat lies in their fascinating biology and their relationship with their environment.

Although jellies have no brains or complex organs, specialized cells and structures allow them to thrive in their dynamic environments. Jellyfish possess an intricate network of nerves covering their bodies. This diffuse nerve net allows them to sense and respond to stimuli coming from any direction, without the use of a centralized brain. While human neurons send information in only one direction, jellyfish neurons can fire in either direction, helping them transfer information rapidly across their entire bodies.

Jellyfish do not have brains, but some possess centralized sensory structures, called "nerve rings." These rings, made up of condensed areas of the nerve net, help initiate movement by triggering contractions in the body. They connect additional sensory structures to the nerve net, including simple eyes that detect light changes in the environment, chemoreceptor cells, and statocysts—organs that help sense balance and relative position in the water.

BODILY MICROBES AS AN INFORMATION LEGACY

The mix of microbes on organismal body surfaces is different for each individual plant or animal. These microbes have personalized effects, ranging from disease cause or prevention, odor production, digestion of nonfood compounds, essential biochemical production, and proper development of the immune system. Some types of microbes are essential to all members of an organismal species. Examples are endophytes in certain grasses, nitrogen-fixing bacteria in legumes, cellulose-digesting microbes in termites and cud-chewing animals, and probiotics in humans and animals. These microbes (called the *microbiome*) are so important that behavioral mechanisms have evolved to assure the passage of specific microbes from one generation to the next. In certain animals, essential microbes are passed through fecal exposure; in mammals, probiotics are introduced through mammary gland secretions.

The effects of these transfers are so profound that offspring may not survive without them. Somatic microbes thus are often as important as the genome and cultural information (memes) as a means to pass information from one generation to the next. Realizing this, the biological engineer should be sure not to impede this process.

Example 4.6.1: Information Content of Micrococcus DNA

There are four bases (A, T, C, and G) in the DNA of the bacterium *Micrococcus phlei*. The probability of occurrence of each of these has been experimentally found to be (Gatlin, 1972)

$$p(\mathrm{A}) = 0.164$$

$$p(\mathrm{T}) = 0.162$$

$$p(\mathrm{C}) = 0.337$$

$$p(\mathrm{G}) = 0.337$$

Calculate the information content of the DNA of the DNA molecule in bits.

Solution

From Equation 4.6.1,

$$I = \sum_i p_i \log_2 p_i$$

$$I = 0.164 \log_2 0.164 + 0.162 \log_2 0.162 + 0.337 \log_2 0.337 + 0.337 \log_2 0.337$$

Since

$$\log_2 x = \frac{(\log_{10} x)}{(\log_{10} 2)} = 3.32 \log_{10} x$$

Then

$$I = 3.32(0.164 \log_{10} 0.164 + 0.162 \log_{10} 0.1620 + 0.337 \log_{10} 0.337 + 0.337 \log_{10} 0.337)$$

$$I = -1.910 \text{ bits}$$

Remark

The negative value indicates that information was extracted from the environment.

Example 4.6.2: Entropy Value of Micrococcus DNA

Calculate the entropy value for the DNA found in Example 4.6.1.

Solution

From Equation 4.6.2,

$$\text{entropy} = -(\text{Boltzmann's constant})(\ln 2)I$$

$$\begin{aligned}\text{entropy} &= -(1.3802\times 10^{-23}\ \text{N}\bullet\text{m/K})(0.693)(-1.910)\\ &= 1.83\times 10^{-3}\ \text{N}\bullet\text{m/K}\end{aligned}$$

Remark

The positive entropy value indicates that the maintenance of the DNA structure of *Micrococcus phlei* has caused disorder in the rest of the universe.

APPLICATIONS AND PREDICTIONS

1. Teachers will contribute information and students will accept information. Because the total amount of information storage for teachers and students together will increase, energy must be expended in the process.
2. Living systems without good memories will be more wasteful thermodynamically than those with good memories.
3. The growth process, where information will be stored at a very rapid rate, will be energy intensive.
4. Distinguishing useful information in the presence of noise will require repetition.
5. The formation of new DNA will increase the entropy of the organism.
6. Weaker signals will require more energy to recognize and process.
7. Environmental stress will require energy for coping; thus, less information can be absorbed during stress.
8. Organisms that adapt to their environments will develop improved information gathering means.
9. Information flows naturally from sources of high concentration to low concentration. A great expenditure of energy will be required to move information from low to high concentration.
10. Information storage and recall will give a survival advantage. Information storage without recall will be nonconsequential.
11. Students who are able to distinguish between important and unimportant information will expend less energy to learn.

4.7 Analog and Digital Signal Processing

> For a giant like Newton, the calculation of π was chickenfeed, and indeed, in his *Method of Fluxions and Infinite Series*, he devotes only a paragraph of four lines to it, apologizing for such a triviality with a *by the way* in parentheses – and then gives its value to 16 decimal places.
>
> —**Petr Beckmann**

Computation and signal processing can be performed in either one of two ways: analog or digital. Analog computation involves an infinite variety of signal levels, and useful information is represented by a value that occurs between high and low extremes. For instance, sound that you hear or light that you see is analog information. You can tell one sound intensity level from another because of its loudness, and there are many different loudness levels that can be discriminated. Many of the body's sensors detect analog signals.

Digital signals occupy discrete levels. In the digital computer world, there are two discrete levels represented by the upper and lower power supply voltages. This is a binary system. When considering DNA, there are four discrete biochemical building blocks. This is a quaternary system. Information in a digital system does not occupy the vast array of possibilities present in an analog system. Instead, information is contained only in the presence or absence of the discrete levels.

Neural communication is digital. Detectable information in a neuron is represented by the presence or absence of an action potential. It's either there or it isn't, and so it's a binary system.

Because many biological sensors detect analog levels, but communication of these levels often occurs by digital neural means, there must be an *analog-to-digital conversion* process in between (other communication, for example hormones, is entirely analog). Through analog-to-digital conversion, the sensed analog signal level is usually converted to a train of pulses the frequency of which corresponds to the input level (sometimes, as with taste, it is the pattern of action potentials that determines the type of taste—sweet, sour, bitter, salt, savory—that is detected).

Digital signal processing is relatively noise free. Because information is recognized to exist only at discrete levels, small differences that occur because of environmental factors (some determined by basic physical fluctuations at the atomic or molecular level) are of no consequence. Thus, a signal level that is within a band of levels is only recognized as the standard level.

Analog processing, on the other hand, can be more efficient as long as it does not have to be particularly precise nor fast. A very precise analog computation requires very sensitive analog components, and there is a limit to how sensitive these can be. Because of thermal noise present nearly everywhere in biological systems, extreme precision must require sufficient time to average random fluctuations. If that time is not available, accuracy can suffer. One means to increase analog computational efficiency is to transform the signal nonlinearly, as, for example, logarithmically. This transformation compresses a multi-order-of-magnitude signal into a relatively small range. This can be done with a few simple nonlinear components the types of which are biologically common.

The same type of processing using digital means would require many digital channels and some fancy digital algorithms. Precision in the digital world translates into numbers of channels for parallel processing or speed for serial processing. Nonlinear transformations are almost entirely performed by analog rather than digital means.

TABLE 4.7.1

Number of Neurons in the Brains of Selected Species

Species	Number of Neurons
Roundworm	302
Pond Snail	10,000
Fruit Fly	100,000
Lobster	100,000
Jumping Spider	600,000
Cockroach	1 million
Honeybee	1 million
Mouse	75 million
Rat	200 million
Octopus	500 million
Cat	1 billion
Chimpanzee	6.7 billion
Human	86 billion

Nature's solution for signal processing is first to process incoming analog information efficiently with specialized analog devices such as eardrums, cochleas, and retinal cells (Sarpeshkar, 2006). The purposes of these steps are to reduce the amounts of data requiring conversion into digital form. For example, the inner ear splits analog sounds into frequency ranges before each range is transmitted to the brain by the auditory nerve. Image processing in the retina detects and locates edges, and this information is transmitted neurally to the brain.

Only after the amount of data has been reduced to conform to the conveyance limits of neurons are the signals converted to digital form. Myelinated neurons can transmit action potentials faster than slower unmyelinated neurons. Thus, in any given amount of time, more information can be conveyed on a myelinated neuron. Analog processing either limits the amount of information to the rate at which it can be transmitted, or additional information is lost.

The neuron itself functions as an analog processor, analog-to-digital converter, and digital signal transmitter. The input side of the neuron has a transmembrane potential that varies either with the incoming sensory stimulus or with the rate of pulses coming from other presynaptic neurons. Analog-to-digital conversion is accomplished by the presence of a threshold voltage above which the action potential is formed. The action potential is the digital form of the signal (Table 4.7.1).

RECOGNITION IN THE BRAIN

How the brain recognizes sensory inputs, builds memories, and recalls past experiences is a topic of much study. The brain apparently has different codes for different types of information (Koch and Marcus, 2014). Some information is represented by neural firing rate, while other information resides in the specific neurons or neural groups that fire. There are differences among different types of information such as new sensations, recognition of familiar entities, or general concepts. Different parts of the brain code information differently. Memories are related to certain synaptic connections. These are all areas of ongoing work and interest.

APPLICATIONS AND PREDICTIONS

1. Living things will use the type of processing that is most efficient for the function that is to be performed.
2. Digital communication lines are less likely to be affected by environmental influences. Thus, neuronal communication systems predominate over other types.
3. Knowing the magnitude of a stimulus is important to formulate a response. Therefore, primary receptors will most likely be analog, with the analog signal somehow being able to be sensed by the organism.
4. Simple creatures will operate almost exclusively in the analog world.

Questions

4.0.1 What place does mathematics have in bioengineering methods?

4.0.2 List the engineering science courses you expect to take and give an assumed relationship to biological engineering for each of these.

4.0.3 State why the study of mathematics contributes to biological engineering understanding.

4.0.4 State why the study of engineering sciences is important for biological engineers.

4.0.5 How does a mathematical model lead to an improved biological engineering prediction?

4.0.6 What is an engineering science and how does it relate to the study of biology?

4.0.7 What are the advantages and disadvantages to applying mathematical analysis to biological systems?

4.0.8 Why is the understanding of control systems important to the study of biological systems?

4.1.1 In the equation,

$$\text{power} = 38.2(\text{speed})^2(\text{grade}) + 9.8(\text{mass})(\text{speed}),$$

where mass is in kg and speed in m/s.

What are the units of power, the coefficient 9.8, and the coefficient 38.2? Grade is a dimensionless fraction.

4.1.2 The quadratic formula is:

$$x = \frac{-b \pm \sqrt{b^2 - 4ac}}{2a}$$

Is this an unequivocal equality, conditional equality, or replacement function?

4.1.3 In Equation 2.5.2 defining enthalpy, the following variables and dimensions are evident:

Enthalpy (F · L)

Heat (F · L)

Work (F · L)

Pressure (F/L^2)

Volume (L^3)

How many independent pi terms can be found? Find one of these.

4.1.4 Describe the process of iteration. Why is iteration only a conditional equality? Give examples of where iteration can be applied.

4.1.5 Add to the list of Applications and Predictions.

4.2.1 What do you think you would have to know in order to know exactly the state of some biological system? Is it possible to know these things with certainty?

4.2.2 What sources of variation are there in a measurement of a biological system that are not directly attributable to the system?

4.2.3 Give examples of biological measurements that are likely to conform to each of the probability distribution functions given in Figure 4.2.1.

4.2.4 Give examples of things that are interesting because they are almost predictable, but not quite.

4.2.5 For the following sets of data, determine the best-fit lines by least squares. Next, graph the data and judge how well the least squares lines represent the function describing the data.

x	y_1	y_2	y_3
10.0	8.04	9.14	7.46
8.0	6.95	8.14	6.77
13.0	7.58	8.74	12.74
9.0	8.81	8.77	7.11
11.0	8.33	9.26	7.81
14.0	9.96	8.10	8.84
6.0	7.24	6.13	6.08
4.0	4.26	3.10	5.39
12.0	10.84	9.13	8.15
7.0	4.82	7.26	6.42
5.0	5.68	4.74	5.73

4.2.6 Is chaotic data random? Why or why not?

4.2.7 Should the strength of a disinfectant be based on the value that is lethal for the average microbe? Why or why not?

4.2.8 Where do statistical methods fit into the scientific method?

4.2.9 Describe a very large experiment. How can this experimental plan be modified to cost less to conduct?

4.2.10 Why should treatment subgroups be carefully considered when analyzing them for statistical significance?

4.2.11 Muscle fatigue can be measured by means of the frequency spectrum, related to standard deviation, of electrical signals recorded from the muscle. The following is a list of values (in arbitrary units) for the myoelectric signal during isokinetic knee extension using the rectus femoris muscle. Compute the standard deviation.

0.35	0.15	0.70	−0.65	0.28	−0.10
−0.25	−0.24	−0.80	0.16	−0.35	−0.03
−0.10	0.15	0.04	0.00	0.04	0.00
−0.10	−0.13	0.50	0.25	0.08	0.06
0.08	−0.23	0.22	0.01	0.08	0.01

4.2.12 In developing a countertop spray cleaner effective against *Salmonella* bacteria, candidate antibiotics are tested on agarose plates inoculated with *Salmonella* bacteria. Single drops (0.1 mL) of 1 M antibiotic solutions are introduced to each plate, and the zones of *Salmonella* inhibition are measured after a standard incubation time. The standard for the experiment is a 70% solution of ethanol in water. Diameters (in cm) for each of 16 replications for each antibiotic are given in the table below. Which antibiotic do you recommend?

Sodium Chlorite	2-Benzyl 4-Chlorophenol	Dehydroabiethylanine	70% Ethanol
1.85	0.95	1.47	1.02
1.58	0.90	1.71	0.80
1.50	1.06	1.25	0.94
1.19	0.86	2.15	0.88
1.87	1.32	2.10	0.85
1.46	1.20	1.57	0.90
1.73	1.08	2.21	0.89
1.58	0.99	2.03	0.76
1.50	1.24	1.63	0.99
1.78	0.93	2.74	0.93
1.76	1.16	1.92	0.85
1.36	1.18	1.87	0.91
0.76	1.07	2.40	0.75
1.02	1.04	0.81	0.81
1.45	1.06	1.58	0.77
1.30	1.27	1.25	0.85

4.2.13 What are the dangers from drawing conclusions from small sample sizes?

4.2.14 How is grouping according to sources of variation related to individualized medicine?

4.2.15 Add to the list of Applications and Predictions.

4.3.1 How are position, velocity, and acceleration related? Write a differential equation relating these three variables.

4.3.2 Why are few biological responses proportional to $e^{t/\tau}$?

4.3.3 Discuss the advantages to a biological system of the exponential response to a sudden change (step input) as given in Figure 4.3.1.

4.3.4 Where, in biological systems, are periodic responses found? Are these the result of intrinsic periodicity or forced by environmental oscillations?

4.3.5 If volume flow rate is $V_o \sin \omega t$, where t = time, ω = frequency, and V_o = magnitude, what is dV/dt? What is the amount of work, as given in Equation 4.3.10?

4.3.6 Add to the list of Applications and Predictions.

4.4.1 List as many bodily sensors as you can. Compare these to sensors in plants. Are they similar?

4.4.2 If you had to replace a natural sensor with one made by humans, how would you do so?

4.4.3 List actuators in addition to muscles. What are the results of their actions?

4.4.4 Would you guess the efficiency of fast twitch fibers to be greater or less than that of slow twitch fibers?

4.4.5 Sometimes cardiac insufficiency is corrected surgically by grafting a piece of skeletal muscle to the heart ventricle. What would you think would be the difficulty with using skeletal muscle in this way? Could skeletal muscle substitute for smooth muscle?

4.4.6 Speculate on why two communication systems are necessary in the body. Why can plants do fine with one?

4.4.7 If resistance, capacity, and inertia of neurons slows and degrades the signal, what mechanisms have evolved to counteract untoward effects?

4.4.8 Distinguish between the sympathetic and parasympathetic nervous systems. Which dominates at rest and which during exercise?

4.4.9 Describe the process forming action potentials. How could you use this knowledge to produce neural prostheses?

4.4.10 Describe the time and space scales of synaptic neural transmission. What challenges do these scales present?

4.4.11 What is meant by feedback and feedforward?

4.4.12 Make a list of examples of open-loop control in biology.

4.4.13 Make a list of apparent feedforward control in biology.

4.4.14 Would you expect adaptive control to be more or less widely present in living things? Why?

4.4.15 Compare neurotransmission times and distances with other biological metrics.

4.4.16 Develop a fuzzy control algorithm to go from your home to a favorite store.

4.4.17 Describe the neural transmission of a signal down one neuron, from that neuron to the next, and down the second neuron.

4.4.18 How would you go about determining events occurring in the interneuronal gap?

4.4.19 Add to the list of Applications and Predictions.

4.5.1 How would optimization be incorporated into an engineering design involving living things?

4.5.2 Compare expected performance of a biological organism to a narrow versus broad optimum.

4.5.3 Add to the list of Applications and Predictions.

4.6.1 What is information?

4.6.2 How is information related to order in biological systems? Are the two synonymous?

4.6.3 Add to the list of Applications and Predictions.

4.7.1 Explain differences between analog and digital signals. Give common examples of each.

4.7.2 Picture a bioreactor. What processes of the bioreactor are treated as analog and which as digital?

4.7.3 What parts of a neuron are analog and what parts are digital?

4.7.4 Add to the list of Applications and Predictions

5

Principles of Biology

> ...The natural world is full of excellent designs that we can learn from....There should be a lot more contact and interaction between good biologists and engineers.
>
> **—Andy McIntosh**

Biology can be a very complex science, and its principles are not always easy to define. Defining life itself is not easy, either, and there has been no totally satisfactory definition of life. As more and more is known about subcellular structures that prey on living things (e.g., prions and viruses), it is clear that the demarcation between living and nonliving things is not well defined.

Living things are complex—they use energy to grow and repair themselves, they respond to external stimuli, and they reproduce. However, inanimate things, like fire, possess similar, if not identical, qualities. Some computer programs seem to be almost living. Contrarily, some living things, like mules, cannot reproduce and others, like seeds, may lie dormant for many years before suitable environmental conditions are present for them to grow.

Living things are not passive players. All living things actively attempt to control their environments to better suit them. If that is impossible, they either become dormant until environmental conditions change or else they die. Sometimes (e.g., the case of the seeds) it is difficult to distinguish between dormancy and death. So, perhaps it is easier to define life as anything that isn't dead. See Section 6.23 (Death) for further explanation.

The study of biology is incomplete without considering the surrounding physical, chemical, and biological environment (Figure 5.0.1), for living things sense environmental attributes, react to them, and cause environmental changes. All of this positions the being to compete and survive better.

From here, we develop the principles of biology:

1. *The primary goal of life is survival and reproduction.* Of these two, reproduction is the ultimate goal because genetic material survives for another generation. Genetic survival extends to the interest an organism shows in the welfare of another organism. In general, the more of a genetic code that two individuals share, the more sacrifice that one would make for the other. Sharing food, defense against predators, and nurturing care are some possible demonstrations of altruism. A parent that shares about 50% of its genes (assuming sexual reproduction) with its offspring will give a lot of nurturing care. Less genetic code in common with other individuals results in less willingness to share. The genes, apparently, look out for their own (see Section 5.4, Selection).
2. *Living things are constantly changing.* Unlike nonliving materials or entities, living things adapt, mature, reproduce, and otherwise react to environmental conditions surrounding them. Some of these changes are patterned and some

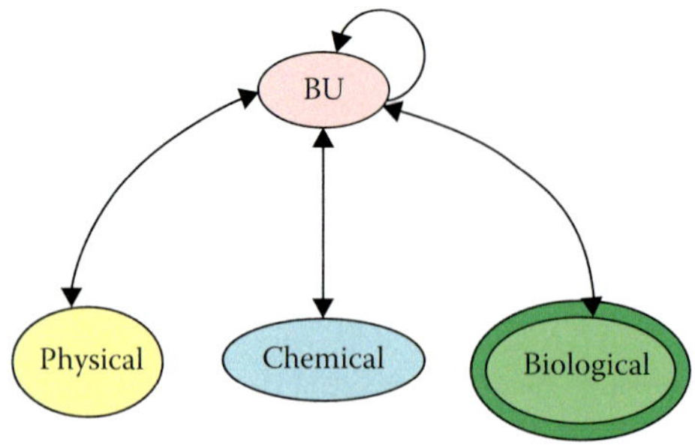

FIGURE 5.0.1
The BU is at least partly defined by and helps to define the surrounding physical, chemical, and biological environments.

are not, but living things are never the same from one time to the next. Changes in living things usually develop gradually.

3. *Long-term changes to a species occur only if there is a reproductive advantage.* This is really a corollary of the first two principles, but it is important to realize when change does not occur. Without a reproductive advantage, there will be no natural selection of a particular genetic code. No matter if some genetic modification seems to confer some advantage, unless there is an actual reproductive advantage, there is no permanent genetic change that occurs. See Section 5.2 (Modularity).
4. *Life is redundant.* There are many redundant features incorporated into living things. This makes life very robust. Because of redundant structures and processes, living things will attempt to face environmental challenges with the most efficient means. When the most efficient means fail, alternatives are used. When redundant features are exhausted, catastrophic failure (death) ensues. See Section 5.2 (Modularity).
5. *Coexistence of species requires that each adapts to a different ecological niche.* Otherwise their competition would be detrimental to one or both, to the extent of disappearance. See Section 5.4 (Selection).
6. *Attributes passed from one generation to the next require an information legacy.* Physical and behavioral attributes may be acquired in response to environmental pressures. However, these cannot automatically be transferred to progeny without some means to do so. Normally the information repository given credit for information transfer from one generation to the next is the genetic code of an organism. Changes in genetic code are passed, usually intact, to the next generation, whereas physical or behavioral changes acquired by an organism that do not involve genetic modification are not passed. Cultural information can also be used to modify physical and behavioral attributes of the next generation, including those individuals not directly in a bloodline. However, this type of information, while broadly more powerful than genes, is not as permanent. See Sections 5.3 (Genes) and 5.4 (Selection).
7. *Each distinguishing biological trait is made valuable by its cost.* Those traits that give survival or reproductive advantages to individuals or groups come at a price. For instance, male birds and mammals often have distinctive coloration, plumage, or other physical characteristics that convey messages of reproductive strength to females. But there are costs associated with these characteristics, either in resources required or vulnerability to threats, which make these characteristics

too costly to fake. Their presences thus remain honest indicators of the messages they are intended to convey. See Section 5.4 (Selection).

8. *An individual is a product of both its genetic code and its environment.* The genetic code is considered to be the basic blueprint of life. However, environment plays a very important role in the expression of genes and in the physical and behavioral attributes acquired by the organism over its life span. Different characteristics can be attributed in different percentages to either genes or environment, but, overall, the environment plays about an equal role with the genes in the development of an organism. See Section 5.3 (Genes).
9. *Life is conservative.* By conservative we mean that living things use those structures and processes already present to achieve its purposes. Each new species does not start anew with no history and no models. Instead, structures and processes already present in ancestral species are modified when needed to new uses. This makes for structural connections among species, and explains why some life processes are accomplished rather indirectly. Similarities among life-forms can be used to predictive advantage. See Sections 5.1 (Function), 5.2 (Modularity), and 5.4 (Selection).
10. *Living things use simple building blocks with complex interactions.* The biochemical bases for metabolic processes, biomaterials, growth, reproduction, and other essential processes are simple ones. There are only just so many combinations of things that are possible. Proteins, for instance, are composed of a small number of amino acids. Physical laws governing the combinations of amino acids are relatively simple. Nevertheless, there are large numbers of proteins and their functions are nearly innumerable because the number of combinations of these relatively simple building blocks is extremely high. And, many of these combinations occur simultaneously to add to the apparent complexity. See Section 5.6 (Biology Simple or Complex).
11. *Extremes are not tolerated well by living things, nor do living things create extreme conditions.* Life will exploit its environment to the maximum extent possible. There is almost no environment on Earth that will not support some form of life; even very hot, very isolated, or very cold environments contain life of some form. However, there are energetic penalties to pay for adaptation to extreme environments. Parasites or predators prey on most organisms, and many of these, in turn, are prey for other organisms. Rivals are controlled, not eliminated. This balance gives rise to an eternal struggle that is never resolved. See Sections 5.4 (Selection) and 5.5 (Hierarchies).

5.1 Form and Function

> Anatomy is to physiology as geography is to history; it describes the theater of events.
>
> —**Jean Fernel**

There is an intimate relation between the form of something living and its function. Let us examine our own bodies to illustrate this. Our ears are meant to gather in sound, so their shapes are like funnels, beginning with the large part of the external ear, and continuing through the ear canal. The outer part of the ear is large in order to focus sound waves toward the smaller ear canal. Animals that depend more on hearing have larger outer ears than those that do not.

Our noses are intended to smell and to condition inspired air (make the inspired air match our body temperature and humidity). Heat is added to air by convection, and convection heat transfer is more effective for larger surface areas and for turbulent airflow. Humidity can be added to the air by the same process of convection. Again, larger surface area and turbulent flow is desired. Smelling a certain chemical in the air stream requires that the chemical come in contact with receptors on the surface of the epithelium lining the inside of the nasal passageway. Unless the air stream is turbulent, chemical molecules could pass through the nasal passage without ever coming in contact with the surface. Thus, to enhance the sense of smell, to add (or remove) heat, and to add (or remove) humidity, air turbulence is desired.

Turbulence is enhanced by a passage with nonuniform cross section and by twists and turns and protrusions. Notice that the nasal cavity has a large surface area, it is of variable cross section, and the path of the air makes a large turn before entering the throat. The large cross-sectional area means that the air velocity slows from its value at the nostrils, increasing the residence time of the air in the cavity and giving it the time to be heated and humidified.

The location of the olfactory receptors is in the place (at the top of the cavity) where it is most likely that incoming chemical molecules will strike the surface. When chemical molecules do contact receptors, they must fit together before a nervous signal can be generated. This requires that the receptor must have a shape complementary to the shape of the chemical in order that the two join together. Each different class of chemicals generally has a different set of receptors sensitive to that chemical class. The form of the nasal passage and the form of the receptors are both determined by their functions.

Let's go farther. The eyes are meant to gather light and to sense an image. Any transparent tissue could allow light to pass, but it takes a lens to focus the light to form an image. Thus, there is a lens in the eye along with transparent tissue. All human tissues require oxygen to live, but the normal way for oxygen to be transported to the cells is through the vasculature. An eye with many capillaries would not be sufficiently transparent to pass a lot of light. Thus, the cornea of the eye has no capillaries. The tissues get their oxygen by diffusion from the air through the tears. Behind the eye is an array of very sensitive light receptors that are nerve cells, most sensitive to electromagnetic radiation in the visual range. Thus, we can see that the form of the eye is determined by its function.

Among other things, the fingers must obtain touch information. Thus, the touch receptors are located close to the skin surface. These receptors are not sensitive to light, as are the receptors in the retina of the eye, but do have some sensitivity to temperature.

If we look at birds, we see that they are equipped with wings shaped to allow them to fly, and they have hollow bones to give strength without inordinate weight. The bones of vertebrates give strength from inside the organisms. The skeleton can also provide external protection, as for insects or arthropods. With an external skeleton, however, the size of the animal is limited by the size of the skeleton. In order to grow, a new skeleton must be formed. This is very energy inefficient for large animals. Thus, internal skeletons are seen for these.

Plants have skeletons, too, but plants are not usually mobile, so they have distributed skeletons. The cell walls of plants are made of rigid cellulose, thus giving plants the ability to withstand forces that would ordinarily drive animals to move to shelter.

There are vestigial organs that retain forms related to previous functions. The human appendix is one of these. The appendix once served as an organ for digestion in ancestral organisms.

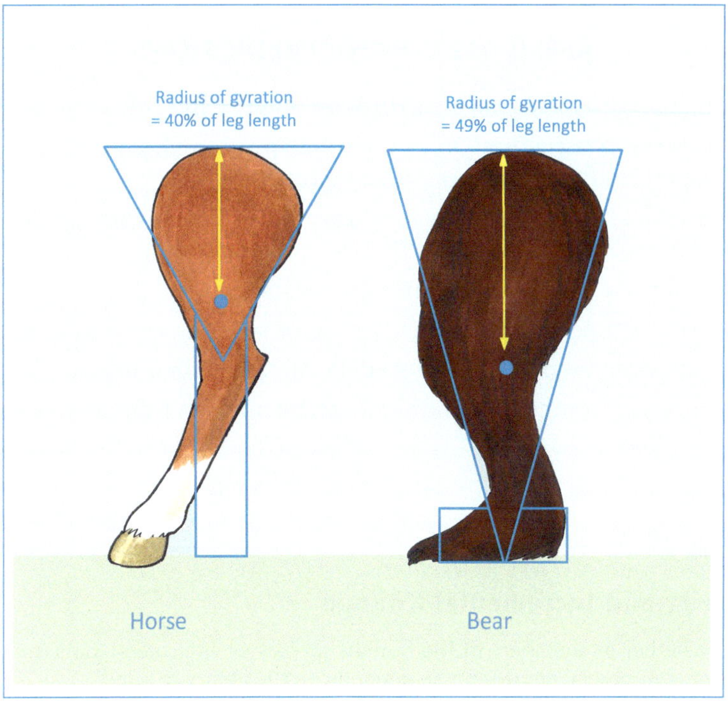

FIGURE 5.1.1
The weight distribution of the hind leg of a horse is compared to the weight distribution of a bear. A horse's weight is much higher on the leg, reducing its inertia (indicated here by the radius of gyration) and allowing faster acceleration and deceleration while running. (Redrawn from Hildebrand, M., *Am. Sci.*, 75, 594–601, 1987.)

Horses are built for speed (Figure 5.1.1). Their legs are elongated relative to their body size, but not too far as to cause undue interference between forelegs and hind legs (as in giraffes). Muscle attachment points are closer to the joints than in other animals, so that a small muscular contraction causes a faster movement of the end of the limbs (Hildebrand, 1987). Ligaments of the hind limb act as elastic bands to store energy upon impact and release it during movement of the leg. Leg bones are as light as possible and still have enough strength to sustain forces incurred during galloping and jumping. The weight of muscles is minimized because the structures of the joints lock the legs into movements in a plane; muscles and tendons are not needed to strengthen the joints in a transverse direction. Everything about the form of a horse's leg is related to its function of generating running speed.

Form and function: They are enough related that function can often be inferred from the form of a part of a living being.

Example 5.1.1: Species That Mimic Each Other

Mimicry, the situation where an unrelated species develops an appearance similar to another species, is sometimes an important strategy in biology. There are survival benefits to the mimicking species if the species being mimicked has some kind of defense against predators. This type of mimicry confers a selective advantage as long as the mimicking species remains scarcer than its model. Even defenseless species may mimic each other if it reduces the chance that any individual may be lost to predators (Vogel, 2010).

APPLICATIONS AND PREDICTIONS

1. The function of a biological structure can often be inferred from its shape.
2. Similar types of structures from different species will have similar shapes and properties if they perform the same function.
3. Similar structures with similar functions in different organisms could have evolved by different paths and still look and act similarly.
4. Ducks' webbed feet mean that they are meant to swim.
5. Organisms adapted to similar environments will possess similar features.
6. Dogs have been bred to have forms determined by their uses.
7. Prehistoric environmental conditions can be inferred from fossil records.

5.2 Modularity and Incremental Change

> Survival, whether as members of the human species or as professionals in a clinical, research, or commercial environment, depends on the ability to adapt.
>
> —**David Dewhurst**

Biological systems are modular. There are a few fairly simple building blocks that are used for multiple purposes, and it is the adaptation of these building blocks for new purposes that makes biology unique.

When environmental conditions change, and a new biological organism is needed, existing organisms change to meet the need. An entirely new organism is not created. Thus, there is a similarity of forms between organisms, and, for the most part, the more closely related the organisms are, the more similarity there is. Thus, the bones and soft tissues are similar among mammals, with strikingly close similarity among the closest related mammals. There is a similarity between mammal hearts and reptile hearts; there is a similarity between fish muscle and amphibian muscle. There is a similarity, too, in enzymes produced and metabolic pathways. Adenosine triphosphate (ATP) is nearly universally used as an energy transfer compound. That means that use of ATP as an energy storage compound evolved very early in the history of life.

Challenged with antibiotics, microbes evolve to new forms capable of immunity. Challenged with insecticides, insect pests evolve to become immune. Adaptation of weeds under herbicide pressure also occurs. In each case, entirely new organisms were not created. Rather, there was an adaptation of existing parts, put together in new ways.

We have become so accustomed to this tendency that we have lost appreciation for the fact that unique functions can be achieved by existing forms assembled in new ways. Thus, each cell in early fetal development can become nerve, bone, or liver; further modification fixes their functions.

It is obvious from their appearances that biological organisms share a close relationship. They share similar morphological features, each with some variation from the next. Indeed, the closer two species are related, the more similar they appear. Among humans, members of the same family often appear very much alike, and can be distinguished from unrelated individuals quite readily (see also Section 6.18.6, Courtship). What is not so obvious is that even at the molecular level, there is a great deal of commonality. All organisms use the same set of bases for their genetic codes. Plants and animals share at least 50% of the same genetic code; fruit flies and humans share 44% of the same genetic code; chickens and humans share 60% of the same genetic code; primates and humans share 99% of the same genetic code; and among humans, there is only a 0.1% variation.

New genes, new organs, new species, and new family members do not appear spontaneously, but are derived from existing genes, organs, species, and family members. The differences are sometimes larger and sometimes smaller, but they are clearly related.

Because of this, it is often easy to guess at the function of an organ in a newly discovered species by knowing the function of a similar organ in familiar species. A stomach is a stomach is a stomach.

If a new form of life is necessary to adjust to a new environment, a new species is not formed from scratch, but the new species evolves from an existing species by incremental change. New functions are found for existing organs; improved function is derived from existing function. The old is not discarded, it is changed. (One might be tempted here to say that the old is improved upon, but improvement is relative; an improvement for one set of environmental pressures may be a worsening for a different set of environmental pressures.) The new does not erupt spontaneously; it is formed from the old.

We can therefore trace lineages by looking back over the incremental changes that are evident from one organism to a previous one. When we do that, it becomes clear that, whether we view a difference as either small or large, any differences we see are not revolutionary but evolutionary. The same gene that controls the process of cell division, and which, when defective, causes cancer in humans, is also responsible for allowing fruits, such as tomatoes, to achieve the unnaturally large sizes we expect from our orchards and gardens. This is an example of identical or similar biological structures having somewhat different outcomes in different species.

We have seen how biological form is related to biological function. That is, geometrical shapes are related to the functions of different parts of biological entities. Incrementality develops the form through many small changes over many generations (Figure 5.2.1), as long as each small change gives the possessor of the change some advantage to survive

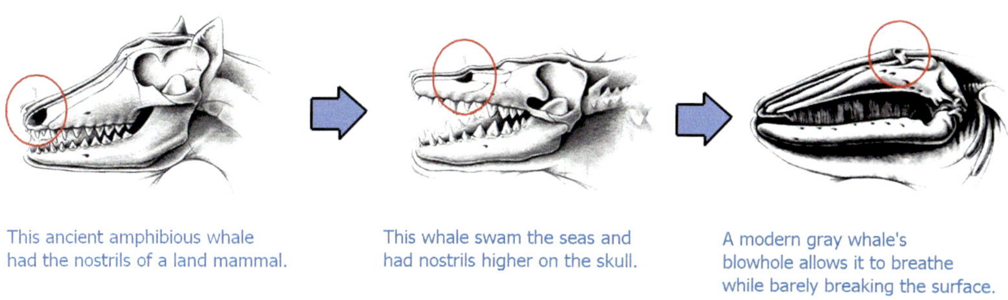

FIGURE 5.2.1
Whales began as land animals that migrated to live in the water in ancient times. As different species spent more time in the water, the blowhole through which they breathed evolutionarily changed location to make it easier to breathe in their aquatic environment. With a smaller energy requirement for breathing, there was more energy that could be expended for growth and reproduction. (Redrawn from Chadwick, D., *Natl. Geogr.*, 200, 65–77, 2001.)

and reproduce. If the incremental change cannot be passed to succeeding generations, and give them survival and reproduction advantages, then the changes are not permanent and must begin anew for each generation (Gould, 1986).

Evolution from one form to another, more adapted, form is unlikely to occur if the intermediate form is less well adapted than the more primitive form from which it starts. Squid are less economical swimmers than salmon, but squid were unlikely to develop fishlike tails that would have made them better swimmers because there does not seem to be any conceivable evolutionary route from a squid to a fishlike animal that would not involve passing through a stage less fit than either (Alexander, 2003).

This is precisely why form and function are so closely related in biological organisms. Not all incremental changes are able to improve the functioning of the organism, but those that do are able to make better use of physical or chemical principles than previous forms. The fittest survive.

Humans have had a profound effect on selection of form and function of species useful or interesting to them. The species include pets, animal companions, livestock, ornamental and agricultural plants, insects, bacteria, yeasts, and other microbes (Pell and Allen, 2015).

Behavior is also incremental. Biological organisms adapt much better to small changes than to large ones; small adjustments are more easily made than large ones. As an example, consider thermoregulatory adjustments in humans. Within a relatively small temperature range, called the thermoneutral zone, humans may regulate their body temperatures by cutaneous vasoconstriction or vasodilation (reducing or increasing blood flow to the skin). A warm skin surface loses more heat to the environment than a cooler skin surface (see Section 2.7, Heat Transfer). This adjustment is small, easily made, and usually not noticed. A large increase or decrease in temperature, however, results in much more noticeable and more difficult changes. The cold elicits shivering, going to the closet for more clothes, and turning up the thermostat.

Hot elicits profuse sweating, removal of clothes, seeking cool spots, drinking extra fluids, and reduction of activity. If the onset of cold or hot is sudden, the person feels extremely uncomfortable. If the onset is slow, such as during the change of seasons, conditions that would feel very uncomfortable for a sudden change can be tolerated well when they occur slowly. This is called thermal acclimation and involves hormonal and other physiological adjustments that occur relatively slowly. Conversely, thermal conditions that would feel comfortable for the unacclimatized may feel uncomfortable for the acclimatized. Biological engineers should be aware of these differences when designing environments for plants, animals, or microbes.

As a second example, consider the case of the common honeybee, *Apis mellifera* and its relation to a parasite mite, *Varroa destructor*. For years, the mite coexisted with the Eastern honeybee, *A. cerana*. It did not disturb the Eastern honeybee all that much because the behavior of the honeybee kept mite populations relatively low; it was an insignificant pest. Sometime in the middle of the twentieth century, *A. mellifera* was introduced into Asia, and *Varroa* underwent a change that made it able to parasitize its new host. Since then, *Varroa* has become a major pest of *A. mellifera*.

The honeybee is slowly adapting to the new threat. There are small pockets of resistance to *Varroa*, and those bees that cannot tolerate *Varroa* are dying. The first defense by beekeepers was to depend on pesticides to control the mites, but the eventual solution will be resistance that develops in the honeybees themselves because of natural selection. As long as this sudden parasitic threat does not result in the catastrophic loss of all honeybees, slow adaptation will allow bees to be able to coexist with the mite. Incremental changes will occur to ensure the survival of this species of honeybee.

Nowhere is evolutionary adaptation more apparent than in the finches of the Galapagos Islands off the western coast of South America. These birds, known as Darwin's finches after the famous biologist who described their similarities and differences, exhibit a wide range of adaptations to local environmental conditions on the islands where they live (Grant and Grant, 2002). Some have developed stronger beaks to crack strong seeds, some have become better adapted to dry conditions rather than rain forest, and some spend more time on the ground whereas others are found in trees. These adaptations have been so profound that interbreeding of the most distant of these birds does not happen; they have developed into entirely different species. By studying these birds, Darwin was able to describe what he stated was the way new species evolve from a common ancestor.

PARSIMONY PRINCIPLE

The *Parsimony Principle in Biology* posits that natural selection operates as simply as possible over time. Therefore, one can judge the traits of an ancestor by looking at traits of present day individual groups. Traits shared by several related groups are even more likely to have been the same or similar to traits possessed by common ancestors (Boehm, 2012). This principle is thus recognition of the modularity of biological physical and behavioral characteristics.

Convergent evolution is the term that describes the result of evolutionary pressures on organisms with different origins. For example, rotifers are small animals that live in soil and in water. Most rotifers have a simple gut lined with cells running from the mouth to the anus. In a number of very small rotifer species, this gut is absent and replaced by a continuous cytoplasmic sac that forms food vacuoles at the mouth end and ejects wastes from the vacuoles at the anal end. This digestive system is exactly like that found in ciliate protozoa, organisms of the same size occupying a similar ecological niche. The small rotifer has apparently replaced its ancestral digestive system with the simpler digestive system of the protozoa because it was subject to the same environmental pressures as was the protozoa. This might indicate that the same genes present in protozoa are also present in rotifers, and are only expressed under the proper set of circumstances.

The degree of similarity between body shapes, biomechanical muscle forces and attachments, and swimming styles of sharks and tunas well illustrates convergent evolution brought about by nearly identical environmental factors (Shadwick, 2005). Although each of these organisms arose at different times from very different predecessors, both have developed constant and fast swimming styles with teardrop-shaped bodies, to streamline water flow, specialized muscle biomechanics to isolate their swimming movement to the tail region, highly specialized gills to supply adequate oxygen amounts, and regions of nearly constant muscle temperature for dependable muscle contractions.

THE EVOLUTION OF HEMOGLOBIN

Four billion years ago, when the earth was young and without life, it is believed that there was no oxygen in the air. The earth's atmosphere contained mostly water vapor, nitrogen, methane, and ammonia (Hardison, 1999). When the first organisms developed about 3.8 billion years ago, these atmospheric constituents were used for food and energy. It seems plausible that these early metabolic reactions were facilitated (or catalyzed) by metals, such as iron and magnesium (Hardison, 1999).

Between 3.3 and 3.5 billion years ago, there appeared cyanobacteria that could convert energy from the sun into chemical energy through photosynthesis. Cyanobacteria removed electrons from hydrogen sulfide present in the atmosphere to produce elemental sulfur and ATP. Photosynthetic bacteria appeared sometime between one and two billion years ago that used water (H_2O) rather than hydrogen sulfide (H_2S) as the chemical substrate. As a result, the earth's environment was remarkably transformed.

The oxygen so produced was released into the atmosphere and gradually came to be its most important constituent. With abundant oxygen, other life forms appeared that could use the highly reactive nature of oxygen to their metabolic advantage. It was not an easy trick: oxygen can be toxic as well as a metabolic enhancer.

In order to manage oxygen presence and availability, some biochemical means had to be employed. In some cases, oxygen had to be transported between cells, in other cases, oxygen had to be bound so that it didn't react when it wasn't supposed to, and in still other cases, oxygen needed to be acquired from places where it was scarce. Sometimes all three functions were required.

The *porphyrins* are cyclic compounds formed from four *pyrrole* rings linked by methane bridges in a ring system (White et al., 1959; Harper, 1963). A characteristic property of porphyrins is the formation of complexes with metal ions such as iron and magnesium.

A particular porphyrin ring containing magnesium is the organic molecule *chlorophyll*. This biochemical is used to help harvest sunlight for use in photosynthesis through the Calvin cycle. Another porphyrin ring containing iron is called *heme*. Heme bound to globin molecules is called *hemoglobin*. Hemoglobin is the protein that binds oxygen in the lungs and gives blood its red color, and performs all required functions listed previously.

Hemoglobin is a very ancient molecule that is present in a huge variety of life forms, although different species do carry different variants of hemoglobin. It is found in animals, plants, protists, and eubacteria, suggesting that the gene for hemoglobin is truly ancient and suggests that they all share a common ancestor very early in organismal evolution (Hardison, 1999).

Example 5.2.1: Cancer Cell Drug Resistance

Cancerous cells are resilient; they can develop effective defense mechanisms to avoid toxic or static effects of chemotherapeutic agents. Although therapeutic drugs may kill the vast majority of cancer cells, those that survive can form the nucleus for drug-resistant cancers. New cancer growth would be impervious to the chemotherapy that was previously effective. Predict the type of cancers that are most likely to be drug resistant.

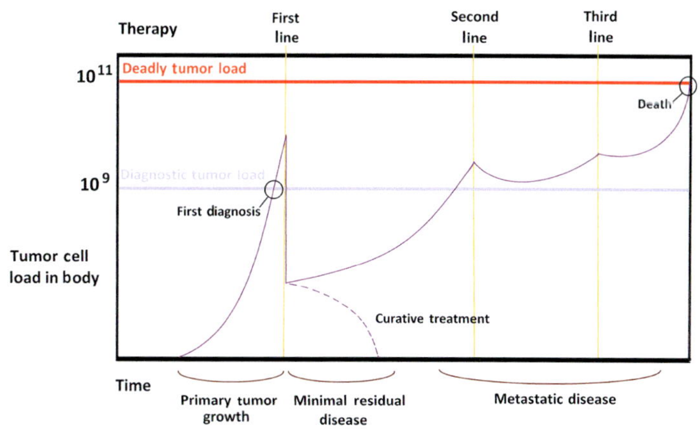

FIGURE 5.2.2
Cancerous tumors grow until they are detected. The first-line therapy knocks out the susceptible tumor cells, but often some survive. These are resistant to the primary therapy drugs, and these resistant cells grow unimpeded until they again reach a detectable number. At that point, alternative therapies are used, and the cycle repeats. Finally, tumor cells reach the point where they cause the death of the tumor host. (Redrawn from Baeuerle, P.A., *Drug Discov. Dev.*, 11, 32–35, 2008.)

Solution

Gastrointestinal tract and kidney tumors are most likely to be drug resistant. These tissues have evolved to tolerate natural cytotoxins in foods.

One important defensive mechanism used by these and other drug-resistant cells is a natural drug pump dependent on a protein molecule called P-glycoprotein resident in the cell membrane. After the cytotoxic drug enters the cell by diffusion, this chemical draws it out again and lowers the intracellular concentration to sublethal concentrations. One solution to this problem is to block the drug pump, either by a second drug or by inserting a gene into the cell that interferes with the drug pump mechanism (Figure 5.2.2).

APPLICATIONS AND PREDICTIONS

1. A majority of the genetic code of insects will match the genetic code of humans.
2. Vestigial organs will be seen at various stages in fetal development.
3. Related species will have similarity of structures.
4. Formation of new species will not be easy to detect because the differences will be small.
5. Microbes in a bioreactor will adapt to new conditions better if the new conditions are imposed slowly rather than suddenly.
6. Permanent changes in the characteristics of a population occur slowly, so they will only adapt to long-term average environmental conditions.

(*Continued*)

APPLICATIONS AND PREDICTIONS (Continued)

7. Adaptations to new environments will usually come from physical or behavioral traits already present to some extent in the population of organisms.
8. Any environmental change that kills the entire population will be too severe to be accommodated. The dinosaurs probably faced this challenge.
9. Human appendixes have no known remaining function, but unless there is a survival cost to retention of appendixes, humans will continue to have them.

5.3 Genetic Basis

> Evolution is blind; technology is mind.
>
> **—Chris Calladine**

It's really the ultimate in biological control. That is the determination of structure and function of the next generation by the previous generation. This is the effect of genetic inheritability.

Genetic material has been called the blueprint of life. It is the material that stores accumulated information about the complexity determined from generations upon generations of adaptations and modifications to environmental pressures.

5.3.1 DNA as the Blueprint

> Instead of the book of life, DNA is more like the scrapbook of life. Sentences, paragraphs, or entire chapters are copied and haphazardly inserted into various parts of our genome. In some people, the same page repeats over and over, while other people don't have that page at all.
>
> **—Steve Olson**

The basic storage code is found in very ancient biochemical molecules called *deoxyribonucleic acids* (DNA). These nucleic acids contain a phosphate group, a five-carbon sugar (deoxyribose), and a nitrogenous base. DNA strands are formed when the phosphate groups and sugars bond covalently to form a backbone, with the nitrogenous bases exposed like ornaments on a Christmas tree (Figure 5.3.1).

The bases are extremely important to DNA. They can either be in the form of a pyrimidine (a six-member ring of carbon and nitrogen) or a purine (a six-member carbon and nitrogen ring attached to a five-member carbon and nitrogen ring). Of the two, the purines are larger.

DNA contains four different bases: the pyrimidines cytosine (C) and thymine (T) and the purines adenine (A) and guanine (G). As it happens, DNA forms a double-helix

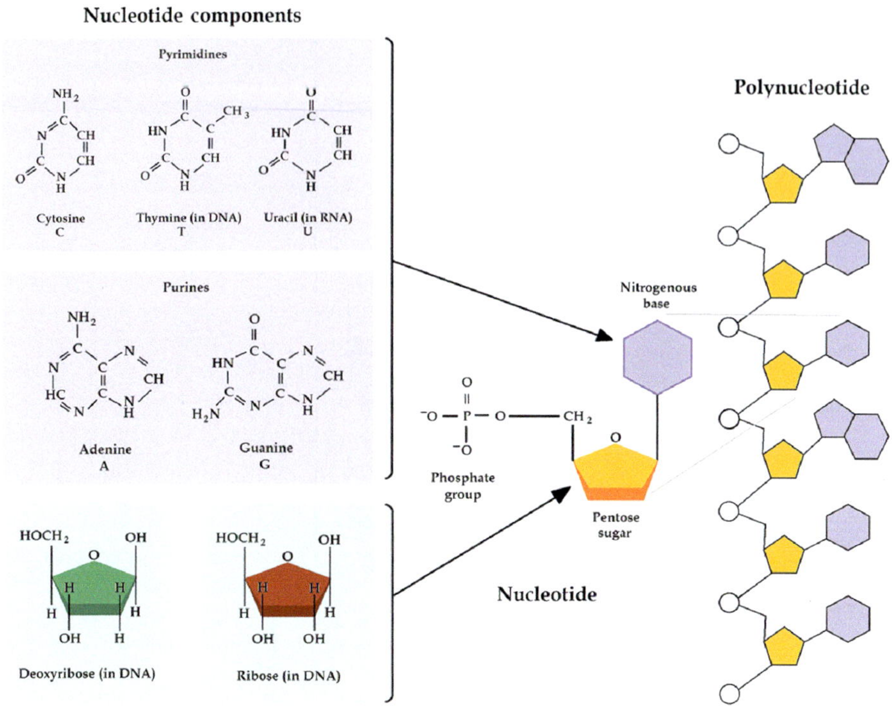

FIGURE 5.3.1
The structure of DNA and RNA. Each unit has a phosphate group (yellow) and pentose sugar (green) to form the backbone. Pyrimidines or purines attach to the sugars to complete the strand. DNA and RNA include slightly different pyrimidines and sugars, but otherwise have the same structure. (Redrawn from Campbell, N.A. et al., *Biology*, 5th ed., Addison Wesley Longman, Menlo Park, CA, 1999.)

structure (Figure 5.3.2), composed of two complementary DNA strands. DNA bases in double-helix form are hidden behind highly charged backbone strings of sugars and phosphates. They are thus protected from the surrounding environment. Hydrogen bonds between the two bases (called a *base pair*) of the complementary strands hold the two strands together. The only possible combinations of bases between the two strands are pairings of purines with pyrimidines:

$$A \leftrightarrow T$$

$$G \leftrightarrow C$$

Thus, if a sequence of bases on one strand is A A G T C, the sequence of bases on the complementary strand must be T T C A G.

DNA is a very stiff molecule, with its double-helix structure and surface charges that repel other DNA surface charges (Babbit, 2011). Because of this, DNA does not curl up unaided into a tight ball inside the cell. Instead, DNA is wrapped spool-like around histone protein cores, called *nucleosomes*, which keep the DNA strands from straightening out into their full 2 m length (in human cells) and disrupting chromosome structure inside the five-micron cell nucleus (Figure 5.3.3). Nucleosome positions with respect to one

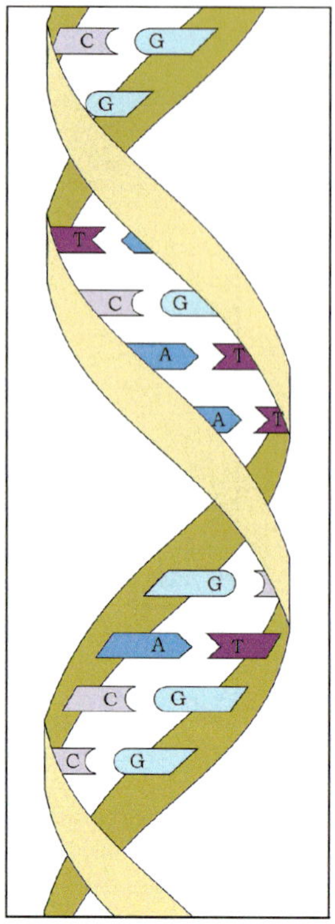

FIGURE 5.3.2
The double helix. The DNA molecule is usually double-stranded, with the sugar-phosphate backbone of the polynucleotides (abbreviated here by blue ribbons) on the outside of the helix. In the interior are pairs of nitrogenous bases, holding the two strands together by hydrogen bonds. (Redrawn from Campbell, N.A. et al., *Biology*, 5th ed., Addison Wesley Longman, Menlo Park, CA, 1999.)

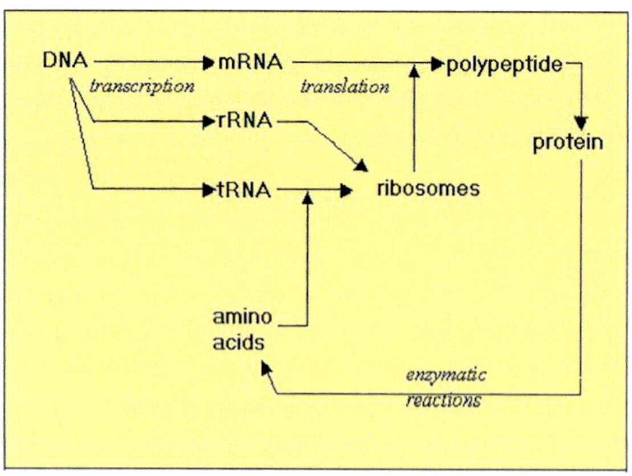

FIGURE 5.3.3
DNA is such a stiff molecule that it would not coil up enough by itself to fit inside the cell nucleus. Instead, it is wrapped around histone proteins to form nucleosomes that are serially located along the chromosome fibers 30 nanometers in diameter. The positions of different genetic materials (genes) in the nucleosomes probably influences genetic expression. Here, a single-stranded RNA molecule is also seen to be transcripted from one side of the DNA double helix. (Redrawn from Babbit, G.A., *Amer. Sci.*, 99, 48–55, 2011.)

another help to determine gene expression within the cell, and these positions appear to have evolved over time by natural selection pressures exerted by the environment. Genes active in the formation of particular proteins are located close to one another in the mass (Lieberman-Aiden et al., 2009).

Methyl groups and other molecules can bind to the helix itself or to its histone protein core and cause the whole assembly to flex. As that happens, some genes are exposed, and others are obscured. The exposed genes are the ones that can be actively expressed.

5.3.2 RNA as the Fabricator

> In nature, technology has already been at work for millions of years.
>
> —**Buckminster Fuller**

The ultimate result of DNA is that proteins are formed regulating the behavior of the cell. DNA is not the means to do this directly, however. Proteins are formed from polypeptides (two or often more amino acids linked together), and these are formed from amino acids that bear a distant relationship to DNA. The organization of this process is found in simplified form in Figure 5.3.4.

Short sections of DNA act as templates for the formation (*transcription*) of messenger RNA (mRNA). RNA is *ribonucleic acid*, which differs from DNA in that the five-carbon sugar is ribose instead of deoxyribose. RNA forms a polymer structure similar to DNA, except that it is single-stranded and the pyrimidine base uracil (U) replaces thymine (T). When RNA is formed from a single DNA strand, base pairs can be

$$A \leftrightarrow U$$

$$G \leftrightarrow C$$

mRNA essentially carries information contained in one gene, which then results in one polypeptide.

Because there are 20 common amino acids and 4 nucleotide bases, 3 bases must be the minimum number to specify each amino acid. If each base coded for one amino acid, only four amino acids, each corresponding to one of the four bases, could be specified. If two bases were necessary for each amino acid, the number of base pair combinations is $4 \times 4 = 16$, still less than the 20 amino acids to choose from. Three base pairs gives $4 \times 4 \times 4 = 64$ possible combinations, a number greater than the required 20. There are some combinations that redundantly specify the same amino acid, and some base pair combinations seem to encode for starting and stopping operations. The combination of the three base pairs required to specify one amino acid is called a *codon*. Whereas one gene is associated with one polypeptide made from several amino acids, many codons comprise one gene. Noncoding sections of DNA are called *introns*, whereas sections of DNA that code for proteins, or portions of proteins, are called *exons*. Exons are said to be *expressed*, but introns are not.

The polypeptide is actually formed outside the cell nucleus as a *ribosome* moves along the mRNA molecule. The ribosome is composed, in part, with ribosomal RNA (rRNA) that helps to align the amino acids with the mRNA codons. Transfer RNA (tRNA) molecules, formed of short strands of RNA, wander through the cytoplasm, linking with specific amino acids floating in the cytoplasm and transport them to the ribosome, where they are

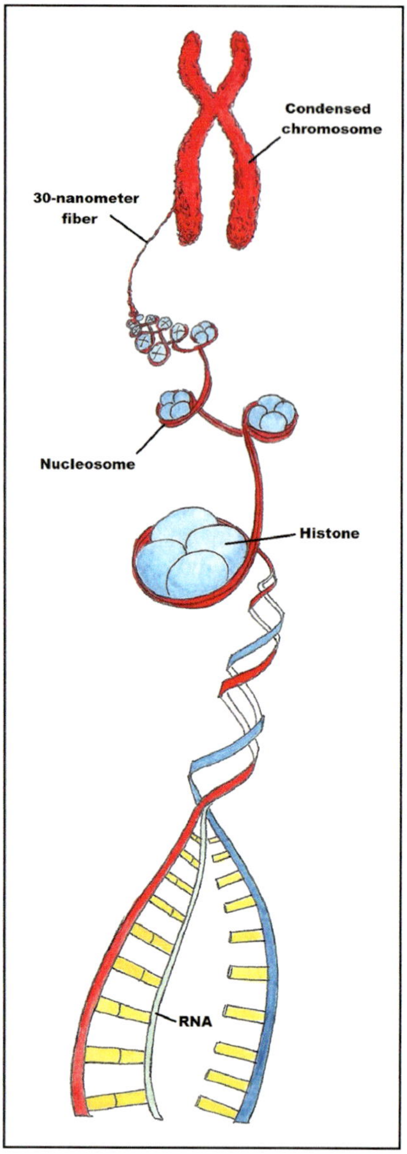

FIGURE 5.3.4
Simplified diagram of the production of proteins from DNA instruction set.

joined to form a polypeptide (Figure 5.3.5). All three types of RNA, and several other types also, are transcribed from different sections of DNA.

As with most biological affairs, the strict association of one base pair with another is not always so. When the association between specific base pairs is relaxed somewhat, several forms of the same tRNA are possible as it folds on itself. This is called *wobble*.

The actions of mRNA and ribosome are similar to a tape playing instructions into a parts-assembly machine. The protein or polypeptide is assembled one amino acid at a time, and, once produced, can act as an enzyme, for example, to create the amino acids

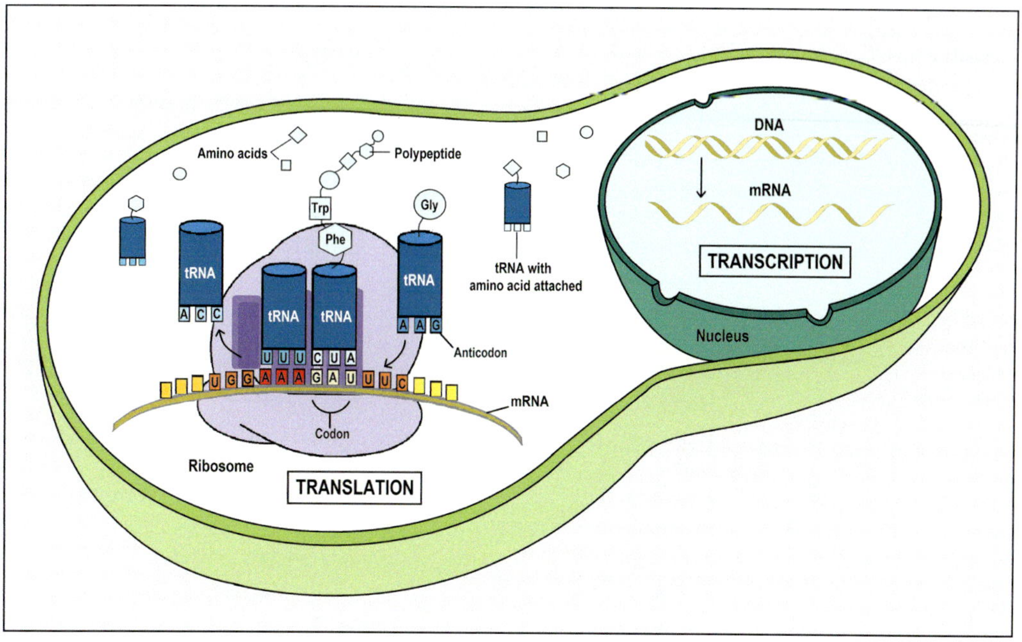

FIGURE 5.3.5
Polypeptides are formed at the ribosome through a process called translation. mRNA acts as a template, and tRNA acts as a scavenger, bringing the correct amino acids to the ribosome in the correct order determined by the mRNA. (Redrawn from Campbell, N.A. et al., *Biology*, 5th ed., Addison Wesley Longman, Menlo Park, CA, 1999.)

required for the next set of proteins. In this way, the sequence of DNA bases is translated into the amino acid sequence of a protein. The feedback loop comprising protein formation, enzymes, amino acid synthesis, and ribosome action indicates that the process can be regulated so that the amount of an enzyme produced depends on the metabolic needs of the cells.

The genes determine the sequence of amino acids in a protein, but they do not, in and of themselves, determine how the protein is folded. Thus, the genes are only the foundation of cell activity, not the entire blueprint.

Only about 1.5%–2% of the human genome contains sequences that code for proteins (Table 5.3.1). This genetically expressed part of the genome, comprising the sum total of all exons coding for a particular protein, is called the *exome*. There are some transcribed strings of RNA that do not produce proteins; these are called noncoding RNA, or ncRNA (or, lincRNA for linear ncRNA). These appear to function by regulating gene expression. One type of ncRNA is called a *riboswitch* that helps to turn genes on (called genetic *up regulation*) and off (called genetic *down regulation*). Riboswitches attach to the genes they regulate and can be sensitive to the levels of compounds targeted by those genes. The genes can then be made to produce more or less of substances regulating the target compound amounts. Long non-coding RNA (lncRNA) is a type of ncRNA that regulates gene activity in the nucleus. As with many regulatory systems in biology, there are mechanisms to turn on a system and mechanisms to turn off the same system and the final system action is given by the difference between these two tendencies; lncRNA, for instance, can upregulate genes (enable them to be expressed), whereas RNAi downregulates genes. Some ncRNAs are important for the expression

TABLE 5.3.1

Correspondence between Amino Acids and DNA Codons from Which They Come

Three Letter Code	Amino Acid	Possible DNA Codons
Ala	Alanine	GCA, GCC, GCG, GCT
Asx	Asparagine or Aspartic acid	AAC, AAT, GAC, GAT
Cys	Cysteine	TGC, TGT
Asp	Aspartic acid	GAC, GAT
Glu	Glutamic acid	GAA, GAG
Phe	Phenylalanine	TTC, TTT
Gly	Glycine	GGA, GGC, GGG, GGT
His	Histidine	CAC, CAT
Ile	Isoleucine	ATA, ATC, ATT
Lys	Lysine	AAA, AAG
Leu	Leucine	CTA, CTC, CTG, CTT, TTA, TTG
Met	Methionine	ATG
Asn	Asparagine	AAC, AAT
Pro	Proline	CCA, CCC, CCG, CCT
Gln	Glutamine	CAA, CAG
Arg	Arginine	AGA, AGG, CGA, CGC, CGG, CGT
Ser	Serine	AGC, AGT, TCA, TCC, TCG, TCT
Thr	Threonine	ACA, ACC, ACG, ACT
Val	Valine	GTA, GTC, GTG, GTT
Trp	Tryptophan	TGG
Tyr	Tyrosine	TAC, TAT
Glx	Glutamine or Glutamic acid	CAA, CAG, GAA, GAG
[a]	stop codon	TAA, TAG, TGA

[a] The three codons (TAA, TAG, and TGA) are termination codons (also called stop codons or nonsense codons) that do not code for amino acids, but signal the end of the mRNA message (with U substituted for T) and provide the "stop" signal for protein synthesis.

of Hox genes in human epithelial cells (Gupta et al., 2010; Williams, 2011). In summary, the ncRNAs control the genes and the genes control the production of the target compound in a multi-level regulation process.

MicroRNA (miRNA) is single-stranded about 22 nucleotides long, and present in plants, animals, and some viruses. Its function is to mediate gene expression. Long non-coding RNA (lncRNA) consists of an RNA string longer than 200 nucleotides, lacking extended open reading frames, and plays a role in epigenetic gene expression.

The *transcriptome* is the translation of genetic information into cellular responses, and includes all RNA molecules mRNA, rRNA, tRNA, and ncRNA produced in a cell. This transcriptome is different between *in vitro* (cell culture) and *in vivo* (in natural context). Different mRNAs are expressed in each, so drugs tested in cell cultures can have different effects in isolated cells compared to cells in their natural environment. Even cells living next to one another, although appearing to be identical, can have different transcriptomes (Stein, 2014).

THE LARGEST GENE

The largest human gene is also the largest gene anywhere in nature. This gene consists of 79 exons, each encoding for a portion of the protein dystrophin. Dystrophin is a necessary muscular protein. Its absence causes the development of a chronic, eventually fatal, condition called Duchenne muscular dystrophy.

5.3.3 Gene Types

Happy is he who gets to know the reasons for things.

—Virgil

DNA material is found in chromosomes. Eukaryotic (higher level cells) chromosomes are each made of a single strand of DNA and several different kinds of proteins (Hale et al., 1995). These chromosomes exist in pairs (said to be *diploid*), except in some lower forms of algae or fungi that have single sets of chromosomes only (said to be *haploid*). Mammalian germ cells are also haploid. Some plants have chromosomes that occur in groups of four (*quadraploid*) or six (*hexaploid*). Prokaryotes (bacteria) have but one chromosome arranged in a circular shape. Diploid cells thus have paired (*homologous*) chromosomes containing identical genetic loci. Both of these genetic locations (called *alleles*) may either have the same DNA base pair sequence or they may be different. If the same, then the cell is described as *homozygous*; if the alleles are different, then the term is *heterozygous*.

Heterozygous cells have two different genes encoding for the same trait, giving the possibility of a conflict between them. Many genes display either *dominance* or *recession*, the dominant gene determining the apparent trait and the recessive gene being hidden except in very nonobvious tests (e.g., by gene-mapping). If both genes are homozygous recessive, then the trait is determined by the recessive gene and becomes apparent.

Crosses between homozygous-dominant and homozygous-recessive individuals (for one particular genetic trait) will give all heterozygous offspring that express the dominant trait (Figure 5.3.6). Crosses between two heterozygous individuals will, on average, yield one-quarter homozygous-dominant offspring, one-quarter homozygous-recessive offspring, and one-half heterozygous offspring (Figure 5.3.6). Three quarters of these offspring will, therefore, exhibit the dominant form of the trait, and only the homozygous-recessive offspring will exhibit the recessive form. Gregor Mendel observed these results with his classic experiments on peas with white (recessive) and purple (dominant) flowers. It was his pioneering work that showed that genetic characteristics did not blend together, but instead retained their essential qualities (Hellman, 1998).

The genetic makeup of a cell is called the *genotype*, and is determined fully when every allele is known. The physical appearance and apparent physiological traits of a cell is called the *phenotype*. In the case of Mendel's flowering peas, there were three separate genotypes resulting from a cross between two heterozygous individuals (purple-purple, purple-white, and white-white); there were, however, only two distinct phenotypes (purple or white).

Two homozygous dominant parents	
AA × AA ↓ AA AA AA AA	100% dominant

One homozygous dominant and one heterozygous dominant parent	
AA × Aa ↓ AA AA Aa Aa	100% dominant

One homozygous dominant and one homozygous recessive parent	
AA × aa ↓ Aa Aa Aa Aa	100% dominant

Two heterozygous dominant parents	
Aa × Aa ↓ AA Aa Aa aa	75% dominant

One heterozygous dominant and one homozygous recessive parent	
Aa × aa ↓ Aa Aa aa aa	50% dominant

Two homozygous recessive parents	
aa × aa ↓ aa aa aa aa	0% dominant

FIGURE 5.3.6
Genetic outcomes from different crosses.

NAMING OF GENES

Discovering and describing a new gene confers with it the privilege of naming that gene.

Some genes carry prosaic names related to their location in the genome, such as *SDCCAG8* in humans (related to a kidney failure disease). Others, however, are named creatively and imaginatively by investigators (Kresge, 2013). Thus, we have, in fruit flies, *couch potato* and *benchwarmer* (related to laziness), *tweek* (quivering), *Piwi* (small testes), *ken* and *barbie* (external genetalia), *kojak* (baldness), *cheapdate* (sensitive to alcohol), *VanGogh* (hair patterns), *fruitless* (sex differences in the fruit fly's brain that make females get drunk faster than males), and *18wheeler* (stripes on larvae). There are others, such as *superman* (flowers with extra stamens), *tiggywinklehedgehog* (neural patterning in Zebra fish), *werewolf* (plants with hairy roots), and *tigger* (human transposon).

5.3.4 Genetic Expression

> The Austrian monk Gregor Mendel was also studying mice, which he bred with the goal of deciphering the inherited traits of coat color. However, to Mendel's conservative bishop, the thought of a monk spending his time with copulating mice seemed inappropriate. He banned the mice, and Mendel set his sights on a less prurient subject for investigation, peas.
>
> **—Terri Peterson Smith**

Mendelian experimental results were extremely simple, much simpler than many other real results. Although there are cases where traits are determined by two independent alleles, one dominant and one recessive, and Mendel happened to have observed these in his experiments, all is not that easy. There are cases of multiple alleles (human blood types are an example). There are cases of incomplete dominance and codominance when intermediate levels of biochemicals are either not sufficient to produce the full effect or when intermediate levels can produce the full effect. There is *pleiotropy*, where one gene can affect an organism in many ways, and there is *epistasis*, where one gene is affected by the presence of other genes. *Antagonistic pleiotropy* is the term applied to a gene that has a strong positive, normally reproductive, effect on the young, but an adverse impact on the old individual. Some organismal traits are determined by multiple genes, a condition known as *polygenic inheritance*. And, finally, there is *genetic linking* between genes located on the same chromosome. It is possible that the effect of a gene can depend upon which parent contributed the gene. Thus, the Mendelian model is one of the simplest of possible genetic models, but it is the place to start (Weiss, 2017).

Not all genes are expressed equally. Expression of genes means that RNA and proteins are formed from the DNA template (Figure 5.3.7). This can be seen by considering the many types of cells in the human body. Each cell contains the same genetic material, but some cells make certain hormones, other cells react to stimuli, and other cells yet have

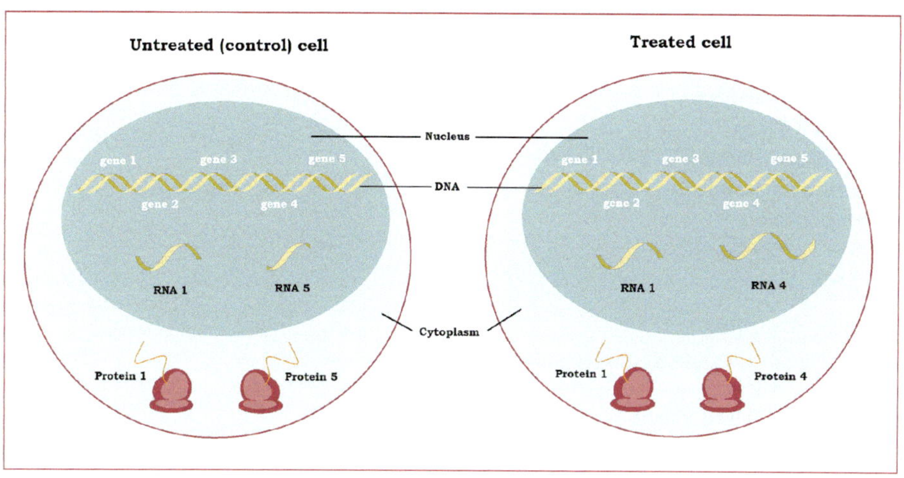

FIGURE 5.3.7
This hypothetical eukaryotic cell has five genes located in the nucleus. Only genes 1 and 5 are active in the control condition. Given a treatment consisting of a different set of environmental conditions (treated cell), genes 1 and 4 become active. RNA from these genes is formed in the nucleus; corresponding proteins are formed in the cytoplasm. (Redrawn from Hamadeh, H. and Afshari, C.A., *Am. Sci.*, 88, 508–515, 2000.)

different structures. Although each cell has the same genotypical prototype, environmental factors change phenotypical properties. Similarly, environmental conditions outside the organism also affect gene expression (Table 5.3.2). Thus, knowledge of the genotype is no guarantee that the phenotype can be predicted.

Very few somatic cell genes are actively expressed. There appears to be something in the cell's cytoplasm that directs genetic expression (Saltus, 2006). Some of this may have to do with *methylation*, wherein a methyl hydrocarbon group (CH_3) replaces a hydrogen atom on a genetic nucleotide base. A gene that is methylated lies dormant.

Fusing human embryonic stem cells with somatic cells can reawaken dormant genes. Sometimes this fusion leads to a double set of chromosomes, but sometimes the adult nuclei completely replace stem cell nuclei (Saltus, 2006).

TABLE 5.3.2

Genes or Environment? Estimates of Relative Contributions of Each to Personality Traits and Physical Conditions Based on Studies of Identical Twins

	Genetic Contribution (%)	Environmental Contribution (%)
Apple fruit size	25	75
Brain Structures		
Corpus callosum (connects brain hemispheres)	95	5
Parietal lobe white matter (logic and visual-spatial skills)	85	15
Temporal lobes (learning and memory)	45	55
Human Diseases		
Age-related macular degeneration	50	50
Asthma	60	40
Breast cancer	27	73
Children's earache	71	29
Crohn's disease	23	77
Depression (men)	29	71
Depression (women)	42	58
Schizophrenia	50	50
Type 1 (juvenile) diabetes	70	30
Type 2 diabetes	10	90
Human Personality		
Chronotype (morning or (evening person)	50	50
Having confidence	51	49
Height preference of mate	89	11
Impulsiveness	49	51
Optimism	50	50
Perception of body weight status	47	53
Procrastination	46	54
Resilience	50	50

(Continued)

TABLE 5.3.2 (*Continued*)

Genes or Environment? Estimates of Relative Contributions of Each to Personality Traits and Physical Conditions Based on Studies of Identical Twins

	Genetic Contribution (%)	Environmental Contribution (%)
Human Attributes		
Autism	70–90	10–30
Blood group	100	0
Blood pressure	55	45
Body composition	80	20
Body weight	75	25
Grip strength	65	35
Hair texture (straight or curly)	90	10
Height	80–90	10–20
Lean body mass	70	30
Musical pitch	76	24
Nightmares	36	74
Obesity	70	30
Voting choices in elections	60	40
Livestock Animals (sheep)		
Birth weight	30	70
Mature body weight	40	60
Multiple births	15	85
Rate of weight gain	30	70

There is not a one-to-one mapping between DNA and mRNA sequences (Hamady et al., 2005). Through a process called *splicing*, certain portions of the RNA transcribed from DNA can be deleted or added. In *alternative splicing*, different pieces of transcribed mRNA are deleted under different circumstances. The result is several different mRNA forms from one DNA master sequence. In *trans-splicing*, mRNA molecules from different DNA sequences are spliced together. The result of this is mRNA from several DNA genes. In fruit flies, for instance, different mRNA forms are produced depending on the sex of the fly.

The phenotype (visible configuration) of the marsh plant *Sagittaria sagittifolia* depends on its environment. As shown in Figure 5.3.8, its leaf forms depend upon the degree to which it is submerged. Nowhere is there a better illustration of the interaction of environment with genetic expression than this.

Although chimps and humans share about 99% of their genetic material, there are large differences between the two species. This is evidence that perhaps it takes more than a genetic blueprint to determine the characteristics of a species, and of an individual within a species.

Most genes have a switch, called a *promoter*, which controls the activity of the gene. Other regulatory elements, called *enhancers*, also are involved. Together, they determine how, when, and if a gene becomes active. Plant genes have been shown to be made ineffective

FIGURE 5.3.8
The marsh plant *S. sagittifolia* as it appears in different conditions. This illustrates that genetic expression is highly dependent upon environmental factors. (Redrawn from West-Eberhard, M.J., *Developmental Plasticity and Evolution*, Oxford University Press, Oxford, UK, 2003.)

by chemically attaching methyl groups to their surfaces (Brown, 2006). This is apparently the way plants regulate gene expression to control cell growth and development. In this way, fundamental genetic sequences are preserved but heritable changes in gene expression are made possible.

The transcription of mRNA from a gene requires the active presence of promoters, enhancers, and transcription factors produced from other genes. Some proteins can act as gene regulators.

Thus, some genes regulate the activity of other genes, and the activities of both are dependent upon environmental influences. Apparently, large differences in physical, intellectual, and emotional characteristics of animals can thus be determined not by changing the presence of certain genes but by whether these genes are used, when they are used, and for how long.

Humans have 23 pairs of chromosomes (or 46 chromosomes) tightly packed together; other species have other numbers of pairs of chromosomes (one copy coming from the female parent and one copy from the male parent). Along each chromosome are found sequences of DNA identified as genes, but there are many more long stretches of DNA for which the functions have not been determined. Some of these stretches transcribe into RNA that does not result in protein formation. Rather, these bits of RNA probably help regulate gene expression. Other parts of the nongene DNA may have other functions such as alignment of genes, coordination of replication, contributing to complex genes overlapping other genetic DNA, direct regulation of gene expression, and emergency genes that function only in extreme situations.

The Hox genes are a set of genes present in nearly all animals from insects to humans, including worms, frogs, chickens, and mice. These genes regulate the animal's basic body configuration, organizing the arrangement from head to tail (Figure 5.3.9). The order of the Hox genes along the chromosome corresponds to the order of body segment

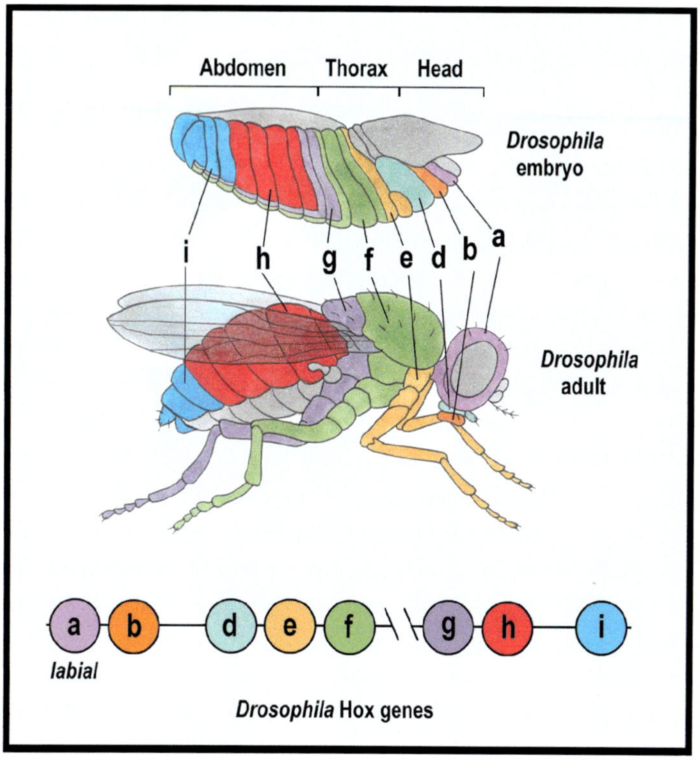

FIGURE 5.3.9
Hox genes in the fruit fly and the portions of the body that they help form. The first gene, known as *labial*, is necessary for the genes to be expressed. The order of the genes on the chromosome corresponds to the order of body part development. (Redrawn from Willmore, K.E., *Amer. Sci.*, 98, 220–227, 2010.)

development, starting from the head and ending at the tail (Willmore, 2010). Hox genes control the actions of other genes by turning them on or off. Thus, although the complement of genes for some species may be the same as for others, it is the action of the Hox genes that enables some of them to be expressed and others to be quiescent, so that the various species turn out entirely different from each other. In this way, differences in a few genes can have profound effects on phenotypical behavior of an organism (Gilad et al., 2006).

The Hoxc8 gene determines the location of the thorax in a developing animal (Figures 5.3.9 and 5.3.10). If this gene is active for a long time during development, then the animal becomes almost all thorax, as is a snake. If the Hoxc8 gene is turned on for a short time, the animal has a short thorax, as with a chicken. An intermediate period of activity results in something like a rat. The snake has all of its vertebrae located in its thorax, the rat has 13 thoracic vertebrae, and the chicken has 7. The same gene is in all these three, but expressed differently.

Variable numbers of a gene can produce a greater or smaller amount of protein important to the body, but a duplicated section of DNA can also disrupt the function of an important gene. Structural differences in the genome (given as duplications, insertions,

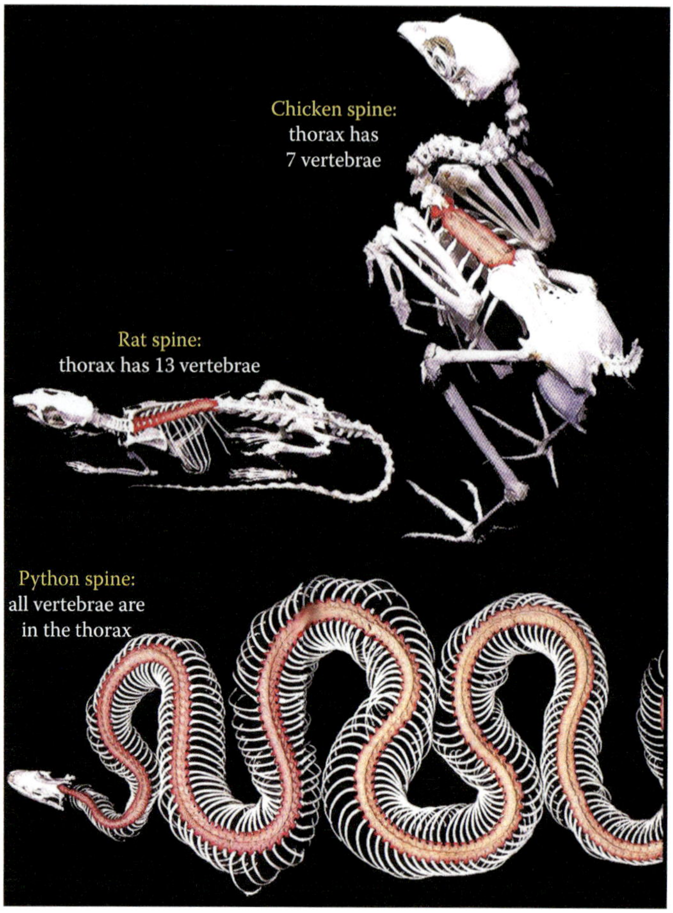

FIGURE 5.3.10
Slightly different promoters lead to big differences in the expression of the same gene for size of the thorax.

deletions, or inversions of genetic material) seem to be as important, or more important, to the causes of genetically linked diseases than are differences in the nucleotides A, T, G, and C (Olson, 2007).

Human cancer cells are overwhelmingly *aneuploid*; that is, some cells have an extra chromosome and some have fewer than normal (Scudellari, 2014). The extra aneuploidy chromosome seems to allow quicker responses to environmental stresses, and also to promote genetic mutations that help the cell to survive. Different parts of cell proteins originate in different chromosomes, so aneuploidy cells have such large amounts of some protein subunits that they overwhelm protein quality control mechanisms within the cell. Aneuploidy also alters the balance of tumor suppressor genes and tumor promoter genes; cancer genomes select for extra chromosomes containing tumor promoter oncogenes and against chromosomes containing extra tumor suppressor genes.

Although it is unusual to produce a large number of proteins from a single gene, one single DSCAM gene has been found to generate a huge number of 38,000 subtle variants of a single protein. The vertebrate immune system can generate a vast array of antibody proteins by shuffling single genes (Davenport, 2008).

There are approximately 15% of the U.S. population who are left-handed. It has been found that handedness is determined by a gene with two alleles, and the allele for right-handedness is dominant. Thus, anyone with at least one copy of this allele is right-handed. People who lack this allele, however, only have a 50–50 chance of becoming left-handed (Brodie, 2004a). Coming from the father, the gene for left-handedness is active; coming from the mother, the gene has no effect. Incidentally, this same allele determines the direction that hair whorls at the back of the head; right-handed people have clockwise spirals. (It is interesting, from an engineering point of view, to note that a bolt that tightens as it is rotated clockwise is called a right-handed thread. Mathematical cross products and vector notations also follow this convention. Thus, after all, there is a sound biological basis for this convention.)

So, what are the external environmental influences on genetic activity? They are not all known, but they include the fetal environment, intellectual and emotional stimulation, and physicochemical environmental factors. All of these, it seems, can influence the activity of genes, and it is well known that there are critical periods during development for the genetic potential to be realized. Everything from ambient temperature to the presence of chemicals to learning to emotional trauma can influence the metabolism of an organism.

Muscle tissue is very plastic. Lack of exercise causes muscle to *remodel* itself into weak muscle, with a smaller amount of tissue, fewer mitochondria, and containing fewer capillaries. Exercise reverses this process to make the muscle stronger and more efficient. Muscular exercise is the trigger to activate genes in the muscle tissue to make this happen (Booth and Neufer, 2005). These genes have wide-ranging effects on health, metabolism, and skeletal muscle strength.

Although it appears that only 10% of the human genome consists of genes that code directly for the formation of proteins, genetic material in the remaining 90% also appears to have useful function. There are short stretches of regulatory DNA that allow regulatory proteins, those involved in gene expression, to bind and influence gene expression through RNA interference. Many of these intermediate sites have DNA that is easily transposed, thus facilitating genetic diversity and species formation.

POLYMERASE CHAIN REACTION

Small samples of DNA are often not large enough to perform identification procedures, such as electrophoresis (see Section 2.11), or other manipulations that have been found useful in disease diagnosis or vaccine development. The Polymerase Chain Reaction (PCR) was perfected to overcome this limitation.

(*Continued*)

POLYMERASE CHAIN REACTION (Continued)

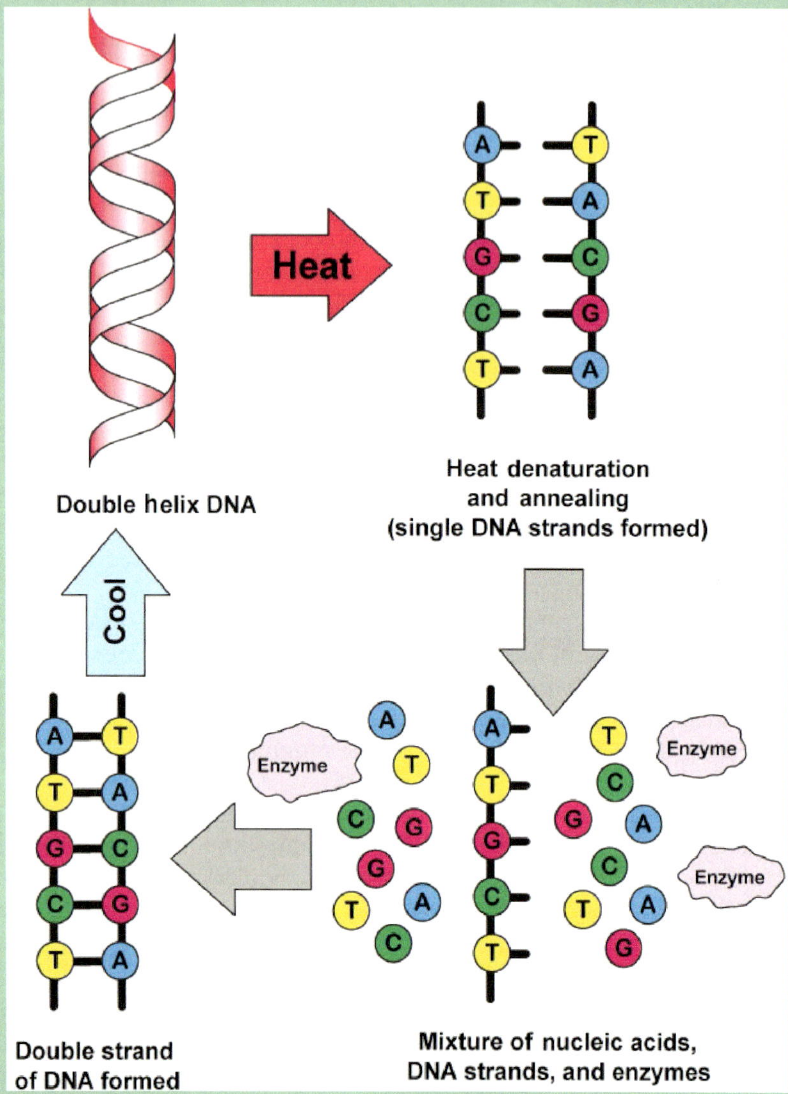

The DNA double helix is first heated so that it denatures and the double helix splits into two separate strands. Newer methods use enzymatic denaturation. *Annealing* (combining of nucleic acids) also occurs during this phase wherein a *primer* (a short segment of RNA) attaches to the DNA strands. Next a mixture of nucleic acids and thermophilic enzymes is added. As the mixture cools, complementary DNA (cDNA), using the original DNA strands as templates, completes the formation of new double-stranded helical DNA. This process duplicates the natural process of DNA replication. This process may be repeated as many times as necessary to multiply the original DNA to the quantity needed.

(*Continued*)

POLYMERASE CHAIN REACTION (Continued)

Refinements of PCR involve reducing the time it takes to produce sufficient DNA for required purposes, reducing error rates, and reducing the amount of sample needed at either the beginning or the end of the PCR process. PCR is normally used to identify only the type of genetic material present in the sample; qPCR can be used to measure, in addition, the amounts of genetic material present in the sample. dPCR stands for digital PCR, which amplifies DNA at the single molecule level. The sample to be amplified is diluted so that only one molecule is present in each detector cell.

The primers must be extremely specific to the template being amplified, and there is a limit to the maximum length of DNA that can be amplified. Any slight contamination with nonintended DNA can result in an incorrect product.

5.3.5 Epigenetics

> Discovery consists of looking at the same thing as everyone else and thinking something different.
>
> **—Albert Szent-Györgyi**

Epigenetics is the term describing gene expression and its long-term effect on phenotype. It is the study of changes in gene activity that do not involve alterations in the genetic code, but are, nevertheless, passed on to at least one, and maybe more, successive generations (Cloud, 2010). Epigenetic markers include methylation, phosphorylation, or acetylation bound at one or more genetic sites. The effects of these markers are to regulate whether the genes to which they are attached are expressed or not. Methyl groups and other molecules can bind to the helix itself or to its histone protein core and cause the whole assembly to flex. As that happens, some genes are exposed and others are obscured. The exposed genes are the ones that can be actively expressed. When a methyl group is added to a gene, it keeps other regulatory molecules from binding to that gene.

Epigenetic markers can incorporate environmental factors into genetic expression of the present generation so that various circumstances such as diet, stress, and prenatal nutrition affect genetic expression in future generations. Feeding B vitamins (methyl donors) to pregnant mice has been demonstrated to overcome genetic tendencies to diabetes and overweight in their offspring (Waterland and Jirtle, 2003). Fruit flies have shown epigenetic effects through at least 13 generations, and round worms through at least 40 generations (Jablonka and Raz, 2009). Even memory in mouse offspring can be improved via epigenetic properties of their parents' genomes (Arai et al., 2009). Epigenetic DNA methylation may also contribute to long-term memory formation in humans (Biergans et al., 2016). Reversible epigenetic changes have been shown to be associated with behavior in honeybees. Young honeybees function as nurses to unemerged brood larva and nurse bees later become foragers. With methylation of a majority of genes these foragers have been made to revert to nurse behavior (Herb et al., 2012).

Whereas epigenetic markers do not permanently change the genomes of the organisms to which they belong, these markers can be used to overcome genetic weaknesses in humans as well as other species.

The chemical switches and markers in the *epigenome* that affect how genes are expressed are important to direct each body cell into the type of cell it needs to become.

These switches assure that the proper set of genes is expressed and that other, interfering genes present in precursor cells are not. The epigenome is vulnerable to disruption by environmental toxins, pollutants, and poor nutrition, thus leading to organic diseases such as various types of cancer.

It has also been found that methylation of the nucleic acid cytosine (the "C" in DNA base units) occurs chronologically in the human genome. Thus, epigenetic changes occur that are correlated with age of the individual (Bocklandt et al., 2011).

5.3.6 RNA Interference

> Thomas Henry Huxley called Richard Owen a "lying Orthognathus Brachycephalic Bimanous Pithecus." Owen charged that Huxley was "nothing but a thorough Archencephalic Primate."
>
> **—Hal Hellman**

RNA is now known to be much more than just an intermediary between DNA and protein. RNA can act as a catalyst (mRNA), a binding site for small molecules (tRNA), a regulator of gene expression (RNAi and siRNA), a structural component of ribosomes (rRNA), and perhaps much more.

As another example of the complexity of genetic operations, some RNAs, along with some destructive enzymes, function by interfering with the expression of some genes. Instead of acting simply as a means to convey genetic information to the ribosomes, some RNAs actually keep other genes from being effective. This process is called *RNA interference,* and often involves double-stranded RNA (iRNA or siRNA, standing for small interfering RNA). RNA interference can be an important tool in *functional genomics*, the term used to describe matching functions with certain genes. RNA interference can be used as a tool to turn off specific genes, and then observations made to see what happens when they are no longer active.

siRNAs, many of which are commercially available, trigger the destruction of targeted mRNA species. MicroRNAs (miRNAs) bind to mRNA and block its translation into protein. miRNAs can also act to regulate transcription pathways and interfere with specific genetic expression. They are evolutionarily conserved and are critical in natural organism development, especially brain development, and in viral infection processes. There are two tags on miRNA molecules, the first being a methyl group that allows the miRNA to be retained and the second a uridine group that signals destruction of the miRNA. In this way, miRNA may sometimes be turned on and sometimes have no effect. Each microRNA molecule can regulate up to hundreds of target protein-encoding genes, and each target gene can be regulated by multiple microRNAs (Stein, 2012). RNA interference is a way to shut down defective genes and can be used therapeutically to treat genetic diseases. RNAi therapies are more specific and less prone to adverse effects than are drug-based therapies.

Interfering iRNA fuses with mRNA to form a double-stranded molecule. This appears to the cell as a virus, and it is subsequently destroyed (Matushansky and Maki, 2005).

There are huge numbers of genetic locations known to encode microRNAs, and each microRNA regulates up to hundreds of different messenger RNAs (mRNAs), and each mRNA is regulated by tens of microRNAs. This makes extremely complex the regulation of the human transcription process (Stein, 2016).

5.3.7 Antisense Molecules and Omics

> Nobody knew health care could be so complicated.
>
> **—Donald J. Trump**

Another genetic regulation mechanism is the presence of *antisense molecules*, complementary *oligonucleotides* (nucleic acid molecules with few (up to 200) bases formed by cleavage of longer nucleic acid segments) that bind to target RNAs or DNAs to inhibit normal functions. An antisense strand contains nucleotides complementary to the target mRNA, thus forming double-stranded RNA (dsRNA) that are ineffective in the translation process to form proteins. dsRNAs are rapidly degraded in the cell.

Single-stranded noncoding RNA (ssncRNA), 19–23 nucleotides long, can regulate gene expression by decreasing levels of target messenger RNA (mRNA). Short hairpin RNA (shRNA) consists of a sense strand, short loop sequence, and antisense strand that interferes with gene expression in the cell in a similar fashion to the action of short interfering RNA (siRNA) (Moriarty, 2012).

Genomics (associated with the genes), proteomics (associated with proteins), or metabolomics (associated with metabolic processes) together are referred to as the field of study of *omics*.

5.3.8 Genetic Variation

> *The* human genome is a misnomer. It's been shown that big changes in DNA—insertions and duplications and deletions and inversions—are extremely common in the population.... These changes play a role in human disease- everything from HIV susceptibility to autism to mental retardation to epilepsy.
>
> **—Evan Eichler**

In the complete genetic sequences of nearly 100 of a wide range of organisms, it has been found that 30%–40% of the genes are the same. About 98.5% of the genomes of humans and chimpanzees are identical (Tiffany-Castiglioni, 2003). There are about 20,000–30,000 human genes, which means that only 2% of the genome functions as genes. The rest has other functions poorly understood at present. It has been estimated that 99.9% of the human genome is identical for all humans; only 0.1% of base pairs are different. Considering those genes common only to all humans, 7% differ from individual to individual (Nesse and Williams, 1994). If there are about 3 billion base pairs in the human genome, then about 3 million of them may differ among individuals.

About 8% of the human genome appears to be remnants of DNA from retroviruses (Ogle and Platt, 2004). At some point in the history of mammals, endogenous retroviruses infected the germ cells of their target animals. Retrovirus genetic material merged with genetic material of their hosts. They are thus passed from one generation to the next. It has been found that this genetic material is essential for proper placental growth and without it the risk of miscarriage of the fetus becomes very high (Dunlap et al., 2006).

One advantage of genetic variation is that the capacity to meet the demands of environmental variations often lies within genetic variants. Through sexual reproduction, new genetic combinations are possible. There is an enormous amount of genetic variability maintained in natural populations. The amount of variability is greater than anyone would expect: average invertebrates are heterozygous at 14.6% of its gene loci, and vertebrates are heterozygous at 5.0%. The cost of maintaining this variability is high. If one or two of these genetic variants has some reproductive advantage, then the rest are at a disadvantage. It would be expected that the reproductively disadvantaged genes should disappear eventually. However, they are apparently maintained at a much higher level than can be explained by random mutation (Powell and Dobzhansky, 1976).

Perhaps some genetic variants are maintained at high levels because of the hybrid vigor exhibited by heterozygotes. Other variants may be useful in different habitats or at different seasons. Still others may have a selective advantage when they are rare but become disadvantageous when they become too frequent (Powell and Dobzhansky, 1976).

There is an estimate that much of the genetic material found in chromosomes does not function as genes; that there are about 20,000–30,000 genes in the human genome. Each gene can result in about 10 proteins in the human (there are only 3 proteins per gene in yeast), so the potential amount of variation is rather large even in the same species.

Bacteria subject to virus (bacteriophage) infections can acquire new DNA sequences identical to sequences in the infecting phages. They then become immune to further infection by the same phages through a mechanism similar to the iRNA mechanism in eukaryotic cells (Baker, 2011; Haurwitz et al., 2010).

GENETIC DIVERSITY REPOSITORIES

Modern agricultural practices have emphasized efficiency (more production from smaller land areas) over other resource issues such as plant nutrients and water. Thus, agricultural crops are highly managed to the extent that small areas in the same field are receiving more or less fertilizer, water, or herbicides for very local weed control.

One invariant in this system is crop variety. Not only are the same crop varieties used on the same farm, but the same varieties are used on many farms in broad localities. Add to that the fact that many farms are now growing only one or two different kinds of crops, and this had led to huge areas devoted to monoculture.

Genetic diversity of food crops has suffered greatly as a result. In the U.S. alone, 75% of crop genetic diversity found 200 years ago has disappeared. In other, less developed countries, the same trends are happening, but they are much more inchoate. Although disappearing fast, locally adopted varieties of many crops still are grown.

There are several threats to monoculture. One is the fact that new diseases or pests may be introduced into an agricultural region and spread rapidly because huge areas of crops may not be resistant. This can result in crop failure of large proportion. Another challenge is to extend crops because of the need for higher production to land not ideally suited. This land may be drier, colder, warmer, or more salty than main crop varieties grown in better locations.

It is highly likely that genes conveying disease resistance or compromised conditions tolerance have existed in plants in the wild or in subsistence agriculture. These

(*Continued*)

GENETIC DIVERSITY REPOSITORIES (Continued)

genes are rapidly being lost as less productive crops are being replaced by crops more productive under intense agricultural conditions.

Gene repositories have been established to collect and store seeds from exotic crops. Scientists gather plant materials from remote areas and bring them to locations where they are stored in climate-controlled conditions. Often this means cryogenically. And, because seeds lose their viability over time, the plants need to be grown every now and again to maintain vitality.

Livestock animal breeders select for specific traits, and, similarly to agricultural field crop monoculture management, seemingly unimportant alleles may be lost as the desirable traits begin to pervade entire animal populations. The result is that animal breeds tend to become more uniform over time.

Genetic variation, however, is important for survival in the cases of devastating diseases, famine, or environmental catastrophe. That is why the U.S. Department of Agriculture has established a National Germplasm Repository in Fort Collins, Colorado. Tissue samples from many livestock breeds have been collected and frozen to await any future need to rescue remaining livestock breeds from annihilation (Myers, 2017).

Gene banks are located in several places in the world. Those located in stable developed countries are being well maintained, but others in more volatile regions are riskier. The Iraqi national seed bank of lentils, rye, barley, and other seeds at Abu Ghraib was destroyed after the 2003 American invasion (Rosner, 2008). Peru's National Agricultural Institute collection of sweet potatoes was robbed by a group of starving homeless people. Afghanistan's seed bank was obliterated during Mujahadin fighting in the 1990s. A typhoon in the Philippines washed away stored varieties of sweet potatoes, taros, and bananas in 2007.

It is important to maintain this genetic legacy for materials to meet as yet unknown future challenges. Once lost, unknown genetic material cannot be (easily) reproduced. At some point in the future, these gene banks may prove to be the salvation of the human population on Earth and elsewhere.

5.3.9 Replication

A book of science is inexhaustible.

—Samuel Johnson

Higher level plants and animals undergo two types of nuclear division as essential steps in genetic reproduction. In *mitosis*, chromosomes are duplicated through a complex process that depends upon the unambiguous linking of DNA base pairs (Starr, 2000). Copies of all pairs of chromosomes are passed on to daughter cells. In *meiosis*, chromosome pairs are separated and passed singly to reproductive cells. When two reproductive cells are joined in the process of *fertilization*, pairs of chromosomes are again formed.

THE CHICKEN OR THE EGG?

Curiosity is, in great and generous minds, the first passion and the last.

—**Samuel Johnson**

This is a philosophical question meant to imply mutual dependence and the difficulty separating steps in a cyclic situation. Asking this question usually ends the discussion, because the question is understood to mean that no resolution of the issue can be found.

Nevertheless, the question can be answered. An organism, such as a chicken, is composed of many cells, each with its own DNA. The DNA in most of these cells is identical to the DNA in all the other cells. The exceptions are the reproductive cells (gametes) that have only one-half the DNA complement of the somatic cells, and damaged somatic cells that may have DNA mutations caused by physical or chemical exposure. Some of these damaged cells may grow uncontrollably, forming a cancer, whereas others may not. In any case, the chicken still remains a chicken, no matter how many cells have DNA mutations.

The egg, however, begins as one cell. A change of DNA in that one cell will be replicated throughout the entire organism that arises from the egg. In other words, the only (natural) way to form a completely different organism is by first affecting the DNA of the egg. The first chicken must have resulted from a DNA change in the egg laid by its reptile or pre-chicken mother. A chicken will always be a chicken, but an egg can develop into something else.

So, which came first: the chicken or the egg? The answer is simple: the egg.

5.3.10 Mutations

Let it also be borne in mind how infinitely complex and close-fitting are the mutual relations of all organic beings to each other and to their physical conditions of life; and consequently what infinitely varied diversities of structure might be of use to each being under changing conditions of life.... If such [variations] do occur, can we doubt (remembering that many more individuals are born than can possibly survive) that individuals having any advantage, however slight, over others, would have the best chance of surviving and of procreating their kind?... This preservation of favorable individual differences and variations, and the destruction of those which are injurious, I have called Natural Selection, or the Survival of the Fittest.

—**Charles Darwin**

During the process of chromosomal manipulation, mistakes sometimes happen. These are called *mutations*. Mutations (genetic changes) occur either spontaneously or they may be induced. Mutations are changes in genetic material that can occur either in nonreproducing (somatic) cells or in reproducing (*germ* cells or *gametes*) cells. Those mutations that occur in somatic cells are not passed to future generations, whereas those that occur in gametes are inherited. The spontaneous mutation rate in humans has been estimated at between 10^{-5} and 10^{-6} mutations per gamete. About 6% of us begin life with at least one brand-new mutation found in neither parent (Nesse and Williams, 1994). The spontaneous mutation rate differs considerably among different organisms, possibly reflecting efficiencies of different DNA repair systems at work (Hale et al., 1995).

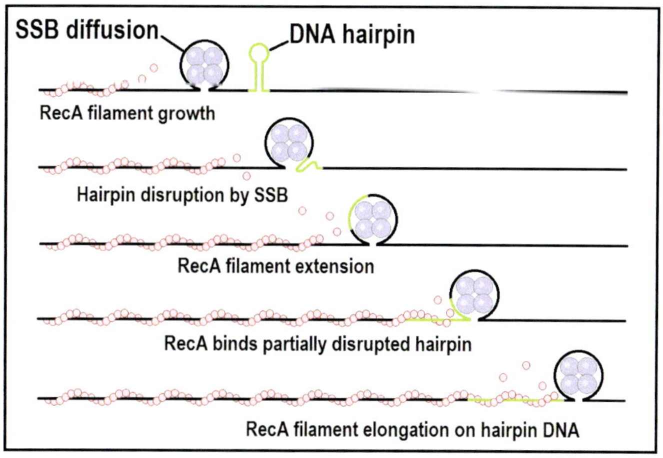

FIGURE 5.3.11
Part of the repair and maintenance operation for single-stranded DNA involves two proteins: single-stranded binding protein (SSB), and Rec A. Double helix DNA unwinds and separates into two single strands during replication. Immediately after separating, each single strand wraps itself around SSB proteins. SSB shuttles back and forth along the strand, fixing small defects known as hairpins. Rec A builds behind SSB and binds to the DNA strand. This action keeps SSB moving along the strand in the right direction and makes the SSB leave the DNA strand at the right time. (Redrawn from Schnabel, J., *HHMI Bulletin*, 23, 45, 2010.)

Structural mutations are classified as (1) *inversion*, where a DNA segment is rearranged in reverse order; (2) *duplication*, where a DNA segment is repeated either next to the original segment or elsewhere along the chromosome; (3) *translocation*, the exchange of segments between nonhomologous chromosomes; and (4) *deletion*, the removal of a segment (Hale et al., 1995). No process as important as genetic duplication can be left without error-correction and repair mechanisms, however (see Section 6.18, Reproduction). There are enzymes that proofread newly created DNA strands and fix mistakes (Figure 5.3.11). That very few of these mutations are not self-corrected pays tribute to the robust system of error detection and correction in the living cell (Drake et al., 1983).

High-energy radiation, and some toxic chemicals are among the environmental *mutagens* that can change the genetic code, and these changes can be felt for many future generations (see Section 6.12, Toxins). In addition, viruses and some bacteria are able to merge their genes with those of host cells and so become agents of genetic change.

During meiosis, when the paired chromosomes are split to form haploid genetic cells, recombination sometimes occurs, which results when some of the DNA in each strand switch places.

Most of the spontaneous mutations in DNA at present appear to be due to deamination changes in the nucleic acid Cytosine, a process that is very sensitive to temperature (increasing roughly 4 times for every 10°C rise in temperature). Life arose when Earth was much warmer than it is now, with the implication that spontaneous DNA mutations at that time occurred at the much higher rate of 4000 times what it is today (Lewis et al., 2016).

Most genetic mutations cause such drastic effects that they are lethal for the host cell. Others are not as drastic, but confer a comparative reproductive disadvantage that may cause this mutation to disappear within a few generations. Others, however, offer clear advantages, and, for these, it can be said that the environment can have profound effects on the biological unit (BU) that inhabit it.

The successful persistence of genetic mutations with favorable effects depends at least in part on where and when the mutations occur (Hawks, 2014). Some mutations arise in places where there is no survival or reproductive advantage to having them, even though they would have huge advantages elsewhere. Other very advantageous mutations happen only after other mutations with advantages only partially as large as the later mutations, but the better mutation doesn't survive because the lesser mutation has largely assumed most of the reproductive advantage that would have happened with the better mutation.

The ways in which genes affect the organism is highly nuanced and more complex than once thought. Mutations in noncoding genes were once thought to have no effect on the organism because they do not directly result in protein formation. These so-called "silent mutations," however, have effects that can be felt in health and disease (Chamary and Hurst, 2009).

In addition to DNA changes, mutations also affect the way the DNA molecule folds into its nucleosome, making some genes more accessible and some less so. The consequence is that different proteins may be formed with different phenotype configurations and characteristics than would otherwise be the case (Eslami-Mossallam et al., 2016).

There is evidence that genetic mutations are not random (Caporale, 2003). Some mutations occur orders of magnitude more often than others, and can even be predicted. Blood pathogens, for instance, can hide from their host's immune system by changing their outside coats. These changes occur by changes in the genes that control coat protein formation. Natural selection has favored biochemical mechanisms that alter just these genes and no others. These genes are located in areas that are bracketed by conserved sequences of bases; these do not change, but the regions between them do.

There are other examples such as cone snails and scorpions that use focused genetic variation to generate new components of their venoms. Human and animal antibody production is also subject to unequal distribution of mutations. Antibodies are composed of two regions: a constant region that links to other body proteins and a variable region that binds to foreign bodies (see Section 6.20.3, Redundancy). The constant region almost never mutates; the variable region, and especially the part that binds to foreign antigens, mutates frequently (Figure 5.3.12). This gives the human or animal the ability to anticipate changes in pathogen surface coat configuration.

Mechanisms that foster selective mutation can be classified as given as follows:

1. Repeats of sequences of bases, such as CCCCC… or AGTCAGTCAGTC…, increase or decrease in length as the two strands of the double helix slip and misalign during DNA copy or repair.
2. Looping of one strand or other of the DNA during copy or repair means that some genetic material may be gained or lost (Figure 5.3.13).
3. Looping of a portion of a DNA strand can cause pairing with itself (Figure 5.3.14). Various bases may then be exchanged by repair enzymes that recognize the arms of the loop as two separate strands.
4. The proteins involved in copying DNA can themselves mutate and change the probability of future mutations at specific sites.

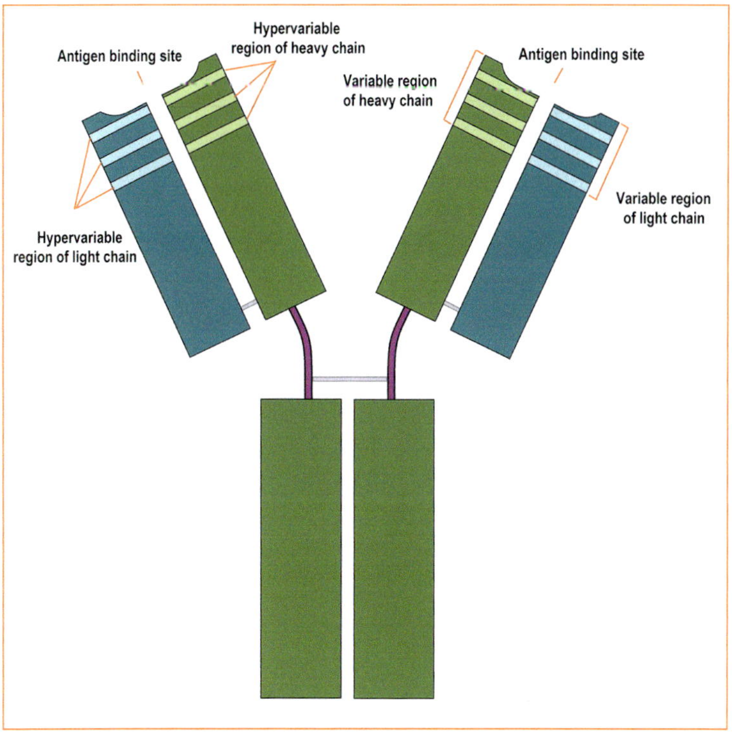

FIGURE 5.3.12
Human antibodies must be immensely diverse to anticipate the unpredictable infections that beset us. The variable regions on the two arms of antibodies include three hypervariable segments (*stripes* at the top) that bind directly to invading antigens. There are four polypeptides joined to form a "Y" shaped molecule shaped specifically to link to a target antigen. These segments mutate much more readily than other parts of antibodies' variable regions (Caporale, 2003). Antibodies are divided into five major classes, IgM, IgG, Iga, IgD, and IgE, based on their constant region structure and immune function.

There may be other mechanisms as well, but it is clear that these genetic changes can have profound effects on phenotypical genetic expression.

The conclusion from this is that there is at least a *second-order selection process* going on. Not only is there natural selection from among mutations that occur that leads to a survival advantage, but there seems to be a selection for those individuals that can mutate their genes in regions that are likely to improve survival. Random mutations are still likely, but nonrandom mutations are more likely, because these give an organism an advantage to survive and reproduce.

Viruses contain RNA or DNA segments that encode for specific proteins encapsulated in a surrounding envelope of protein, lipid bilayer, and external proteins that protrude from the viral surface (Figure 5.3.15). The surface proteins bind to surface receptors of the host cell (Figure 5.3.16), whereby the viral RNA or DNA strands move into the cell nucleus. Viral RNA or DNA strands encode mRNAs that ultimately produce new virus particles. Other viral surface proteins enable the newly created viruses to separate from the host cell and invade other cells.

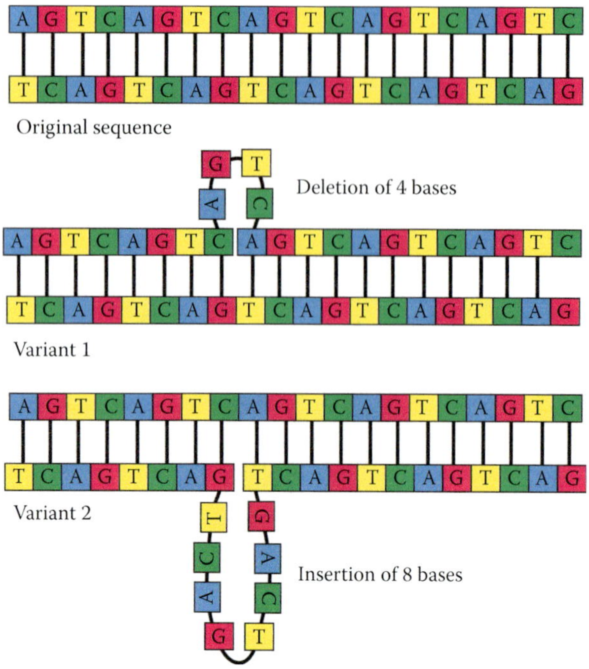

FIGURE 5.3.13
One strand of repetitive DNA can slip during DNA copying or repair, leading to a mispairing that changes the length of the repeat. If the original strand loops out while being copied, a portion of the repeat is deleted. If the new strand loops out, the repeat becomes longer in the next generation. (Redrawn from Caporale, L.H., *Am. Sci.*, 91, 234–241, 2003.)

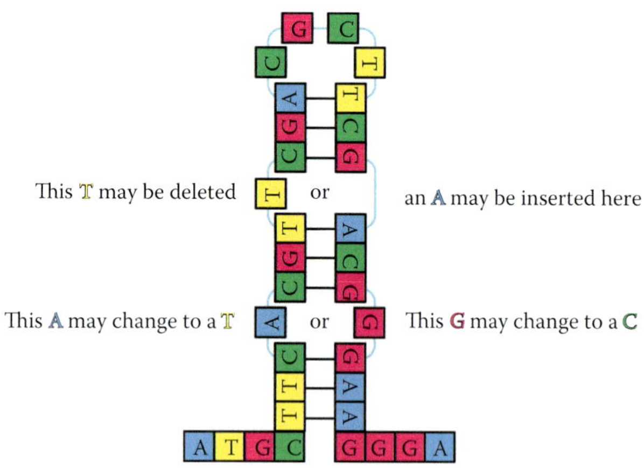

FIGURE 5.3.14
When a DNA segment forms a loop and pairs with itself, repair mechanisms can correct apparent mismatches. This leads to selective DNA mutations. (Redrawn from Caporale, L.H., *Am. Sci.*, 91, 234–241, 2003.)

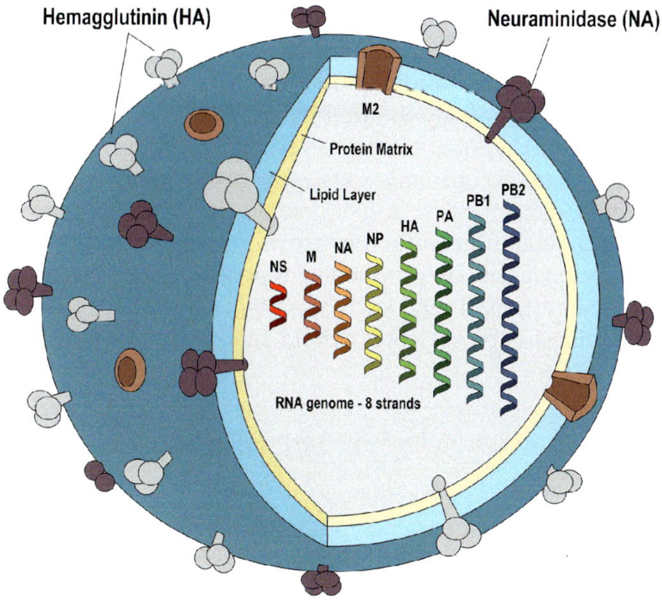

FIGURE 5.3.15
Diagram of a Type A influenza virus. These viruses are spherical with a diameter of about 10^{-7} m. Surface proteins hemagglutinin (HA) and neuraminidase (NA) enable the virus to enter and leave host cells. Different forms of HA and NA give the influenza virus its unique designated identification (e.g., H1N3 or H7N5). Inside the viral coating are eight strands of RNA that code for 10 proteins, including HA and NA. The ion-channel protein M2 forms a conduit through which small molecules, such as water, can pass through the impermeable outer coating. (Redrawn from Webster, R.G. and Walker, E.J., *Am. Sci.*, 91, 122–129, 2003.)

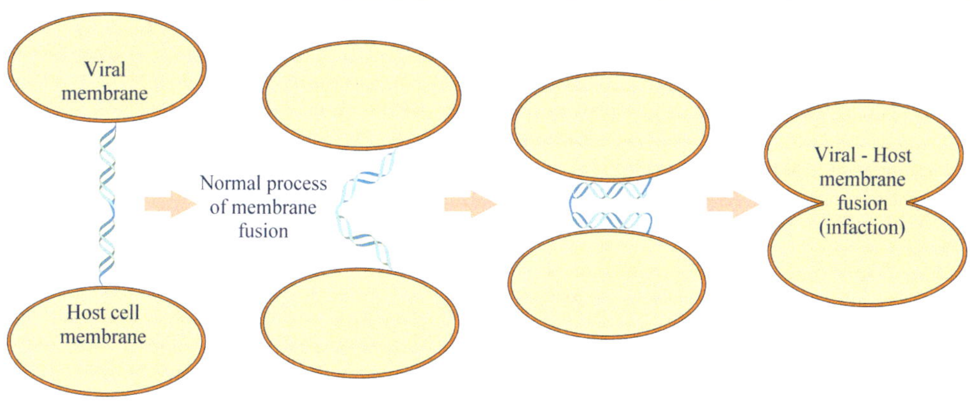

FIGURE 5.3.16
Most viruses invade cells by first attaching to the cell membrane with viral fusion proteins. Their outer layer containing a lipid membrane fuses with the lipid-containing membrane of a host cell (Goforth, 2009). Viral fusion proteins are configured as an alpha helix with a hinge. As the catalytic hinge closes, the two membranes are brought close together until they fuse. At that point, the virus contents enter the cell (Dutton, 2006). Not all viruses use this means.

In viruses that contain RNA and not DNA, replication of viral genetic material is not subject to the same error-checking and repair mechanisms as is cellular DNA. Thus, mutations in viral RNA are relatively more common, and viruses, should they be able to survive the mutations, can evolve rather quickly to circumvent cellular or therapeutic drug antiviral countermeasures. Antiviral defenses oftentimes recognize viral surface proteins as foreign bodies, and deal with them accordingly. *Cytokines* such as interferon and tumor necrosis factor are used as the nonspecific first line of defense that does not require previous exposure to the virus. Immunoglobulin antibodies are produced specifically reacting to a certain viral surface configuration; antibody production depends on previous exposure. Some viral RNA mutations result in different viral surface proteins, and these changes can be enough to make ineffective antibodies produced to defend against the previous form of the virus.

AMES TEST FOR MUTAGENICITY

There are many toxic substances in the environment, some natural and some man-made. One issue of interest when a new material is identified is whether or not it is capable of changing the cellular genetic code. If so, then it is a likely cancer-causing substance. So, the search for *carcinogens* resolves into a search for *mutagens*.

The Ames test is extremely simple and fast. A strain of bacteria is cultured that is unable to grow on a minimal medium devoid of histidine. When mixed with the suspected mutagen, some of the bacteria may mutate into a bacterial strain capable of growth on the minimal growth medium. Presence of bacterial colonies growing on plates containing the minimal medium indicates that the substance is, indeed, capable of mutagenicity.

5.3.11 RNA Correcting DNA

> Why do some controversies resolve satisfactorily, while others seem to continue on and on? In the latter case, the science itself may be recalcitrant, just plain slow to develop. As a result, competing ideas go back and forth. More often, there is some subtle or not-so-subtle question of beliefs or values that underlies the whole debate.
>
> **—Hal Hellman**

One possible additional function of RNA is to correct abnormalities that appear in DNA. It has been found that certain plants with mutated genes could revert to the unmutated DNA forms possessed by the second previous generation. It has been speculated that the mechanism by which this happens is a cache of RNA that can be used to correct harmful DNA mutations (Pennisi, 2005). RNA involved in genetic modification during cell replication is known as *reverse transcription* (Gage and Muotri, 2012), and is somewhat similar to the way a retrovirus reproduces in a healthy target cell.

5.3.12 Mitochondrial and Chloroplast DNA

> Microbiology has undergone an explosion of discovery…into realms that are as bizarre as anything appearing…in novels.
>
> **—Joan Slonczewski**

Mitochondria in the cell carry their own DNA separate from that in the cell nucleus. Sperm of all animal species contain their mitochondria in their tails, Ih are separated from their heads at the moment of fertilization. Thus, the newly fertilized egg contains mitochondrial DNA from the female parent only. Tracing relatives through the female lineage is thus relatively simple. Mitochondrial DNA (mtDNA) suffers a higher rate of mutation compared to nuclear DNA due to factors such as proximity to highly reactive free radicals, lack of histone protection, and poor fidelity of mtDNA replication and repair. Because of this, differences in mtDNA can be used to study relationships among population groups. Knowing the rate at which spontaneous genetic mutation occurs can be used to estimate the ages of different species.

DNA in the chloroplasts of plants is transmitted from one generation to the next through the egg (Juniper, 2007), similarly to the DNA of mitochondria. The DNA for both of these structures is circular, as it is in prokaryotes. Thus, for higher level plants that depend on photosynthesis to fix carbon from Iic carbon dioxide (and thus contain chloroplasts), there are three independent DNA stores in the cell: in the chloroplast, in the mitochondria, and in the nucleus.

DNA INHERITANCE

Female inheritability can be traced through mitochondrial DNA because mitochondria come only from egg cells. Thus, lineage can be traced by comparing mitochondrial DNA from one generation to the next. The number of generations, and thus the time between two relatives, can be estimated by assuming a certain rate of DNA mutation and looking at DNA differences.

Male inheritability can be traced through the Y sex chromosome. Each male somatic cell contains two sex chromosomes, an X and a Y (Figure). Females have two X's.

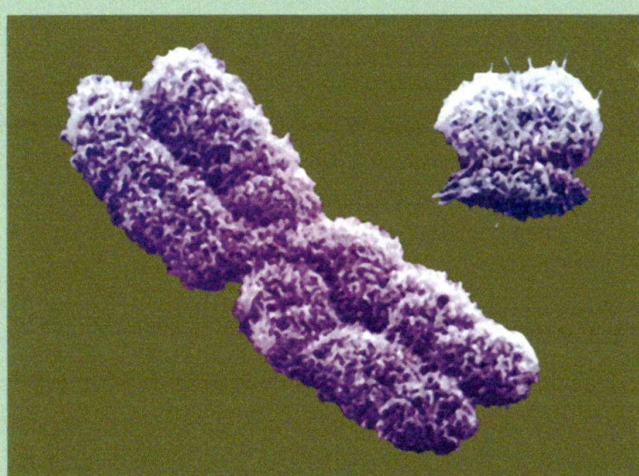

Images of X and Y sex chromosomes. The Y chromosome is much shorter than the X chromosome, and is carried only by males.

When pairs of chromosomes split to form germ cells by the process of meiosis, male germ cells (sperm) carry either an X chromosome or a Y chromosome. Female germ cells (eggs) carry only X chromosomes. Thus, the only way for a zygote (offspring) to inherit a Y chromosome is from the male parent.

Because of this, genes appearing on the Y chromosome can be compared similarly to comparisons of mitochondrial DNA, and reflect male-only lineage. Time between particular generations can be estimated by again assuming a normal rate of DNA mutation.

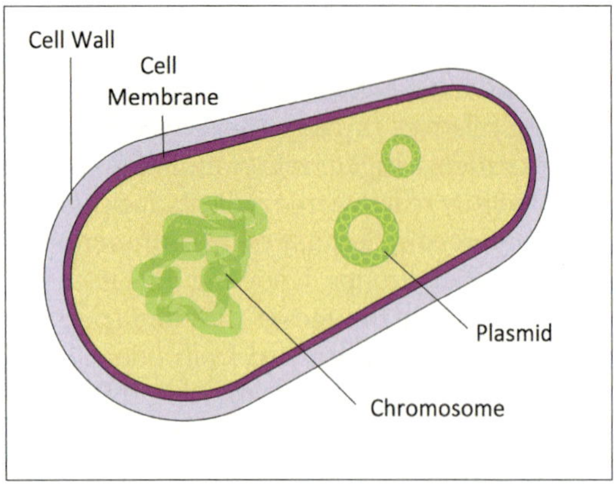

FIGURE 5.3.17
The plasmid is a packet of supplemental DNA material found outside the bacterial chromosome. Plasmids often contain DNA molecules that serve specific purposes supplemental to chromosomal DNA. Genes for specific antibiotic resistance are found in plasmids, and genes conferring virulence are also found there. Exchange of plasmid DNA among bacteria is relatively easy. (Redrawn from Amábile-Cuevas, C.F., *Am. Sci.*, 91, 138–149, 2003.)

5.3.13 Plasmid DNA

> Bacteria, far from being opportunistic loners, are highly social creatures that incessantly chatter among themselves, with the hosts they infect, and even with other species of bacteria….
>
> **—Paul Raeburn**

Bacteria contain some extrachromosomal genetic material in the form of a small, circular, self-replicating DNA molecule called a *plasmid* (Figure 5.3.17). Certain plasmids can be incorporated reversibly into the cell's chromosome. The plasmid contains a small number of genes (designated as *pDNA* for plasmid DNA) that supplement chromosomal genes; they are not required for the survival or reproduction of the bacterium under normal conditions. They do offer advantages to the bacterium in a stressful environment, however. The F (for fertility) plasmid facilitates genetic recombination involving other bacteria that may be required to survive in a changed environment that can no longer sustain existing bacterial strains (Campbell et al., 1999).

Parts of plasmids inside common bacteria are being used commercially to supply genes for specific purposes, such as to make a mouse's brain cells react to light, or to turn on each fruit fly gene one at a time, or to make certain types of proteins. These plasmid-containing bacteria can be ordered off-the-shelf and grown locally to produce the amount of genetic materials required for gene-editing purposes (Regalado, 2016b).

Example 5.3.1: Searching for the Causes of Autism

Autism is a mental condition where people, mostly children, crave routine, have trouble communicating, and don't understand intuitive social rules. Some autistics barely speak, while others have some very large vocabularies. Some are riotously overstimulated, while others are isolated and withdrawn.

It is thought that the causes for autistic behavior are a combination of genetic makeup and environmental factors. Although the right genes must be present, not all children with these genes develop autism. Speculate on possible environmental causes.

Solution

There has been found a genetic variation present in 47% of the U.S. population that codes for a protein active in the brain, gastrointestinal tract, and immune system. This gene is associated with autism. It is likely that there is a genetic predisposition to the condition that is triggered by unknown environmental events.

Because autism develops in the very young, only those happenings occurring in early development can be the trigger. After children are born, they may be placed in an incubator; they may be placed on a ventilator; they can be held much or little; they may be breast fed or not. Autistics seem to develop for all of these conditions.

More generally, drops of silver nitrate or other silver compounds are placed in the eyes of newborns to prevent infectious conjunctivitis (*opthalmia neonatorum*). Vaccines are given shortly after birth. It has been speculated that vaccines could contribute to autism, although this has been discredited.

Before birth, immune system abnormalities of the mother may contribute to autism by interfering with the timing of fetal brain development. There is evidence of a link between autism and viral infection of the mother during pregnancy (see Section 6.22.8, Mind–Body Interactions). Because autistic children exhibit extremely male brain characteristics, others are focusing on prenatal testosterone exposure.

Interestingly, high rates of autism seem to run in families of physicists and engineers. These are professions that require focus on details and not on language or social skills. There has also been found a positive correlation between autism incidence and precipitation rates as an environmental trigger for the condition (Waldman et al., 2008).

Example 5.3.2: High-Energy Radiation

High-energy radiation can damage DNA and cause mutations. How can this same radiation be used to advantage?

Solution

High-energy radiation can be used intentionally to induce DNA damage in bacteria and viruses present in and on food. When high-energy photons strike the electrons of irradiated food, they send them flying in all directions. Some of these electrons strike the relatively huge double-stranded DNA of bacteria. Atoms become ionized and molecular bonds are disrupted. Breaking both DNA strands kills the organism; breaking one strand weakens the bacterium and renders it sterile. Targeting viruses is a little more difficult because they are smaller than bacteria and contain only short pieces of genetic material packed inside a protein coat. Thus, high-energy beams can be used to sterilize food to make it safer and store longer.

Example 5.3.3: Crossbreeding Tigers

The well-known illusionist team of Siegfried Fischbacher and Roy Horn had been performing together at the Mirage Casino in Las Vegas for more than 34 years. In their act, they worked with extremely rare white tigers, making them seem to disappear and appear at other times. This duo had become interested in propagating these tigers in order to increase their numbers. What strategy should they use in order to do this?

Solution

If the white tigers were mated among themselves, they would soon become too inbred and display all the weaknesses (and the strengths) of animals that are nearly homozygous. Although offspring of cross-matings between white tigers and yellow tigers will not display the desired white coloring (white being a recessive gene), these heterozygous tigers can subsequently be bred among themselves, with about one-quarter of the offspring expected to be white. The result is that the selected white tigers will be stronger because their genetic material will contain genes from a wider range of animals. Because many undesirable genetic traits are recessive, even one copy of a desired gene will avoid genetic weaknesses.

Marker-assisted gene selection is used in animal or plant breeding by associating a desired trait with easily assessed physical or behavioral indicators (markers) rather than searching for exact genes governing the trait. That way, the desired trait can be bred into a plant or animal as long as they exhibit the marker.

Crossbreeding is used often to add genetic strength to a breed, especially one that has limited numbers to start with. Crossbreeding is also used to develop new varieties of plants; the crossbreeding usually involves parents that display extremes of the trait of interest.

Example 5.3.4: Genetic Causes of Alcoholism

Marilyn Vos Savant authors a column for the weekly *Parade* magazine, in which she answers mostly intellectual questions from readers. The following appeared in the October 12, 2003, issue of *Parade*:

> Why do you doubt the idea that certain people are genetically prone to alcoholism?
>
> **—J. T., New York**

One reason is that alcohol doesn't exist in nature. Instead, alcohol is a creation of mankind: Our genes don't know about it. Another reason is that about 80% of alcoholics are male. Yet no one suggests that problem genes are sex-linked, such as male-pattern baldness. So I am concerned that, in an effort to remove the stigma of alcoholism from individuals and to blame their genes instead, we are stigmatizing whole families and ethnic groups. In my opinion, that's far worse.

Response

Permanent changes in the genetic code of an organism, either the addition of new capability or the removal of an old liability, must come about as a result of a survival and reproductive advantage to the organism. The speed at which the change happens would be related to the degree of the reproductive advantage.

If humans had developed a genetic response to alcohol, there would have had to be some good reason why alcoholics were more procreative than nonalcoholics. It is hard to imagine this happening, so it is not likely that humans would ever develop a genetic proclivity toward alcoholism.

Alcohol and similar biochemicals have been present in nature for a lot longer than humankind. So, if there is a genetic component to alcoholism, it is likely that the genes were present in more primitive organisms. Because humans share approximately 30%–40% of their DNA with all other organisms, it is likely that these vestigial genes could still reside in the human genome. If there were no strong reproductive disadvantages to the presence of these genes, they would probably be maintained.

Biologists have been searching for such a gene, and there is evidence of the gene in a worm *Caenorhabditis elegans* (Davies et al., 2003). Whether this gene is responsible for alcoholism, and whether it has been passed on to humans is still an open question. It could very well be that the gene influences the metabolism of a whole class of biochemicals, not just alcohol, and the intoxicating effects of alcohol are just an inadvertent side effect to the major function of the gene. Sickle cell anemia, for instance, is a genetic disease that confers a survival advantage to those who live in malaria-infested regions; the long-term effects of sickle cell are inadvertent.

Something like alcoholism can have complex causes, as we know. To attribute alcoholism to no genetic causes would be too simplistic (Curry, 2017). If there are genes involved in alcoholism, they may influence such diverse attributes as metabolism, personality, and sensory perception, all of which are determined in part by genetic causes. Thus, there are probably many genetic components to alcoholism.

For these reasons, the answer that was given to the question was not complete and wasn't based on sound reasoning and knowledge about the workings of the genes.

Example 5.3.5: Biochips for Disease Detection

Disease detection is simple and fast with biochips that contain up to several hundred small wells (of the order of 10 μm in diameter) etched into a substrate carrier. In each well is a drop of gel containing a DNA segment, protein, peptide, or antibody that tailors each drop to recognize a specific biological agent or biochemical signature substance. These drops are placed in known locations on the biochip so that when a reaction takes place between the test reagent and the detected sample, the sample can be identified by its location. A biochip scanner makes this process almost automatic.

Many biochips use dyes that fluoresce when illuminated. Different color dyes can be used to determine, not only whether a reaction has taken place, but specific details of sample reactants. Other biochips may work with small magnetic beads.

Detector substances in biochips can be tailored to identify diseases of human, veterinary, or horticultural importance, and samples can be processed in just a few minutes. Viral or bacterial genetic material can be detected with DNA fragments complementary to the candidates for detection; when the target DNA is present, it links with the DNA affixed to the biochip, and the resulting hybrid can be detected by fluorescence.

Other test wells may contain DNA strands (called *aptamers*) that stick only to specific proteins. These can then detect a single protein out of a mixture of many proteins.

APPLICATIONS AND PREDICTIONS

1. Complete knowledge of the genotype will never be able to be used to predict the phenotype.
2. Family resemblance will be determined by genes.
3. Not all traits possessed by an individual will be genetically determined. Thus, the process of selection for a particular trait must start with determination of its inheritability.
4. Certain behaviors will be linked to appearances.
5. A new breed of dog will be able to be established through genetic selection.
6. Most random mutations will be detrimental. Many will be deadly. A small few will be advantageous.
7. The accumulated successes of survival will be manifested in the genes.
8. Nothing will be as simple as you would like.
9. Knowledge about genetic mechanisms will be important for the development of new products and tests.
10. There will be vestigial genetic material present in the genome that has no present purpose.
11. Genetic approaches will cure many diseases and provide new vaccines.
12. Inbreeding will result in a higher than normal expression of recessive traits.

5.4 Competition and Selection

> Finding Toumai man, the oldest hominid, in Chad fits well with the theory of punctuated equilibrium developed by paleontologists Niles Eldridge and Stephan Jay Gould [the theory explains why new species, rather than evolving gradually over millions of years, seem to suddenly appear in the fossil record, punctuating long periods of species stability, or equilibrium]. . . .People derided the theory, calling it evolution by jerks. Gould's famous retort was that the alternative theory is evolution by creeps.
>
> **—David J. Melvin**

There is perhaps no process more important for anyone working with living systems to be familiar with than the process of competition and selection. All BU are in competition with each other and with different types of BU. An environment without BU is unnatural, and will soon contain BU as the opportunity arises. It is difficult to sterilize packages containing food, medical instruments, or even enclosures for humans with weakened immune systems, but it is very easy for these environments to be colonized by BU at just the slightest opportunity.

The vast amount of variation of BU and the adaptability of BU ensures that there will be a BU to thrive in all but the harshest environments. Even there, BU may take forms such as spores, seeds, or hibernation to survive the worst conditions imaginable until these conditions ameliorate, and growth can again take place. There is hardly a place that does not contain BU in some form (see also Sections 3.4, Acids and Bases, 6.5, Heat Sources, 6.15, BU Alterations, and 6.20, Control).

BU will grow and reproduce to the extent allowed by the environment. As long as sufficient resources are available, the only limit to growth will be time. With sufficient time, all available resources will be used by BU. These resources include chemical substrates, light, heat, or space. Negative resources useful to some BU are lack of toxins, lack of heat (or cooling), and lack of light.

With unchecked growth, it should be apparent that BU will expand limitlessly. Other BU will also tend to do the same thing. There are many cases where growth of one type of BU can enhance the growth of another type (for instance, growth of nitrogen-fixing bacteria can enhance growth of legumes, or growth of humans enhances the growth of human immunodeficiency viruses [HIV]). There are also many cases where growth of one type of BU limits the resources available to another type. Competition between these two types is the natural result.

Competition is a natural part of biology. When two or more BU could use the same resources, but each BU limits the opportunities for growth and reproduction of the others, then the competition becomes severe. The result is that some BU may thrive, some may barely survive, and some may die. As Gillespie (2005) has put it, "Ecologists who have examined communities over time have shown that [introduction of new species] may be stochastic, but deterministic processes can dictate the community's set of species at equilibrium. However, communities are complex, dynamic systems, so many factors decide whether stochastic forces dominate deterministic ones or vice versa.... Early on, species just pile in, with nothing to stop any one species from arriving and existing in a community. Over time, the early colonists 'settle down' during which competition may bump out certain species. After this competitive jostling, the remaining species are not a random bunch of those that arrived, but rather a set that is more tightly co-adapted—they've figured out how to live together. So any species can get into a community to start with, but only those that form cohesive co-adapted sets remain."

Each BU is a carrier for its genetic material, and can be considered to be the vehicle at present to deliver genetic material to the future. Thus, progress in biology travels in one direction only, from the past to the future. Genetic material survives only by being transported from the past into the future. If genetic material is lost or changed in some way in its journey from the past into the future, it cannot be recovered in its original form, and so is lost forever. In this way, time can be considered to be an effort variable.

CHEATING GENES

Certain genes are able to out-compete their homologues in an interesting way (Ganetsky, 2000). When certain chromosome pairs split during meiosis, genes of one type and its corresponding gene on the other chromosome find themselves in different gametes. The gene for red-eyed fruit flies (dominant gene) in certain males can eliminate the gene for white-eyed fruit flies (recessive) by killing sperm with the recessive gene. Thus, a cross between a heterozygous male and a female homozygous for the recessive gene should yield 50% offspring recessive homozygous (white eyes) and 50% offspring heterozygous (red eyes).

Not all crosses end this way. Some of the red-eye genes can destroy white-eye sperm and the results are 100% heterozygous offspring (red eyes).

Thus, when a BU dies as a result of competition, not only does that BU fail to survive, but all potential future generations also fail to survive. This has enormous consequences. It means that

Those BU that are best adapted to their environment will stand the best chance of populating that environment.

Future generations will be well adapted to that environment because those less well adapted will be crowded out.

Competition among all future BU will become even stronger.

Genetic material contained by BU that do not survive will be forever lost.

Ie the future is so vast, and the number of generations represented by the future is so numerous, it does not take much of a competitive advantage to have a huge ultimate effect on a population. A competitive advantage of 0.0001% can become 100% in 693,148 generations. So, when we talk about natural selection, we usually do it in terms of reproductive advantage and genetic survival rather than the immediate effects on one generation only.

DARWIN'S LEGACY

Charles Darwin has been recognized as the person most responsible for the recognition of biological evolution and continual creation of new forms of life. Before Darwin, biological scientists had sought purpose and meaning in the relationship among organisms. Darwin asserted that the order was the purpose; that relationships themselves were derived from history (Gould, 1986). Form and function of organisms and their parts resulted from adaptations of things that were already there, modified under the influence of environmental pressures. That's why Darwin's ideas were so novel; he removed the order and purpose from the creation of an individual, and placed it on the environment to which the individual responded.

Selection pressures lead to evolution of biological forms better adapted to their environments. (Courtesy of Beetle Bailey, King Features Syndicate, New York.)

Perhaps now we can understand why intermediate forms of BU (the so-called "missing links") are not easy to find. If we suspect that two types of BU are related through natural selection, then the intermediate BU between these two forms would probably have been lost completely due to their competitive disadvantages. They would no longer exist, and their genetic material would survive only in changed form.

Changes in environment change selective pressures and change reproductive potentials. These environmental changes can be physical, chemical, cultural, legal, or may take other forms. We can easily see the effects of physical selection pressures on the animals that populate tropical contrasted to arctic regions (see Section 2.7, Heat Transfer). Tropical animals tend to be lanky with long limbs to help lose heat. Arctic animals are stocky with short limbs; their low surface area-to-volume ratios conserve heat. Chemical pressures have led to bacteria populating ocean vents that use sulfur compounds as sources of energy; other bacteria would find these extreme environmental conditions too harsh for survival. Cultural reproductive advantage can be seen among groups of humans less likely to use birth control methods than other groups. As long as higher birth rates are not accompanied by lower survival rates, groups not using birth control methods will continue to increase disproportionately to other groups. Legal changes can also affect selection pressures.

SELFISH GENES

According to Dawkins (1976), genes are impersonal replicators, dedicated to multiplying as widely as possible. This idea considers the organism as the means to pass genes from one generation to the next. Aggression and selfishness are natural attributes of genes locked in a competition to dominate. Altruism, love, and generosity would only be expected to be shown toward those other individuals who share the same genes, and the closer the relationship, the more care and attention would be lavished on the relative. This is an interesting idea, and seems to have some merit in the animal kingdom (see box on Hamilton's Rule).

Humankind, however, does not always act this way. There is care and concern for even total strangers. There is even care and concern for other species, especially those kept as pets. This is a cultural attribute.

There are selfish genes that introduce themselves into both mosquito X and Y chromosomes and thereafter cause infertility of their offspring pose the possibility to eradicate completely mosquitoes that carry malaria disease (Regalado, 2016a).

Perhaps the paradigm for selfish genes are genes called *transposons*, or "jumping genes." The transposon encodes a protein that cuts the transposon DNA free of its place in the chromosome and then reinserts it in another unrelated place in the genome. DNA repair mechanisms of the cell then mend the hole at the original transposon position by recreating the transposon nucleic acid sequence. After this, the cell has two copies of the transposon rather than one. This is an example of competition even among different genes within the same cell (Gould et al., 2006). Transposons make up about 50% of the human genome (Burt and Trivers, 2006), and make up a large part of the so-called "junk DNA" that appears to have no purpose except to replicate itself.

Transposable segments of DNA are present in all organisms, and were found in maize (corn) many years ago. Transposable elements comprise about 85% of maize DNA and affect the color of corn kernels (Figure). The maize genome with ten chromosomes and 2.5 billion base pairs is almost as big as the human genome (ASABE, 2010).

Because of their activities, transposons have been called "genetic parasites." They do serve a useful purpose sometimes. Jumping genes are one cause of new species formation in cichlid fishes of Africa (Meyer, 2015). In the fungal disease called

(*Continued*)

SELFISH GENES (Continued)

powdery mildew, transposons disguise the pathogen by changing the genomic structure determining target molecules that identify it to the host plant (Spanu et al., 2010). Transposons appear to be important in animal, including human, brains and brain activity (Gage and Muotri, 2012). It has been speculated that the genetic variation that results from transposon insertion makes at least some brains better able to confront environmental challenges and improve species survival.

The conflict among genetic elements is sharpened by the existence of *gamete killer* genes in certain fungi. The gamete killers are a series of tightly linked genes (genes that nearly always are replicated together, just as certain sex-linked genes are tightly-linked) that code for a toxin that is formed or deposited in all spores produced by the fungus. In the spores that contain the gamete killers, an antitoxin is made. Thus, spores that contain the gamete killer genes survive but spores without the genes are extirpated. The result is that gamete killers end up in 100% of the viable spores.

All is not clearly one-sided, however, because there are other genes that confer resistance to gamete killer genes. The tide of battle turns around when these genes enter the gene pool.

Transposons are responsible for the different colors of kernels in these corn ears.

Most of our food that we eat today, the food that we call "natural," is really evolved food (Palumbi, 2001). Seedless oranges, pink grapefruit, Idaho potatoes, sweet corn and popcorn, Angus beef, and huge Thanksgiving turkeys are all among these (Ferrari and Chi, 1998). Our different breeds of dogs and cats are also products of not-so-natural selection that illustrates how reproduction can be used to alter characteristics of living organisms.

Not all characteristics are subject to successful selection pressure. There must be natural variation, differences in reproduction, and inheritability of the trait in order for selection to produce results. Natural selection requires not only that there be genetic variation but also that the genes be expressed (Mulcahy and Mulcahy, 1987). Without the

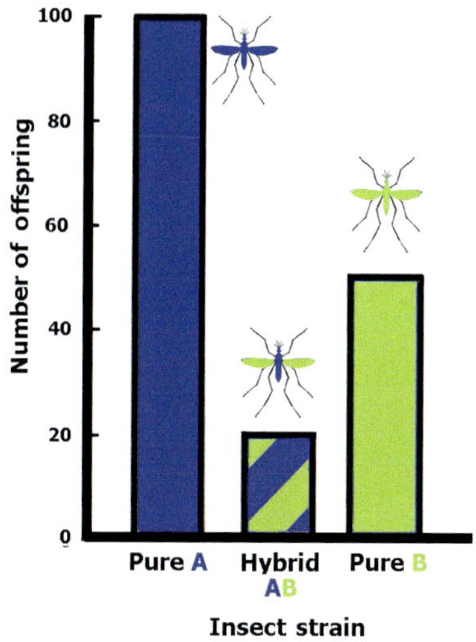

FIGURE 5.4.1
Underdominance describes a condition when a mating between two strains (A and B) results in offspring (AB) less fit than either parent. This mechanism has been proposed as a means to eliminate malaria-carrying mosquitoes from the environment. (Redrawn from Gould, F. et al., *Am. Sci.*, 94, 238–246, 2006.)

latter, the genetic code present in a BU is simply irrelevant. Those traits that are not genetically determined cannot be selected for. Predation can select for different traits than would otherwise be chosen. An example of this is the color of male guppy fish: females breed preferentially with the most colorful ones, but predators can most easily locate and eat the most colorful ones. Thus, male guppies in upland streams where there are few predators are more colorful than guppies downstream where predators abound (Palumbi, 2001).

Underdominance is the term used to describe a counterintuitive competitive situation where the more fit group of interbreeding strains does not result in the fitter group surviving (Figure 5.4.1). This can happen when these conditions prevail (Gould et al., 2006):

1. There are two (or more) distinct genetic groups, one of which has higher numbers of surviving progeny.
2. The two groups crossbreed.
3. The progeny of the crossbreeding are less able to survive than either of the two pure-bred (homozygous) groups.

The result is that the total population begins with dominance by the more fit genetic group: their numbers are greater than the group with less fit genes. It is more likely that the less fit individuals mate with more fit individuals than with individuals similar to themselves. The progeny of this mating have a small likelihood of survival. Some of the better fit individuals will mate with the less fit individuals. This progeny will also not survive in great numbers. Eventually, the proportions of the two subpopulations equalize. At this point, interbreeding is very likely, and few progenies survive. The total breeding population may then collapse.

HAMILTON'S RULE

William D. Hamilton was a theoretical evolutionary biologist who, among other contributions, expounded on altruistic actions performed by individuals for others (Hamilton, 1964a, 1964b). Each of these actions is considered to have a cost in resources that could be used to support reproductive success. On the other hand, each action by a benefactor could enhance the resources available to the recipient to boost the recipient's chance of reproductive success. So, each action has a cost to some genes, and a benefit to other genes. Hamilton's Rule states that the altruistic action will be performed if the cost to the benefactor (C) is less than the fitness benefit to the recipient (B) times the probable degree of genetic relatedness (R) between the two:

$$B \bullet R > C$$

The probable degree of genetic relatedness is a number, less than or equal to 1.0, which expresses the probability that two individuals share the exact same genes. The closer the relationship between two individuals, the larger the degree of relatedness.

Table of Probable Degree of Genetic Relatedness

Relationship	R
Parent-offspring	½ (0.5)
Grandparent-grandchild	¼ (0.25)
Great grandparent-great grandchild	1/8 (0.125)
Identical twins	1.0
Full siblings	½ (0.5)
Half siblings—one parent the same out of two	¼ (0.25)
First cousins	1/8 (0.125)
Second cousins	1/32 (0.03125)
Full hymenopteran sisters—ants, termites, bees, wasps, etc.	¾ (0.75)
Full siblings, sex linked genes	¼ (0.25)

Hamilton's Rule has been identified as leading to the field of *sociobiology*, which posits a genetic basis for all actions among individuals (Dawkins, 1989). According to sociobiology, all genes are in competition with one another to reproduce to the maximum extent possible.

Biodiversity is an essential element of a natural selection process in the face of continuous competition. Biodiversity is valuable for two reasons:

1. It leads to a greater abundance of the species in question because different populations can exploit different habitats and resources in unique ways.
2. It fosters enhanced long-term stability by spreading the risk and providing redundancy in the face of unpredictable catastrophes.

Responses to environmental pressures are often gradual and incremental, resulting from slight changes in genetic expression from one generation to the next. There have appeared in the fossil record, however, periods of time when changes seemed to have occurred suddenly, more widespread, and of much greater magnitude than at other times. This effect still happens in the present, when microbes well adapted to one species suddenly jump to another species, and then evolve quickly to adapt to their new host. This seems to have been the case for the influenza virus that chronically infects birds, but has the ability to move to swine, and from there to humans. It is also the case for the HIV that was apparently a mild infectious agent of monkeys (simian immunodeficiency virus, or SIV).

The quote at the beginning of this section illustrates these two different views of evolution. One group posits that the evolution of new life-forms is gradual and incremental ("evolution by creeps"). They look for small changes that sometimes make it difficult to distinguish a newer life-form from a previous one. In this view, interbreeding between a prior species and an emerging species is possible, but there is no interbreeding among species genetically far enough distant from one another.

At the other end of the spectrum are the *punctuated equilibrium* evolutionists, those who point to fossil evidence that demonstrates epochs in Earth's history when new species appeared explosively fast ("evolution by jerks"). These proponents argue that differences between older and newer species are so vast that they are clearly distinguishable.

Technological progress can be said to encompass both of these properties. A particular product may evolve gradually for a while, with incremental improvements appearing at regular or irregular time intervals. Automobile models, medical devices, light sources, and food products demonstrate the occurrence of incremental changes.

Nevertheless, every now and again there is a technological paradigm shift. Substituting automatic transmissions for manual shifting, the introduction of nuclear magnetic resonance (NMR, now called MRI) imaging, light-emitting diodes instead of incandescent bulbs, and quick-freezing of food each ushered a new technological era that will see more slow improvements until the next big breakthroughs. Thus, there is a parallel between advances made in human technology and the biological world: slow increments punctuated by sudden and large changes that seem to come from out of nowhere.

Those who deal with living systems need to understand that these systems are not static. They will change in response to the things done to them. Just because BU are of a certain form and function now, does not guarantee that they will be the same after changes are imposed. Whether we are considering new laws to protect the environment, the implantation of an artificial heart in a human, or cooling a bioreactor, the BU involved will change. If the change we make is permanent, and affects the reproductive potential of the BU involved, then the BU can change very dramatically. We must keep this in mind as we extend technological control over our domain.

MEMES

An intergenerational information legacy can use other means besides genes to determine physical and behavioral qualities. This alternative takes the form of cultural information passed from one generation to the next. Simple packets of cultural information are called *memes*, analogously to the simple packets of physical information called genes.

Culture has been likened to the genome, and individual ideas (memes) to genes. Memes are now in a competition to survive, and evolution of ideas (and knowledge)

(*Continued*)

MEMES (Continued)

can be much faster than genetic evolution. This may lead to a new paradigm for survivability and improvement of performance. It is unclear how these new evolutionary rules may affect future life, but Palumbi (2001) suggests that they may interact, with ideas, and the knowledge that comes from the accumulation of ideas, forming the basis for future evolution of humans and other species with which we come into contact. Perhaps this can be said to have happened already through the products of plant breeding, agricultural selection, and domesticated animal improvement. Humans have had inadvertent evolutionary effects on the disease-causing microbes that they had hoped to subdue and on plants and animals whose territories they have invaded. Add to this mix the new knowledge of biotechnology and the ability to directly select those genes that will not only survive but also come to dominate the planet. Ideas may have their own evolutionary course to follow, but they certainly have been a powerful force for natural selection in more physical ways.

The examples of individual adaptability directly dependent upon cultural information are many. Humans, for instance, learned how to domesticate animals and plants for their own purposes, and thus begat agriculture. Those who had the best agricultural knowledge were the ones who had the best chance of surviving and reproducing. This knowledge, when passed from one generation to the next, clearly gave reproductive advantage to that human line. Thus, information became a determinant of survival alternative to genetic makeup. Some animals do the same by teaching their youngsters about tools, plants safe to eat, avoidance of predators, and migratory patterns (see box on Mother Bear Man, Section 6.22.7, Social Infrastructure).

For instance, adult meerkats (small members of the mongoose family) teach their pups how to catch and eat dangerous prey. Adults bring dead scorpions, lizards, and spiders to very young offspring. As the pups get older, they are given prey disabled by, for example, biting off the stinger of a scorpion. The pups are eventually taught how to handle normal live prey (Thornton and McAuliffe, 2006).

Sheep and goats do not innately know the difference between poisonous and nonpoisonous plants. They learn about this through the social interactions of the herd. Young goats learn about edible forages by imitating older goats, and eating what they eat when they eat it (NIAA, 2007). Learned behaviors among animals include (Dugatin and Hasenjar, 2015):

1. Chimps and crows use tools.
2. Parrots solve problems with logic.
3. Elephants disable electric fences by dropping large rocks on them.
4. Dolphins learn to cooperate with humans to catch fish.

Information can be passed biochemically among microbes of the same or different species. This chemical communication confers competitive advantages to those most sensitive and responsive to chemical signals coming from other organisms. Antibiotic resistance, for instance, can be passed from one organism to another through physical or chemical means, and can become a permanent part of the progeny of that organism (Dorit, 2008). Insects and plants have similar means of information exchange that

(*Continued*)

MEMES (Continued)

can permanently change responses of subsequent generations. These exchanges may somewhat generalize the concept of memes, but the result is the same: memes allow for information to be passed from one generation to the next without directly involving genetic information storage.

This book is an example of how information can modify behavior and thus lead to better survivability through successful applications of knowledge. Understanding of biology results in control of living things, and the control is meant to benefit (in some way) human beings.

Although it is easy to see how knowledge can affect behavior, knowledge passed from one generation to the next can also change physical characteristics of the recipients of that knowledge. Examples of this are male circumcision among Jews, foot-binding of Chinese girls, neck-elongation among some African women, plastic surgery, and human selection of cultivars of fruits, vegetables, grains, and flowers.

Among species for which cultural information can be taught and learned, the notion that memes have become powerful forces for survivability rivaling, or even overshadowing, genes is not very far-fetched. It seems likely that the future of the human race will depend on its ability to discover new information and to transfer that information to others. This is the ultimate response to new environmental challenges.

Example 5.4.1: Natural Selection from Genetic Variation

Color of the moth *Biston betularia* was profoundly influenced by the rise of industry in Britain. Moth collections, spanning more than 50 years, record the rise in frequency of darker colored moths as the deposition of soot on surfaces visited by the moths made them darker. Darker colored moths were thus better protected against bird predation. Although darker genotypes had existed at a low prevalence in Briton before the Industrial Revolution, the darker genotype came to dominate genes for lighter colors after the Industrial Revolution (Koehn and Hilbish, 1987).

APPLICATIONS AND PREDICTIONS

1. Organismal changes will occur most rapidly where competition is most severe, environmental selection pressures are the greatest, and generation time is the shortest.
2. An organism that can adapt easily to environmental conditions will not evolve. Likewise, an organism that cannot survive in an environment will not evolve.
3. Evolution will not apply to an individual organism, but to a progenitor and progeny.
4. A mother will be more likely to care naturally for her own children than to care for her stepchildren.
5. Competition will be present for all biological systems.

(*Continued*)

APPLICATIONS AND PREDICTIONS (Continued)

6. Social groups with cultural information passed from one generation to the next will have a competitive advantage over social groups without that cultural information. Knowledge about the preservation and storage of food is one such piece of cultural information. Proper sanitation and medical care information is another example. Still another is information about the use of tools in hunting, domestic activities, and fighting.
7. Spatial or temporal isolation is necessary for the formation of a new species.
8. Symbiosis will improve competitive advantage for the species involved; cooperation among organs in a body will improve the survival of all.
9. Humans have changed selection pressures for many other species; all species modify selection pressures for other species.

5.5 Biological Hierarchies

EPIDERMIS, n. The thin integument which lies immediately outside the skin and immediately inside the dirt.

—Ambrose Bierce

Those who deal with biological systems might offer that, beginning with the simplest BU and classifying to the most complex BU, a biological hierarchy could be constructed as

1. Cell
2. Tissue
3. Organ
4. System
5. Organism
6. Population or colony
7. Biome
8. Ecosystem

We see in this section that the cell is distinguished as the basic BU. Any group of cells of similar structure that performs a specific function is called a *tissue*. Examples of these are muscle tissues in animals and phloem tissues in plants. A multicellular structural or functional unit to perform a specific role of an animal or plant, which may be composed of different tissues, is known as an *organ*. Examples are the liver, a leaf, or an eye. A *system* is a functional unit made up of correlated and semi-independent parts. Examples of these are vascular or digestive systems. An *organism* is any living thing, be it animal, plant, microorganism, or other (depending on the classification scheme, which will be avoided in this text). A *population* is a group of similar individual organisms inhabiting a particular locality or region. A *biome* is a major regional community of organisms defined by the habitat and determined by the interaction of the substrate, climate, fauna, and flora. An *ecosystem* is a collection that includes all the biotic organisms and abiotic components of the total environment.

From the earlier definitions, one can see that there is a certain increase in complexity from the cell to the ecosystem. However, this increase in complexity is largely definitional. Each of these levels can be considered to be a BU, and, as such, has similar responses to its external and internal environments. For instance, consider competition and cooperation. Two cells can compete, and this is basic to nearly all of life and to the natural selection process. If these cells cooperate in a certain way, they can form a tissue. Likewise two biological systems can cooperate for mutual support of an organism, or they can compete as when the digestive system and the muscles compete for blood flow in an exercising animal.

If we consider each of these as BU, then the responses of these BU can be studied in general and applied in the context in which they arise. Thus, biological responses in context (BRIC) form the basic building blocks for the study of predictive biology. Part III is devoted to developing the BRIC concept.

Ecology or bionomics is the study of plants and animals in relation to their total environment. The techniques and methods are more powerful than this, however, because they can be related to all BU. Studying the BRIC of a tissue BU is the same as studying the BRIC of a population BU. Thus, the approach here will be largely the same as the application of ecology to all BU, no matter what the level.

There are thus certain hierarchies in biology, and these hierarchies are often hardly more than conceptual in nature. There are overlaps such as when a cell can be an organism or a part of a tissue, and there are different ways to classify the same BU. However, many of the same principles apply no matter what the classification scheme, and no matter what level in the particular scheme is being considered.

5.5.1 The Cell

> ...an embryo is the result of the union of a live human egg and a live human sperm. If either is dead, no viable embryo is produced. Therefore, life does not begin at conception, at the transition between embryo and fetus, or at birth. It continues.
>
> **—John Majka**

The cell is the basic BU in a way similar to the atom as the basic unit of physics and chemistry. Particles smaller than the atom exist, but chemistry, as we know it, does not exist below the atomic level, and the laws of physics are completely changed at that scale. At levels lower than the cell, we do not have biology, but rather have chemistry. This is an example of an *emergent* property where the properties and actions of the whole entity are greater than the sum of its parts.

Because the cell exhibits all the properties and actions that we would call living, the cell is considered to be the basic biological entity. All combinations of cells will also be considered to be living (Figure 5.5.1).

There are two basic types of living cells. *Prokaryotes* are the most primitive of organisms, and this includes all bacteria (Figure 5.5.2). They have no true cell nucleus, nor do they have internal organelles (specialized regions separated by membranes to perform certain functions such as food storage or energy production). They have a single chromosome made up of nucleic acid only, and they reproduce by binary fission (see Section 6.17, Competition). They are small, where diffusion does not severely hamper cell function. Many of these functions are performed within or along the cell membrane.

FIGURE 5.5.1
One combination of cells considered to be a living unit.

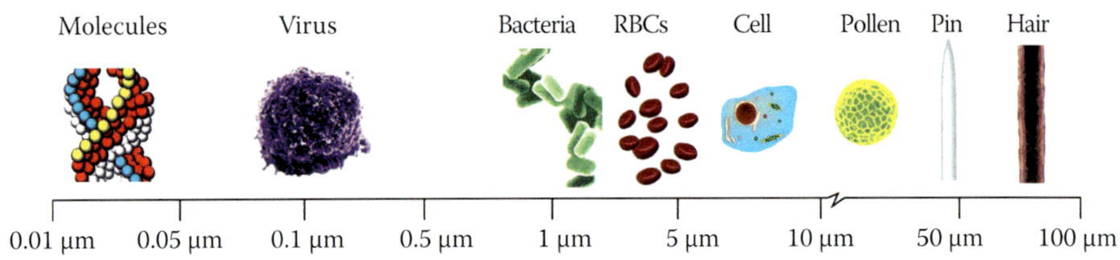

FIGURE 5.5.2
Comparative diagram of different particle sizes showing where typical bacteria fit into the size scheme.

Eukaryotes are distinguished by the fact that they have a membrane-bound nucleus with several chromosomes made of nucleic acids complexed with protein (see Section 5.3, Genetic Basis). They have an internal structure including organelles such as Golgi apparatus, endoplasmic reticulum, lysosomes, and mitochondria (Table 5.5.1). Their relatively large sizes might incur materials transport limited by diffusion. Thus, they have specialized transport proteins and vesicles to move materials and store them at sites throughout the cell. Cell division is by mitosis (asexual) and meiosis (sexual). All plants, fungi, and animals contain eukaryotic cells.

Outside the cell is a membrane. The structure of the cell membrane (Figure 5.5.3) is believed to be a bimolecular lipid layer covered on both sides by protein coats (Schneck, 1990). The lipids cause the membrane to exclude water and polar molecules. The membrane behaves as if it contains pores slightly larger than a urea molecule (5–10 Å), so small molecules can pass freely from one side to the other. The outer surface of the cell is replete with linkage sites for external biomolecules (see Section 4.4.1, Sensors) and is very active in transport of chemicals (see Section 6.19.2, Chemical Stimuli) and in detection of foreign bodies. The cell membrane is populated with proteins that perform several functions. Some proteins act as attachment sites for other molecules, and these can be very specific. When a mating molecule attaches to these receptors, some action is usually triggered in the cell, from the formation of an action potential in a neuron in response to a neurotransmitter to the formation of reactive proteins. Other proteins included in the cell membrane are shaped in such a way that they form channels (or tunnels) through which small ions or molecules may pass. Depending on the shape of the channel, its size, and the surface charges inside, the channel may be a good gatekeeper for specific molecular species.

TABLE 5.5.1

Summary of Typical Components of Different Types of Cells

		Prokaryotic	Eukaryotic			
Cell Component	**Function**	**Archaebacteria, Eubacteria**	**Protistans**	**Fungi**	**Plants**	**Animals**
Cell wall	Protection, structural support	Some	Some	All	All	None
Central vacuole	Increasing cell surface area; storage	None	None	Some	All	None
Chloroplast	Photosynthesis; some starch storage	None	Some	None	All	None
Cytoskeleton	Cell shape; internal organization; basis of cell movement and, in many cells, locomotion	None	Some	Some	Some	All
DNA	Encoding of hereditary information	All	All	All	All	All
Endoplasmic reticulum	Initial modification of many of the newly forming polypeptide chains of proteins; lipid synthesis	None	All	All	All	All
Flagellum, bacterial	Locomotion through fluid surroundings	Some	None	None	None	None
Flagellum or cilium with 9 + 2 microtubular array	Locomotion through or motion within fluid surroundings	None	Some	Some	Some	All
Golgi body	Final modification of proteins, lipids; sorting and packaging them for use inside cell or for export	None	All	All	All	All
Lysosome	Intracellular digestion	None	All	Some	Some	All
Mitochondrion	ATP formation	None	All	All	All	All
Nucleolus	Assembly of subunits of ribosomes	None	All	All	All	All
Nucleus	Physical separation and organization of DNA	None	All	All	All	All
Photosynthetic pigment	Light-energy conversion	Some	Some	None	All	None
Plasma membrane	Control of substances moving into and out of cell	All	All	All	All	All
Ribosome	Protein synthesis	All	All	All	All	All
RNA	Transcription, translation of DNA messages into polypeptide chains of specific proteins	All	All	All	All	All

Source: Starr, C., *Biology: Concepts and Applications*, Brooks/Cole, Stamford, CT, 2000.

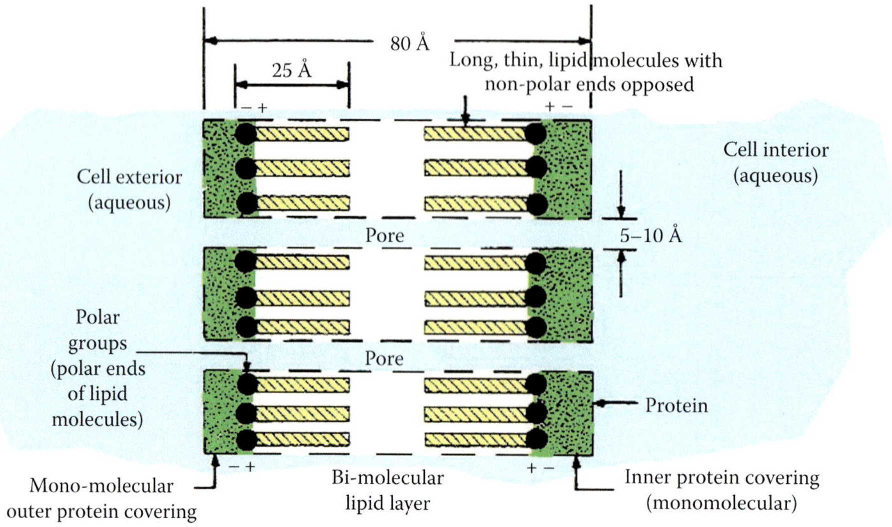

FIGURE 5.5.3
Schematic representation of the bi-lipid configuration of the cell membrane. Membrane phospholipids in a water medium naturally align themselves into double layers held together by no other bond than hydrophobia (Williams, 2009). Membranes are very dynamic, studded with proteins that act as ligands, sensors, and both passive and active channels for ions and small molecules.

Within the cell there is a gel-like substance composed of water, proteins, sugars, lipids, ions, and complex organic compounds. Various organelles, many enclosed within their own membranes, perform functions such as storage, metabolism, regulation, and so on. Together these substances and structures perform the complex functions we characterize as a living cell. The inside of a cell is packed full of ions, molecules, and inclusions (see Figure 3.7.7).

The cellular cytoskeleton gives the cell some structure and rigidity. Actin filaments that, among other things, form parts of the cellular cytoskeleton prefer to polymerize at one end and depolymerize at the other. Hence, one end grows and the other shortens. Actin filaments thus stream from one end location to the opposite end, forming a one-way transport system for molecules linked to the filaments (Brodie, 2004b).

Although the cell is the basic unit of living organisms, cells are not always completely autonomous and separate units. Fusion of cells appears to be commonplace (Ogle and Platt, 2004). Skeletal and cardiac muscles, and some liver tissue, are composed of giant cells with multiple nuclei that appear to be fused from multiple precursor cells. These are termed *syncytia*.

Cell fusion within the same organism, and even between cells from different species, is possible. If the nuclei fuse and contain DNA from both precursor cells, the resulting hybrid is called a *synkaryon*. If the nuclei remain separate within the cell, the hybrid is called a *heterokaryon*. In both cases, DNA from both precursor cells exerts an influence over proteins and other complex molecules fabricated by the cell.

Fused hybrids possibly have a major role in the differentiation of fetal cells during early development. Fusion also might allow a mature cell whose location and function are well established, to induce an immature cell, a stem cell for instance, to assume the function of the mature cell. This may well be the mechanism that regulates the ability of stem cells to differentiate into tissues that need augmentation. Fused hybrids may also have a survival advantage by introducing superior DNA from one species into another.

Embryonic stem cells are useful if they retain the ability to transmutate into many other types of cells. This *pluripotency* becomes compromised the longer a cell line is maintained in culture. Therefore, frequently freezing vials of stem cells as the line is expanded helps to maintain cells from validated batches, and is crucial to success with a stem cell line.

EPI-, MESO-, ENDO-, AND ALL THOSE KINDS OF CELLS

Epithelial cells comprise closely packed monolayers that separate different tissue compartments. One side of the epithelium faces the outside of the body, or the side occupied by the external environment. The other side faces the inside of the body, or the side containing the blood and extracellular fluids. The skin is made of epithelial cells, as are the lumens of the gastrointestinal tract, the kidneys, the lungs, and the urinary bladder (Putnam, 1995). Simple epithelium is one cell thick, and compound epithelium is several cells thick. The cells usually cover connective tissue, and are held together by a cementing substance to form a sheet (Hale et al., 1995). Their shapes give rise to descriptive names of columnar, cubical, and squamous (flat).

Epithelia sometimes have a secretory function (in the liver), but otherwise function as a barrier between the external environment and the internal environment. They also selectively transport substances between compartments. To enhance this task, the outward-facing membrane, called the *apical* membrane, is often covered by *microvilli* (small, finger-like protrusions that increase surface area for absorption). The inward-facing membrane, called the *basolateral* membrane, has a different lipid and protein composition from the apical membrane. Because of these specialized structures, and the tight junctions between cells, the only substances that normally enter the body from the outside are those transported through the cells.

Endothelial cells are a single layer of flattened epithelial cells lining a tube such as the heart, blood vessels, and lymph vessels of vertebrates. *Mesothelium* cells are similar to epithelial cells, but they line the inside of the body cavity. The *endoderm* is a single layer of tissue found outside the vascular layer in many *angiosperm* plants.

Ectoderm (or ectoblast) is the germ cell layer lying outside of the developing embryo that eventually gives rise to epidermis, nervous tissues, and *nephridia* (a primitive tubular excretory organ present in many invertebrates such as the earthworm). *Endoderm* is the embryological germ layer in animals that develops into the gut and its associated organs. *Mesoderm* is the layer of embryonic cells lying between ectoderm and endoderm in all higher animals that forms the muscles, blood system, connective tissues, kidney, the skin dermis, and the axial skeleton. The limbs of some animals, such as newts, can be regrown if cut off. The tissue from which the limb arises comes from the mesoderm at the middle layer of the embryo (Tickle, 1981). The ectoderm, endoderm, and mesoderm are all embryonic cells that develop into epithelium, endothelium, and mesothelium, respectively.

The *epidermis* is a designation of location rather than type of cells. The epidermal layer of vertebrates, which is the outer layer of skin, is usually made up of stratified epithelium with an outer layer of dead cells and an inner layer of growing and dividing cells. The invertebrate epidermis is normally one cell thick and often forms a protective cuticle (Hale et al., 1995). The epidermis of plants is a one-cell thick tissue that surrounds young roots, stems, and leaves. The epidermal cells (not epithelium) of stems and leaves secrete a cuticle (a protective layer of protein or lipids).

(*Continued*)

EPI-, MESO-, ENDO-, AND ALL THOSE KINDS OF CELLS (Continued)

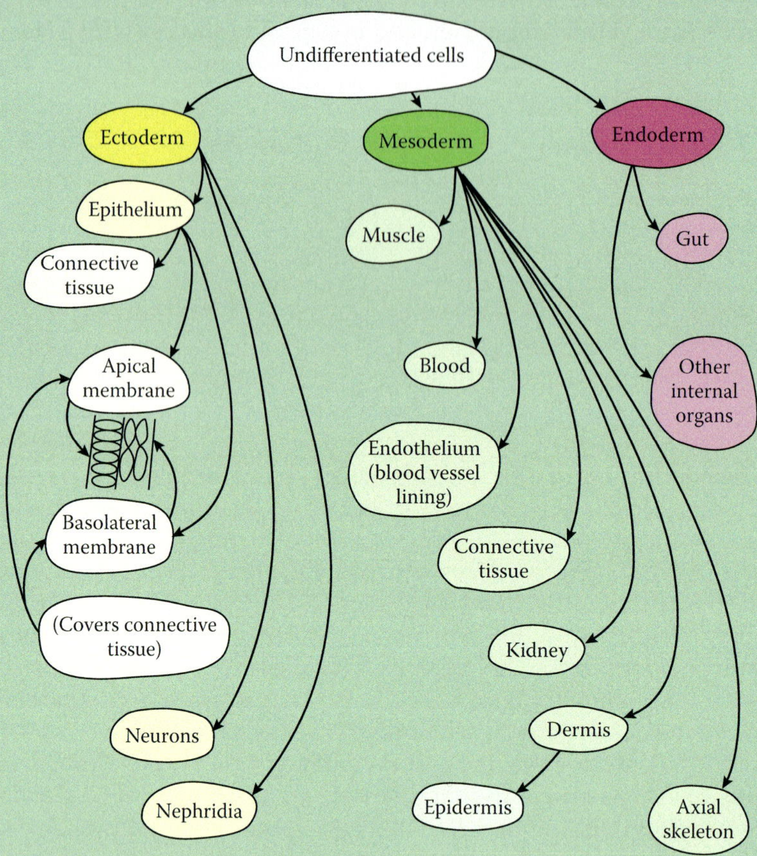

Cell types resulting from undifferentiated fetal cells.

5.5.2 What Is Life?

> If you call a tail a leg, how many legs has a dog? Five? No; calling a tail a leg doesn't make it a leg.
>
> **—Abraham Lincoln**

Defining the basic living unit as the cell begs the question about what life really is. This definition is more than just an academic exercise if we are to search for life outside the Earth, called *exobiology*. In the next chapter of this book there will be descriptions of organisms living in some extreme environments of the Earth. There are extremophiles, such as *Shewanella*, found in very harsh environments that grow electrically conductive proteins from their outer membranes to exchange electrons directly with metal surfaces; there are others (*Borrelia burgdorferi*, the cause of Lyme disease) that use manganese instead of iron in essential enzymes (most other organisms require iron); there are extremophiles in regions of high salt concentrations (called *osmophiles*), extreme water pressures (blobfish and devil worms), and able to live in space without either water or air (*Tardigrades*). These forms of life make it difficult to determine if life is encountered on other heavenly bodies (Powell, 2015).

Even these organisms could be classified as alien or highly unusual. Life outside the Earth might be expected to be even more unusual. Will they be classified as living or not?

The separation between living and nonliving is not sharp. Most viruses are little more than RNA enclosed in a protein coat. Yet, they reproduce in the right environment of a host cell. Prions are bits of protein that become active within cells, and also reproduce. Neither of these carries on all the functions that we would call "life," but they do exhibit some living properties.

So what are these properties? They are numerous and depend on who is doing the listing. Some examples are as follows:

1. Life is (Starr, 2000)
    - An outcome of ancient events by Ih nonliving matter—atoms and molecules—became assembled into the first living cells as a way of capturing and using energy and raw materials.
    - A way of capturing and using energy and raw materials.
    - A way of sensing and responding to changes in the environment.
    - A capacity to reproduce, grow, develop and change over generations.
2. Life is (Hazen, 1999)
    - Highly complex chemical systems
    - Composed of cells
    - Able to obtain and use energy
    - Able to reproduce using the same genetic mechanism
    - Able to grow and develop
    - Able to respond to changes in the external environment while maintaining a relatively constant internal environment
3. Life is a condition characterized by (Campbell et al., 1999)
    - Order and complex organization
    - Reproduction
    - Growth and development
    - Utilization of energy for its own purposes
    - Responses to environmental stimuli
    - Maintenance of relatively steady internal environment
    - Evolutionary adaptation
4. Life is (Webster's New World Dictionary, in Anbar, 2001)
    - That property or quality of plants and animals that distinguishes them from inorganic matter or dead organisms; specifically, the cellular biochemical activity or processes of an organism, characterized by the Ingestion of nutrients, the storage and use of energy, the excretion of wastes, growth, reproduction, etc.

5. Life is (NASA, in Anbar, 2001)
   - A self-sustained chemical system capable of undergoing Darwinian evolution.
6. Life is (Anbar, 2001)
   - *A* process *that spontaneously organizes matter to higher levels of complexity and then maintains that complexity in potentially destructive environments.... We may search for living systems or for tangible products of life, but not for life itself.*

Rasmussen et al. (2004) have given sufficient properties of life as localized molecular assemblages that regenerate, replicate, and build new functionality through evolution. Three essential functions to life are as follows:

1. A genetic system for transmission of hereditary information.
2. A metabolic system for extracting energy and materials from the environment.
3. A containment system to maintain separation from the surroundings.

They did not, however, consider sensing and reactive functions that often characterize living things.

Stec (2004), however, brought a chemical perspective to the issue of the definition of life. Stec maintains that living things must contain chemicals (usually proteins) with alternate forms at the same or nearly the same energy levels to allow the living system to adapt. If one form of the protein clearly predominates (i.e., its energy level is much lower than those of alternate forms), then the system is not adaptable, and appears to be more physical than biological. It takes a protein system with many possible forms at nearly equal energy levels to exhibit the adaptability characteristic of living matter.

Each of these descriptions begs the question, "What is life?" enough so that none can be called a definition, although the one by Anbar (#6) probably comes close. Rather, most of these are descriptions of a combination of attributes. Taken together, they describe a living cell and combinations of cells, but they do not describe subcellular components. Hence, the cell is considered to be the basic unit of life.

5.5.3 Synthetic Biology

> Evolution prefers short-term survival at the expense of long-term function.
>
> —**J. Douglas Bremmer**

There are many reasons for trying to define what constitutes life. One of these is the attempt to create forms that have many of the same characteristics of life, including reproduction, information storage, complex behavior, and others. These artificial life-forms have never existed before and may be like no other living thing.

Some are attempting to create life from the bottom up. They are locating the molecular machinery for their protocell on the outside, where a membrane is not needed. A clump of hydrophobic fatty acid molecules glues the protocell together as a structure called a micelle (Stroh, 2005).

Genetic material for the protocell will be supplied by peptide nucleic acid, or PNA, Ih has the same double-helix structure and the same four chemical bases as DNA, but has a peptide backbone. A light-sensitive molecule will be able to provide the energy to convert precursor molecules into new fatty acids and PNA molecules.

Newly created fatty acids will be incorporated into existing micelles, making them larger and larger. At some point they become unstable and split into two, as a simple form of binary fission. Artificial nucleic acid base pairs have been produced that have been found to be accepted as part of cellular DNA (Malyshev et al., 2014; Thayer and Ellefson, 2014). Further work along these lines could result in a completely new synthetic biology.

A top-down approach to artificial life is also being attempted. One team starts with a simple 517 gene organism called *Mycoplasma genitalium*, and pares away as many genes as possible while still maintaining a semblance of life. As many as 215 genes may be unnecessary. A substitute genome is to be constructed from scratch and will require 300,000 chemical bases. If successful, the artificial life-form could be loaded with genes to perform useful functions.

Enough is now known about genetic expression that synthetic genes can be fabricated to produce a protein in a specific organism like *E. coli*. Algorithms have been developed to design synthetic genes to express target proteins optimally (May, 2010). These proteins can be used in pharmaceutical research.

Synthetic biologists change the behavior of a cell by designing and rewiring the complex genetic foundation inside. They utilize existing biological (mostly genetic) parts, fitting them together to transform cells into micromachines capable of performing whatever is their designed function. This may range from building cells that move especially fast, to cells that produce desirable compounds, to cells that turn color in the presence of ultraviolet light. This is true genetic engineering, where a set of standard genetic parts are created and characterized, and Ih can then be combined with other standard genetic parts to implement intended purposes.

Mathematical simulation modeling of single cell activities, including metabolic energy and nutrient flows, has largely been accomplished for the small organism *Mycoplasma genitalium*, and could lead to computer models of more complex cells (Covert, 2014). These models could be useful for everything from predicting the outcomes of biological experiments to designing new organisms with desirable properties.

STORING DATA WITH DNA

Digital data have historically been stored on various electrical or magnetic technologies. The vast amount of binary data needing to be stored has been increasing faster and faster, and outstripping the capacities of current data storage methods.

DNA molecules are very compact in size and durable enough that they are used in living things to store hereditable information used as a template for future generations. Using DNA molecules as data storage media could replace huge storage facilities with small test tubes containing multiple copies of DNA.

Digital files are presently stored in binary (base 2) sets of 1s and 0s. There are four different bases that make up DNA (A, C, G, and T), thus making DNA base 4. So,

1. The first step in converting a binary file into DNA is to select pairs of 1s and 0s and translate each pair of 1s and 0s into one of the DNA base molecules. For instance, 00 could be assigned to A, 01 to G, 11 to C, and 10 to T.
2. The entire binary file is then translated into a sequence of base molecules according to the assignment in step 1, giving a map of a long DNA strand composed of all four bases.

(*Continued*)

STORING DATA WITH DNA (Continued)

3. The DNA map is then used by a synthetic biology scheme to manufacture the real strand of DNA according to specification.
4. Multiple copies of the DNA strand can be made for storage.
5. A test tube containing the DNA molecules can be chilled and stored until needed.
6. To read the DNA molecule and convert back into a binary file, a standard DNA sequencing technique is used to identify the DNA code of the stored molecule.
7. The DNA code is then back-translated into the original binary file.

This whole process could be automated and made to be very efficient, especially for very large binary data files (Strickland, 2016b).

5.5.4 Ecology on Micro- and Macroscales

> With man gone, will there be hope for gorilla? With gorilla gone, will there be hope for man?
>
> —**Daniel Quinn**

When a biological engineer looks at BU responses to various environmental stimuli (or BRICs, as we have called them), it is best to think about the myriad of interactions as an ecological system. This applies no matter at Ih hierarchical level the BU happens to reside.

Ecology as popularly defined is the study of communities of organisms and how they all fit together (May and Seger, 1986). Yet this can be considered to be macroecology, or ecology on a large scale. In essence, all BU are subject to environmental influences, and every cell affects the flora and fauna around it. This is microecology, or ecology on a scale that can resolve to the smallest possible level, even if there is no identifiable biological component present. Without microecology, biochemicals could not be produced through biotechnology, biological homeostasis could not be maintained within a transplanted liver, and your radio would not be able to deliver music to your ear.

A colony or other type of social structure acts in a coordinated way Ie information flows both horizontally (at the same level) and vertically (between levels). The same is true within a cell, within a tissue, or within an organism. The social structure thus acts as an entity unto itself with independent and identifiable organization, actions, and input–output relations. Taken another step farther, the definition of BU extends to symbiotic relationships, parasite–host pairs, and predators with prey. Indeed, each of these has predictive physical and behavioral responses to environmental stimuli (Grene, 1987). These will be considered further in Part III of this text.

HUMAN ECOLOGY SYSTEM

Each of us is composed of roughly 100 trillion (10^{14}) cells, but only 10 trillion (10^{13}, or 10%) of these are human. The other 90 trillion are bacteria, parasites, fungi, and other small creatures (Buckman, 2003; Dorit, 2008). On every square centimeter of human skin there are 10^7 microbes.

Humans provide a complete and varied ecosystem, and like other ecosystems, theirs is balanced in the sense that various species cooperate and compete, with no species dominating. Included in the bacteria of note are the common *Staphylococcus aureus*, the microbes that can cause deadly *Staph* infections. *S. aureus* is present in great numbers on the skin, but they are kept in check by other bacteria and viruses that compete and limit their populations.

Escherichia coli is a common intestinal bacterial parasite that is normally harmless unless the ecosystem of the gastrointestinal tract becomes upset in some way. It is this balance that is the key to human health. The indiscriminate use of antibiotics and disinfectants can upset this balance and give competitive advantage to dangerous microbes. Their use can also keep the human immune system unchallenged and thus much more vulnerable to infection than if full immunities were allowed to develop. Related to this is accumulating evidence that allergies are more likely to develop in children who are kept too clean and their environments too sterile. It is the balance of nature to which we have accommodated. To believe that we can conquer or control all other species is very egotistical. We have allies in our struggle to survive, and it is not a good idea to ignore or harm them.

5.5.5 Food Pyramid

Time flies like an arrow. Fruit flies like a banana.

—Groucho Marx

Energy is captured from primary sources for use by BU by primary producers. These may be plants that convert solar energy into energy-rich carbon compounds such as glucose, cyanobacteria that do the same, or thermophilic microbes that use sulfur compounds from deep ocean vents as their primary energy sources (Figure 5.5.4). These BU are called *autotrophs* because they are not dependent on other sources of organic substrates to manufacture their own organic requirements; they use inorganic sources. *Heterotrophs* obtain their basic organic materials from the environment; they may dine directly on autotrophs or obtain organic wastes attributable to other biological sources.

There are four general classifications of organisms made by combining energy and carbon sources (Figure 5.5.5):

1. *Photoautotrophs*: These use light as the energy source and carbon dioxide as the source of carbon.
2. *Photoheterotrophs*: Light is the source of energy, and carbon comes from organic compounds such as alcohols, fatty acids, other organic acids, or carbohydrates.

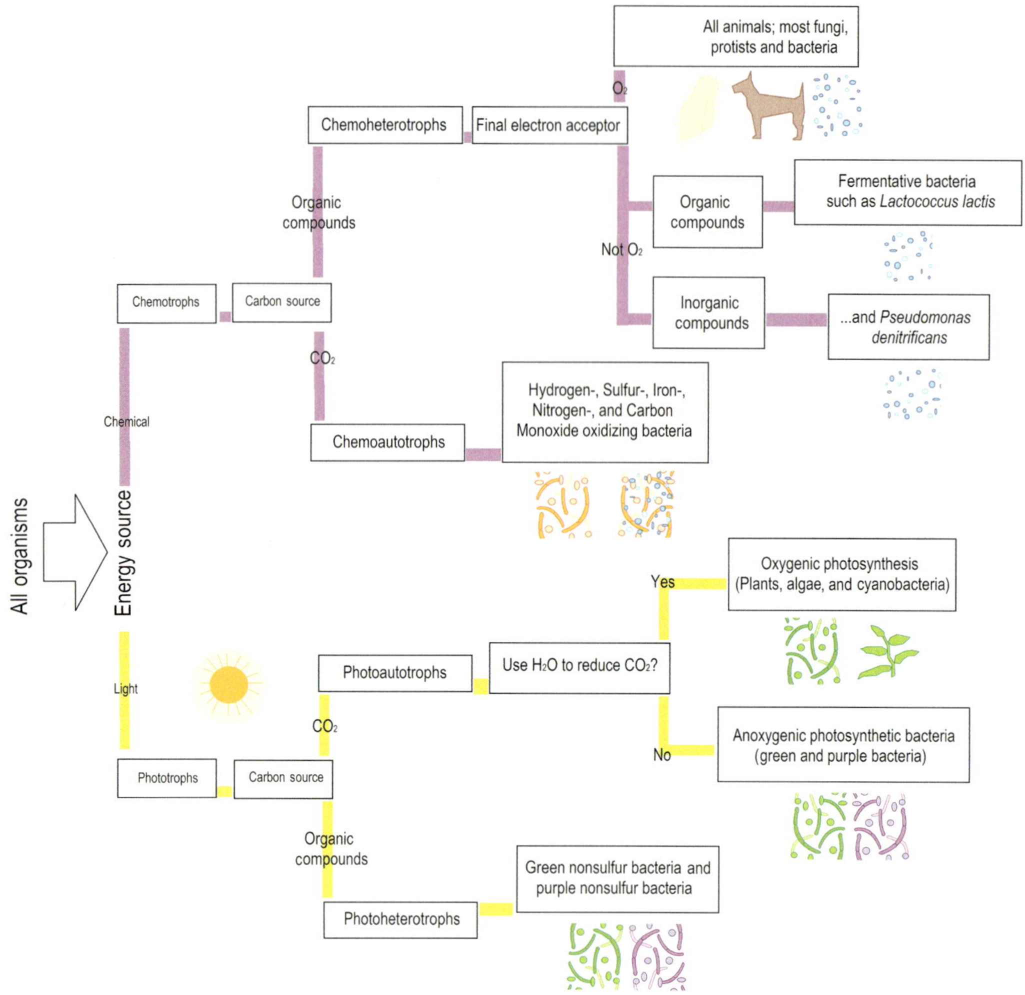

FIGURE 5.5.4
Classification scheme for organisms by energy and carbon sources. (Redrawn from Tortora, G.J. et al., *Microbiology: An Introduction*, Addison Wesley Longman, San Francisco, CA, 2001.)

3. *Chemoautotrophs*: These use electrons from reduced inorganic compounds as their energy source and carbon dioxide as their source of carbon. Inorganic compounds can include hydrogen sulfide (H_2S), elemental sulfur (S), ammonia (NH_3), nitrite ions (NO_2^-), hydrogen (H), and ferrous ions (Fe^{-2}).
4. *Chemoheterotrophs*: Both energy and carbon come from the same organic compound, such as glucose. These organisms are medically important, but some can be beneficial for bioremediation.

Once the primary producers fabricate organic compounds from inorganic sources, they can become sources of food for other heterotrophs. The ideal order of feeding is called the food pyramid (or food chain, or trophic levels), and is classically described

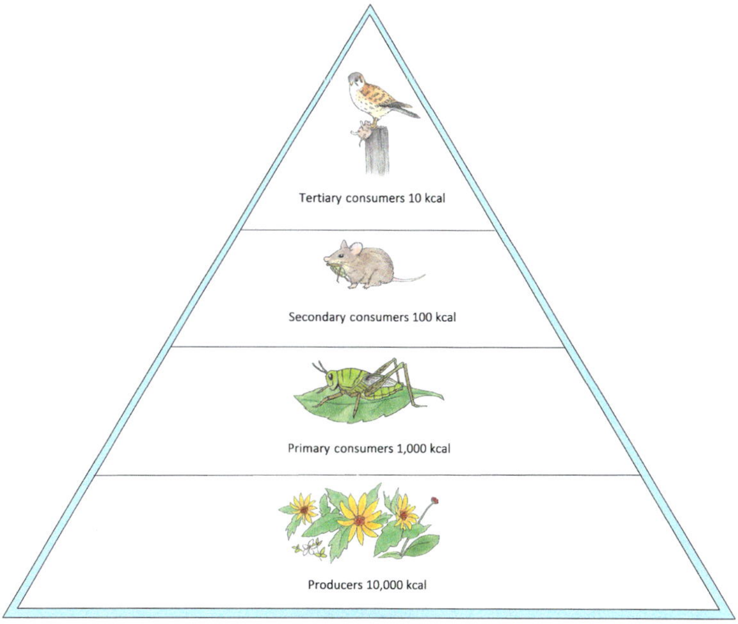

FIGURE 5.5.5
The food pyramid, also called trophic levels. Energy is transferred from the bottom to the top with an efficiency between levels of about 10%. (Redrawn from Fried, G.H. and Hademenos, G.J., *Theory and Problems of Biology, Schaum's Outlines*, McGraw-Hill, New York, 1999.)

as herbivores grazing on plants, carnivores eating the *herbivores*, and higher-level *carnivores* eating the lower-level carnivores. This is not the way it always happens, but gives the idea.

Energy levels are degraded at every step of a process (Section 2.4, Work and Energy). That is, there is always some inefficiency in converting energy from one form to another, and some energy is unrecoverable (see Section 2.4.3, Efficiency). Because of this, there is lost energy (about 90%) at each step on the food chain. Consequently, whether we consider numbers of BU, total biomass, or energy equivalence, the bottom of the food chain always is larger than the top. That is, there are more autotrophs than grazers, more grazers than lower-level carnivores, and more lower-level carnivores than higher-level carnivores. Diagrammatically, these statements describe a pyramid shape (Figure 5.5.5).

Trophic levels in the sea were originally defined as discrete steps describing the food chain (Pauly et al., 2000). For instance, tiny zooplankton (second level) feed on phytoplankton (first level) (see Figure 5.5.6). Many marine creatures don't feed exclusively on the level just below. Instead, like anchovies (level two), they may feed on organisms from several lower levels. Depending on the relative amounts of their diets, these creatures can be assigned a fractional trophic level number. Anchovies have thus been assigned to level 2.2. Humans who fish for anchovies would then be assigned to level 3.2, one above anchovies.

The difficulty comes when a lower trophic level is fished nearly to extinction. Then the entire ecosystem above this trophic level collapses.

Incidentally, humans are not always at the top of the food chain. Ask any shark.

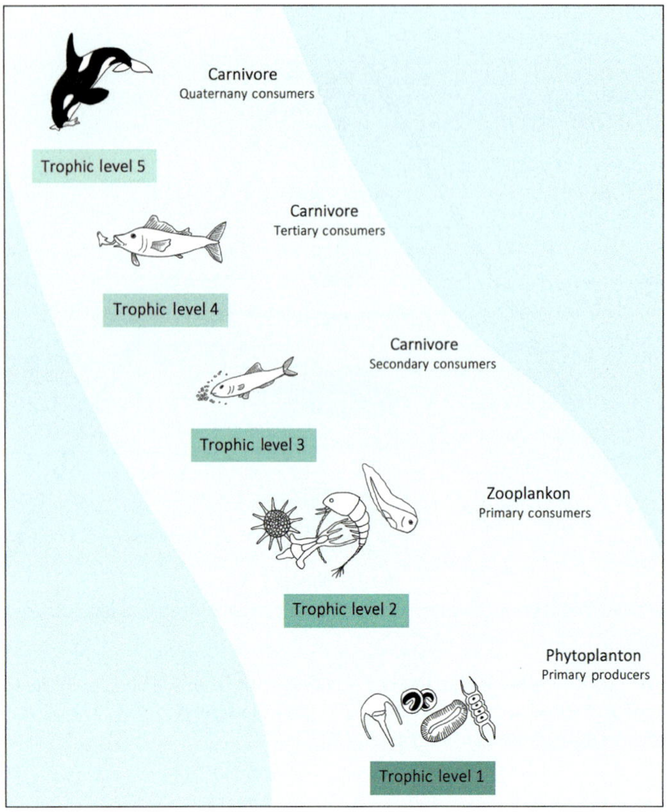

FIGURE 5.5.6
Trophic levels were initially defined to include only discrete steps (left). Organic detritus and microscopic plants (phytoplankton) occupy the first trophic level. Tiny zooplankton, which feed on phytoplankton, reside at the second level. Creatures that eat zooplankton sit at the third level, and so forth. But many marine creatures feed from multiple trophic levels and so could not be fit into this classic scheme. Thus the modern approach allows the assignment of trophic level to span a continuum rather than forcing it to take on integral values. Marine biologists would, for example, assign the anchovy, which supplements its main diet of phytoplankton with some zooplankton, to a trophic level of about 2.2; people fishing for anchovies (and eating a diet of only these small fish) would then be assigned a trophic level of 3.2 (right). (Redrawn from Pauly, D. et al., *Am. Sci.*, 88, 46–51, 2000.)

Example 5.5.1: Chimeric Antigen Receptor T-Cells

Adaptive *immunotherapy* is being developed as a weapon to fight cancer. One way that cancer cells differ from normal cells is that cancer cells have abnormal antigens on their surfaces. The body's immune system T-cells can be modified to recognize cancer cell abnormalities and destroy the cancer cells. Chimeric antigen receptors (CARs) use the variable antibody regions spliced into T-cells, which then recognize the cancer antigens. These CAR-modified T-cells are inserted into the body and can then reproduce and remain in there, ready to fight cancer cells wherever and whenever they occur. Synthetic biology techniques can be helpful to develop CAR antibody T-cell expression (Doyle, 2015).

APPLICATIONS AND PREDICTIONS

1. Colonies of ants and human livers will have similar responses to environmental challenges.
2. Cells are necessary for more complex life-forms.
3. Life requires sources of energy and nutrients to develop, grow, and reproduce.
4. All life is connected.
5. Certain nutrients and biochemicals will become more concentrated in BU at higher trophic levels.
6. Some living things are difficult to distinguish from nonliving things.
7. Nutrients and energy are recycled.
8. Trophic levels of societies will be higher than the trophic levels of individuals within the societies.
9. Higher trophic level organisms depend for their survival on lower trophic level organisms.

5.6 Is Biology Complex or Simple?

> There are known knowns; there are things we know we know. We also know there are known unknowns; that is to say, we know there are some things we do not know. But there are also unknown unknowns—the ones we don't know we don't know.
>
> **—Donald Rumsfeld**

There is an aura surrounding biology that attributes great complexity (Trefil and Hazan, 1998) that even tends to the metaphysicality of a special life force. The huge numbers of biochemicals within the cells, the way in which they interact in strangely efficient ways, the apparent self-organization, and the emergent properties of higher biological levels reinforce the notion that we are dealing with entities that somehow suspend the limits of physics and chemistry. Complexity, some say, is the key to understanding biology (Capra, 1996; West, 2006) and this very notion makes it nearly impossible to completely understand biological foundations.

Yet, each biological property that is studied turns out to be surprisingly simple. Whether it is the attachment of the gecko's feet to vertical surfaces or the height to which water rises in the xylem tubes of the giant Sequoia, what was once not understood and so thought to be very complex turns out to be based on very simple principles. And because the principles are simple, the effects are robust.

Could it be that biology is not so complex, but instead is an accumulation of a vast number of very simple outcomes based upon simple and robust modes of action? It is indeed possible, because, at least in many instances thus far, mechanisms of biological action that have been discovered and consequently understood have not required that any new principles or modes of action be postulated. The study of biology has not revolutionized the fields of physics and chemistry. Rather, it has reaffirmed them.

Engineering predictions about biological phenomena are usually based upon (mostly empirical) mathematical equations. The usefulness of these equations depends upon the accuracy of their predictions. Sometimes, however, biological behavior is too complex to be described in simple mathematical form.

The biological beings that are the subjects of these equations, however, are not able to understand mathematics, and, indeed, have been operating successfully for many generations. Thus, there must be a simpler basis for biological activity than the submission to equations that require a high degree of generalized intelligence and many years to understand.

Whether it becomes useful to identify these simple rules, the robustness of biological activity must depend almost solely on simple processes. The simplest, we know, is the gene.

There is a problem with the burgeoning and overwhelming aggregation of information being learned about genetic structure, placement, and function. Such a vast information aggregate cannot be easily classified and presented, although the multidisciplinary approach called *systems biology* attempts to understand how individual cellular components are organized, function, and, especially, interact within live cells. Manipulation of genetic knowledge is one domain of bioinformatics, but the ideal pictorial representation of genetic structure and function has not yet been demonstrated. Is there a genetic analog to the periodic table of chemical elements? One can look at the table and immediately infer generalizations about chemical properties. Will there be possible such a representation for genetic elements? The search continues.

In many natural environments, from biological cells to clusters of galaxies, complex geometric forms can appear spontaneously and propagate into other forms under very simple conditions (Madore and Freedman, 1987). The Belousou–Zhabotinskii chemical reaction, for instance, develops into waves of chemical activity propagating through a receptive liquid medium. These can be seen in a series of photographs (Figure 5.6.1). Other structures can be formed spontaneously in biological gels under relatively simple conditions. The conditions under which these structures are formed are just beginning to be discovered, but the implication is that subcellular biological structure may be thermodynamically preordained.

In a very lucid book on the interior conditions of cells, Pollack (2001) takes issue with many of the assumptions of cellular structure and organization. Integrity of the cellular membrane is not required for proper functioning of the cell, and the number of active transport mechanisms thought to be required to maintain chemical equilibrium inside the cell is much too high to be realistic. Instead, Pollack asserts that there is a basic intracellular gel maintained by the interactions of polar water molecules and surface charges on the actin filaments of the cytoskeleton (Figure 3.7.7). He has explained many cellular actions based upon this structure. Again, the basic functional mechanisms are extremely simple and do not require levels of complexity previously thought necessary to explain cellular activity.

If there is complexity in biological systems, it is most apparent in the relationships among simple elements. In Figure 5.6.2 is shown a diagram of the interactions among contributors

FIGURE 5.6.1
The Belousou–Zhabotinskii reaction produces complex geometric forms without human intervention. (Redrawn from Madore, B.F. and Freedman, W.L., *Am. Sci.*, 75, 252–259, 1987. With permission.)

to a model of the limits of human work performance while wearing a respirator mask (Johnson and Dooly, 1995). This diagram has been described as looking like a "plate of spaghetti" with its interconnecting lines between elements of the model. Each element is fathomable and modelable, and therefore relatively simple, but the overall appearance seems to be complex.

Biological materials, also, show this same feature. There are a very limited number of macromolecules available to produce all the materials present in biological systems (Section 3.6.4, Macromolecules). However, the appearance of an infinite variety of biomaterials is due more to the interconnections of this small number of building blocks rather than to a large number of precursors.

Even within the cell, regulatory mechanisms appear to be complex, but can be seen to be composed of a myriad of overlapping and complementary mechanistic pathways.

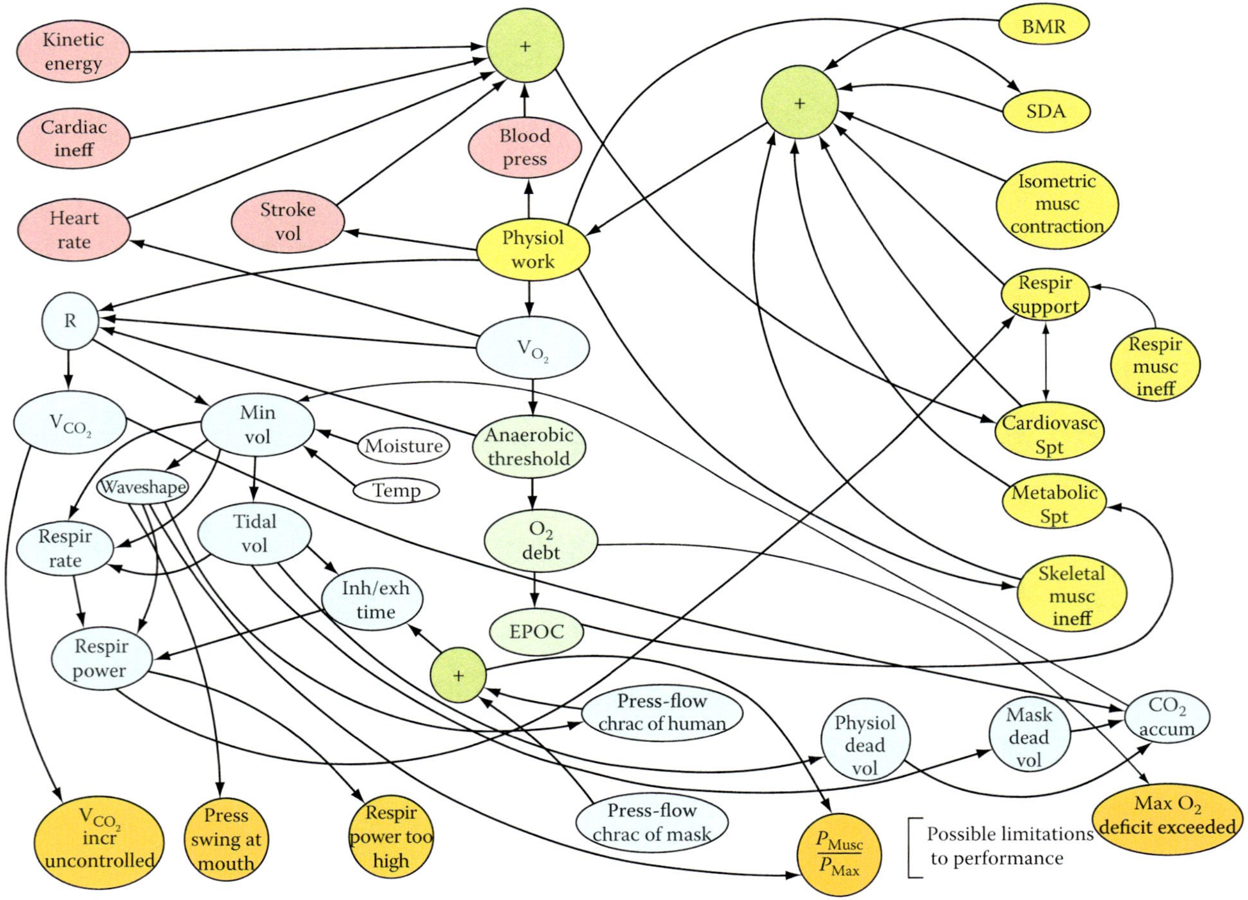

FIGURE 5.6.2
Schematic overview of a model to predict human performance. The appearance is that of a complex pattern involving simple elements. (Redrawn from Johnson, A.T. and Dooly, C.R., Design of respiratory protective masks to improve human performance, In *The Biomedical Engineers Handbook*, J.D. Bronzino (Ed.), CRC Press, Boca Raton, FL, pp. 2321–2334, 1995.)

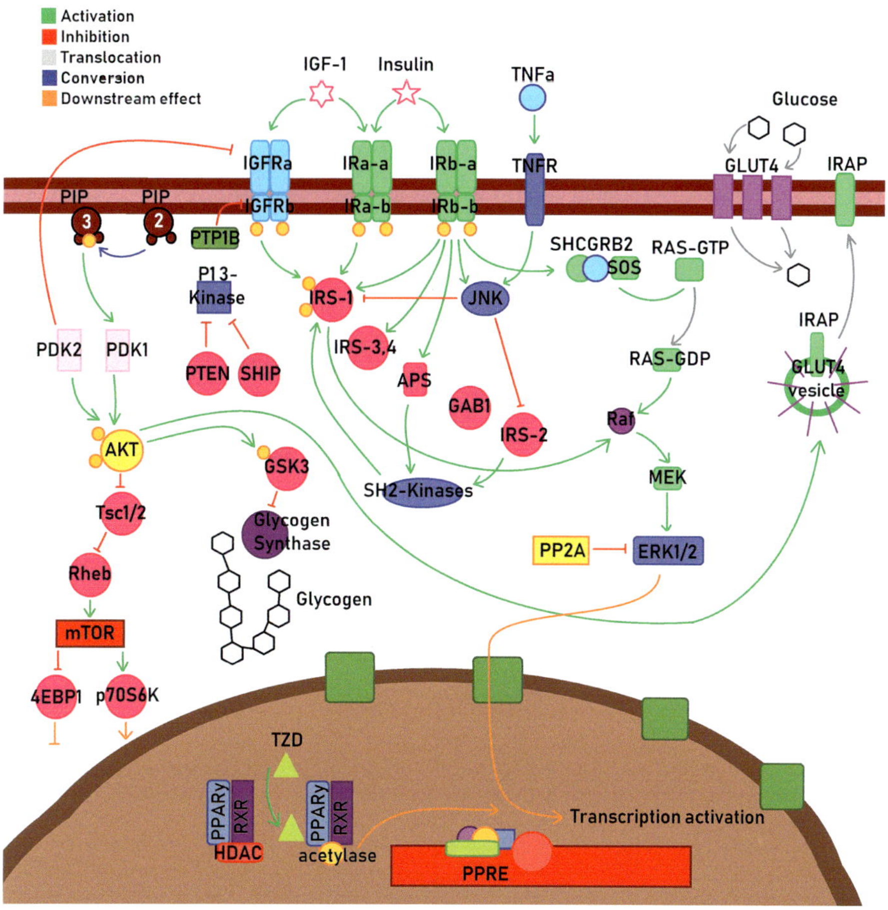

FIGURE 5.6.3

Insulin acts on the cell to enhance glucose permeability of the cell membrane. This diagram also shows the insulin-like growth factor (IGF-1) that promotes cell growth. Biochemical diagrams such as this appear to exhibit biological complexity.

In Figure 5.6.3 is shown a diagram of the action of insulin on glucose entry into a cell, and in Figure 5.6.4 is shown the action of the p53 tumor suppression gene. The former (insulin) illustrates action from outside the cell, whereas the latter (p53 gene) shows action from deep within the cell.

Reductionists (those who study the most fundamental scientific states) often believe that all properties of more complex states can be explained by their fundamental laws. All science could then be classified as either: (1) related to the discovery of fundamental laws or (2) explaining natural phenomena in terms of known fundamental laws (Anderson, 1972). Biological scientists, however, talk about emergent properties, or characteristics of a higher-level biological organization that cannot be deduced simply by adding the characteristics of its component parts. In other words, it may be that there is no unique set of fundamental laws that can be used to explain all phenomena at all hierarchical levels. At each new level, there are new laws, concepts, and generalizations that are necessary to apply to that level. Thus, they say that biology is not just applied chemistry.

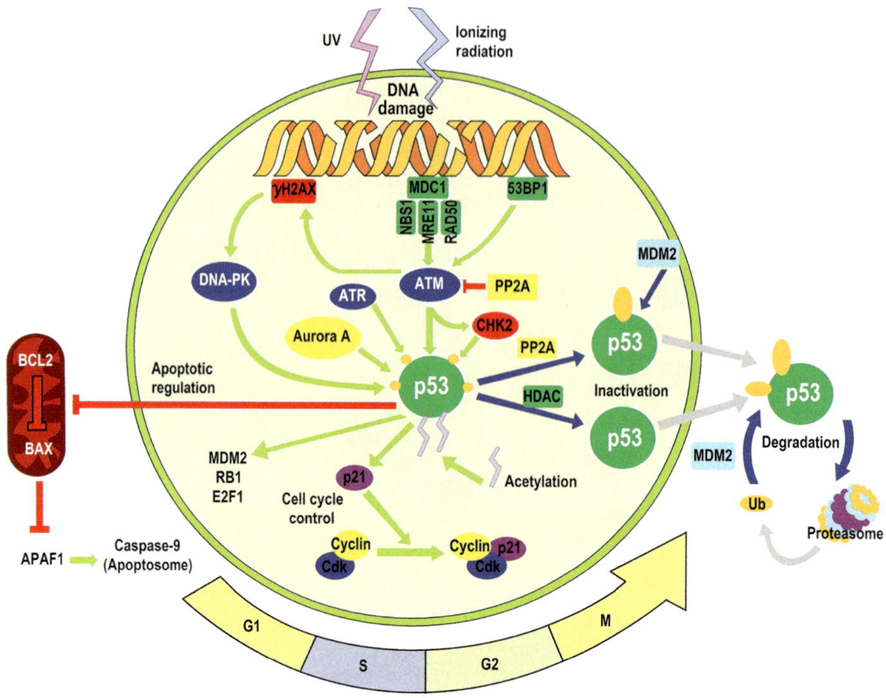

FIGURE 5.6.4
The p53 tumor-suppression gene acts from within the cell nucleus to destroy or correct cells with mutations that lead to uncontrolled proliferation. The mechanisms involved can appear to be complex because of the multiple steps in the process.

The view taken in this text sidesteps the reductionist and emergent properties views. As engineers or technologists who are concerned more with the application and use of biology than with its explanation, it is sufficient to become familiar with generalizations that apply across all hierarchical levels, those principles that apply at certain levels only, and exceptions to both that modify the biological behavior expected when challenged with a certain set of environmental conditions.

Thus, the view here is that basic biological activity is based upon very simple and robust physical and chemical principles. The complexity of biology comes in the multitude of different activities that work together to achieve the common goals of survival and replication.

Summary of scientific principles related to the understanding of biological systems:

1. There are places with higher potential and places with lower potential. Things move from higher to lower potential.
2. The maintenance of order requires energy.
3. What goes in but doesn't come out is stored inside.
4. Different forms of energy can be used to perform mechanical work.
5. The transfer of something from one place to another depends directly on the surface area and is inversely proportional to the distance between the two places.
6. Mechanical strength depends on geometrical configuration, the amount of material present, and properties of the material.

7. Unbalanced mechanical forces cause acceleration.
8. Heat is the ultimate nonspecific form of energy.
9. Hydrostatic pressures are equal in all directions.
10. Flowing fluids require energy to overcome resistance.
11. There is a periodicity of properties of elements, and these properties are related to numbers of electrons and their energy states in elemental atoms.
12. Elements can combine to form compounds with different properties.
13. Molecular configurations determine usages.
14. Chemical reactions occur spontaneously when they yield energy to the environment.
15. Reaction rates depend upon reactant concentrations, temperatures, and pressures.
16. Intermediate reactions are most important to living things.
17. There is an element of randomness in biology.
18. Appropriate responses require control systems, and each of these needs sensors, actuators, processing, and information pathways.
19. Optimization conserves resources.
20. Information implies order.
21. The primary goal of life is survival and reproduction.
22. Living things are constantly changing.
23. Long-term changes to a species occur only if there is a reproductive advantage.
24. Life is redundant.
25. Coexistence of species requires that each adapts to a different ecological niche.
26. Attributes passed from one generation to the next require an information legacy.
27. Each distinguishing biological trait is made valuable by its cost.
28. An individual is a product of both its genetic code and its environment.
29. Life is conservative.
30. Living things use simple building blocks with complex interactions.
31. Extremes are not tolerated well by living things, nor do living things create extreme conditions.

Example 5.6.1: How Wolves Saved Yellowstone (Ward, 2010)

Wolves historically inhabited Yellowstone Park in the western US, but were exterminated in the early 1900s to protect livestock herds bordering the area. This set off an ecological chain of events that ended up drying streams, creeks, marshes, and springs.

Without wolves, native elk overpopulated Yellowstone. They chewed and overgrazed willow and aspen seedlings while they were still small. Without replacements, the population of these trees declined dramatically. Beavers, which normally ate these trees and used them for building dams and lodges, fell into decline. Without the beavers to build dams and ponds, wetlands disappeared, and so did the natural habitats for insects, amphibians, fish, birds, and plants. Water runoff that was no longer slowed and distributed by ponds, rushed down streams and was soon lost to the uplands.

Yellowstone's overgrazed river banks eroded and silted spawning beds for fish. Amphibians lost precious shade, and nearly disappeared. Birds that once thrived in the lushness of Yellowstone soon flew on by. The area dried up, and turned arid. Springs that had been recharged by pond water seeping into the ground no longer flowed, and drought became common.

Wolves were reintroduced into Yellowstone in 1995–1996, and immediately began to have a beneficial ecological effect. They killed fat elk, thinning the herd and improving the genetic quality of the survivors. Without so many elk, the trees regrew. With the trees came beavers and water to recharge springs and create habitats. Wolves had regulated the Yellowstone ecosystem from the top down, and restored ecological balance.

Questions

5.0.1 Formulate additional principles of biology and justify them.

5.1.1 Choose a biological example, and show how form is related to function.

5.1.2 How does the relation between form and function affect a biological engineering design?

5.1.3 What characteristics of a plant are most important for designs involving it?

5.1.4 Give examples where the forms and functions of these BU influence engineering designs involving them:

- Slithering animal
- Plant
- Animal that communicates verbally
- Bacteria
- Virus
- Intestinal wall cell
- Muscle cell
- Flying insect
- Finfish

5.1.5 Add to the list of Applications and Predictions.

5.2.1 Describe how you would infer relationships among organisms from their forms.

5.2.2 Describe the conceptual design of a new biological function by combining elements from different biological forms.

5.2.3 What happens to intermediate biological forms once better adapted forms have been developed?

5.2.4 Why is it important to know about normal living conditions when providing for a new habitat for living things?

5.2.5 Contrast the expected optimal environments for hepatic cells compared to dermal cells.

5.2.6 Development of resistance to a new pathogen, parasite, or biochemical challenge requires what?

5.2.7 What is convergent evolution and explain why it might occur.

5.2.8 Add to the list of Applications and Predictions.

5.3.1 Why is it that the sequences of bases on both DNA strands can be determined if the base sequence on one strand is known?

5.3.2 What are the relationships among DNA, a gene, a codon, amino acids, RNA, and proteins? Why is a codon considered to have a minimum length of three bases?

5.3.3 Compare the process of development of a computer operating system (like DOS or Windows) with the operation of protein formation.

5.3.4 Explain why the dominant–recessive genetic model is much simpler than most genetic trait determinations.

5.3.5 Speculate on the difficulty in identifying the relationship between a gene and its effect.

5.3.6 Why is it surprising that genetic variability is maintained in an organism? What is the advantage to maintaining genes that result in individuals less suited to their environments than others?

5.3.7 Of what importance is PCR? Think of some unique uses for PCR.

5.3.8 Discuss the chicken or egg question.

5.3.9 If sufficient knowledge about the genetic makeup of an individual is known, what factors would be important to determine if individualized medicines would or would not be made?

5.3.10 What is the role of mutation in genetic progress? What factors determine whether mutations occur and survive? How can these be used in biological engineering designs?

5.3.11 Why can it not be said that genetic mutations occur randomly?

5.3.12 Discuss the simple elegance of the Ames test.

5.3.13 Why is it important to know about extranuclear genetic material?

5.3.14 List specific factors influencing gene expression.

5.3.15 Of what use is the fact that chloroplast and mitochondrial DNA is transmitted separately from nuclear DNA?

5.3.16 Describe the function and use of biochips.

5.3.17 Look up the results of attaching the different genetic markers to the DNA nucleotides and describe the results.

5.3.18 Why are epigenetic changes important in the proper functioning of a living organism, such as a human or a plant?

5.3.19 How can epigenetic changes be used to advantage in biological engineering projects?

5.3.20 Describe the sequence of events in an RNA interference process? Give possible biological engineering uses for RNA interference action.

5.3.21 Add to the list of Applications and Predictions.

5.4.1 Why is competition necessary for biological selection?

5.4.2 Name an Earthly environment where no life would be expected to exist.

5.4.3 The fact that some sort of microbe will grow almost everywhere means what to a biological engineer?

5.4.4 If less-competitive species are constantly being lost, does that necessarily mean that competition for a species becomes greater in the future? Why?

5.4.5 Discuss the theory that the closer the relationship, the more interest one individual shows in another. How does this relate to biological engineering designs?

5.4.6 Is there a sudden advance in technology that can be anticipated in biology? What is it?

5.4.7 How can the idea of memes be used to advantage by biological engineers?

5.4.8 Will evolution always result in a new species? Why?

5.4.9 Give examples of evolution that does not result in genetic changes. How can these be used by biological engineers?

5.4.10 Describe the competition among individual genes or groups of genes.

5.4.11 How are the reproductive strategies of transposon genes and gamete killer genes similar and how are they different?

5.4.12 Add to the list of Applications and Predictions.

5.5.1 List similarities and differences among different hierarchical levels in biology.

5.5.2 List ways in which eukaryotes differ from prokaryotes. Why are these differences important?

5.5.3 Which definition of life do you prefer? Why do you think it best, and what could be improved?

5.5.4 How would you define life?

5.5.5 Why is it important for a biological engineer to be able to distinguish between living and nonliving forms?

5.5.6 If all of life is considered to be a component of an ecological system, what implications does this have for the ultimate outcomes of engineering uses of biology?

5.5.7 Describe ways in which the ecological balance of nature can be used as a potent tool for biological engineering design. How can a balance be incorporated and nurtured?

5.5.8 Where would you place humans on the trophic level scale if all human food is considered? Why?

5.5.9 How does synthetic biology differ from conventional biology?

5.5.10 How would you go about counting the many microscopic organisms living on and in the human body?

5.5.11 Add to the list of Applications and Predictions.

5.6.1 Give an example where a biological characteristic can be attributed to a multitude of simple building blocks.

5.6.2 Is biology simple or complex? Why?

5.6.3 Is the summary of scientific principles sufficient? Is there need for others? Are some not necessary?

Section III

Responses of Living Systems

6

Biological Responses in Context

> The human body turns out to be both fragile and robust. Like all products of organic evolution, it is a bundle of compromises, each of which offers an advantage, but often at the price of susceptibility to disease. These susceptibilities cannot be eliminated by any duration of natural selection, for it is the very power of natural selection that created them.
>
> **—R. M. Nesse and G. C. Williams**

Biological units (BU) do not exist in isolation. There are relations with a physical environment, with a chemical environment, and with other BUs. These other BU may be at the same hierarchical level as the target BU, as between a vegetable plant and a weed in a garden, or as between two hepatic cells in the liver. They may also be at different hierarchical levels, as with an ape in a forest ecosystem.

No matter what the arrangement, the interactions are many (see Section 5.6, Is Biology Simple or Complex). Each BU affects and is affected by both its biological environment and its physicochemical environment (Figure 6.0.1). The presence of other BU in the same physicochemical environment changes the physicochemical environment for the target BU, so that environmental conditions are constantly changing. Equilibrium is never reached (we saw that in Section 2.6, Entropy), and a true steady-state, where changes over time are nil, is rare and not long-lasting when it does occur. Thus, being able to predict typical responses is about all that can be expected.

Biological responses in context (BRIC) are meant to draw attention to the fact that each BU is not isolated, and that a BU that adapts one way to one type of environment can adapt

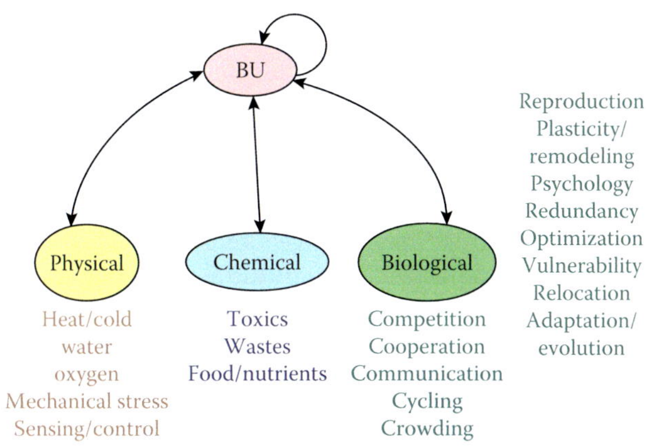

FIGURE 6.0.1
Many interactions with the environment are possible. The presence of other BU can affect a target BU directly or indirectly through their environmental responses.

another way to a different environment (as long as the environment can be sensed in some way; thus communications are very important).

Said aloud, the term BRIC sounds like the word "brick." Bricks are used as the basis for constructing complex buildings; BRICs can be used as the basis for constructing design solutions to engineering problems utilizing living things.

Life is remarkably robust, and to each BRIC there are some exceptions. Various forms of life are turning up everywhere (Lemonick and Dorfman, 2002; Levin, 2002):

- In anoxic underground pools of crude oil
- In hot, sulfurous ocean vents
- In brine pools five times saltier than the ocean
- In places with toxic levels of heavy metals, acids, and radiation
- In solid stone
- In the cold temperatures and high pressures under Antarctic ice

There are extremophiles (genus *Shewanella*) found in very harsh environments that grow electrically conductive proteins from their outer membranes to exchange electrons directly with metal surfaces (Powell, 2015). There are others that use manganese instead of iron in essential enzymes (to most other organisms, iron is required). There are extremophiles in regions of high salt concentrations (Dead Sea bacteria and other osmophiles), extreme water pressures (blobfish and devil worms) and some able to survive in space without air or water (tardigrades).

These extremophiles are not typical of the life forms that make up the majority normally encountered by biological engineers. But, they serve as a reminder that life can exist in very harsh conditions, and that there are organisms with competitive advantage to inhabit very unlikely spots. In fact, it is an enzyme from one of these extremophiles (*Thermus aquaticus*, living in the hot springs of Yellowstone National Park in the United States) that is used in the *polymerase chain reaction* (PCR) to amplify the small samples of DNA (see Section 5.3.4, Genetic Expression).

Thus, the reader is cautioned that these BRICs, while useful, are not inviolate. They should be considered to be general guidelines, not absolute principles.

Ecology is the study of the relationships between organisms and the environment. As normally considered, the term ecology usually applies to *macroecology*, wherein the environment consists of the biophysicochemical surroundings comprising whole organisms and complex influences both on and by these organisms. Similar responses are seen in *microecology*, wherein the environment can consist of suborganismal units as well as total organisms. Hence, no distinctions will be drawn among BU levels (Figure 6.0.2).

There is a surprising similarity across all levels of biology. In many ways, an environmental wetland can act the same as the human kidney. Conditions for soil bacteria are very similar to conditions for bacteria in the human gut.

Generally, the same responses can be expected no matter whether it is macroecology or microecology that is the object of attention. Therefore, as given in Section 5.5 (Hierarchies), a BU will be expected to act as a BU no matter at what level it is located.

The term "ecosystems" does not always need to mean a group of plants and animals sharing the same space and which have interdependent relations. A definable ecosystem could just as well be the interior of a human or animal body. For instance, long-term studies of the internal parasites of rabbits have found that there is a marked relationship among

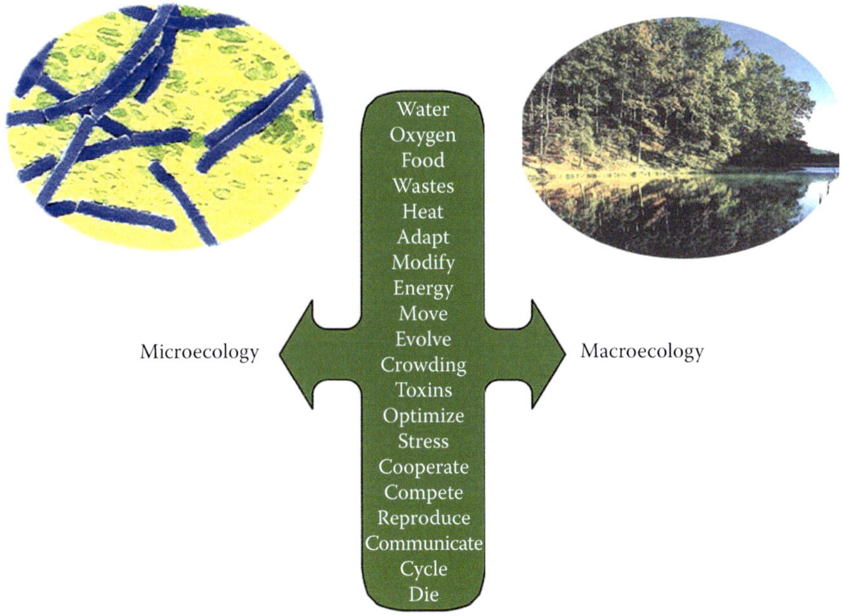

FIGURE 6.0.2
The same needs and responses are important even at widely differing BU levels.

the populations of different parasites. If a drug is given to suppress one type of parasite, then the populations of the other types may increase greatly. The net result is that the total population of parasites may actually increase as a result of administering the drug.

Likewise, it has been hypothesized that metastasized cancers spread throughout a human body, growing tumors in numerous locations. In at least some circumstances, one tumor dominates the others, keeping the others in check by inhibiting their growth. Removing the dominant tumor by surgery then leads to the appearance of other tumors in other parts of the body.

In both of these cases, the ecosystem involves the interactions of each biological unit with its physical, chemical, and biological environment within definable confines (most ecosystems are not totally closed to outside influences, so the definition of ecosystem boundaries is somewhat arbitrary.)

6.1 Biological Units Need Water

> We never know the worth of water till the well is dry.
>
> —**English Proverb**

There is no other ingredient more essential to BU than water. All life as we know it, is water based. Water is essential to life because it is a unique solvent that forms the medium in which to transport many other substances, both organic and inorganic, ionized and nonionized (Section 3.2.4, Water).

6.1.1 Water Has Unique Properties

> In wine, there is wisdom. In beer, there is strength. In water, there is bacteria.
>
> —**Dave Auerbach**

Water is not just simple H_2O, but a unique and complicated material with distinct and varied chemical properties (Buswell and Rodebush, 1976). Water is denser as a liquid than as a solid, thus allowing life to be maintained beneath the frozen surface of the bodies of water. This property also causes the death of BU when water internal to the BU freezes, expands, and mechanically disrupts the structure of the BU.

Water also has a high heat capacity, or specific heat. That means that water requires a relatively large amount of heat to raise its temperature, and a large amount of heat must be extracted to cool it. This property helps BU to maintain its thermal equilibrium despite sometimes rapid changes in environmental temperature. Many heat transfer processes are described by exponentially changing temperature with time (Sections 2.7, Heat Transfer, and 4.2.3, Random Data); water produces a relatively long thermal time constant.

The strong hydrogen bonds in water explain why it has such high melting and boiling points. Again, the large amount of heat required to change the state of water aids the maintenance of a state of equilibrium for the BU.

In an ordinary unassociated liquid such as benzene, the molecules flow by each other by sliding around one another. In water, because of the hydrogen bonds that must be broken before any flow can occur, the motion is rolling rather than sliding. This property gives water a relatively low viscosity and makes the pumping of water by BU (e.g., the flow of blood in the heart, the flow of water in the hydra, or the movement of water in plants) much less energy intensive than it could otherwise be.

Cellular BU contain a soup of proteins, salts, and other molecules and ions in a water matrix. The water often forms hydrates by surrounding these molecules with water molecules that collect into crystalline cages. This has several consequences: first, it separates ions from each other and allows them to remain in chemically active forms; second, it facilitates the movement of molecules with irregular forms that normally would catch on each other when they came in contact.

6.1.2 Water Surrounding Biological Units

> Everywhere water is a thing of beauty, gleaming in the dewdrop; singing in the summer rain; shining in the ice-gems till the leaves all seem to turn to living jewels; spreading a golden veil over the setting sun; or a white gauze around the midnight moon.
>
> —**John Ballantine Gough**

Many BU are bathed in water or a water-based medium. Most tissue or organ BU are supplied with this medium by other systems in the organism. The brain, for instance, is bathed in cerebrospinal fluid originating from the blood plasma. Population or biome BU are supplied with water from the Earth's environment, such as from a river or from rain.

There are some BU that must struggle to acquire enough water to survive. These are desert BU, and they cannot waste precious water. Desert plants have a very impenetrable surface cuticle that conserves water. Stomata in their leaves close during midday, when water stress is highest. Desert animals excrete very concentrated urine, and move to shady locations during hot periods.

Tardigrades, also called water bears, are tiny animals found all over the world, and can tolerate extreme drying. They can survive in the extreme vacuum of space for days.

Because water is one product of carbohydrate aerobic metabolism (Section 3.9, Energy Compounds), water that is produced as a metabolic byproduct can supply the water needs of BU that produce the water, or the needs of nearby BU that are close enough to use excess moisture. One reason that spoilage of foods or grains is so difficult to stop once it begins is that the moisture that results from metabolism modifies and enhances the environment for the growth of additional spoilage microbes.

6.1.3 Water Balance

The thirsty earth soaks up the rain,
And drinks, and gapes for drink again.
The plants suck in the earth, and are
With constant drinking fresh and fair.

—Abraham Cowley

Water, like all other substances that cross the boundary of a BU, must obey the basic balance:

$$\begin{aligned}&\text{Rate of water in} - \text{rate of water out} + \text{rate of water generated}\\&\qquad\qquad\qquad\qquad - \text{rate of water consumed}\\&= \text{rate of water storage}\end{aligned} \tag{6.1.1}$$

6.1.4 Biological Units Barriers to Water Movement

Continual dropping wears away a stone.

—Lucretius

Water can enter a BU due to a concentration difference (diffusion) or because of pumping (convection). Because the water concentration of the atmosphere is much less than the water concentration internal to a BU, there must be an effective water barrier between the organismal BU and its atmospheric environment. Plants have their cuticle and animals have their skin. Water can enter through this barrier when the organism is submersed, but only to a limited degree. Any disruption of this barrier, for example by scraping a plant, or by burning human skin, can lead to life-threatening water balance problems.

Surface lipids in the form of waxes have much to do with limiting drying rates to values below lethality. Plants have a water barrier composed of cutin, embedded waxes and pectin. Insects and other arthropods have embedded wax in their outer chitin layer. Amphibians have skin that is very permeable to water movement in order to absorb water from their environments. When spending time in the sun or in the ground, however, they either spread exuded lipids on their skins, or else form an impermeable skin layer. Reptiles use keratin as their principal water barrier. Mammals have a thin lipid film that covers their outer skin layer, the stratum corneum (Hadley, 1980).

BU in an aquatic (including interstitial fluid) environment do not usually have such a barrier. As long as the osmotic concentration of solutes inside the BU is the same as the osmotic concentration of the solutes outside the BU, then a water balance can be maintained

(Section 2.8.2, Osmosis). Small excesses or deficiencies of water can be corrected easily and without notice. If, however, the surrounding fluid changes suddenly to a different solute concentration, then the BU may not be able to survive. Water leaving a BU to a surrounding concentrated solution may cause desiccation of the BU. Water entering a BU from a surrounding dilute solution can cause bursting. Rigid plant cell walls offer some protection against bursting. Sea water has a salt concentration of about 3.45 g/100 mL; human blood plasma has a salt concentration of about 0.9 g/100 mL. These concentrations of fluid surrounding BU largely define concentrations inside the BU.

Example 6.1.1: Monitoring Almond Tree Trunk Diameter

Almond trees in arid California require irrigation water to survive and produce crops. However, with the increased competition for water supplies because of continuing urban development, there is pressure to decrease the water ration for agriculture. Tree water status has been found to be able to be used for tree-based irrigation scheduling, but requires trips to the field and significant labor if frequent readings are needed. Suggest an alternative.

Solution

A linear variable differential transformer (LVDT) is a very sensitive electronic transducer that makes linear distance measurements very accurately. It consists of a rod inside a coil. The rod is placed against the tree trunk and its position inside the coil is converted into an electrical signal (Figure 6.1.1). When LVDTs were mounted on almond tree trunks, trunk diameters could be recorded automatically every 30 s and average values transmitted to a laboratory computer by a modem and cellular phone (Goldhamer et al., 2003). Trunk diameters varied by up to 0.5 mm from morning to noon (Figure 6.1.2). When indicated by the measurements, irrigation water could be supplied either manually or automatically.

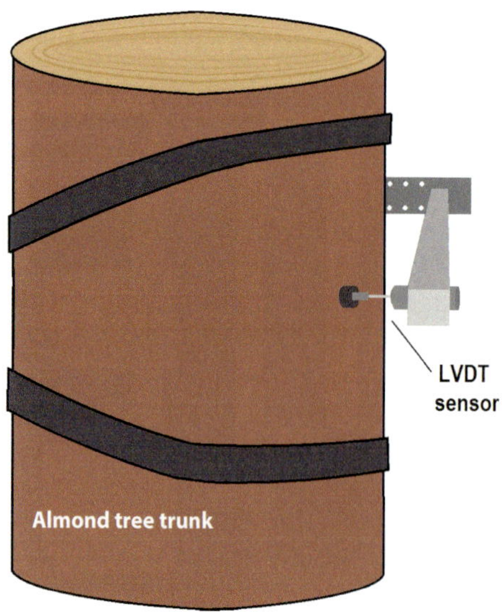

FIGURE 6.1.1
Linear variable differential transformers continuously monitored the diameters of tree trunks. (Redrawn from Goldhamer, D.A. et al., *Calif. Agric.*, 57, 138–144, 2003.)

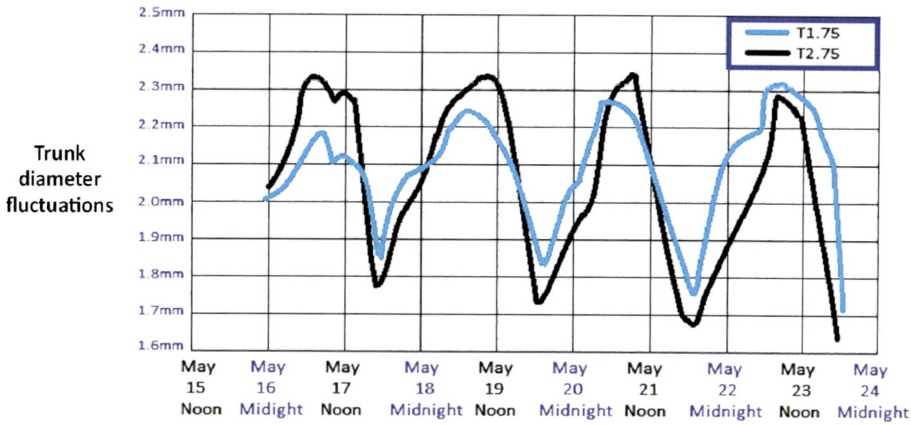

FIGURE 6.1.2
Continuous recordings of trunk diameter fluctuations for two irrigation treatments (T1.75 and T2.75). (Redrawn from Goldhamer, D.A. et al., *Calif. Agric.*, 57, 138–144, 2003.)

APPLICATIONS AND PREDICTIONS

1. Killing bacteria or molds in walls or ventilation systems removes sources of moisture.
2. Freezing will usually kill BU. The exception will be if large water crystals do not form and disrupt cell structure.
3. Immersion in a water bath will ameliorate temperature fluctuations.
4. Nearly all BU that are not surrounded by water will have a protective layer to keep them from drying.
5. Water will help deliver nutrients and remove waste.
6. Burn victims will be in danger of dehydration.
7. Contaminated water will negatively affect an ecosystem.
8. Water can transport pathogens.
9. BU adapted to cold temperatures will generally have higher solute concentrations that freeze at lower temperatures.
10. Aquatic BU will generally have less well-developed adaptations to temperature changes than will terrestrial BU. Similarly, organs within the body of a homeotherm will function well only over a narrow temperature range.
11. The presence of water on other planets means that life is possible there.

6.2 Biological Units Need the Right Amount of Oxygen

> Roses are red
> Violets are blue
> Without your lungs
> Your blood would be too.
>
> —**Susan Ott**

Oxygen is needed for aerobic metabolism. Oxygen is usually required for the final metabolic stages that form water and carbon dioxide (from carbohydrates and lipids) and various nitrogen compounds (from some lipids and proteins). Oxygen accepts electrons from another element in order to complete its outer electron shell.

6.2.1 Anaerobes and Facultative Anaerobes

> And 'tis my faith, that every flower
> Enjoys the air it breathes.
>
> —**William Wordsworth**

All BU do not need oxygen. *Anaerobic* metabolism can proceed in the absence of oxygen to form intermediate products that can supply some of the energy needs of the BU. Some bacteria, including some of the most primitive life forms, cannot survive in the presence of oxygen, and are called anaerobic organisms. Those that can survive and function in either aerobic or anaerobic conditions are termed *facultative* (which actually means "adaptable"). Those BU, such as yeasts, which are facultative but normally aerobic are called *facultative anaerobes.*

6.2.2 Oxidative Metabolism

> We see only what we know.
>
> —**Johann von Goethe**

Oxidative metabolism generates much more energy than does nonoxidative (or anaerobic) metabolism. The oxidative *catabolism* (metabolism in which a compound is broken into simpler compounds to obtain energy) of glucose yields 18 times as much ATP as does nonoxidative glycolysis. The oxidation of the fatty acid palmitate generates 64.5 times as much ATP as does nonoxidative metabolism (Brooks et al., 1996). Thus, there is a large energy advantage to the BU to complete its metabolism in the presence of oxygen.

6.2.3 Oxygen Delivery

> We should concern ourselves with the impacts of technology, not just its development.
>
> —**Raj Tonnash**

The metabolic rates of various BU are different, and so are the oxygen requirements. Tissues such as the brain or organs such as the liver have extremely high oxygen requirements. The oxygen needs of other BU may depend strongly on temperature and nutrient availability. Hypoxia at 37°C that results in human unconsciousness in 20 s and death in 4–5 min may not be lethal in an hour if the body temperature is reduced by 20°C. This is the reason why surgery on vital organs is normally performed during hypothermia.

The repair of damage increases the need for oxygen. Thus, humans or animals that have undergone physically traumatic events have noticeable increases in oxygen needs. Hospital ventilators may have to be adjusted accordingly. Biomes in which there are high levels of decaying organic matter suffer from low dissolved oxygen levels in the water. Fish that require at least 4 ppm of oxygen dissolved in the water may suffocate if the decaying organic matter reduces the dissolved oxygen to nearly zero.

It is the partial pressure of oxygen in the atmosphere or dissolved in water that is important to sustain life. The atmosphere at high elevations contains the same percentage of oxygen (21%) as does the atmosphere at sea level. However, very low oxygen partial pressures cannot sustain life.

The human lung contains water vapor at a partial pressure of 47 mmHg. This is determined solely by the vapor pressure of water at a human body temperature of 37°C. In addition, the lung contains carbon dioxide at 35–45 mmHg. Oxygen must be available, then, at least at 100 mmHg in order to sustain some human activity (Billings, 1973a, 1973b). This is equivalent to the oxygen content of the atmosphere at 15,000 m elevation.

Anoxia is the term used to describe lack of oxygen. *Hypoxia* is low oxygen level. Anoxia in humans results in immediate unconsciousness, convulsions, and paralysis. Hypoxia may have less profound effects, depending on the level. Hypoxic effects become worse as the duration increases.

Temperature can also have a profound effect on oxygen dissolved in water. The solubility of gases in water decreases as temperature increases (see Section 3.7, Water Chemistry). So, aquatic life at higher temperatures must be able to extract oxygen that is more rarified than at lower temperatures. Higher temperatures have the same effect on aquatic species as higher altitudes have on air-breathing species.

Much of the complexity of higher-level BU is derived from the need to deliver oxygen to all component parts. The circulatory systems in animals, with special oxygen-carrying compounds, internal stomatal structures in plant leaves, tracheae in insects, and plant root hairs are all examples of these. Because the diffusion of oxygen through the water medium surrounding most cells is so slow, the distances from oxygen supply to each individual cell must be very small.

6.2.4 Too Much Oxygen

> Every path has its puddle.
>
> **—English Proverb**

Animal respiration is usually less sensitive to oxygen lack than it is to the presence of carbon dioxide. There are both carbon dioxide and oxygen chemoreceptors in the body, but the carbon dioxide-sensitive receptors have a greater effect on respiratory adjustments to exercise and atmospheric changes. In the presence of hyperbaric oxygen, however, the response to carbon dioxide may be eliminated entirely. When this happens, continuously produced carbon dioxide can build up to the point where cranial vasodilation can cause coma, convulsions, and even death. In addition, hyperbaric oxygen can cause pulmonary irritation, congestion, exudation, and edema. Central nervous systems are affected by the inactivation of certain enzymes as a direct result of the high levels of oxygen in the blood. Very high dissolved oxygen concentrations in ecosystems is also toxic to some cells.

Example 6.2.1: Modeling of Composting

Composting is the controlled biological decomposition of organic wastes by microbial activity (Cundiff and Mankin, 2003). Compost is free of unpleasant odors, does not harbor pathogens or pests, can be stored for long periods of time, and has nutrient value usable by plants.

Satisfactory composting requires a carbon to nitrogen composition of about 30:1, moisture in the range of 50%–60%, a pH of about 6.5–8.0, a temperature in the range of 55°C–60°C, and enough oxygen to allow microbes to metabolize the biological substrate aerobically. Less oxygen than this permits anaerobic microbes to grow and produce unpleasant odors.

Temperature, moisture, and oxygen availability are all interdependent. Greater rates of air movement through the compost pile make more oxygen available, but dry and cool the pile. With more oxygen, microbial activity may increase and this increases both temperature and moisture production. Hence, a complete analysis of the compost process includes temperature, water, energy and substrate in addition to oxygen.

A balance equation for oxygen becomes (Cundiff and Mankin, 2003)

rate of change of oxygen uptake = (volume flow rate of air)
× (O_2 concentration difference between air entering and leaving the pile)

or,

$$\frac{dm_{O_2}}{dt} = \dot{V}(c_{O_{2i}} - c_{O_{2o}})$$

where:

m_{O_2} is the mass of O_2 used by the bed (kg/h)
$\dot{V}$ is the air flow rate (m^3/h)
c_{O_2} is the concentration of O_2 (kg/m^3)
i, *o* denotes incoming and outgoing conditions

Because oxygen concentration changes as air flows through the subsequent layers of compost, the bed is usually analyzed as a series of layers that each has a nearly uniform set of conditions. For each layer, then, an oxygen balance gives

(rate of change of oxygen usage) = (oxygen consumption per unit mass of substrate)
× (rate of change of substrate concentration)
× (volume of layer) + (air flow rate)(O_2 concentration difference between air entering the layer and leaving the layer)

or,

$$\frac{dm_{O_{2j}}}{dt} = M_{O_2} V_L \frac{dc_s}{dt} + \dot{V}(c_{O_{2j-1}} - c_{O_{2j}})$$

where:

M_{O2} is the oxygen consumption per unit mass of substrate (kg_{O2}/kg_{sub})
V_L is the volume of layer (m^3)
c_s is the substrate concentration (kg_{sub}/m^3)
j is the layer designator

Dividing this last equation by layer volume gives oxygen in terms of concentration. The minimum oxygen concentration for satisfactory composting is 5% (mol fraction), which corresponds approximately to 0.07 kg O_2/m^3 compost.

APPLICATIONS AND PREDICTIONS

1. Animals that are sealed in enclosures without oxygen will die.
2. Plant roots in packed soil, where oxygen cannot contact them, will die.
3. Microbes used to bioremediate toxic chemical spills may need additional air to be blown through the soil (called sparging) in order to metabolize the chemical.
4. Air (including oxygen) introduced to a container where anaerobic microbes are growing will favor the growth of aerobic microbes, which will then probably out-compete the anaerobes.
5. People drowning in icy water will be able to be resuscitated after a longer time than people drowning in warm water.
6. Humans and animals suffering severe physical injuries will need more oxygen and nutrition than normal in order to heal.
7. Fish and other aquatic life will require water to be oxygenated.
8. Hyperbaric oxygen will be dangerous to life.
9. Exercise will require more oxygen than rest.
10. Increasing the oxygen transport rate of bioreactors will make them more efficient.
11. Algal blooms can damage an ecosystem by consuming a great deal of oxygen.
12. Fetus hemoglobin will have greater affinity for oxygen than will the hemoglobin of the mother.
13. Because oxygen is less soluble in warm water compared to cold, fish in warm water will have a more difficult time acquiring oxygen than fish in cold water.
14. Hemoglobin controls oxygen supply and toxicity.

6.3 Biological Units Need Food and Nutrients

What we feed cells, and how we feed them, has been responsible for most of this [cell culture] progress.

—**Florian Wurm**

It may seem to be an obvious requirement that BU need food and nutrients, but too often the obvious is overlooked. In Spallanzani's time (the 1700s), it was thought obvious that life arose spontaneously, that beetles and wasps were generated in cow dung, and that mice were a product of Nile River mud (de Kruif, 1926). Through a careful series of experiments in which he sterilized the contents of flasks, sealing some and merely covering others, Spallanzani proved that microbes would grow only when they had access to the contents by means of the atmosphere. His sealed flasks grew no microbes.

Similarly, nutrition is required of all BU, and not just any nutrition, but the proper nutrition for that particular type of BU. Anderson and Underwood (1976) described an area in Australia that was a desert despite adequate rainfall. When small amounts of zinc and copper were applied to the land, plants flourished. Other plants are particularly sensitive to molybdenum levels and sheep particularly need cobalt.

Having a stable food supply is important at all biological levels. Humans, of course, had stabilized their food supply when they changed from hunting and gathering to agriculture. But versions of farming, or investing in a crop, are practiced by simpler organisms as well. There are fungus-growing ants that cultivate fungi inside their nests; they harvest leaves, feed the leaves to the fungi, and eat the fungi (see Section 6.16.4, Communal Benefit). There are damselfish that care for algal plots, removing unwanted algal types to benefit the types that they prefer to eat. There are solitary amoebas that engulf and keep intact bacteria inside themselves, searching for and feeding nutrients to the bacteria to help them survive and grow. At some point, the bacteria are released to reproduce and become a food supply for the amoebas (Burke, 2014).

6.3.1 Essential Elements

> Joseph Louis Gay-Lussac, most noted French chemist of his generation, was the natural person to consult about any odd substance. One of his fellow scientists, experimenting with ashes from burnt seaweed, had isolated a strange grayish-black solid. He had no idea what it might be or whether it would prove of any use. So he sent a specimen to the Sorbonne.
>
> Gay-Lussac studied the queer stuff and concluded it to be a new element. At ordinary temperatures he found it to behave in staid fashion. But when he heated a quantity to 185 degrees, it changed to a strange blue-violet vapor. There was a striking resemblance between the color of the odd gas and that of the chemist's favorite flower, the violet.
>
> It was customary to base scientific names upon Greek words. So the Frenchman took *iode*, classical name for the common violet, and bestowed it upon his discovery. Sir Humphrey Davy, experimenting with iode, found it to have many valuable properties. Its flower-based name modified to iodine, the new chemical became a standard weapon in the war against bacteria.
>
> **—Webb B. Garrison**

There are at least 24 elements essential for life (Tables 6.3.1 and 6.3.2). These must be present and available from the environment in order for BU to survive and flourish. Some are present, but not readily available. Phosphorus, for example, is required for plant growth and is absorbed in its dissolved inorganic (usually H_2PO_4) form by plant roots. This form of the phosphate ion is more readily soluble at low pH (more acidic) than at higher pH values.

So, the phosphate may be present but not available. Similar availability issues are there for the trace elements iron, manganese, zinc, copper, and boron, which can be deficient in alkaline soils. As we will see in Section 6.6 (Adaptations), some plant roots and rhizosphere bacteria secrete acids into the soil solution, which raises the solubility of phosphorus and other elements in their immediate neighborhood.

TABLE 6.3.1

Elements Essential for Life

Element	Symbol	Atomic Number	Comments
Hydrogen	H	1	Required for water and organic compounds
Helium	He	2	Inert and unused
Lithium	Li	3	Probably unused
Beryllium	Be	4	Probably unused; toxic
Boron	B	5	Essential in some plants; function unknown
Carbon	C	6	Required for organic compounds
Nitrogen	N	7	Required for many organic compounds
Oxygen	O	8	Required for water and organic compounds
Fluorine	F	9	Growth factor in rats; possible constituent of teeth and bone
Neon	Ne	10	Inert and unused
Sodium	Na	11	Principal extracellular cation
Magnesium	Mg	12	Required for activity of many enzymes in chlorophyll
Aluminum	Al	13	Essentiality under study
Silicon	Si	14	Possible structural unit of diatoms; recently shown to be essential in chicks
Phosphorus	P	15	Essential for biochemical synthesis and energy transfer
Sulfur	S	16	Required for proteins and other biological compounds
Chlorine	Cl	17	Principal cellular and extracellular anion
Argon	A	18	Inert and unused
Potassium	K	19	Principal cellular cation
Calcium	Ca	20	Major component of bone; required for some enzymes
Scandium	Sc	21	Probably unused
Titanium	Ti	22	Probably unused
Vanadium	V	23	Essential in lower plants, certain marine animals, and rats
Chromium	Cr	24	Essential in higher animals; related to action of insulin
Manganese	Mn	25	Required for activity of several enzymes
Iron	Fe	26	Most important transition metal ion; essential for hemoglobin and many enzymes
Cobalt	Co	27	Required for activity of several enzymes; in vitamin B_{12}
Nickel	Ni	28	Essentiality under study
Copper	Cu	29	Essential in oxidative and other enzymes and hemocyanin
Zinc	Zn	30	Required for activity of many enzymes
Gallium	Ga	31	Probably unused
Germanium	Ge	32	Probably unused
Arsenic	As	33	Probably unused; toxic
Selenium	Se	34	Essential for liver function
Molybdenum	Mo	42	Required for activity of several enzymes
Tin	Sn	50	Essential in rats; function unknown
Iodine	I	53	Essential constituent of the thyroid hormones

TABLE 6.3.2

Nutrients Necessary for Plant Growth Media

Chemical Symbol	Element Name	Functions
B	Boron	Essential in cell division, Flower formation
Ca	Calcium	Component of rigid cell walls slows aging
Cu	Copper	Required to form pollen, lignins in woody plants
Fe	Iron	Releases energy stored in starches
K	Potassium	Drives photosynthesis, prevents wilting
Mg	Magnesium	Component of chlorophyll molecules
Mn	Manganese	Releases oxygen from water during photosynthesis
Mo	Molybdenum	Aids in the absorption of nitrogen
N	Nitrogen	Part of proteins, enzymes, DNA, RNA, and more
P	Phosphorus	Helps store end transport energy
S	Sulfur	Used in amino acids that form proteins
Zn	Zinc	Needed to build chlorophyll for photosynthesis

HOW EVOLUTION SHAPED NUTRIENT NEEDS

(excerpted from Lane, 2005)

Some mineral requirements for sheep are expressed as a percentage of the diet—calcium, phosphorus, magnesium, potassium, sodium, chlorine, and sulfur. These elements are often called the *macro* minerals, because they are needed in large—macro—amounts.

The other minerals are expressed in *ppm* (parts per million). These are only needed in small amounts and are called the *micro* minerals. Iron, manganese, and zinc are required in the range of 20–50 ppm. Copper is required at slightly lower levels, in the range of 7–13 ppm. But three minerals—cobalt, selenium, and iodine—are only required at 0.10–0.30 ppm—very, very tiny amounts. This is intriguing.

Why? Because the three minerals that animals need in the smallest amounts are the same minerals that plants don't need at all.

How interesting. And I don't think that this is an accident. Evolution may be random in the broadest sense, but on the molecular level, evolution is rigorous and unforgiving. It has forced our livestock to use rare resources in the most efficient way.

Plant cells and animal cells share most of the same metabolic systems, and with a few exceptions, plants require the same array of nutrients as animals. Most of this overlap occurs in the macrominerals and some microminerals. But plants also have their own special needs for elements that our livestock cannot use, like boron, which is involved in fiber metabolism, and molybdenum, which is used in nitrogen metabolism and nitrogen fixation. But what about the three minerals that plants don't need—iodine, cobalt, and selenium?

(*Continued*)

HOW EVOLUTION SHAPED NUTRIENT NEEDS (Continued)

Well, animals use iodine almost exclusively in thyroid hormones. Since these hormones are critical in controlling metabolic rate and heat production, plants don't have much need for them or their iodine.

In animals, cobalt only occurs as a component of vitamin B-12 (properly called *cyanocobalamin*). All animals use vitamin B-12 in amino acid metabolism and red blood cell formation, and ruminants especially use it for converting rumen fermentation products into glucose in the liver. But again, plants do not have red blood cells, rumens, or livers, and thus have no need for either vitamin B-12 or any other molecules that contain cobalt.

Selenium is used by animals primarily in cellular antioxidant systems and also a few specialized enzymes. Most plants contain no metabolic systems that include selenium. But remember the Periodic Table? Elements in the same column of this table have similar chemical characteristics. Look at selenium. Directly above selenium in the Periodic Table, in the same column, is *sulfur*. Sulfur, of course, is an integral atom of the two essential amino acids methionine and cysteine, and plants require relatively large amounts of sulfur. Since selenium chemically resembles sulfur, plants don't differentiate them very well. Plants will absorb selenium from the soil along with sulfur and incorporate it into amino acids to form *selenomethionine* and *selenocysteine*. The plants then insert these amino acids into various proteins instead of the regular sulfur-containing amino acids. Although those proteins may not function as well, these aberrant amino acids do no real harm because their levels are so low.

Iodine, cobalt, and selenium show up in most forages because they come along for the ride through the roots. Although plants don't need these elements, they generally do no harm, and they are absorbed in small amounts if they occur in the soil. Animals, on the other hand, require all three minerals for critical metabolic systems. Evolution ruthlessly selected animals that could survive on low levels of these minerals. Animals that genetically required higher levels tended to leave the gene pool early.

6.3.2 Food and Nutrients for Energy and Essential Biochemicals

> A poet once said, "The whole universe is in a glass of wine." We will probably never know in what sense he meant that, for poets do not write to be understood.
>
> —**Richard Feynman**

In general, there are two purposes to food and nutrition: (1) to meet energy needs and (2) to supply essential biochemicals for life functions. Both of these requirements must be met, and the specific nutrients to satisfy them are sometimes very exacting.

All life needs a source of energy to fuel metabolic processes. Most of this comes through carbon compounds called carbohydrates that are hexoses or polymers of hexoses (the most important of which are glucose, galactose, and fructose). Fats can also act as concentrated energy sources. Carbohydrates have the general chemical formula $(CH_2O)_x$, which, when oxidized, changes the carbon to carbon dioxide and liberates the water; fats have smaller amounts of hydrogen relative to the number of carbon atoms, and much smaller amounts of oxygen. Thus, when fats are oxidized, energy is liberated not only from the production of carbon dioxide, but also from the oxidation of hydrogen to form water.

Autotrophic BU are more dependent upon their physicochemical environments to supply essential elements than are heterotrophs, but, of course, autotrophs are less dependent upon nutrient uptake to supply their energy needs. Herbivores can usually obtain most of their nutrient requirements from the plants they eat, although nutritional requirements of plants and animals differ somewhat, and this may lead to nutritional deficiencies in the animals.

Animals fed a fat-free diet do not grow, develop skin and kidney lesions, and become infertile. These symptoms can be reversed by feeding linolenic acid, linoleic acid, and arachidonic acid, which are called essential fatty acids.

Proteins can also be used as energy sources. When glycogen levels in the liver are low, amino acids that form the products of protein digestion are deaminated and catabolic processes form energy. High levels of circulating glucose depress this process and spare the protein for other uses.

Amino acids are used by the body to form proteins, hormones, and enzymes. Transamination reactions can convert one amino acid into another to meet immediate needs. However, just as there are essential fatty acids, there are also essential amino acids. These amino acids cannot be synthesized in the body and must come from external sources. Humans require phenylalanine, valine, tryptophan, threonine, lysine, leucine, isoleucine, and methionine as essential amino acids. All other amino acids in the body can be synthesized at rates sufficient to meet body needs. If any one of the amino acids necessary to synthesize a particular protein is not available, then the other amino acids that would have gone into the protein are deaminated, and their excess nitrogen is excreted as urea (Ganong, 1963).

From this, it should be clear that what an animal is fed influences the composition of its waste. If any closed-cycle BU, such as a biome is to be formed, then the normal cycle of wastes from some biome constituents feeding other constituents may be disrupted without proper nutritional availability.

Vitamins are any organic dietary substance necessary for life, health, and growth that do not function by supplying energy. They usually function as coenzymes. Vitamins for one species may not be vitamins for another. Only humans, monkeys, and guinea pigs lack the ability to synthesize ascorbic acid (vitamin C), for instance. Indeed, bacteria in the gut synthesize some essential vitamins, which are absorbed in amounts sufficient to meet daily requirements. The administration of antibiotics for a long period of time could thus result in a vitamin deficiency of the bacterial host.

Vitamin D can be produced in the skin when exposed to the ultraviolet rays of the sun. It was once thought that the only effect of a vitamin D deficiency was a childhood disease called rickets, where the bones grew weak and malformed. Lately, however, vitamin D has been found to be very important in the adult years (Park, 2010). It has been found that vitamin D dramatically reduces the risk of cancers by suppressing cell growth and blood vessel formation. Vitamin D controls the release of stress hormones that lead to high blood pressure and inflammation. Because of this, it can reduce incidence of heart disease. The vitamin has also been found to protect against autoimmune diseases, such as multiple sclerosis, lupus, and rheumatoid arthritis. It may also help to reduce depression by promoting the release of the mood enhancing hormone serotonin. Perhaps related, vitamin D helps to alleviate back pain.

CODEPENDENCE OF FOOD AND GENES

Diet, and the ability to adapt to eating a variety of foods, may have been a factor in the rapid evolution of our own *Homo* species as climates changed dramatically and shifted available plants and animals in prehistoric times (deMenocal, 2014). There is a connection between cultural aspects of diet and evolution (memes and genes). The switch in hominids from plant foragers to meat eaters was associated with larger body sizes in human ancestors because the greater amounts of energy available in meat could support greater energy requirements of larger body sizes. Larger bodies, in turn, allowed for more successful hunting.

At the same time, brain sizes were also increasing. Brain tissue is highly energy intensive, so more concentrated energy sources were needed to support them. Larger brains also resulted in more successful food acquisition. Larger bodies and larger brains were determined by genetic changes accompanying diet adjustments.

The shift to a larger portion of carbohydrates in early humans required genetic adaptations to produce the enzyme salivary amylase. This enzyme begins the process of converting starches to sugar in the digestive system.

Lactose is the sugar found in milk. In order to be digestible, lactose must be biochemically altered by the enzyme lactase. Lactase had been normally present only in babies and young children, but not in adults. When adult humans began to use milk as a dietary source of energy and nutrition, they had to develop the genetic ability to maintain lactase production later in life. Adult humans from geographical regions where milk was not commonly drunk have not developed the ability to produce lactase, and are thus lactose intolerant.

Human diets are still changing, and providing pressures for evolutionary modifications. Similar pressures are present in all biological organisms, from bacterial to higher-level plants and animals. These pressures demonstrate the adaptability of all life (Arjamaa and Vuorisalo, 2010).

6.3.3 Microbes Assist Digestion

Simplicity is the ultimate sophistication.

—Leonardo Da Vinci

Microbes (bacteria, protozoa, algae, and fungi) are extremely important to the nutrition of animals, insects, and plants. Microbes in the soil decompose complex organic matter into simpler forms that can be readily absorbed by plant roots. This is especially true for nitrogen and phosphorus compounds. Some microbes, called *mycorrhizal fungi*, are necessary to pine species and other conifers to form root structures that allow these species to absorb water and nutrients from the soil. Some soil microbes degrade harmful pesticide chemicals residing in the soil that might otherwise accumulate and kill or depress growth of plants growing in those soils.

Microbes living in the mammalian gut, called *microflora* and collectively called the *microbiome*, are important to supply certain nutrients that cannot be synthesized by the animal or human. There are an estimated 40 trillion (40×10^{12}) bacteria that live in the human gastrointestinal system. They either extract nutrients from undigested foods or aid in the digestion of otherwise indigestible compounds. Dietary fiber, lactose, and sugar alcohols are examples of non-starch polysaccharides that escape digestion in the upper part of the intestine. Microflora ferment these foods into short-chain fatty acids that are readily absorbed in the colon.

There are between 300 and 1000 different microflora species in the human gut that collectively add over 1 kg to body mass. These microbes are first introduced to babies through breast milk and more can be added through ingestion of fermented milk (yogurt), and uncooked fruits and vegetables. One of the most beneficial microbe types belongs to the genus *Lactobacillus*, found in yogurt.

Taking antibiotics kills beneficial as well as disease organisms, and it is important to reestablish thriving microflora colonies after discontinuing the drug. Antibiotics reduce or kill gut microbes that normally compete against the establishment of intestinal pathogens. This gives the pathogens a chance to grow and thrive. Overuse of antibiotics can actually make the user sicker in the long run (Raymann et al., 2017).

Other ingested substances that kill gut microbes are alcoholic beverages, chlorinated water, and food preservatives.

Herbivores with rumens are particularly dependent on microbes to digest cellulosic materials into sugars, fatty acids, and vitamins. Termites also rely on the same types of microbes for their nutritional requirements. Without these microbes, the animals or insects would die of starvation.

Bacteria and yeasts in honey bee saliva ferment the tough outer shell of pollen grains and partially digest them for bee food. Without these microbes, pollen would be indigestible; with them, pollen becomes the major protein source for honey bees.

THE REALLY BIG DINOSAURS HAD REALLY BIG APPETITES

Dinosaurs lived on Earth between 65 and 230 million years ago (Lane, 2010). The sauropods were the largest class of these, and were much larger than any animal of our time. The most massive of these was the *Argentinosaurus*, longer than 100 feet and weighing 100 tons. The largest African elephant weighs only a tenth as much.

Argentosauruses were herbivores that ate leaves, ferns, and shrubs. They likely maintained in their digestive systems populations of fiber-fermenting microbes to help digest this plant material, similar to herbivorous animals of today. If they ate only 1%–1.5% of their body weights in a day, then they consumed 2000–3000 lb. of dry matter each day. Plant material was probably about 75% water, so these dinosaurs needed to consume more than 8000 lb. of leaves per day. They had to keep moving just to find enough forage to eat.

Fiber fermentation does not happen efficiently at low temperatures. In order for the fiber-digesting microbes to survive and perform as needed, they must be kept at 60°F or above. Thus, the dinosaur probably had some body temperature regulation mechanism uncharacteristic of cold-blooded animals of today.

6.3.4 Synthetic Growth Media

Bread! Bread! Bread! Everywhere and every way. You could bake it, sop it, drizzle it, dunk it, smother it, or stuff with it. You could eat it at the beginning of a meal or as the last delicious drop to finish a good one off. But no matter what you did with it, no real meal, it seems, was complete without it.

—Lorraine Johnson-Coleman

BU bathed in a water-based medium, (and this includes cells in a bioreactor, tissues and organs in an organism, or microbes in the ocean), must have their nutritional needs met by the surrounding medium. Whether the medium is synthesized by humans (as in a bioreactor) or is naturally occurring (as with circulating blood), it must contain at least those foods and nutrients already described. To meet energy needs, animal blood usually contains glucose. Bioreactor growth medium may contain sucrose, glucose, dextrin, fructose, methanol, ethanol, or methane as carbon sources. Plant tissues may contain sucrose, glucose, fructose, or others. There are also circulating fatty acids, amino acids, and vitamins specific to the BU of interest. Synthetic media may also need to contain natural compounds whose chemical composition is not completely known. Hormones are normally part of internal media. Substances, such as insulin and thyroxine, regulate animal BU metabolism, whereas auxins and cytokinin regulate plant growth. The absence of any of these essential substances can lead to death of the BU, and, if not death, then its malfunction. That is why many synthetic growth media are formulated based on trial and error experiments, and that is one major obstacle to the formulation of human artificial blood.

Example 6.3.1: Inside Food Development Labs

Food meant for human consumption must taste good, be attractive, have the correct texture, and be nutritious. What factors must be considered in the development of new foods?

Solution

The modern trend is toward food consumed outside the home. As of 2003, fully 25% of all meals were consumed in restaurants and two-thirds of home meals were prepared elsewhere. Foods are big business, and the development of new foods goes well beyond food chemistry and nutritive value.

Where will the food be eaten? If it is take-out, it must be able to be eaten in automobiles, on park benches, or in the office. Thus, it cannot be too messy, and it must be firm enough to remain whole during eating. To appeal, it probably must be somewhat sweet without stickiness. Thus, the sweetness must come from inside.

It must be able to be heat processed to eliminate microbes without disintegration or, worse, without destroying desirable odor and nutrition. If the food is to contain artificial flavors, they must contain the essence of the original.

Different cultures expect different tastes. Those from equatorial regions expect spiciness. Those from northern regions might expect higher caloric content. American children prefer 60% more flavor in foods than adults do. Some people are traditionally more inclined to be lactose-intolerant than others.

In addition to the four other types of taste sensation, bitter, sweet, sour, and savory (or umami), it is likely that there are taste receptors for detecting fat. Fat in foods enhances flavor, and adds moistness. Low-fat foods must include substances that compensate for the effects of fats that are no longer there.

Even pet food must appeal both to the pets and to the human buyers of the food. Cats like acidic foods and prefer a glassy texture. Dogs prefer sweet or salty food with both rough and creamy textures. Cats like thinner morsels, while dogs like amorphously shaped dense pieces. Different aged pets need different nutritional compositions. And, it all has to appeal to the buyer.

In order to develop a new food, much more must be considered than just the food itself.

APPLICATIONS AND PREDICTIONS

1. All living things will require sources of nitrogen, potassium, sodium, and phosphorus. When considering the needs of animals, the nitrogen source is usually called "protein."
2. The lack of micronutrients and trace elements will stunt growth, hinder reproduction, and cause an unthrifty appearance.
3. Humans lacking vitamins will develop nutritional diseases.
4. Animals and plants lacking vitamins or trace elements will be more susceptible to contagious diseases than those that have the proper nutrition.
5. Growing microbes in a bioreactor, or in the soil, will need carbon sources for energy.
6. Composting will require carbon as well as nitrogen.
7. Different species will have different nutrient needs.
8. To reduce pollution from animal wastes, change the type of food they are fed.
9. BU exposed to powerful antibiotics may require vitamin supplements.
10. BU will adapt their lifestyles to the availability of nutrients and food.
11. The shortage of specific nutrients will be felt in structures that have the most need for that nutrient.
12. More carbohydrate than fat is needed to supply energy needs.
13. Nursing mothers will require extra nutrition.
14. Severely injured patients will require extra nutrition.
15. Intestinal microbes are an important part of animal health.

6.4 Biological Units Become Ill in the Presence of Wastes

> It is not possible to adequately protect the health of our nation without addressing infectious-disease problems that occur elsewhere in the world.
>
> **—Scott F. Dowell and Alexandra M. Leavitt**

Wastes are produced from metabolic processes necessary to maintain life. Many of these wastes are toxic to the organisms that produced them. Aerobic metabolism produces carbon dioxide and water. While water is not a toxic substance, carbon dioxide is toxic at high concentrations. Depending on the concentration and time of exposure, carbon dioxide may cause perception changes or unconsciousness in humans (Figure 6.4.1). Because of this, carbon dioxide must be continuously removed from the atmosphere containing humans or higher-level animals. Closed-loop rebreathing systems are common in space vehicles, undersea vehicles, and diving equipment (Figure 6.4.2).

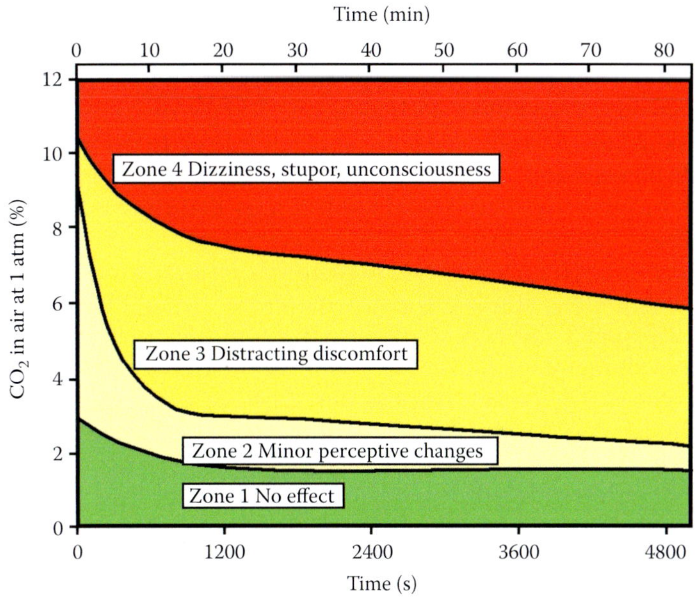

FIGURE 6.4.1
Symptoms common to most subjects exposed for various times to carbon dioxide–air mixtures at 1 atm pressure. (Redrawn from Billings, C.E., Atmosphere, In *Bioastronautics Data Book*, J. F. Parker Jr. and V. R. West (Eds.), NASA, SP-3006 U.S. Government Printing Office, Washington, DC, pp. 35–63, 1973b.)

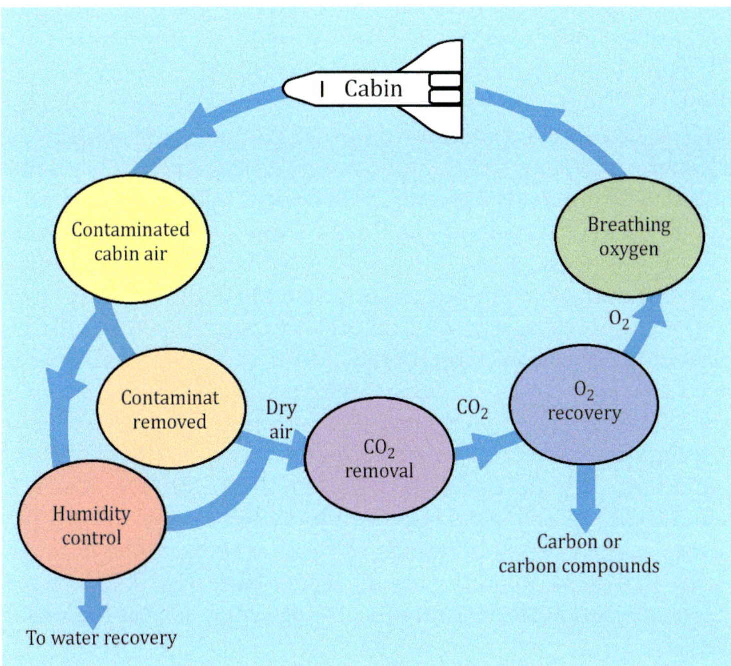

FIGURE 6.4.2
Schematic atmosphere control system for space vehicle application. (Redrawn from Jones, W.L. and Ingelfinger, A.L., Atmosphere control, In *Bioastronautics Data Book*, J. F. Parker Jr. and V. R. West (Eds.), NASA, SP-3006 U.S. Government Printing Office, Washington, DC, pp. 807–845, 1973.)

Although water is not toxic, the high atmospheric concentrations of water vapor encourage the growth of undesirable or dangerous microorganisms on surfaces where condensation occurs. Thus, excess water vapor should also be removed from the atmosphere.

Products of anaerobic respiration, such as lactate and pyruvate can be toxic to animals. Alcohol, as a product of anaerobic yeast metabolism is toxic to the yeast at concentrations approaching 15% by volume.

Liquid wastes (urine) frequently contain substances that can be toxic to the organism that produced them. The products of protein metabolism are nitrogen compounds that may have ill effects if not removed. Acids in urine can irritate skin.

Solid wastes contain disease-causing microorganisms. Again, they must be removed, or diseases may easily spread. The removal of decaying or rotting tissues from an organism can easily cause disease if maintained in contact with the organism they came from. Infectious diseases such as cholera, dysentery, typhoid, and diarrhea were largely abolished as major causes of death in the United States when treatments to drinking water eliminated infection sources (Bell, 2010).

The amount of waste produced depends on the amount of food ingested, and this, in turn, depends on body mass and activity levels (see Section 7.4.11, Food and Waste). The biological engineer designing closed or semi-closed environments for BU must account for waste removal. Indeed, it is well known that sanitation is one of the leading contributors to health maintenance among the BU of all kinds and trophic levels.

HUMAN DISEASES HAVE HAD PROFOUND EFFECTS

Disease has had a profound effect on the affairs of peoples. Smallpox, for one, requires susceptible hosts in order to survive; without unprotected humans, the virus would soon be extinguished. And so it was that the American people during the Revolutionary War period became just that group of people; smallpox became an epidemic. The British used smallpox infected civilians and soldiers as a weapon against the Americans (Fenn, 2001), and George Washington had to take the calculated risk of inoculating his troops in 1777, knowing full well that some would die. As the epidemic waxed and waned, the Revolutionary War took turns one way and the next.

It should come as no surprise that environmental conditions and human disease are closely linked. Wet years encourage insect vectors, and so West Nile Virus, Hanta Virus, and bubonic plague are more prevalent during those periods. The string of relatively mild winters may be linked to the spread of malaria into Virginia and Maryland in 2002. Researchers have found that a bumper year for acorns may be followed by a high incidence of Lyme disease: deer gather in oak forests to feed on the acorns, and large numbers of deer ticks drop from the deer to the forest floor. There they lay their eggs in the leaf litter. White-footed mice are also attracted to the acorns, and they carry large populations of Lyme disease microorganisms in their blood. The baby ticks feed on mouse blood, become infected, and then are poised to attach themselves again to deer or humans. As a result, acorns promote the spread of Lyme disease.

(*Continued*)

HUMAN DISEASES HAVE HAD PROFOUND EFFECTS (Continued)

A cartoon as recently as 1883 urged New York City to remain vigilant against infectious diseases. (Courtesy of the Granger Collection, New York.)

Many human diseases are spread through contaminated wastes. Cholera, typhus, typhoid, and bubonic plague are among these.

Nutmeg was considered to be precious in the seventeenth century and it was thought to protect against plague. The Dutch and the English fought over the control of nutmeg commerce, and the Dutch prevailed militarily. With the Treaty of Breda (1667), the Second Anglo-Dutch War was ended, and the Dutch secured a monopoly on nutmeg. The English were forced to concede a remote East Indian island where nutmeg was grown, but received, in return, a lightly inhabited New World island called Manhattan (LeCouteur and Burreson, 2003).

APPLICATIONS AND PREDICTIONS

1. Waste from one species will be waste also to related species, but may be food for distantly related species. Thus, one way to remove waste will be to feed it to another type of BU (as in composting).
2. Carbon dioxide will affect the anxiety levels of humans.
3. The quality of food will affect the amount of waste produced. Food with a large indigestible component will produce more waste than totally digestible food.
4. Animals that eat fruit will produce less solid waste than animals that eat hay.

6.5 Biological Units Need Heat Sources and Sinks

The reward of a thing well done is to have done it.

—**Ralph Waldo Emerson**

We have already seen in Section 2.4 (Energy) that heat is a special form of energy that enables many other life processes. BU are caught in the middle of needing some heat, but not too much; of needing a temperature intermediate between the phase change of ice to liquid water and the temperature where proteins, including enzymes, lose their functionality. In spore form, some BU can survive these extremes, but most BU cannot survive unless their temperatures are somewhere in the physiological temperature range defined earlier. Thus, we see that BU need heat sources from which to extract heat and BU need heat sinks to accept extra heat.

6.5.1 Heat Sources

What a dreadful hot weather we have! It keeps one in a continued state of inelegance.

—**Jane Austen**

The sun is the ultimate heat source for most BU. The sun bathes the Earth in an average of 1400 N m/(s m^2), which is a lot of energy. Some of this is absorbed in the atmosphere, some is reflected, and some is absorbed by plants to drive the chemical conversion of carbon dioxide into organic sugars. Much of the sun's energy is used to maintain the Earth at a temperature necessary for life.

There are BU that thrive in otherwise inhospitable environments such as hot springs. These BU (called thermophiles) use geothermal energy as their heat source (Hoffmann, 2001).

6.5.2 Removing Excess Heat

> [My vision is] to create a world where science and technology are celebrated...where young people dream of becoming science and technology heroes.
>
> **—Dean L. Kamen**

All BU produce heat as a result of metabolic processes, and so the major problem faced by many BU is the removal of excess heat. *BU that are not provided with the ability to rid themselves of waste heat will eventually die.* We have already seen in Section 2.7 (Heat Transfer) that BU may lose heat by conduction, convection, radiation, and evaporation. Of these, conduction is not usually significant because BU are bathed in air or water fluids, and conduction heat transfer is subsumed by convection heat transfer. Radiation heat loss is not usually significant for heat loss (although it is for heat gain from the sun) either, because the BU is not at a temperature very much different from its surroundings. Radiation heat loss to a clear, cold night-time sky can sometimes be significant, causing frost damage to plants, but the air must also be still enough that the plants do not gain heat from the air by convection.

Convection heat loss may or may not be significant. Tissue and organ BU depend upon the convection of the blood flow to maintain a constant and uniform internal thermal environment. We often think of blood as heating the tissues through which it flows, but a heat balance requires that for the blood to heat some tissues it must cool others. The brain and liver, as two of the highest producers of heat in the body, are organs for which blood cooling is especially important.

Cattle adapted to colder climates store fat under the skin and between muscle fibers. This fat is poorly vascularized and acts as insulation against heat loss. Hot-weather adapted cattle, however, store most of their fat in the gut, around body organs, and in humps on their backs. These cattle can lose much more heat to the environment than if they had the insulating fat layer underneath their skin.

Convection is also important for microbes bathed by a water-based medium and for plants in the air. Each has such a small capacity to generate heat compared to the capacity to transfer heat by convection that they exist at nearly the same temperature as their surrounding fluids. The temperatures of biomes and ecosystems can also be considered to be determined by the temperature of the air and water around them.

That leaves evaporation as the one variable means of heat removal available to many BU surrounded by air. Water, with its unusually high latent heat of evaporation of 2.447×10^6 N m/kg, can remove a large amount of heat with only a modest loss of mass. Plants reduce the surface temperatures of their leaves by 1°C–2°C as water evaporates through stomata on their undersurfaces (stomata of water plants with floating leaves are on the upper surfaces). Some animals pant and other animals sweat, and the amount of panting or sweating is regulated to remove the required amount of heat.

As the water content of the air approaches saturation, evaporative heat loss becomes less efficient. When the air can hold no more water vapor, further evaporation is impossible. When providing ventilation systems for human or animal BU, the possibility of air saturation must be taken into account because the combination of high temperature and high humidity, especially if air movement is limited, can easily be fatal. The environment inside an impermeable protective suit, inside a livestock housing facility, or inside a modern apartment building when the air conditioning shuts off can reach this lethal state. People and animals have died as a result. Evaporation, of course, is not available as a heat loss mechanism to BU bathed in water.

COOLING CHICKENS

Chickens raised commercially for broilers are raised in flocks of up to 100,000 or more in a single poultry house. Temperatures inside these houses sometimes rise to intolerable levels during the heat of summer, and many birds die as a result.

One proposed means to reduce the temperature of ventilation air entering a poultry house is to cool it through evaporation of water. As the water evaporates the air temperature falls. This scheme has been proposed by engineers who attempt to cool the birds and eliminate bird mortality.

The difficulty with this scheme is that birds also use evaporation by panting to cool themselves. If the inlet air is cooled by evaporation, water vapor is also added to the air. The extra moisture in the air makes it more difficult for individual birds to cool themselves, as would still be necessary if the air temperature is not cooled at or below thermoneutral. In this case, the cooled air can gain heat not only from the birds themselves, but also from hot surfaces of the poultry house. Thus, this scheme of cooling air actually can make matters worse rather than better.

6.5.3 Moving to a Better Neighborhood

Do not go where there is a path—go instead where there is no path and leave a trail.

—Ralph Waldo Emerson

Locomotion is a behavior that is important to regulate temperature. Moving to someplace warmer or cooler can take advantage of local climatic conditions more suitable to the BU than the average. *Ectothermic* animals typically seek shelter from cold night-time temperatures and then move to bask in the warm morning sun to raise body temperature. When body temperature has been raised sufficiently, the animal may move to the shade in order not to overheat. *Endothermic* animals may exhibit similar movements, except that they have, in addition, the ability to regulate body temperature in other ways as well. Humans have taken this a step further by artificially producing cool environments in the heat and warm environments in the cold. Thus, they have moved the environments to themselves rather than move themselves to friendlier environments.

Although movement to choose environmental conditions is thought to be limited to animals, other BU take advantage of this mode of thermoregulation on a longer time scale. Populations and biomes find climates in which to live where they can exploit the resources available, compete with other BUs, and reproduce satisfactorily. The ranges of these BU are limited by the ability to exploit, compete, and reproduce. When climates change, as they do over a long time scale, the ranges of these BU change. Plants, animals, and microbes can then be found in locations where they never appeared before. The range of the California Redwood tree (*Sequoia sempervirens*) and relatives covered much of the northern hemisphere 140 million years ago. Today, the range is limited to a few locations in the upper northwestern United States. Similarly, the ranges of certain hardwood trees are extending northward, and those of conifers are retreating, as the modern-day climate warms.

This climatic change is not just global. Microclimates around and in houses, dams, and other structures erected by humans have allowed species survival in places where they have not been seen before. Some disease organisms, such as *Legionella pneumophila* (the cause of Legionnaires' disease) thrive in the cool, moist environments of air conditioning and cooling tower systems.

Other bacteria, fungi, and viruses have been found to live in the favorable thermal environments of houses. Plants, birds, insects, fish, and mammals have been able to extend their ranges because human structures provide shelter from harsh climates. The biological engineer should expect that the construction of a new structure will change the biological environment of its surroundings, and give new opportunities for microecological systems to develop.

LIOLAEMUS LIZARDS

Liolaemus multiformis is an unusual lizard that thrives in the cold environment of the Andes Mountains at 4800 m and above. Temperatures are cold year-round, with morning temperatures falling as low as –5°C. The lizard spends the night in burrows, where it cools less than it would in the open. Nevertheless, the lizard's body temperature may still fall to as low as 2.5°C.

The lizard emerges from its burrow early each morning and basks in the sunlight, lying on a mat of plant material to insulate itself from the cold stones underneath. It orients its back toward the sun to increase its radiation heat gain. Pressing itself against the plant material underneath reduces its exposure to wind and convection. The dark color of the lizard also enhances its ability to capture radiant energy.

The result of this is that the body temperature of the lizard rises to 33°C and is maintained at that level despite a surrounding air temperature of 1.5°C (Molles, 1999).

6.5.4 The Best Thermal Conditions

> Jupiter has loaded us with a couple of wallets: the one, filled with our own vices, he has placed at our back; the other, heavy with those of others, he has hung before.
>
> —**Phaedrus**

For the most part, BU have adapted to thermal environments in which they evolved. Thus, homeothermic animal tissue and organ BU need a constant temperature to be provided, and their proper functioning depends upon thermal (as well as chemical) stability. Lowering their temperatures exponentially reduces their metabolism and oxygen and other environmental demands. Cooling is used to advantage in the transport of organs slated for transplantation. Similarly, microbial BU that spoil food cannot grow and reproduce unchecked if cooled below their preferred temperatures. However, there are BU that have acclimatized to extremely hot or extremely cold conditions. These BU will thrive in conditions where other BU will not; it is thus imperative that biological engineers expect that unusual BU may appear in unusual environments, especially those environments that are maintained for long periods of time. By "long," we mean an environment maintained over many generations of the BU.

BU that have developed in constant temperature conditions often do not perform well if the temperature is allowed to vary. Other BU, however, have adjusted to these variations and cannot grow and reproduce without them. Deciduous trees, for instance, have developed a cycle of losing their leaves in the winter and re-growing new leaves in the spring. Fruit trees must experience a number of hours below 7°C (called the *chilling requirement*, usually in the range of 600–1000 h) in order to blossom correctly. Ectothermic animals, including insects, have modified their behaviors to use temperature variations to their advantage. If the BU under consideration has evolved under a certain set of environmental conditions, then it should not be expected that the BU will thrive unless those conditions are present.

Lastly, let us consider the consequences of an environment that provides either little heat or too much. If too little, then the water in the BU will likely freeze. Water crystals disrupt the physical structures of BU, and the BU dies. Some BU (plants, fungi, and microbes) have developed the ability to form dry spores that can survive low temperatures for the long periods of time, and others (fish) have developed natural antifreeze (glycoproteins) solutions to keep bodily water from freezing until water surrounding them has frozen first. The external ice layer then rises to the top of the body of water and insulates the water from further heat loss.

Too much heat can have the same lethal effect. With too much heat, proteins and enzymes become denatured and ineffective. Without these, normal metabolism stops, and the BU dies.

Hence, it can be seen that the thermal environment of a BU must be maintained at a level to sustain life. In some cases, for example in the preservation of meat, fish, or produce, a cold thermal environment is maintained to keep the BU alive, but growth and reproduction are of no consequence. In other cases, when growth and reproduction are important, as in a bioreactor, the thermal environment will be different, and perhaps more exacting.

6.5.5 Too Hot

> No amount of experimentation can ever prove me right; a single experiment can prove me wrong.
>
> **—Albert Einstein**

Excessive temperatures can pose a threat to life. At high temperatures, critical enzymes become denatured and metabolism goes awry. Neural circuits misfire, causing seizures and loss of thermoregulation. Intercellular clefts open and allow microbes access to physical body interiors. Cognition, dexterity, and coordination suffer. Human decision-making becomes impaired; plants just bake.

Responses of living organisms to excessive temperatures include:

1. Moving to cooler locations, if possible.
2. Enhancing water evaporation.
3. Increasing heat loss surface area.
4. Making use of convective or conductive heat loss.
5. Increasing surface temperature to increase heat loss.
6. Shedding or adding surface insulation, depending on radiation exposure.
7. Reducing metabolic heat production.

So, important physical parameters in the heat relate to air temperature, air humidity levels, and radiant energy. Each of these directly affects heat gain or loss.

There is a metabolic advantage to high body temperatures. Higher temperatures allow for faster chemical reactions, quicker responses, and higher signals-to-noise in sensors. Movement becomes sustainable. Honeybees, for instance, must tense their flight muscles on cold mornings to raise their temperatures and allow them to contract fast enough to sustain flight. If they did not perform this pre-flight ritual, they would fall from the air and die.

The disadvantage of high body temperature is the proximity to the lethal temperature limit. Thus, it is imperative that living organisms be able to confine body temperature increases to the safe range.

The issue of human activity in the heat is so critical that indices have been developed to estimate the ability for people to continue work during hot exposure (Moran and Pandolf, 1999). Some of these are calculated from physical measurements only (temperatures, humidity, wind speed, and radiation), and some use physiological data also (heart rate and rectal temperatures). When these heat stress indices reach threshold values, the exact value depending on the index used, employers in some countries are legally obligated to ameliorate the stress, either by stopping or modifying work schedules or by providing cooling strategies.

6.5.6 Too Cold

Simplicity is the ultimate sophistication.

—**Leonardo Da Vinci**

Cold temperatures can pose a threat if they are very low and convective heat loss is very high. Under these conditions, flesh can freeze. If this is a continuing problem, then the organism has likely developed strategies to lessen the threat. Deciduous plants, for instance, lose their leaves and greatly decrease their water dependences. Some arctic poikilotherms circulate glycoproteins throughout their bodies to act as natural antifreeze. Other homeotherms develop thick insulating furs or feather coats. For humans able to move freely and dress warmly, cold temperatures are mostly of interest from a comfort standpoint.

Weather bureaus in various countries have empirically developed correlations of physical weather factors with comfort, and often report these as wind-chill temperatures. The wind-chill temperature is influenced largely by convective heat loss caused by air moving across bare skin. The wind-chill temperature is supposed to be the temperature equivalent to that with the sensation of bare skin under minimal wind speed (usually taken to be 1.8 meter/second) conditions.

Example 6.5.1: Storing Platelets

Platelets are made in the bone marrow and play a central role in forming blood clots. Aggressive cancer chemotherapy can cause the bone marrow to cease functioning, and hence can halt platelet production. As a consequence, platelets must be transfused into these patients.

Once they are separated from whole blood, platelets are very fragile. If they are refrigerated, they undergo a chemical change that makes them the target of macrophage attacks in the patient's blood. For this reason, platelets are stored at room temperature and become useless after 5 days. Predict means to overcome this conundrum.

Solution

If the problem is platelet deterioration when not refrigerated, then the solution might be to slow or arrest this deterioration. If refrigeration is to be considered, then the problem is how to keep the macrophages from attacking the platelets. Either the macrophages could be disabled, or the platelets could be modified to prevent them from being recognized as foreign objects. The solution to the problem depends on how the problem is defined.

In this case, it has been found that adding a small amount of the sugar galactose to the platelets masks another type of sugar on their surfaces. It is this other sugar that the macrophages target. Thus, the platelets can be refrigerated and remain viable for up to 12 days.

Example 6.5.2: Heat Stress in Cattle

Cattle, too, are susceptible to heat stress (Figure 6.5.1). This stress can be accompanied by diminished appetite, reduced growth rate, compromised disease resistance, or death (McGinnis, 2010). People who raise cattle want none of these.

Cattle, like humans, use sweating to remove heat. Some animals are more able to control body temperature in this way, and some are not. In order to identify individual animals that may require targeted management of their thermal environment, agricultural engineers developed index values that evaluate animals susceptible to heat stress (Brown-Brandl, 2008).

The index they came up with relates to respiration rate of cattle in the heat (above 80°F):

Breathing rate index = 2.83 × air temperature (°F) + 0.58 × relative humidity (fractional) + 0.76 × wind speed (min/h) + 0.039 × solar radiation (Watts/m^2) – 196.4

If no solar radiation data are available, use 1000 W/m^2.

A score	
<90	cattle are fine
90–110	cattle are on alert
110–130	cattle in danger
>130	emergency

Example 6.5.3: Wind Chill Temperature

The U.S. National Weather Service provides wind-chill equivalent temperatures as charts of equivalent temperature with ambient temperature and wind speed. Values in these charts are based on the formula:

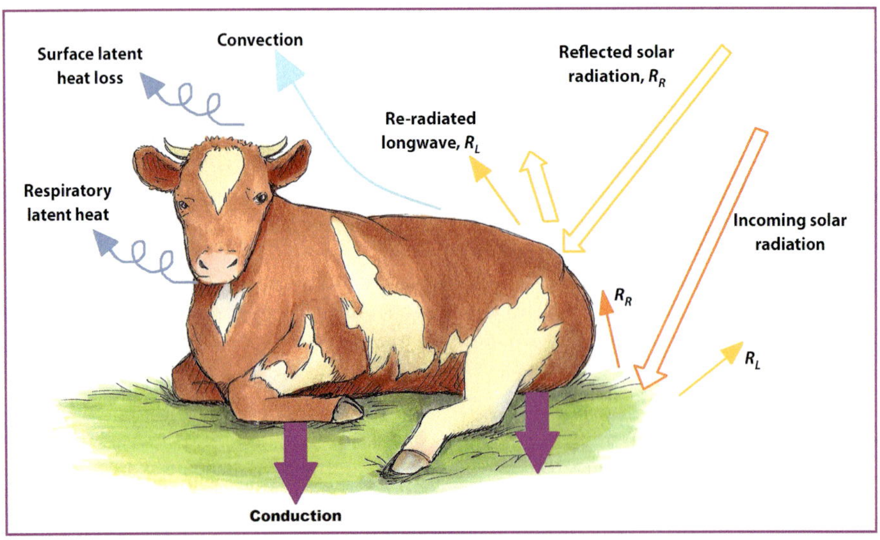

FIGURE 6.5.1
Heat transfer modes for a bovine outside without shade. (Redrawn from Brown-Brandl, T.M., *CAB Reviews: Perspectives in Agriculture, Veterinary Science, Nutrition and Natural Resources*, 3, 2008.)

$$T_{\text{wind chill}} = 35.74 - 0.6215\,T_{\text{ambient}} - 35.75\text{ wind speed}^{0.16}$$
$$+\,0.4275\,(T_{\text{ambient}})(\text{wind speed})^{0.16}$$

Temperatures are in °F and wind speeds are in min/h. This equation is limited to temperatures at or below 50 °F and wind speeds above 3.0 min/h.

APPLICATIONS AND PREDICTIONS

1. Freezing large tissues, organs, or organisms will usually kill them because of the formation of ice crystals.
2. Some microbial biochemical conversions will occur faster if heat is added.
3. Wine must be cooled during fermentation to remove excess metabolic heat.
4. Northern apples will not grow well, and probably will not develop fruit, in Florida.
5. Spores and seeds introduced into the soil from the past generations of plants, fungi, and microbes will sprout once conditions are favorable. To remove all seeds requires repeated sprouting and killing through periodic cultivation.
6. BU adapted to unusual environments will be found in those environments.
7. Complex microbial, plant, and animal behaviors will develop over time to compensate for environmental challenges.
8. BU in colder climates will have well adapted heat maintenance and production; BU in warmer climates will be expected to lose heat more efficiently.
9. Pathogens can be killed with high temperatures.
10. Plant, animal, and microbial adaptations to temperature extremes can yield useful solutions for engineering designs.
11. Campers are sometimes surprised by insects, snakes, or mice in their warm sleeping bags.
12. Heating in the cold or cooling in the heat will require the BU to expend energy.

6.6 Biological Units Adapt to Their Environments

However dumb they seem to be
Camels usually appeal to me.
Camels seldom need be fed
They store up fat and plan ahead.

—Nancy Gilchrist

The interaction between living things and the environment is not static. BU adapt to their environments and also modify their environments to better suit their needs. In this section, we consider only the first: how BU adapt to their environments to

1. Survive,
2. Make the better use of resources,
3. Reduce environmentally induced stress.

Survival is paramount in biology, but the survival of an individual is not nearly as important as the survival of the genes that are carried by the individual. Thus, we can consider two types of adaptation:

1. Relatively short-term adaptation.
2. Relatively long-term adaptation that operates over many generations.

There is no absolute time scale here, because "short-term" and "long-term" refer to times relative to generational time. If the adaptation occurs within the lifespan of a particular BU, then it is short-term. However, adaptation does occur over many BU generations, and this adaptation is usually called *evolution*. Evolution will be discussed in the next section.

Adapting to the environment takes many forms, and usually allows the adapting BU to perform its function, and thus compete against other BU, better than without adaptation. There are many adaptation modes, and we will discuss some of these.

CAMELS AND CACTI

(excerpted from Molles, 1999)

On the surface, camels and saguaro cacti appear entirely different. If you look deeper into their biology, however, you find that they take very similar approaches to balancing their water budgets. Both the camel and the saguaro cactus acquire massive amounts of water when water is available, store water, and conserve water.

The camel can go for long periods in intense desert heat without drinking, up to 6–8 days in conditions that would kill a person within a day. During this time, the animal survives on the water stored in its tissues and can withstand water losses of up to 20% of its body weight without harm. For humans, a loss of about 10% to 12% is near the fatal limit. When the camel has the opportunity, it can drink and store prodigious quantities of water, up to one-third of its body weight at a time.

Between opportunities to drink, the camel is a master of water conservation. One way it conserves body water is by reducing its rate of heat gain.... The camel faces into the sun, reducing the body surface it exposes to direct sunlight. In addition, its thick hair insulates it from the intense desert sun, and rather than sweating sufficiently to keep its body temperature down, the camel allows its body temperature to rise by up to 7°C. This reduces the temperature difference between the camel and the environment and so decreases the rate of additional heating. Reduced heating translates into reduced water loss by evaporation.

The bulbous nose on a camel has the important function of moisture conservation. Moisture in the exhaled breath is reabsorbed into the camel's body; the nose is constructed with large interior surface area to facilitate moisture conservation.

The saguaro cactus takes a similar approach. The trunk and arms of the plant act as organs in which the cactus can store large quantities of water. During droughts, the saguaro draws on these stored reserves and so can endure long periods

(*Continued*)

CAMELS AND CACTI (Continued)

without water. When it rains, the saguaro, like a camel at an oasis, can ingest great quantities of water but instead of drinking, the saguaro gets its water through its dense network of shallow roots. These roots extend out in a roughly circular pattern to a distance approximately equal to the height of the cactus. For a 15 m tall saguaro this means a root coverage of over 700 m^2 of soil.

The saguaro also reduces its rate of evaporative water loss in several ways. First, like other cactus, it keeps its stomata closed during the day when transpiration losses would be highest. In the absence of transpiration, in full sun, the internal temperature of the saguaro rises to over 50°C, which is among the highest temperatures recorded in plants. However, as we noted for the camel, higher body temperature can be an advantage because it reduces the rate of additional heating. The saguaro's rate of heating is also reduced by the shape and orientation of its trunk and arms. At midday, when the potential for heating is greatest, the saguaro exposes mainly the tips of its arms and trunk to direct sunlight. However, the tips of the saguaro's arms and trunk are insulated by a layer of plant hairs and a thick tangle of spines, which reflect sunlight and shade the growing tips of the cactus.

The parallel approaches to desert living seen in saguaro cactus and camels are outlined in the figure (Molles, 1999).

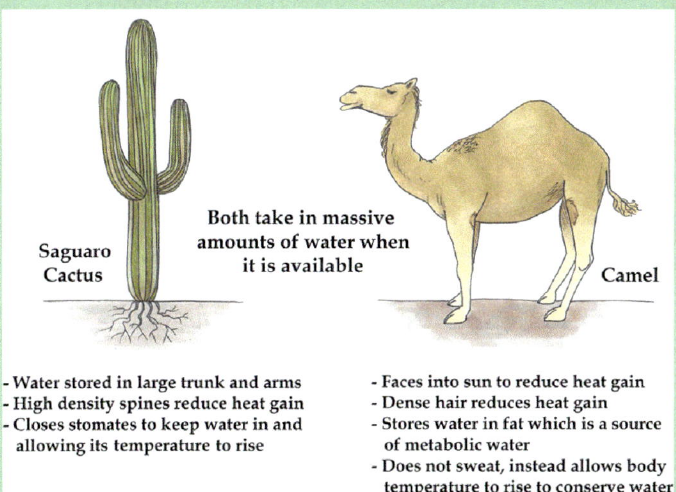

Desert plants and animals use similar adaptive strategies. (Redrawn from Molles, M.C., *Ecology: Concepts and Applications*, McGraw-Hill, New York, 1999.)

6.6.1 Cells and Microbes

The genes tell us what could happen in the cell;
Messenger RNA tells us what might happen;
Proteins tell us what is happening.

—**Catherine Fenselau**

In another instance of adaptation, cellular metabolism can change depending on the availability of certain substrates. A very common example of this is the use of anaerobic metabolism in muscle cells when sufficient oxygen is not available (Section 3.10, Energy-Rich Compounds). The product of this metabolic pathway is lactic acid rather than carbon dioxide and water. When oxygen is again available, the lactic acid is used to reformulate glucose and some lactic acid is metabolized to water and carbon dioxide.

Microorganisms that produce but one metabolic product are called *homofermentative*, whereas those that produce many different products are called *heterofermentative* (Nielsen and Villadsen, 1994). Heterofermentative microbes can be manipulated in bioreactors to produce more economically valuable products by controlling the availability of metabolic substrates, and therefore determining the metabolic pathways taken. This adaptability confers a competitive advantage to the microorganism because it allows the microbe to utilize available resources to grow and reproduce in environments to which it is not completely well adapted.

6.6.2 Hypertension

> I write long epigrams, you yourself write nothing. Yours are shorter.
>
> **—Martial**

There is an adaptation of internal tissues and organs to the abnormal physiological states of the body, and this can lead to complications when trying to combat the underlying disease. For instance, hypertension is higher than normal blood pressure. In many instances, hypertension is caused by the constriction of arterioles and their increased resistance to blood flow. The activation of the sympathetic nervous system or circulating hormones (rennin, angiotensin, vasopressin) can cause this constriction.

Hypertension is usually a temporary adjustment to emotional stress or trauma, but after sustained hypertension, the condition perpetuates itself, becoming *essential hypertension*. The left ventricle of the heart, challenged by the need to pump blood against pressure higher than normal, develops a thickened muscular wall, called *cardiac hypertrophy*. The extra muscular mass requires additional oxygen, and any decrease in coronary blood flow has more serious consequences in hypertensive patients than in normal individuals. Myocardial infarction (muscle cells die due to lack of oxygen) is therefore more of a threat in hypertensive patients than in normal ones. Even with the implantation of a replacement heart, the underlying causes of the heart failure have not been eliminated, and the cycle is likely to repeat.

Although it may seem that the previous example is not an adaptation that improves survival, consider the alternative. If the heart does not develop hypertrophy, it cannot pump adequate amounts of blood through the vasculature, and, at the very least, the person cannot exert himself or herself. Every small physical challenge thus becomes an emergency that the person may not survive.

6.6.3 Color Changes

> Traditional is just something that was contemporary in the past. Things always evolve.
>
> **—Russ Barenberg**

We have all seen the color adaptations of the chameleon, which changes body color to match that of the surface upon which it rests. Ermine is the name for a weasel that has brown fur in summer and white fur in winter; this provides better camouflage. Some insects, as well, can change their colors to better match their backgrounds.

Brightly colored insects are often toxic to predators. The bright orange Monarch butterfly carries alkaloid toxins that it gets from eating milkweed leaves. The Viceroy butterfly looks almost exactly like the Monarch, but carries no toxins. Birds mistake the Viceroy for the Monarch, and that protects the Viceroy. However, it jeopardizes the Monarch because eating a nontoxic Viceroy would encourage birds to try a Monarch. Thus, we would expect the number of Viceroy butterflies to be a small fraction of the number of Monarch butterflies.

Mimicry is common in biology, especially in insects, when one species adopts the color, habits, or structure of another, less vulnerable, species usually to enhance chances of survival (Nelson, 2014). When one species (the model) is either distasteful or toxic to predators, and another nontoxic species (the mimic) assumes the appearance of the model, it is called *Batesian mimicry*. When both model and mimic appear similar and are both distasteful, it is called *Mullerian mimicry*. Predators learn to avoid eating both species based on appearance.

6.6.4 Adaptations to Light

> A man gazing at the stars is proverbially at the mercy of the puddles in the road.
>
> —**Alexander Smith**

Plants adapt to light levels. Seedlings face the sun because sunlight inhibits the action of growth hormones called *auxins*, thus fostering more growth on the side of the plant away from the sun. Sunflowers face the sun all day by turning themselves as the sun moves. Plants growing at high altitudes, where sunlight is stronger, are often shorter than plants growing at lower altitudes. Move the high-altitude plants to lower altitudes and they grow larger.

Animals are also sensitive to light levels. Birds tend to lay their eggs and raise young during the spring when the duration of light is increasing. Light levels during this time apparently signal to the bird that it is entering the season when food will be abundant, and the survival of the young will be favored. Chickens can be manipulated to lay more eggs in the autumn if given extra hours of electric light.

Humans, also, react to light. Seasonal Affective Disorder (SAD) occurs in many people during the autumn and winter as light levels decline. SAD causes emotional distress and depression among those who suffer this condition. It is not known exactly what the function of SAD might be, but it may have to do with a vestigial arousal to migration to warmer climates in the winter. Exposure to high intensity light alleviates SAD.

6.6.5 Other Adaptations

> Reading maketh a full man; conference a ready man; and writing an exact man.
>
> —**Francis Bacon**

When BU from one environment seize an opportunity to move to another, different environment (perhaps to escape draconian competition), the adaptability of BU becomes well illustrated (Figure 6.6.1). Such is the case for sea snakes, reptiles that have forsaken their native terrestrial environment for a marine environment. These species have had to develop nostril valves to exclude water, respiratory and metabolic adaptations to allow

FIGURE 6.6.1
There was a succession of life forms after an undersea eruption. First came microbial mats forming near ocean vents. Small tubeworms appeared within a year, and these were soon replaced by larger tubeworms. Eight years after the eruption, mussels encroached and replaced the tubeworms. Eventually, a whole new ecosystem had developed.

intermittent breathing, salt glands to excrete excess salt in their bodies, and circulatory adaptations to match their styles of breathing (Heatwole, 1978). The fact that these snakes are able to live their entire lives in water demonstrates the remarkable ability of BU to adapt to new environmental demands.

The interconnectedness of biological systems can be illustrated by the plight of native foxes in the Channel Islands National Park off southern California (NGS, 2002). Several of the islands have been overrun by pigs brought to the islands years ago by farmers, and native bald eagles were decimated by pesticides in the 1940s and 1950s. Mainland golden eagles began cruising over the islands to prey on baby pigs, but soon they learned that the tiny island foxes (smaller than house cats) were very easy to kill. Without the foxes to prey on them, the pigs flourished. Removing the pigs, reintroducing bald eagles (which don't eat foxes), and restocking the island foxes is being attempted to restore ecological balance (and reverse the adaptations that had occurred naturally).

There are many other examples of adaptation, from the kind and amount of food ingested in relation to food availability, to BU movement to avoid danger. In other sections, we will consider the evolution and co-adaptation of several BU in close proximity. In all cases, it should be clear that BU are not passive puppets that must be satisfied with whatever they receive from their surroundings; they are all active players that change themselves to better exploit that which is available.

FETAL INFLUENCES LAST A LIFETIME

We pass more milestones before birth than at any other time in our lives. It's a no-brainer that unless you pass those milestones correctly, you're going to enjoy less good health for a lifetime, says Peter Nathanielsz, at the Cornell University Laboratory for Pregnancy and Prenatal Research (Tregaskis, 2002). The world inside the womb is much less isolated from the outside environment than might be expected. Consequently, prenatal biology is influenced very much by the conditions experienced by the mother. If, for instance, the mother has barely adequate food, the fetus develops a thrifty metabolism because it is likely that food will not be any more plentiful after birth. A thrifty metabolism is a good way to adapt to famine, but it

(Continued)

FETAL INFLUENCES LAST A LIFETIME (Continued)

can also lead to overweight, atherosclerosis, and heart disease if heavily processed, high-fat, low nutrient foods are abundant. Maternal nutrition modulates DNA methylation (epigenetic changes) and shapes the phenotype of the offspring, thus leading to the era of environmental epigenomics (Jirtle and Skinner, 2007).

A fetus with inadequate oxygen supply protects its brain by directing less blood to internal organs such as the liver and pancreas. Later on, these undersized organs can have inadequate capacity to protect against high cholesterol or diabetes.

Toxins, either inhaled or ingested, can also reach the fetus. Large molecules in tobacco smoke damage the placenta, and this hampers the ability to convey nutrients from the mother to the baby. As a consequence, babies born to smokers, or to non-smokers living in polluted air, are on average, a half-pound lighter than babies born to mothers breathing nonpolluted air.

Some toxins are accumulated during the lifetime of the mother, and released to the fetus during pregnancy. Lead, for instance, is stored in the mother's bones and teeth, only to be released during the sixth month of human pregnancy when the fetal skeleton hardens. Fat-soluble organic pollutants (such as polychlorinated biphenyls, or PCB's) become stored in the mother's fat tissue and are mobilized when it is time to produce milk for the baby. This PCB-laced fat forms the myelin sheathes surrounding the baby's neurons.

Pregnancy also has permanent effects on the mother. In addition to fetal cells that continue to circulate in her bloodstream after the baby is born, there are permanent developmental structural changes in the mother's brain, especially in reduction of gray matter in areas subserving social cognition (Hoekzema et al., 2016).

Prenatal stress of female sheep (ewes) was found to affect adversely the brains of lambs (Coulon et al., 2013). Researchers observed structural changes in brain neurons associated with emotion, behavior, and awareness. Similar results have been found in experiments with mice and rats (SRC, 2013). Prenatal stress in humans can result in childhood development of depression, post-traumatic stress disorder, and schizophrenia.

Ewes fed a diet enriched with polyunsaturated fats for one month prior to conception have a significantly higher chance of giving birth to male offspring (Green et al., 2008). It has been speculated that diets high in polyunsaturated fats would be expected to be eaten by healthy and well-fed females. In a wild herd with many females dominated by a few males, it would be genetically advantageous to increase the number of males if the herd were well-fed. If the ewes were poorly fed, more female offspring would be more likely to pass the mother's genes to future offspring.

Prenatal exposure of human babies to testosterone has been shown to predispose them to analyze and explore a system, to extract underlying rules of behavior, and to construct various mechanisms (Baron-Cohen, 2005). Conversely, prenatal exposure to high levels of estrogen results in a sympathetic and empathetic outlook centered on people's feelings. These differences are found in newborns as well as in adults, and seem to be the basis for typical psychological differences between the sexes. Also, women who have a male twin are less likely to marry and have children than women with a female twin, perhaps because of the fetal exposure to testosterone (Lumma et al., 2007). A *freemartin* is a female twin of a male animal that has a small, underdeveloped

(*Continued*)

FETAL INFLUENCES LAST A LIFETIME (Continued)

reproductive tract due to prenatal exposure to the hormone testosterone. A freemartin is almost always infertile.

Birth order can be a strong environmental factor for offspring development. Eldest human children, as a general rule, take on some parenting roles for their siblings; teaching, guiding, caring, and disciplining younger brothers and sisters. As adults, they tend to achieve education, wealth, and status. Because they do not have the same access to their

parents as do their older siblings, middle children are often more independent and influenced more by friends than are others in the family. Youngest children, who are treated much differently from their older brothers and sisters, tend to please, be funny, and provoke others (Kluger, 2007).

Even animals and plants seem to confer special advantages on their offspring depending on birth order. Egrets hatch during successive days, and, if resources are scarce, the latter hatchings are allowed to die. Fruit trees tend to set many more fruits than they can support, so a good many of them never mature.

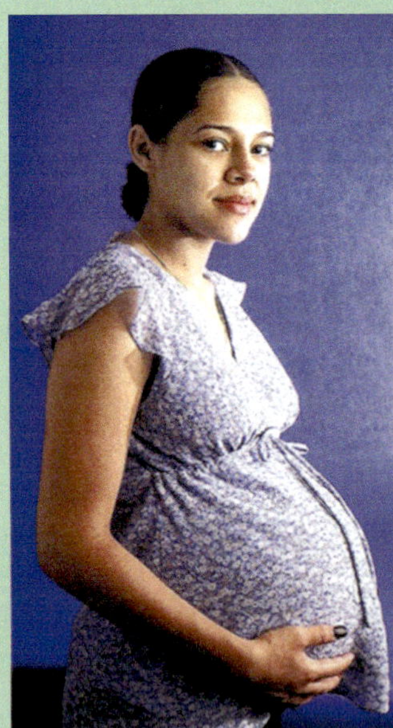

The health of the baby depends on the health of the mother. Sufficient foliate (a B vitamin) eaten by the mother during pregnancy protects against spina bifida (neural tube defect in the spine), anencephaly (failure for the brain to develop properly), cardiovascular disease, certain types of cancer, and some mental health problems in the child (Redrawn from Cena, E.-R. et al., *Calif. Agric.*, 61(2), 85–89, 2007). (Copyright © 2007 The Regents of the University of California. Used by permission.)

The fetal period is the time when the most drastic changes occur in the formation of tissues and life patterns. Pollutants, pesticides, hormones, nutrients, and diseases of the mother are buffered, but not isolated, from the fetus. Many of the health problems of life can be traced, at least partly, to the environment during the first nine months of life.

Example 6.6.1: Ecology of Piney Run Lake

This 100-ha lake, surrounded by 180 ha of forests and wetlands in Sykesville, Maryland, was at the center of a contentious debate over how best to obtain sorely needed drinking water for fast-growing South Carroll, the county's most populous area.

Carroll wants to tap the lake to meet its water needs, but 1999 photos of the man-made lake, when the county had to lower the water level by about 1.5 m to repair docks, show a muddy and lifeless expanse of shoreline devoid of plants and animals. Explain what happened to the plants and animals, and predict the effect on the lake of using it as a water supply for the local community.

Solution

When the water level drops rapidly, plants do not have time to adjust, and so they die. Animals that depend on the plants die also. Plants no longer supply oxygen to the water, and rotting organic matter from dead organisms uses additional oxygen. The quality of the water also suffers. Using the lake as a water source will probably result in lowered lake levels during dry periods, and the lake will not be able to recover in between these periods. Over a very long time, a new lake ecology will probably emerge tolerant of these new conditions.

Example 6.6.2: Microbes in Salt Solution

Three identical batches of bacteria are to be cultured in three different strengths of salt solution. The first is very dilute, and has no noticeable immediate effect on the bacteria. The second is extremely concentrated, and immediately kills the bacteria. The third is of an intermediate concentration, and kills about half of the exposed bacteria. Predict the long-term consequences of each of the salt solution concentrations on this type of bacteria.

Solution

The dilute solution poses no challenge, so will have no long-term effect. The concentrated solution kills all bacteria, and so the bacteria cannot adjust to the challenge. The intermediate concentration will select for salt-tolerant bacteria, and, over the long term, all bacteria will be able to survive in that concentration.

Example 6.6.3: Varroa Mites on Honeybees

Since the 1980s, *Varroa destructor* mites have been a serious pest of honeybees (*Apis mellifera*) in the United States. The female adult mite enters the honeycomb cell containing the bee larva and lays her eggs. When the eggs hatch, the juvenile mites suck the blood of the developing bee larva and weaken the larva. Although these mites will infest worker (female) bee larvae, they prefer drone (male) bee larvae. One means that has been tried to reduce mite populations in the hive is to kill and remove drone larvae before they emerge as adults. This is an effective means to reduce mite populations. (Also, the hives can function well without as many drones.) Predict the eventual effect on the mites of continually removing drone larvae.

Solution

Mites that infest worker brood will be selected for, and the mite problem will become worse because mites will weaken or kill necessary worker bees.

Example 6.6.4: Allergy Epidemic

The incidences of allergies are increasing greatly in recent years in the United States. The manifestations of allergic reactions range from asthma to food allergies. Speculate on causes for the recent rise in allergic reactions.

Solution

There may be many causes. One of these would be that people are being exposed to many more allergens than they were in the past. Another would be that the genes leading to allergic reactions are more widespread than they were in previous generations. This second speculation is not likely to be the cause, however, because it is unlikely that a genetic change would be so rapid in the population. The first is definitely true: we are being exposed to many more materials and chemicals than in the past. However, allergies to certain foods such as wheat, soy, peanuts, and shellfish are increasing rapidly, and these products are unlikely to have changed significantly, recently.

Another possibility could be an unintended consequence of cleanliness. Many systems of the body function best when challenged: muscles and intellect are among these. The immune system may work the same way. From what is known about the immune system, we know that it responds best to a microbe that it has previously encountered. If there are few microbes that challenge the immune system, would it not be possible that the system does not develop properly and that it begins to malfunction by reacting to common proteins present in pollen, dust, mold, food, and even in one's own body? It is indeed possible that the widespread use of disinfectants and medicines, and isolation from dirt, disease, and other critters in early years could portend an immune system that develops incorrectly. We do know that chronic exposure to pathogens or parasites leads to a certain tolerance in animals. It has also been found that human children who received antibiotics in their first 6 months were 1.5 times as likely to develop allergies and more than twice as likely to develop asthma as children who didn't get the drugs. Children raised in homes with pets were less likely to develop asthma and allergies. The relation between cleanliness and immunities is called the *hygiene hypothesis.*

There are other possible reasons for the recent epidemic of allergic reactions, but if people deny themselves the biological challenges to which their forebears were exposed, then they may not develop the abilities to fight what needs to be fought and to coexist with those things that pose no real danger.

Example 6.6.5: Genetic Expression of Fruit Fly Larvae Hairs

The tiny hairs that exist on the surface of newly hatched fruit fly larvae are called *trichomes*. The pattern of trichomes is governed by a gene called *shavenbaby*, which is influenced by at least six other enhancer DNA sections, some of which are far removed from the shavenbaby gene. Mutations in these enhancers can produce different trichome patterns in different species (McGregor et al., 2010; Michalowski, 2010).

These different patterns do not show up when fruit flies are reared in the laboratory with comfortable temperatures. The only time larvae have been found to have different trichome patterns is when the flies are raised in more natural temperature conditions of hot days and cold nights. Environmental conditions can have a direct bearing on genetic expression.

Example 6.6.6: 2D:4D Finger Ratio Is an Indication of Prenatal Testosterone Exposure

The ratio of the length of the second (index) finger to the length of the fourth (ring) finger on the right hand is called the 2D:4D ratio (Figure 6.6.2). This ratio has been shown to

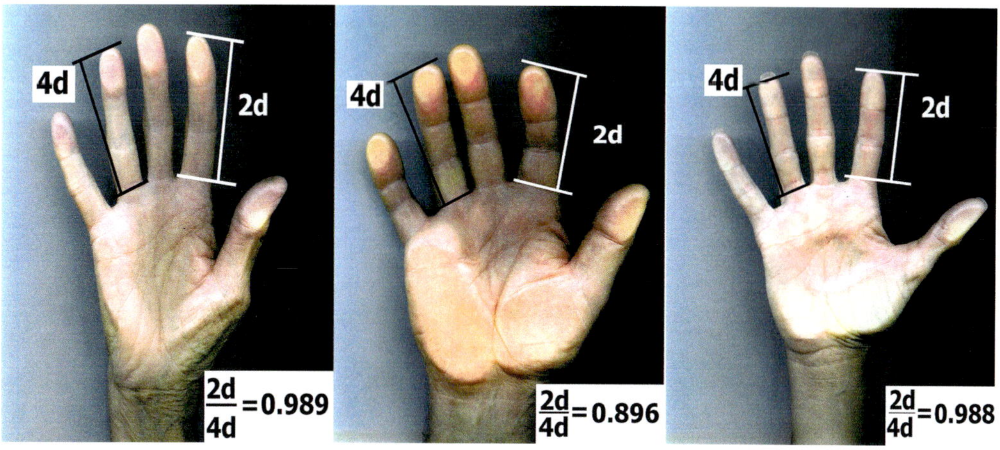

FIGURE 6.6.2
Impression of one male (center) and two female hands (outsides) illustrating the 2D:4D ratio in each. The male ratio is much lower than the two female ratios, indicating a higher level of prenatal testosterone in the male.

be different for male and female humans (Veale et al., 2010) and animals (Zheng and Cohn, 2011). It is sexually dimorphic, with male 2D:4D ratios (≈ 0.95) lower than those for women (≈ 1.0). The 2D:4D ratio appears to be influenced by prenatal exposure to androgenic hormones in the womb.

Lower, more masculine, finger ratio appears to be correlated to attitudes of gratitude (Misiak et al., 2015), more aggression or hostility (Bailey and Hurd, 2005b; McIntyre et al., 2007), increased spatial ability and sensation-seeking in men (Fink et al., 2006; Schwerdtfeger and Heer, 2008), lower pain perception (Schwerdtfeger and Heer, 2008), reduced male depression (Bailey and Hurd, 2005a), greater agreeableness and less quarrelsomeness of men with women (Moskowitz et al., 2015), sexual attractiveness in men (Bogaert et al., 2009), sperm count and testosterone levels in men (Manning et al., 1998), more preschool disruptive behavior in young girls but not boys (Roberts and Martel, 2013), faster skiing times in men and women (Manning, 2002), delayed menarche in girls (Matchock, 2008), and a variety of other physical and psychological traits

Higher, more feminine, finger ratios appear to be correlated to depression severity in women but not men (Smedley et al., 2014), personality traits in females (Fink et al., 2004), and estrogen and prolactin levels in women (Manning et al., 1998).

Finger ratios have been found to be related to career and occupational interests (Hell and Päßler, 2011; Weis et al., 2007; Nye and Orel, 2015), and handwriting style for women but not men (Beech and Mackintosh, 2005).

Animal models have indicated that androgenic steroid (especially, testosterone) exposures before birth might influence their sexual orientations, and that might also be true for adult humans. The 2D:4D finger ratio was used as an indicator of prenatal androgen presence, and evidence was found that homosexuality in both sexes is more likely with higher prenatal androgen exposure (Williams et al., 2000). Mothers previously pregnant with prior sons appear to have higher levels of testosterone in their bodies, thus altering the fetal development of subsequent children and increasing the likelihood of homosexuality in adulthood.

APPLICATIONS AND PREDICTIONS

1. Animals fed an unlimited diet will get fat.
2. Ringworm, when treated with medicine that does not kill it, will change its appearance.
3. Muscles used to lift heavy loads will grow; unused muscles will atrophy. The use of lifting aids and support belts will weaken muscles.
4. Urine of animals in locations with an abundance of water will be very dilute; urine of animals in locations with a scarcity of water will be very concentrated.
5. Digestion in animals with an abundance of food will not be as efficient as digestion in starving animals.
6. Microbes not killed by antibiotic compounds will become immune to them.
7. Adaptable BU will better survive environmental changes.
8. Brightly colored BU are often toxic when eaten.
9. Different organs in the body will adapt to each other to perform intended functions most efficiently.
10. Practice makes perfect.
11. Behavior-modifying drugs often must be taken in increasing amounts to give the same effect.
12. Astronauts exposed to microgravity for a long time will experience decreases in bone and muscle strengths.
13. Parasites adapt to the environments of their hosts.

6.7 Biological Units Modify Their Environments

An engineer ultimately designs for a customer, an artist for an audience.

—Greg Hull

Environmental interactions with BU are not just one-way. BU are adept at modifying their environments to suit their purposes, and one should never assume that the environment will be the same after BU have been introduced. We probably all know the weathering effect that certain plants have on rocks and soils. These plants extract required minerals from the rocks by secreting acids that dissolve the rocks. Over time, the rocks will diminish and disappear.

Some plants can alter the immediate environment around their roots by exuding materials that change soil acidity. This enhances, by three or more orders of magnitude, the solubility of such essential elements such as iron. These plants have a finely tuned biofeedback control that enhances soil fertility precisely at the root surface without altering the rest of the soil. This, then, minimizes the energy cost to the plant of soil modification (Olsen et al., 1981). Root crops and rhizosphere bacteria secrete acids into the soil solution to raise the solubility of solid-phase phosphates (highly pH dependent) in their immediate neighborhoods (Russell, 1961).

At the other end of the scale, trees are responsible for the removal of carbon dioxide and the addition of oxygen to the atmosphere. It is widely believed that all of the oxygen in the Earth's atmosphere is the result of photosynthetic activity by primitive organisms, and that the atmosphere before photosynthesis was much different than it is today.

For a very broad and global example of the way living things can enhance the hospitality of their environments, consider the *Gaia theory*. This theory is a view of the Earth that sees it as a self-regulating system due to the interactions of living things and their physical and chemical environments (Lovelock, 2006). The goal of these interactions is the regulation of the surface conditions of the Earth to make them most favorable for contemporary life. Because of this huge, multichannel feedback system, Earth's climate is much more moderate than it would be if it were lifeless. Its temperatures are less extreme, its atmospheric oxygen is more abundant, and its fecundity is greatly enhanced because of the presence of life.

Humans have had an inordinately large effect on the environment, both on a global scale (indeed, even beyond our own planet) and on a local scale. Human activities have leveled mountains, brought water to deserts, increased the expanses of deserts, changed the courses of rivers, rolled back the ocean, diminished huge lakes, changed the composition of the atmosphere, warmed the entire planet, lit the darkness of night, and caused the Earth to glow with electromagnetic emissions. Humans have changed the biota of the planet so much that only the deepest parts of the oceans can be said to be untouched. Humans have transformed wild areas full of biodiversity into expanses devoted to agriculture and, if not monoculture, at least restricted culture.

Although humans with the machines that they have invented have caused the most dramatic environmental change, other species are not innocent in this respect. Giraffes and other herbivores in the African Savanna are thought to limit the spread of acacia trees and other woody plants. The result is that grasslands remain as grasslands rather than revert to forest. Elephants, too, tear down trees and open up the ground to light, allowing new trees to sprout and grow.

Honeybees, ants, and termites condition the air in their nests, cooling it when it is warm and warming it when too cold. In the hot summer, bees can be seen bunched at the entrance to their hive, all facing outward and fanning air into the hive with their wings. They also gather water from streams, ponds, and puddles to carry back to the hive to evaporate and cool the hive. In winter, they cluster together, conserving the small amount of body heat they generate in order to maintain their larval brood at 30°C–32°C even on the coldest days. Termites air-condition their nests to control both temperature and humidity.

Probably one of the most intricate examples of environmental modification is the hormonal system inside the bodies of humans and animals. Hormones are biochemicals produced at certain sites, usually endocrine organs, which are transported by the blood to remote sites, where they elicit particular responses. Some hormones are very specific in action and some have more widespread effects.

Insulin is a hormone produced by the islets of Langerhans. These are isolated collections of cells scattered throughout the pancreas. The insulin they secrete passes through their cell membranes into the bloodstream, where it pervades the body. Insulin acts on muscle tissue, adipose tissue, some eye tissues, and the pituitary gland to increase glucose uptake. Glucose uptake is not influenced by insulin in the brain, kidney, intestinal mucosa, and red blood cells. In the liver, insulin helps to maintain blood glucose levels by forming glycogen during times of blood glucose excess and by inducing glycolysis when blood glucose levels are low. Thus, insulin helps to maintain homeostasis internal to the BU. Other circulating hormones act to do similar things.

Although the air in the lungs and the food in the gut are not strictly inside the body, they are in the regions of closely controlled environment. The air is warmed (usually) and moisturized (usually) as it passes through the mouth or nasal tissues. Inside the lungs, there is a considerable volume of air (1.2 L for normal adult humans), called the *residual volume*, which remains in the lungs even with the most forceful exhalation. The presence of this air tends to dampen extreme compositional variations that would normally accompany the inhalation and exhalation parts of the breathing cycle. The result is that air in the lungs is controlled to eliminate extremes that would inhibit lung function.

In the gut, food first passes into the esophagus, then to the stomach, small intestine, and large intestine. Herbivores may have several stomachs, each with a different function, and gizzards are found in birds.

Even in the simplest digestive system, environmental conditions must be closely controlled. The process of digestion (*catabolism*) is an energy-consuming process wherein digestive enzymes are formed, food is broken into simple compounds, and the compounds are then absorbed. If the environment were not closely controlled, then enzymes could be lost to the environment and more would have to be produced, the enzymes would not be as effective at different temperatures or pH values, and digested food would go elsewhere. Thus, although food passing through the body does not strictly enter the body until it is absorbed through the lining of the digestive system, it is essentially captured by the body until released.

One should never forget that a BU will affect its environment. This has serious implications for materials that are brought into contact with BU, the ways in which BU are controlled, interactions among BU, and the abilities for BU to exploit substrates that are assumed untouchable. The biological engineer must anticipate possible ways in which BU can modify their environments, and act accordingly.

THE PRIME DIRECTIVE

There is a real possibility that microbes from Earth would find conditions on another heavenly body suitable for them to live. If so, is it ethical for planetary probes from Earth to infect the other world? NASA personnel say "no". Just as in the TV program "Star Trek" the prime directive was to prevent interference with alien societies, especially those less advanced than the one that operated the starship "Enterprise"; there should be no contamination of other places in space. Yet, in reality, there is

(*Continued*)

THE PRIME DIRECTIVE (Continued)

the possibility that some errant microbe might hitchhike its way on a probe from Earth. So, to what degree should hardware from Earth be sterilized? What level of assurance is acceptable (Greenberg and Tufts, 2001)?

APPLICATIONS AND PREDICTIONS

1. No materials will be biologically completely inert. Materials implanted into the body will be corroded and materials on which microbes grow will be degraded over time.
2. The modification of the environment by one type of BU will affect another BU.
3. Environments containing BU can never be assumed to be static.
4. People will create homelike atmospheres when they go on vacation.
5. The interior environment of a BU is almost always completely controlled and difficult for humans to change.
6. The modification of environments by BU will require energy. To maximize growth or reproduction, an optimum environment should be provided.
7. If BU cannot modify their environments, they may attempt to migrate to a more suitable environment, as do Monarch butterflies.
8. BU can fight other pathogenic BU by changing their environment, as in fever.
9. When humans wash their hands, they affect the environment of microbes living on them.

6.8 Adaptations Require Extra Energy and Resources

Experience is the name everyone gives to their mistakes.

—Oscar Wilde

Some of the biological principles stated in this book are probably obvious; others may be considered to be unimportant. This one, "adaptations require energy" could probably be regarded as subordinate to "BU adapt to their environment." Nonetheless, because the fact that adaptations require energy and resources is not always completely obvious, and does not always immediately come to mind, we have raised this response as a special matter for consideration.

Genetically speaking, the first order of business for BU is to reproduce—as much and as often as possible—so that the genetic material included in the BU is spread as widely as possible and has the best chance of surviving over time. Although reproduction implies that the BU of which we speak is an organism, there are symbiotic BU (for example, lichens) that reproduce, and the functions of nonorganismal BU (for example, tissues and organs) are meant to support the organism so that it can reproduce. Any energy or resources directed away from reproduction are wasted from a genetic standpoint unless these resources somehow enhance reproductive potential. Thus, there is the possibility that resources addressed to other BU needs may be resolved as reproductively positive, negative, or neutral.

A few examples are in order. The first involves bioreactors filled with cells cultivated to produce economically important enzymes. In order to maximize enzyme production, overproducing strains of organisms are used (Shuler and Kargi, 1992). These BU must be fed extra nutrients and energy in order to overcome the inefficiency of producing more of these enzymes than they actually need.

These BU are not actively multiplying when they produce the excess enzymes. Actively multiplying BU (the *growth phase*) do not usually produce economically useful amounts of any product; their energies are directed to increasing their numbers and to individual growth. Most useful bioreactor products come from the *stationary phase*, which is artificially prolonged by manipulating the growth medium for the bioreactor (Figure 6.8.1).

Rapidly growing plant tissues usually contain less toxin than stable or slowly growing structures (Nesse and Williams, 1994). Toxins are one means for a plant to defend itself against herbivore grazing. The energy spent on growth cannot be used to produce toxins without diminishing the rate of growth as long as the rate of energy utilization by the plant is limited by some means.

Hearts that must pump blood against high resistance vessels develop extra-thick walls. This is an adaptation to specific environmental conditions that uses extra energy to produce myocardial proteins and then extra energy to keep the heart beating.

Conditional strategies for survival tend to be more successful than fixed strategies for survival (Hutchinson, 1981). An example of this is found in a large number of aquatic bugs and beetles. Not all members of a species can fly. The flightless form must save a good deal of energy by not providing for resting metabolism of large wing muscles. This form is more successful at reproduction when food is scarce.

The flying form has a chance of survival if the habitat dries up. The number of flying insects is dramatically determined by the environmental temperature, so that flying forms appear in the late summer when the danger of desiccation is greatest. We have here either a case of the same genes being present, but not always allowed to be active, or the case of the environmental temperature that determines the genetic makeup of the species. It is more

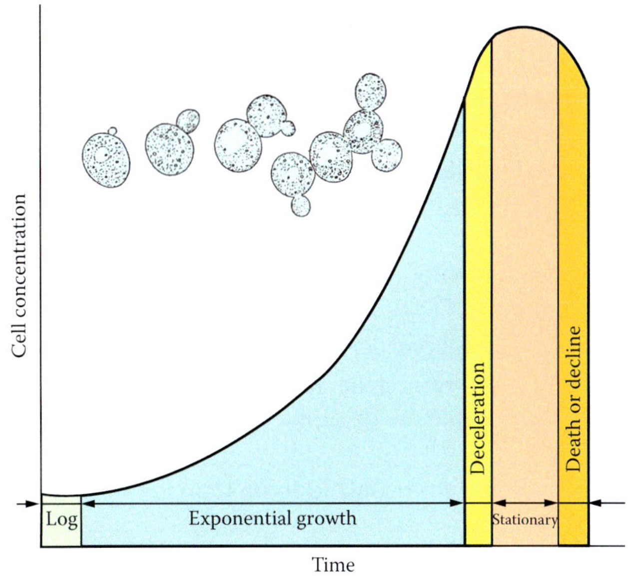

FIGURE 6.8.1
Typical stages of microbial growth.

likely that the former is true than the latter, so one can infer that the phenotype is only stochastically related to the genotype.

Taking advantage of the fact that the cost of adaptation slows growth and reproduction can lead to surprisingly practical results. For example, there is a parasitic mite, *Varroa destructor*, which causes heavy losses in honeybees. There are several different miticides available for beekeepers to use to kill the mites. If one such miticide kills 99.9% of the mites, selection pressure ensures that resistant mites will dominate the population in a few generations. Therefore, this particular miticide would soon be useless.

However, if a less effective miticide is used, it could be effective for a much longer period. If a miticide was used that kills 70% of the mites each time it is used, but resistant mites reproduce only 50% as fast as nonresistant (susceptible) mites, then the number of resistant mites and susceptible mites present in the beehive would be the same in 30 days. Resistant mites would not dominate the mite population, and this miticide could be used forever. The mite kill at each use would not be complete, but it would be sufficient (Oliver, 2007).

As stated previously, BU are very adaptable, and can find a means to grow and reproduce in very different environmental conditions. However, chemical and physical changes each take their toll by limiting the ultimate reproductive potential that would have existed if no adaptation were necessary in the first place.

Example 6.8.1: Disease Immunity in Feral Sheep

A good example of the trade-off between adaptation and reproduction can be found when comparing domesticated sheep to wild sheep living on the Scottish island of Hirta (Martin and Coon, 2010). The feral sheep have been found to be much more disease-resistant than are domesticated sheep. Blood levels of antibodies are much higher in the free-living sheep population, which helps adult females survive harsh winters. However, it has also been found that individual sheep with higher antibody levels have fewer offspring. Resources used to adapt reduce those available for reproduction.

APPLICATIONS AND PREDICTIONS

1. Plants that must resist wind forces will not produce as many seeds as those that do not need to resist wind.
2. Animals in the wild will not be expected to reproduce as readily as captive animals raised in a benign environment. Thus, captive animals will require contraceptives.
3. Isolated microbes growing on growth medium will grow faster than microbes in natural situations.
4. Good nutrition will cause precocious puberty in humans.
5. Cultural and societal stress will act as a birth control method. Males subordinate to the dominant or alpha male will not be as virile as the dominant male.
6. More energy is required for people to swim than for a fish to swim.

6.9 Biological Units, If Possible, Move to Friendlier Environments

Technology is neither good nor bad, nor is it neutral.

—Melvin Kranzberg

One characteristic of BU is that they attempt to control as much as possible of their internal and external environments (see Sections 6.7, Environmental Modification and 6.20, Control). We have seen how BU use energy to overcome tendencies to physical and biological disorder inside their own cells, organs, bodies, and biomes. We have even seen that the Earth tends to regulate itself through physical, chemical, and biological means to maintain a stable environment suitable for life. This regulation extends outward from a BU in an attempt to make the environment a friendly place that supplies all needed resources, removes all unwanted wastes, and assists in the defense against predators and competitors.

Sometimes, it is not possible to bring that kind of environment to the BU, or sometimes, it is not possible to maintain it. In such cases, the BU may move to locations that are more tenable.

What comes to mind initially is probably an organism capable of locomotion. If there is not enough food in the neighborhood, then the organism (an insect, arthropod, fish, bird, reptile, amphibian, mammal, or other) moves to where the food is. If water is not drinkable, then the organism moves to find potable water. If the temperature is too high, then the organism moves to a cooler location. If there is too much reproductive competition, then the organism moves to establish an individual defendable territory. All these, and more, are likely responses to environmental situations that are not within the capability of the BU to control satisfactorily.

Whole BU populations may also move. Honeybees swarm and fly to establish new hives when crowded. Microbes migrate in the soil to exploit new food sources, as when oil is spilled in a local area. Earthworms retreat from dry or cold conditions. Even plant populations move when conditions change from those that can easily sustain growth to those that cannot. Even the most sessile of organisms usually have a mobile phase during which they can adjust to new environmental conditions by moving.

Sub-organismal BU are tied very much to the organism in order to respond to environmental needs. In that case, the whole organism must move to escape the source of discomfort. If external vibrations are so great that internal organs may be damaged, the organism usually moves to a quieter spot. If noise is so great that the inner ears may be damaged, then the organism responds by moving. If a finger touches a hot surface, the whole arm moves away. However, if the threat cannot be sensed, or if other organismal needs are more important than those of the tissues, the organism will not move, and the tissues will have to cope as best as they can. Therefore, an animal may tolerate a hot environment by sweating or panting because sufficient food, water, and other needs are met adequately by the environment. Water evaporated as sweat or saliva comes from the blood plasma, and so the blood tissue must sacrifice for the sake of the entire organism.

There are some peripatetic tissues; however, that do seek out friendlier environments within organismal constraints. Many of these are pathological in nature: metastasizing cancer cells, syphilis bacteria and rabies viruses that seek out neural tissue, hepatitis viruses that locate in the liver, and bovine spongiform encephalopathy (BSE, or mad cow disease) prions that invade brain tissue. Others are not as harmful. Fetal development is made possible because the fertilized egg moves to a location specialized for its support. Stem cells and cells from the islets of Langerhans migrate to the locations of greatest effectiveness after being injected into the body.

Cancers are likely to metastasize, not as single cells, but as clusters of cells. These travel through the bloodstream as a group until they reach a location where they establish a new tumor. As they travel, they exchange signaling molecules that help to sustain the cluster. Single cancer cells in the bloodstream are likely to die (Callier, 2016).

Cell actin cytoskeletons do more than just stabilize the integrity of the cell; they are also instrumental in cell locomotion, extending by branching into cell protrusions and dissolving in the parts of the cell following behind (Lamar, 2012). The cell probes possible pathways for movement by growing and branching the actin cytoskeleton. At the same time, actin filaments at the tail end disappear. The cell can respond to surfaces over which it travels, needing some traction to move, but not so much stickiness that it cannot escape. The cell can also respond to chemical signals from other cells; this is the means by which *keratocytes* (highly mobile cells found in the basal layer of epidermis) move to injured skin areas to repair them.

Those, such as biological engineers, who wish to manipulate biological systems must keep in mind that BU are not passive. Unless constrained, they will move from less friendly to more friendly environments; unless friendly environments are provided, they will move elsewhere. If constrained in an environment that does not satisfy all its needs, the BU will exhibit signs of overcrowding (see Section 6.11, Crowding), inadequacy, deprivation, and even death.

PRESERVATION AND EXTINCTION

In the study of biogeography and the mathematical analysis of wildlife reserves, Mazumdar (1989) has examined relationships among species, area, and extinction rates. Design of animal and plant reserves cannot happen haphazardly if extinction is to be minimized or avoided altogether. Reserves must be designed to allow maximum mobility within a reserve and between reserves. The figure illustrates the principles involved in this design:

1. A large reserve is better than a smaller reserve as it holds more species at equilibrium and has lower extinction rates.
2. Given a certain total area available for a reserve, the reserve should be divided into as few disjoint pieces as possible.
3. If the available area must be broken into several smaller reserves, then they should be as close together as possible to increase migration rates between the reserves.
4. If there are several smaller reserves they should be grouped equidistantly from each other to enable the highest possible immigration rates between them. Placed in a line, the reserves on either end will have smaller immigration levels between them.
5. Several smaller reserves could benefit from being connected with strips of protected habitat enabling more dispersion between reserves.
6. Any given reserve should be as close to circular as possible as this minimizes dispersal distances within the reserve. An elongated reserve may suffer local extinctions at the ends.

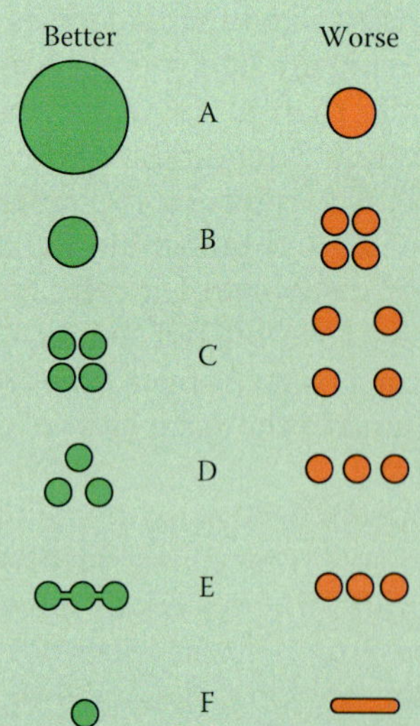

Illustration of better and worse habitat areas for species preservation. (Redrawn from Mazumdar, J., *An Introduction to Mathematical Physiology and Biology*, Cambridge University Press, New York, 1989.)

(Continued)

PRESERVATION AND EXTINCTION (Continued)

Some readers may not be concerned with preservation, but instead with extinction. At the microbial level, some may be faced with eliminating bacterial or fungal growths. From a public health standpoint, controlling a disease outbreak may be the goal. Preservation of a particular ecology may require that an aggressive non-native species be eradicated. No matter what the need, elimination of an interloper requires the opposite principles from preservation. To use a common phrase, "divide and conquer."

In the biological realm, however, nothing is as simple as it may seem. Although fragmentation of range leads to extinction, it is also true that isolation is necessary for new species to become established. Thus, the current trend towards fractionating habitats could also favor species development. Levin and Levin (2002) presented historical data showing that among both plants and animals, the rates of extinction of species and establishment of new species are linked, with higher extinction rates corresponding to higher speciation rates (figure). It is not clear how much time is required to establish new species, or how that time relates to the time required for extinction of species. Clearly, if all individuals of a certain species disappear before they can adapt to new environmental conditions, then a new species cannot evolve.

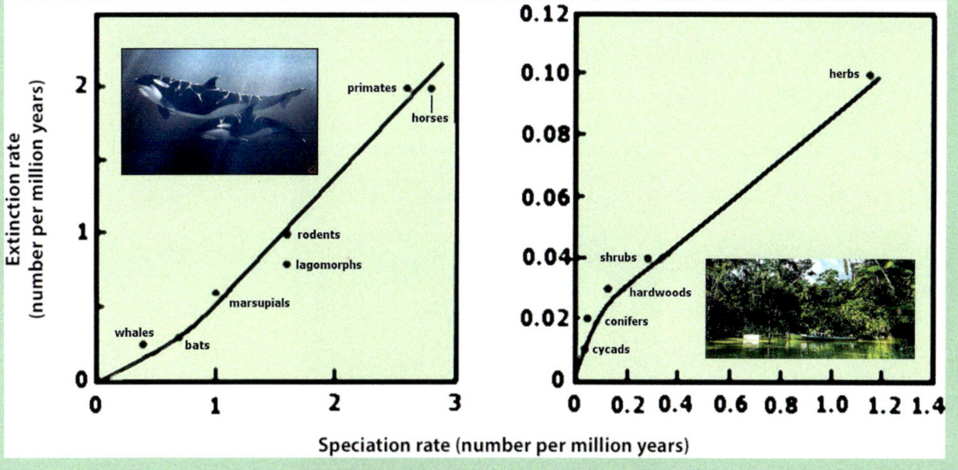

Although speciation and extinction rates vary widely in magnitude, they are related for both plants and animals. (Redrawn from Levin, P.S. and Levin, D.A., *Am. Sci.*, 90, 6–8, 2002.)

Example 6.9.1: Wild Animals on Display

An open-range animal display shows off wild animals in simulated natural habitat. The display will be very popular if free-ranging animals can always be seen by sightseers. Among the animals is a lion, and it would be very dramatic if he could always be seen lying on a rock atop a small rise. Suggest biologically compatible ways for this to happen.

Solution

The lion will be most likely to lie on the rock if it is not too hard, and is cooled in the summer and heated in the winter. After a while, the lion will develop the habit of lying on the rock. Hence, the resting site will be made to look like a rock, but will really be a locale for resting and comfort.

Remark

This is the solution tried by people at Disney's Animal Kingdom, and it works well.

Example 6.9.2: Slithering Snakes

Snakes and some lizards move with body undulations that depend on friction between their bodies and surfaces underneath. As the snake wiggles, it produces reaction forces both normal and axial to its body segments (Figure 6.9.1). The sum of the forward normal forces must exceed the backward axial forces in order to propel the snake forward.

On their undersides, snakes have overlapping scales that snag on the ground more in the backwards direction than in the forward direction (*frictional anisotropy*). Measurements made of friction coefficients of a milk snake on a cloth surface were 0.10 in the forward direction, 0.14 in the backward direction, and 0.20 in the sideways direction (Goldman and Hu, 2010). This helps snakes turn body undulations into forward motion.

To reduce friction even further, snakes raise parts of their bodies off the ground when moving. This dynamic load balancing can increase speed by 35% and efficiency by 50%. Snakes slithering on land use a similar amount of energy as a legged organism of the same weight.

FIGURE 6.9.1
As the snake slithers, its forward reaction forces normal to its body propels it forward; axial reaction forces hold it back.

APPLICATIONS AND PREDICTIONS

1. Killing unwanted organisms by making an environment that cannot sustain life will only succeed if the organisms cannot move elsewhere. Otherwise, the area could become infected again.
2. Animals tagged in infancy will be found elsewhere as adults.
3. Pregnancy can be avoided by blocking the movement of the fertilized egg to the uterus.
4. Weeds can be eliminated by confining the seeds to a small area and killing seeds as they germinate in that area.
5. Diseases may be able to be prevented if the causative organisms cannot reach their preferred locations.
6. You will catch more flies with honey than with vinegar.
7. Looking for humans in the summer? Try air-conditioned buildings or the beach.
8. Catching fish by net will be more successful in places where there is an abundance of food; catching fish on a rod will be more successful where food is scarce.
9. Birds will migrate south in the winter unless they can meet their needs in the north.
10. Worms are forced to the surface during a heavy rain.
11. Predators go where the prey are, not vice versa.
12. If you feed a stray cat once, it will probably return.

6.10 Biological Units Evolve under Environmental Pressures

> In nature, there is a functional purpose and reason for the shape and color of every living thing.
>
> —**Lee Harrisburger**

Genes will be expressed only if the environmental conditions are favorable. Thus, the presence of a gene, or set of genes, will not completely determine the appearance or behavior of the BU. If the environment is favorable, then the particular behavior or appearance coded by that gene will be manifested. Without the gene, however, the same environment will not likely lead to the same kind of behavior or appearance. The exception would be due to redundant genetic control.

Genes will survive best if they adapt, or allow adaptation. Thus, we would expect that in the presence of a variable environment, the BU that survive best are those that are more adaptable. If the environment is relatively stable, then this adaptability is not necessary.

Various environmental conditions that are sure to induce evolutionary changes are the same ones that induce individual adaptations on a short timescale: things such as temperature, availability of food and water, light, and competition. The difference is that those BU that adapt best to new environments are also those that are expected to be more successful at reproduction. Over the course of many generations, the genes from those BU that demonstrate better harmony with their environments will predominate.

It was shown previously that adaptations require extra energy and resources that could otherwise be used for growth and reproduction, thus reducing growth and reproduction. Adaptation, as a term used in this book, is meant to mean a nonpermanent, non-genetic, change in a BU. Before the discovery of epigenetic gene silencing, it was easy to distinguish between adaptation and evolution. The latter involved a change in the genetic code whereas the former did not. However, epigenetic changes can persist for many generations without a fundamental genetic change. Therefore, the feature most distinguishing between adaptation and evolution must then be that adaptation reduces growth and reproduction, whereas evolution leads to BU more successful at growth and reproduction.

For evolution to occur, there must be

1. Pre-existing genetic variation or some means for genetic changes (mutations) to occur
2. An environmental condition that remains over the course of many generations
3. A differential reproductive advantage of the responses of some genes compared to others

With no variation, there would be no competition allowing eventual genetic dominance; all genes would be alike, and all would be equally as suited (or not) to prevailing environmental conditions. Fortunately, there is genetic variation.

Environmental conditions can interact with individual genes to influence whether or not they become active. In this way, the environment can directly influence the development of a BU. Thus, individual cells differentiate into specialized cells, whole organisms acquire individual traits, and ecological systems include mixes of flora and fauna best suited to the exploitation of their environs.

SPECIES FORMATION

At some point, when portions of a population are isolated from one another, their genetic codes begin to diverge due to mutations and natural selection, allowing them to survive and reproduce better in their individual environmental conditions. There are two theories about the emergence of a new species. The first is that two species diverge from a common ancestor; the second is that a new species emerges as gradual changes take place in a previous species, making the evolved species significantly different from its previous form. In both cases, the new species would not be able to reproduce with other species.

Barriers that isolate a population and lead to a new species formation (Shermer, 2010; Reisch, 2016) are:

1. Geographic (such as a mountain range, desert, ocean, or river)
2. Morphological (changes in coloration, body type, or reproductive organs)
3. Behavioral (a change in breeding season, mating calls, or courtship actions)

Environmental factors may allow an individual to acquire particular traits, but, unless these are able to change the genetic code, they cannot be passed directly to subsequent generations.

The environmental conditions leading to the natural selection of certain genetic variants over others must remain stable for many generations. Genes that give a slight reproductive advantage to, but one or two generations will not overpower other genes in that time. Thus, genetic variants that best survive are those that are well suited to long-term environmental conditions and adapt to short-term conditions.

During most of the evolutionary continuum for humans, environmental conditions were much different than they are today. Things like air-conditioning, a safe and reliable food supply, clean water, and crowded cities have existed for only a short epochal time. Humans in particular have changed environmental conditions for themselves and for other BU, and this change was relatively recent, at least for themselves and for other species that reproduce at a slow rate. Because of this, genes such as those causing sickle-cell anemia have a reproductive advantage by protecting against malaria (the gene causes a deficiency of the enzyme glucose-6-phosphate-dehydrogenase and the red blood cells burst when the malaria parasite uses oxygen in red blood cells (Nesse and Williams, 1994)). The gene persists today despite the fact that malaria can be managed in other ways because the history of malaria control is only recent. Other genes such as the ones for Tay-Sachs disease (protects against tuberculosis), and the DR3 gene that causes childhood-onset diabetes (decreases miscarriage rate) are perpetuated even under recent environmental conditions (Nesse and Williams, 1994).

Genetic diseases are relatively rare, in part because most are caused by recessive genes that cause little trouble except in individuals who inherit two copies (Nesse and Williams, 1994). This is a very rare event because the recessive gene must come from both parents, and, if either of the parents is homozygous recessive, that parent will not likely have the same reproductive success as any parent with the dominant gene (Figure 5.3.5).

The shorter the life span, the faster evolution will occur. Thus, the constant and pervasive survival pressure of antibiotics, discovered by Alexander Fleming in 1929, have produced an environment wherein bacteria have developed immunities. In the Oregon Veterans' Administration Hospital, the rate of antibiotic resistance of staphylococcus bacteria went from less than 5% to over 80% in a single year (Nesse and Williams, 1994).

Antibiotics were originally produced as natural defense agents against bacterial infections by molds and fungi, also as a result of natural selection in the presence of the environmental pressure of infection. Why then hadn't bacteria developed resistance to antibiotics, given that molds and fungi have been around for extremely long periods of time? The answer is probably two-fold: first, there probably was some resistance developed, but because these bacteria also existed where there were no molds and fungi, the environmental selection pressure to pass on resistance genes was very low (remember that the genetic mutations of resistant bacteria can cause them to become nonresistant), and, second, no one noticed the resistant bacteria because they did not victimize humans.

Selection pressure has been effective in agriculture, where humans have selected plants and animals over many generations to exhibit more desirable traits. Thus, cultivated tomatoes are much different from wild varieties, Holstein cows give much more milk than their progenitors, dogs have many different characteristics from the wolves they came from (see Trut (1999) for an interesting account of selection of foxes for domesticity), and Red Delicious apples are the predominant variety grown in North America. From the standpoint of the genes involved, each of these cases represents a reproductive success.

It might be expected that natural selection should influence the responses of insects to echo location signals of bats, and that is found to be the case (Fenton and Fullard, 1981). Flying moths receiving weak impulses from a pair of auditory neurons, indicating a distant bat, turn and fly away from the bat. The strong stimulation of the auditory neurons causes the moth to fold its wings and dive to the ground.

Inadvertent ecological selection is today favoring certain traits and species and acts against others. Many ecologists (Levin and Schiewe, 2001) are decrying the modern loss of species due to habitat change; at the same time, other species such as deer, woodchucks, wild rabbits, and dandelions are thriving. Human activities are changing environmental pressures on all Earthly species (and soon perhaps other worlds—see box, Section 6.7). Living organisms are bound to change as a result.

That evolution allows organisms to adapt to their environments is embodied in the *Island Rule*. Large animals isolated on small islands with limited resources tend to become smaller over time. Contrarily, small animals that face reduced predator threats tend to become larger than relatives living in large territories.

Cooperation and competition lead to some interesting views concerning evolution. Darwinian evolution involves a more or less random genetic mutation that can then survive or not depending on its ability to confer a competitive advantage to the organism carrying it. Ben-Jacob et al. (1997) present the hypothesis that mutations in the cooperative colonies of similar bacteria are directed toward enhancing the colony, although these same mutations give no particular advantage to any specific bacterium within the colony. In this way, bacterial colonies growing in media with limited food resources develop genetic strategies for increasing their utilization of available resources and for seeking out new resources. Patterns of colony growth appear to be genetically modulated and do not involve the movement of established cells. The authors conclude that in this cooperative way, evolution changes from a system that depends on luck to one that can be controlled, at least in part.

Kessler and Hill (1997) have shown that unconscious cooperation among freely swimming organisms can produce similar benefits to those observed by Ben-Jacob et al. (1997). Although interactions among bacteria are not direct, by some means they form complementary swimming patterns that result in bioconvection and the better mixing of nutrients and oxygen.

The biological environment is always changing (Bak and Paczuski, 1997). Groups of organisms form ecological units that interact in many ways, often cooperating for the good of the group. Each set of organisms in this group evolves together, changing from generation to generation. Where one generation for one organism overlaps many generations for another organism, the environment may be said to be unchanging for the one with the shortest generation time. However, even there, the adaptations make evolution a cooperative venture. Random single mutations and natural selection can eventually reach a state of adaptation that is locally optimal. Further advances must come from many coordinated mutations.

It appears that gradual evolution is only part of the story of adaptation (Bak and Paczuski, 1997). Also possible are catastrophic events that completely overwhelm genetic adaptation. These events may be controlled by humans (for instance) or uncontrolled. The use of an autoclave on bacterial contaminants is a locally catastrophic event for the microbes that is under the control of humans. A comet striking the Earth and obliterating the dinosaurs is an uncontrolled catastrophe. Gradual evolution works best for unchanging environments and has little direct influence on biotic life in the face of a catastrophe.

DIRECTED EVOLUTION AS A DESIGN TECHNIQUE

Evolutionary principles can be used as a design paradigm. Begin with a set of specifications for performance of the final product and a means to quantify progress toward the specification goal. Next, define the starting point or configuration of the product. Then make random changes in the product while measuring how much closer to the final specification is each permutation. If the change improves the product, keep it. If the change does not, then discard it and try again.

This system can be computerized, and has been used to create hundreds of inventions (Brown, 2012; Keats, 2006). This system automatically mutates a single detail of the product and distributes product characteristics to two parent products. It then mates the two parents to produce offspring. If one child is closer to the ideal than the others, it alone is retained, and its siblings are eliminated. The process of mutation, distribution, mating, and selection repeats until the product comes close enough to the specifications to be acceptable, or until no further improvement is possible (see Figure).

Directed evolution can be used at an entirely different scale to improve enzyme function (Chica et al., 2005). Röthlisberger et al. (2008) used directed evolution to improve the efficiency of their artificial enzyme created by computational modeling. Their enzyme was designed to remove a hydrogen ion from a carbon atom as part of an organic compound. No such enzyme existed in nature. In order to respond to evolutionary methods, there must be some amount of functionality at the start. That's where they designed their first prototype molecule using computer modeling. They then induced random mutagenesis in the designed enzyme molecules and selected those that demonstrated improved functionality. After seven evolution replicates, the functionality had improved by 200 times.

Much of technology develops in a similar fashion, except, perhaps, the step of deliberately introducing random changes is not explicitly taken. Technology often starts with an idea that is developed using engineering principles. Thereafter, technological progress is made empirically, which means that improvement information comes from testing or use of the product. Empirical improvements derive quite often from trial-and-error, serendipitous, or practical observations, and not from application of basic engineering principles.

There is an analogy that can be drawn among the processes of technological innovation and development, the scientific method, and biological evolution and natural selection. Each of these involves a recurring loop with a starting point, testing, and discrimination against unsuccessful variants. The results are better products or processes, better information, or more successful living beings.

An artificial intelligence variant of the directed evolution design technique is called *reinforcement learning*. In reinforcement learning, computers improve their skills by practice and experience. During training sessions, software instructions are changes a little at a time with each attempt at achieving a goal. Whichever variants are more successful than the prior code are kept; those less successful are discarded. In this manner, computers can learn how to do things that programmers could not teach them (Knight, 2017).

(*Continued*)

DIRECTED EVOLUTION AS A DESIGN TECHNIQUE (Continued)

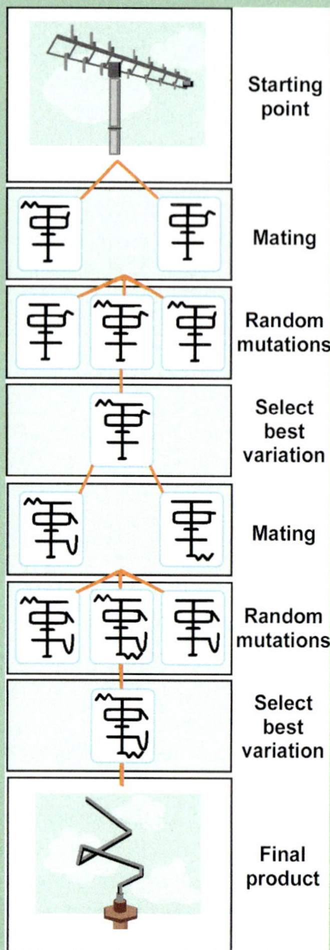

Evolutionary principles were used to select an improved electromagnetic wave antenna starting from a conventional design and ending with a device looking like a bent paper clip (Keats, J., *Pop. Sci.*, 268(5), 66–72, 92, 2006). A similar procedure has been used by civil engineers to design better buildings (Kicinger, R. and Arciszewski, T., *Am. Sci.*, 95, 502–508, 2007.)

Example 6.10.1: Selection of Pink Salmon

Predict the effect on a pink salmon population of paying fishermen by weight instead of by the piece.

Solution

Selling by the piece results in catching fish of all sizes; selling by weight prompts the use of nets that capture only the largest fish. Because all pink salmon return to spawn after their second year, the largest fish are the ones with the fastest growth rates. Size selection by fishers promotes the survival chances of the smaller fish and thereby favors genes for slower growth. Therefore, the genetic makeup of the pink salmon population will be altered toward that for slow growth rates. The longer this pressure continues, the more effect it will have.

Remark

This is exactly what happened in Canada when the scheme for payment to fishermen was changed in the late 1940s. A decline of more than 30% in average body weight of spawning pink salmon was found in 30 years.

Example 6.10.2: Microbes Respond to External Pressures

An alternative to recombinant DNA methods with microorganisms (see Section 8.2.3, Biotechnology) is growth under selective environmental pressure. Microbes have very short generation times and high populations in small spaces. If it is desired to produce a microbe with particular characteristics, then that microbe may result from growth in an environment that selects for those particular capabilities. This is especially true if the genes giving those capabilities are already present, but some enhancement is needed. This technique has been used to produce algae capable of increased production of biofuels and to produce microbes to remediate certain environmental pollutants.

APPLICATIONS AND PREDICTIONS

1. Pesticides will not work forever.
2. Animals continually bred and raised in a zoo environment will eventually diverge from what they would have been in the wild. They will become better adapted to close quarters, moderate environment, abundant food, and the presence of people. They will lose the ability to hunt, to adapt to the seasons, and to eat intermittently.
3. Some microbes, plants, and animals will become extinct due to the proliferation of humans; others will flourish.
4. New and improved fruits, vegetables, and ornamental plants will be developed by selective pressure.
5. Acquired skills will not be genetically passed on to your children.
6. Fleas will evolve faster than dogs.
7. Microbes will develop resistance to widely used disinfectants.
8. BU that can use alternative nutrients will survive better than those that can't in the face of nutrient shortages.

6.11 Crowding of Biological Units Produces Stress

Why do people question?
To question is to doubt.
Why do people doubt?
To doubt is to find the truth.
What is the truth?
That's a good question.

—Ted Leaptrott

The biological engineer who designs or adapts space for the confinement of organisms must do so with care, because there are important implications in the way this is done. Economic considerations would dictate that the largest possible number of organisms be enclosed in the available space. Yet, there are untoward consequences that can be encountered. Depending on the species, there may be social, reproductive, health, or developmental effects to be accounted for.

6.11.1 Antisocial Behavior

> Merely as an observer of natural phenomena, I am fascinated by my own personal appearance.... As a matter of fact, my upper lip is pretty fascinating by itself, in a bizarre sort of way.
>
> **—Robert Benchley**

Aggression among animals serves the purpose of survival (Southwick, 1970). Aggression against predators defends against being eaten. Aggression against members of the social class helps to establish the fittest individuals for reproductive success. The more aggressive animals become the dominant class, and the most aggressive is the preeminent individual. This animal has access to the most food and the most sexual partners. Hierarchies established in this way are stable and effective.

When these hierarchies are disturbed, new social orders must be established. When two small groups merge, the disturbance is short-lived, and the stress is limited. With crowding, however, the number of individuals involved is large, and the possibility of establishing a new, stable social order diminishes. In this case, emotional distress continues unabated.

Extreme aggression, when the animals are not in a crowded environment, can lead to contrary results. Male birds vocalize their intentions to mate and protect their territories through their songs (see Section 6.19.1, Acoustic Stimuli). Extremely aggressive birds sing extremely aggressive songs, as indicated by certain pitches and cadences. The most aggressive birds may be set upon by other males acting in concert to kill, injure, or drive away the super-aggressive males. In this case, the males must be aggressive to a certain extent, but not to a degree that they can't defend.

Laboratory studies on rats have shown that rats raised under crowded conditions exhibited perverse and antisocial behavior (Booth, 1976). The incidence of aggressive episodes increased markedly, which tended to stress the entire community. Those rats that were the hardiest avoided stress by establishing territories in areas that were easiest to protect. Some of the rats engaged in indiscriminate sexual behavior, while other rats became very passive, avoiding sexual encounters, play activities, and aggression. Still others went berserk and attacked almost all others. Social hyperactivity was common. Maternal care suffered, and a great many offspring did not survive to adulthood. Tolerance for new animals declined markedly.

6.11.2 Crowding in Humans

> ...engineers, whose job it is to harness nature, are required to take action, while scientists, whose task is to understand nature, are not.
>
> **—Lyle Feisel**

Research results on human crowding are much less clear, perhaps because of the complex nature of the human social environment. Crowded household and neighborhood conditions have been found to have little, if any effect on human health (Booth, 1976). However, crowded household conditions were positively related to the incidence and severity of stress ailments in males, but not in females. While some studies have shown that household density inversely affects reproduction rates, others have shown no such results (Booth, 1976). Crowded household conditions seem to have a small adverse effect on the physical and intellectual development of children. Children living in congested households are shorter, less heavy, and are sicker than their uncrowded counterparts. Crowded children are not as advanced in school achievement compared to their age peers, and are more often seen by school authorities. Crowding has more health and physical development effect on males, first born, and children over ten than on females, higher birth order, and younger children. Crowding affects the school performance of females more than males.

Crowded household and neighborhood conditions have little or no effect on neighborhood participation patterns. Some withdrawal symptoms have been observed in females affected by crowding. However, the more crowded a city, the less likely it is that people would help others in need (Levine, 2003).

The previous study results were obtained comparing conditions within the city, but did not compare urban against rural living (Booth, 1976). More dramatic differences might have been obtained if the social environments were more varied.

Green space can improve a child's attention span. Children who move to new homes with lots of room around them have been found to have gains in cognitive function.

6.11.3 Personal Space

> Scientific trail blazers are routinely treated roughly.... Apparently contempt is viewed as a perfectly normal and appropriate response to anyone who thinks outside the box.
>
> **—Richard Greenberg**

Crowding in humans is a multidimensional concept incorporating physical, social, and personal variables (Insel and Lindgren, 1978). People tend to be territorial, and consider an area around themselves as personal space. This *personal space* buffer zone was found to be within 0.93 m of the man for violent prisoners or 0.45 m for nonviolent prisoners (Insel and Lindgren, 1978). Personal space is dependent upon culture: Germans expand the space, and Mediterranean people contract it (Insel and Lindgren, 1978). When this space is violated, then aggressive behavior ensues. Even outside this space, physiological stress increases and blood catecholamines (biochemicals that sustain the "fight or flight" reaction) also increase.

The human personal space is made up of cubical cubits about 46 cm on a side (at least for British and Americans). When standing, a person is about 4 cubits high (Figure 6.11.1), stacked one on top of the other (Scheflen and Ashcroft, 1976). When sitting in a chair, the four cubits include one to the front (Figure 6.11.2). This is the space that is considered to be violated if another person comes too close. Unless the violator is accepted by the person in the space (as for a lover, for instance), the person owning the space will be uncomfortable, defensive, and try to flee. Particularly aggressive individuals may intentionally violate someone's personal space in order to dominate that person. Personal superiors (for example, a parent) can be more aggressive, dominating another's personal space (for example, a child) than the other way around.

FIGURE 6.11.1
A standing human occupies a personal space of four stacked cubits. (Redrawn from Scheflen, A.E. and Ashcroft, N., *Human Territories: How We Behave in Space-Time*, Prentice-Hall, Englewood Cliffs, NJ, 1976.)

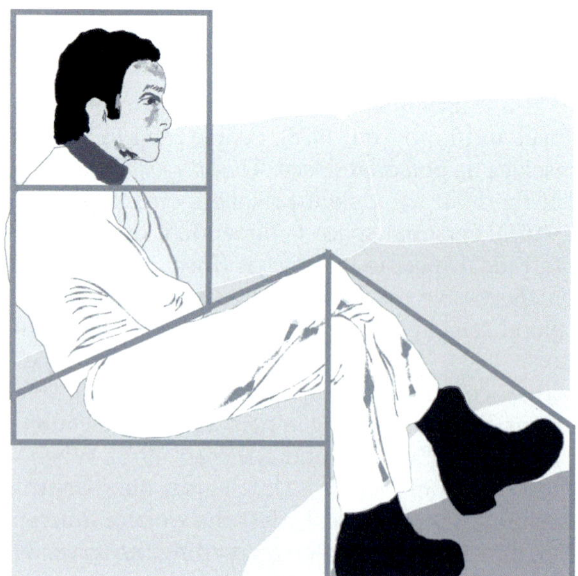

FIGURE 6.11.2
A sitting human occupies a personal space of four cubits, but in a bent shape. (Redrawn from Scheflen, A.E. and Ashcroft, N., *Human Territories: How We Behave in Space-Time*, Prentice-Hall, Englewood Cliffs, NJ, 1976.)

Even caterpillars are territorial. They scrape their teeth on leaves, and the resulting vibrations deter rivals.

The design of enclosures for animals or humans should account for the potential stress induced by crowding (Aiello and Baum, 1979). There is a need to provide a means for individuals to relieve the stress of close proximity. Room dimensions should be appropriately sized. Males need more personal space than females, and children need more space than adults. Indoor space requirements are greater than outdoor space requirements, perhaps because of the perception of limited escape routes inside. Spaces must appear to be controllable, defensible, and personal. They should be of light color or well-lit (at least for humans) because such spaces are perceived as less crowded and friendlier.

Spaces designed for human occupancy should support the characteristics of the occupants as well as the function of the space. Mental hospitals, for instance, need intimate spaces where patients can be alone if they wish (Sommer, 1969). These patients who are rarely physically isolated inside the institution can often be by themselves in parks or adjacent grounds during the day. Another example is school classrooms designed for learning. Increased distance among students fosters less interaction, and physical position (as at the head of a table) confers status on the child occupying that position (Sommer, 1969). Comfortable interaction may be required during part of the school day but not during others, so the physical seating arrangement must be able to be changed to reflect this difference. The study of these personal interactions is called *small group ecology*.

SPACES OCCUPIED BY HUMANS

Human territoriality is evident in this list of facilities. Their designs need to incorporate personal spaces for the occupants.

Schools	Stores
Airports	Homes
Churches	Apartment buildings
Dormitories	Bars
Prisons and Jails	Restaurants
Hospitals	Nurseries
Office Buildings	Parks
Libraries	Bathrooms

6.11.4 Sensory Overload

Try to learn something about everything and everything about something.

—**Thomas H. Huxley**

Sensory overload affects some people (Heller, 2002). Some people cut the tags from the inside of their clothes because the pricks or tickles they cause are intolerable. Others may stick a piece of paper over a blinking computer monitor icon. Or, some may run their air conditioners in November to drown out the sound of a car alarm on the street that never seems to stop. Add the sound of a dripping faucet, or nails on a chalkboard, or whispers in the next office cubicle. Each of these may overwhelm the tolerance of an individual and cause them to be either deeply disturbed or defensive in their actions.

Children raised in inner-city environments are constantly subjected to loud street noises and, as a result, have difficulties concentrating on school work, reading, and learning. Sleeping can be disrupted, and prevent sufficient deep sleep. Some adapt by learning to ignore external noises, but others quickly fall behind their peers in school and are handicapped for much of the rest of their lives.

Premature infants are very sensitive to light, sound, and touch before their nervous systems mature. When overstimulated by any of these senses, the infants may stop breathing and turn blue due to lack of oxygen.

Environmental noise can have both physical and developmental effects on both humans and animals (Cohen et al., 1981). Research has indicated three effects of high intensity noise. First is a narrowing of focus of attention to decrease the amount of information to be processed. This may enhance the performance of simple tasks but be detrimental for complex tasks. Second is a feeling of loss of control over the environment. This leads to depressed mood and lowered initiative. Third is physical arousal leading to long-term increased blood pressure and hormonal secretion. For every 10 dB increase in noise, systolic (heart contraction) blood pressure has been found to increase by an average of 10 mmHg. Diastolic blood pressure (during heart relaxation) increased 13 mmHg for each 10 dB increase in noise. Chronic exposure to noise, especially intermittent loud noises, can interfere with learning, behavioral responses, and physical health.

Experts suspect that animals they are trying to breed can also be distracted by external stimuli such as sound or sight. Serious attempts at reproducing these species often require isolation in comforting surroundings. The design of housing, working, health care, and growing facilities should consider the sensitivities of the individuals resident there.

6.11.5 Animal Spaces and Animal Welfare

> He has done like Orbaneja, the painter of Ubeda;...when he had scrawled out a misshapen cock, was forced to write underneath in Gothic letters, "This is a cock."
>
> **—Miguel de Cervantes**

Modern farming methods can easily disrupt animal social behavior (Mench and van Tienhoven, 1986). Overcrowding is common, single sex or uniform age groups prevent normal social contacts, and parent–offspring bonds are either disrupted or prevented altogether. The five basic freedoms proposed for animals raised in a production environment are the freedom to

1. Turn around
2. Groom
3. Get up
4. Lie down
5. Stretch limbs freely

When these are violated, physiological stress may result, although this is often difficult to prove.

Animals are inevitably kept by humans for many purposes, including for food, for companionship, and for other animal products. Given that, there is concern about ethical treatment of these animals. The World Organization for Animal Health has a Terrestrial Animal Health Code for animal welfare that states five freedoms:

1. Freedom from hunger and malnutrition, through ready access to fresh water and a diet sufficient to maintain full health and vigor.
2. Freedom from discomfort, through provision of an appropriately sheltered and comfortable environment.
3. Freedom from pain, injury, and disease, by prevention or rapid diagnosis and treatment.
4. Freedom to express normal patterns of behavior, through provision of sufficient space, suitable facilities, and company of the animal's own kind.
5. Freedom from fear and distress, ensuring conditions and treatments imposed avoid mental suffering.

Animal living and handling spaces should be designed with these provisions in mind.

6.11.6 Crowding and Disease

Your three best doctors are faith, time, and patience.

—Chinese Proverb

The spread of diseases among crowded BU is easier than among more isolated BU. Pathogens that are passed from one individual to another are less likely to be virulent than pathogens that require a vector or alternate host, under normal conditions. That is because individual-to-individual contact cannot happen as readily if the individuals are so sick that they cannot move around.

The crowding of individuals modifies this characteristic. Very virulent pathogens can spread from one individual to another if the second cannot escape the first. Thus, whereas natural selection in the wild would have been biased against the extremely virulent forms of disease, confinement of large numbers of individuals would select for the most virulent forms.

Crowding also exposes animals and plants to parasites. Cliff swallows in large breeding colonies suffer from high infestations of swallow bugs that feed on the blood of adults and brood alike (Brown, 2010). The infestation is worse for large bird colonies. Similarly, plants that occupy spaces with high population densities are more likely to attract animals that prefer to graze on them than plants of the same species sparsely located.

Agriculture is tending toward concentrated animal production practices. More than 100,000 chickens are housed in the same building, up to 10,000 hogs are raised together, and feedlots can contain 60,000 or more of cattle. When disease strikes these facilities, it can be devastating. Diseases such as avian influenza, cattle shipping fever, and hoof-and-mouth disease cause large economic losses and health concerns for humans.

6.11.7 Densities in the Wild

If we can really understand the problem, the answer will come out of it, because the answer is not separate from the problem.

—Jiddu Krishnamurti

In the wild, there is a natural population *carrying capacity* for an area. When food, nesting sites, materials, or other essentials become scarce, reproduction slows or ceases, disease incidence increases, or weakened individuals are caught and eaten by predators. Thus, there seem to be mechanisms for the regulation of stable population sizes. Crowding upsets at least some of these mechanisms.

As animals grow from childhood to maturity, they may migrate from their family units to find their own territories where others of the same species do not live. If there are others in the new region, competition for that space can become fierce. Eventually, the natural density of animals reaches a sustainable value.

For plants, too, there is a population density limit. For many plants in temperate climates, the maximum density is determined by access to light. Overcrowded plants tend to be tall and spindly, with small leaves. These plants have poor chances to survive additional environmental insults such as drought or insects (see Section 7.6.3, Population Densities).

Plants cannot move once they are established. As they age, they increase greatly in size. They begin life as small seedlings, and may grow to become trees as large as the *Sequoia gigantea*. For these organisms, a process of self-thinning occurs as they grow. The more robust individuals grow faster than the less robust, and crowd them out. The less fit individuals die.

In desert areas, water is often the limiting resource, and the greater distance between plants results from water nonavailability. Attempting to grow desert plants at greater densities than this will result in the death of many of them.

Microbes and body cells, too, have populations limited to certain numbers depending on nutrient or space availability (see Section 7.6.3). There is active communication among single cells, signaling when growth and reproduction is appropriate and when it is not. Under normal circumstances, when the carrying capacity of the environment has been exceeded, the number of cells dying exceeds the number of new cells.

Example 6.11.1: Vegetative Crowding Leads to Severe Forest Fires

Throughout history and before human dense settlements, periodic fires reduced or eliminated plant undergrowth as flammable tinder that burns with severe intensity. The results were fires that burned with low to moderate intensity and ecosystems adapted to these periodic fires.

As a consequence of human interventions, the natural cycle of low intensity fires was eliminated. When forest fires did occur, they subsequently became tragically severe, burning hotter and over a wider range, leaving few, if any, survivor trees.

Forest management also resulted in the harvesting of large trees, the ones best able to survive forest fires, and the dense growth of small trees and undergrowth (Winford et al., 2015). Vegetation crowding with the wet winters and dry summers of California subjected all trees to the risk of mortality from moisture stress and insect pests. As a result, there was even more flammable material accumulated in the forest to burn with severe intensity.

Hand thinning of small trees and underbrush in these forests reduces the vertical continuity of fuels and increases the spacing between tress that remain. During thinning operations, trees that are left can be selected to be those most likely to be fire tolerant. Low-intensity fires rarely kill the large trees that contribute most of the seeds for forest regeneration.

FIGURE 6.11.3
The forest on the left shows vegetative crowding with an abundance of flammable material that promotes intense forest fires. Trees in the forest on the right are more likely to survive a wildfire when it occurs. (Redrawn from Winford, E.M. et al., *Calif. Agric.*, 69, 150–155, 2015. Copyright © 2007, The Regents of the University of California. With permission.)

The results of these practices are forests better able to survive burning, ecosystems with less shrub regeneration and higher native species diversity, and reduced carbon emissions from subsequent wildfires (Figure 6.11.3). As an added benefit, the woody biomass removed during thinning can be chipped and used as a biofuel (Tittmann, 2015).

APPLICATIONS AND PREDICTIONS

1. Human students wanting to isolate themselves at a table will gravitate toward a position at the end of the table near a wall. Students wanting to discourage others from using the table will sit in the center of the table near an aisle.
2. Humans wanting privacy will face away from the door; humans wanting to defend the room will face the door.
3. There will be a smaller number of elephants than mice in any given area.
4. The density of trees in a forest will decrease as the forest matures.
5. Disease pressure on the occupants of a confined space will be higher than in an open area. Because of this, more drugs and antibiotics will be used. Antibiotic-resistant microbes will be most likely to develop in hospitals.
6. Homes with vaulted ceilings in comparison to a 2.5 m ceiling give a feeling of openness instead of constraint.
7. Crowding affects eating habits.
8. Dorm rooms for males should be designed differently from dorm rooms for females.
9. The merging of social groups permanently disturbs social order.
10. People living in areas of high noise levels will be less alert than people living in quieter areas.

6.12 Biological Units Are Affected by Chemical Stresses

> The difference between a deadly poison and a life-saving medicine can be very small; in fact, it is sometimes merely a question of dosage.
>
> **—R. E. Schulter**

Toxins are present in many forms. There are natural inorganic toxins, natural organic toxins, and, more important in modern life, synthetic toxic compounds.

Toxins usually interfere with metabolic processes. Targets of toxins are enzymes, metallic *cofactors* (substances that are essential for the catalytic activity of enzymes, binding to the enzyme only during the reaction), and *coenzymes* (organic cofactor molecules smaller than proteins that bond with the enzyme while the reaction is being catalyzed, and are not altered or consumed by the reaction). Hence, toxins are most effective where enzymatic presence is highest in the cell, such as the nucleus, mitochondria, lysosomes, endoplasmic reticulum, and plasma membrane (Reeves, 1981).

Most compounds entering the living organism are subject to metabolic transformations. If they have no nutritive value, they are called *xenobiotics* (foreign compounds). The metabolism of xenobiotics is intended to reduce their toxicity or facilitate excretion.

6.12.1 Toxicity

> Theory guides. Experiment decides.
>
> **—Anonymous**

The following always holds for substances foreign to the organism: increased molecular size is associated with increased molecular polarity, which is associated with increased ionization, which increases excretability by the vertebrate kidney, which gives decreased toxicity. Thus, larger molecular size means less toxicity (Reeves, 1981). Part of the reason that higher molecular mass compounds are generally less toxic than lower molecular mass compounds can be attributed to two traits: (1) greater mass compounds are generally less chemically reactive than lower mass compounds, and (2) greater mass compounds are physically larger and so cannot pass as easily through the permeable membranes of capillaries, cells, and organelles. Larger molecules are more likely to carry a net charge and can be more easily filtered by the kidneys.

There is an effective exception to the rule that smaller molecules are more toxic than larger molecules. In the case of enzymes, molecule activity depends on the configuration of the entire molecule. One part of the enzyme molecule may conform to one type of substrate and another part of the enzyme molecule fits with another reactant. Bringing those two substances together on the surface of the enzyme allows them to combine much more readily than they would otherwise. Splitting the enzyme molecule destroys its ability to catalyze its intended reaction. Therefore, breaking an enzyme molecule into smaller pieces makes each piece just as ineffective as denaturation.

Enzymes are not usually classified as toxins. However, antibiotics, many of which act directly or indirectly enzymatically, are toxic to bacterial cells. Cleaving antibiotic molecules removes their toxic effects. Beta-lactamases in certain antibiotic-resistant bacteria, for example, cleave penicillin molecules and render them ineffective as antibiotics.

Many nonpolar molecules (e.g., methane and ethane) are biologically inert and require no detoxification. But, if a nonpolar molecule is biologically active (e.g., benzene and carbon tetrachloride), it is not easily detoxified (Reeves, 1981).

There are four general classes of biotransformations available to deal with toxic substances:

1. *Oxidation*. This is a common metabolic response, and involves the addition of oxygen to the molecule or the removal of electrons from an ion. The rate of oxidation is limited, however, so continued ingestion can result in blood accumulation. Alcohol oxidation, for instance, occurs only at 4–8 g/h.
2. *Reduction*. This is the change in a molecule by loss of oxygen, addition of hydrogen, or gain of electrons. This is not a common means of detoxification, because the loss of oxygen reduces the molecular mass of the compound (see above).
3. *Degradation*. Some compounds may require cleavage before further metabolism. The results of degradation are often more toxic than the original compound (see above).
4. *Conjugation*. Combining the toxin with common substances such as amino acids or carbohydrates makes molecules larger and detoxifies them. This is the most important detoxification mechanism (Reeves, 1981).

Chemotherapy drugs are typically small enough that they can move through the body by means of the bloodstream. Thus, all organs are exposed to these toxic substances. Likewise, once inside tumors, their stays are relatively short because they are small and mobile. Macromolecular drug carriers can be linked to chemotherapeutic drugs to reduce their toxicity to nontarget tissues and to lengthen their residence time within tumors. If the macromolecule is soluble, then the drug does not have to be mixed with noxious substances to make it more readily carried in the blood.

BITTER BODYGUARDS

Bitter taste receptors serve a useful purpose in the tongue, where they guard against ingestion of toxic substances that usually have a bitter taste. Identical bitter taste receptors are also located in the nose, heart, lungs, intestines, and other organs of the body, but are not neurally connected to perceptive regions of the brain and so do not contribute to taste. Instead, these bitter receptors serve a protective purpose by detecting chemicals emitted by dangerous gram-negative bacteria and responding with the release of nitric oxide that kills nearby bacteria (Lee and Cohen, 2016). Nitric oxide also stimulates hair-like cilia in the respiratory system to beat vigorously to expel the bacteria. Sensors sensitive to sweetness, also located throughout the body, release calcium that dampens the effect of nitric oxide to provide feedback and prevent an overreaction to the bacteria. People who are particularly sensitive to bitter flavors in food have more active antibacterial activity from their bitter receptors than do people who are not so sensitive.

6.12.2 Dose–Response

> All things are poison, and nothing is without poison; only the dose permits something not to be poisonous.
>
> **Paracelsus (500 years ago)**

Toxic effects depend upon dose. Two typical dose–response curves are given in Figure 6.12.1 (Timbrell, 1995). Plotted is the percent response from none (0%) to maximal (100%) against the dose on a log scale. Compound A can be seen to have an immediate effect. There is no dose of compound A that has no effect; consequently, there is no safe dose for compound A.

Compound B has a different shaped response. There is a range of dosages that result in no response or a response lower than a threshold amount; consequently, there are safe dosages for compound B.

Cellular responses to a mutagen are diagrammed in Figure 6.12.2. There are many possible outcomes, all related to dosage.

One point of extreme interest with any toxin is whether or not there is a safe dose for that toxin. The answer to this question is very important technically and economically because toxins that have no safe dose must be eliminated completely for perfect safety. A zero-level reference is never totally achievable because compliance depends on the detection sensitivity of instruments, and that usually continues to increase. Thus, a toxin thought to be totally eliminated in the present may be found in the future because monitoring instruments have changed.

To make the issue even more complicated, toxic effects can change depending on the method of administration and on the species used. Most dose–response information about toxins have been obtained from species other than humans, and this always opens the possibility that the other species will not respond exactly as humans respond. If the dose–response data is to be applied to a nonhuman species, then necessary experiments can be conducted to obtain the required information. If the target species is humans, however, then the results will nearly always have some uncertainty associated with them.

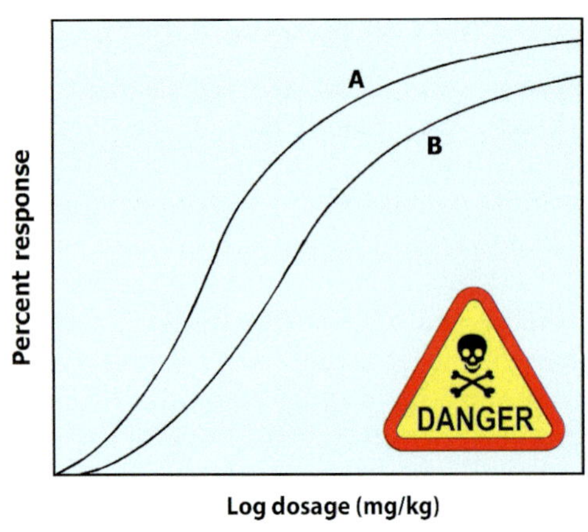

FIGURE 6.12.1
Two typical dose–response curves. Compound A has no threshold value.

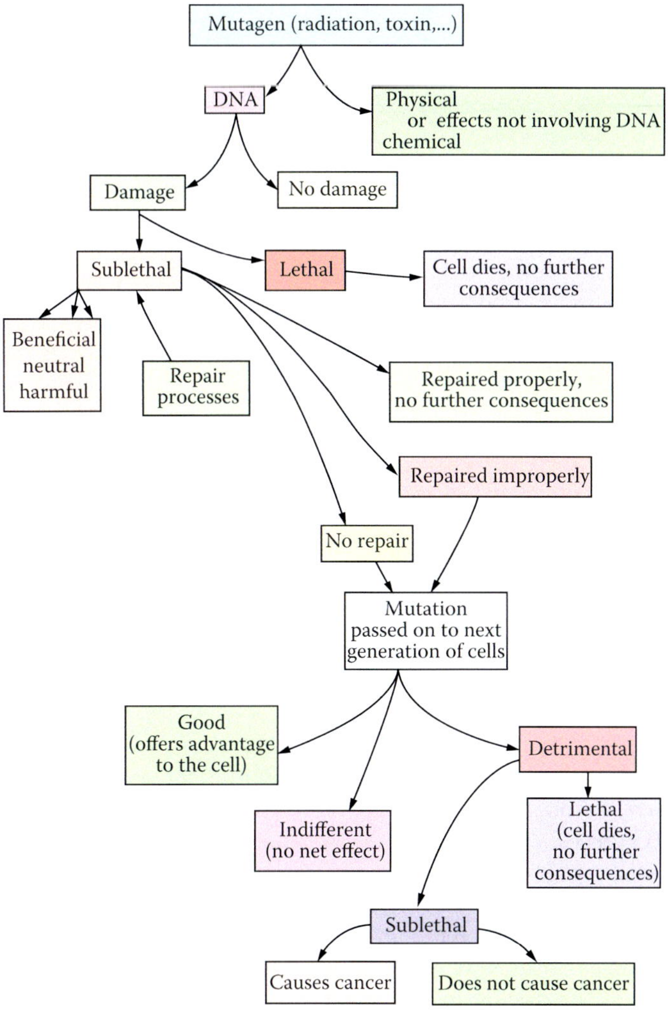

FIGURE 6.12.2
Diagram of responses of a cell to a toxic substance that can cause cancer.

One means to compare the toxicities of different compounds is to compare the dosages that result in the death of 50% of those exposed. These dosages are called the LD_{50}. Representative LD_{50} information is given in Table 6.12.1.

There are two competing hypotheses relating response to dose (Karam, 2003). The first is the linear, no-threshold (LNT) hypothesis that suggests that all exposure is potentially harmful and that the risk of harmful effects is directly proportional to the dose received. This means that the risk of toxic effects can be extrapolated from higher levels down to levels that are vanishingly small (use of the word "risk" here means that measurements are made statistically, and that the only certainty is a probability of occurrence, see Section 4.2.4, Statistics). The LNT hypothesis has been used to calculate expected cancer rates from extremely small radiation exposures. For example, if the risk from a given radiation exposure is five additional cancer deaths for every 10,000 person-rem, then exposing 10,000 people to 1 rem (a unit of radiation exposure) each should result in

TABLE 6.12.1
Lethal Dosages for Common Substances

Compound	LD_{50} (mg/kg body mass)
Acetaminophen	186
Arsenic trioxide	15
Aspirin	160
Benadryl	19
Botulinum toxin	0.004
Caffeine	140
Codeine	7
Cyanide	4
Ethanol	7,000
Lead	450
Nicotine	53
Phenobarbital	20
Salt (sodium chloride)	12,000
Sugar (glucose)	30,000
Valium	47
Water	114,000

an extra five cancer deaths among those people. Using LNT, exposing one million people to 10 millirems each should also lead to five additional cancer deaths. This is like saying that, if a 1000 kg rock will crush someone, throwing a million 1 g rocks at a million different people will crush someone. Clearly, the calculation may be conservative but not necessarily realistic.

The second hypothesis states that there is a threshold below which no harm accrues, and, further, that lower doses may even be beneficial. When a toxin at high levels is beneficial at low levels, it is called *hormesis*. Substances exhibiting hormesis are, for example, vitamin D, selenium, aspirin, table salt, water, and, for plants, fertilizer. There may even be a hormetic effect for low-level radiation because this might provide the stimulus for the heightened activity of DNA repair mechanisms (Karam, 2003).

Taken singly, toxicity of a substance is usually related to dose. However, cumulative synergistic effects of sub-acute levels of several toxins acting together can enhance the toxicity of each contributor. The total exposure of toxic chemicals, especially in cancer studies, is called the *exposome* (Traynor et al., 2016).

6.12.3 High Doses

> A chief criterion for the selection of a correct hypothesis…seems to be the criterion of beauty, simplicity, or elegance.
>
> —**Murray Gell-Mann**

Nutrients can also have toxic effects in large dosages (Figure 6.12.3). The figure shows that at low levels, increasing the amount of nutrients results in increased growth. Then follows a range of nutrient amounts where the nutrient no longer limits growth, and growth does

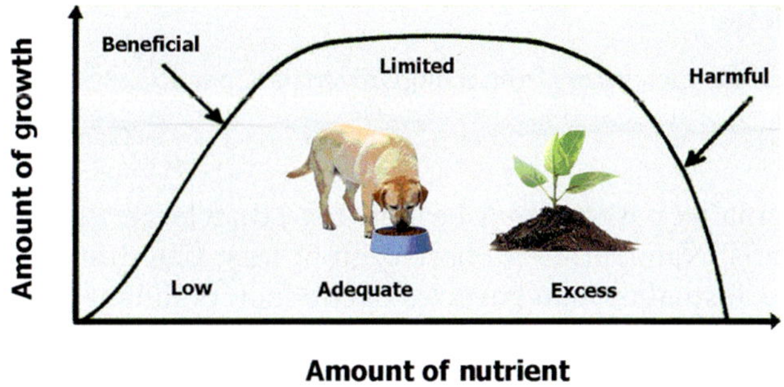

FIGURE 6.12.3
Nutrients can also be toxic in large amounts. (Redrawn from Russell, E.W., *Soil Conditions and Plant Growth*, John Wiley & Sons, New York, 1961.)

not depend upon the amount of supplied nutrient. At very high nutrient levels, there is a deleterious effect that adversely affects growth (Russell, 1961).

An example of a stimulant that is also a toxicant is the pesticide strychnine. This alkaloid works by neutralizing glycine, an amino acid responsible for transmitting inhibitory nerve impulses to muscles. Without inhibition, the muscles contract but do not relax. Skeletal muscles become hyperexcited and contract continuously. Limbs become stiff and respiration stops. Animals poisoned with strychnine suffocate and die. This poison is toxic to fish, insects, mammals, and birds.

Some chemicals in the environment act as estrogen mimics: they have effects similar to the female hormone estrogen when animals at many different levels are exposed to them. These effects include the promotion of certain types of cancers, sexual dysfunctions, genital defects, obesity, attention-deficit disorders, and the reduction of male secondary sexual characteristics (Hinterthuer, 2008).

Bisphenol-A is frequently used in certain types of plastics and is one of these estrogen mimics. But, unlike more conventional toxics, hormones, and hormone mimics, are sometimes not more toxic at higher doses. The body's response to high doses of hormones is sometimes to stop responding.

6.12.4 Metabolic Wastes

> Carbon dioxide emitted into the air has a 200-year half-life in the carbon cycle.
>
> —**Eric Loewen**

Metabolic products themselves are often toxic to the BU that produced them. Thus, alcohol is toxic to yeast cells, carbon dioxide is toxic to humans, and ammonia is toxic to birds (see Section 6.4, Wastes). It is best to remember that metabolic products need to be removed from BU in cultures, bioreactors, greenhouses, barns, and hospitals in order that the BU not be affected by their own toxic byproducts.

Wastes may also contain other substances with toxic environmental effects. Manure deposited from cattle fed antibiotics drastically disturbs the bacterial and fungal populations in surrounding soils. Microbes showed higher levels of stress, and many microbes appeared with antibiotic resistance. Perturbations in microbial populations can drastically change biological properties and fertility of the soil (Wepking et al., 2017).

6.12.5 Nanoparticles

The nanohyperbole meter runs from nanopanic to nanopanacea.

—**George M. Whitesides**

As a general rule, smaller particles may be more toxic than larger ones (Adams, 2007; Karn and Matthews, 2007). Nanoparticles, those with at least one dimension between 1 and 100 nm, can be much smaller than particles of common pollutants, and all are manufactured. Nanoparticles can be used to deliver medicines, for sun blockers, to strengthen materials, as antimicrobials, as environmental cleansers, and in electronics, energy and agricultural materials, and in coloring agents. However, they also may cause physical harm to those who contact them (Wu, 2010; Sadik, 2017). Nanoparticles have very high ratios of surface area to volume. This can affect their chemical reactivity, electrical conductivity, melting point, and color. Their chemical and physical properties are size-dependent, and so can be manipulated for particular purposes. Even shapes of nanoparticles can affect their interactions with living cells.

Engineered nanomaterials have high surface catalytic activity. Some nanomaterials can pass through cell membranes to directly interact with cellular components and cause adverse biological effects. Inhaled nanoparticles can become distributed in most organs, including the brain. They can cause cellular oxidative stress, which leads to genotoxicity (Barrie et al., 2017). Nanoparticles can also be used to kill cancer cells after they are specifically taken up by the cancer cells and become lethal when irradiated by light.

Nanoparticles that have penetrated the cell interact with free radicals in several ways. Metal-based nanoparticles can combine chemically with hydrogen peroxide naturally occurring within the cell to produce hydroxyl free radicals. Other nanoparticles attract free radicals in the cell and concentrate them in certain locations. Free radicals can damage DNA and cause mutagenesis.

The reason that some nanoscale materials become a lot more reactive than the same materials at either the macroscale or atomic scale is that there is a greatly increased fraction of atoms exposed on a nanoscale particle surface, and these atoms become sites for chemical binding. Smaller particles at the atomic scale have only few atoms with a very limited capacity to accommodate linking to functional chemical groups. Likewise, a macroscale particle has most of its atoms hidden in its interior, away from the surface where chemical bonding takes place (Murashov and Howard, 2015).

Nanoparticles have properties not exhibited by larger particles made of the same materials. Bulk gold, for instance, is inert; nanogold catalyzes chemical reactions. Some pollutant nanoparticles can penetrate the skin and enter the bloodstream. Carbon nanotubes in the lungs of mice and rats trigger areas of inflammation. Nanoparticles can pass through the nose and into the brain through the blood-brain barrier membrane. Nanoparticles passing into the brains of bass via the gills trigger an enzymatic reaction called oxidative stress. Nanoparticles are too small to settle in water or air, and can be toxic to smaller animals. Aluminum oxide nanoparticles slow plant growth. If bacteria ingest nanoparticles, they could accumulate up the food chain.

6.12.6 Toxins Used as Defenses

> We are engineers, and we should remember two things: first, there are diminishing returns in trying to get past the 90% point instead of just doing the job, and second, we can always work to change the constraints.
>
> **—David J. Dewhurst**

The biological effects of toxins include damage to an organ system, disruption of a biochemical process, or disturbance of an enzyme activity (Schiefer et al., 1997). Many plants and animals purposefully use toxins to protect themselves against predators and competitors. Thousands of chemicals have been isolated from plant tissues, and many of these serve to defend the plants. *Alkaloids* are basic organic compounds containing nitrogen that have poisonous and medicinal properties. Examples include nicotine, quinine, morphine, and cocaine (Hale et al., 1995). *Terpenes* are unsaturated hydrocarbons of plant resins and oils that are toxins and feeding deterrents to many herbivores (Taiz and Zeiger, 1998). *Phenolics* are derivatives of carbolic acid with antimicrobial activity. The phenolic compound lignin is one of the main constituents of plant cell walls (along with cellulose) that cannot be digested by herbivores without help from internal microbes. Lignin and cellulose may serve, among other things, as a first line of defense against indiscriminate herbivore browsing. Living trees resist beetle attacks by increasing the flow of potentially toxic resin, as well as through the other mechanisms of resistance (Birch, 1978).

SEARCHING FOR NEW ANTIBIOTICS

The key to successful control of pathogens is redundancy, simultaneously using several different mechanisms to kill or weaken target microbes. Most antibiotics, however, are used alone with the disadvantage that resistance can develop over time. Thus, there is a constant battle to find new antibiotics to cure microbially based human, animal, and plant diseases.

Prior searches for effective antibiotics have centered on natural products appearing in soil or aquatic microbes, plants, and animals. Fungi and bacteria produce natural antibiotics to protect themselves from infectious agents. These can be the sources of many of the antibiotic drugs that we humans use to fight disease. This is how the antibiotic penicillin was discovered. These same fungi and bacteria also produce compounds that make them immune to their own antibiotics.

It has been found that the human microbiome can be a rich source of antibiotic compounds (Fischer, 2016). The body is teeming with microbes with the ability to produce antibiotic compounds as they struggle to compete against other microbes that have the potential to dominate their internal or external environments at the expense of more benign organisms. These antibiotic compounds comprise a host of small molecules, including thiopeptides, and ribosomal and nonribosomal peptides, which

(*Continued*)

SEARCHING FOR NEW ANTIBIOTICS (Continued)

enable these microbes to survive challenges from other, less friendly bacteria, viruses, yeasts, and fungi. Researchers with the ability to quickly scan the genomes of bodily microbes, and to identify those genes encoding for specific peptides, can discover new antibiotic compounds with an efficiently not achievable previously. *Pharmacogenomics* is the name given to the use of genetic information to forecast drug effects in patients.

An example of the possibilities when using this search technique was a study in which about 25% of microbes isolated from men's beards produced antibiotics capable of killing other bacteria, including MRSA (Methicillin-Resistant Staphylococcus aureus). Growing a beard may protect men from bacteria and their infections (Wakeam et al., 2014).

Another example is maple syrup. Maple syrup, produced by boiling down the sap from sugar maple trees, is a rich source of phenolic compounds. Phenolics produced by plants can be medically beneficial to humans Maple syrup extract has been found to enhance the effectiveness of antibiotics against bacteria (Vimal et al., 2015).

Strategies to control infections are nearly as old as human history. A remedy given in a 1000 years old English document was discovered to be effective against stubborn microbial infections (Harrison et al., 2015), and has been shown likely to be more effective than modern antibiotics for treating the very difficult to control MRSA infection. The remedy consists of a mixture made from wine, garlic, cow bile, and copper. When tested against microbes in the lab, this concoction killed up to 99.9% of the bacteria it was tested against.

Each of these classes of compounds is used by plants as protection against predators. Tropical plants seem to contain more toxic alkaloids than do temperate species (Molles, 1999). Despite this, tropical herbivores remove approximately 11%–48% of leaf biomass in tropical forests, while in temperate forests they remove only about 8%. This higher level of grazing in tropical areas has apparently produced more intense selection pressure for plants with chemical defenses (see Section 6.17.1, Plants and Herbivores).

Some animals, as well, use toxins to defend themselves. Some poisonous toads and frogs can synthesize toxins. Others acquire their chemical defenses from plants they eat. The Monarch butterfly is probably the best example of this. The larvae eat milkweed leaves and the toxins are retained in their bodies. The toxins are still there after metamorphosis from larva to butterfly. Birds that eat Monarch butterflies regurgitate their prey and quickly learn to avoid others of the same species.

Many toxic substances have a bitter or bad taste, and taste, of all the senses, has a close learned association with illness. Animals, when presented with a new food, will only nibble at the substance. If they subsequently develop illness, they will avoid the taste of that food forever. This behavior protects them from dying as a result of poisoning. Likewise, if humans eat some food, and become ill afterwards, then they likely develop a distaste and an aversion to that food, whether the food was the actual cause of the illness or not. This close association between taste and illness is learned rapidly and can last a lifetime. Sight or sound does not have the same response (Davidson and Riley, 2015), but emotional state can affect the perception of food taste (Noel and Dando, 2015). Some food providers know about this link. Sights, smells, and sounds can affect perception of flavor, and this has led to targeted programming of background music in eating establishments to promote certain kinds of foods (Spence, 2017).

EVOLUTION AND CYANIDE TOLERANCE

(excerpted from Lane, 2005)

You may have heard about plants with the potential for cyanide toxicity, such as young sorghum-sudangrass, apple seeds, and wild cherry. You may have also heard that very low levels of cyanide are usually not dangerous because animals can detoxify these molecules.

Have you ever considered *why* animals—livestock and humans—have the ability to detoxify cyanide?

When cyanide is absorbed across the gut wall, the blood carries it to the liver and other cells, where a specialized enzyme system converts cyanide into the non-toxic molecule *thiocyanate*, which the kidneys then dump into the urine for excretion. But enzyme systems are expensive. It costs energy and nutrients to manufacture all these enzymes, and evolution is notoriously frugal about overspending. So why do animals spend nutritional capital just to build extra metabolic equipment that they would rarely use?

Because cyanide is actually quite common in the plant world. Cyanide discourages certain animals like slugs, weevils, and grasshoppers, and plants use cyanide compounds as weapons in their ongoing biochemical warfare to stay alive. The toxic plant species are well-known, but a surprising number of other species can contain low levels of cyanide, including forages like white clover and common vetch and even some human foods like cassava and lima beans.

But cyanide is extremely toxic to animals. In animal cells, cyanide irreversibly binds to a critical enzyme called *cytochrome oxidase*, which controls the last step in the extraction of energy from carbohydrates using oxygen. Cyanide is so devastating because it completely blocks the main energy-producing sequence in a cell. That's why animals who die of cyanide toxicity have blood that's bright cherry red—because their cells cannot use the oxygen in the blood, and thus the hemoglobin cannot get rid of it and stays bright red.

Thinking about nutritional evolution, animals like to eat. If animals didn't have mechanisms for detoxifying cyanide, a lot of animals wouldn't survive very long. This includes grazers, browsers, and species like omnivorous humans who eat plants as well as animals. Eons of evolution therefore exerted relentless pressure. Genetic lines that developed the equipment for living with low levels of cyanide did a better job of reproducing than genetic lines that lacked this equipment. Maybe not enough equipment to allow us to consume cyanide truffles, but enough to keep us alive when we encounter the occasional molecule.

6.12.7 Toxin Tolerances

A moment's insight is sometimes worth a life's experience.

—Oliver Wendell Holmes, Sr.

Tannins (phenolic compounds) in acorns disrupt digestion, and make raw acorns unable to be eaten. Over the years, some groups of people have been able to develop means to deal with this toxin. The prehistoric residents of California mixed unprocessed acorn meal with a certain kind of red clay to make bread. The clay bound enough of the tannin to make the bread palatable. Other groups boiled the acorns to extract the tannin (Nesse and Williams, 1996). This illustrates the point that for every biological measure of defense, there is a BU somewhere with a countermeasure. Coping with new environmental chemicals may involve the induction of new enzymes or the modification of existing ones. Over time, there is a gradual elimination of those incapable of adjusting.

It is reasonable to surmise, therefore, that toxins to which BU have been exposed over many generations can be accommodated. It is new environmental toxins that should cause concern.

POISONOUS TO PETS

There are some common foods that humans eat that are dangerous to their pets. These include:

- Bread dough—may swell and produce alcohol in the stomach of dogs and cats
- Onions—cause hemolytic anemia (lysis of red blood cells) in dogs and cats
- Grapes and raisins—cause kidney failure of dogs and cats
- Avocado—produces cardiac tissue damage, respiratory distress, and mammary gland damage in dogs, horses, rabbits, fish, and birds
- Chocolate and cocoa—induces irregular heartbeat, irritates the gastrointestinal tract, and triggers epileptic seizures in dogs
- Coffee grounds—similar to chocolate

There are a number of common chemicals that can poison your pets. These include:

- Flea powder with permethrin (for dogs)
- Antifreeze with ethylene glycol (tastes sweet)

There are medicines for humans that cannot be given to pets. These include:

- Aspirin (causes risk of bleeding)
- Anti-inflammatory drugs, ibuprofen, naproxen, acetaminophen (toxic)

And there are many common plants that can be poisonous, too, if eaten. These include:

- Aloe Vera
- Daffodil
- Poinsettia
- Tomato leaves

and many others (ASPCA, 2002).

Embryonic and fetal tissues may be harmed by lower concentrations of toxins than are adult tissues (Nesse and Williams, 1996). As in Figure 6.12.4, toxin vulnerability is greatest around the first trimester of human pregnancy.

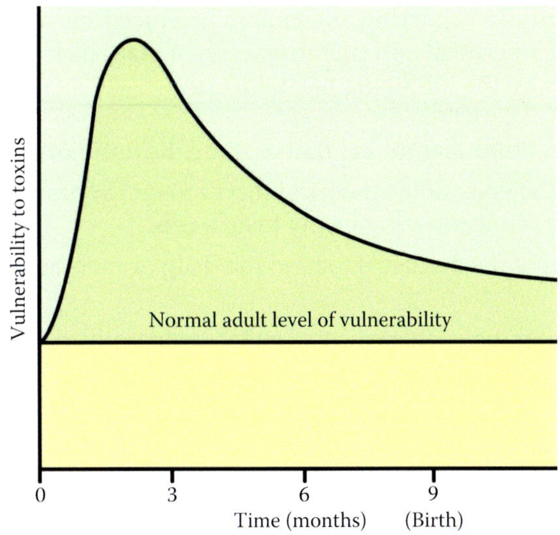

FIGURE 6.12.4
Vulnerability to toxins in the unborn. (Redrawn from Nesse, R.M. and Williams, G.C., *Why We Get Sick: The New Science of Darwinian Medicine*, Vintage Books, New York, 1994.)

To illustrate the extreme specificity of interactions between toxins and their consumers, which sometimes may occur, consider the *Veratrum californicum* plant. If a female sheep consumes this plant on day fourteen of her pregnancy, her lambs may be born with *cyclopia*, having one eye in the middle of the forehead. She can graze this plant before and after day 14 with no ill effects (Gessert, 2003).

It has been hypothesized, therefore, that the nausea that accompanies early pregnancy (morning sickness) in humans, cows, and perhaps other mammals is meant to protect the fetus from ingested toxins (Nesse and Williams, 1994). Strong tasting foods often contain toxic substances already touched upon (alkaloids, terpenes, and phenolics), and these are particularly nauseating to newly pregnant women. Because the extra energy demands made by the fetus at this stage in pregnancy are not great, there is little cost to protecting the unborn child by restricting food intake and potential exposure to food-borne toxins.

The same water barrier present on the outside of an organism that lives in the air (see Section 6.1) also can protect against the absorption of toxins from the environment. The integrity of this barrier layer is complex and must be maintained to be effective. The use of soap and water for the emergency cleanup of skin exposed to chemicals might actually increase skin absorption in many instances (Cummins, 2004). Solvents are sometimes employed to move certain chemicals past this barrier layer. For example, Dimethyl Sulfone (DMSO) can be used to carry drugs across the skin for dermal administration, or other solvents can be used to administer herbicides to plants.

Solubility differences can lead to some interesting results. People who eat high fat diets have increased lipid levels in their blood. Solvents in the environment can endanger health and often comprise nonpolar liquid and vapor molecules. These are differentially more soluble in lipids than in plasma. Thus, it has been reported that people exposed to environmental solvents who also eat high fat diets can carry elevated solvent levels in their blood (Cummins, 2004).

Antibiotics are generally toxic to microbes (although the general term "antibiotic" can be used as well for substances to combat nonmicrobial pathogens and others). Antibiotics are often based upon naturally occurring substances produced by some BU to combat other BU, and often disrupt microbial cell membranes or metabolism. Microbial immunity to antibiotics can be built by the following mechanisms:

1. Changes in the cellular membrane that exclude the antibiotic.
2. Improving the efficiency of biochemical mechanisms to pump the drug out of the cell and reduce its concentration below toxic levels.
3. The development of mechanisms to bind the drug or metabolize it into something less toxic.
4. Changes in enzymatic pathways that either overproduce enzymes that are the targets of the drug or change to alternate metabolic enzymes (Figure 6.12.5).

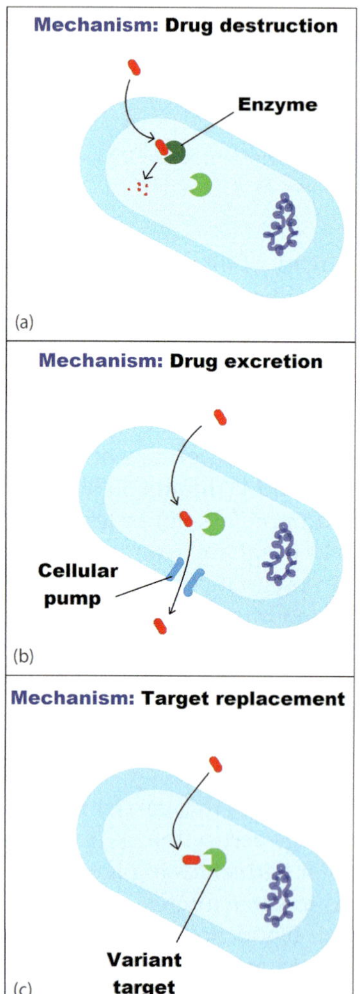

FIGURE 6.12.5
The three common forms of antibiotic resistance development are: (a) deployment of an enzyme that destroys or disables the antibiotic drug, (b) pumping the drug back through the cell wall before it can be effective, and (c) replacement of the target protein with a variant that is not affected by the drug.

THE FIRST BIOLOGICAL THERAPY

The first biologically based drug therapy was very expensive when first used around 1940. When used clinically, the drug was transported in armored trucks with police escort. The drug was even recovered from the urine of patients to whom it had been administered because that procedure was less expensive than manufacturing it anew. The name of that drug is penicillin, which is now cheaper than the glass vials in which it is sold (Amábile-Cuevas, 2010).

MYSTERIOUS FOAL DEATHS IN KENTUCKY

The spring of 2001 will be remembered for a long time in the horse country of Kentucky. That was the spring that baby horses (foals) died mysterious deaths.

There were precious few clues to the deaths. There were no poisons in the water, nor in the grain eaten by the mothers (mares). The grass was normal, and the air was normal. No unusual pesticides were sprayed to anyone's knowledge. All anyone really knew was that the $1 billion-dollar Kentucky horse industry was reeling.

The answer finally turned out to be a rare coincidence of unusual weather and a cyclical peak in caterpillar reproduction. The population of tent caterpillars reached a 10-year high and they munched on wild cherry leaves to gorge their hunger.

Wild cherry leaves contain substances that can form naturally occurring cyanide. All livestock owners know about wild black cherry leaves that are safe for animals to eat when they are green and again when they are completely dry. But in the wilted stage they can be deadly.

Unusual weather in March and April increased the potency of young cherry leaves. However, mares rarely had the opportunity to eat cherry leaves. Horse owners would have eliminated them from pastures long ago.

This is where the hungry caterpillars came into the picture. They ate the cherry leaves and migrated to other sites, carrying the toxic chemicals with them. Mares ingested the abnormally abundant insect larvae or their feces from grass or water tanks, poisoning their babies in the womb.

(*Continued*)

MYSTERIOUS FOAL DEATHS IN KENTUCKY (Continued)

This incident emphasizes a broad imperative for humans to understand better the dynamics of nature, and to use that understanding to craft a solution to problems such as this. Mystery solved: now should we spray chemical pesticides, saw down all remaining wild cherry trees, or what?

6.12.8 Toxin Concentration

> Our individual and collective failure to comprehend and act on the connectedness of things is pervasive, systemic, and threatens our health and long-term prosperity.
>
> **—David Orr**

As toxins are passed from one trophic level to the next in the food chain, they are often concentrated. Many of these materials accumulate in fat tissue and are not detoxified or eliminated from the bodies of those ingesting them. Many more organisms must be eaten than are doing the eating at each trophic level (Figure 6.12.6). Thus, toxins from many BU accumulate in the body of each higher trophic level BU. Eventually, chemicals that were at relatively benign levels at lower trophic stages reach harmful levels at higher stages. Such was the case with dichlorodiphenyltrichloroethane (DDT), which is a very effective insecticide. Unfortunately, it concentrates to such levels that birds of prey do not deposit enough calcium in the shells of their eggs. When they attempt to incubate their eggs, the eggs break, and drastically lower the birds' reproductive rates.

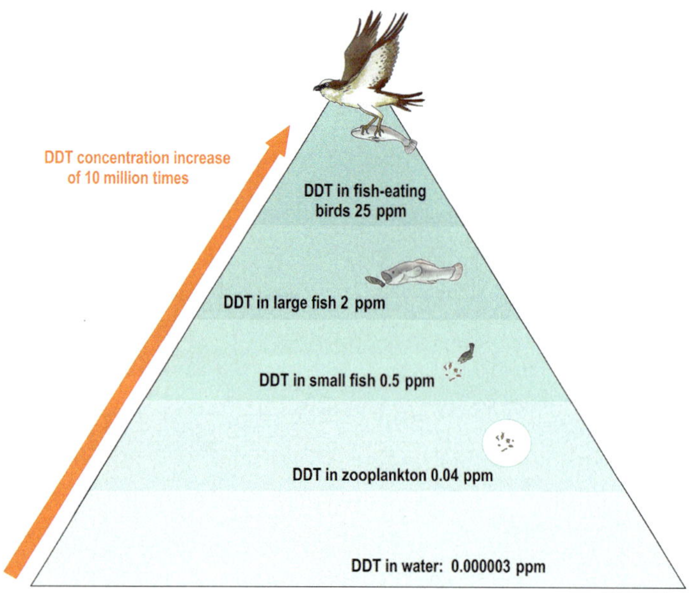

FIGURE 6.12.6
Concentration of DDT in the food chain. (Redrawn from Campbell, N.A. et al., *Biology*, 5th ed., Addison Wesley Longman, Menlo Park, CA, 1999.)

TABLE 6.12.2

Mercury in Seafood Tends to Move Up the Food Chain

	Mercury Concentration (ppm)[a]	
Species	**Average**	**Range**
Swordfish	1.00	0.65–3.73
King mackerel	1.00	0.10–1.67
Shark	0.96	0.05–4.45
Tuna (fresh or frozen)	0.32	ND[b] to 1.3
Pollack	0.20	ND to 0.78
Tuna (canned)	0.17	ND to 0.75
Catfish	0.07	ND to 0.31
Salmon (fresh or canned)	ND	ND to 0.18
Shrimp	ND	ND

[a] Parts per million.

[b] Not detectable.

Elemental mercury, an inactive by-product of plastic production, was once routinely dumped into rivers and the sea in insoluble form. Bacteria, however, converted the muddy waste into methyl mercury. This very toxic compound has accumulated in the bodies of water-dwelling organisms and in the bodies of humans who have eaten contaminated fish (Table 6.12.2). At present, most mercury pollution comes through deposition of airborne residues from coal-burning power plants. This mercury falls on rivers, lakes, and forests, and accumulates up the food chain. Mercury compounds are highly potent neurotoxins that impact the function and development of the central nervous system in both people and wildlife.

Toxins that produce the most difficulty for humans and for other BU are those of the most recent origin. As they concentrate at higher trophic levels, they provide selection pressure to accommodate to them, but unfortunately this accommodation includes the loss of individual lives of those who cannot adjust.

Perfluorochemicals (PFCs) are a class of man-made substances consisting of long-chain carbon-fluorine bonds that persist in the body and in the environment for many, many years. They are of concern because they can interfere with functioning of the immune system (Schmidt, 2017).

Drugs and other biochemicals used by humans do not disappear once used. Whether by excretion or by washing, many more of them end up in wastewater than was originally realized. Municipal wastewater treatment is designed to deal with solids and microbes, but not most biochemicals. Thus, substances such as antimicrobials, hormones, medicines, and detergents pass through treatment intact and find their ways into the environment either in treated water effluent or municipal solid waste. The water is returned to streams or ground water and the solids are often spread on the land (Brodie, 2007a). Traces of amphetamines from urban waste water have been found in area streams. They had many extensive effects, including stunting algae photosynthesis, changing stream biofilm bacterial populations, and affecting insect growth (Lee et al., 2016). Antibiotic residues in wastewater can lead to antibiotic resistance in pathogenic bacteria.

What makes this a cause for concern is that these biochemicals can be concentrated in the process of municipal wastewater treatment. Many of them are fat soluble, and so can accumulate up the food chain. The presence of antibacterials and microbes in the same mix of wastewater could easily lead to the formation of the resistant strains of bacteria. There is also concern that people are being exposed to medicines originally taken by others. This problem will only get worse as population growth puts pressure on clean water sources, available land, the increased use of drugs, medicines, and disinfectants, and greater reliance on recycling.

6.12.9 Endocrine Disruption

> Humans have evolved rapidly as a dominant species because of three characteristics: the desire to know, the ability to question, and both desire and ability to adapt the environment to suit ourselves.
>
> **—Evelyn Tiffany-Castiglioni**

A chemical that either mimics or blocks the action of natural hormones is called an *endocrine disrupter*. Most of the known disrupters modify either the production of the female sex hormone *estrogen* or the abilities of cells to respond to estrogen. Far fewer compounds interfere with the production and function of male sex hormones (*androgens*, including *testosterone*). There are naturally occurring phytochemicals that act this way, usually enhancing estrogen production (for example, soybeans produce such products). By far the most troublesome and potent endocrine disrupters are man-made compounds functioning as pesticides, plastics, and industrial chemicals.

Endocrine disrupters are present in much of our environment, and the threats of these compounds seem to be worsening as more and more new biochemicals are being used. Many of these compounds are present in drinking water; it has been found that wastewater treatment does not remove many of them, and low concentrations of medicines, pesticides, food preservatives, and industrial chemicals have been discovered in ground water and municipal water.

All of the endocrine disrupting compounds have not been identified, and determining which compounds have this effect is a daunting task. There are nearly 90,000 chemicals that need to be tested, but common chemical structural characteristics have not been found to make the identification easier. To make matters worse, endocrine disrupters may have opposite effects at low and high doses, or may be most potent at low doses (Vandenbergh, 2003).

These compounds can interfere with normal animal development or reproduction. They have been implicated in problems occurring in marine snails, frogs, alligators, and people, among others.

A common estrogen mimic called bisphenol A is found in the lining of food and beverage containers and in dental resins. It leaches into human food and into the surrounding environment. Female mice pups, whose mothers were fed bisphenol A during pregnancy, were found to have accelerated puberty onset. The effects of this compound were modulated by their fetal exposure to uterine testosterone (Vandenbergh, 2003).

The pesticide DDT is also an endocrine disrupter, and concentrates in predator birds at the highest levels of the food pyramid. DDT causes eggshell thinning and other developmental abnormalities. Eggs in the nest break, and reproduction suffers.

The type of play characterizing juvenile animals varies between males and females, with males generally showing more aggressive physical behavior. Prenatal exposure to testosterone affects this type of play, and litter females situated next to males in the uterus are more likely than females next to other females to engage in male behavior. Contrarily, males with higher estrogen exposure are more likely to behave like females. Endocrine disrupters, especially during pregnancy, can have a large effect on animal behavior throughout their life spans.

Example 6.12.1: Dose–Response Extrapolation

Dose–response curves for toxic chemicals are almost always generated from animals thought to have responses similar to humans. High chemical concentrations are often used because these result in definitive responses in reasonably short times. Predict how these responses can be extrapolated to the low concentrations normally encountered in the human environment.

Solution

No prediction can be made. Aside from the transfer of response information from an animal to a human, not enough is known about chemical effects at low concentrations. Some chemicals may continue to be harmful, others may be beneficial at low concentrations. In addition, differences in individual responses will become relatively more important at low concentrations. The result of all these uncertainties is that one cannot say for sure what will happen.

Example 6.12.2: Antimicrobial Plastics

Antimicrobial compounds can be incorporated within plastics that are then molded into useful shapes. Instead of wearing off, as surface applications do with time, part wear just exposes new compound and renews the ability of the plastic to kill microbes on its surface (Alder, 2002). Antimicrobial plastics come in a variety of rod, plate, tube, and thin sheet shapes, and can be used for medical and food applications.

Nonlethal antimicrobial compounds, or those that are less than 100% effective in killing bacteria, would select for microbes resistant to these chemicals. Therefore, it is necessary to select compounds with multiple lethal mechanisms. Metals, especially, and compounds containing metals are good candidates for this application. Silver compounds, for instance, have been used for more than 100 years to control bacteria, mold, mildew, and fungus in medical and food applications. Silver nitrate, for instance, has been used in the eyes of newborn children to protect against disease transmission. Because silver disrupts the cell walls, cell metabolism, and DNA replication, it is not likely that microbial resistance will develop. Thus, incorporating silver compounds in the plastic from which parts are made can be an effective means to combat pathogenic microorganisms.

Example 6.12.3: Why *Bacillus thuringiensis* Toxin Isn't Always Deadly

Bacillus thuringiensis (Bt) is a microbe that produces a toxin deadly to several kinds of caterpillars that ingest it. Bt toxin causes mature caterpillar gut cells to swell, burst, and die. Normally, if the Bt concentration is high enough, the caterpillar dies. However, a lower Bt concentration allows surviving cells to emit cytokines (see Section 6.20.3, Immune System), which signal gut stem cells to multiply and rapidly form new mature gut cells. If more new cells can be produced than are killed, the caterpillar survives (ASAE, 2001).

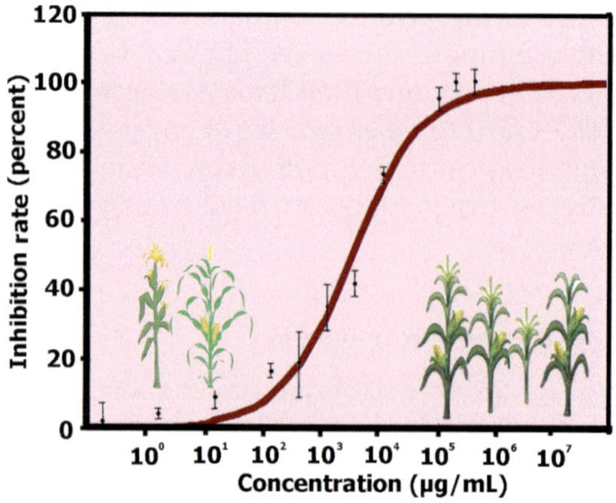

FIGURE 6.12.7
Dose–response curve for the fungicide polyoxin applied to the maize pathogen *Bipolaris maydis*.

Example 6.12.4: Chemical Fungicides to Control Fungal Infections in Corn

Sometimes chemical toxicity is sought, especially when it is selectivity toxic to pathogenic organisms that damage crops. Chemical fungicides are used separately or together in combination to control to maize fungal pathogen *Bipolaris maydis* (southern corn leaf blight). Wang et al. (2015) have determined the fungicide effectiveness response curves of several common fungicides. The results for one of these, polyoxin, appears in Figure 6.12.7, and displays the typical dose–response curve expected of a chemical compound.

APPLICATIONS AND PREDICTIONS

1. Many plants will contain toxins in their stems and leaves to protect themselves. These same plants, which depend upon animals to eat their fruits and spread their seeds, will have no toxins in their fruits.
2. Only those parts of the plant that require protection will contain toxins. Thus, the tomato has no toxin in its fruit and the potato has no toxin in its tuber despite the presence of toxins in their leaves.
3. The more toxin produced, the slower plant growth will be.
4. Toxins will be found everywhere.
5. The most conspicuous plants and animals will contain the most toxins.
6. Animals that can defend themselves by aggression, fighting, or flight will not be toxic when eaten.
7. Small biochemical compounds not found in nature will be the most difficult to detoxify.
8. BU will contain more toxins during life stages when they are most vulnerable to predation.
9. Natural toxins can be used to advantage by other BU that acquire them.

(Continued)

APPLICATIONS AND PREDICTIONS (Continued)

10. The same toxin that kills a baby can be tolerated by an adult.
11. Koala bears have special detoxifying bacteria in their gut that enable the bears to eat toxic eucalyptus leaves.
12. Drinking municipal water may have adverse health effects.

6.13 Biological Units Respond to Mechanical Stresses

Trees are like giant brooms
Sweeping the sky.
When they get old and worn
Nothing is left but the handle.

—Shelia Holloman

BU live in a physical world that challenges them in many ways. There are many stresses incurred from the flow of fluids around and through BU, for instance, and the biological engineer must know about these and the consequences of exceeding limits to resisting them. For instance, cells in proximity to a flowing fluid are subject to shearing. As the fluid flows past these cells, friction tends to distort cell shapes by elongating them along the direction of flow. Endothelial cells inside blood vessels grow in such a way that they can better resist shear effects. If shear rate increases too rapidly, the cells cannot adjust, and they may be torn apart. Cells in bioreactors have been known to be destroyed by shear stresses exceeding 20 dynes/cm^2.

6.13.1 Sedimentation and Clotting

Strength is born in the deep silence of long-suffering hearts, not amid joy.

—Felicia Hemans

Regions of low flow promote deposition or clotting. Sediment and other suspended solids carried along by flowing fluids will be deposited where the fluid stagnates (Johnson, 1999). Thus, a stream that flows rapidly through hilly terrain and picks up a lot of sediment on its rush downhill will drop its load in the lake or marsh into which it flows. High flow rates can suspend a lot of material. Turbulent flow, where there is a lot of mixing and churning, is especially good for moving even very heavy objects. Slowly moving fluids cannot sustain these objects, and they fall. Hence, sedimentation tends to fill lakes, marshes, and river locations where the velocity is particularly low. This not only means that basins will fill with sediment; it also means that benthic plants and animals can be completely covered by unusual sedimentation events. Without access to light, the plants may die and change the ecology of the region.

The clotting system of the blood is not entirely like sedimentation, because it involves biochemical reactions that form clotting compounds (Figure 6.13.1). However, there are similarities, and the one most like sedimentation is that clotting occurs in the regions of low

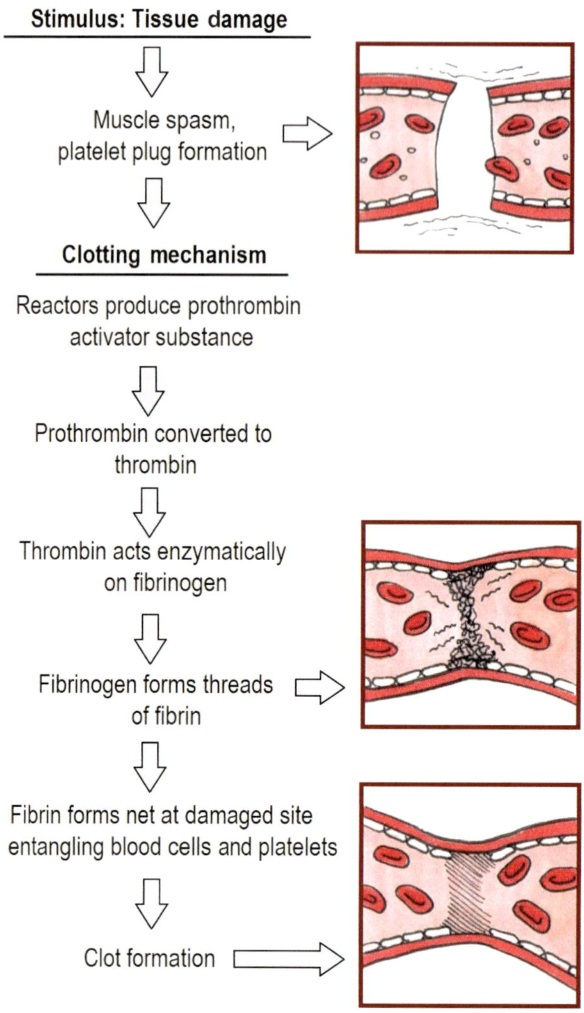

FIGURE 6.13.1
Blood clotting occurs when tissue is damaged or when blood stagnates. Reactions cause rod-shaped plasma proteins called fibrinogens to stick together as long, insoluble threads. These adhere to exposed collagen, forming a net that traps blood cells and platelets. The entire mass of fibers and cells is the clot. (Redrawn from Starr, C., *Biology: Concepts and Applications*, Brooks/Cole, Stamford, CT, 2000.)

blood flow. These are not normally a problem, because *thrombus* (a blood clot) formation is a dynamic process that involves clot formation and dissolution. However, blood that pools in the leg veins during long periods of inactivity can form thrombi. When clots break loose from their points of attachment inside the veins, they can be carried by the blood to other vessels in the body smaller than they are. The result is that the tissue downstream from the occlusion (blockage) receives inadequate oxygen or glucose and dies. Even this may not be life threatening except when that tissue is in the lungs, the heart, or in the brain.

One difficulty with the implantation of mechanical hearts, artificial valves, or blood vessels is that there may be regions of blood stagnation inside these artificial organs. Thrombi can form there, and have been known to cause life-threatening incidents in organ replacement patients. The difficulty is compounded by the fact that many early biomaterials were sensed as foreign by the body's immune system and it attempted to cover the materials with clot-like depositions.

High flow rates have high amounts of kinetic (moving) energy and low potential (pressure) energy; low flow rates are just the opposite. From an energy balance (see Section 2.2, Balances) in the fluid, when a flow stream accelerates, its pressure must decrease. It has been observed that *atherosclerotic plaque* (mostly lipid material) is deposited in blood vessels where the pressure is lowest. Plaque occupies part of the lumen of the vessel and reduces the area through which blood can flow. In order for the same volume of blood to flow through the partially occluded vessel in any given amount of time, its velocity must increase. When that happens, it accelerates, and pressure decreases. Reduced pressure is just the condition that caused the plaque formation in the first place, so the situation can easily get worse.

6.13.2 Strengthening and Stiffening

> Man is but a reed, the weakest in nature, but he is a thinking reed.
>
> **—Blaise Pascal**

High mechanical stresses require strength. Nothing illustrates this better than trees growing where the wind blows consistently from the same direction. The trees are usually bent to reduce their exposure to the wind, and have gnarled and thickened trunks and branches. Seeing trees growing like this should indicate to the observer that light structures would not survive intact for long in such a place. In fact, the estimates of historical wind velocities have been obtained from the study of the shapes of trees in the area.

The biological energy necessary to resist wind damage would otherwise be spent on growth and reproduction. Plants that withstand the pressure of wind are often stunted and without many offspring.

Human bone strength has been found to be related directly to intermittent applied forces, similar to (not surprisingly!) forces generated during walking or running. Osteoporosis (weakening of the bones due to the resorption of bone material) in the elderly can cause bones to become brittle and break easily. Exercise is one way to reduce the effects of reduced bone density.

The integrity of certain types of bacteria depends on cell walls that they construct. The antibiotic penicillin was effective because it interfered with this construction process (Palumbi, 2001). From the Law of Laplace (Section 2.9.4, Circulatory System):

$$\text{pressure} = \frac{2(\text{wall tensile stress})(\text{wall thickness})}{(\text{wall radius})} \tag{2.9.6}$$

It can be seen that a bacterium can resist higher pressures for thicker walls, smaller size, or walls that are made of stronger materials. Bacteria not round shaped have a radius of infinity, and cannot theoretically resist any pressure, so one reason that bacteria are shaped as spheres and cylinders is that they must be in order not to implode or explode. Cylindrically shaped bacteria must have some curvature to the parts of their walls that are straightest, and the walls must be thicker at those parts. The use of stronger materials for their walls is presumably not an option because bacteria utilizing weaker wall materials would have been eliminated from the gene pool long ago.

The Law of Laplace can also be used without the 2 in the numerator for cylinders. The pressure can also be considered the highest pressure that can be resisted from inside out. Thus, we see that smaller blood vessels are stronger than larger ones, all else being equal.

But all else is not equal: capillaries must have thin and weak walls to promote the diffusion of O_2, CO_2, glucose, and other materials to and from the surrounding tissue. They can have thin and weak walls, and still not burst from internal blood pressure, as long as they are small enough. Which they are.

6.13.3 Critical Shear Stress

> Human genius amazes because it is a mystery. If science could explain how genius came to be, the wonder would be gone.
>
> —**Nathaniel M. Campbell**

Individual cells react to mechanical stress in a number of ways. Cells that normally adhere to surfaces and to other cells have more difficulty doing so at high levels of shear stress. Endothelial cells tend to distort and to align their long axes in the direction of the chronic shear. A shear stress of 20 dynes/cm^2 for 48 h was found to cause alignment parallel to the flow in aortic endothelium, but caused alignment perpendicular to the flow in the aortic valve. It is not known if the differences in alignment are due to different phenotypes.

Shear stress induces larger amounts of elastin in cells and inhibits tissue calcification. Occludin, a transmembrane protein that forms tight junctions between cells, and is the main contributor to the *blood–brain barrier* (an obstacle to free passage of complex molecules into the central nervous system), is present in lesser amounts at high shear stress values (about 10% less at 20–30 dynes/cm^2).

Shear stress also causes a transient structured disruption of the cell membrane, and increases membrane permeability to Ca^{++} and other ions (Serbest et al., 2002). This temporary disruption of the cell membrane allows many types of external materials, including nucleic acids, to enter the cell (Bradley, 2014). It can also increase lipid peroxidation. Shear stresses in the range of 8–14 dynes/cm^2 begin to damage cells in human vasculature. The percentages of neuronal cells killed *in vitro* varies linearly with the shear stress (Figure 6.13.2) and also linearly with the rate of shear stress application (Serbest et al., 2002).

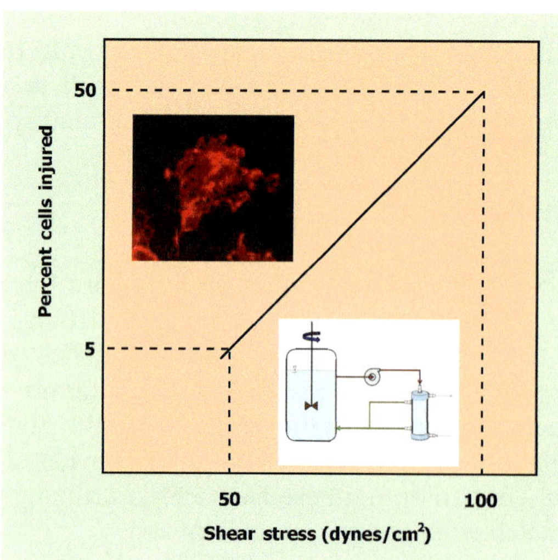

FIGURE 6.13.2
Neuronal cell injury increases linearly as shear stress increases.

Mechanobiology is the area of study that relates cell behavior to its physical surroundings. Different physical forces exterior to a stem cell, for instance, can cause the cell to morph into various cell types (Piccolo, 2014), or cause differentiation of somatic cells subject to different forces.

6.13.4 Stem Cell Substrates

> The hardest part (of the research) is that it's done outside under natural conditions that are changing all the time.
>
> **—William Manning**

Stem cells, those undifferentiated cells that can become almost any type of cell in the body, are apparently affected by the stiffness and the texture of the surfaces upon which they are grown (Figure 6.13.3). Stem cells grown on the stiffest matrix became bone cells (Brodie, 2007b). The softest surfaces resulted in nerve cells. Stem cells grown on a medium-soft substrate formed muscle cells. The differentiation of stem cells into these types of somatic cells was determined by physical responses instead of the chemical media that they were grown on.

Also, the ability of stem cells to become any type of cell (their *pluripotency*) is maintained best if they are grown on a surface lined with ridges. The scale of the ridges can vary from nanometers to micrometers without effect (Brodie, 2007b).

Somatic and stem cells grown outside the body prefer three-dimensional matrix and basement membrane growing environments to two-dimensional flat plates or Petri dishes (Saltus, 2010b). Cells nurtured in 3-D environments experience more normal physiological, biochemical, metabolic, and physical conditions, and, in turn, function more like the types of cells they began as. The microenvironment of the cell plays a large role in directing its growth and shaping its behavior.

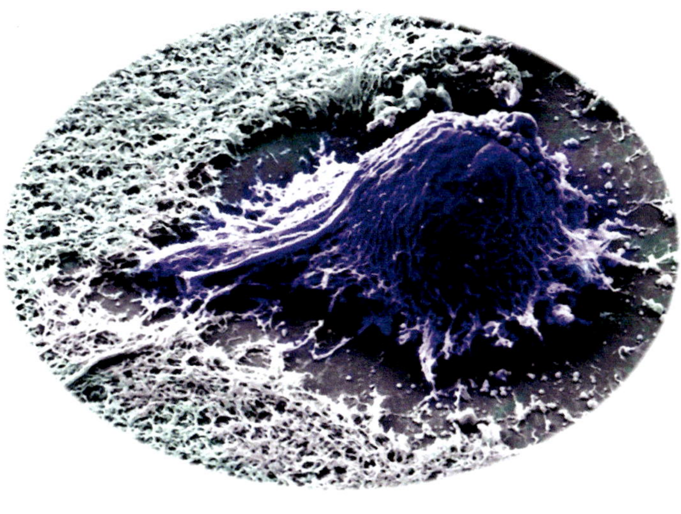

FIGURE 6.13.3
Human embryonic stem cell.

BIOPRINTING

Using technology similar to that in a common inkjet printer, bioprinting uses materials to produce three dimensional scaffolds for tissues and to insert cells of the correct types within these scaffolds to form organs, parts of organs, or patches to repair organs. These bio-printed parts can be made from tissue building materials and have the advantages that:

1. They can be custom-made for a particular patient,
2. They can avoid immune system rejection,
3. They can supplement or replace limited organ bank stores, and
4. They can form standardized tissues for drug and other types of testing.
5. They can form three-dimensional matrices for natural cell or tissue growth.

Testing drug toxicity and efficacy on animals is expensive, possibly unethical, and does not always duplicate effects seen in humans. These difficulties can be avoided when bioprinting is used to provide living tissue for *in vitro* testing and drug screening.

6.13.5 Whole Plant Responses

> If anything, living systems consistently violate all of the criteria for reducibility. The number of elements that compose any living system – an ecosystem, an organism, an organ, or a cell – is enormous. In living systems, the specific identities of these parts matter. Unlike chemistry, for instance, in which an electron in a lithium atom is identical to an electron in a gold atom, all proteins in a cell are not equivalent or interchangeable. Each protein is the result of its own evolutionary trajectory. We understand and exploit their similarities, but their differences matter just as much. Perhaps most importantly, the relations between the components of living systems are complex, context-dependent, and weak.
>
> **—Robert L. Dorit**

Physical stresses impose limits or otherwise affect cellular responses, as was presented in the previous sections. When cells join to form multicellular organisms, the possible range of physical effects increases greatly. Not only can individual cells respond to mechanical stresses, but the integrity of the whole organism can be compromised when intercellular connections become stressed beyond limits.

Plants, for instance, can be challenged by high winds, hail, or herbivore grazing. Plants that are habitually eaten sometimes develop tougher tissues that are less palatable than rapidly growing stems and leaves. This is in addition to the chemical responses discussed elsewhere in this text.

When mechanical stresses exceed strength limits, limbs may break, or the plant may snap off at ground level. These stresses may take the form of bending, tension, compression, shear, or torsion. Mature wood has maximum stress levels in the range of 5000–1000 pounds per square inch (psi). Green (immature) wood may tolerate a greater degree of bending than can mature wood, but is more likely to tear off (weaker in tension and bending). Plants constantly challenged by mechanical stresses that do not exceed physical limits often develop extra tissue to help cope with chronic mechanical stress. Although this extra tissue does not significantly alter the ultimate strength stress that can be tolerated, it does add bulk that when multiplied by stress results in larger forces that can be resisted:

$$\text{Stress} \times \text{area} = \text{force} \tag{6.13.1}$$

This is analogous to the myocardial hypertrophy that develops in a human heart resisting high blood pressure.

Wood is a composite material made from linear polymer cellulose fibers embedded in a *hemicellulose* matrix. *Lignin* acts as a glue to hold the fibers together. The *grain* of the wood is the direction parallel to the cellulose fibers. Wood is stronger in this direction than in any other.

Some trees grow more slowly than others, and these produce wood that is denser than faster-growing species. More dense wood translates into stronger wood. Dried wood is also stronger than green wood. The values in Table 6.13.1 are for dried wood. These values should be degraded by at least 50% for wood still part of a tree.

TABLE 6.13.1

Strengths of Some Wood Species

Wood Species	Specific Gravity	Compressive Strength (psi)	Bending Strength (psi)
Ash	0.60	7,410	15,000
Aspen	0.38	4,250	8,400
Basswood	0.37	4,730	8,700
Beech	0.64	7,300	14,900
Birch, Yellow	0.62	8,170	16,600
Butternut	0.38	5,110	8,100
Cedar, Aromatic Red	0.47	6,020	8,800
Cedar, Western Red	0.32	4,560	7,500
Cherry	0.50	7,110	12,300
Chestnut	0.43	5,320	8,600
Cypress	0.46	6,360	10,600
Elm	0.50	5,520	11,800
Fir, Douglas	0.49	7,230	12,400
Hemlock	0.45	7,200	11,300
Hickory	0.72	9,210	20,200
Lauan	0.40	7,360	12,700
Mahogany, African	0.42	6,460	10,700
Maple, Hard	0.63	7,830	15,800
Maple, Soft	0.54	6,540	13,400
Oak, Red	0.63	6,760	14,300
Oak, White	0.68	7,440	15,200
Pine, White	0.35	4,800	8,600
Pine, Yellow	0.59	8,470	14,500
Poplar	0.42	5,540	10,100
Redwood	0.35	5,220	7,900

(*Continued*)

TABLE 6.13.1 (*Continued*)

Strengths of Some Wood Species

Wood Species	Specific Gravity	Compressive Strength (psi)	Bending Strength (psi)
Sassafras	0.46	4,760	9,000
Spruce, Sitka	0.40	5,610	10,200
Sycamore	0.49	5,380	10,000
Teak	0.55	8,410	14,600
Walnut	0.55	7,580	14,600

6.13.6 Bodies in Motion

There are no things man was not meant to know.

Michael Kurland

As long as human beings are using only their own bodily capabilities to move or perform work, then they most likely easily remain within tolerable limits of strength, comfort, and tissue failure. However, when they use modern tools, devices, or modern transportation modes, humans can experience discomfort, injury, or even death when physical limits are exceeded.

Dealing with a sentient being goes beyond the mechanical limits of physical failure discussed in the previous section on plants. There are also issues of comfort, psychological damage, and chronic health effects to consider.

Rapid accelerations and decelerations of the human body can cause many problems. The maximum deceleration ever recorded for a human was 46 times the acceleration due to gravity (9.8 m^2/s), also called 46 g's (Ward, 2011). At this rate of deceleration, the human body was distorted, with some organs shifting position. The eyes, for instance, displace from their eye sockets. Some representative accelerations for different events appear in Table 6.13.2.

Rapid accelerations cause large forces, and these can be enough to cause breakage. Without tissue cushioning and elastic energy storage in the muscles and tendons, a jump of less than ½ meter would be enough to break a femur (Johnson, 2007). Over time, such forces would result in extra strength added to the bone.

TABLE 6.13.2

Acceleration and the Human Body

Acceleration (g's)	Event
2.9	Sneezing
3	Space Shuttle flight
3.5	Coughing
3.6	Crowd jostle
3–4	Roller coaster, maximum
4.1	Slap on back
4–6	Induces blackout in fighter planes
6	Maximum gravitational force exerted during suborbital launch
8	Fighter pilot, brief maximum

The maximum tolerable acceleration depends both on the direction of the force and the type of restraint used. A seat belt or shoulder harness helps to protect against horizontal accelerations, for instance (front-to-back inertial force, called *Gx*, is the most tolerable, but can peel skin back; sideways inertial force, called *Gy*, is rare, but can move or dislodge organs). Forward thrust horizontal accelerations are tolerated better than reverse-thrust acceleration. Both the brain and the eyes may be damaged with horizontal forces. Accelerations of 70–100 g's during car crashes can tear the pulmonary artery from the heart.

Vertically upward acceleration, called *Gz*, drives the blood down to the feet. This deprives the eyes and the brain of needed oxygenation. Induced tunnel vision, loss of consciousness, or even death can result. The remedy for fighter pilots, who commonly experience up to 4–6 g's vertically, is a "g-suit" that applies external pressure to the legs to push blood back up again. Vertically downward acceleration does just the opposite, forcing blood to the eyes and the retina, causing a "red-out" of vision.

It is important to maintain comfort for humans. Commercial standards in the U.S. limit acceleration to 1.5 g and deceleration to 1 g (Ward, 2011). These are applied to various modes of transportation. Commercial airplanes can accelerate passengers by no more than 1.6 g's, and banking angles on turns can be no more than 30°. Horizontal movement can be very uncomfortable also. These motions are tolerated better in the prone position than in the seated position.

The natural frequency of the human viscera is about 6–8 cycles per second. Vibration in that range can be extremely uncomfortable and must be avoided. Rocking back and forth, as riding on an animal, is more tolerable at higher frequencies (a horse) than at lower frequencies (a camel). This rocking can produce a condition called *motion sickness*.

Exposure to whole body vibration for a long time is one of the leading risk factors for low back pain disorders in humans. This poses potential problems for professional bus and truck drivers, among others. Most of the vibration these groups encounter is in the vertical direction (*z*-axis) rather than either horizontal direction (*x*- or *y*-axis). Different seat designs can help ameliorate the whole-body vibration experienced by these groups (Blood et al., 2015).

Chain saw vibrations of 30 g's or more have been recorded on the hands and arms (Cundiff and Suggs, 1976), and these exceed international standards for tolerable vibration (Bingham et al., 1992). Other hand-held devices can also vibrate beyond tolerances. One common consequence of excessive hand vibration is constriction of the peripheral arteries shutting off the blood supply to the hands and fingers. This occupational condition, called *Raynaud's Syndrome*, can become permanent and result in loss of feeling, dexterity, and, in extreme cases, entire hand function.

WOODPECKERS DO NOT SUFFER CONCUSSIONS

One interesting example of accommodation to mechanical shock is the woodpecker. This bird pounds into wood to find insects hidden underneath the surface. Although their heads decelerate at 1200 times the acceleration due to gravity, their shock-absorbing structures consisting of elastic beaks, musculotendinous attachments of the tongue and throat muscles (called *hyoids*), skulls with a minimum of cerebrospinal fluid, and partially spongy skulls that attenuate rapid accelerations protect their brains from injury (Yoon and Park, 2011). Brain injury is caused by a combination of acceleration and duration resulting in internal shear and excessive tensile forces (Alexander, 2015). Studying the woodpecker can potentially benefit humans who suffer concussions, most notably from violent sports participation.

Example 6.13.1: Plants Affected by Human Stroking

A study of plants in Pennsylvania has shown that scientists' attention can, by itself, alter the way insects feed on plant leaves (ASAE, 2001). A team of ecologists marked 605 plants in 12 plots for study and visited half of the plots weekly. When they visited them, they stroked the plants once each from base to tip to imitate typical contact made when a scientist studies plants in the field.

One species experienced higher insect leaf loss when visited compared to the unvisited control. Several others experienced less leaf loss when visited. Some kinds of plants showed no visitation effects. These results showed that the act of conducting an experiment can, itself, alter experimental results.

As to why this happens, the reasons are open to speculation. Touching a plant may cause changes in its structure or leaf toughness. It may also trigger the release of chemicals attractive to insects. Trampling surrounding vegetation could also alter the target plants' growth and visibility to insects.

Example 6.13.2: Wind Load on a Small Poplar Tree

Assume a small poplar tree as diagrammed in Figure 6.13.4. The wind blows against the tree at 30 miles per hour. Will the tree trunk break?

Solution

Determining the answer to this problem requires knowledge of statics and strength of materials. Wind load (F) in pounds can be calculated from:

$$F = A \times P \times C_d$$

where:

A is the area presented to the wind in ft^2
P is the wind pressure in lb/ft^2
C_d is the drag coefficient, unitless

$$P = 0.00256\ v^2$$

where:

v = wind speed in miles/hour
$P = 00256\ (30)^2 = 2.304\ lb/ft^2$

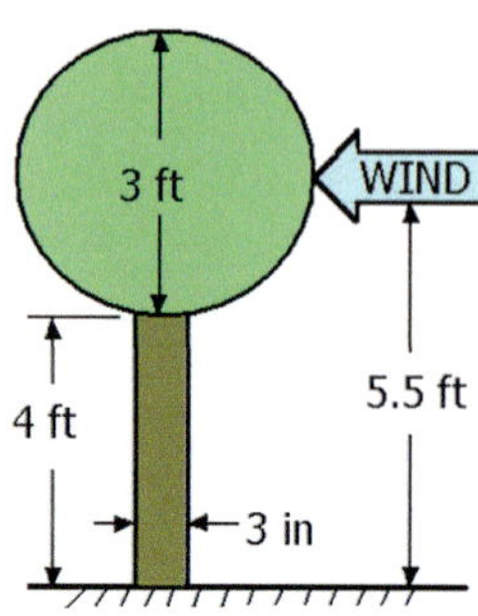

FIGURE 6.13.4
Simplified diagram of the poplar tree in Example 6.3.2.

Assuming the tree top is spherical in shape, and relatively non-porous (opaque) to the wind,

$$A = \pi\ d^2/4 = 7.069\ ft^2$$

For a brisk wind, $C_d \approx 0.4$ for a spherical shape (Johnson, 1991). So,

$$F = 7.069\ ft^2 \times 2.304\ lb/ft^2 \times 0.4 = 6.51\ lb$$

The wind force can be considered to act at the center of the tree top. Thus, the moment (M) it causes at ground level is:

$$M = F \times L = 6.51\ lb \times (4\ ft + 1.5\ ft) = 35.8\ ft{\cdot}lb$$

From strengths of materials, the maximum bending stress on the trunk at ground level is:

$$\sigma_{max} = M \times c\ /\ I$$

where:

σ_{max} is the maximum bending stress in lb/ft^2
M is the moment produced by a force acting over a distance in ft · lb
c is the maximum distance from the axis of rotation in ft
I is the moment of inertia of the circular cross-section in ft^4

For the circular cross-section of the trunk (Eshbach and Souders, 1975),

$$I = \pi\ d^4/64 = \pi \times (3\ in/[12\ in/ft])^4/64 = 1.917 \times 10^{-4}\ ft^4$$

Thus,

$$\sigma_{max} = 35.8\ ft \cdot lb \times (1.5\ in/[12\ in/ft])/1.917 \times 10^{-4}\ ft^4 = 23{,}360\ lb/ft^2$$

From Table 6.13.1, The maximum bending stress for poplar wood is 10,100 lb/in^2, or 1,454,400 lb/ft^2. This is over 60 times higher than the wind-induced maximum bending stress. Hence, even the green wood of the tree trunk should be able to withstand the 30 mi/h wind without breaking.

Example 6.13.3: Entertainment Engineering for Fun and Excitement

Entertainment of visitors to highly technical shows and theme parks combines engineering with thrill solicitation. Shows such as the Cirque du Soleil bring extreme human gymnastic and artistic expression together with mechanical technology. The result is awe-inspiring.

Theme parks combine technology with interesting settings to elicit excitement from ride participants. These must convey a sense of danger without being dangerous.

There is a good deal of biomechanical engineering in entertainment engineering. Capabilities of humans to be accelerated, thrown about, and psychologically challenged must be known and limits strictly observed. Yet, within these limits, a great deal of latitude results in unique experiences for participants. Entertainment engineering academic programs have been established to instruct students in the specialized knowledge applicable to the entertainment industry (Creighton, 2005).

Virtual reality with sophisticated eyepieces and accelerometers can immerse the wearer in a very realistic virtual environment. These apparatuses can be used for realistic entertainment as well as for training exercises.

Example 6.13.4: Enhancing Safety and Performance with Sports Engineering

Combining mechanical engineering, materials science, and applied physiology, sports engineering applies modern technology to sports equipment and sports measurements. Women tennis players, for instance, are four times more likely to injure their ankles than men (Grose, 2004). Realizing this, sports engineers can design better shoes for women tennis players; such shoes would differ from men's.

Sports engineers can also tackle designs of tennis racquets, golf clubs, rowing oars, and baseball bats. Of special importance is protective gear to minimize injury while skiing, playing baseball, football, or hockey, or even riding a bicycle. Sports engineers need to know the limits to human strengths and endurance. Along with knowledge of materials and mechanical design, they can make sports participation safer and more rewarding.

Example 6.13.5: Drug Delivery Using Cell-Penetrating Peptides

Delivering large molecules for medicinal purposes into cells is impeded by the bilayer cell membrane that repels hydrophilic compounds. For this reason, cell membrane penetration techniques have been developed that include viral vectors and membrane disruption. These methods, however, can have significant toxic and immunogenetic side effects.

Cell penetrating peptides (CPPs) can be used to evade the membrane barrier. These peptides are either: (1) positively charged amino acids, (2) hydrophobic, or (3) amphipathic (both hydrophilic and lipophilic). When they contact and attach to the membrane, they are ingested through endocytosis, forming peptide-inclusive vesicles and being transported into the interior of the cell (Figure 6.13.5). Once there, they are released and free to conjugate with other biochemicals inside the cell (Thundimadathil, 2013).

Therapeutically significant biomolecules can be attached to these CPPs and be ingested into the cell at the same time.

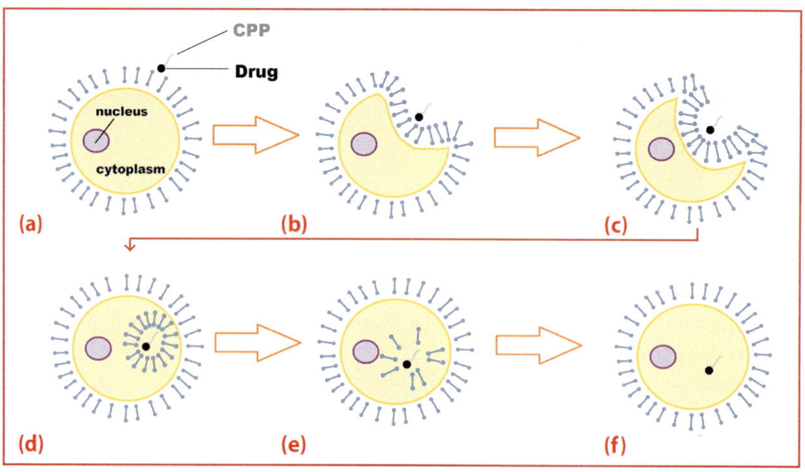

FIGURE 6.13.5
Steps in cell penetration of a cell-penetrating peptide (CPP) through endocytosis: (a) membrane interaction, (b) endocytosis, (c) vesicle formation, (d) endosome in cell, (e) endosomal release of the CPP, (f) CPP conjugates with other biomolecules inside the cell. (Redrawn from Thundimadathil, J., *Drug Discov. Dev.*, 16, 22–23, 2013.)

APPLICATIONS AND PREDICTIONS

1. Cylinders and spheres will be found to be common biological shapes.
2. Clots will form in the regions of stagnant blood flow.
3. If organisms can withstand mechanical stresses, they will have thicker body parts.
4. The larger the surface area, the less pressure can generally be withstood.
5. Anchored plants will not grow in fast-moving streams.
6. The human spinal cord is mechanically strong enough to withstand torsion and compression mechanical stresses.
7. Two strategies are to bend with applied forces or to resist them. Seaweed bends, but a mollusk resists.
8. Paralyzed individuals will need to be exercised to prevent blood stagnation.
9. Guy wires to stabilize a tree need to have sheaths where they contact the tree to better transmit forces to the tree without damage.
10. Clotting is a problem when stents are used.
11. Strength can be influenced either by shape or size. Lightweight tools trade off one for the other.

6.14 Optimization Is Used to Save Energy and Nutrient Resources

> A man who works with his hands is a laborer; a man who works with his hands and his brain is a craftsman; but a man who works with his hands and his brain and his heart is an artist.
>
> **—Louis Nizer**

Competition in the biological world is usually so great that there is no room for inefficiency. At every turn, in all environments, energy must be conserved where possible because extra energy means better defenses, or better growth, or better ability to garner resources. In all cases, these lead to better reproductive success.

6.14.1 Reproductive Advantage

> [Pregnant fruit flies] spent considerable time searching their environment for a suitable spot to lay their eggs. After all, selecting an appropriate site to lay its eggs is presumably the ultimate decision a fly mother has to make, as the consequences of such decisions are likely to have significant impact on the reproductive success of the species.
>
> **—Rebecca Yang**

Think of the case where there are two BU: one is able to perform the same functions as the second, but it can do so with one half the energy requirements. If each BU is an animal, which is most likely to be caught and eaten by a predator? If each BU is a predator, which is most likely to chase and catch an elusive prey, or which is most likely to survive long times between kills? If each BU is a plant, which is most likely to be able to outgrow competitors and grazing herbivores? If each BU is a microbe, which is most likely to inhabit a region with limited nutrient availability? If each BU is a bodily tissue, which is more likely to confer to the entire organism a reproductive advantage? If each BU is an entire ecosystem, which is most likely to thrive and expand into new territory?

In each of these cases, the answer is clearly that the BU with the energy advantage is the winner of the competition. Only in the instance where competition is at a minimum, say, for example, for the first species in a virgin environment, will there be little primary reason to reduce energy costs. However, as soon as the second species arrives, or even as soon as the number of individuals of the first species increases to the point that they force significant intraspecific competition, there will be an advantage to those individuals that can make more efficient use of resources. Thus, there is a tendency to minimize dependence upon the environmental sources of energy and nutrients.

Natural selection is a powerful force leading to the evolution of living things (see Sections 5.4, Competition, and 6.10, Evolution). We can, for instance, see evidence of *convergent evolution* (where organs and tissues with different origins form identical final forms) and *allometric relationships* (scaling of different forms and functions among species, see Chapter 7, Scaling). There is enough competitive pressure in biology that the benefit-to-cost ratio of almost every biological function must be optimized. A hummingbird, for example, needs enough strength in its wings and energy to be supplied to its wing muscles to hover. There is no advantage to be gained with excess wing strength, and so the benefit-to-cost ratio changes abruptly after sufficient strength is satisfied. On the other hand, animals that jump have a survival advantage if they can jump farther, faster, or higher. There is an overhead cost of supporting larger muscles or bones needed for better jumping, but the benefit-to-cost ratio changes gradually for jumping. The biologically optimal solution for these cases would be expected to be different for each. Thus:

> Each biological form, function, and action has a cost and a benefit. Survival considerations demand that the benefits outweigh the costs.

6.14.2 Locomotion

> Books are the bees which carry the quickening pollen from one to another mind.
>
> **—James Russell Lowell**

Optimization (see Section 4.5, Optimization) is a concept that often depends on two resource-consuming processes. One of these increases with some meaningful variable, and one of these decreases with the same variable. Total resource expenditure is given as the sum of these two variables, and, since one increases and the other decreases, there is usually a point where the sum of the two is a minimum.

Take locomotion, for instance. Human walking, which is one form of locomotion, has been found to have a rate of energy expenditure that depends on walking speed (Dean, 1965; Milsum, 1966; Johnson, 2007):

$$\text{Rate of energy usage} = a + b(\text{walking speed})^2 \tag{6.14.1}$$

Dividing by walking speed gives average power per unit speed:

$$\text{Average power} = \frac{a}{(\text{speed})} + b(\text{speed}) \tag{6.14.2a}$$

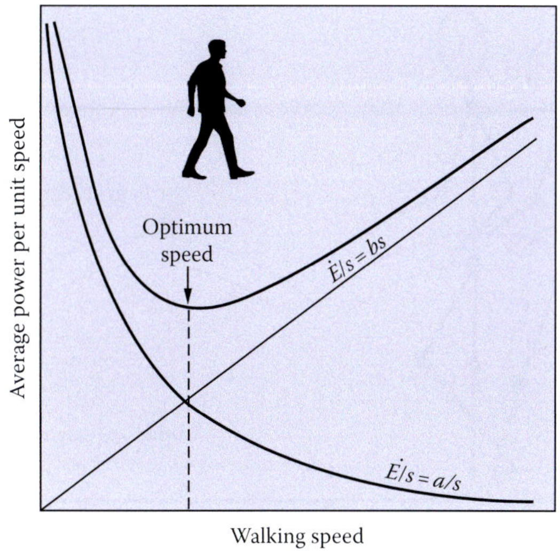

FIGURE 6.14.1
The average power per unit speed for walking has a minimum at a particular speed. (Redrawn from Johnson, A.T., *Biomechanics and Exercise Physiology: Quantitative Modeling*, Taylor & Francis, Boca Raton, FL, 2007.)

or symbolically

$$\frac{\dot{E}}{s} = \frac{a}{s} + bs \tag{6.14.2b}$$

This represents average power with two components, one linearly increasing with speed and one hyperbolically decreasing (Figure 6.14.1). The average power of walking is a minimum where the sum of the two components is a minimum, or, the minimum average power can be found by taking the derivative of E/s and setting the derivative to zero:

$$\frac{d}{ds}\left(\frac{E}{s}\right) = \frac{-a}{s^2} + b = 0 \tag{6.14.3}$$

$$s_{opt} = \sqrt{\frac{a}{b}} \tag{6.14.4}$$

where
s_{opt} is the optimum speed, m/s.

Human walking is energy intensive because it involves the raising and lowering of the body's center of gravity as the body moves over the extended leg and then as the leg extends backward while the other leg swings forward (Figure 6.14.2). Also, there is a time when both feet are on the ground and pushing against one another. Eliminating or minimizing the vertical movement of the center of gravity or the time during which the feet are simultaneously exerting forces in opposite directions can increase the efficiency of walking.

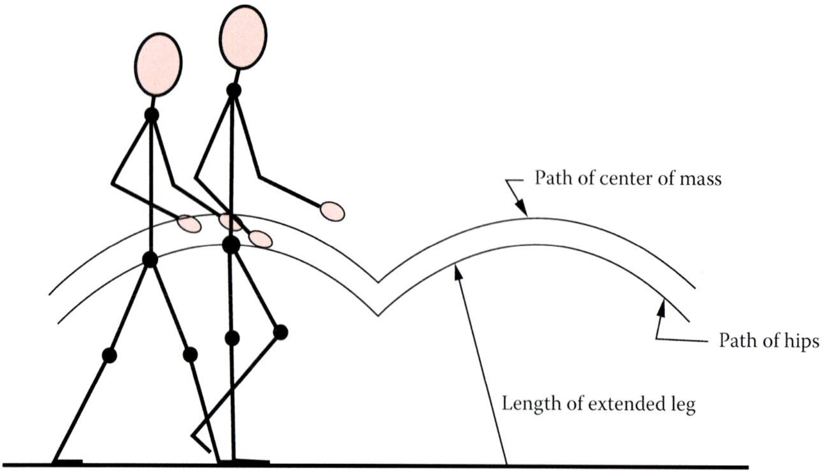

FIGURE 6.14.2
Raising and lowering of the body's center of gravity during walking contributes to walking efficiency. (Redrawn from Johnson, A.T., *Biomechanics and Exercise Physiology: Quantitative Modeling*, Taylor & Francis, Boca Raton, FL, 2007.)

Running accomplishes at least some of these goals. During running, only one foot is on the ground at a time. The body does not move vertically as much as it would during walking because each step involves a small leap to the next position. Energy absorbed during the falling stage of each step is stored in the muscles and sinews and recovered during the pushing stage of the next step.

Some animals have other means to reduce the energy inefficiency of locomotion. Animals such as centipedes have so many legs that their bodies do not rise and fall during locomotion. Snakes crawl without the benefit of legs, but their locomotion must overcome the friction created between the ground surface and their skin. Crocodiles use a side-to-side waddling motion to move their legs forward to propel themselves. Fishes' bodies contain a special swim bladder that allows them to maintain a vertical position in the water without muscular effort. It has been estimated that a 1% improvement in the efficiency of a swimming fish can be expected to make 3% more energy available for growth and reproduction (Alexander, 2003). Birds have wings that convert forward motion into lift to improve efficiency. Humans even use bicycles to propel themselves forward without the rising and falling of the body, and so decrease energy expenditure (Figure 6.14.3).

Walking and running are two possible means of locomotion for humans. Walking is less energy intensive for adult humans than running at speeds of about 2.5 m/s or less (Johnson, 2007); running uses less energy than walking at higher speeds. Walking on the level at 1.3 m/s requires about 185–370 N·m/s, whereas running on the level at 4.5 m/s requires about 223–488 N·m/s (Johnson, 2007). The transition from walking to running in adult humans appears to be made at the same point at which it is more energy efficient to run than to walk.

Quadrupedal animals also use the gallop as a locomotion means. Galloping involves bending movements of the back that briefly store leg kinetic energy fluctuations as elastic energy, and this contributes to overall efficiency. In addition to the walking–running transitional point, there is also one for the switch from running to galloping.

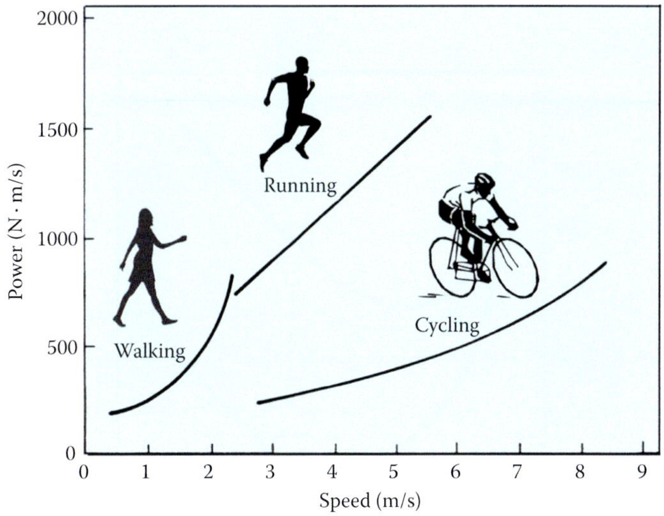

FIGURE 6.14.3
Cycling is more energy efficient than walking or running, despite the extra weight of the bicycle because the body's center of gravity stays at a particular level. (Redrawn from Johnson, A.T., *Biomechanics and Exercise Physiology: Quantitative Modeling*, Taylor & Francis, Boca Raton, FL, 2007.)

6.14.3 Breathing and Heart Rate

> It would be futile to accomplish with a greater number of things what can be accomplished with fewer.
>
> **—William of Ockham**

Breathing is another organismal process that appears to be optimized. Human breathing at rest consumes approximately 1%–2% of total oxygen consumption of the body, whereas during exercise breathing may consume 8%–10% or higher. Oxygen consumed by the diaphragm and other muscles involved in breathing cannot be used by the skeletal muscles in the legs to escape predators, so it seems apparent that breathing should be accomplished in such a way that the exercise oxygen consumption of breathing is minimized. And, indeed, it is.

Breathing at normal frequencies is dominated by resistance (see Section 2.1, Effort and Flow) located in the airways and lung tissue, and the compliance (also see Section 2.1) of the lung tissues. As respiration rate increases, so does the rate of airflow and so does the pressure required to push air through the respiratory resistance. Work rate depends on both pressure and flow rate, so the rate of work required to overcome resistance increases nonlinearly with frequency.

The pressure required to store energy in a compliance depends upon the volume stored. At higher respiration rates, the lungs are not required to fill as much to deliver the same amount of oxygen to the tissues. Thus, compliance pressure, and, consequently work rate, is nearly inverse in magnitude to the frequency increase.

The result that we have is work rate composed of two components, one of which increases with frequency and the other of which decreases with frequency (Figure 6.14.4). There is an optimum breathing rate that minimizes work rate, and most published data indicates that people and some other animals breathe at a rate corresponding to the minimum.

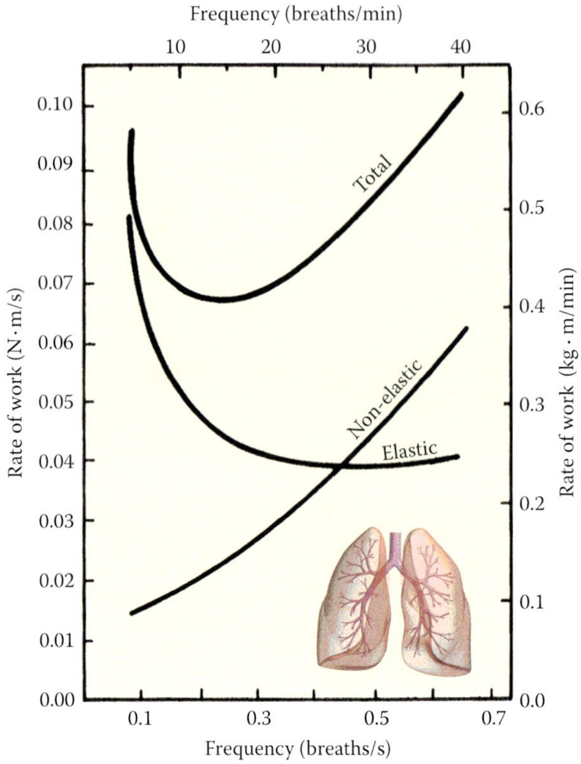

FIGURE 6.14.4
The work of breathing is the sum of two components: that due to resistance and another due to compliance. The sum reaches a minimum at some particular frequency. (Redrawn from Johnson, A.T., *Biomechanics and Exercise Physiology: Quantitative Modeling*, Taylor & Francis, Boca Raton, FL, 2007.)

That is why we breathe faster during exercise: the minimum work rate frequency moves higher as we inhale more air. Not only that, but airway caliber, airflow waveshape, the ratio of inhalation time to exhalation time, and lung midposition appear to be adjusted to reduce energy expenditure.

Cardiac parameters related to blood pumping may also be optimized to reduce energy use. Model results appear to be consistent with an optimization hypothesis (Figure 6.14.7).

6.14.4 Genetic Variability

> Biology has traditionally had more success when driven by good data rather than by theory.
>
> —**Robert H. Carlson**

It was mentioned in Section 5.3.8 (Genetic Variation) that there is genetic variation within a species that cannot be easily explained. The principle of survival of the fittest (natural selection) should result in the elimination of all but the most survival successful genes. This means that genes not optimum for survival in a competitive environment should not persist, no matter how small their disadvantage. However, a few of them remain, and they give the species the possibility that, should the environment change, there would be genes already present that could be better able to allow the species to adapt correctly.

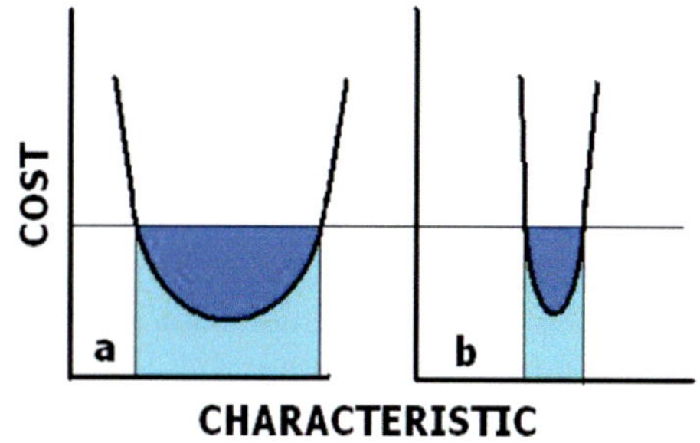

FIGURE 6.14.5
Illustration of broad (a) optimum and narrow (b) optimum. For the same level of cost, the broad optimum allows a greater range of the optimized characteristic. Biological systems seem to prefer broad optima.

If this is considered to be an optimization problem, then the unexplained genetic variation could be a consequence of the broad optima that characterize biological systems. Optima can be broad or narrow (Figure 6.14.5). Narrow optima are very selective, and don't tolerate much variation before the cost of locating at a nonoptimal point becomes too high to be sustained. Broad optima can still have the same optimum, but the costs of deviation from the exact optimum point do not rise significantly quickly. Biological systems seem to have broad optima.

Therefore, genetic variation could be explained by the fact that carrying nonoptimal genes does not turn out to be too expensive for the species as long as the result of those genes being present does not differ too much from the results of the fittest genes. As with all other biological optima, this means of genetic optimization turns out to be energetically less expensive (and maybe more likely for species survival) than an optimum that confers too much advantage to the best genes.

The broad genetic optimum can be a consequence of several properties characteristic of living things. The first is the ability of a living system to sense its environment and to communicate in various ways. The second is the ability to respond to the immediate environment that results in a mathematically chaotic system. That is, the end result is dependent on the pathway it takes to get there. The third property is the appearance of alternate forms that differ very little in energy levels. As a result of these properties, it is easy for biological systems to have varying forms with very little additional energy costs, but which have been determined by the history of the organism. In the case of the genome, each individual's genes are determined by history of the individual and its predecessors, but there persists genetic variability within a population because of somewhat different histories. The genes, once in place, cannot be routinely changed (despite some possibility of random mutations or of epigenetic expression modifications). In the case of optima for breathing rate, heart rate, and stride length, responses can vary with time.

6.14.5 Ecological Optimization

> He who waits to do a great deal of good at once will never do anything.
>
> —**Samuel Johnson**

Likewise, there is an optimization that occurs within an organism to conserve scarce nutrients. Iron, for instance, is recycled from worn-out red blood cells to be used by new red blood cells. The kidney reabsorbs glucose, bicarbonate, some sodium (depending on intake), potassium, and chloride in order to conserve them and reuse them. Desert animals save water by excreting a very concentrated urine.

Entire ecosystems are also models of efficiency. The populations of species that cohabitate these ecosystems complement and supplement each other. Waste from some species is used as nutrition by others; some plants shade others and cool them, and keep them from drying; large trees protect less hardy species from the wind; nutrients that would ordinarily be lost to the ecosystem are stored in the bodies of organisms and recovered when they die.

While this is not exactly the same as optimization that lends itself to mathematical modeling, it is optimization nevertheless. If a more efficient BU is introduced, it likely has a competitive advantage over existing BU, and the less efficient BU is soon displaced. In this way, biological systems are constantly improving their utilization of scarce resources.

6.14.6 Mode of Action

> One idea is that there are two components in the brain—a teacher and a tinkerer. The tinkerer is constantly adjusting things, which produces the background noise; the teacher goes back and fixes or optimizes the changes.
>
> **—Sebastian Seung**

Of course, the body doesn't act as a computer dedicated to calculating optimum settings for the muscles and tissues. Optimization and its calculation are mathematical abstractions. Instead, it appears that mechanoreceptors (to measure stretch or force) and chemoreceptors (to measure blood O_2 or CO_2) are monitored for their outputs (Figure 6.14.6).

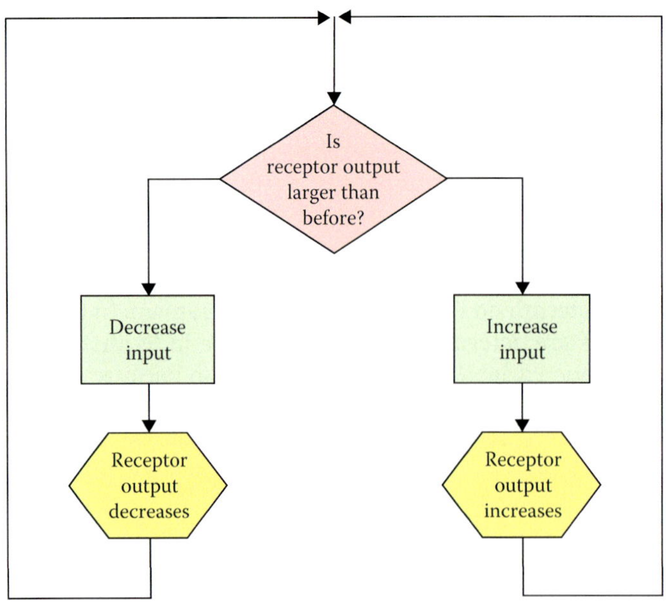

FIGURE 6.14.6
Trial and error means to find an optimum. This scheme only requires tracking whether receptor outputs are becoming smaller or larger. If the receptor output decreases, then a change is made to increase its output. If the receptor output increases, then a change is made to increase its output. No complicated mathematics is required.

Small differences in neural signals to the tissues result in small changes in receptor outputs. If the oxygen content of the blood, for instance, decreases, then a different neural signal is sent until the oxygen content increases. And then a larger signal in the same direction can be sent to see if oxygen content rises still farther. If it does, then an even larger signal is sent. If the oxygen content falls, then a smaller signal is sent. By this means, the optimum point can be reached by trial and error, and the optimum point can be tracked as it changes because the neural signal is always being modified by a small amount.

Evidence of the trial-and-error approach is given by variations that occur around the optimal point. It has been already discussed that the walking stride appears to be optimized to reduce the energy expenditure for locomotion. The time for each stride, however, varies by about 6% for adults and 8.5% for children (Chau and Parker, 2004). The frequency of breathing also varies around the optimum value (Figure 6.14.7), thus allowing the optimum breathing frequency to track changes in input conditions such as oxygen consumption.

Heart rate variability has been used as an indicator of vagal nerve control. Heart rate variability appears to give the cardiovascular controller precise information for optimum blood pressure control (Figure 6.14.8).

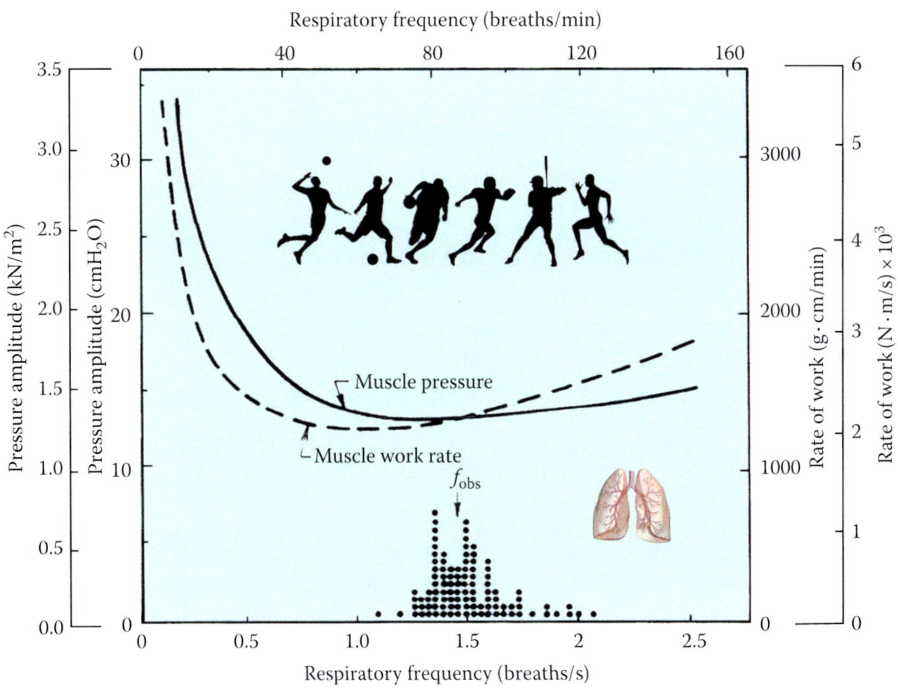

FIGURE 6.14.7
Observed human breathing frequency is not controlled at the exactly optimal value. Rather, frequency varies around the optimal values. Such variation allows better tracking of changes in breathing demand. (Redrawn from Johnson, A.T., *Biomechanics and Exercise Physiology: Quantitative Modeling*, Taylor & Francis, Boca Raton, FL, 2007.)

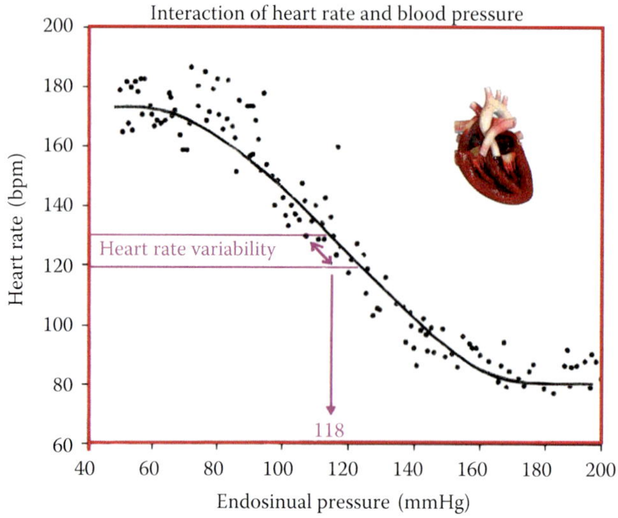

FIGURE 6.14.8
Variation in heart rate appears to be designed to determine empirically the maximum sensitivity for blood pressure control. (Redrawn from Moser, M. et al., *IEEE Eng. Med. Biol. Soc. Mag.*, 27, 29–37, 2008.)

APPLICATIONS AND PREDICTIONS

1. People will learn to perform repetitive muscular motions in the most efficient way possible.
2. Left to themselves, inefficient BU will not survive.
3. To design a new physical process, check first to see if there is a similar process that occurs in nature. Chances are the natural process will use the most efficient means under the circumstances.
4. Desert plants and animals will have developed efficient means to conserve water.
5. Breathing during exercise will be at a faster rate than at rest.
6. Hibernation is a means to conserve resources.
7. The shape of plants is optimized to collect solar energy in the presence of other environmental constraints such as temperature and water availability.
8. Eskimos and polar bears hunt seals in similar fashion because it is most efficiently done that way.
9. BU with specialized genes will not survive environmental changes.

6.15 Biological Units Alter Themselves to Protect against Harsh Environments

The best way to predict the future is to invent it.

—**Alan Kay**

BU are very adaptable, and can tolerate a wide range of environmental conditions. There are limits, however, and exceeding these results in BU changing forms or states to be able to survive particularly harsh conditions. Of environmental qualities to which the BU must adjust, those of temperature, water availability, and food scarcity are the three that seem to be the most likely to cause major BU accommodations.

These alternative forms and states are of great consequence. Inactive states in animals must be recognized and incorporated into designs if success is to be expected. Inactive forms for bacteria can eventually cause food spoilage or human diseases. Inactive forms for plants are useful means for reproduction. Inactive organs during trauma must be corrected before healing can begin. The biological engineer must then be prepared to deal with these altered states to produce successful endeavors involving BU.

6.15.1 Torpor, Hibernation, and Estivation

Now let us suppose that such a vessel is divided into two portions, A and B, by a division in which there is a small hole, and that a being, who can see individual molecules, opens and closes this hole, so as to allow only the swifter molecules to pass from A to B....He will thus, without expenditure of work, raise the temperature of B...in contradiction to the second law of thermodynamics.

—**James Clerk Maxwell**

Animals change metabolic states in response to temperatures either too high or too low, or the nonavailability of adequate food. Hummingbirds, for instance, have very high metabolic rates (see Section 7.4.4, Metabolism) that must be satisfied by eating roughly 2/3 times their own weights of nectar and insects per day. When hummingbirds arrive at breeding or wintering sites before flowers are abundant, when flowers they visit have decreased nectar production, or when their feeding is reduced by storms, hummingbirds switch to a state called *torpor* (Molles, 1999). In torpor (Figure 6.15.1), metabolism, heart rate, respiration rate, and body temperature are all lowered in order to save energy and allow the animal to survive for relatively short duration environmental extremes (Campbell et al., 1999).

Many small mammals and birds with high metabolic rates exhibit a daily period of torpor adapted to their feeding patterns. This includes most bats and shrews that feed at night and revert to states of torpor during the day (Campbell et al., 1999). Chickadees in cold northern forests may drop their body temperatures as much as 10°C at night. This enables them to survive on energy stored in their tissues.

Torpor appears to be controlled by an internal biological clock. Even if food is made available at all hours to a shrew, it still experiences torpor (Campbell et al., 1999). The need for sleep, and the concomitant slight decrease in body temperature, may be evidence of vestigial torpor in humans.

Hibernation is long-term torpor during which the body temperature is lowered in winter when ambient temperatures are low, and food is scarce. As the days before winter shorten,

FIGURE 6.15.1
When food availability is low, hummingbirds switch to a state of torpor. (Redrawn from Molles, M.C. Jr., *Ecology: Concepts and Applications*, McGraw-Hill, New York, 1999.)

some animals will eat huge quantities of food before hibernating. Ground squirrels, for example, more than double their body weight in a month of gorging (Campbell et al., 1999). The body temperature of hibernating ground squirrels may drop to just above freezing, or 2°C (Molles, 1999). The metabolic rates of hibernating marmots (short-legged rodents also called groundhogs or woodchucks) may fall to only 3% of active levels (Molles, 1999). When hibernating, the woodchuck's heart rate drops from a normal 80 beats/min to 4 or 5/min. Its body temperature falls from 37°C to −3°C, below freezing and the lowest known body temperature of any living mammal.

The altered metabolic state during the summer is called *estivation*. Characterized by slow metabolism and inactivity, estivation allows an animal to survive long periods of high temperatures and scarce water supplies. The metabolic rates of estivating long-neck turtles may fall to 28% of their normal metabolic rates (Molles, 1999).

Both hibernation and estivation appear to be induced by changes in the length of daylight. Artificial lighting can sometimes alter tendencies to hibernation or estivation.

Even if they do not react as drastically to extreme environmental temperatures, many insects and animals become inactive during at least part of the day. During cold nights, snakes, lizards, and other small animals hide in burrows where temperatures do not fall as low as outside. Insects such as the predatory tiger beetle hide in the shade (Figure 6.15.2) during the day when surface temperatures on the black sandy beaches of New Zealand, where it lives, reach 70°C (Molles, 1999).

Animals with reduced metabolism still produce metabolic wastes, although at a rate far lower than they would at higher metabolic rates. Cellular wastes would build up to poisonous levels (see Sections 6.4, Wastes and 6.12, Chemical Stress) if allowed to accumulate over a long time. Hibernating animals solve this problem by returning to a semi-wakeful state every few weeks, during which time their body temperatures ascend to normal. Their cells flush their wastes, which the animal then excretes. In any engineering design

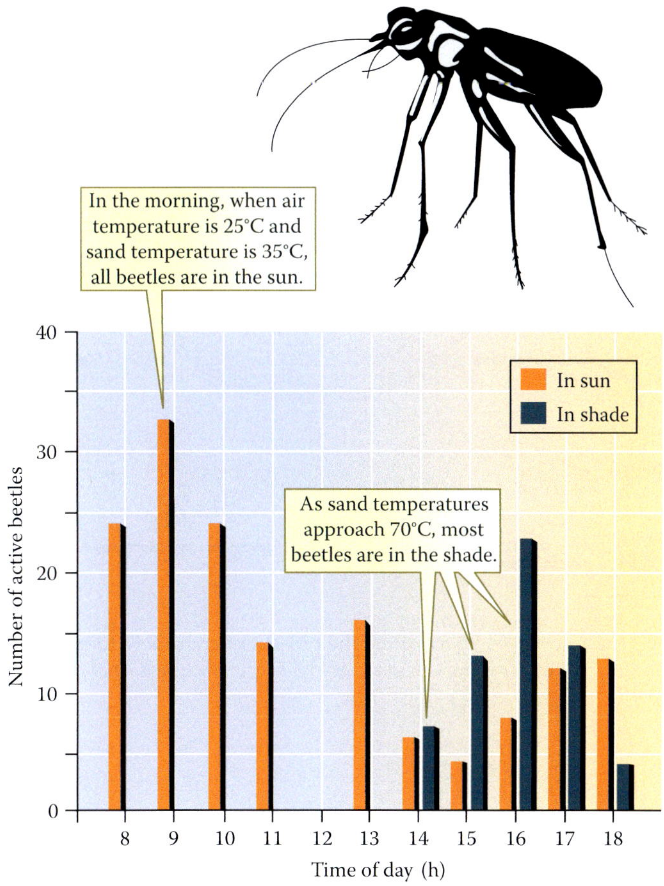

FIGURE 6.15.2
Tiger beetles seek out environmental temperatures that cause the least stress. (Redrawn from Molles, M.C. Jr., *Ecology: Concepts and Applications*, McGraw-Hill, New York, 1999.)

dealing with metabolically altered states, including artificially induced hibernation for medical or space travel purposes, provision must be made to allow these wastes to be voided, both at the cellular and at the organismal levels.

6.15.2 Endospores

> A competitive world offers two possibilities. You can lose, or, if you want to win, you can change.
>
> **—Henry Ford**

Actively metabolizing microbes are described as *vegetative*. When essential nutrients are depleted, some prokaryotic microbes form specialized inactive cells called *endospores*. These are highly durable dehydrated cells with thick walls and additional layers. They are formed inside vegetative cells by the process of *sporulation* or *sporogenesis* (Tortora et al., 2001).

Endospores contain a newly replicated bacterial chromosome and a small portion of cytoplasm inside a double-layered membrane, all enclosed within a thick spore coat

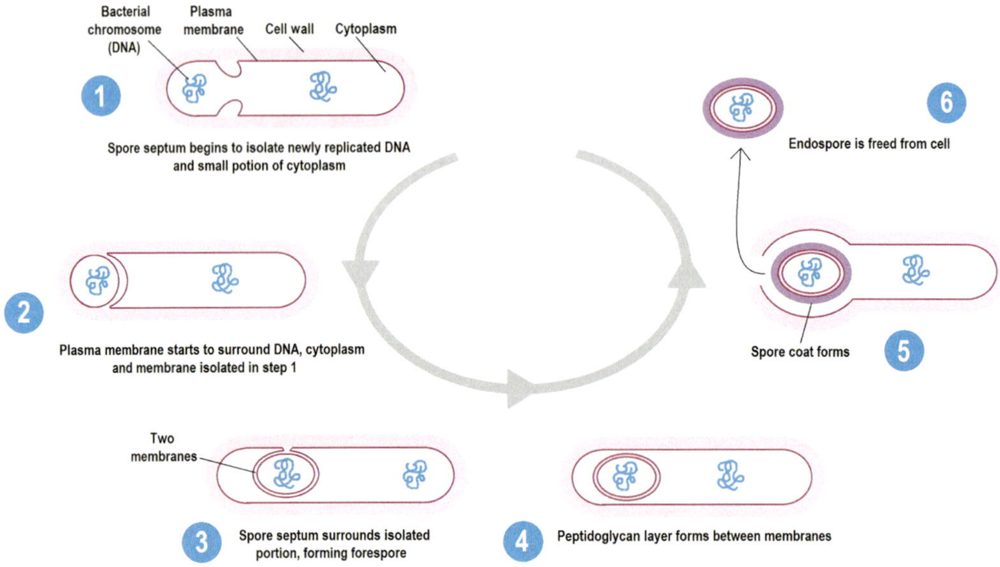

FIGURE 6.15.3
Formation of a bacterial endospore. (a) Sporulation, the process of endospore formation. (b) An endospore in *Clostridium difficile*. (Redrawn from Tortora, G.J. et al., *Microbiology: An Introduction*, Addison Wesley Longman, San Francisco, CA, 2001.)

of protein (Figure 6.15.3). This coat makes the endospore resistant to many harsh chemicals (Tortora et al., 2001). Most of the water present in the endospore is eliminated, and endospores do not carry out metabolic reactions.

Endospores can remain dormant for thousands of years. This durability and the fact that some very virulent bacteria produce endospores is the reason that they are so dangerous. They are resistant to processes that normally kill vegetative cells.

The germination of the endospore into its vegetative form is triggered by physical or chemical damage to the protein coat. Enclosed enzymes then break down the layers surrounding the endospore. Water enters, and metabolism resumes.

Sporulation is not a reproductive process. One vegetative cell forms one spore, and one spore forms one vegetative cell. There is an extremely small chance that any given spore will return to its vegetative state.

Two genera of bacteria form endospores. The first is the genus *Clostridium*, an obligate anaerobe. *C. tetani* causes the disease tetanus; *C. botulinum* causes botulism; *C. perfringens* causes gas gangrene and foodborne diarrhea. The second genus is *Bacillus*, which includes *B. anthracis*, causing anthrax, *B. thuringiensis*, a bacterial pathogen, and *B. cereus*, which can cause a form of food poisoning (Tortora et al., 2001).

Processing food to ensure safety requires the reduction of viable endospores to a negligible level. Because it is impossible to ensure that all spores are completely eliminated, a standard reduction rate of 10^{-12} has been accepted. That means that one surviving endospore for each 10^{12} cans processed is acceptable. Through a combination of temperature and time, thermal processes for food sterilization can be designed (Teixeira, 1992; Johnson, 1999). Similar considerations are important for the sterilization of medical devices and bioreactor growth media.

ANTHRAX THROUGH THE MAIL

Terrorist attacks on the World Trade Center Buildings in New York City on September 11, 2001 were soon followed by letters addressed to the headquarters of a tabloid publisher in Boca Raton, Florida; NBC News in New York City; Microsoft, Inc. in Reno, Nevada; and the Congressional Office Building, Washington, DC. In each of these letters was a little white powder, which testing confirmed contained endospores of *B. anthracis*.

Because of its easy availability and the durability of its spores, *B. anthracis* has often been mentioned as a tool for bioterrorists. Anthrax has, at times, been an important disease of livestock in many parts of the world. It wasn't until 1876 that the German Robert Koch proved that anthrax was caused by a microbe that can change form into one that is extremely inert and hardy (de Kruif, 1926). That was the reason that sheep could contract anthrax in fields that had not seen sheep in many years.

Before 1996, it was appallingly easy to buy anthrax samples from one of the 500 culture collections kept by universities, governments, or private companies. For about $50, anyone could buy anthrax cultures and turn them into bioweapons.

In 1995, however, Aryan Nations member Larry Wayne Harris purchased plague bacteria from the Virginia company American Type Culture Collection and stored the vials in the glove compartment of his car. When he was caught, Congress finally paid attention and passed legislation in 1996 restricting access to deadly microbes including anthrax, Ebola, smallpox, and yellow fever. Since the day of the terrorist attacks on New York City, it is almost impossible to obtain any of these microbes legally (Lemonick, 2001).

When exposed, people can contract the disease in several different forms. Inhaled spores are the deadliest, and need the most powerful antibiotics administered early. The skin form of the disease, from spores that enter the body through small cuts or abrasions in the skin are much less serious and easier to treat.

Anthrax endospores are only 1–5 μm in diameter. It takes an average of 8,000–10,000 inhaled spores to kill an average person (LD_{50}), although inhaling only 100 spores may kill some people. Once they have begun to grow, the anthrax bacteria produce a toxin that is the real cause of death.

In the twentieth century, 18 Americans had fatally inhaled anthrax spores. The victims include a San Francisco woman who played bongo drums made from infected skins. Others included gardeners who handled fertilizer made with ground bone from infected animals (Park, 2001). Goat and sheep skins from the Mideast are a small but persistent source of anthrax in the U.S. Five more people died of anthrax in the U.S. in 2001 as a result of the anthrax-laden letters they handled.

6.15.3 Seeds and Spores

> Happiness held is the seed; happiness shared is the flower.
>
> —**Samuel Johnson**

Plants also have alternate forms for dealing with harsh environments. These usually take the form of seeds and spores.

To reword an old quip, a plant is a seed's way of making another seed. Seeds are reproductive forms that each contain an *embryo* and *endosperm* tissue to supply nutrients.

There is a hard-outer protective shell called the *seed coat*. *Gymnosperm* plants produce rudimentary seeds from male and female cones, and the plants are called conifers. *Angiosperms* are flowering plants that also produce grains and fruits. Angiosperms are divided into *dicotyledons* and *monocotyledons* depending on the number of seed leaves (or cotyledons) present in the embryo.

Unlike microbial endospores, seeds are not completely metabolically quiescent. Like endospores, however, seeds do not grow unless they are situated in a favorable environment.

During seed maturation, the embryo dries and becomes relatively inactive, and remains so during its dormant phase. Seeds may remain viable (able to grow into a mature plant) for many years. Some seeds lose viability after a few years. Others may remain viable for hundreds of years; seeds of *Canna compacta* can live for at least 600 years (Taiz and Zeiger, 1998); *Cassia multijuga* seeds were found to be viable after 160 years, and seeds of *Verbascum blattaria* remained viable for 100 years. Storage conditions will affect seed viability.

In order for a seed to germinate (defined as resumption of embryo growth), water and oxygen must be available, and the temperature must be suitable. Optimum temperatures to germinate are different for different seeds. Some seed coats are so tough that germination does not occur until the coat is damaged or removed (Taiz and Zeiger, 1998). In coat-imposed dormancy, the seed coat may prevent water uptake, provide mechanical constraint to the growth of the embryo, interfere with oxygen permeability, retain inhibitor chemicals contained within the seed, or produce inhibitor chemicals (Taiz and Zeiger, 1998).

In other species (i.e., European hazel, European ash, and peach trees), the cotyledons can exert an inhibitory effect on germination growth. If the cotyledons of a peach embryo are removed at an early stage of development, the plant shifts from extremely slow growth and dwarf size to normal growth (Taiz and Zeiger, 1998). It is thought that embryo growth inhibition is due to the presence of abscisic acid (ABA), a growth inhibitor, as well as to gibberellic acid (GA).

Spores are produced by slime molds, by fungi, and by vascular plants, the largest group of which are ferns. Spores are smaller and less complex than seeds: many are single reproductive cells that may be as small as 1 μm in diameter (Simpson et al., 1957). Fungal spores may be either sexually or asexually produced, depending on the stability of the environment and the part of the life cycle in which the fungus finds itself. Airborne spores are ubiquitous, and have even been found more than 160 km above the Earth (Campbell et al., 1999).

Spores are reproductive bodies and are usually produced when conditions are favorable for growth. Like seeds, however, spores are able to withstand environmental stress until such time as environmental conditions are favorable for growth.

6.15.4 Plant Responses

> We are survival machines—robot vehicles blindly programmed to preserve the selfish molecules known as genes.
>
> **—Richard Dawkins**

Vegetative plants produce a natural stress hormone, abscisic acid (ABA) to help them survive drought by curtailing growth. Plants exposed to environmental stress, such as drought, can adjust their responses so that they respond better to the next episode of drought. Plants that have not been prestressed wilted faster than those that had survived a previous period of dryness. This adaptation has a limited memory time, however. After a certain period of adequate moisture, the drought tolerance can be lost (Ding et al., 2012). There are soil microbes that may also help to enhance plant drought tolerance.

There are plant structures that are meant to enable the plant to survive through the periods of unfavorable conditions. There are bulbs, tubers, and corms that enable a plant to survive cold winters or dry summers by using stored water and nutrients. The storage of body fat and water in animal bodies serves the same purpose. Although these are not alternative forms or states, they are enhanced while the environment is mild to be used when the environment is harsh.

6.15.5 Response to Hemorrhage

> [Competition] requires us to be tough-minded, never hard-hearted.
>
> **—John Kerry**

In another example of a part of an organism that changes its state in the face of brutal environmental conditions, we consider the effects of hemorrhage on the human body. A great loss of blood, even bleeding into internal tissues, can lower cardiac output to the point where blood pressure falls dramatically. In response, the blood vessels constrict to increase resistance (and thus blood pressure), and to reduce blood stored in them. The veins, in particular, are normally blood storage vessels, and these constrict to move blood back to the heart faster. The contraction of the spleen discharges more blood into the circulation. The heart begins to beat faster. The levels of circulating hormones increase dramatically. Among these are vasopressin, or antidiuretic hormone (ADH), glucocorticoids, aldosterone, erythropoietin, and catecholamines. The first three of these directly affect the kidney, and it essentially stops producing urine (Ganong, 1963; Ruch and Patton, 1966). This altered state of the kidney helps the body cope with the trauma of severe hemorrhage.

6.15.6 Psychological Trauma

> There is no exercise better for the heart than reaching down and lifting people up.
>
> **—John Andrew Holmer**

Humans who have experienced some severe psychological traumatic event may even develop an altered psychological state to deal with the event. The sudden death of a loved one, involvement in a serious car accident, or combat experience can produce a traumatic reaction that may include a denial state characterized by sleep disturbances, amnesia, fatigue, and headaches (Smith, 1998). *Post-traumatic stress disorder* (PTSD), also called "shell shock" or "battle fatigue" is common to those who have had to face extreme distress in military or civilian life. PTSD is characterized by negative emotional reactions beyond coping.

It has been found that the high levels of the stress hormone adrenaline is associated with the formation of trauma memory, and that people in whom adrenaline production has been blocked suffer less post-traumatic stress disorder than those who produce the hormone. Thus, quadriplegics (those with the spinal cord severed above the arms) suffer less stress disorder than do paraplegics (those with spinal cords severed above the lower limbs only); the neural connection between the brain and adrenal glands is interrupted for quadriplegics (Lemonick, 2007).

Example 6.15.1: Using Anaerobes to Combat Cancer

Endospores are not all bad. Researchers have injected spores of *Clostridium novyi* into animals with cancerous tumors. The spores do not germinate unless the environment is anaerobic, which it is in oxygen-starved tumors. After the spores germinated, the bacteria consumed cancerous cells from the inside of the tumor. Tumors either disappeared or shrank dramatically. The bacteria were engineered to lack their toxin-producing gene, and so were otherwise safe. Not only anaerobic bacteria, but also other viruses and bacteria, engineered to exploit the unique characteristics of tumors, are being harnessed to combat cancer.

Example 6.15.2: White Nose Syndrome in Bats

Since 2006, millions of bats in the eastern U.S. have died from a fungal disease called White Nose Syndrome. The disease attacks bats while they hibernate in caves in the winter. The disease irritates the bats and wakes them from their deep sleep. This increases their metabolism to nearly summertime levels. In winter, there is no food available, and the bats die of starvation. The disease is so pervasive that all the bats in the affected area die out. They have long generation times and low reproductive rates, so the threat is very serious.

As might be expected from a biological perspective, however, not all bats have been killed by this serious disease. Some bats are therefore developing resistance to this disease (Langwig et al., 2016).

Bats are important ecologically because of the prodigious numbers of insects that they eat. One hundred fifty bats can eat millions of potentially damaging insects in a year.

APPLICATIONS AND PREDICTIONS

1. Endospores are relatively easy to destroy after they have transformed into vegetative cells.
2. Endospores are so small that simple dust masks will not protect against them.
3. The chances of any specific endospore to re-sprout into vegetative form are extremely small.
4. Seeds may need special treatment in order to germinate.
5. Altered states will make the organism much less destructible, but also much less recognizable.
6. Human blood pressure must be restored quickly in order to enhance chances to survive traumatic injury.
7. It will be difficult to protect against aerial contamination by spores.
8. Buffers in the blood protect against pH change.
9. Starving animals will survive on stored fat and protein.

6.16 Biological Units Cooperate with Other Biological Units

A man must live like a great, bright flame and burn as brightly as he can.

—Boris Yeltsin

One possible interaction BU have with other dissimilar BU is cooperation. There can be no better example of this than the cooperation of all the tissues and organs of the body to form an effectively functioning whole organism. In animals, the gills or lungs gather oxygen, the heart circulates blood, the muscles locomote, the mouth ingests food, the gut digests food, the excretory system eliminates waste, the liver processes biochemicals, various endocrine organs help to maintain homeostasis, the skin protects the interior of the organism, and the nervous system coordinates workings of all of these. In plants, the root hairs absorb water and nutrients from the soil, the xylem and phloem transport materials from one place to another, the leaves act as the chemical factories for the plant, and cells in various locations emit chemicals that act to communicate with itself and other plants. The individual cells have sacrificed the ability to reproduce as individuals for the survival of the whole group of cells.

Although one might be tempted to yawn and question the importance of this type of cooperation, there are instances when this cooperation breaks down. Diseases such as cancer, lupus, and multiple sclerosis represent states of noncooperation. There are other times when certain organs or tissues cannot uphold their ends of the bargain (diabetes, congestive heart failure, schizophrenia, or amyotrophic lateral sclerosis), and the entire organism dies because the cooperative effort is not upheld.

Cooperation, it appears, evolved in life amid a highly competitive environment. Despite the direct disadvantages that helping others confers on an individual, there are several ways that cooperative tendencies could survive and grow through evolution in a population (Nowak, 2012):

1. Direct reciprocity. Doing something positive for another and the other does something positive for the first.
2. Special selection. Clusters of cooperating individuals exist amid highly competitive environments, as in neighbors helping other neighbors.
3. Genetically related. Helping kin promotes survival of related genes.
4. Indirect reciprocity. Cooperating with another individual of higher stature reflects back on the one of lower stature and confers rewards.
5. Greater good. Cooperating with other individuals of a group helps the whole group to survive and prosper.

Many animals live in social networks and cooperate together for mutual benefit. Indeed, cooperation may be a more powerful determinant of evolutionary change than is competition (Dugatin and Hasenjar, 2015).

Human cooperative association has been found to be more important than competitive self-interest. Research has shown that team members were chosen by people for their generosity and character (cooperation) rather than for the material advantages that they could bring to the partnership (competition). The conclusion is that humans look beyond concrete outcomes even if the material costs to the individual doing the choosing are significant (Hackel et al., 2015).

6.16.1 Symbiosis

> No man really becomes a fool until he stops asking questions.
>
> —**Charles P. Steinmetz**

Symbiosis (or mutualism) is a relationship between dissimilar organisms in which both partners benefit (Hale et al., 1995). An example is the symbiosis between the hermit crab *Pagarus* and the sea anemone *Adamsia palliata*. The anemone attaches to the crab's shell and obtains food scraps from the crab. The crab is camouflaged by the anemone and defended by its stinging cells.

The lichen is a composite organism formed by the symbiotic association of a green alga or a cyanobacterium and a fungus. The fungus gains oxygen and carbohydrates from the photosynthetic alga or cyanobacterium. The alga or cyanobacterium gains water, carbon dioxide, and mineral salts from the fungus. The fungus also provides protection from desiccation. Lichens are very common on trees and rocks in cooler unpolluted areas.

Endophytes are microscopic fungi that live inside other plants (Lane, 2006). In return for a safe haven and supply of needed nutrients supplied by the plant, endophytes produce various biochemicals that help the plant survive and persist (Khan et al., 2012). Two important forage plants infected with endophytes are tall fescue and perennial ryegrass. Endophytes in these grasses secrete a wide range of compounds, including toxic alkaloids that suppress insect attacks and can also be toxic to grazing animals. After the ingestion of endophyte-containing grasses, animals can exhibit symptoms of malaise, reduced vigor, and neural incoordination. The result is that the plants grow better and the predators (insects and grazing animals) are handicapped.

Probably, the ultimate case of symbiosis involves cellular inclusions called *mitochondria* and *chloroplasts*. Mitochondria are small cylindrical bodies within eukaryotic cells that function as the chemical powerhouses of these cells (see Sections 5.3.12, Mitochondria, and 5.5.1, The Cell). It is in the mitochondrion that ATP is formed through the biochemical reactions of the Krebs cycle (see Section 3.10, Energy Compounds). Mitochondria are self-replicating within the cell and their numbers increase as cellular energy needs increase. They contain their own DNA (mtDNA) separate from the DNA in the cell nucleus.

Chloroplasts are lens-shaped organelles in higher photosynthetic algae and plants. their own DNA that governs replication of chloroplastic proteins. Chloroplasts are the sites for photosynthesis and contain pigments, including chlorophyll.

It is believed that mitochondria and chloroplasts evolved from prokaryotes that became residents within larger host cells. Thus, biochemical mechanisms to exploit the environment to produce energy and nutrients for the earliest living cells are today part of all living things.

CHIMERA

There is a rare condition called *chimera* (named after a mythical creature having a lion's head, goat's body, and dragon's tail) wherein two or more non-identical fetuses fuse at an early stage in the womb. The body of the resulting person (or animal) is composed of non-identical cells with different DNA. This means that the germ cells and the somatic cells can have different DNA. A DNA match between parent and child could then show that they are not directly related. Chimera are also found in plants.

In the culture of embryonic stem cells, chimeric germline animals are produced and grown. Cell fusion techniques are used in this process.

It is not uncommon that some fetal cells remain with the mother after pregnancy and some of her cells remain with the fetus. These errant cells, contributing to *microchimerism*, can eventually reside in almost all organs and tissues of the host. Chimeric cells can also arise between twins, and from blood and marrow transfusions. They can sometimes cause immunity problems, but other times assist the immune system.

(*Continued*)

CHIMERA (Continued)

The general concept of chimera can be considered to be applicable to the sexual fertilization process between male and female gametes. The resulting zygote contains genes from both parents. Symbiotic relationships that develop between two dissimilar and unrelated species could also be called chimera. It has also been hypothesized that the larval forms of various animals are the result of hybridization between species (Williamson and Vickers, 2007), the larval forms being added to the adult forms through interspecies cross-fertilization. The resulting two sets of genes are then expressed sequentially rather than at the same time.

Chimera are also present in some forms of immunotherapy, when T-cells are modified in order to be able to recognize cancer cells as alien to the body (see Section 6.20.3, Redundancy).

THE NITROGEN FIXING DANCE

(Excerpted from Lane, 2009)

During the growing season, a most extraordinary phenomenon occurs every hour in our fields—nitrogen fixation. Modest legume plants with their little root nodules quietly extract nitrogen gas from the air and convert this nitrogen into compounds that plants use to create proteins. Although we take this biological process for granted, it's so critical that human society might not exist without it. There's really nothing modest about it at all.

Actually, this story has two amazing parts. The first part is the general process of how plants "fix" nitrogen—which means capturing nitrogen from the atmosphere. The second part of the story is how free-living bacteria infect a willing plant and together create a highly specialized nitrogen-fixing factory in the roots.

Pulling nitrogen from the air is no trivial matter. Although nitrogen is 78% of our atmosphere, this nitrogen occurs as the stable N_2 molecule. The two nitrogen atoms are bound together with a very strong triple bond so strong that if we want to break this bond ourselves, we must use the *Haber Process*, a heavy-duty industrial procedure involving very high temperatures, high pressures, and metal catalysts. In fact, we rely on the Haber Process to produce the 100 million tons of synthetic nitrogen fertilizer used on farms each year.

In contrast, legume plants quietly fix nitrogen in their roots. No loud industrial clanging, no risk of high-pressure explosions or toxic fumes. Instead, in the tiny dark spaces in the soil, specialized gram-negative rod bacteria called *Rhizobia* (there are a number of related genera) infect the plant roots during early root development. These bacteria combine with plant tissue to form a highly organized root mass called a *nodule*. The bacteria contain an enzyme known as *nitrogenase*. This is the actual molecular complex that uses plant energy—originally captured in the leaves during photosynthesis—to break the N_2 triple bond, add hydrogen atoms, and convert the free N to ammonia (NH_3). The ammonia is then quickly converted to other

(*Continued*)

THE NITROGEN FIXING DANCE (Continued)

biologically useful nitrogen compounds. This is a classic *symbiotic* arrangement, where two species form a partnership in which both gain. The bacteria gain nutrients from the plant and a secure place to flourish and reproduce, while the host legume plant gains nitrogen that gives it a competitive edge in a harsh world where nitrogen is often in short supply.

The nitrogenase enzyme, however, deserves a little closer inspection. This is the guts of the fixation machinery. Nitrogenase consists of two large metalloproteins (proteins that contain metal): *dinitrogenase reductase* which contains iron, and *dinitrogenase* which contains iron and molybdenum. These metalloproteins are two of the most complex metal-protein arrangements known. And we can see why—as they intimately work in tandem to fix nitrogen, they routinely do something that the industrial Haber Process only achieves with high pressures and temperatures.

Because nitrogenase contains molybdenum, legumes require molybdenum to capture nitrogen. Therefore, if they are expected to add nitrogen to the soil, fields of clover or alfalfa must have more molybdenum than grass fields. But molybdenum is a micronutrient, so only a little is needed—maybe only a few grams per acre. The application rate depends on specific soil characteristics.

One characteristic of nitrogenase is that it is irreversibly inactivated by oxygen—which sets up an interesting metabolic conundrum. Root cells and Rhizobia require oxygen to survive, but oxygen will also shut down the nitrogen-fixing molecules that are their reason for existing. How can this be solved? Well, legumes have evolved an elegant solution: *leghemoglobin*. This compound is built much like the hemoglobin in our own blood, and it does much the same thing—hold and transport oxygen. The nodule cells synthesize leghemoglobin. Although oxygen can permeate through the nodule's outside shell, the leghemoglobin captures these oxygen atoms, holds them, and keeps them away from the nitrogenase enzyme. But at the same time, it transports enough oxygen to root cells and bacteria to allow them to respire properly.

And like our own hemoglobin, leghemoglobin is red when it's loaded with oxygen. So if you take a healthy nodule and carefully split it open, you can see a distinct pinkish color. That's the leghemoglobin.

Now for the rest of the story—nodulation—the formation of root nodules. If anything, this is even more remarkable than the chemistry of fixation.

It starts with a young legume seedling just beginning to send out roots. These roots release *flavonoid* compounds into the surrounding soil. If the right species of Rhizobia is present and detects these flavonoids, its *nod* gene (probably for nodulation) goes into action and produces a species-specific *Nod factor*, which then binds to surface receptors on the root epidermis cell.

Then the root epidermis cell begins to bulge outwards, forming a microscopic *root hair* that extends outwards and pushes into the soil. Calcium ions stream from the interior of the epidermis cell to the tip of the lengthening root hair. The Rhizobia bacterium then attaches itself tightly to the side of the root hair. As the root hair continues to grow, the bacterial Nod factor causes the root hair to change its direction of growth. Instead of continuing to grow outward, the tip of the root hair turns and grows back on itself, forming a tight clamp that looks a little like a bobby pin—trapping the

(*Continued*)

THE NITROGEN FIXING DANCE (Continued)

bacterium between the two parts of this clamp. Now sandwiched between two root hair cell walls, the bacterium reproduces and grows into a tiny microcolony. Then this bacterial colony projects a living microtubule—called an *infection thread*—down through the center of the root hair. This infection thread extends downward, working its way into deeper layers of cells. Parts of the infection thread fuse with the cell walls of some root cells. Bacteria populate the infection thread and then move into the nodule cells.

This is the beginning of a proto-nodule—an intermingled blend of bacterial and legume material. The Rhizobia then differentiate into *bacteroids*, which lie inside the nodule cells, surrounded by plant cell cytoplasm. These bacteroids synthesize nitrogenase, and the plant synthesizes leghemoglobin and other molecules and also provides nutrients to the bacteroids. The proto-nodule grows larger and differentiates into a well-organized protuberance on the root. There is now a fully functioning nodule.

Communication...recognition...attachment...growth...involvement...fusion... mutual benefit...nitrogen—a system for capturing atmospheric nitrogen that evolved piece-by-piece over millions of years. It's the epitome of species cooperation. The legume and the Rhizobia engage in a delicate duet, move for move, increasing complexity, almost like a dance.

6.16.2 Coevolution

> Nature uses extraordinarily ingenious techniques to avoid conflict and competition, and cooperation is extraordinarily widespread throughout all of nature.
>
> **—Juan Samaranch**

A little less intimate than symbiosis is the cooperation exhibited by coevolution. In this instance, two or more species have developed a mutual dependence that is very profound, even essential. Usually, this mutual dependence involves the forms and functions of physical features of one species that match the complementary forms and functions of another species. In other cases, one species has modified its behavior to match the complementary behavior of another species. The protozoan that causes malaria, *Plasmodium*, has developed a cooperative arrangement with the *Anopheles* mosquito, wherein the mosquito ingests *Plasmodium* from an infected individual and transmits the disease to another healthy individual during her next feeding. The mosquito is said to be the *vector* for the disease.

One of the most common cooperative associations is the one between nitrogen-fixing bacteria and plant hosts. Some plant roots contain nodules that contain the bacteria, protecting them from competition from other soil-borne bacteria. The plants supply the bacteria with photosynthate, and the bacteria supply the plant with nitrogen in a form that it can use (ammonium ions, nitrate ions, or amino acids). Other plants nurture nitrogen-fixing bacteria in close proximity, but outside their roots by excreting organic photosynthate for the bacteria. These cooperative relationships are very important economically and nutritionally (Brill, 1979).

COOPERATION BETWEEN HIPPOS AND FISH

It has long been known that fish and hippos are constant symbiotic companions. Fish clean hippos and are in turn nourished by the algae, parasites, and dead skin scraped from the hides of their hosts. The amazing thing, however, is that certain fish specialize in cleaning specific body parts. The carp *Labeo* is the main cleaner, using its wide rasping mouth to scour a hippo's hide. *Barbus* feeds on dung and cleans the cracks in the soles of the feet. Small *cichlids* graze around the tail bristles. And tiny *Garra* clean out the wounds.

Hippos deliberately splay their toes and spread their legs to provide easy access to the fish. The hippos even visit places where fish congregate in order to solicit cleanings.

The hippopotamus depends on fish to clean its body.

6.16.3 Plant Reproduction

I think that I shall never see a poem lovely as a tree.

—Joyce Kilmer

Nowhere is biological cooperation more apparent than for reproduction of higher level plants. Flowering plants did not appear before there were insects to pollinate them.

Some plants are licentious and make no effort to control where their pollen goes. Their flowers tend to be flat and wide, easily reached by all creatures and able to spread pollen through the air (see Section 5.1, Form and Function). Other plants are choosier.

Carelessly scattering pollen to the wind is a risky strategy in a tropical rain forest where leaves can block the pollen from reaching its goal. Only about 10% of Hawaiian species uses this strategy. Most tropical plants rely on birds, bats, bees, or bugs as partners in their reproduction to make sure that their pollen ends up in the right place. These plants entice their partners with pollen (a protein source) or nectar (a carbohydrate) to visit them.

The goal is to coat the pollinators' heads, backs, feet, or chests with golden, sticky pollen, which they carry to the next feeding spot.

Flowers meant to attract birds are brightly colored and with little, if any, smell because birds have a poor sense of smell but a refined color vision. They especially like red. These flowers are also found on strong stems that offer the birds a perch. Flies like meaty, rotten smells, and bats and moths that fly at night like white flowers that are easily seen in the dark.

Bees have a great sense of smell, but their vision extends into the ultraviolet range. Flowers meant to attract bees are sweetly scented and soft in color. They are attractive under ultraviolet light, and have guides leading to their nectar sources. The yellow stripe (beard) on an iris's throat is an example of this (Figure 6.16.1).

Honey bees of several species are used throughout the world for pollination of crops. Honey bee societies rank among the most complex of all cooperative insect societies, with such advanced features as strong dimorphism between queen and worker, elaborate division of labor according to age, precise control of nest temperature, and a remarkable system of communication based on dance language (Seeley, 1983). When honey bees discover good pollen and nectar sources, they return to the hive and recruit their sister bees to visit the same sources (see Section 6.19, Communication). Thus, cooperation among bees in the same hive greatly benefits them all.

Bumblebees do not recruit additional workers to flowers found to have a lot of nectar. Instead, individual bumblebees learn to specialize in one kind of flower, and the whole hive benefits collectively from each bee specializing in a different kind of flower (Figure 6.16.2).

In isolated locations where pollinators and their plant partners can coevolve, this mutual dependence has reached extreme specialization. Conditions on most of the world's tropical islands have been so isolated that plant defenses did not have to be highly developed.

FIGURE 6.16.1
The yellow beards on these iris flowers function by pointing the way to the nectar and pollen sources in the center of the flowers.

FIGURE 6.16.2
The individual bumblebee specializes in visiting certain types of blooms, and the whole hive benefits from cooperation among these bees. (Redrawn from Heinrich, B., *Am. Sci.*, 64, 384–395, 1976.)

No voracious deer or cattle ate these plants and no hooved animals trampled them. Instead, they concentrated their evolutionary energies on improving their reproductive processes. Over time, the plant and its winged companion evolved together to become better suited to each other and more incompatible with any other partner. One example is the *Trematolobelia singularis* native to Hawaii. It has flowers borne in clusters that have a distinctive curve that exactly fits the beak of the Hawaiian nectar-eating bird called the I'iwi. The bird's red beak and the plant's red flower are perfectly matched (Figure 6.16.3).

Clearing of Hawaiian forests has greatly diminished the population of I'iwi birds. Because its reproduction is so dependent on the I'iwi, the *Trematolobelia* is also an endangered species (Dewar, 2001).

FIGURE 6.16.3
The I'iwi bird's bill and the *Trematolobelia* flower have evolved together. (Redrawn from Dewar, H., Dangerous liaisons in Hawaii, *The Baltimore Sun*, 22 July, p. 2A, 2001.)

GUANACASTE LOVES EQUUS

The Guanacaste tree (*Enterdobium cyclocarpum*) is a large member of the pea family that grows in tropical dry forest in Costa Rica. This tree produces up to 5000 disk-shaped fruit per year, which fall to the ground when ripe. In prehistoric times, large herbivores such as ground sloths, camels, and horses roamed this area, ate the fruit, and dispersed the seeds. However, all these large animals became extinct about 10,000 years ago, leaving few large animals to consume the fruits and disperse the seeds.

About 500 years ago, Europeans introduced horses and cattle, which ate the fruits of the Guanacaste tree and dispersed its seeds around the countryside. Faced with the task of restoring tropical dry forest, ecologist Daniel Janzen incorporated horses into his management plan. Restoration would be accelerated because this and other trees would be able to reproduce according to the cooperative method that had evolved long ago (Molles, 1999).

6.16.4 Communal Benefit

> As a thinker and planner the ant is the equal of any savage race of men; as a self-educated specialist in several arts she is the superior of any savage race of men; and in one or two high mental qualities she is above the reach of any man, savage or civilized.
>
> —**Mark Twain**

Nest-weaving ants also provide an example of extreme cooperation for communal benefit. These nests are constructed from leaves. When a worker ant succeeds in folding a portion of a suitable leaf, other nearby workers join in the effort. They line up in a row and pull together. If the gap that remains to be closed is longer than a single ant's body, they form a living chain by holding one another's waists and pulling together (Figure 6.16.4).

FIGURE 6.16.4
When a single ant cannot bridge the gap between two leaves to be used in a nest, the worker ants arrange themselves in chains and pull together to close the gap. (Redrawn from Hölldobler, B. and Wilson, E.O., *Am. Sci.*, 71, 490–499, 1983.)

The combined force of these chains can be very large. When the leaves have been maneuvered into a suitable tent-like configuration, workers carry larvae from the interiors of existing nests and use them as sources of silk to bind the leaves together (Hölldobler and Wilson, 1983).

Various social insects, microbes, birds, and mammals have developed distinctive structures in which to live cooperatively. In addition to the example given earlier, bird nests differ according to species, ant hills are characteristic of the species, bee hives vary depending on the race, and termites of different kinds have much different habitat structures. It appears that the architecture for each of these is genetically determined and subject to the same evolutionary pressures as all other features of living organisms (Dunn, 2014).

Biofilms are complex, multilayered, multispecies consortia of microbes. These aggregations form sticky and persistent coatings on surfaces. Biofilms protect bacteria growing in them by slowing the diffusion of toxic substances, by trapping water, and by providing environments that allow microbes to thrive together where they might perish separately. Such close contact among species may be conducive to horizontal gene transfer through plasmid sharing.

Hepatocyte cells in the liver have been likened to ants in a colony (Tiffany-Castiglioni, 2004). Most ants are workers who perform multiple tasks, almost as interchangeable units. Hepatocytes perform multiple tasks as interchangeable units. Unlike the cells of the small intestine that do many things with very specialized cells, the liver does many things (such as absorb nutrients and ammonia from the blood, store vitamins, release glucose, make blood proteins, hormones, cholesterol, and bile, and detoxify poisons) with just one kind of hepatocyte cell. This is just another example of the parallelism that exists among BU.

The ecology of bacteria in the human gut has been called the *human microbiome* (Williams, 2010). These 10^{11} microbes in residence are different for each person. They stimulate immune responses, help detoxify food compounds, enhance new blood vessel growth, allow proper tissue development, produce vitamins, and transform undigestible compounds into useful forms. It's the balance of microbial populations that keeps individuals healthy or sick. Disrupting normal bacterial populations can lead to asthma, allergies, obesity, and weakened immune systems. Gut bacteria have a large influence on the success of oral vaccines (Huda et al., 2014; Courage, 2015).

Cooperation can aid individuals in the struggle to survive and reproduce. Grouping of individuals into herds, packs, schools, flocks, and swarms helps to obtain resources, avoid enemies, and find a suitable mate (Berryman, 1999). There is both intraspecific and interspecific cooperation that can lead to cooperative defense, hunting, and resource allocations. Many of these cooperative adaptations have evolved over time because more independent individuals have been less likely to survive and reproduce than more dependent individuals.

The hormone oxytocin enhances the tendency for an individual to trust others (Zak, 2008). Like oxytocin, the hormone vasopressin also has a strong effect on social and bonding behaviors in mammals and birds.

In many cases, group activity leads to efficiencies that can best be achieved through larger size; there are, of course, times when larger sized groups are counterproductive, and this leads to an optimum size for prevailing conditions. The biological engineer should always look for those particulars that support cooperation, and those that oppose it, in order to understand the degree of cooperation that exists in any given biological system.

A dramatic example of common benefit over individual benefit involves the programmed cell death (*apoptosis*) of single-cell organisms (Durand et al., 2011; Youngsteadt, 2011). Should an individual single-cell organism be stressed to the point that it cannot survive, it commits suicide, actively expending energy to shrink, disassemble its own DNA, and release beneficial chemicals into the surrounding environment. These substances stimulate surrounding cells, probably having the same DNA or DNA similar to the suicidal cells, to grow robustly, even more effectively than would fresh nutrient broth or leftovers from unprogrammed cell death (*necrosis*). Necrotic substances contain toxics that can sometimes poison their neighbors. Therefore, it appears that some cells that would have trouble surviving turn their demise into a benefit for related neighboring cells.

Stable as a cooperative society is, there is still an opportunity for cheaters to be rewarded. Examples of this abound, from rogue males mating with females belonging to the harem of dominant males to viral parasites using resources produced by other viruses necessary to the reproduction of both (Turner, 2005). Cheaters usually exist in small numbers on the periphery of a stable population, but can predominate if the right evolutionary rewards prevail. Game theory (Maynard Smith, 1982) can be used to predict eventual outcomes (Figure 6.16.5).

Social cooperation among groups of humans and among other animals has had profound effects on survival and reproduction over time (Stix, 2014), but, in addition, the expectations for behavioral responses of individuals within these groups can have large implications for successful biological engineering designs, influencing, for example, complexity of operation, meaningful symbology, placement of modules, and number of operations required. Consider, for example, piloting an airplane with its heads-up displays rich in symbols for different gauges and operations, and that requires a copilot to perform tasks supplemental to those of the pilot. Such a highly sophisticated and cooperative operation

	Cooperator	Cheater
Cooperator	Reward	Sucker's payoff
Cheater	Temptation to cheat	Punishment

FIGURE 6.16.5
The payoff matrix for an encounter between two individuals. If both are cooperators, they receive a mutual reward. A cooperator who meets a cheater loses something to the cheater (sucker's payoff). A cheater who meets a cooperator is tempted to cheat. When two cheaters meet, nothing is gained and punishment may ensue. (Redrawn from Turner, P.E., *Am. Sci.*, 93, 428–435, 2005.)

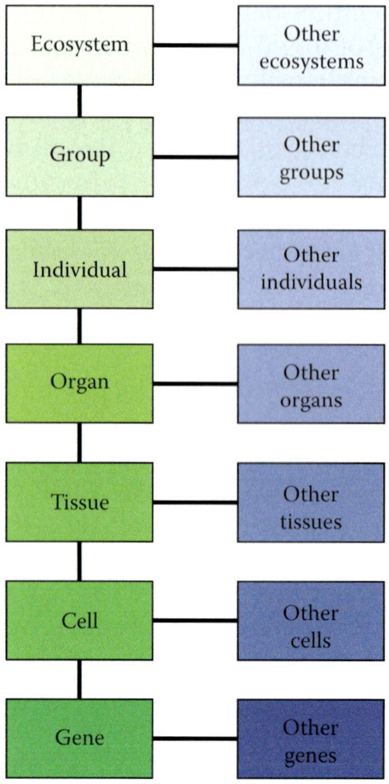

FIGURE 6.16.6
Evolution appears to operate at many different biological levels, from populations and ecosystems to the genome. The MLS concept relates evolution at each level to competitive interests at the levels below.

would be impossible without the design assumptions of intelligence, experience, and willing cooperation expected of adult human beings.

One concept of evolution is the multilevel selection (MLS) theory. The heart of this concept is that natural selection operates at many different BU levels, from the genetic to the ecosystem level (Figure 6.16.6). The higher levels provide the environment for the levels below. Adaptation at any level requires a corresponding process of selection at that level, with the fittest BU eventually dominating others. However, natural selection at that level tends to be undermined by selection at lower levels (Wilson and Wilson, 2008). Hence, natural selection at the group (or ecosystem) level can favor the survival and reproduction of certain groups over other groups, but individual selfish cheaters within that group can benefit more than individuals who give full support to the group instead of furthering their own individual needs. Likewise, natural selection at the individual level can favor the prevalence of a certain set of genes carried by the individual, but cheating genes within the genome can derive even more competitive benefits. Thus, the evolutionary process can exist at multiple levels and its success at each level depends on the suppression of those that take unfair advantage of others at that level (Buston and Wong, 2014). Humans, it seems, have successfully survived as a species because there is a predominant tendency toward cooperation and relatively few noncooperators (cheaters).

ANTS AND ANT FARMING

One of the most important developments in human civilization is the practice of sustainable agriculture. But humans were not the first to do this. Ants have been doing it for more than 50 million years. It has helped leaf-cutter ants become dominant herbivores and one of the most successful social insects.

Leaf-cutter ants have developed a system to keep their gardens pest-free, an impressive feat that has evaded even human agriculturists. Leaf-cutter ants put their freshly cut leaves in gardens where they grow a special fungus that they eat. New material is continuously incorporated into the gardens to grow the fungus and old material is removed by the ants placed in special refuse dumps away from the colony. The ants also care for their gardens; when a microbial pest is detected by the worker ants, there is an immediate flurry of activity as ants begin to comb through the garden. When they find the pathogenic microbes, the ants pull them out and discard them into their refuse dumps.

Because the ant gardens are maintained in soil chambers, they are routinely exposed to a number of potential pathogens that could infect and overtake the gardens. Sometimes ant colonies are overcome by fungal pathogens and the colony dies. Scientists have shown that a specialized microfungal pathogen attacks the gardens of the fungus-growing ants. These fungi directly attack and kill the crop fungus, and overrun the garden in a fashion similar to the way weeds and pests can ruin human gardens.

Some worker ants were observed to have a white wax-like substance across their bodies. When observed under a microscope, these coverings were found not to be wax, but bacteria. These bacteria are part of the group *actinobacteria*, which produce more than 80% of the antibiotics used by humans. The bacteria produce antifungal compounds that stop the microfungal pathogen from attacking the garden. This discovery was the first demonstrated example of an animal, other than humans, using bacteria to produce antibiotics to deal with pathogens. The bacteria appear to be specially suited to inhibit the pathogenic fungi that infect the ants' fungus garden.

The interactions among the ants, their fungus crop, and the bacteria is known as *mutualism*. A mutualism is established when both members of the interaction benefit from the relationship. In the ant-fungus relationship, the ants get necessary food from the fungus; the fungus receives a continuous supply of growing material, protection from the environment, and protection from disease-causing pests.

The bacteria supply antibiotics for the ants' use; bacteria are fed and protected by the ants. Many species of fungus-growing ants have evolved special crypts on their bodies where the bacteria live and grow. Scientists believe that the ants feed the bacteria through glands connected to these crypts. The bacteria are protected against the intense competition they would face if they lived in other environments, such as the soil.

The tight association between ant, bacteria, and pathogens sometimes results in pathogens winning. This interplay has been described as a chemical arms race between the bacteria and the fungus, with one side beating the other as new compounds evolve.

6.16.5 Inadvertent Benefit

The quality of mercy is not strained
It droppeth as the gentle rain from heaven upon the place beneath.

—**William Shakespeare**

The interactions among biological organisms can take some interesting forms and show results in inadvertent cooperation. Cavity spot is a soil-borne disease of carrots that can be controlled by the fungicide mefenoxam (Farrar et al., 2002). The pathogen remains highly sensitive to mefenoxam, but when the fungicide is used in certain fields for a long time, populations of soil microorganisms increase that degrade it. The cavity spot fungus becomes a problem in these fields, not because it has developed resistance to the fungicide, but because the activity of other microorganisms benefit the fungus by removing the compound used for control. This represents the other side of bioremediation (see Section 8.2.1, Hybrid Systems).

As another example, consider plant roots; they are leaky. In order for plants to absorb nutrients from the soil, the roots must exude many chemicals, and, these, in turn, provide food for the surrounding soil microbial community.

Example 6.16.1: Soil Microorganisms Interact

Predict the effect of soil-borne microorganisms on the resistance to the fungicide mefenoxam by the cavity spot fungus.

Solution

The microorganisms can reduce the level of mefenoxam in the soil to levels sublethal to the fungus. Thus, there will be the opportunity for those fungi resistant to the fungicide to dominate the fungi population through natural selection. As long as the challenge to a population of organisms is not lethal to all members, the population can over time develop resistance to the challenge.

Example 6.16.2: Costly Signaling Theory of Ritual

Cooperation within a group often conveys survival advantages on its members. Benefits are extended to all group members equally or according to need. Costs of group membership can be considerable, often involving personal sacrifice and hardship. Examples of such groups are communes and various forms of religious groups among humans. Animals, too, form herds wherein individual freedoms are subserved to the well-being of the herd.

Benefits to group members can be considerable, as long as all are equally dedicated. The presence of too many freeloaders, however, dilutes these benefits, and may completely overwhelm them. In this case, there must be some outward demonstration of dedication to the group.

Behavior that confers survival benefits on its members can be studied in the context of evolution theory (Sosis, 2004). Why are there prohibitions against certain kinds of foods or drinks (coffee, alcohol, meat, etc.)? Why are burnt offerings of perfectly good food made? Why are bodies mutilated in rituals of personal religious sacrifice? Why are uncomfortable clothes worn to demonstrate unity with other members of the group? The answer is related to the costly signaling theory of ritual.

This theory proposes that ritual is meant to signal to others the level of dedication kept by the individual performing the ritual. The reason why many rituals are so extreme (as, for example, self-mutilation or self-immolation) is that the display must be too costly to fake. A hungry person must be dedicated in the extreme to sacrifice perfectly good food, and a sensitive person must be extremely dedicated to endure the pain of physical abuse. These rituals cannot be faked, because they are too costly not to have deep meaning to the person performing them.

From the costly signaling theory of ritual, one can predict that groups that impose the greatest demands on their members will elicit the highest levels of devotion and commitment (Sosis, 2004). Thus, churches that require the most from their members have experienced the greatest rates of growth. It is these groups that can more easily attain their collective goals compared to groups with less-committed members.

Over the course of many generations, the rewards for commitment to common group goals have been the greater reproductive success. Certainly, unless the group intends to be self-destructive (as the Peoples Temple in Jonestown, Guyana in 1978), or celibate (as the Shakers in England, New York, and New England, 1700s to the present), a group of like-minded people is more likely to survive the challenges of life than are lone individuals. Hence, natural selection can also explain social behavior as well as physical attributes.

Example 6.16.3: Infant Formula Probiotics

Probiotics are live microbial dietary supplements that benefit the consumer. Probiotics are often normal flora found growing in or on the bodies of healthy individuals. They often serve important roles in protection against harmful microbes through *competitive inhibition*, or by producing nutrients necessary for health maintenance.

The early colonization of a child's intestines with beneficial bacteria occurs during breast feeding. Bacterial genera such as *Streptococci*, *Lactobacilli*, and *Bifidobacteria* are found in breast milk and become established in the intestines of the child. There they help protect the child against disease.

If a mother is not capable of breast feeding, or if she chooses against breast feeding, then it takes longer for the child to incorporate these microbes. Some infant formula companies, therefore, seek to add some of these bacterial strains into their products.

Optimal growth media are used to culture these probiotics in bioreactors. When they have reached sufficient population densities, they are introduced into infant formulas prior to sale.

Example 6.16.4: Probiotic Treatment for Diabetes

Intestinal epithelial cells secrete insulin when stimulated by either of two proteins, GLP-1 or PDX-1 (Duan et al., 2008). The first of these causes secretion of insulin in response to blood glucose levels, and, with the second, insulin is secreted irrespective of blood glucose concentration. A common probiotic strain of *E. coli* was engineered to secrete either of these two proteins. It was found that ingestion of these bacteria could be a promising treatment for Type 1 (insulin-dependent) diabetes. A pancreas is not necessary for insulin secretion.

Example 6.16.5: Orcas Cooperate for the Benefit of All

Orcas (large dolphins also called "killer whales") also have a cohesive social structure that enables them to care for each other and to hunt more successfully than they would as individuals (Figure 6.16.7). When coordinating with each other during a hunt, they communicate constantly with calls, whistles, and echolocation. Different orca groups, called *pods*, have even been seen to care for a handicapped orca youngster, feeding him and protecting him from harm (Morrell, 2015).

FIGURE 6.16.7
Orcas have developed cooperation to a high degree.

APPLICATIONS AND PREDICTIONS

1. Some species are highly dependent on others for reproduction and survival. Noting the forms of reproductive parts of plants can be used to speculate on the partner species.
2. All functions of an organ must be known before it can be replaced by an artificial organ.
3. Isolation of one or two species will likely not satisfy all the needs of those species.
4. Pollination within a greenhouse will require the introduction of a pollinator species.
5. The continued use of antibiotics can lead to vitamin deficiencies because bacteria in the gut that normally produce vitamins will be killed.
6. Termites without cellulose-digesting bacteria cannot survive. Antibiotics will kill termites.
7. Wolf packs help the survival of the group.
8. The key to designs involving living things is an understanding of dynamic relationships.
9. Species diversity helps maintain cooperative relationships.
10. Parasites will not harm their hosts to any great extent.
11. Species that do not cooperate with others will become extinct. Likewise, removing a cooperating partner from a cooperative relationship will disadvantage or doom the remaining species.

6.17 Biological Units Compete with Other Biological Units

> I always try to skate where the puck is going to be, not where it is.
>
> **—Wayne Gretzky**

The biological world is a very competitive place, with BU of all kinds attempting to use environmental resources to their exclusive advantage, and to the disadvantage of others. Competition is what drives BU to adapt to their environments and competition is what eventually selects for those genes that have a reproductive advantage.

By competition, we mean every type of contest from aggression and defense to simple rivalry. This is the struggle to survive, and, when survival is no longer the issue, to dominate. The world of living things is much harsher than we care to admit from our human vantage point where we have largely achieved dominance and exhibit a modicum of civilized rules. Indeed, our struggles pale in comparison to those of many BU, where survival through the next day is not assured. And humans have extended their influence to shelter other favored BU from the harsh realities of exploitation. In effect, humans have become competitors to predators of our favored species, causing the predators to struggle against not only their natural enemies, for which they have developed defenses and strategies, but also to struggle against humans, for which they are ill-equipped.

6.17.1 Plants and Herbivores

> Life is a constant oscillation between the sharp horns of dilemmas.
>
> **—H. L. Mencken**

Perhaps, the most basic case of exploitation is *herbivory*. Herbivores consume live plant material but do not usually kill the plants they feed on. According to the food pyramid (see Section 5.5.5, Food Pyramid), herbivores transform autotrophic plant material into the first level of (usually) animal tissue, from which many other BU derive their nutrition. Herbivores, then, are the first consumers in the traditional food pyramid. Plant eaters outnumber meat eaters in nearly all ecosystems.

If herbivores are the aggressors in this interaction, then we would expect some defensive behavior on the part of the plants. And indeed there is. Some plants have developed toxins that either (1) alter the taste of the plants, (2) make the plants less palatable than other target plants, or (3) affect the growth or survival of the herbivore. Fescue is a grass of temperate climates often found in fields, meadows, and pastures. Fescue often contains an *endophyte* (a fungus or bacterium living entirely within a plant and that may parasitize the plant) toxic to livestock. Animals eating the fescue often don't grow as fast as others, and they may die. Thus, this is one means for the fescue to defend against exploitation by herbivores.

Nicotine is an alkaloid found in tobacco plants that have insecticidal properties. It is likely that nicotine is produced by tobacco to defend against insects that would otherwise feed on the plant.

Some plants have developed physical defenses such as thorns and burrs. Other plants have leaves with stiff hairs or tough skins to discourage grazing. However, plant fruits that require animals to eat them to scatter the seeds do not have bad taste, toxins, or thorns when ripe. The persimmon tree, *Diospyros virginiana*, has fruit that is astringent and unpalatable until fully ripe, but sweet and attractive when ripe. The chestnut tree, *Castanea dentata*, has nuts covered with a very prickly hull until they ripen fully. The May apple, *Podophyllum peltatum*, has fruits that are toxic until they ripen, at which time they have a lemony flavor. It is obvious that these plants have strategies to eliminate exploitation of their fruits, except at the time when it is advantageous for the plants to be exploited.

This check and balance system has at least one more aspect: herbivores influence the distribution and abundance of their host plants. Grazing animals clearly prefer some types of plants over others, and this can be seen in late summer pastures where grasses and legumes will be eaten to the ground, but New England asters (*Aster novae-angliae*), multiflora roses (*Rosa multiflora*), and musk thistles (*Carduus nutans*) are standing tall. These plants have developed a competitive advantage over neighboring plants, and the herbivores have played a role in the ecology of the area.

Allelopathy is the term used for plants that produce chemical compounds harmful to nearby competitor plants. Such plants have a growth and reproduction advantage without competition for resources. Two plants exhibiting allelopathy are black walnut trees and canola (rape).

The herbivorous insect *Helicopsyche borealis* inhabits streams across most of North America (Molles, 1999). Larval *Helicopsyche* graze on algae and bacteria that grow on exposed surfaces of submerged stones. As the larvae grow through the summer and fall, they attain densities of over 4000 individuals per square meter and represent 25% of the total biomass of *benthic* animals (those animals living in water). At that density, *Helicopsyche* do not only reduce their food supply, but also deplete it.

6.17.2 Predators

> The ability to learn faster than your competitors may be the only sustainable competitive advantage.
>
> **—Arie de Geus**

There is a difference between herbivores and predators. While both exploit target species, *predators* kill and consume other organisms, whereas *herbivores* do not usually kill their targets. *Parasites* live in the tissues of their hosts, but do not usually kill them. *Pathogens* induce diseases that may or may not be virulent.

In their natural habitats, plants and organisms that exploit them usually achieve a dynamic balance, or cyclic equilibrium. The density of the target species may become low and highly scattered, thus making it difficult for the predator to find enough individuals upon which to grow. When the population of predators falls, as it will when the opportunity to grow and reproduce is severely limited, then the target population can expand. This continues until the predator species again increases its numbers (see Section 6.21, Cycles).

Ungulates (plant-eating, cud-chewing animals such as sheep, goats, and elk) modify their behaviors where the risk of predation is high (Howery and DeLiberto, 2005). They avoid higher quality forages with high risk in favor of lower quality forages with lower risk. Many form herds instead of remaining as individuals, and those animals on the periphery of the herd spend more time watching alertly than do animals in the center. They can also form a common nursery, where the young are protected by stronger adults. Social cooperation increases survival of the whole group.

There are many notable examples of species introduced into alien ecosystems where natural predators did not exist. The water hyacinth (*Eichhornia crassipes*) in Florida, kudzu

(*Pueraria thunbergiana*) in the southeastern United States, the star thistle (*Centaurea calcitrapa*) in California, and purple loosestrife (*Lythrum salicaria*) in the northeastern United States are examples of these. In many of these instances, the alien plants spread widely with little to stop them. Native plants with their natural enemies were at a disadvantage compared to the introduced aliens.

In the mid-1800s, a prickly pear cactus (*Opuntia stricta*) was introduced to Australia as an ornamental plant, but, as with many other introduced plants, escaped cultivation and became established in the wild (Molles, 1999). The plant spread quickly, and covered over 20 million hectares by the late 1920s. As it spread to over 24 million hectares by 1930, it was clearly out of hand. This plant had grown densely and made the land unfit for other purposes. Nowhere in its native range of North America does the cactus reach the densities seen in Australia. Without natural enemies, there was very little to check its growth.

After a search of its range in North America, biologists eventually discovered several insect species that attack the cactus and could be used in its growth control. The most effective of these is a moth *Cactoblastis cactorum* (who says biologist don't have a sense of humor?). Female *Cactoblastis* moths deposit eggs on the cactus pads. When 70–90 eggs hatch, the larvae burrow into the cactus and feed on the flesh inside. As they burrow, they introduce fungi and bacteria that also attack internal tissues. Cactus tissues quickly turn to mush.

With so much food available, the *Cactoblastis* spread rapidly, causing the collapse of whole thickets of cacti. In 2 years, this assault by *Cactoblastis* reduced the density of *Opuntia* cactus from 12,000 per hectare to 27 per hectare and the area covered from 24 million hectares to a few thousand hectares (Molles, 1999). The two populations, *Opuntia* and *Cactoblastis* now coexist in the cyclic equilibrium previously described, and demonstrate the dynamic nature of biological populations.

SPIDERS AND THE WEB OF LIFE

Take the case of the spider. Four hundred million years ago, spiders used their silk to weave a hiding place. But then insects developed wings and the spider began to develop aerial webs to catch them as food. About one-third of the 35,000 known spider species weave orb webs (the standard kind with spokes and spirals) and another third weave sheet webs, cobwebs, and other types of webs to catch insects.

With no vision and a limited nervous system, the spider makes some rather complicated calculations and decisions about the size of the space to be filled, how much silk is available, and where to attach the web. Spiders are not automatons that repeatedly make the same webs, but are flexible and smart at what they do.

Some prey, like mosquitoes, fly very tentatively, with their forelegs out. As soon as they touch a web, they retreat. Spiders have developed orb webs in a concave shape with a spring line running straight back from the hub to winch the web into a cocked position. This web springs out to follow the elusive prey and catch them anyway.

There is a small slender wasp about three quarters of an inch long that parasitizes the brightly colored orchard spider *Plesimeta argyra*. The wasp stabs its stinger into the mouth of the spider, which renders the spider still. Then the wasp curls her hind end under and deposits her egg in the abdomen of the spider. When the spider wakes in ten or fifteen minutes, it resumes its normal life, but carries its own killer inside its body.

(*Continued*)

SPIDERS AND THE WEB OF LIFE (Continued)

The wasp larva hatches in a couple of days and makes little holes in the spider's cuticle to suck its blood. Spiders feed on their prey, oblivious to the larva feeding, in turn, on its own blood. For the first week or two the spider continues to build its orb webs several times a day, but then the wasp larva suddenly takes control of the spider's mind.

In the middle of the night, the spider goes back and forth up to 40 times on the same few spokes of its web. Normally the spider waits until dawn to make its usual orb web. It is during daylight that the insects it eats usually fly.

But the larva is not interested in the health of the spider at this point. The spider makes a very strong web and stops and sits in the middle, waiting to die. The larva finishes it off by sucking the remaining juices from the spider and drops the carcass to the ground. Later, the wasp larva will spin its cocoon and remain suspended by the web until the adult wasp emerges sometime after dawn. Then, it would find a mate and a spider to attack, and renew the entire cycle (Conniff and Murawski, 2001).

Biological cycles, such as this, are sometimes incredibly complex. There are other parasites that assume behavioral control over the actions of their hosts so that those hosts may become agents for the reproduction of the parasites: a male sheep crab is changed by a parasitic barnacle into a female so that the barnacle may hatch in the crab; a house cricket can be made to commit suicide in a body of water by a parasitic worm that emerges as the cricket drowns; a female wasp can sting a ladybug and leave behind a single egg that then hatches, eats the ladybug from the inside out, forms a cocoon, and is protected from enemies by the ladybug; a flatworm burrows into a bullfrog tadpole and causes the tadpole to develop deformed limbs, making it easy prey for birds, and, and, finally, the flatworm reproduces inside the bird; Amazonian ants are penetrated by a fungus that enters its brain causes the ant to climb a nearby tree and fasten itself to a leaf where the fungal stalks burst forth and rain spores on other ants below (Maure et al., 2011; Zimmer, 2014).

6.17.3 Parasites

Parasitism is the most popular animal lifestyle on the planet.

—Kevin Lafferty

There are many parasites in nature, and these are organisms that have become specialized to obtain nutrients from inside a host organism, grow, reproduce, and spread to other hosts by leaving the original host. Sometimes, there is more than one host involved in the life cycle.

Whereas the predator–prey competition provides the more exciting conflict, the parasite–host relationship provides the more interesting examples of adaptation. For instance, *Acanthocephalans* (a spiny-headed worm) that infects an amphipod host (a crustacean) causes the amphipod to change its behavior from light avoidance to light seeking when the *Acanthocephalans* reaches a life stage that is capable of infecting a vertebrate host. When that life stage is reached, the amphipod swims to the water surface, where it is likely eaten by feeding ducks, beavers, and muskrats. Amphipods that are not infected usually hide near the bottom of ponds and lakes; amphipods infected with *Acanthocephalans* do not change their behavior until the parasite is in the most infectious stage. When eaten, the *Acanthocephalans* spends the rest of its life in the body of its alternate host, preparing to reproduce and spread, once again, to amphipods (Molles, 1999).

Mycoplasmas (not to be confused with mycobacteria, a normal sized bacterium) are exceptional parasites descended from bacteria that had lost their cell walls and shed much of their genomes and metabolic capacities. They became highly dependent upon host environments and biomolecules (including cholesterol for their cell membranes), and, in the process, became the smallest (less than 1 μm in diameter) and simplest free-living organisms. They have extremely small genomes ranging from 450 to 1,000 million *Daltons* in size (a Dalton is a number equal to the molecular weight of the molecule), which makes them an obvious choice to be the starting point for synthetic biology (Morowitz, 2011).

It is not in the best interest of the parasite to kill its host or enfeeble the host enough to diminish the reproductive potential of the parasite. Those parasites that are too virulent do not reproduce, so, clearly, genes for that parasite would not be perpetuated. When either a parasite or pathogen jumps from a habitual host species to a new species, it is often very virulent, because the new host has not developed an immune strategy to deal with the new threat. Many host individuals die or are at least incapacitated. Over time, this does not suit the parasite or pathogen any more than it does the host, and, as a consequence, the parasite or pathogen becomes less virulent to the new host. It may evolve into a chronic rather than acute condition. This pattern has been followed by the spread of Simian Immunodeficiency Virus (SIV) from monkeys to become Human Immunodeficient Virus (HIV), Severe Acute Respiratory Syndrome (SARS) from bats and cats to humans, and strains of influenza virus from birds to swine to humans. The same pattern has been observed in other species (e.g., varroa destructor from the Asian honey bee, *Apis cerana,* to the European honey bee, *Apis mellifera*).

6.17.4 Pathogens

> Competitors take bad breaks and use them to drive themselves just that much harder. Quitters take bad breaks and use them as reasons to give up.
>
> **—Nancy Lopez**

Pathogens cause diseases important in humans, their pets, their food, their ornamentals, and in other remote species, usually in that order of importance. Diseases are usually accompanied by symptoms, and it is the nature of these symptoms that has been reinterpreted in light of evolutionary medicine (Nesse and Williams, 1994). Fever, for instance, is a resetting of the internal thermoregulatory mechanism that increases metabolic rate, increases antibody production, and quickens disease-fighting mechanisms. As a reaction to infection, fever has positive consequences. So, why doesn't the body maintain its temperature at 40°C all the time instead of at 37°C? This is because the maintenance of 40°C uses more energy and is too close to lethality to be worth the risk.

Pain and irritation are also defense mechanisms against disease. The itch of a mosquito bite leads to slapping and perhaps killing the insect that transmits malaria and heart worms. Feeling vaguely ill leads to inactivity that likely favors immunological defenses and repair of damaged tissues (Nesse and Williams, 1994).

Vomiting rids the stomach of bacteria and toxins. The distress of nausea discourages us from eating more of the same kind of apparently tainted food. The nausea of early pregnancy is likely an adaptation meant to protect the fetus during the time that it is most sensitive to the effects of circulating toxins (see Section 6.12, Toxins). During the first trimester, the energy burden on the mother is not so great to offset the protection to the fetus offered by withholding potentially dangerous food (Nesse and Williams, 1994).

Other means of expulsion benefit both the host and pathogen. Nasal discharge cleans the nasal passages of pathogens and, at the same time, disseminates the pathogen. Diarrhea performs the same function for the intestine. Patients who were treated to alleviate these symptoms were found to suffer from the diseases up to twice as long as those who let the diseases run their natural course (Nesse and Williams, 1994).

Diseases caused by bacteria are often more threatening when caused by gram-negative, rather than gram-positive, bacteria. The cell walls of gram-negative bacteria are often toxic, and their cell membranes are highly protective and antibiotic-resistant.

Many pathogens have evolved to be specific to one or several specific host(s). It has long been considered that pathogenic viruses are exclusively found in their preferred host. Plant viruses have never before been found to spread from plants to animals. However, the tobacco ringspot virus (TRSV) has been seen to infect honey bees, a representative of the animal kingdom (Li et al., 2014) in the first known shift of a plant virus to an animal.

Fungi and bacteria that produce natural antibiotics (the sources of many of the antibiotic drugs we use) also produce compounds that make them immune to these same antibiotics.

Coevolution is at work for pathogens and hosts as well as it is for cooperation among BU (see Section 6.16, Cooperation). When hosts develop better defenses, pathogens usually develop means to overcome the defenses. Biologists have named this the Red Queen Principle, after Lewis Carroll's Red Queen. She explained to Alice, "Now, here, you see, it takes all the running you can do, just to keep in the same place" (Nesse and Williams, 1994). This expresses the dynamicism of biological competition.

Certain competitors against humans can be eliminated by technology. Humans have already been the cause for elimination of many plant and animal species, and not all of these happened in modern times. It is when humans appear at the top of the food chain that they have succeeded best in this regard.

But humans are not always the hunters. To sharks, *Streptococci*, and *Schistosoma*, humans are the hunted species, and their technological advantages are of limited value. Especially when the predator has a short reproductive cycle and can adapt very quickly to things like antibiotics, we see that any technological advances are only temporary. The most sophisticated technology is still within the human body, and, in many circumstances, we need to realize that we are up to the competition.

Example 6.17.1: Humans against Food Microbes

Spices in food affect more than the taste. Some spices are powerful inhibitors of bacterial growth and appear to have been used because they keep food from spoiling. There is a positive correlation between the mean average temperature of a country and the number of spices used in food recipes in that country (Sherman and Flaxman, 2001). This might be expected because food in hot countries such as India can spoil faster than food in cold countries such as Sweden. Meat-based recipes use more spices than vegetable-based recipes.

Spices that are particularly good for inhibiting bacterial growth are onion, garlic, chili pepper, bay, cinnamon, cloves, thyme, cumin, and allspice. Much poorer are lemon-lime, ginger, paprika, and celery. Pepper, parsley, coriander, nutmeg, and mustard are intermediate.

Example 6.17.2: New Corn Pest

The Western Bean Cutworm rarely caused economic problems in corn until recently. The insect first appeared in Colorado, Wyoming, and Idaho, and has been moving eastward into Iowa, Kansas, Minnesota, Nebraska, and South Dakota. Entomologists theorize that the Western Bean Cutworm is taking advantage of the opportunity afforded by the widespread planting of Bt (*Bacillus thurengiensis*) corn. Bt corn is a genetically modified corn containing a gene that results in a protein that kills the major insect pest of field corn, the European Corn Borer. Eliminating the corn borer suppressed significant competition to the Western Bean Cutworm, and allowed the cutworm to expand well beyond its original range. This is an example of the resiliency of living things, and the opportunism of competitor species when their competition is eliminated. It is also an example of the unintended consequences that result from trying to control nature.

Example 6.17.3: Fighting Aflatoxin Naturally

Wheat and other grains grown in a wet year often play host to a common fungus called *Aspergillus flavus*. This fungus produces a poison called aflatoxin that can result in sickness or even death in humans and animals who eat the grain. Aflatoxin is a known carcinogen. Grains containing aflatoxin are difficult (and costly) to detect, and cannot be separated from grain free of the poison. Thus, if even a tiny bit of grain is suspected of containing aflatoxin, the whole load must be thrown away. Suggest a means to protect the grain from developing aflatoxin (Wolfshohl, 2003).

Solution

Seed grain can be inoculated with a strain of *A. flavus* (AF36) that doesn't produce aflatoxin. In a method called *competitive exclusion*, the purpose of AF36 is to grow where native *A. flavus* strains would grow. Competition between the population of AF36 and aflatoxin producers becomes tipped heavily away from the aflatoxin producers. It is hoped that AF36 can be registered as a protective *biopesticide* in wheat, cotton, corn, and peanuts, among others.

Remark

Biopesticides are used elsewhere, as well. Certain bacteria can be sprayed on strawberries and other fruit crops to protect them against late frosts. Other bacteria, fungi, and yeasts are used to control diseases of stored fruits. Beneficial bacteria, called *probiotics*, can be fed to humans and animals to relieve the symptoms of diseases and maladies. Probiotics can produce useful biochemicals (such as vitamin K in the human gut), attack harmful viruses and bacteria through natural antibiotics, and crowd out harmful microbes. Competitive exclusion is a powerful weapon that has only begun to be used to advantage.

Example 6.17.4: Immunomodulation to Treat Autoimmune Diseases

Autoimmune diseases include allergies, asthma, rheumatoid arthritis, multiple sclerosis, diabetes, and inflammatory bowel disease. These are nasty diseases, sometimes fatal, caused when the body's immune system becomes hypersensitive and begins to attack parts of its own body.

Parasitic worms (called *helminths*) are able to subtly dampen immune response so that they can live for long periods of time in the human body. Some of these creatures can cause very unpleasant outcomes themselves. *Schistosoma*, for instance, burrows under the skin and moves to the intestines or bladder, gorges on blood cells, and causes fevers, blindness, or liver damage.

Some of the less damaging helminths (Figure 6.17.1) can be given as treatment for autoimmune diseases. Ingesting worm eggs can result in significant benefit, wherein

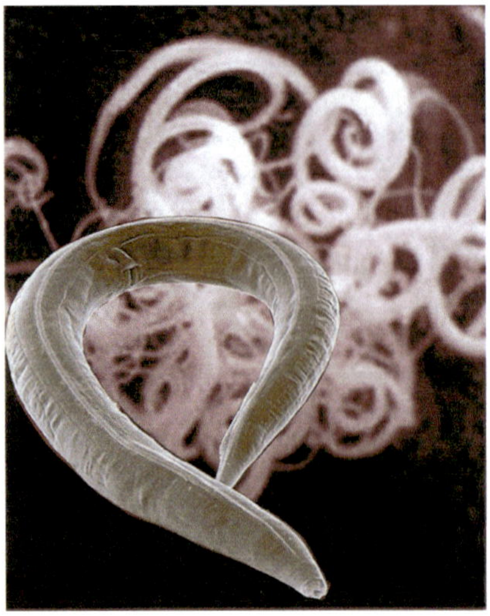

FIGURE 6.17.1
Helminths, or parasitic worms, have the ability to calm the host immune system.

diseases such as multiple sclerosis or inflammatory bowel disease become substantially arrested. Searches are underway to find how the worms achieve their immunomodulation. Then drugs can be developed to mimic these means.

Medical scientists say that it is no accident that autoimmune diseases among humans have escalated coincidentally with a sharp drop in parasitic diseases. At one time, helminths may have helped humans avoid some of these diseases.

Example 6.17.5: Cat Parasite Has Unusual Neural Effects

Toxoplasma gondii is a parasitic microbe of cats with unusual characteristics (Shepherd, 2010). Cats come in contact with the microbe when they eat an infected mouse or bird. They don't usually get sick, but the parasite reproduces in the cat's gut, forming eggs (or *oocysts*) that develop and are shed in the cat's feces.

When another warm-blooded animal, such as a mouse, ingests the oocysts through contact with the cat's feces, the oocysts enter its gut. There they release cells that migrate mostly to the muscles and brain, forming cysts to protect themselves from the mouse's immune system.

But *T. gondii* must get back to into a cat to reproduce again. In order to do that, it needs the mouse to be eaten by the cat. So, it changes mouse behavior to induce it to take risks, including an attraction to the smell of cat urine. This makes mice vulnerable to attack by cats.

T. gondii has two unusual genes that contain instructions for an enzyme that makes the neurotransmitter dopamine. This is likely the mechanism that *T. gondii* uses to change normal mouse avoidance behavior. It may also explain a statistical link between *T. gondii* infection of humans and human schizophrenia. Some anti-schizophrenic drugs are dopamine antagonists.

Example 6.17.6: Pathogens in Produce

There is always a risk of eating produce contaminated with pathogenic microbes. This risk is made larger by the produce preparation and distribution system that serves a

FIGURE 6.17.2
Leaf lettuce growing in the field, where it could be exposed to pathogenic bacteria.

large portion of the consuming public and so that any defect in the safety of the produce is spread to a large number of consumers. There is also the greater chance that a larger number of food handlers come in contact with the food before it is sent out.

Even before then, however, the proliferation of antimicrobial and antibiotic chemicals used on the produce as it grows may be partly to blame for produce contamination (Figure 6.17.2). These chemicals have the greatest effect on the weakest microbes, and eliminate them from competing with stronger and more pathogenic microbes. Without this check on their growth, pathogenic bacteria can grow quickly to toxic levels (Fonseca and Ravishakar, 2007).

APPLICATIONS AND PREDICTIONS

1. If BU have limited access to energy, then natural selection will favor BU that are more effective at acquiring energy.
2. The losing competitor will die.
3. Plants that do not grow as fast as they can will eventually be shaded by those that do.
4. Competitors not genetically related will compete most fiercely; relatives will not compete as seriously.
5. When resources are limited in the extreme, cooperation turns to competition.
6. Competition is the best way to control an unwanted population.
7. Taller trees beget taller giraffes.
8. Gram negative bacteria are usually to be avoided.

6.18 Biological Units Reproduce

> The hand of God is to be found in the very first molecule that was compelled to soak up energy and resources in order to replicate itself.
>
> **—Steve Schreiner**

Reproduction is the most basic of activities of living systems (and even nonliving systems if we consider subcellular units such as viruses). The reproduction and evolution of RNA molecules in the test tube have been observed (Schuster et al., 1997). It is through reproduction that genetic material is perpetuated, and, in that sense, genes can be immortal as long as they are not mutated into different genes.

6.18.1 Asexual Reproduction

> For years, scientists have dreamt of making robots that can self-reproduce. Someday, such a machine could be sent to explore a distant planet, where it could clone itself.
>
> **—Neil Greenfieldboyce**

There are several classes of reproduction, and we start with *asexual* vs. *sexual* reproduction. Asexual reproduction is the creation of new individuals whose genes all come from one parent without the fusion of egg and sperm (Campbell et al., 1999). That means that each offspring of the process is a genetic clone of the parent.

Bacteria normally divide and reproduce by asexual reproduction in a process called *binary fission*. Bacteria are prokaryotes, and, as such are relatively simple BU. They carry most of their genes on a single chromosome attached to the plasma membrane. When a single bacterial cell divides, it first replicates its chromosome and attaches it to a different membrane site. When the bacterium has grown to about twice its normal size, the plasma membrane grows inward between the two chromosomes and divides the parent cell into two daughter cells, each with a complete genome.

There are interactions, such as competition and adaptation, that indirectly impinge on the reproductive process. If external resources are just adequate, cellular reproduction balances with cellular death, and net reproduction is nil. If excess external resources are present, net reproduction increases beyond that necessary to maintain a static population.

6.18.2 Exchange of Bacterial Genes

> Biodiversity [is defined] not only as the variety of living organisms, but also as how these organisms organize themselves (structure) and how they interact with each other (function).
>
> **—J. Michael Scott**

Prokaryotes may exchange genetic material in three ways: by (1) transformation, (2) conjugation, and (3) transduction. *Transformation* is the ability of a bacterium to incorporate naked DNA from its environment. This DNA probably came from cells that died and released their contents. *Conjugation* is the direct transfer of genetic material from one cell to another when the cells are temporarily joined. The transfer only occurs in one direction. *Transduction* occurs when viruses (called phages) infect bacteria and carry bacterial genes from one host to another (*transfection* is placing nucleic acids into cells without using viruses). Each of these processes is common enough that bacteria can acquire pathogenicity from other bacteria, antibiotic resistance can move from species to species, and there is concern that certain genes can jump spontaneously from a *genetically modified organism* (GMO) to a related species in the wild.

Horizontal Gene Transfers (HGT) occur in other species besides bacteria. Parasitic plants that establish direct connections with their plant hosts obtain genes directly from their hosts. Mitochondrial gene transfers are very common, and nuclear gene transfer happens often.

6.18.3 Somatic Cell Reproduction

> Plato having defined man to be a two-legged animal without feathers, Diogenes plucked a cock and brought it to the Academy, and said, "This is Plato's man." On which account this addition was made to the definition, "with broad flat nails."
>
> **—Diogenes Laertius**

There is another common form of asexual reproduction that is clearly important to all multicellular organisms, and that is reproduction of *somatal* (or body) cells. These cells do not form *gametes* (or germ cells), and die when the organism dies. However, they reproduce in order to support growth or heal wounds.

Somatal cell reproduction does not happen without interaction from other cells and tissues. Growth factors are proteins released by certain body cells that stimulate other cells to divide. Thus we find that platelet-derived growth factor (PDGF) is made by blood cells called *platelets* in response to a wound. Connective tissue cells called *fibroblasts* are stimulated to divide by PDGF. The result is that the wound begins to heal (Campbell et al., 1999).

There are many other growth factors present in BU, and they regulate growth of cells, some quite distant from the emanating tissues.

Density-dependent inhibition is a term describing the phenomenon in which crowded cells stop dividing (Campbell et al., 1999). When cells divide to the point that they touch each other and fill the available space, they automatically stop dividing. When cells die and cause a hole to form, adjacent cells begin to divide until they again fill the space. This occurrence is apparently due to inadequate amounts of nutrients and growth factors when the cells are crowded (Campbell et al., 1999).

There is also an *anchorage dependence* that requires that cells must be attached to a substratum in order to divide. Freely floating somatal cells are thus inhibited from dividing.

For each somatal cell division, an equal copy of genetic material is passed from the mother cell to the daughters. This complete genetic reproduction comes about through the process of *mitosis*. During mitosis, eukaryotic cells duplicate both sets of chromosomes before dividing.

The human body is made of about ten trillion cells (Nesse and Williams, 1994), (see Section 5.5.4, Ecology). Among all the actively dividing cells in the body, there are bound to be mistakes made in the genetic codes of some of them, and so there must be mechanisms to detect genetic mistakes and either (1) correct errors, or (2) destroy the cell with defective genes (*apoptosis*). If these cells are allowed to continue with defective genomes, they can become tumorous or cancerous. There are many tumor suppressive and tumor-supportive (oncogenes) in the genome, and it is the cumulative effect of these two types that determines if cell growth is limited or not. The immune system continuously eliminates tiny tumors in a process called *immune editing*, until some tumor cells develop the ability to turn off the immune system.

A great deal of research is being performed to elucidate mechanisms of genetic monitoring and correction. A great deal of attention has been placed on the *p53 gene*, called the "guardian angel of the genome" (Campbell et al., 1999). This gene halts cell division, mediates genetic repair, or kills cells with irreparable DNA. There are at least three different ways that the p53 gene prevents a cell from passing mutations due to DNA damage. Whenever the p53 gene is missing or damaged, cancer may ensue.

There are other tumor-suppressor genes in the body, and there can be tumor-enhancing genes called *oncogenes*. Oncogenes arise from damage to the genes that code for proteins that stimulate normal cell growth and division.

CANCER IN HUMANS

With the complexity and redundancy built into biological systems, it is difficult for malfunction, and when something does go wrong, it is because there were a series of challenges that overcame the many obstacles against malfunctions. Cancer is preceded by about a half-dozen genetic changes, all of which must occur before uncontrolled cell growth takes place.

In the first place, there are proto-oncogenes that promote normal cell growth in the genetic codes of all individuals. These proto-oncogenes change into *oncogenes* (cancer-causing genes) either through translocation of genetic material or faulty replication of the gene during mitosis, or by a point mutation within the gene. Because oncogenes are normally dominant, these changes need only occur on one *allele* (or on one of the pairs of genes on one of the paired chromosomes).

There are also tumor-suppressor genes that produce proteins to: (1) repair damaged DNA, (2) control the adhesion of cells to each other or to an extracellular substrate matrix, or (3) inhibit the cell cycle of growth. In order for cancer to develop, tumor-suppressor genes must be made ineffective. Mutated tumor-suppressor genes are usually recessive, so both copies must be mutated. Additionally, many malignant tumors require that the gene to produce telomerase be activated. Therefore, with all these obstacles, cancer is relatively rare in the young, and only increases with age because of the greater probability for compounding of mutations as time goes by.

There are approximately 10^{13} cells in the human body, and the average life (Nesse and Williams, 1994) of skin and blood cells is about 3 weeks (some cells, of course, do not reproduce and are with the body throughout the entire lifetime). That means that in a 75-year lifespan there are about 10^{16} new cells produced.

The average diameter of human somatic cells is (Simpson et al., 1957): 24×10^{-6} m. The average volume per cell is (assuming spherical cells):

$$V_{cell} = \frac{4}{3}\pi r^3 = \frac{4}{3}\pi(1.2\times 10^{-5}\ \text{m})^3 = 7.2\times 10^{-15}\ \text{m}^3.$$

The average density of the human body is (Johnson, 1999):

$$\rho = 1050\ \text{kg/m}^3$$

(*Continued*)

CANCER IN HUMANS (Continued)

The average mass per cell is:

$$m_{cell} = \rho V = (1050\ kg/m^3)(5.24 \times 10^{-16}\ m^3) = 7.6 \times 10^{-12}\ kg/cell.$$

Thus, for a 70 kg man, the number of cells in the body is:

$$\text{number of cells} = \frac{\text{total mass}}{\text{mass/cell}} = \frac{70\ \text{kg}}{7.6 \times 10^{-12}\ \text{kg/cell}} = 9.2 \times 10^{12}\ \text{cells}.$$

The average lifespan of a cell is about (Nesse and Williams, 1994):

$$t_{cell} = 5\ \text{weeks/generation}$$

The average lifespan in weeks of a 75-year-old is:

$$t_{life} = (75\ \text{years})(52\ \text{weeks/year}) = 3900\ \text{weeks}.$$

The number of cellular generations is thus:

$$\text{number of generations} = \frac{3900\ \text{wks}}{5\ \text{wks/generation}} = 780$$

The number of cellular reproductions is, therefore:

$$\text{number of reproductions} = \left(\frac{\#\text{generations}}{\text{cell}}\right)(\#\text{cells}) = (780)\left(9.2 \times 10^{12}\right)$$

$$= 7.2 \times 10^{15}$$

With this number of cellular replications, the incidence of cancer in the U.S. population (1 in 250 for all ages, 1 in 33 for those aged 75) is more understandable. It also illustrates why the body's defense against cancer (and other diseases) must be so complicated.

6.18.4 Telomeres

Biology is not destiny. Will is destiny.

—Arline B. Curtiss

It has been noticed that somatal cell lines cultured *in vitro* reproduce about 20–50 times and then the cells die. This has prompted biologists to speculate that there is a natural end to multicellular BU in vivo. One reason for this may be events in the process of cellular mitosis, as chromosomes are split and replicated. Replication requires an involved series of steps, and includes RNA priming, DNA primase, DNA polymerase, and DNA ligase in an intricate set of maneuvers. These steps, however, are not able to replicate the end of the

DNA strand. Errors in the replication process can be corrected by an elaborate and elegant set of repair enzymes present in the cells. These enzymes, however, cannot correct certain errors or omissions at the ends of the chromosomes. The result is that the chromosome shortens during each somatal cell division.

Losing meaningful genetic material at the end of the chromosome would be disastrous for the cell, so it appears that the cell begins life with an amount of genetic material called a *telomere* at the ends of its chromosomes. This material is a buffer against losing functional genetic material during mitosis. Human telomeres consist of the sequence ...TTAGGG... repeated thousands of times. The length of human telomeres is about 15,000 base pairs at conception, 10,000 base pairs at birth, and the rate of loss is about 50 base pairs a year after birth. At about 5000 base pairs, human cells have largely lost the ability to divide (Hooper, 2011). Telomere sequences in other species vary from one to another, but usually consist of repetitions of 6–10 base sequences.

As the cell divides again and again, this telomere is shortened, and, when it is gone completely, the cell line dies (Figure 6.18.1). Laboratory cultured cells, not immortal stem cells or cancer cells, can divide about 50–70 times before they die; this number of divisions is known as the *Hayflick Limit* (Hooper, 2011). Prokaryotes avoid this problem by containing DNA in a circular macromolecule that has no ends.

Shortened telomeres have been associated with age-related health like heart disease, obesity, diabetes, cancer, and mental decline. Chronic mental depression is associated with shortened telomeres, with the most severely depressed having the shortest telomeres (Verhoeven et al., 2014).

There is an enzyme called *telomerase* that functions by restoring the length of the telomeres. It is present in germ line cells, which enables a new individual to begin life with a full potential life span.

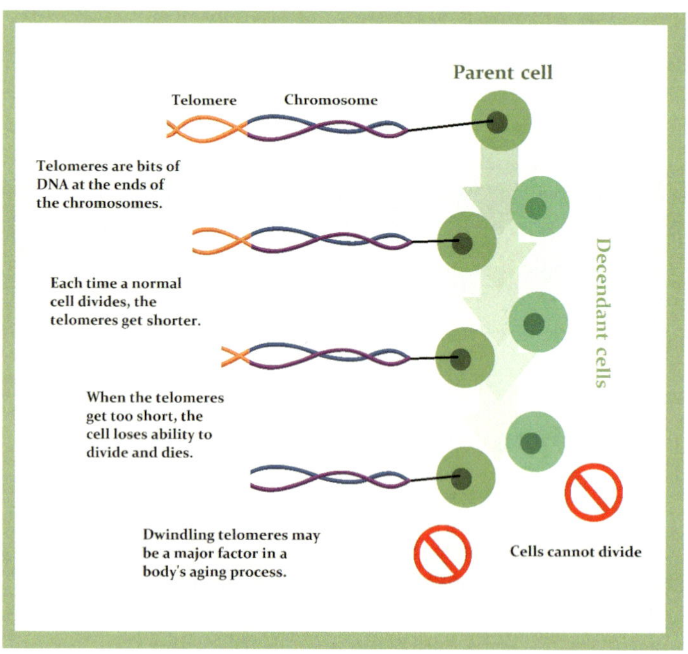

FIGURE 6.18.1
Telomeres are bits of genetic material whose function is to guard against errors in useful genes.

Telomerase has been found in cancer cells (Campbell et al., 1999), which has the effect of making them immortal. Thus, we see that cancer requires about a half-dozen changes to occur in the DNA. At least one active oncogene must appear, and several tumor-suppressive genes must become inactive. Density-dependent inhibition and anchorage dependence must be overcome, and the gene to produce telomerase must be activated. Because some of these are recessive rather than dominant genes, both copies of DNA material must be changed. Therefore, cancer is much less prevalent than it would be without these safeguards. Cancer then becomes more a disease of the aged because, over time, cellular DNA can be damaged by exposure to high-energy electromagnetic radiation, by viral infections, or by naturally occurring mutations.

Telomere length is a primary biomarker for cellular aging, and several influences on it have been found. Telomere length and telomerase activity have been shown to be affected by various environmental factors such as oxidative (free radical) stress, psychological stress, and socioeconomic status (Ludlow et al., 2008). A moderate human lifetime physical activity level of 991–2340 kcal/week was observed to be related to significantly longer telomere length than either very low (less than 990 kcal/week) or very high (greater than 3541 kcal/week) physical activity levels (Ludlow et al., 2008). Different human genomes were associated, as expected, with different telomere lengths. These results indicate that reduced stress and lifetime physical activity can help to protect humans from cellular aging effects leading to cardiovascular disease, insulin resistance, hypertension, and premature death.

HeLa CELLS

Cancer cells differ from healthy cells in that they grow unchecked by normal cell limits and they can live forever. The HeLa cells are a line of immortal cervical cancer cells biopsied from Henrietta Lacks, a cancer patient at Johns Hopkins University in 1951. They are amazingly resilient and robust. As such, they have been shared among many researchers worldwide and used for research on cancer, Acquired Immunodeficiency Syndrome (AIDS), radiation, toxic substances, and gene mapping. This cell line is so hardy, prolific, and overpowering that it is a contaminant in perhaps 10%–20% of all research cell lines.

HeLa cells differ from normal human somatic cells because of the presence of powerful telomerase that prohibits the shortening of the telomeres during cell division. These cells also have 82 chromosomes, unlike the normal human number of 46. The additional chromosomes apparently came from Human Papilloma Viruses (HPV) that fused with the original cancer cells. Although some variation in the HeLa cell genome has been reported when cultured under different environmental conditions, the HeLa genome has been remarkably stable over the course of hundreds of cell divisions.

The Hayflick Limit is the number of times that normal human somatic cells can divide before their telomeres are completely consumed. This number is a value somewhere between 40 and 60. When the Hayflick limit is reached, senescence sets in, eventually leading to cell death (apoptosis). The HeLa cell line has shown no evidence of a Hayflick Limit.

6.18.5 Sexual Reproduction

Birds do it, bees do it.
Even educated fleas do it.
Let's do it, let's fall in love.

—Cole Porter

It is easier to see the interdependence of BU in sexual reproduction. We normally think of sexual reproduction as the union of a male *gamete* (or germ cell) with a female gamete to produce a *zygote* (a fertilized egg). Male gametes are usually called sperm in animals and pollen in plants. Female gametes are called eggs in both cases. Gametes are *haploid* because they each contain but one of the set of chromosomes found in mature individuals. Zygotes are *diploid* because they contain chromosomes in pairs, just as they will when they mature into adults.

Why there is sexual reproduction at all is open to speculation. Certainly, sexual reproduction increases the potential for genetic diversity, and there is conferred an advantage to populations of organisms that must adapt to new environments by genetic diversity. If environmental conditions were static, then there would be an enormous advantage to *parthenogenesis* (development from an egg without sperm fertilization). A parthenogenic individual who could guarantee that all the genes she carried would be passed on to all her offspring has a large advantage over a sexual individual who can only pass half her genes to her offspring. However, if the world was populated by only one genotype, then a change in temperature, moisture availability, food resources, or pathogens could wipe out the entire population.

This situation occurs in agriculture, which tends to cultivate those varieties of crops that perform best under a certain set of environmental circumstances. The longer those specific circumstances prevail, the more monocultural agriculture becomes. Every so often, however, a new disease shows up or there is a climate change that causes panic throughout the agricultural community. This has happened with corn that was planted in the United States in the 1970s. A new corn disease threatened almost all the cultivated corn until a gene was found in a wild corn that was quickly incorporated into the corn genome to halt the disaster in subsequent years. The Irish potato famine of the 1870s was due to the almost complete dependence of the Irish people on the potato crop for their food supply. When potatoes were decimated by disease, the Irish people faced starvation or migration. Nature has apparently experienced enough of these disasters that genetic diversity is valued above temporary efficiency.

Sexual reproduction in animals requires the union of a sperm and egg of the same species (one distinction between species is the inability to breed). Both sperm and egg must be sufficiently mature in order to form a successful zygote (Figure 6.18.2). This usually requires an extremely complex pattern of physical and behavioral activities that are different, but complementary, in both male and female.

Sexual activity is regulated by *hormones*, chemicals produced in very small quantities that elicit particular responses in other parts of the body. *Androgens* are the principal male sex hormones, of which *testosterone* is the most important. These hormones regulate primary sex characteristics such as the development of sperm and sperm delivery systems. They also are responsible for secondary sex characteristics that are characteristic of male animals, including the sex drive, lower voices in mammals, singing of birds, and croaking by frogs.

Unlike the male, who must be ready to supply sperm whenever the female's egg is receptive, the female produces mature eggs only on a limited basis. The result is that successful reproduction does not require coordination of two complex cycles, just one (the female), the male being receptive nearly all the time. This leads to a typical difference in male–female

FIGURE 6.18.2
Sperm surrounding an egg cell at the moment of fertilization.

reproductive behaviors: the female typically exhibits a wider range of receptivity for *copulation* than do males.

Nevertheless, there are many examples in nature where one sex competes regularly for mates. Whether the competitive sex is male or female does not seem to depend as much on the relative sizes and numbers of *gametes* (specialized sex cells) produced by each sex, as it does on the total investment each sex makes in the rearing of offspring (Thornhill and Gwynne, 1986). Female mammals are almost exclusively the sex that invests more in producing and raising offspring and, therefore, are sought competitively by males intending to mate. Males of these species tend to be the larger and more aggressive sex. In some insects, fish, amphibians, and birds, however, the male is the sex that invests more in the next generation. Some males care for the *zygotes* (fertilized eggs) during gestation, some provide extra food for the female during this period, and some guard the fertilized female and prevent harm to come to her. In this case, the female is often the larger and more brightly colored sex who competes for sexual attention from the male. Monogamous species tend to have nearly equal investment by both sexes and little differential competition.

The most important female sex hormones are a class called *estrogens*. These are produced by the vertebrate female ovary and maintain the secondary sexual characteristics of the female. Estrogen also has a primary role in maintaining the female reproductive system (see Section 6.21.6, Monthly Cycles).

There is a complex interplay of different hormones in the female, and this is often unique to a particular species or class of species. For instance, humans and many other primates undergo a *menstrual* cycle, wherein the egg is matured, and the uterine lining is prepared for implantation of the fertilized egg. Hormones participating in this process are gonadotropin-releasing hormone (GnRH), follicle-stimulating hormone (FSH), luteinizing hormone (LH), estrogens, and progesterone. Others, such as oxytocin and prostaglandins, play a role in the birth process. Other mammals have an *estrus* cycle different from the menstrual cycle. Birds, amphibians, and insects undergo different processes. The common characteristic of these cycles, however, is the preparation of the egg (or eggs) for fertilization.

When the female is receptive, she signals this to the male through chemical and behavioral means. Chemicals, called *pheromones*, are used to communicate between organisms of (usually) the same species (see Section 6.19.2, Chemical Stimuli). Some pheromones may convey nonsexual messages (fear, defense, aggression). Pheromones released into the air are low molecular weight (and, hence volatile) organic chemicals that elicit a very sensitive response. One molecule of gypsy moth pheromone out of 10^{17} molecules of other gases will evoke a response in target males. The males, in turn, give off aphrodisiac odors.

Then follows a set of complicated behavioral responses (called courtship). The apparent functions of these are to (1) assure each partner that the other can produce the best possible offspring with the greatest chance of survival, and (2) to coordinate copulation, or coitus. For many species, a particular pattern of behavior is required for completion of the sex act, and any disruption of this behavioral pattern will not lead to successful fertilization.

Low light levels in deep ocean depths of 500 m or more can challenge animals that live there to find food and also to find mates. Bioluminescence is one way to signal the presence of an individual looking to reproduce (McClain, 2010). Also of importance in the quest to find a mate is the sense of smell. Small male anglerfish seek out females by smell. When a male contacts a much larger female, enzymes fuse his mouth to her body. Eventually all of his organs except his sperm-producing gonads atrophy, and he becomes a lasting source of sperm (McClain, 2010).

SILPHIUM

The ancient Greeks and Romans used a species of wild fennel as a safe and effective female contraceptive. This plant, known as *siliphion* to the Greeks and *silphium* to the Romans, became the economic staple of the North African Greek city-state of Cyrene. The plant was highly prized and worth more than its weight in silver. Due to the insatiable demands for this plant, silphium became extinct about fifteen hundred years ago (Plotkin, 2000).

Fennel is a relative of silphium. Silphium may have looked something like this.

6.18.6 Courtship

> Everything is incredible, if you can skin off the crust of obviousness our habit put on it.
>
> **—Aldous Huxley**

There certainly is no more elaborate communication than the elaborate courtship behaviors of many animals. Some female birds signal their availability to males by assuming enticing positions. The males, in turn, respond by strutting or fluffing. There may follow a set of behaviors that appear quite ritualistic, involving move and countermove, until, when completed, copulation takes place. Although minutely and intricately wrought, these behaviors are repeated each time males mate with females of the same species. These movements are apparently intended to communicate to the opposite sex both availability and fertility.

Courtship behavior is important also for insects. The flashing of fireflies, or lightening bugs (*Coleoptera*) is part of a courtship ritual. Flying males produce bioluminescent signals from their abdomens when the protein *luciferin* is oxidized by ATP in the presence of the enzyme *luciferase* (Hale et al., 1995). The rate and pattern of light pulses is characteristic of a particular species. On the ground, females respond with bioluminescent signals of their own. When males and females answer each other appropriately, mating and fertilization occur, although there may be total darkness otherwise.

The courtship behavior of the three-spined stickleback fish (*Gasterosteus aculeatus*) has been studied and described in detail (Lewis and Gower, 1980; Campbell et al., 1999). Parental care in this species is given mainly by the male, who constructs a tunnel nest at the bottom of the river. Although fiercely territorial, the appearance of a *gravid* (an egg-carrying or pregnant) female with her swollen belly inhibits his aggressive tendencies. He then begins a zigzag dance that attracts the female (Figure 6.18.3). Soon they are swimming in coordinated fashion. He leads, and she follows until they reach the nest. If this courtship has gone well, she accepts his bid and enters the nest. His trembling and nuzzling of her tail stimulates her to spawn, and then he immediately enters the nest and deposits sperm on the eggs. Once she no longer exhibits a swollen belly, he aggressively drives the female from the area.

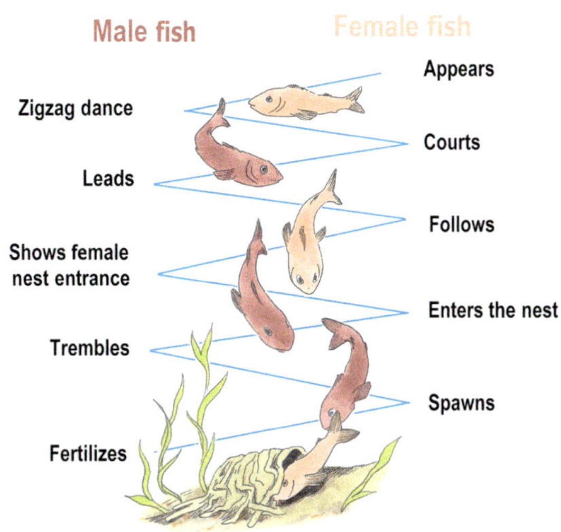

FIGURE 6.18.3
Coordinated activity during the courtship of the spined stickleback fish. (Redrawn from Lewis, D.B. and Gower, D.M. *Biology of Communication*, John Wiley & Sons, New York, 1980.)

Spiders use many types of signals in their mating rituals. Each different species behaves differently from the others, as would be expected based on the fact that mating with a different species would be nonreproductive and, hence, a waste of resources. Male web-spinning spiders attract the attention of females by shaking their webs. They may display body areas of bright colors or engage in elaborate and tentative movements to distinguish themselves from prey. To fail in this quest is to be eaten, which is a high price to pay for failure. Even successful males are often eaten after fertilization occurs.

In all these rituals, the objective is successful mating and reproduction. Males and females assess each other for strength, health, and reproductive suitability. If at any time, one or the other partner does not perform in a satisfactory way, the courtship routine may be terminated, and partners go off in search of other mates (or, in the case of the male spiders, they may be eaten).

[I was struggling for words to use in the preceding paragraph. Phrases like "does not meet expectations" or "is judged to be unsatisfactory" kept coming to mind. But, we are not talking here about creatures that "expect" or "judge," and even if it could be shown that they are thinking critters, these courtship routines are largely outside the realm of rational thought. If you doubt that, look at the pages of the newspaper that announces human weddings and engagements. Notice the many couples where the man and woman have faces that look similar. What you will see will amaze you, as it did me when it was first brought to my attention. As evidence of this observation of physical similarities, a study of married couples in the U.S. show that paired couples share a higher share of genetic material than would randomized pairings (Domingue et al., 2014).]

There is evidence that vertebrate and invertebrate mating adults of both sexes prefer certain mating partners over other candidates (Purdy, 2005). When the special partners are allowed to mate, the number of offspring is usually smaller, but the survivability of the offspring is higher than if mating occurs between nonpreferred partners. Contrary to the concept that the partner with the greater investment in rearing the offspring exhibits more selectivity in choosing a mate, both partners appear to have some mate selection discrimination, and a tendency to mate with several individuals. This results in genetic variation in offspring and greater chances that at least some of the offspring survive life's challenges.

The concept that animals may seek mating partners that fit a particular image extends to dragonflies (Ackerman, 2006). One species of dragonflies, known as bluets, has females that come in two colors, blue and green. Males are all blue. Why females of two colors should be maintained in the population is not known, but it has been speculated that males reared by blue females prefer blue females with which to mate. Males raised by green females mate preferentially with green females. This suggests that portions of male dragonflies' sexual behaviors are learned.

WHY ARE HUMAN BABIES BORN AT NINE MONTHS?

Human babies are born after a gestation period of nine months despite the fact that their brains and cognitive development would require gestation times of 18–21 months to be comparable to the relative level of development of chimpanzee births. Why this 9 month gestation period is relatively so short is not completely known. One theory is that human mothers are limited at 9 months to a metabolic rate that can no longer supply the needs of both the fetus and the mother. Thus, the babies are born and derive most of their metabolic needs outside the body of the mother (Wong, 2012; Dunsworth et al., 2012).

FORM AND FUNCTION FOR BIRD NESTS

Bird nest structures differ among different species, and are largely genetically determined, but each nest must serve, more or less, four different purposes:

1. Provide warmth or shade for hatchlings,
2. Attract mates,
3. Prevent parasite infestations,
4. Protect against predators.

Thus, nests are either larger or smaller, made of some materials or others, and shaped in particular ways, depending on the evolutionary needs of the individual and species (Zuckerman, 2015).

Birds' nests take different forms depending on species needs.

6.18.7 External or Internal Fertilization

> The world is an uncertain and changing place, to which humans and animals respond by considering the potential reward and cost of different options and estimating the odds of success before committing to a choice.
>
> —**Richard Saltus**

Some animals fertilize their gametes externally. In this case, both eggs and sperm are released into the surroundings, and fertilization occurs when sperm encounter eggs. Because external fertilization requires an environment where the young can develop without heat stress or desiccation, it occurs almost exclusively in moist places. Externally fertilized eggs that also develop in the water are a food resource for other animals, so there is a good chance that they will never develop into mature individuals. Because of this, the eggs and sperm released into the environment are usually relatively numerous.

Species with internal fertilization usually produce fewer zygotes, but provide more parental protection than species with external fertilization. There is a greater resource investment in each offspring with internal fertilization, so far fewer can be produced. With fewer numbers, and more expensive offspring, greater protection is required to assure an adequate survival rate.

The amount of parental care given to the offspring is generally inversely related to the number of offspring. Animals higher on the food chain tend to have fewer offspring, sometimes numbering less than 10 per generation rather than thousands per generation for some lower species. Parental care can dramatically improve survival rates, so thousands of offspring are unnecessary for the transmission of genes to the next generation.

6.18.8 Hermaphrodites

None of us is normal…no one has the perfect genome.

—Evan Eichler

There are certain species that may have trouble meeting others of their same kind. Sessile animals, burrowing animals, or internal parasites may have this problem of meeting a member of the opposite sex. *Hermaphrodites* are individuals with both male and female reproductive systems. Although these individuals could potentially fertilize themselves, most must mate with another member of the same species. Because both individuals can assume both male and female roles, this doubles the chances of encountering someone with which to mate (Figure 6.18.4).

In *sequential hermaphroditism*, an individual reverses its sex during its lifetime. In various reef fish species, sex reversal is associated with age and size (Campbell et al., 1999). These fish live in harems consisting of a single male and several females. When the male dies, the largest female in the group changes sex and becomes the new male. The largest fish would be able to protect the group better than a smaller member, and so confers on the group better reproductive advantage (Warner, 1984).

Oysters release both eggs and sperm into surrounding waters for external fertilization. As explained earlier, this requires greater numbers of eggs and sperm to be released than if oysters had been internal fertilizers. Eggs are larger than sperm, and require more energy to produce. Thus, larger female oysters have a reproductive advantage because they have the strength to produce more eggs than smaller oysters. Oysters change sex from male (when smaller) to female (when larger), probably to produce more eggs and boost reproduction.

Approximately 0.2%–2% of human live births can be classified as hermaphrodites or intersexuals. Some of these contain some cells with XX (female) and some with XY (male)

FIGURE 6.18.4
Earthworms have both male and female reproductive organs to improve their chances of reproductive success. (Redrawn from Simmons, K., Evolution ecology and biodiversity, Lab Manual Online, www.kentsimmons.uwinnipeg.ca/16com05/16labman05/1b6pg2_files/earthworm2.bmp, 2006.)

chromosomes, possibly as chimera (see box, Section 6.16, Cooperation). Some of these have other unusual chromosomal patterns such as a single X chromosome (called Turner syndrome). Some may also have responded to fetal levels of estrogen or testosterone (see Sections 6.6.5, Adaptations, and 6.12.9, Endocrine Disruption). Intersexual children have the sex organs of both sexes, which don't often develop normally.

6.18.9 Plant Reproduction

Flowers are sunshine, food and medicine to the soul.

—**Luther Burbank**

Sexual reproduction among plants occurs in a number of ways with a lot of similarity between them. All schemes involve a stage of *meiosis* to form haploid cells and a fertilization stage where male and female cells combine. There may be additional mitosis stages to reproduce male and female *gametophytes* before they are ready to combine. Flowering plants that dominate the landscape have male gametes called *pollen* and female gametes called *ovules*. When fertilized (or *pollinated*), the ovule develops into a seed containing an embryo and a supply of nutrients.

Some flowers possess both male and female parts. They are called *perfect flowers*. *Imperfect flowers* have either one or the other, but not both.

Pollen grains are formed on male flower parts called *stamens*. The male gametes in pollen are encapsulated in several tough outer polymer layers to protect against drying and physical damage. There is a very small opening in the pollen grain, called a *micropore*, through which the gametes can leave when they land on the female flower structure, called a *pistil*, the top part of which is called a *stigma* (Figure 6.18.5). They then grow into the pistil to fertilize the ovule.

There are two popular mechanisms for transferring pollen from its source to the ovule (see Section 6.16.3, Plant Reproduction). The first is similar to external fertilization in animals in that pollen is released into the air and haphazardly falls on many surfaces, including the *stigma* of a flower where it can pollinate the ovule. In order for this strategy to be successful, a huge number of pollen grains must be released by each plant. One ragweed plant can launch a million pollen grains a day (Ackerman, 2001).

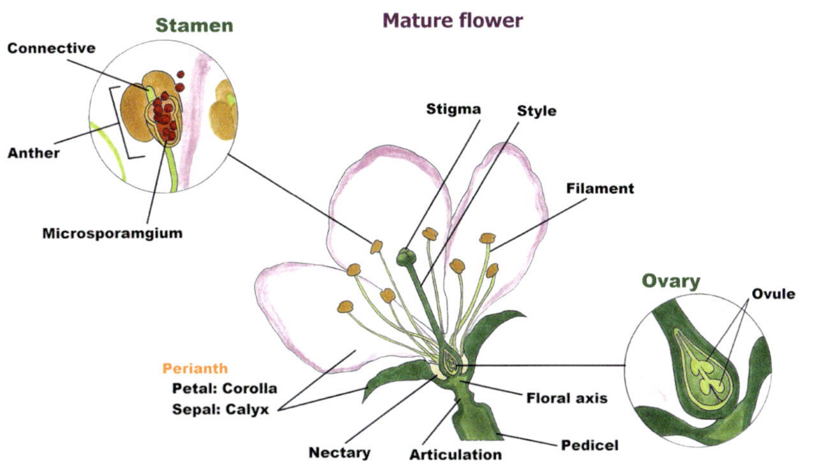

FIGURE 6.18.5
The reproductive parts of an angiosperm flower. (Redrawn from Wikipedia, Marianna Ruiz, 2011.)

The other strategy is to cooperate with animals that can carry pollen grains (usually inadvertently) from their sites of release to the stigma. Brightly colored flowers and those with intricate shapes usually operate in this way. Birds and insects seek nectar secreted in the throat of the flower, and encounter sticky pollen grains along the way. These hitchhike on feathers or hairs to the next flower, where they can pollinate the ovule there. There are rejection mechanisms that keep many of these flowers from pollinating themselves.

Plants sometimes also have difficulty with fertilization by cross-pollination. There may be no other plants around to provide pollen, as at the edge of the geographical range of the species, or neighboring plants may still be many years away from sexual maturity. There are genes in some plants that control fertilization by interfering with self-pollination. The absence of one of these genes can allow the plant to become self-fertile under some environmental circumstances (Fessenden, 2010).

Some insects, especially bees, collect pollen as a protein source for raising their young. After they bring it back to their nests or hives, the pollen is partially digested to make the protein nutritionally available. Perhaps as a reproductive strategy, plants that depend upon insects for pollination have more protein in their pollen than plants that depend upon the wind for pollination.

The type of pollinator that normally pollinates a plant can have consequences on the evolution of plant traits and flower characteristics. Plants respond to their preferred pollinators by becoming more attractive to them. Plants pollinated by bumblebees became larger and had more fragrant flowers with a greater ultraviolet color component. Plants pollinated by hoverflies became smaller, their flowers were less fragrant, and they tended to self-pollinate more (Gervasi and Schiestl, 2017).

Once fertilized, the ovule develops into a seed (see Section 6.15.3, Seeds). The surrounding ovary tissue may expand to form a fruit protecting the seed and attracting animals to disperse the seeds (see Section 6.12.4, Wastes).

Seeds are miniature plants (or *embryos*) with all plant parts included. Packed inside the seed coating are starch and other stored foods to support growth when conditions are favorable. Within its hard coat, the seed may remain viable for years. Viability, however, decreases with storage time.

Seeds that contact water absorb some, and this triggers enzyme production to convert starch into sugar. Metabolic changes induce the embryo to resume growth. Roots grow into the soil and the shoot tip breaks into the light. The plant is then on its way to eventually make another seed.

Plants contain *meristematic tissues* of dividing, undifferentiated cells that can sustain or renew growth indefinitely. Plants also contain *parenchymal cells* throughout the plant that can divide and differentiate into specialized cells in various parts of the plant. Thus, detached parts of some plants can develop into whole new plants by *vegetative reproduction*.

Example 6.18.1 Shipping Animals

Each year nearly 13,000 animals arrive at North American zoos and aquariums. Of those, about 5000 are born there. The rest, according to the International Species Information System, are transported by plane, truck, or boat from across town or from remote locations across the world (Davis, 2004). Some 5000 species are protected under an agreement called the Convention on International Trade in Endangered Species of Wild Fauna and Fiona (CITES). With a conservation mission, zoos and aquariums have begun to produce more wild animals than they consume. More than 90% of mammals and 70% of birds on display were born in captivity.

To guard against inbreeding, maintain genetic diversity, and achieve sustainable population growth, computer-based Species Survival Plans (SSPs) have been developed to track the whereabouts of all potential breeding stock at accredited zoos (Example 5.3.3). When the time comes to begin the breeding process, air freight is a popular means to move animals quickly. Cockatoos, barramundi, chimpanzees, turtles, lizards, and snakes are all shipped in this way. Including pets, Delta Air Lines alone handled 40,500 animal shipments in 2003.

Animal welfare is paramount. Some animals are more easily shipped at night, others during the day. Some Antarctic birds require their own refrigerated chartered cargo flights. Little nectar feeders such as hummingbirds, fruit bats, and shrews have fast metabolisms that need constant replenishment. Hanging feeders in the crates isn't practical, so other means to feed must be innovated.

Shipping animals is a challenge for specialists who need to know what the animals can tolerate and what conditions they require. The International Air Transport Association annually issues a 400-page book of regulations for shipping, with specifics on crate sizes, crate materials, ventilation, humidity, and temperature. When their trips are completed and they have arrived at their destinations, the animals are ready for assimilation into their new homes.

APPLICATIONS AND PREDICTIONS

1. Animals and plants that release their offspring or seeds into their environment at an immature stage will produce more offspring than animals and plants that nurture their offspring.
2. Species with internal fertilization will usually produce fewer zygotes than those with external fertilization.
3. Successful sexual reproduction will require that mature sperm encounter fertile eggs of the same species.
4. Species that have difficulty encountering another of the same species will likely depend on hermaphroditism for sexual reproduction.
5. Among hermaphroditic species, protogynous hermaphroditism or protandrous hermaphroditism will depend on which confers a size advantage.
6. Asexual reproduction will be favored by a successful species in a stable environment to increase numbers rapidly.
7. The reproductive process is extremely complex, and its success will depend on the interactions of many different kinds of internal and external stimuli.
8. Genetic variability will be maintained by mutation in species reproducing asexually.
9. Looking at the shape and color of a flower can indicate the type of pollinator that visits.

6.19 Biological Units Coordinate Activities through Communication

There are many ways for organisms to probe the external world. Some smell it, others listen to it, many see it. Each species therefore lives in its own unique sensory world of which other species may be partially or totally unaware.

—Richard Axel

BU do not exist as isolated units, but instead share their environment with others, both similar and dissimilar. Sometimes relations with other BU are friendly and cooperative, and sometimes they are hostile or competitive. In all cases, however, BU must be aware of the actions of surrounding BU in order to be able to react appropriately (see Figure 6.0.1). Imagine a BU unable to determine the proximity of a predator. This BU would live an extremely short life. Thus, there is strong selective pressure to develop sophisticated sensing abilities. Going even further, however, a real survival advantage belongs to the BU that not only can sense the actions taken by other BU, but to anticipate future actions and adjust accordingly (Yoerg, 2001). Hence, sometimes extremely complex behaviors have evolved that appear to be quite intelligent. But oftentimes, these intelligences are very specific and not able to adapt to new circumstances as they arise. We all know about squirrels: they collect nuts during the abundance of autumn and hide them by burying them in the ground. In the winter, they remember the locations of enough of these nuts to be able to retrieve them and eat them to sustain themselves. There is a kind of intelligence there that puts ours to shame. However, squirrels are doubtless unable to visualize using a stick to help bury the nuts or to scrape the snow from on top of the ground. Their intelligence is quite mindless because it is exquisitely refined to perform one task extremely well; it does not adapt well to other similar tasks.

Communication is important as a means of coordination, and, as such, is used by microbes, plants, animals, and tissues and organs. Organismal BU use communication to meet basic needs: food, security, and reproduction are among these. They use a sophisticated set of stimuli and responses for these purposes. The more social the BU, the more elaborate are the communications.

Von Uexlcüll suggested in 1934 that each species inhabits a sensory world uniquely evolved to meet its needs (Dyer and Gould, 1983). Thus honey bees see colors, but these colors include ultraviolet and lack red. Also, bats hear sounds, but they hear very high frequencies that humans cannot. Sharks and other fishes are sensitive to electric fields, and some birds can tell direction from lines of force in the Earth's magnetic field (Gould, 1980). Sensory reality to each of these animals is much different from what we humans would perceive.

Nonorganismal BU use communications to support the needs of the organism. Their sensing and responses are parts of overall biological control systems that maintain the ability of an organism to act as one coordinated whole rather than a disorganized assemblage of independent parts (see Section 6.16, Cooperation).

Sensation is but one part of communication. In order to be complete, communication is sensation that invites response, and that, in turn, may elicit an additional response from the original initiator. Communication can take one of four forms (Nelson et al., 1970):

1. Acoustic
2. Chemical
3. Tactile
4. Visual

Each of these will be considered.

6.19.1 Acoustic Stimuli

> There are extremely few examples where we really know that the nervous system is doing from sensory input to a behavior. We can map them out in simple reflexes, like an animal's escape response, but what we'd really like to understand is the steps by which information is transformed and integrated all the way through.
>
> —**Cornelia Bargmann**

Acoustic signals are used by a variety of organisms. Insects, such as crickets (*Gryllidae*), vibrate various parts of their bodies to make specific signals (see Section 3.10). Male woodpeckers of some species drum on hollow objects during their courtship of females. The buzzing of honeybees from outside the hive can be used by some beekeepers to judge the condition of the bees inside: a loud buzzing may indicate a hive with an ineffective queen, whereas a quiet hive is a healthy hive.

Vocally produced acoustic signals are also used by some vertebrates. Humans, of course, use talking as a means to communicate. Languages have developed in ways that allow humans to express ideas important in all aspects of life. The *Sapir-Whorf Hypothesis* posits that the way we see the world is shaped by language; the structure of a language determines a native speaker's perception and categorization of experience.

Other vertebrates use vocalizations to communicate to their peers. There has been shown a relationship between type of hen activity and vocalization (Stone et al., 1984). Similarly, different dog vocal sounds have been associated with marking of territory, various wants, recognition of danger, request for attention, and aggression (Houpt and Wolski, 1982). Pig vocalizations for various husbandry practices have been recorded and found to correlate to states of the animals (Xin et al., 1989). Pig sounds were found to be correlated better to stress level than was the physiological measurement of respiration rate (White et al., 1995). In this way, animals can be assessed to determine whether the conditions in closed confinement are too stressful to satisfy animal rights activists or to produce the most economical meat.

HUMAN LANGUAGE ACCORDING TO CHOMSKY

Early studies of languages focused on differences and similarities among them, with the idea that languages were acquired and learned. The ideas that they conveyed were innate, but the means to communicate them to others were learned. The study of language was tied closely to anthropology (Lerer, 1998).

Chomsky (1965) proposed a different concept, one that turned traditional ideas inside-out. Chomsky promoted the idea that language is innate; the capacity to communicate is not learned, but is hard-wired into the human brain. A model of communication exists within each child at birth.

What must be learned are the words that make the model effective, and that allow others to comprehend the meaning of the speech. Thus, human communication is likened to the ability to see and hear, in that no one must acquire these skills, but interpretation of the meaning of what is seen and heard must be learned.

Animals, too, seem to have a limited ability to communicate without acquisition from their peers of the meanings of different sounds. Thus, there is a basic similarity between human and animal speech. Given the presumed difference in general intelligence between humans and animals, it probably makes more sense for humans to learn meanings of animal utterances rather than attempt to teach human speech patterns to animals.

Human language, sentence structure, and grammar have been found to provide biomarker clues for several neural and mental disorders, such as schizophrenia (Bedi et al., 2015), Parkinson's disease (Ibáñez et al., 2012), and Amyotrophic Lateral Sclerosis (Ash et al., 2015). Computer algorithms have been written to make these diagnoses.

Verbal communication among humans (speaking) and birds (singing) is learned very early in life when their brains have an extraordinary ability to focus on sounds made by adults. Social interaction, however, is required for this learning to occur. Without conscious contact with live speakers of the language early in life, the human babies do not learn their languages and birds do not learn their songs. Autistic children often are not capable of realizing this type of social interaction, and so are impeded in their language development (Kuhl, 2015).

Bird songs are a type of acoustic communication that have the following functions:

1. They convey that the singer is a virile male with a defended territory
2. They reduce unnecessary fighting with other males who are made aware of the boundaries of the territory
3. They attract breeding females
4. They help bring females into breeding condition

Others say that birds sing for four good reasons:

1. Sex
2. Real estate
3. Who's boss?
4. What's for dinner?

Bird songs are largely hereditary, with little variation within a species. Because of this, bird types can be identified by song even though visual sightings are not made.

Birds can also alert others to danger through alarm calls. Distress calls of starlings will cause a flock to take flight even if the calls come from a recording. Crows and ravens (*Corvus*) have a large repertoire of different calls that are used among birds to convey messages. These birds can be taught to form human words, and there is some indication that they can associate words with their meanings.

Polychlorinated biphenyls (PCBs) are synthetic chemical pollutants that have been used to remove heat from electrical components and have also been widely dumped into the environment. PCBs have been demonstrated to have detrimental toxic and developmental effects on humans and wildlife. A study of two bird species, black-capped chickadees and song sparrows, has shown higher total blood PCB concentrations in regions of the Hudson River valley where environmental PCB contamination has historically been higher. High blood PCB concentrations were shown to affect bird songs adversely, thus interfering with avian intraspecies communications necessary for satisfactory reproduction (DeLeon et al., 2013).

Vervet monkeys (small monkeys common throughout Africa) communicate among themselves about various predators. They have different alarm calls for leopards, eagles, and pythons, their chief predators. When the alarm call is made for leopards nearby, the monkeys run to the nearest trees and climb to the smaller branches where they are safe. The eagle alarm causes vervets to scan the sky and dive into the underbrush to avoid being seen from the air. The python, or snake, call causes monkeys to stop and alertly determine where the snake is located. Once located, the vervets pounce on the snake and drive it off. Lions, hyenas, and cheetahs as a group also have an alarm call (Shumaker, 2010).

Vervets fighting other vervets and getting the worst of the fight sometimes give the leopard call as a ruse to save themselves from further beatings. Once the call is given, all vervets stop fighting and display their defensive behavior.

FREQUENCY CONTENTS OF SOUNDS

Communication within a species requires both a talker and a listener, and interactive communication requires that these two interchange roles every now and again. Interspecies communication often does not rely upon changing roles, because the relationship of the species to one another is often that of predator and prey.

Not all species hear sounds in the same frequency ranges. Indeed, there are some with twice the frequency range of humans.

Frequency of emitted sounds can depend on the purpose of the emissions and the environment in which they are normally emitted. There is a large advantage for bats to emit very high frequency sounds because the wavelengths of such sounds are small; this results in sounds that readily reflect from nearby objects and have high spatial resolution. Reliable reflection only comes from objects that appear large compared to the wavelengths.

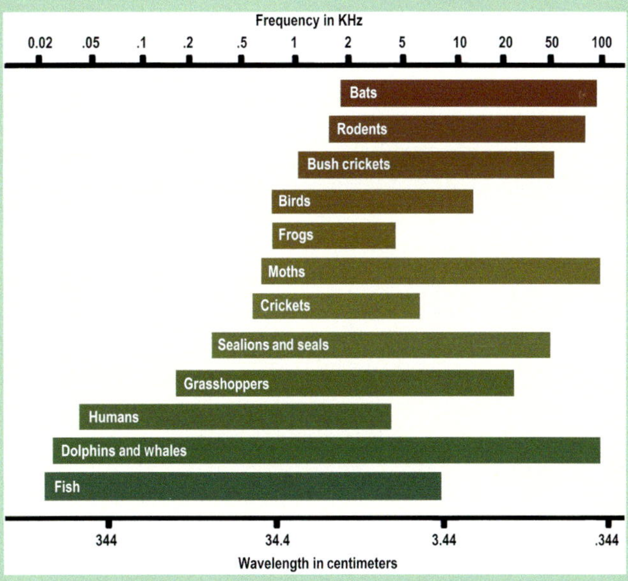

Approximate limits of hearing in various animals. (Redrawn from Lewis, D.B. and Gower, D.M., *Biology of Communication*, John Wiley & Sons, New York, 1980.)

(Continued)

FREQUENCY CONTENTS OF SOUNDS (Continued)

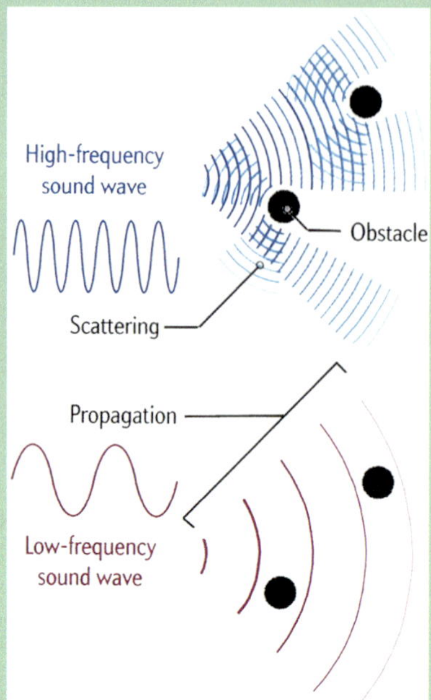

Sound frequencies with wavelengths long with respect to the sizes of obstacles can be propagated around the obstacles without significant attenuation. High frequency sound waves, with short wavelengths are scattered and reflected by obstacles. These effects are similar for electromagnetic waves used for electronic communications.

Very low frequency sounds are used for communication between members of larger species. It takes a large animal (such as an elephant, rhinoceros, whale, or cassowary) to produce the mechanical power necessary to send low frequencies over long distances. These sounds are used by large animals to coordinate activities without converging on scarce resources. Elephants, for instance, produce rumbling noises between 5 and 30 cps (Hz), and these can fill an area up to 300 square kilometers (Ross, 2004). These sounds are not reflected by commonly sized objects, so are not attenuated by surrounding vegetation.

Very low frequency sounds also have a psychological effect. Tigers produce an 18-cycles per second (cps) component in their roar that induces a feeling of terror in humans and paralyzes prey for up to 10 seconds (Ross, 2004).

Deep sounds are usually warnings; higher pitched sounds are usually conciliatory (Friend, 2004). Dominant individuals produce deeper sounds; subservient individuals accommodate their sounds to dominant ones.

Higher frequency sounds require much less energy to transmit than do lower frequency sounds, and thus convey more information per unit energy than do lower frequency sounds. Vowels in human speech are lower frequency sounds; consonant sounds are formed from high frequencies. Acoustic filters are more easily produced to diminish high frequency consonants than to attenuate low frequency vowels.

(Continued)

FREQUENCY CONTENTS OF SOUNDS (Continued)

Casowaries are ancient land birds that now live in the forests of New Guinea. They can produce calls as low as 23 cps.

6.19.2 Chemical Stimuli

> When one looks at the macroscopic structure of living systems, what one is really seeing is the cooperative activity of cells of different types communicating by different molecules.
>
> **—Douglas A. Lauffenburger**

Chemical signaling often involves the use of *pheromones*, volatile chemicals released into the environment, usually as sex attractants. They are extremely specific and affect only others of the same species. The concentrations of pheromones that elicit a response are extremely small. Glypure, the gypsy moth sex attractant pheromone, has a male threshold value of about 1.25×10^{-6} kg/m^3 in the air (Johnson, 1999). Female gypsy moths release the glypure into the air and males follow the scent to the female. Artificially synthesized insect pheromones are used in two ways: (1) to trap insect pests, and (2) to overwhelm and confuse insects to disrupt mating.

Pheromones are important for higher level animals, as well. Dogs, cats, foxes, and other territorial mammals (Peters and Mech, 1975) use urine to mark the boundaries of their territories. Females use urine to attract males when they are in estrus (see Section 6.18, Reproduction). Whether pheromones play an aphrodisiac role for humans is still an open question. Human male sweat and other body odors are apparently attractive to the opposite sex, especially at certain times during the menstrual cycle.

To think that one chemical compound would be the sole pheromone constituent is to underestimate biological complexity. From a quality assurance standpoint, reliance on the appearance of one chemical for a function as critical as reproduction would place a species in a very vulnerable position to predators who could easily mimic a single chemical. Instead, pheromones may consist of three or more chemical compounds, mixed in correct proportions (Birch, 1978). Some of these compounds may come from other environmental sources such as plant hosts. And, in addition, the effectiveness of pheromones in the natural environment may be enhanced by visual shapes, colors, and textures corresponding to natural plant targets of insects. Chemical communication thus exists as a system rather than as an isolated unit.

The human nose, it has been estimated, can distinguish among 1.7 trillion different odors (Bushdid et al., 2014). Odors are important signals for human memory, and humans

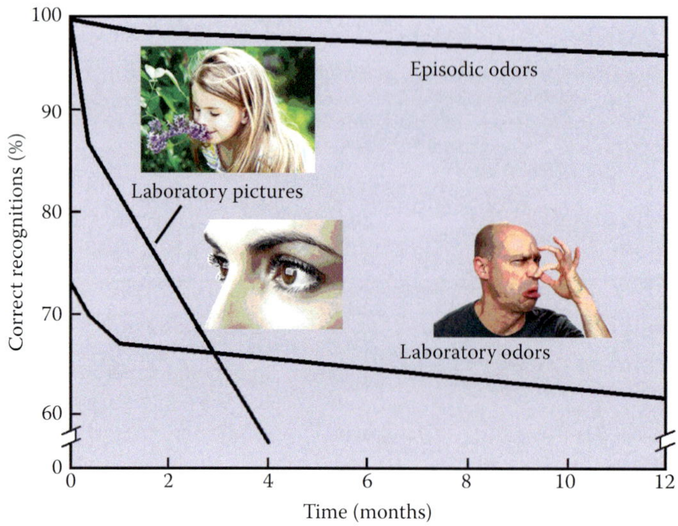

FIGURE 6.19.1
Odor memory is very strong, especially if associated with significant life episodes. Odor memory is retained for a long time, but is not accompanied by correct recognition. In contrast, pictures presented to subjects are remembered extremely well for only a very short time. (Redrawn from Engen, T., *Am. Sci.*, 75, 497–503, 1987.)

can vividly remember episodes in their lives upon smelling the same odors again. This memory lasts for an extremely long time, but the association with verbal descriptors is very weak (Engen, 1987). Presenting subjects with odors and telling them the words associated with these odors can be remembered for months (Figure 6.19.1), but only a few of these odors are remembered for that long. Odors associated with significant real-life experiences can be remembered very accurately almost indefinitely.

Chemicals can also be used to locate food. Fire ants, for instance, lay odor trails to a food source by exuding pheromones as they touch their stingers to the ground (Johnson, 1999). The denser the food source, the stronger is the pheromone signal laid by returning workers. As the pheromone evaporates, the trail disappears. Old trails no longer used disappear below threshold level in 2 min.

Honeybees (*Apis mellifera*) have a dance that is thought to communicate the location of a food source. If the food is near the hive, the discoverer bee performs a round dance in the hive, telling the other bees that there is a food source nearby. When the food is farther from the hive, the bee performs a wagging dance in the shape of a modified figure-eight pattern (Figure 6.19.2). The angle of the center portion of the figure-eight indicates the angle between the food source and the sun. While the bee is moving along the central part, it waggles (or moves abdomen from side to side). The rate at which it waggles indicates the distance to the source, with closer distances being given by faster movements. This dance can either be performed horizontally on the hive bottom, or vertically on the combs (Nelson et al., 1970).

There is controversy associated with the meaning of this dance. Some argue that the dance by itself is not the way honeybees locate the new source of food. They say that other honeybees take the opportunity afforded by the dance to sniff the odor on the body hairs and regurgitated food of the dancing worker and they follow the odor back to its source. The other possibility is that the dance allows recruited worker bees to locate the general area of the food, and they follow the odor to the exact location (Wenner, 1998). In any case,

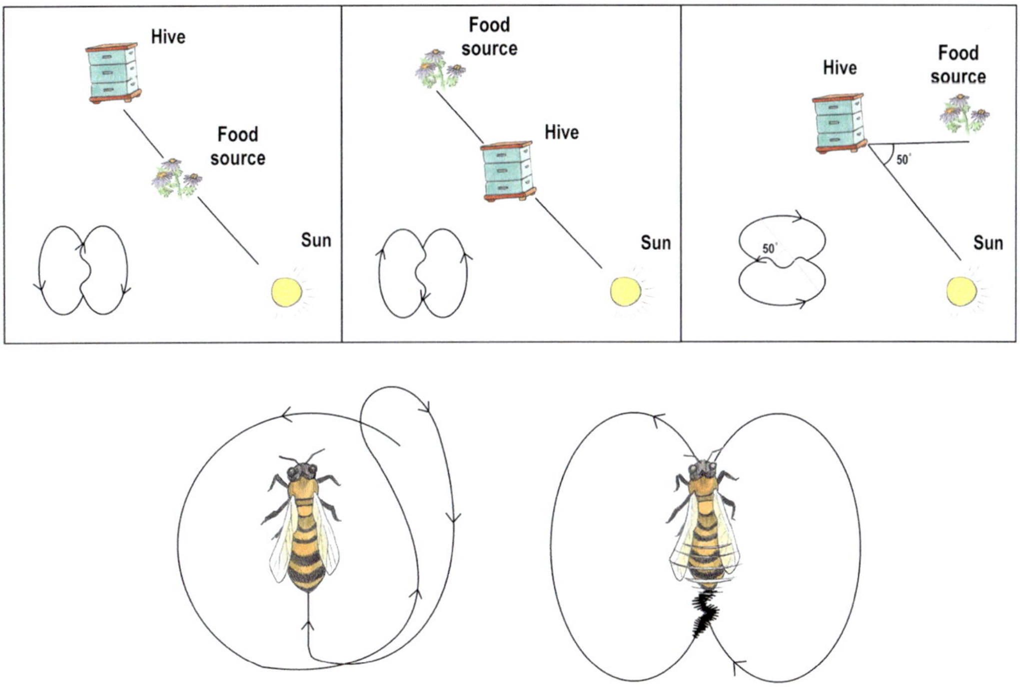

FIGURE 6.19.2
Communication by dances in honeybees. (Redrawn from Nelson, G.E. et al., *Fundamental Concepts of Biology*, John Wiley & Sons, New York, 1970.)

odor probably plays a big part in food location, although air pollutants have been found to destroy enough flower hydrocarbon emissions to reduce the distance a pollinator can detect a flower from several kilometers to 200 m (McFrederick et al., 2008).

The queen honeybee emits a substance from her mandibular gland that controls the hive. This substance is passed to each worker at the rate of about 0.1 μg per day (Lewis and Gower, 1980), and has three separate effects:

1. It allows the workers to know that the queen is healthy, so they will not rear another queen in the hive.
2. It prevents the workers' (all females) ovaries from developing, so they won't lay eggs.
3. It acts as a sexual attractant to the drone (male) during her nuptial flight (she becomes fertilized in flight), but not when she is in the hive.

Similar substances are probably present in all social insect colonies.

There is a genus of water molds called *Achlya* that uses chemicals transmitted between male and female to synchronize their reproduction. There are four chemicals involved and each causes a specific behavior to occur in the proper sequence.

When their food supply is exhausted, slime mold amoebae send out a 2 s pulsed release of cyclic adenosine monophosphate (cAMP). Low concentrations of cAMP (see Section 4.4.1, Sensors) cause a clumping of cells, whereas high concentrations produce both the attraction of cells (chemotaxis) and the release of their own pulses of cAMP upon

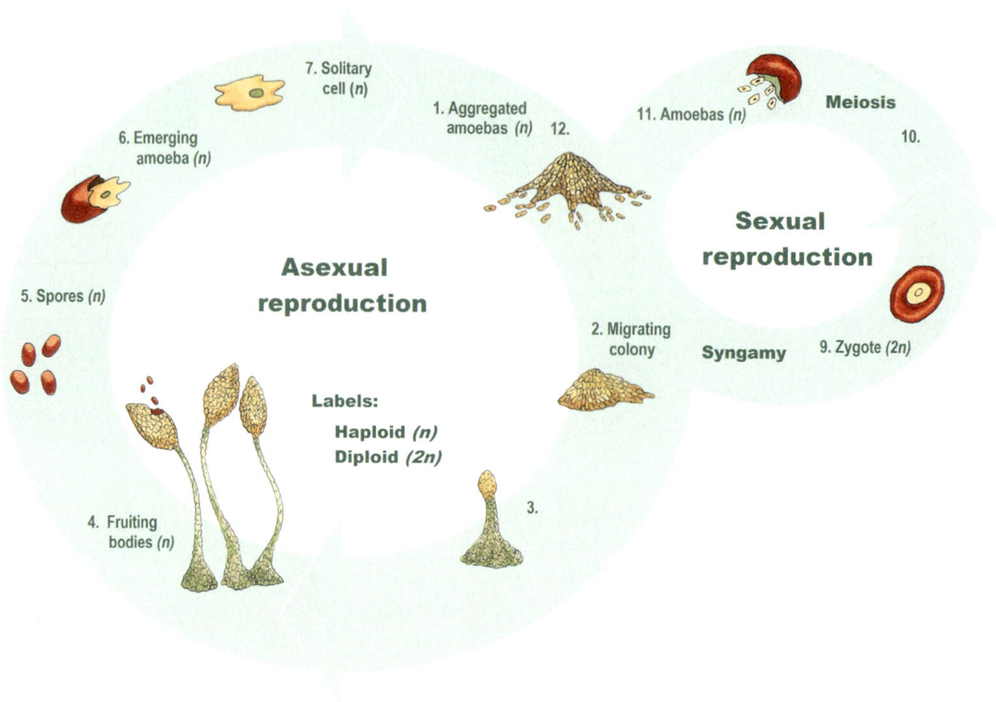

FIGURE 6.19.3
The life cycle of a cellular mold. (Redrawn from Campbell, N.A. et al., *Biology*, 5th ed., Addison Wesley Longman, Menlo Park, CA, 1999.)

receiving the first pulse. Following this release, there is a refractory period of at least 2 min during Ih the amoeba is not sensitive to further pulses. This guarantees that the amoeba cannot be stimulated by its own signal or that reflected back from its neighbors (Lewis and Gower, 1980). The aggregated amoebae migrate for a short distance as a single unit, and then form a fruiting body to form asexual spores (Figure 6.19.3).

Tobacco plants emit a chemical vapor to give airborne warning to neighboring plants about attacking viruses. When attacked by mosaic virus, tobacco produces salicylic acid in defense (Park et al., 2007). Some of that becomes methyl salicylate, Ih evaporates. Airborne methyl salicylate stimulates defense mechanisms of neighboring healthy plants (Johnson, 1999).

Many plants emit volatile compounds specific to the type of insect that is attacking it. These compounds are apparently triggered by the saliva of the insect. These volatiles, in turn, are attractive to the parasites and predators of the attacking insects. There are other volatile compounds that plants emit depending on soil condition or different stresses. Some plants emit methane, some absorb methane. Gamma-aminobutyric Acid (GABA) acts as a stress-signaling hormone in plants as well as in animals (Ramesh et al., 2015).

It has been found that one type of annual plant (the Great Lakes sea rocket) can recognize the difference between its own kind and others, perhaps by the volatile chemicals they all release. When placed in pots surrounded with unrelated plants, the sea rocket will divert resources into root growth in order to bolster its competition with the others. Among sibling plants, however, it exhibits no such behavior (Dudley and File, 2008).

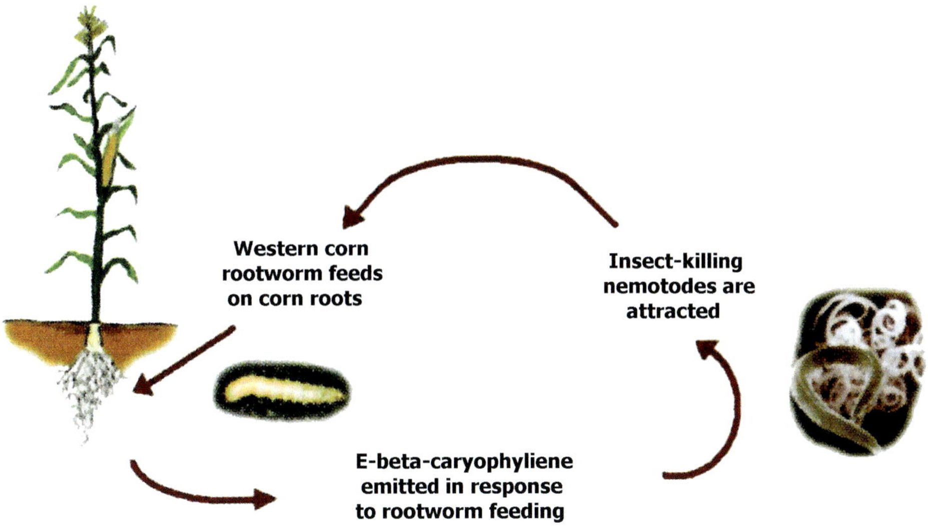

FIGURE 6.19.4
A natural feedback loop helps to defend corn plants against predator grubs.

Gene expression profiles for plants are different for plants exposed to different bacterial strains. If these can be detected, then plants could be used as environmental biosensors by humans.

Corn root worm larvae feed on root hairs and bore into maize roots. Left unchecked, the plants take up less water and nutrients and collapse. Corn plants, however, emit a chemical, E-beta caryophyllene, when attacked by the root worm. This, in turn, attracts beneficial insect-killing nematodes to control the root worm population (Glover, 2009). Feedback loops such as this one, which includes natural insect pest controls, can be manipulated and utilized for environmentally friendly insecticide applications (Figure 6.19.4).

In a series of experiments, Karban and Shiojin (2009) demonstrated that plants not only communicate with one another by airborne chemical cues, but that they can also distinguish between chemicals emitted by parts of their own selves from chemicals emitted by others. These chemical cues can be used to protect plants against further insect and herbivore damage (Phillips, 2009).

Masting is the term used to describe reproductive synchronization among individual plants of some species (Koenig and Knops, 2005). Those species that mast, such as oak trees, completely forego reproduction for some years and produce an overabundance of seeds in another. This appears to have the advantages of

1. Over satiating predators who eat the seeds,
2. Controlling predator populations,
3. Assisting efficiency of pollination through releases of saturation levels of pollen into the atmosphere during reproductive years.

Exactly how masting synchrony is maintained is not known, but it does extend spatially to distances of thousands of kilometers (Koenig and Knops, 2005).

The most important communication medium among body organs in humans and animals is chemical. Organs and tissues must perform their functions despite physical separation from other organs and tissues with related functions. The liver, for instance, is not part of the intestine, but must interact closely with the products of intestinal digestion. Communication between these two organs is through the blood stream, as it is for most organs. Circulating hormones allow organs to control each other's activities. Although some organ control is by means of the nervous system, the ultimate neural action is also chemical.

Communication among cells is largely through chemical means. Neurotransmitters (see Section 6.22.3, Neurotransmitters) emitted by neurons cause depolarization of target cells. Hormones secreted by specialized cells can have profound effects on other cells. Small ions can pass between cells, thus enabling surrounding cells to share information about the states of certain cells. A solution of small molecules in water flows through the plasmodesmata (threads of cytoplasm that pass through cell walls and join the cytoplasm of adjacent cells) between plant cells. To share more complex or fat-soluble molecules, or even RNA, long, thin filaments called *tunneling nanotubes*, or *pili*, (Figure 6.19.5) bridge between cells and can transport membranous packets (*endosomes*) of chemicals from the interior of one cell to Ir (Brodie, 2004b). Plasmids may be transferred between bacterial cells through pili and confer traits from one cell to the next. This is thought to be the means that antibiotic resistance passes from one group of bacteria to another.

Cells adjacent to one another and have communication channels, called *gap junctions*, which allow simple substances to pass directly from one cell to another through pores in their neighboring membranes (Figure 6.19.6). This sharing of chemical information helps cells to maintain a healthy coordination of activity (Laird et al., 2015).

Bacteria have an amazing chemical communications system that they use to detect the presence of similar and dissimilar bacteria. This has been termed *quorum sensing*, and it works like this: each bacterium produces certain chemicals (called *autoinducers*) and releases them into its environment. As the population grows, the chemicals accumulate until they reach threshold levels. The bacteria can sense the point at which the threshold is reached, and then, in concert, they all begin to act in certain coordinated ways.

The ability to act together has advantages for bacterial survival. For instance, pathogenic bacteria invading the body do not produce harmful toxin until they reach a critical population. If they didn't wait until their numbers had reached a certain level, they would

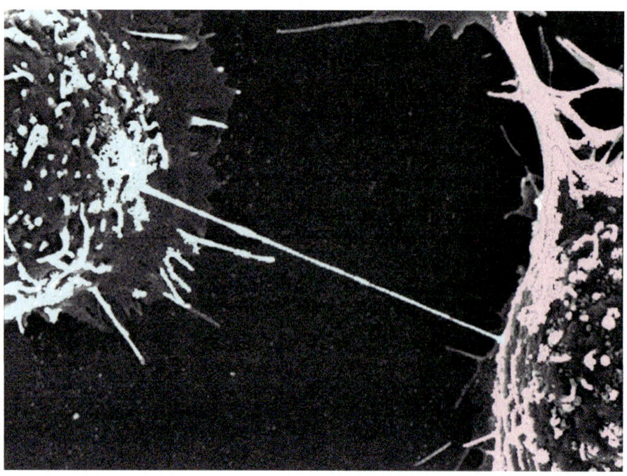

FIGURE 6.19.5
Tunneling microtubes transport chemicals one-way between cells. (Redrawn from Brodie, C., *Am. Sci.*, 92, 415–416, 2004b.)

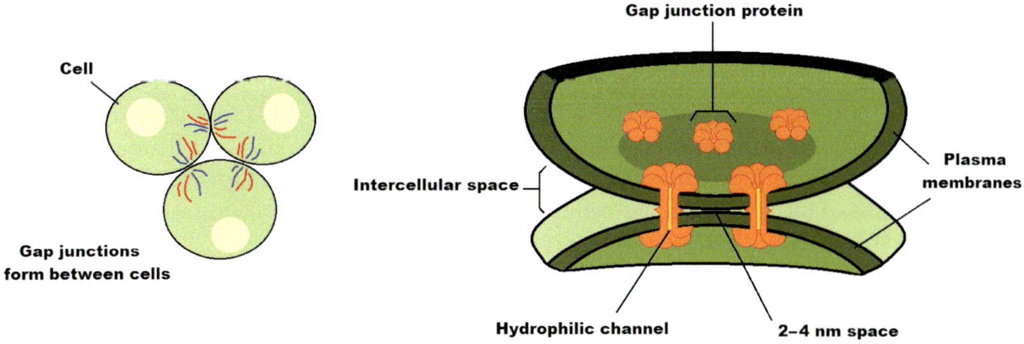

FIGURE 6.19.6
Gap junctions form paths through which simple chemicals can flow between cells, bypassing their outer membranes.

be easily detected by the immune system and readily destroyed. Once a critical population level is reach, and all bacteria together begin producing toxin, then the immune system is more likely to be overwhelmed.

The organism causing cholera acts in the same way. It lives in contaminated water, and enters the body when this water is drunk. It subsequently attaches itself to the intestine wall and immediately begins producing toxin. There it thrives and multiplies better than it would in a pool of standing water. When a critical population threshold is reached, all bacteria cease producing toxin and instead produce an enzyme to detach themselves from the intestinal wall. By this time, enough toxin is present to cause diarrhea, and the bacteria are expelled into the environment.

Quorum sensing also coordinates functions among different types of bacteria constituting a biofilm. There are chemicals that bacteria use to sense the presences of others of the same kind, and there are chemicals that are used to detect bacteria of other species. Bacteria seem to be able to smell each other by sensing ammonia as a byproduct of bacterial metabolism (Nijland and Burgess, 2010). This was found in physically isolated cultures, and smelling their neighbors allows them to find resources or brace for competition. Thus, bacteria possess sophisticated chemical information systems that give them abilities to act in ways appropriate to their survival (Raeburn, 2007).

Populations of bacteria can appear to be much more antibiotic resistant than expected. That is because some bacteria in the population will have resistance and some will not. Those resistant bacteria challenged with antibiotics emit a compound called *indole* that signals to surrounding non-resistant bacteria to expel antibiotic and to change chemical pathways to neutralize toxins produced inside the cell by the antibiotic (Williams, 2010; Lee et al., 2010).

Bacterial colonies can also communicate with each other electrically by passing charged ions from one to another through ion channels in their outer membranes, in some ways similar to communicate among neurons. This helps to coordinate actions and responses for the benefit of the entire colony (Prindle et al., 2015).

In addition to its role as an energy-storage medium, ATP also plays an important role in communications (Khakh and Burnstock, 2009). ATP serves as a molecular signal affecting cell behavior. ATP has been present since early in the history of life, and its signaling actions have broad influences on physiological responses. ATP has been shown to be released by neurons at neuromuscular junctions to cause muscles to contract. ATP can be released by a presynaptic nerve along with other neurotransmitters into the synaptic cleft and subsequently depolarize postsynaptic neurons. ATP or its unphosphorylated products ADP, AMP, or adenosine alone, act on brain stem networks controlling breathing, heart rhythm, and gastrointestinal function. ATP and its co-transmitter acetylcholine play a large role in information processed by the retina and optic nerve. Inner ear cochlea cells and sensory nerves in the taste buds have ATP receptors. ATP as a communicator substance is also involved with sleep, memory learning, movement, psychological disorders, pain sensing, release of nitric oxide to relax blood vessels, blood clotting, bone strength, wound healing, and immune inflammation response.

OXYTOCIN MAKES LOVE

Oxytocin is the hormone that influences social attachment and compassion in both humans and animals. Oxytocin circulating in a mother's blood immediately after birth facilitates mother-to-baby bonding and stimulates breast milk production. Oxytocin is also present in the bodies of males, and makes them more emotionally aware and socially outgoing. It has been proposed that administering oxytocin to autistic people could help alleviate symptoms of emotional indifference, isolation, and obsessive repetitive behaviors (Rosenwald, 2008).

6.19.3 Touch Stimuli

> Pain is part of being alive, and we need to learn that. Pain does not last forever, nor is it necessarily unbearable, and we need to be taught that.
>
> **—Rabbi Harold Kushner**

Touch receptors are present in all animals, and are important in such activities as obstacle avoidance, fighting, and copulating. Many animals use tactile signals in their courtship behavior. Male turtles often stroke or scratch the female during courtship; field cricket males repeatedly touch females with their antennae.

Touch is extremely important for humans and their pets. Stroking pets has been found to have a calming effect on both the human and the pet. Touch is also important in the bonding process between mother and child. Touch also forms the sensory basis for Braille, the means of spelling words by a system of raised dots on a flat surface to enable the sightless to read.

Touch is one of the first sensations to be developed after birth, and seems to be important in classification of one's own physiological states and interpersonal relationships later in life (Linden, 2015). Skin-to-skin tactile sensation has been found to be beneficial for physiologic and neurologic maturation of newborn premature infants (Feldman et al., 2014). This intervention, called *Kangaroo Care*, features maternal to newborn contact for at least one hour per day, and results in improved physiological and cognitive function lasting at least ten years in the life of the child.

Positive physiological responses to manual massage of both preterm low-birth-weight and full-term healthy babies include a strengthened immune system, increased weight gain, improved circulation, better digestion, and better behavior. All this improves survival rate of premature infants; massaged infants are typically released from the hospital up to six days sooner than non-massaged infants. These same benefits are promised when a bioengineered Neonatal Automated Physiotherapy device is used to provide gentle tactile stimulation for infants (Karlin, 2010).

Human touch, especially as sensed through the hands, associates with relevant mental concepts. For instance, physically touching a warm object promotes interpersonal warmth (Williams and Bargh, 2008). Having heavy objects in one's hands associates with importance and seriousness in other people and other matters (Ackerman et al., 2010). Handling rough objects forms an impression associated with difficulty and harshness (Ackerman et al., 2010). Hard objects made others appear stricter, more stable, less emotional, and less flexible (Ackerman et al., 2010). Thus, we have the expressions that associate touch with personal attributes, such as:

1. warm-hearted
2. weighty thoughts
3. rough day
4. coarse language
5. hard-hearted
6. rock-solid
7. cold personality
8. keep in touch
9. smooth operator

Physical pain also has a psychological analog in depression or rejection (as exemplified by "I'm crushed" or "I'm hurt by that remark"). Acetaminophen (Tylenol) has been found both to relieve physical pain and, at the same time, social or moral conflicts that tend to cause depression (Stix, 2010).

These observations have biological engineering implications. The type of packaging may critically determine the attitude of customers toward the product inside. The design of human-occupied spaces may depend upon the use of the space and the impressions that are to be fostered. Manipulations of the human environment can be made easier if the correct tactile choices are made.

It is likely, too, that these results extend to the animal kingdom. Classical studies by Harlow(1958) on maternal-infant bonding in macaque monkeys demonstrated that infants preferred to stay close to a cloth surrogate mother warmed with a 100 W light bulb rather than a bare wire surrogate mother with a baby milk bottle as a source of food. Harlow's conclusion was that contact comfort was very important to the monkey infants compared to meeting nourishment needs. Monkeys raised with the warm cloth mother developed relatively normal social skills, in contrast to infants with only wire mothers.

Physical intimacy (touch) is important also at the cellular level. The cell membrane has many receptors, each with a specific function. Many of these provide the means for substances to be transferred into or out of the cell; the cell membrane is otherwise nearly impervious to transmembrane movement of larger ions or compounds. These receptors can also act as portals through which viruses access the cell interior. The Transferrin Receptor 1, for instance, which usually brings the element iron into the cell, can also act as access for the deadly Machupo virus to enter the cell (Vastag, 2010).

Viruses contacting and entering cells illustrate a process similar in certain ways to the intricate courtship rituals of advanced animals. In both cases, there has to be a coordination of advances and responses: communication, if you will. When the virus first approaches the cell, glycoproteins on the cell surface (acting like small bristles) and glycoproteins on the viral surface must spread apart. This allows the virus to contact the cell membrane, where it binds. After attaching, the viral envelope membrane fuses to the cell membrane. After combining with the cell membrane, the virus inserts its genetic RNA material into the cell, where it hijacks cellular mechanisms to reproduce. *Defensins* are compounds on the cell surface that can prevent viruses from reaching the cell membrane. They do so by crosswise binding of the glycoprotein bristles, so the bristles are prevented from spreading apart and giving access of the virus to the cell membrane surface (Figure 5.3.13).

Surprisingly, also, plants may be able to sense sound vibrations. In one study, plants grew roots toward gurgling water flowing in a tube (Gagliano et al., 2017), and, in another study, plants were shown to distinguish between caterpillar chewing sounds and wind vibrations (Appel and Cocroft, 2014).

6.19.4 Visual Stimuli

> The human features and countenance, although composed of but some ten parts or little more, are so fashioned that among so many thousands of men there are no two in existence who cannot be distinguished from one Ir.
>
> **—Pliny the Elder**

Visual communication is an important mode for birds and primates, as well as certain other species. Birds communicate their romantic intentions by showy displays of their colorful plumage. Contrarily, raising the hairs on the back of a dog or wolf and baring the teeth is a warning of anger, defense, and aggression, not only toward others of its kind but also to other animals. Fluffing the feathers of birds serves the same purpose.

Nestlings convey the message of hunger to their avian parents by the noises they make and by the postures that they assume: necks extended, heads up, beaks open. Some animals show appeasement to their foes by lying on their backs and exposing their abdomens. Ducks provide the same message by repeatedly bowing their heads down and to the side in the presence of an aggressor. The stomping of the foot on the ground is the way sheep convey a warning for others to stay clear.

Humans have raised the art of visual communication to new heights. There are a host of body language symbols that convey very specific messages (Scheflen, 1976). Hands on hips, for instance, may convey the idea of questioning, disbelief, or interrogation, depending on the facial signals that accompany hand placement; arms crossed on the chest usually signify separateness and noninvolvement. A large number of hand signals are used to convey meanings (and they may be different in different cultures). People have even invented a new set of visual signals for computer messages. These, called *emoticons*, use combinations of standard typewriter symbols to suggest facial configurations. They work Ie we are familiar with reflections of a person's frame of mind in facial appearance.

Humans have the ability to imagine visual objects, even without being able to see them. The blind have considerable innate pictorial abilities that they can use, not only to visualize in their minds, but also to draw these objects on paper (Kennedy, 1983). This ability to place objects in the space around them can be used to move about in a world inhabited by the sighted.

People with Alzheimer's disease have impaired visual senses. They have hard time distinguishing colors and color patterns. Some contrasting colors look like elevation changes, patterned rugs appear as uneven terrain. Small colored tiles look like objects to be picked up. Sensory overload is present with the simplest things. Lighting changes throughout the day and dark shadows can be confusing. Biological engineers should be sure that essential things are easy to find.

SEEING INSIDE US

Being able to see what is inside a human body, or an animal, a plant, or even an inanimate object, allows us to know what is there, how it is placed, and whether it is functioning properly. Based upon this information, decisions are made regarding medical procedures, anatomical relationships, or localized regions of activity.

Before modern technology, the most obvious means to see inside a living thing was to cut it open and expose its internal parts. That is still done for exploratory surgery (although this is rapidly diminishing) and post-mortem autopsies. However, opening up a living human, animal, or plant is crude, dangerous, and unnecessary. Today there is a selection of imaging technologies, each with its particular advantages and uses.

One of the simplest is the *endoscope*, which is a relatively long (about 1 m), thin tube containing optical fibers. A very powerful light at the proximal end of the tube sends light through the optical fibers to shine on the surroundings at the distal end of the tube. Other fibers receive reflected light in the form of an image and carry the image back to a video camera at the proximal end. The image is usually displayed on a video monitor, although an optical eyepiece can also be used.

In practice, the endoscope is threaded into a channel, perhaps the nose, mouth, anus, vein or abdominal cavity through either a natural opening of the body or a small slit cut through tissue. The endoscope has been instrumental in allowing surgical procedures to proceed without large incisions (arthroscopic).

Ultrasonic imaging uses high frequency pressure waves in the range of 1 MHz. These waves are generated by a *piezoelectric crystal*, a device that contracts and expands when a time-varying voltage is applied to it. The resulting mechanical waves are focused into soft tissue, where they are propagated deep within. At the interface between one kind of tissue and another, some of the wave energy is reflected back to the source.

The same piezoelectric device that generated the ultrasound wave can convert pressure waves into an electrical signal. So, the time it takes for the wave to travel through tissue to the interface and return to the source is proportional to twice the thickness of the tissue. An image can be formed by focusing the wave in different directions, and this image corresponds to places where different kinds of tissue come together. In addition, ultrasound may be used with the *Doppler effect* to measure flow rates of blood and other fluids as long as they contain inclusions of some kind, such as red blood cells, to reflect ultrasound waves.

Acoustic angiography is related to ultrasound imaging, but uses a relatively lower frequency transmit element and a higher frequency receive element to produce high-resolution images of the microvasculature *in vivo* (Gessner et al., 2013). *Electrocardiography* is the process to produce an image of the heart using ultrasound.

Computed tomography (or CT) uses the variable absorption of x-rays by different tissues to visualize structures within the body. An x-ray source is needed, and

(*Continued*)

SEEING INSIDE US (Continued)

an image is formed by obtaining absorption data from many individual points as the x-ray is focused in different directions. A single CT image is the product of thousands of individual measurements made as the source encircles the body. Unlike ultrasound, the source and detector are not the same device. CT has good spatial resolution, but not always for imaging soft tissues, and the x-rays it uses can ionize tissues and cause health risks.

Magnetic resonance imaging (MRI) relies on powerful magnets to align nuclei of hydrogen atoms in body tissues. Most of these are in water molecules. When the magnet is switched off, the nuclei return to their normal unaligned states and release radio frequency energy. The frequency of these waves provides a measure of the hydrogen concentration of the tissue. Bone, fat, muscle, etc. can be identified. A very fine-grained map can be produced by accumulating many individual measurements. Once stored in the computer, MRI scans can be used to form a virtual three dimensional model of the body. MRI provides good spatial resolution, but is an expensive procedure.

MRI detects individual molecular responses to intense magnetic fields, and uses spatial offsets (slices) to form three-dimensional images of the portion of the body being imaged. Due to differential magnetic susceptibility (the ability to support a magnetic field internally) of oxygenated hemoglobin (diamagnetic) and deoxygenated hemoglobin (paramagnetic), brain activity can be detected using fMRI (functional MRI). Neuronal activity requires the use of oxygen, and thus reduces hemoglobin oxygenation. This change can be detected and neuronal activity inferred (Mumford and Nichols, 2006; Pekar, 2006). fMRI can also be used to detect changes in blood flow throughout regions of the body. In order to detect rapid changes in blood oxygenation, images from fMRI must be faster than regular anatomical MRI; fMRI images consequently have lower resolution.

Diffusion Tensor Imaging (DTI) uses a different type of magnetic field and programs to display vectors of water diffusion. DTI is particularly good at distinguishing neural white matter from gray matter because water largely diffuses radially inside and from the axon in gray matter, but is confined to diffuse axially in white matter (Fields, 2008).

Going beyond mere structure is achieved with *Positron-emission tomography* (PET). PET can be used to detect metabolic and other biochemical events by detecting photon energy emitted as various short-lived radionuclides decay The emission of a positron from carbon-11, nitrogen-13, oxygen-15, fluorine-18, or other inhaled or injected tracer compounds, is in the form of two photons emitted 180 degrees apart. This fact allows the source of the photons to be spatially localized, and, as with CT and MRI, a completed image is constructed from many individual measurements. PET has been used to determine what happens in the brain during learning, to detect tumors with intense metabolism, and to illuminate cerebral blood flow (see Section 8.2.2, Neural Engineering).

(*Continued*)

SEEING INSIDE US (Continued)

Single-photon emission computed tomography (SPECT, or SPET) is a nuclear medicine tomographic imaging technique using gamma rays. The technique requires that a gamma-emitting radioisotope, usually Technetium 99m, be injected into the bloodstream. The radioisotope is often attached to a specific ligand to create a radioligand, which binds to specific types of tissues. These can be imaged by means of a gamma camera.

Optical Coherence Tomography (OCT) is similar to ultrasound, except that light, rather than acoustic vibrations, are used as the signal source. Like ultrasound, an image can be composed from scans of light reflected from tissue interfacial boundaries (echo location). Rate of flow can be measured using phase shift information. OCT has better resolution than MRI or ultrasound, and there is no ionizing radiation or special tissue sample preparation involved.

Fluorescence spectroscopy is used to identify and locate different organic compounds in a tissue. A light source is used to energize electrons in the tissue. Certain compounds fluoresce at particular light frequencies. Those that do not fluoresce naturally can be chemically labeled by attaching chemical functional groups that do fluoresce. The fluorescent molecules can be identified by comparing signals with standard samples of known composition. *Absorption spectroscopy* is similar except that it measures absorbed light wavelengths rather than emitted wavelengths.

Near infrared spectroscopy (NIRS) uses non-ionizing radiation that can penetrate deeper than visible light, but not as deeply as fMRI. NIRS uses pulses of infrared light to excite different tissues. Chemical bonds in molecules that make up the tissue differentially absorb certain wavelengths. By measuring the absorption fractions of different infrared wavelengths as they pass through tissues, the types of tissues present can be identified. This method can be used widely in animal and plant tissues, foods, and inanimate objects. NIRS is particularly useful for detecting blood hemoglobin levels. It can be useful for foods, pharmaceutical powders, and products of respiration. There is at present a minimum compound concentration that can be reliably detected with NIR.

Electron tomography (ET) is similar to electron microscopy, except that multiple images are captured at different angles. The result is a three-dimensional picture with very high resolution. ET takes place in a vacuum, so the sample cannot be living.

Fluorescence resonance energy transfer (FRET) can be used to determine the locations of individual molecules. When two light-absorbing molecules lie close to one another, they can pass absorbed energy between them. The efficiency of this energy transfer depends precisely on the distance between donor and acceptor molecules. Measuring luminosities of each molecule can determine the distance between them. FRET has been used to track molecular movements along a DNA strand (Schnabel, 2010).

In *Laser Ablation Imaging* (LAI), short-duration high-power laser pulses are used to vaporize thin sections of a sample as images are taken of each level. These images

(*Continued*)

SEEING INSIDE US (Continued)

are then combined to produce a full three-dimensional view of the anatomy of the sample (Clabby, 2014). LAI is very quick, easy, and accurate. LAI destroys the samples it analyses, so it cannot be used with live specimens. This technique, nevertheless, could be found valuable for determining the anatomical structure of plants, insects, and other items to be screened.

Optogenetics is a technique wherein a gene is inserted inside neurons that then produce a light-sensitive protein (called an *opsin*). The opsin is sensitive to a particular color of light introduced through a fiber optic thread. The result is that the neuron can either be excited or depressed by shining a light on it. Optogenetics is used to study neural circuits in the brain (Palmer, 2014).

To make neurons light up when they fire in the living brain, the neurons are seeded with fluorescent proteins that glow in the presence of calcium ions that become abundant when the cell fires. Benign viruses may be used to carry the fluorescent proteins inside the neurons, or genetic modification of the neurons can compel them to produce the proteins on their own. Infrared laser energy focused on the brain primes the proteins to fluoresce whenever the neuron the proteins to fluoresce whenever the neuron fires. That way, activity of each individual neuron can be recorded (Waldrop, 2017).

Fluorescence and *Bioluminescence* are two optical methods used to visualize cells in live animals. Fluorescence requires fluorescent proteins to be incorporated into the cells, often tumor cells, both metabolically active and inactive, and the proteins to be excited by an external light source. Bioluminescence can be used to monitor active cells and requires interaction of luciferase with luciferin and ATP. These methods are used to visualize the effects of drugs on cancers in animals.

Combinations of these imaging modalities are proving useful. PET-CT and PET-MR can locate tissues, such as cancer with high metabolic rates. *Photoacoustic tomography* uses light pulses to heat target tissues or cells, and the heat reflects back to the transducer with less scattering than with light.

Microwave Sensing and Imaging (MSI) uses electromagnetic waves at microwave frequencies (hundreds of megahertz to tens of gigahertz). Its use requires a source antenna and receiving antenna. It detects different tissue electrical properties and can distinguish between healthy and pathological tissues (Chandra et al., 2015).

Raman spectroscopy is a spectroscopic technique where monochromatic light, usually from a laser source, is absorbed by the sample and then reemitted at different frequencies, depending on the identities of the target molecules. A spectrograph of the emitted light can be used to identify the chemical composition of a sample without damaging or destroying it. Hence, Raman spectroscopy can be used on living tissue. Raman spectroscopy can also help detect cancer cells in real time, as during surgery. Infra-red light is shined on the cells and the resulting emission spectrum is characteristic of proteins found in cancer cells. The differences between normal cells and cancer cells are subtle, however, so the spectrum is transformed into auditory signals that are easier to detect than are the visual signals (Cowen, 2015).

(*Continued*)

SEEING INSIDE US (Continued)

Serial Femtosecond Crystallography (SFX) uses extremely fast X-ray pulses to visualize proteins as they change structure during reaction processes. For instance, a light-responsive protein has been visualized using SFX as it responded to a light stimulus (Fromme and Spence, 2017).

Three dimensional images of living cells can safely be obtained using *Brillouin microscopy* (Scarcelli et al., 2015; Antonacci and Braakman, 2016). This method is based on the interaction of light with spontaneous acoustic phonons in the gigahertz ultrasonic frequency range. The refractive index of tissue changes under deformation caused by the acoustic signal, and this change is detected with refracted light. This method can be used to visualize internal structures too small to be seen with visible light wavelengths and can also measure the stiffness of subcellular structures.

Some small transparent creatures can be observed through a microscope to elucidate life processes in living organisms. But, light images are distorted when passing through intervening biological tissues. Astronomers have solved a similar problem in the way the atmosphere distorts light emanating from distant heavenly bodies. Wang et al., (2014) have used a similar technique to clarify and sharpen images from nerve cells deep in the brains of zebrafish.

6.19.5 Others

> If I didn't believe it with my own mind, I never would have seen it.
>
> **—Unknown**

Nonverbal communication with animals is also a possibility. Mice, like humans, express pain through facial expressions (Dove, 2010). Knowing this, laboratory scientists can assure that animals under their care do not suffer unnecessarily. The same characteristic can also lead to improved pain-relief drugs for humans.

Pet owners usually agree that their animals can tell from non-verbal cues when there will be a change in normal routines; they can tell when their owners will go on vacation, take them to see the veterinarian, or give them baths. Communications with animals can happen at several levels, and there is opportunity for biological engineers to extend communication devices for human hearing or seeing impaired to connect to animals as well.

Not all sensation leads to communication, and not all actions are taken because a message was received by the actor. There are means to sense information from the environment that are in addition to the four modes discussed earlier. These include temperature, magnetism, and electrical fields.

These, however, are not part of communication. Sensing is an important part of control (see Section 4.4.1, Sensors), and has been discussed there. What the biological engineer should remember, nevertheless, is that communication is a normal part of life, and nothing the engineer does should interfere with this communication unless it is explicitly determined to be a design goal.

PERCEPTION OF STIMULI

An interesting example of how knowing about sensory mechanisms in an animal's body can help to solve other recognition problems is given by Young's Principle (Erickson, 1984). Young began by trying to explain the perception of all the many colors in the visible spectrum. After all, we don't have a specific receptor for each color, yet we perceive a large number of individual hues. Similarly, we don't have tactile sensors in every small area of the skin, yet we can localize tactile stimuli to very small areas. Other examples abound: there are a limited number of taste receptor types, yet many perceived tastes; we have but two ears, yet we can localize the source of a sound to a particular point in space. Thus, the problem is to understand how a limited number of sensors can be used to discriminate among a much larger number of choices.

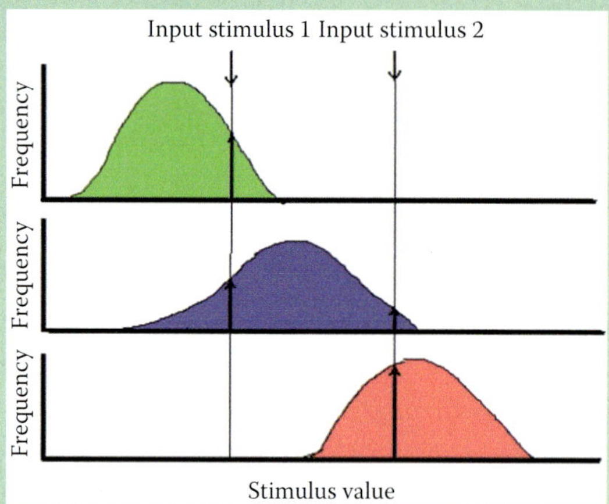

An example of three receptors with different stimulus sensitivities. Input stimulus 1 can be discriminated from input stimulus 2 based on the combination of output frequencies from the three receptors. Any particular input stimulus will have a unique combination of frequencies. For instance, the human eye can distinguish 2.3 million colors despite only three types of retinal color sensors (called *cone cells*).

The answer to this question apparently comes from the sensitivity of the receptors themselves. Each receptor has a broad range of sensitivity, being most sensitive to one particular input value and less sensitive to other input values, with the degree of sensitivity becoming smaller as the difference from the most sensitive input value becomes greater. With several of these receptors, each with a different maximum sensitivity perception can be related to the combination of outputs from the different receptors. Thus, we may see different colors that do not correspond to the maximum sensitivities of any of the color receptors. Tactile localization could occur as long as the output frequencies of the skin touch sensors vary with distance from the receptor location.

This means of sensory discrimination has practical value when applied to other types of problems. Location of a military target can be based on a combination of electronic emitters and receivers. Global positioning (GPS) can be ascertained similarly

(*Continued*)

PERCEPTION OF STIMULI (Continued)

by comparing transmission times from various transmitting satellites to a receiver on Earth. Other triangulation schemes use the same principle.

In other cases, the eyes can be fooled into perceiving colors not really there, and stereo sound can be formed from a virtual sonic environment coming from speakers with different acoustic delays.

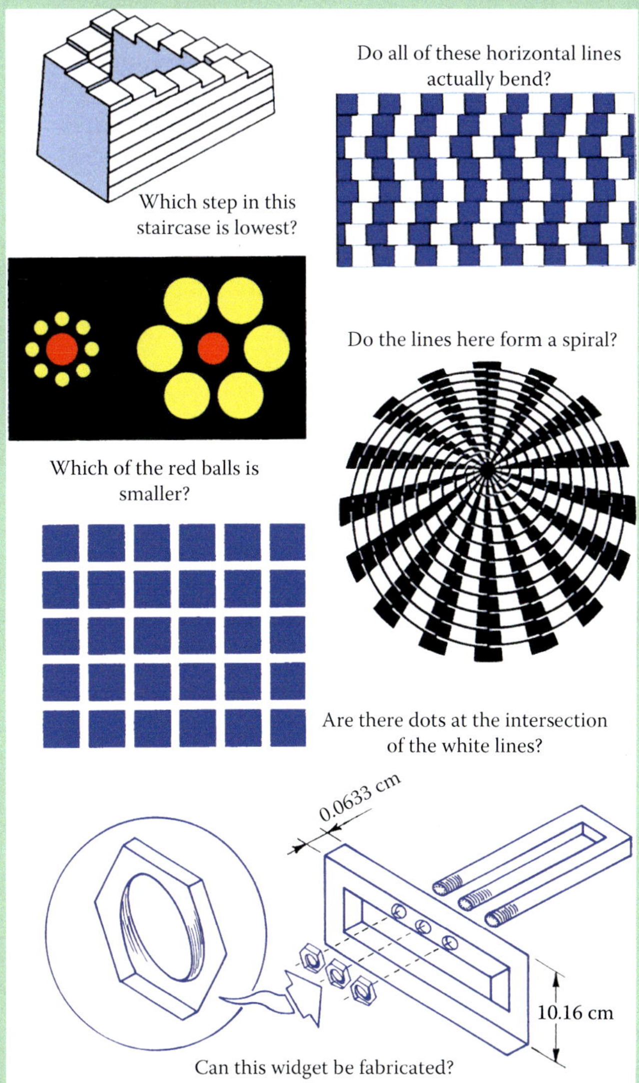

The mind can be fooled with visual stimuli.

6.19.6 Just Noticeable Difference

> There can be no doubt, that the difference between the mind of the lowest man and that of the highest animal is immense.... Nevertheless, the difference in mind..., great as it is, certainly is one of degree and not of kind.
>
> **—Charles Darwin**

The perception of added stimulus intensity in the presence of an existing stimulus intensity was the subject of work by Weber, Fechner, and Stevens (Smith, 1998). Weber found that the smallest detectable difference in intensity was a fraction of the intensity already present. This finding has been called *Weber's Law*, and has been shown to be at least approximately true for stimuli as diverse as light, sound, and the discrimination of heaviness (see Table 6.19.1) for moderate intensities.

Fechner found that Weber's law is not true for extreme sensations, so he postulated a logarithmic relationship between the sensation felt and the physical stimulus intensity:

$$\text{sensation} = \text{k} \cdot \log(\text{intensity}) + \text{constant} \tag{6.19.1}$$

This is called the *Weber–Fechner Law*.

Stevens found that Fechner's formulation did not adequately predict sensation for all senses, so he postulated *Steven's Power Law*:

$$\text{sensation} = \text{k} \cdot (\text{intensity})^n \tag{6.19.2}$$

The importance of all this is that human judgment is not absolute, but relative. The judgment of what one sees or feels depends upon the present condition in which one finds oneself.

Example 6.19.1: Training Animals to Come When Called

Animals on display in the daytime in an outdoor environment are to be returned to their cages at night for safety, feeding, and health care. How can this be accomplished, and what difficulties can be anticipated with this method?

Solution

Higher level animals can be conditioned to offer a particular behavior in response to an initially unrelated stimulus. So it was that the people at Disney's Animal Kingdom in Orlando, Florida trained the animals on display during the day to return to their nighttime cages when they heard a whistle blow. However, the whistle sound used was the same for many animals, and, after long and arduous training periods with

TABLE 6.19.1

Weber Fraction for Moderate Intensities

Sense	Sensation	Smallest Detectable Fraction
Vision	Brightness	1/60
Hearing	Tone intensity	1/10
Pressure	Cutaneous pressure	1/7
Smell	Odor	1/4
Kinesthesis	Weight	1/30
Pain	Temperature of skin	1/30
Resistance	Breathing difficulty	1/30

each animal group, it was found that either too many animals or the wrong animals came running the first few times the whistles were blown. The animals had to be reconditioned using a unique sound for each distinct animal group. Humans can often get animals to respond the way they want them to, but, in this case, the humans did not give sufficient thought beforehand.

Example 6.19.2: Siren's Song

The Mediterranean fruit fly (*Ceratitis capitata*) damages more than 250 different types of fruit. Strategies for halting the spread of the Mediterranean fruit fly ("medfly") include total eradication from expansion territories by aerial insecticide sprays and release of sterile males. Public pressure restrains the extensive use of pesticides and forces increased development of environmentally friendly alternatives. Medfly females are the ones that cause direct fruit damage, and so they are the main target for control. Female attraction can be used for population monitoring and control. Suggest a means to do this.

Solution

The search for medfly detection and control techniques focuses on mating behavior and sexual communication. These can be highly selective and efficient.

Medfly males produce three distinct sounds as part of their sexual communication ritual. Their calling song is a low amplitude vibration at 350 cps, and can be the attractant sought (Mizrach et al., 2004). This sound, when artificially produced by a sound speaker, was found to attract a large proportion of test females. Females thus attracted could be killed electrically or by other means.

Remark

Sexual rituals are normally very complex, with several different components. Utilizing sound by itself to attract females will likely select against those females that ignore components other than sound. Thus, it is expected that over time, the attractiveness of the electronically produced sound will become less and less.

Example 6.19.3: Ground Squirrels Warn Rattlesnakes

Rattlesnakes in the western United States hunt with infrared sensors located in their noses and mouths. These detect the presence of bodies at a different temperature from their surroundings. As a countermeasure, California ground squirrels have the ability to heat up their tails when threatened by rattlesnakes (but not other kinds of snakes). The function of this action is apparently a form of infrared communication to warn rattlesnakes away (Allen, 2004).

Example 6.19.4: Pheromones for Your Cat

Sara was the name of a cat owned by Andrea Sachs (Sachs, 2010). Sara had come nine years ago from an animal shelter. She was a gentle cat, but skittish, afraid of new situations and strange people, a proverbial scaredy-cat.

Sammy was an aggressive and competitive cat which Andrea brought home two months ago from the shelter. Once Sammy moved in, he and Sara fought over and over until she would run to the nearest safe hideout. This continual fighting disrupted the peace and quiet of the household.

Desperate to get some sleep, Andrea consulted her veterinarian, who gave her pheromone collars and electric plug-in pheromone diffusers. The diffusers emitted a chemical that mimics a pheromone that cats leave behind when they rub their cheeks on furniture or people; the collar had a pheromone that mother cats emit while nursing. Both calm agitated cats.

Pheromones are known to have behavioral effects on many kinds of animals, from alarm alerts to soothing siblings. Besides cats, pheromone products are also available for dogs and humans.

APPLICATIONS AND PREDICTIONS

1. Successful rearing of wild species in captivity will require normal courtship rituals to be followed. This may require space, maneuverability, light levels, low background noise, air currents, and other physicochemical environmental characteristics.
2. Animal stress can be monitored in some species by the sounds that they make.
3. Honeybees can be trapped by luring them with queen substance.
4. Humans need to touch and be touched. Pets can satisfy this need.
5. Humans have extended coordination of their movements by technology such as Global Positioning Satellites.
6. Food is a powerful attractant.
7. Odor is a powerful memory stimulant.
8. Conversations are friendlier with a warm beverage rather than cold.
9. Biofilms are extremely hard to eradicate.

6.20 Biological Units Maintain Stability with Exquisite Control

> Failure is the preamble to success. Most first efforts don't work. If you persist, you'll eventually figure it out.
>
> **—Thomas Fogarty**

Internal stability is of paramount importance to BU. Achieving the required balanced state with the challenge of sometimes hostile environments is not an easy task. It requires complex and elaborate control mechanisms that have evolved over many, many generations. These control mechanisms are able to maintain *homeostasis*, which is defined as the stabile state of equilibrium among different but interdependent elements and subsystems of an organism. As previously noted, the use of the word equilibrium is not meant in a global, thermodynamic sense, but only as a term describing stability in a local region. True equilibrium comes about only after a living organism dies and its chemical elements disperse.

We have seen that this exquisite control may extend beyond the organism to the surrounding environment. Through this means, the organism attempts to influence environmental effects in its favor. Plants do this by adjusting the pH of the soil surrounding their roots to make nutrients more available; animals do this by building nests to insulate themselves from harsh temperatures; and microbes do this by locomotion to areas of abundant food.

6.20.1 The Senses

> If assemblages of neurons cannot be viewed as the building blocks of consciousness, then consciousness must be a primary principle.
>
> **—Robert G. Tabor**

The basics of control systems have been explained in Section 4.4. As explained there, all control systems start with the ability to sense an environmental attribute. If it can't be sensed, it can't be controlled. Hence, the ability to sense the environment was developed very early in the history of life. We usually talk of five human senses:

Touch
Smell
Vision
Hearing
Taste

Taste and smell are both chemical senses (Tables 6.20.1 and 6.20.2), touch and hearing are mechanical, and vision is electromagnetic in nature. To this list, we could perhaps add the

TABLE 6.20.1
Some Olfactory Thresholds

Substance	ppm
Acetic acid	0.21
Acetone	100
Ammonia	47
Anise	0.20
Banana oil (isoamyl acetate)	0.22
Benzine	4.7
Butyric acid	0.001
Camphor	120
Ethanol	10
Lemon	0.22
Peppermint	0.22
Sulfur dioxide	0.47

TABLE 6.20.2
Some Taste Thresholds

Taste	Concentration and Substance
Sour	0.00009 M hydrochloric acid (pH = 4.0)
Salt	0.01 M sodium chloride
Bitter	0.00000016 M strychnine hydrochloride 0.000008 M quinine
Sweet	0.08 M glucose 0.01 M sucrose 0.000023 M saccharin
Umami	0.0007 M glutamate

senses of balance, somatic condition, blood chemical composition, and temperature. Some other animals also sense heat (mosquito, pit viper), magnetic fields (many birds), and electrostatic fields (sharks). Sensing magnetic fields is called *magnetoception*, and involves the presence of small bits of the mineral magnetite located in subcellular organelles called a *magnetosomes*.

Many specialized receptors are actually sensitive to several different kinds of primary signals. The hot, burning sensation we sometimes get from intense mechanical pain is because cutaneous thermoreceptors (those that sense skin temperature) are also somewhat sensitive to mechanical stimulation. Even more common are the flashes of light we sometimes see when rubbing our closed eyes. The photoreceptors in our eyes are also sensitive to mechanical stimulation. The primary signal to which receptors are most sensitive is called the *adequate stimulus*, and this is the stimulus that is used in biological control. We can see; however, that sensing is imperfect, and may be fooled to some degree.

Receptors that link to the nervous system are often free nerve endings (modified neural dendrites) that can be depolarized by the adequate stimulus. The neuron of which the receptor is part repetitively discharges, or forms action potentials. As the receptor depolarizes, it affects the discharge frequency of the neuron, often (but not always) increasing frequency as the intensity of the adequate stimulus increases (Figure 6.20.1). It is most common for the neural discharge frequency to convey receptor information to the central nervous system (CNS). However, taste receptors in the tongue also change discharge pattern with the type of chemical compound tasted.

Receptors are distributed unevenly in the body. Touch receptors and cutaneous thermoreceptors are concentrated in the hands, feet, and face. Somatic proprioceptors are concentrated in the tendons, joints, and muscle connections. Light receptors are located in the retinas of the eyes, and there they have an uneven distribution. Sound receptors are located in the ears. This uneven distribution provides a weighting of different areas of the body, some more important to control than others.

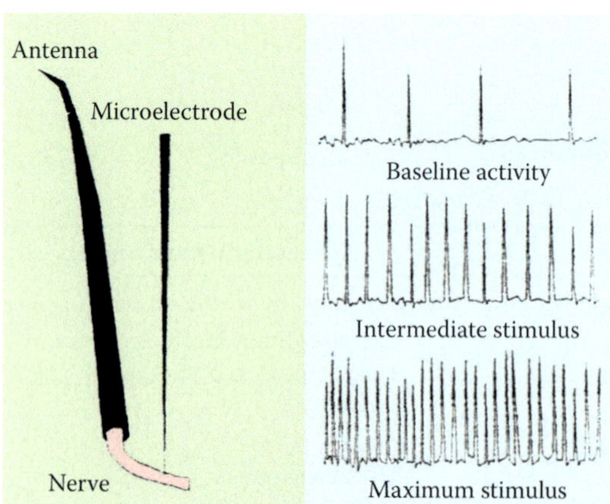

FIGURE 6.20.1
As the intensity of the stimulus increases, so does the frequency of neuronal discharge. Here, the frequency output of an insect antenna is shown for several stimulus levels. (Redrawn from Buerk, D.G., *Biosensors: Theory and Applications*, Technomic Publishing, Lancaster, PA, 1993.)

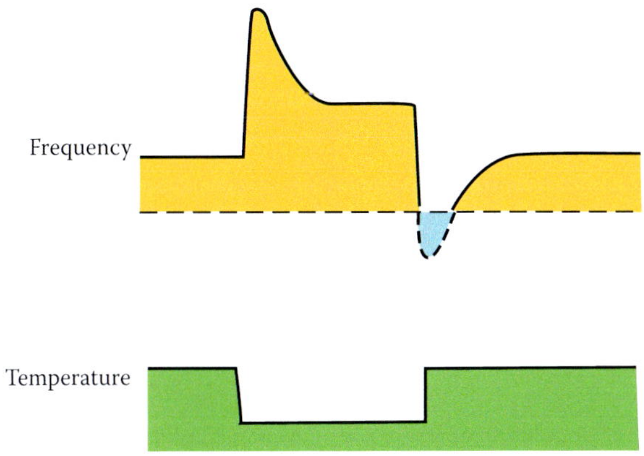

FIGURE 6.20.2
Cold receptor response. When temperature is lowered, receptor output frequency increases. There is an initial overshoot in frequency, which disappears after a short time. When temperature is increased, steady-state frequency of the cold receptor again decreases. The transient response is negative this time, even causing the receptor to remain silent for a period of time corresponding to the time the frequency would have been negative. Larger temperature steps cause larger transient and steady-state responses. Recovery to steady-state is exponential in shape. (Redrawn from Johnson, A.T., *Biomechanics and Exercise Physiology: Quantitative Modeling*, Taylor & Francis, Boca Raton, FL, 2007.)

Neurons that carry signals to the CNS are called *afferent*, but those that carry signals from the CNS are called *efferent*. An afferent nerve may have connections to many receptors, and some processing may occur at the local level. Local processing is especially important in the eyes.

Many receptors are not only sensitive to the level of the adequate stimulus, but also to the time rate of change of the stimulus. Thus, a stimulus that increases with time gives a discharge rate higher than expected based just on the stimulus level, but a stimulus that decreases with time gives a lower than expected discharge rate. In the event of a sudden decrease in stimulus level, virtual negative frequencies are found to occur (Figure 6.20.2).

Receptors also show *adaptation* to the signal level impinging upon them. With adaptation, the firing rate of associated neurons decreases with time for sustained input signals. Because of adaptation, it takes a while to see clearly in the dark after having been in the light, persistent odors are no longer smelled, and we become unaware of the clothes touching our bodies. Adaptation allows us to attend to our constantly changing environment without having to keep track of the parts of our surroundings that are invariant. Adaptation is another form of rate sensitivity, and those receptors that give output only for changing inputs are ones for which rate sensitivity is much higher than level sensitivity.

This rate sensitivity is usually conveyed throughout the entire control system to the effector organs. Rate sensitivity tends to stabilize operation of a feedback control system because it imparts the ability to anticipate changes and to make corrections before a full response is needed.

Ecological systems do not have this anticipatory quality. Instead, it takes time for populations of predators to increase in numbers following increases in the numbers of their prey. Instead of anticipating changes in the prey population, predator numbers follow prey numbers. There is an additional delay in the decline of prey numbers as prey, which have accumulated during periods of predator scarcity, are caught and eaten. These delays lead to the overshooting instability that can be seen when plotting predator and prey numbers over time (Figure 6.20.3).

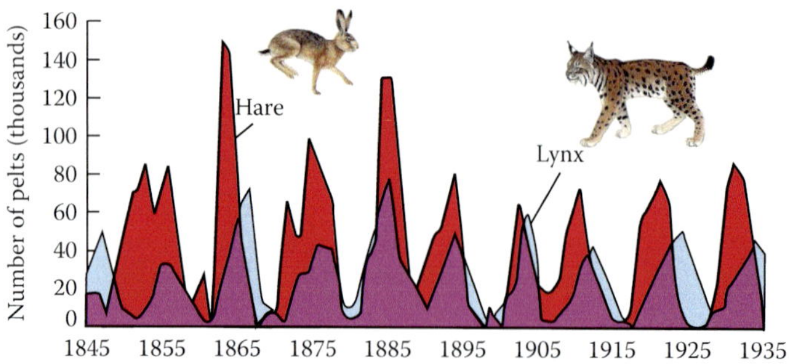

FIGURE 6.20.3
One-hundred-year record of population cycles of the snowshoe hare (*Lepus americanus*) and the Canada lynx (*Lynx canadensis*), based on pelt records of the Hudson's Bay Company in Canada. Lack of anticipation in predator–prey systems lead to unstable population oscillations. (Redrawn from Gotelli, N.J., *A Primer of Ecology*, Sinauer Associates, Sunderland, MA, 1998.)

6.20.2 Controllers

> A robot may not injure a human being, or, through inaction, allow a human being to come to harm.
>
> —**Isaac Asimov**

Biological control responses are rarely simple. There is often a great deal of processing that takes place in the nervous systems of animals. At the very least, for complex animals, there is a spatial summation of inputs from many of the same type of receptors at different locations around the body. There may also be added the effects of additional inputs. For example, hemorrhage can decrease blood pressure, which is sensed and at least partially corrected by increasing the resistance of peripheral blood vessels. This response can be influenced also by information about the partial pressures of oxygen and carbon dioxide in the blood.

For responses that are needed to meet emergencies, the CNS processing of the signals is minimal. The simplest response is the reflex, in which afferent *sensory neurons* are connected through one synapse to efferent *motor neurons*. The signal transmission delay at each synapse is 0.5 meter/second or more (Ganong, 1963), and more complex control tasks require processing through many synapses. The reflex thus saves time to complete a simple act in a minimal amount of time.

The kind of control that determines its output based on information from its inputs is feedback control (see Section 4.4, Control). Much biological control is feedback control. The output from the controller is directly related to the amount that sensory information differs from a desired set point (proportional control). And, in addition, the output is usually related directly to the rate of change of sensory information (proportional plus derivative control), as we have already discussed. There is almost always a nonlinear relationship between output and input, especially over a wide range of inputs. Most biological control responses demonstrate very high sensitivity (output divided by input) in the center of its range, with lessening sensitivity toward the extremes. In this way, the control system will maintain a high degree of control unless the system becomes overloaded and catastrophic failure ensues.

Higher animals can learn to produce adequate control without dependence upon sensory input. This is called *open-loop control* (see Section 4.4, Control), and it happens with highly repetitive responses (such as stepping, eating, and blinking). Advantages of open-loop control are that responses are faster and do not take attention away from other tasks. As long as no differences are required from patterned responses, open-loop control may

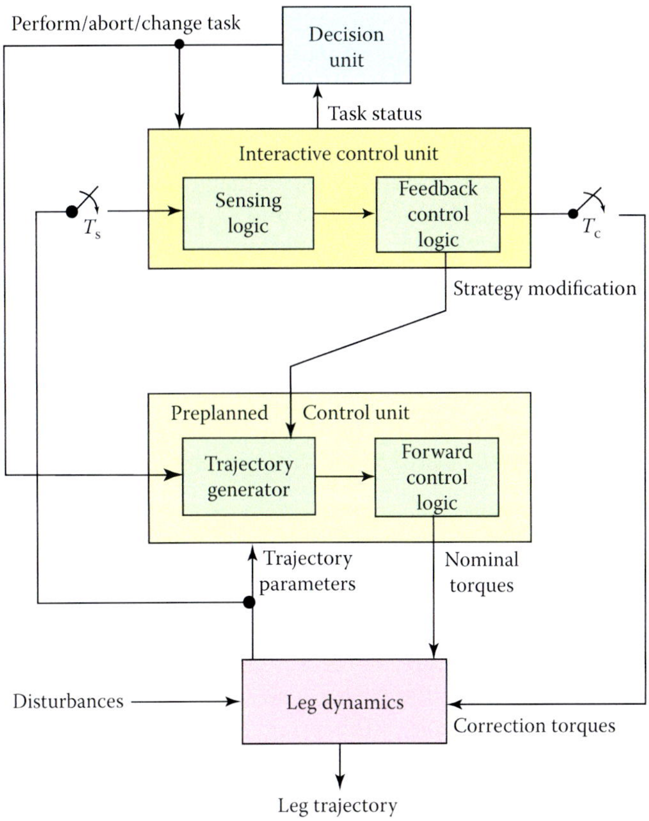

FIGURE 6.20.4
Hierarchical control of a stepping motion. The decision to perform the motion results in a trial trajectory given to the leg. Sampled data from the leg are sent to the interactive control unit, where it is determined whether or not to send correction torques to the leg to overcome unforeseen disturbances. (Redrawn from Johnson, A.T., *Biomechanics and Exercise Physiology: Quantitative Modeling*, Taylor & Francis, Boca Raton, FL, 2007.)

be used successfully. When differences do occur (for example, spotting an obstacle while walking, or encountering a hard inclusion while eating food), the system can return to feedback control with its concomitant attention requirement (Figure 6.20.4). Behaviorists have used this tendency to pattern responses whenever they apply *conditioning* to animals.

There is an intracellular feedback loop that involves the X sex chromosome. Females have two of X chromosomes, but males have but one. Genes on the X chromosome help control everything from filtering blood to repairing DNA sequences. Two copies of these genes can produce too many proteins and can be disastrous.

There are mechanisms inside the cell that can count the number of X chromosomes present and shut down extras. That's called *X chromosome inactivation*, and it protects against too much X gene activity (Leslie, 2008).

In Section 4.4 was mentioned the additional mode of feedforward control. There are several places where biological control responses have the appearance of feedforward control. One of these is control of breathing during exercise. It is known that increasing the level of inhaled carbon dioxide at rest stimulates both the depth and frequency of breathing with the purpose of removing excess CO_2 from the respiratory system. There is, however, a small portion of this CO_2 that is not exhaled, and there is a small but measurable increase in CO_2 dissolved in the blood (Johnson, 2007).

Carbon dioxide is produced as a byproduct of exercise metabolism, and breathing is stimulated during exercise to remove this CO_2 excess. In this case, however, all the metabolically produced CO_2 is removed and there is no measurable increase of CO_2 dissolved in the blood. Furthermore, inhaled CO_2 during exercise results in an increase of dissolved CO_2 just as it did at rest. This difference between responses to inhaled CO_2 and metabolically produced CO_2 has baffled scientists and engineers for many years. Exercise apparently recruits different respiratory mechanisms to compensate for CO_2 produced internally. Some (Whipp, 1981) have speculated that this is evidence for feedforward control that adjusts breathing to remove all of the metabolically produced CO_2 without resorting to CO_2 sensed in the blood.

6.20.3 Redundancy

> One would think that if we put 10 copies of that gene [that lead to a certain protein being produced] in [a cell], we should get 10 times as much protein produced. But, it has been found that often the more copies of the gene you put in, the less protein you get out....
>
> **—Douglas A. Lauffenburger**

Because the consequences of control failure are so severe, there is often a built-in redundancy, so that if the primary system fails, there is a secondary system to keep some semblance of balance. This secondary system is often the vestigial remains of a more primitive system present in lower life forms, and does not exert control as precisely as does the primary system. An example of this is cutaneous sweating in paraplegic humans. The human thermoregulatory system is controlled quite well by the hypothalamus in the human midbrain. In response to thermal challenges, signals are fed to sweat glands through a neural pathway in the spinal cord. Individuals with severed spinal cords cannot receive the message to begin sweating below the disjunction, but it has been found that they sweat anyway—perhaps not as precisely in amount related to heat challenge, but they do sweat.

Many cellular metabolic pathways also show redundancy. If, for some reason, a critical metabolic substrate is in short supply, there often are alternative metabolic pathways to allow the cell to survive the shortage. These alternative mechanisms are usually not as efficient, or they may produce unusable metabolic wastes, so these pathways are not normally used by the cell. Anaerobic metabolism of glucose is an example of an alternative metabolic mechanism when sufficient oxygen is not available (see Section 3.10, Energy-Rich Compounds).

The manipulation of cellular metabolic pathways in bioreactors is the aim of *metabolic engineering*. Cells are used that are capable of producing useful biochemicals. However, they often don't produce them in economical quantities. Changing the genetic machinery of the cell or changing the composition of the bioreactor growth medium can enhance by several hundred percent the amounts of useful products.

An example of biological redundancy, as well as adaptability of the brain, is given by recovery from stroke (Azari and Seitz, 2000). Stroke is the term used when parts of the brain are damaged, either by physical disruption when an *aneurism* bursts or by *ischemia* when blood flow is cut off. Recovery from a stroke may be rapid or slow, depending on the extent of damage. If the damaged area is not too pervasive, within-system pathways can adapt to assume primary responsibility for the function. These within-system pathways may only be engaged by normal adults when they first learn a task. Thereafter, they play a supporting role in the undamaged brain. When the primary pathways are damaged, these alternate routes may need to relearn how to perform the task. But, because there is a general familiarity with the function, recovery may only take a few weeks.

If, however, there is complete damage to a neural system, then the brain must sort out how to compensate for the damaged neurons. It can often do so by *adaptive plasticity*, recruiting neural networks not usually involved in performance of the lost function. This process takes a longer time in order to learn anew how to perform the task.

IMMUNE SYSTEM AS A MODEL OF ULTRA-REDUNDANCY

The immunological response to a foreign invader is a cascade of events that begins when special sentinel cells, prestationed throughout the body, alert the immune system. Some of these are mast cells that release histamine that makes nearby capillaries leaky. Plasma pours out to flood the area (causing swelling) and other defender cells pass through the capillary walls to join the fight. Another group of sentinel cells is called macrophages; these release chemicals called cytokines, which signal for reinforcements. Wave after wave of specialized cells coming from all parts of the body are directed into the fight. They bring with them special weapons of war called antibodies that knock out the invader. The battle continues as long as necessary to win the fight. When it is all over there is the rubble, debris, and detritus normally seen on a battlefield. Then the war must stop and the process of healing commence.

At the heart of the immune response is the ability to distinguish between one's self and others. There are marker molecules on the surfaces of all cell bodies that serve this purpose. Any molecule capable of triggering an immune response is known as an *antigen*.

The huge number of possible molecular structures belonging to antigens from which the body needs protection is a recognition problem of great magnitude. The ability to recognize these chemicals resides in a vast array of natural protein receptors, found in all vertebrates, called *antibodies*, or *immunoglobulins* (Talmage, 1979). There are probably more than a million different kinds of immunoglobulins (or Igs) in each individual, and each of these is specific to one foreign molecular structure. Insects are among the number of animals that do not produce antibodies as part of their immune responses. Instead, they depend on RNA interference mechanisms to defend against foreign invaders.

The function of an antibody is to be a link between a foreign antigen and a standard immune defense (Yelton and Scharff, 1980). In this way, the number of defenses can be relatively small, yet the few defenders there are can be used with a wide range of antigens. The antibodies have two ends: one end attaches to the body's defenses and the other end attaches to the antigen. Because the number of antigens is so huge, then that end must have a huge number of variants.

Antibodies are protein molecules shaped to form a "Y." The stem of the Y is the standard end that links the antibody to other immune defenses. This is called the constant region, and is the same for all antibodies of the same class. The arms of the Y vary greatly from one antibody to another. This is called the variable region, and is shaped to fit a specific antigen as a key fits a lock.

There are five classes of human antibodies (Table). Each has a specific purpose, but is involved with linking between foreign bodies and the body's defensive cells.

(*Continued*)

IMMUNE SYSTEM AS A MODEL OF ULTRA-REDUNDANCY (Continued)

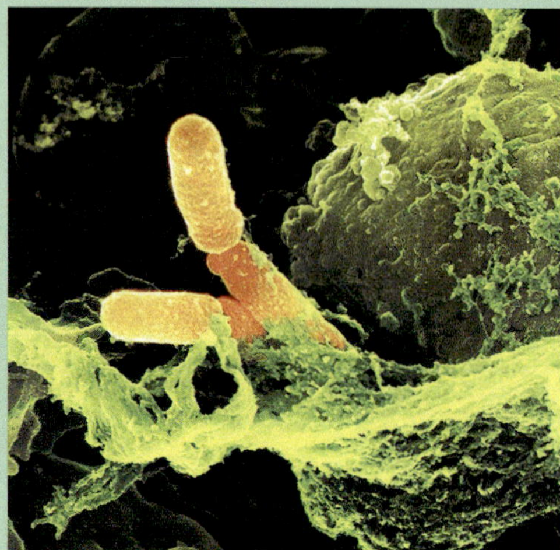

A phagocyte devours invading bacteria.

Different Classes of Immunoglobulins

Class	Function	Found In
IgA	guards body entrances	body fluids, tears, saliva, respiratory and gastric secretions
IgD	regulates B cell activation	B cell membranes
IgE	triggers inflammation and allergies	attaches to mast cells and basophils
IgG	protects against bacteria, viruses, and toxins in blood and lymph	blood and tissue fluids
IgM	kills bacteria	Blood

Once linked to an antibody, there are several possible ways the antigen is dealt with. The first is *neutralization*, where the antibody covers up the antigen and shrouds it enough to keep it from attaching to receptor sites on antigen target cells. The second is *opsonization*, where the bound antigens enhance ingestion by phagocyte cells. The third is *complement fixation*, where one of 20 circulating serum proteins called *complement* forms a complex with the antigen and ruptures the cell membrane of the bacterium or virus that contains the antigen.

Cells of the immune system serve many different purposes, and they support each other to form complex loops. *Phagocytes*, or *white blood cells*, ingest bacteria, viruses, and foreign particles, and digest them when possible (Figure). This often kills the phagocytes, which accumulate to form the pus at the site of an infection. *Lymphocytes*

(*Continued*)

IMMUNE SYSTEM AS A MODEL OF ULTRA-REDUNDANCY (Continued)

produce antibodies and kill target cells with lethal chemicals. There are several different types of each class of cells.

Circulating biochemicals also perform supporting functions. *Complement* proteins circulate in the blood in inactive form. When an antigen is encountered, usually linked to an antibody, a cascade of reactions begins involving the complement proteins. At the end of this process is a protein cylinder that punctures the invader cell membrane and dooms the target cell.

Cytokines, also called *Interleukins*, are messenger molecules that bind to target cells, promote cell growth, encourage all activation, and destroy cancer cells.

Cellular Defenses of the Immune System

Phagocytes	devour cells and particles
Monocytes	in blood
Macrophages	scavengers in tissue, activate T cells
Neutrophils	most common type in blood, contain potent chemicals
Granulocytes	associated with inflammatory response, contain granules of potent chemicals
Eosinophils	
Basophils	in blood
Mast cells	in tissue
Lymphocytes	
B cells (plasma cells)	produce plasma cells to secrete antibodies, one specific type for each B cell
T cells	
Regulatory T cells	
Helper T cells	activate B cells, other T cells
Other T cells	suppress immune cells
Cytotoxic T cells	remove infected body cells, and cancer cells, reject tissue and organ graphs
Natural Killer (NK) Cells	kills target cells with lethal chemicals

Chemical Defenses of the Immune System

Antibodies	link with antigens
Complement	binds to antibody and antigen, kills target cells
Cytokines (Interleukins)	potent chemical messengers, bind to target cells, promote cell growth and activation, destroy cancer cells
Lymphokines	secreted by T cells and B cells
Monokines	secreted by monocytes and macrophages

(*Continued*)

IMMUNE SYSTEM AS A MODEL OF ULTRA-REDUNDANCY (Continued)

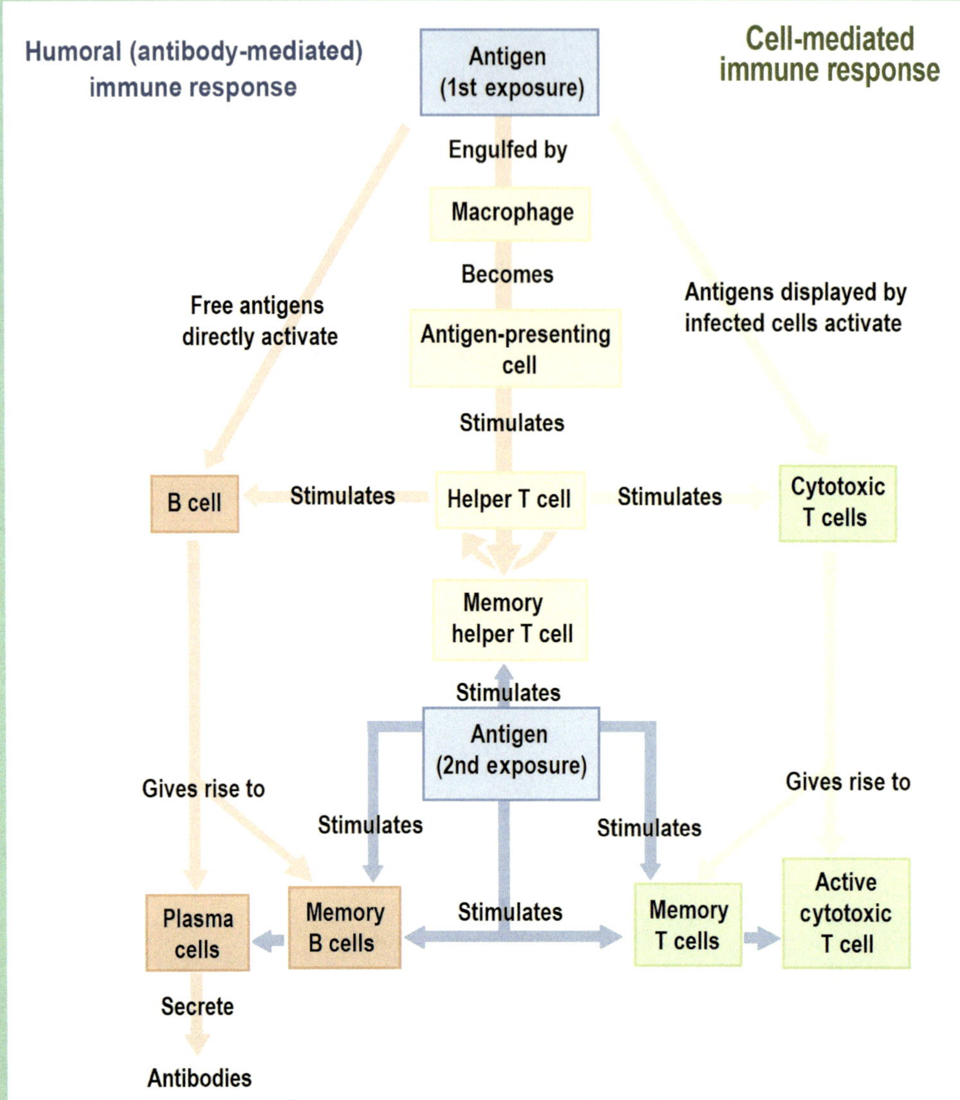

Diagram of the human immune response. Both chemical and cellular defenses are recruited. (Redrawn from Campbell, N. A. et al., *Biology*, 5th ed., Addison Wesley Longman, Menlo Park, CA, 1999.)

There is also within a cell a mechanism to screen all RNA molecules to detect foreign particles or cells that have evaded immune system protections (Ashe et al., 2012; Shirayama et al., 2012; Williams, 2013a). This mechanism particularly protects against viral infections that incorporate viral RNA within the genetic duplicating mechanisms of the host cell. This system uses a type of RNA, called *Piwi-interacting*

(*Continued*)

IMMUNE SYSTEM AS A MODEL OF ULTRA-REDUNDANCY (Continued)

RNA (piRNA) to check if the foreign RNA has been seen before. If it has not, then it is recognized as "non-self" and epigenetically silenced (RNAe, or RNA-induced epigenetic silencing).

Microbes attempting to enter the body must first get past the skin and mucous membranes. These pose a physical barrier and are rich in scavenger cells and Ig A antibodies. Next, there are the nonspecific defenses that attack all invaders. Patrolling scavenger cells, complement, and other enzymes and chemicals form this second line of defense. Then, infectious agents must face specific antibodies and cell defenses. If the body has been assaulted before by the same kind of antigen, then the B cell is ready to produce plasma cells that, in turn, manufacture large quantities of a specific antibody. These antigens activate B cells to recognize them upon the first exposure (Figure).

T cells are recruited. These are very powerful cells that could kill any kind of cells of the body if not carefully controlled. Thus, there is a complicated arrangement to assure that T cells act only on precise targets at close range. T cells do not mature to their full lethal capabilities until a macrophage has ingested a cell and produced cytokines that allow the T cell to mature.

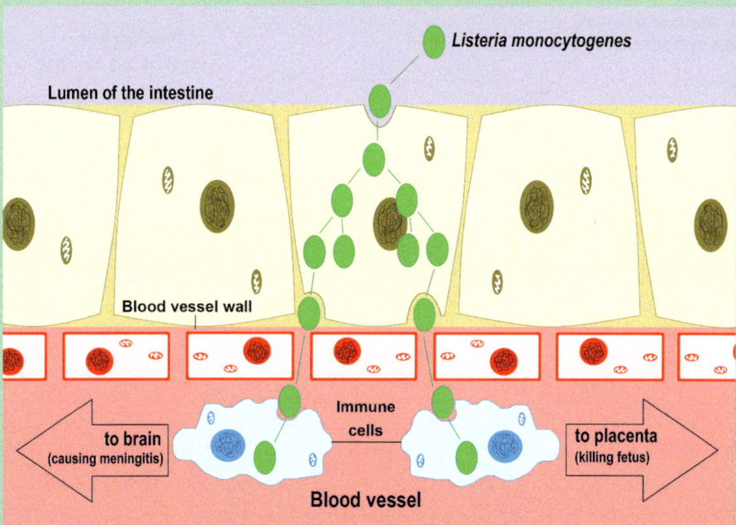

Listeria bacteria that breach the intestinal lumen are capable of life-threatening infections. (Redrawn from Gulbins, E. and Lang, F., *Am. Sci.*, 89, 406–413, 2001.)

Proper development of the immune system depends strongly on early exposures in the digestive tract (Brandtzaeg, 2007). The nascent immune system must be conditioned to distinguish between those foreign substances that are to be tolerated, and those to be considered dangerous enough to fight. Early exposure to the right kinds of microorganisms is important in this process. Among other things, gut

(*Continued*)

IMMUNE SYSTEM AS A MODEL OF ULTRA-REDUNDANCY (Continued)

microbes help to control leakage in the intestinal lining and in the blood-brain barrier (Schmidt, 2015). In the first moments of life after birth, a baby's mucosal surfaces in the lung and gut begin to be exposed to all kinds of microbes, and these are coated with antibodies IgA in order to prevent them from breaching the barrier between outside and inside the body. The source of this IgA is either inside the body of the baby or from breast milk. Beneficial microbes to which the baby is exposed (such as lactic acid bacteria—see Figure) stimulate cells that produce IgA (rather than IgE) and regulatory T cells that suppress effector T cells (which produce inflammatory responses). Without this early tuning of the immune system, regulatory T cells are not produced, and IgE antibodies are produced. The result is an overly aggressive immune system allergic to many foods and environmental antigens (see hygiene hypothesis).

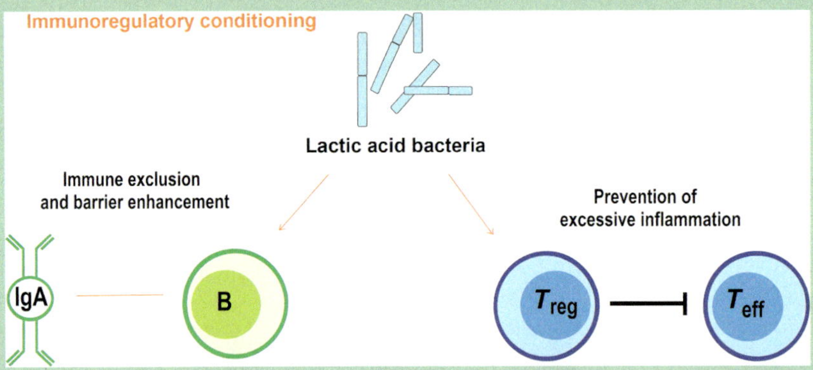

Lactobacilli and other beneficial microbes to which the baby is exposed early in life help tune the immune system by stimulating B cells (that produce IgA) and increasing the numbers of regulatory T cells (that suppress effector T cell activity). Without this early exposure, the immune system treats many foods as dangerous. (Redrawn from Brandtzaeg, P., *Am. Sci.*, 95, 28–35, 2007.)

Newborn ruminant mammals have very low protease activity in their digestive tract, which means that ingested proteins, including antibodies, are not degraded. Specialized receptors on small intestinal epithelial cells bind to antibodies and incorporate them into the body by endocytosis. This allows antibodies appearing in mother's first milk, called *colostrum*, to be passed to their offspring. Six to 12 hours after birth, the newborns lose the ability to absorb antibodies directly from the mother (Stewart, 2011).

If bacteria such as *Listeria monocytogenes* are ingested with contaminated food, they eventually reach the intestine. There the bacteria penetrate the epithelial cells forming the intestinal lining. They can multiply inside the epithelial host cells and are released into the underlying tissue, which are less tightly connected. When they gain access to the blood stream, they can penetrate white blood cells, which spread infection throughout the body (Figure).

(*Continued*)

IMMUNE SYSTEM AS A MODEL OF ULTRA-REDUNDANCY (Continued)

One vulnerable point of entry of microbes into the body is at the interface between the teeth and surrounding tissue. There is normally a tight seal between tooth and tissue, and antibody-rich saliva bathes the area constantly. Gum disease, however, can compromise this seal and allow bacteria access to the blood stream and interior organs. Gum disease has been linked to other disease conditions such as those in the heart, pancreas, and brain.

The immune response of an individual requires the presence of blood or lymph to deliver antibodies to the site and free passage of antibodies from their source to the location of the xenobiotic substance. Two types of tissue grafts can be made without the usual immune response (Rodger and Drake, 1987). The first is the cornea of the eye, which has no blood supply, and the second is cartilage, which separates foreign cells from antibody exposure with an impenetrable barrier.

The chromosomes of every vertebrate species contain a set of linked genes known as the *major histocompatibility complex* (MHC). Two animals with vastly different MHC genes will reject a xenograft very quickly. If the MHC genes differ by a little, then a tissue from one will be only slowly rejected by the other (Talmage, 1979).

Vaccination prepares bodily defenses and stimulates the production of antibodies by the lymphocytes. When the animal encounters an antigen for the first time, it responds with a small *Primary Response*. If it encounters the antigen again soon after, the so-called *Secondary Response* is much larger, much quicker, and more prolonged.

It takes time for the body to manufacture antibodies. If the body's lymphocytes have not encountered the antigen before, the body can take 2–3 weeks to generate enough antibodies to protect against the disease. This is because the lymphocytes must divide and multiply themselves after they have been exposed to the antigen before they can produce large quantities of antibody. This is what happens with the first dose of vaccine. After the first exposure to an antigen, the amount of antibody produced by lymphocytes can be relatively small. The levels of antibodies produced might only offer protection for a few weeks rather than months.

If the body is exposed again to an antigen such as a vaccine or bacteria relatively soon after the Primary Response has occurred, the antigen will encounter many lymphocytes that have been previously exposed to the antigen. These lymphocytes "remember" how to make an antibody to fight that antigen, so the response is more refined, and a far greater quantity of antibodies is produced in a much shorter time (Secondary Response). The Secondary Response can produce protective levels of antibody in hours or days rather than weeks. Antibodies attached to antigens inactivate the invaders. The antigen-antibody complex then attracts scavenger cells that destroy the antigens and so prevent disease.

There is growing evidence that an immune system unchallenged by pathogens and parasites can become dysfunctional to the point where autoimmune diseases develop. Parasitic worms (*helminths*) seem to be able to protect against allergies, asthma, diabetes, multiple sclerosis, and inflammatory bowel disease (Wickelgren, 2004). People and mice who had developed such diseases were helped when given helminth eggs so that parasitic worms became established in their bodies. Children in locations where parasitic worms are endemic were found to be much less sensitive

(*Continued*)

IMMUNE SYSTEM AS A MODEL OF ULTRA-REDUNDANCY (Continued)

to dust mites and other allergens than were children where helminths were not normally present. Although detailed mechanisms are not known for immune system malfunction, they seem to involve T-cells and the ability to recognize antigens.

Sometimes it is hard to turn off an immune response. Chronic inflammation has been implicated in many autoimmune diseases such as rheumatoid arthritis, lupus, multiple sclerosis, diabetes, asthma, and unstable atherosclerosis. Diabetes can be caused when insulin-making beta cells are destroyed by autoreactive T cells. Other diseases such as Alzheimer's and cancer may have their origins in chronic inflammation.

Immunotherapy is being developed to use the body's own immune system to fight diseases, especially cancers. Cancers have surface proteins somewhat different from those present on the surfaces of healthy body cells. One type of immunotherapy modifies a person's own T-cells to recognize these differences and tag them for destruction by the immune system. Immunotherapy can also be used to enhance effectiveness of vaccines against other diseases.

Immunotherapy can be dangerous if it results in drastic immune responses to all potential threats, whether they be actual threats or not. An overactive immune system can lead to organ dysfunction or failure, and death. Gene regulating switches can turn immune responses on and off with administration of an ingested pill, eliminating unwanted immunotherapy activities (Kozubek, 2016).

A molecule produced by the liver in response to an inflammatory signal is called *C-reactive protein* (CRP). The concentration of CRP in the blood quickly rises from less than 10 mg/L to more than 1000 mg/L during a severe bacterial infection. The chronic level of 10 mg/L may signal chronic inflammation somewhere in the body, and this may lead to chronic disease. Anti-inflammatory drugs (such as aspirin) can be effective in lowering the effects of inflammatory diseases.

PLANT IMMUNE RESPONSES TO DISEASE THREATS

Plants, too, have multiple responses to combat diseases caused by microbes. These include:

1. Stomata close when certain proteins near the stomata detect signature molecules in bacterial membranes or flagella.
2. Infected cells die (e.g., apple scab immunity works in this fashion).
3. Nutrients necessary to bacterial growth are removed from open spaces inside the stomata.
4. Biochemicals are produced inside the leaf that disable pathogenic compounds.
5. Plant cell death compounds are activated by high humidity when infection is most likely to happen.

Plants, too, have microbiomes, collections of microbes living in, on, and around the plants. These microbes apparently assist with plant metabolism, nutrient uptake, disease resistance, and reproduction (Gould, 2012; Wagner et al., 2014).

6.20.4 Antagonistic Action

> Give me a lever long enough, and a prop strong enough, and I can single-handedly move the world.
>
> **—Archimedes**

The heart is innervated by both *sympathetic* and *parasympathetic* nerve fibers of the *autonomic nervous system*. Although the heart can generate its own heartbeat independently of nervous control, stress, exercise, and physical trauma make it advantageous to adjust cardiac contraction to meet the needs at the time. Thus, the cardiovascular control system (Figure 6.20.5), which is located in the brain, controls the contractility of the *myocardium* (the muscle of the heart), and produces both *inotrophic* (force of contraction) and *chronotrophic* (rate of contraction) effects.

The same cardiovascular control system regulates blood distribution and blood pressure by affecting the small *arterioles* of the peripheral blood vasculature. The entrance to each of these vessels is surrounded by a *sphincter* muscle (a ring of involuntary muscle that surrounds the arteriolar aperture) with sympathetic, and in some cases, parasympathetic, nerve fibers. The sphincter is usually contracted. When the signal comes for the muscle to relax, the neuron produces nitric oxide at the neuromuscular junction, and this gas relaxes the sphincter. When the sphincter muscle expands, it increases the area through which blood flows and decreases its resistance. With decreased resistance, blood pressure falls.

Thirst is also determined by antagonistic action. There are cells in the brain that, when activated, lead to drinking. Another group of cells causes drinking to cease. These cells are located outside the blood-brain barrier where they can sense the electrolyte balance of bodily fluids (Oka et al., 2015).

These examples illustrate an important control strategy in living beings. The presence of both sympathetic and parasympathetic innervation means that there will be active control in both excitatory and relaxation directions. Skeletal muscles, too, are often

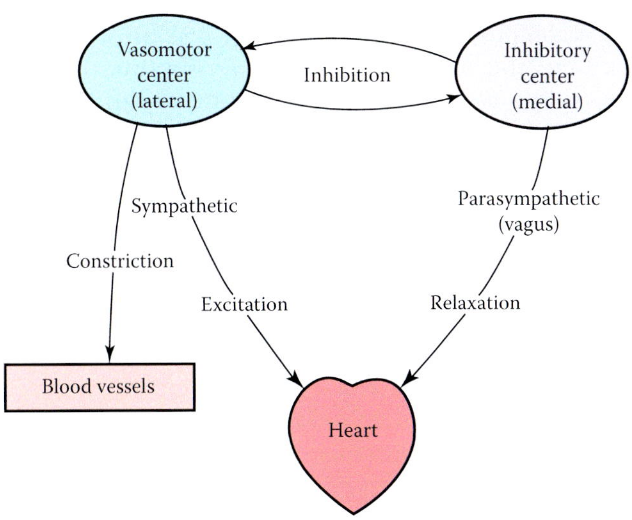

FIGURE 6.20.5
Basic diagram of the cardiovascular controller. Innervation by sympathetic and parasympathetic nerves allows antagonistic actions. The many inputs to the controller are not shown. (Redrawn from Johnson, A.T., *Biomechanics and Exercise Physiology: Quantitative Modeling*, Taylor & Francis, Boca Raton, FL, 2007.)

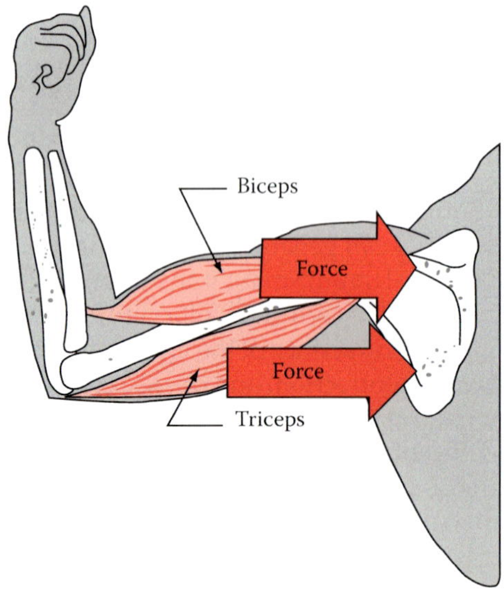

FIGURE 6.20.6
Antagonistic control of movements occurs as a result of two or more active muscles pulling in opposite directions. The result is that the movement can be made more precisely than if only one active muscle was involved.

found in pairs. One muscle (an *extensor*) actively moves a joint to result in extending or straightening a limb. The opposed muscle (a *flexor*) moves the same joint to bend the limb. This combination of muscles is said to be *antagonistic*. Movement of the limb to any position can be made precise because it is the difference between antagonistic muscle contractions that determines the final position (Figure 6.20.6). The smaller and more precise the movement (as with my fingers in writing this manuscript with pen), the more important is antagonistic control.

Control could be accomplished by regulation in one direction only. This could be accomplished by accumulating restorative force at the same time that a position is changed. A good illustration of this is regulation of breathing at rest. With no extra exertion, inhalation is active and exhalation is passive. When the lungs inflate, there is an extra amount of work that the inspiratory muscles must expend in order to overcome the elastic recoil of the lung tissues. During passive exhalation, this work is recovered while the air is expelled from the respiratory system, just as air escapes an inflated balloon. This system of breathing can work because the demands on the system are not too great, and there is little penalty for imprecise control. As a consequence, any two breaths during resting inhalation do not trace the same pattern.

There is a second penalty to this type of control: it is slow. Passive exhalation cannot proceed any faster than allowed by the recoil pressure and airway resistance. Flow rate is exponential in shape, which means initially high and slowing with time (see Section 4.3.3, Exponential Responses), with a time constant given by the product of capacity (or *compliance*) and resistance of the system.

Imprecise and slow control cannot be tolerated during the demands of exercise. Excess CO_2 must be removed at a rate high enough that it doesn't poison the body. Exercise breathing control becomes active in both inhalation and exhalation directions.

The penalty to be paid for antagonistic control action is energy inefficiency. Maintaining active control in both directions means that both sets of muscles will, at least in part, be working against each other. This wastes energy (see Section 6.14, Optimization), but in the hierarchy of priorities, loss of control is more wasteful.

We see this antagonistic action again in humoral control. Glucose is a major fuel for cellular respiration and a key source of carbon for synthesis of other organic molecules. The availability of glucose can vary greatly, depending on digestion, metabolic level, and stress. It is important to maintain a nearly constant level of glucose in the blood for proper metabolic functioning of the body.

The concentration of glucose in human blood is maintained at about 90 mg/100 mL (or 0.9 kg/m^3) by two pancreatic hormones: insulin and glucagon. Insulin lowers blood glucose levels by stimulating nearly all body cells (except those in the brain) to assimilate and consume blood glucose. It also slows the conversion of liver glycogen (an insoluble storage form of glucose) into glucose and inhibits the transformation of amino acids and fatty acids into glucose. Glucagon raises blood glucose levels by signaling liver cells to increase glycogen hydrolysis and to convert amino acids and fatty acids to glucose. The glucose is then released into the blood (Figure 6.20.7). The combination of these two hormones, both of which may be present in the blood at the same time, results in exquisite glucose control.

A good example of antagonistic control in the human body is the Mechanostat Theory to describe regulation of bone strength (Foutz, 1996). This theory was proposed by Frost (1987), and depicts the mechanostat as working like a thermostat that can either add bone or delete bone as needed.

Normal bone loading exists in a physiological range of mechanical strain, where the Mechanostat will not activate and therefore will not trigger bone geometric adaptation. Two set-points, an upper strain level and a lower strain level, bound this physiological range and control the type of bone adaptation needed.

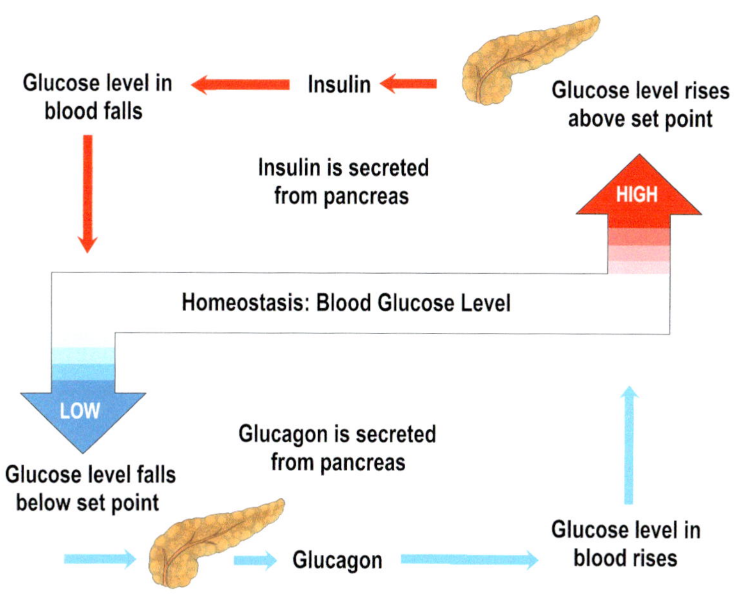

FIGURE 6.20.7
The glucose control system. Glucagon and insulin act antagonistically to maintain blood glucose level very precisely. (Redrawn from Campbell, N.A. et al., *Biology*, 5th ed., Addison Wesley Longman, Menlo Park, CA, 1999.)

When abnormal overloading occurs and the bone experiences a strain exceeding the upper set-point, the Mechanostat activates and recruits cellular activity for bone formation. The adaptive change will occur until the added bone mass increases stiffness and lowers the loading strain below the upper set-point. At this time, the Mechanostat switches off the cellular activity.

Similarly, when abnormal underloading causes bone strain to be less than the lower set-point, the Mechanostat activates and recruits cellular activity for bone resorption. The reduction in bone mass makes the skeleton more fragile, causing bone strain to increase.

The reduction will continue until strain increases above the lower set-point, allowing the Mechanostat to turn off the appropriate processes. Figure 6.20.8 provides a diagrammatic explanation of the Mechanostat concept. The theory presumes that the upper and lower set-points are genetically preset. For humans, published studies place the upper set-point at approximately 3000 microstrain and the lower set-point at approximately 500 microstrain.

Thirty percent of mature bones are destroyed and regenerated every year. These two processes are balanced so that the same amount of bone is reabsorbed as is regenerated. Thus, mature bones stay the same length and thickness.

The set-point concept of the Mechanostat allows this theory to describe bone adaptation from agents (hormones, diseases, drugs) that produce no mechanical stimulus. These agents can alter the location of one or both set-points, causing the Mechanostat to activate, although the bone experiences physiologically normal mechanical strain.

For example, a physician wishes to activate bone formation in an elderly patient, but a brittle bone condition will not allow the patient's skeletal system to experience loads more than 1500 microstrain. Under normal conditions for humans, this strain level is less than the upper set-point (3000 microstrain) and therefore the Mechanostat will not activate cellular processes. The physician gives the patient a drug that causes the upper strain set-point to fall to 1000 microstrain. Now the physiologically normal

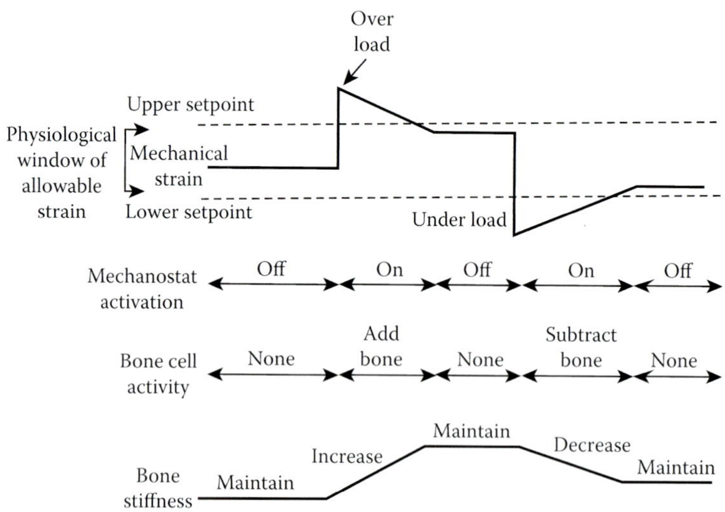

FIGURE 6.20.8
The Mechanostat has two set-points: one for adding bone mass and another for bone reabsorption. Between the set-points, no active regulation takes place. The set-points may be changed by externally administered drugs or by levels of circulating hormones. (Redrawn from Foutz, T., *IBE Newsl.*, 1, 2–3, 1996.)

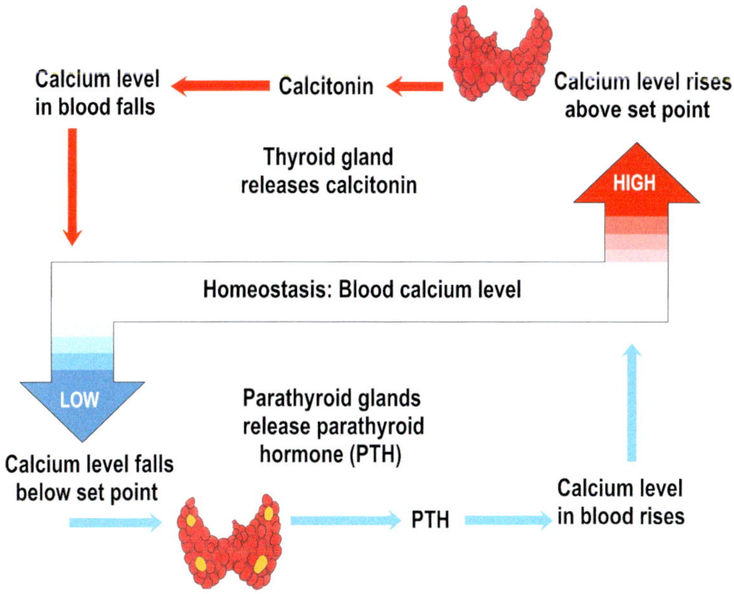

FIGURE 6.20.9
The blood calcium control system. Calcitonin and parathyroid hormones act antagonistically to regulate blood calcium level precisely. (Redrawn from Campbell, N.A. et al., *Biology*, 5th ed., Addison Wesley Longman, Menlo Park, CA, 1999.)

1500 microstrain exceeds the upper set-point, activating the Mechanostat and triggering cellular activities for bone production.

Antagonistic hormone control pervades living systems. Blood calcium level in mammals is regulated by the action of two hormones with opposing effects: *calcitonin* produced in the thyroid gland that reduces calcium and *parathyroid hormone* (PTH) that raises its level (Figure 6.20.9). Insect development is controlled by the hormones *ecdysone* and *juvenile hormone* (JH); ecdysone stimulates molting of the exoskeleton and growth, but the presence of JH hinders development of pupa and adult stages; only when JH levels wane does the insect progress to the next life-form stage.

Plants contain a pigment called *phytochrome* that exists in two forms: P_{660}, which is the inactive form, and P_{730}, the active form. P_{660} absorbs the red light of daytime and is converted to P_{730}; P_{730} absorbs the infra-red radiation of nighttime, and is converted to P_{660}. P_{730} predominates in the plant during the day and P_{660} predominates at night. These two different forms are thought to be involved in the phenomenon of *photoperiodism* in plants (Figure 6.20.10). Photoperiodism is the term that describes plant sensitivity to day length for flowering, fruit-ripening, and leaf-dropping.

Antagonistic action has also been found to be important in embryo development. Two developmental proteins play important roles: one (bone morphogenic protein, or BMP) causes *dorsal* (back) cells to differentiate into neural cells, whereas the other (antidorsalizing morphogenetic protein, or ADMP) causes *ventral* (belly) cells to become epidermal cells. There are two inverse gradients of these proteins, with BMP concentrations high dorsally and low ventrally, and ADMP concentrations high ventrally and low dorsally. The balance between these two proteins is apparently what determines the ultimate cell type outcome in an embryo (Meredith, 2006).

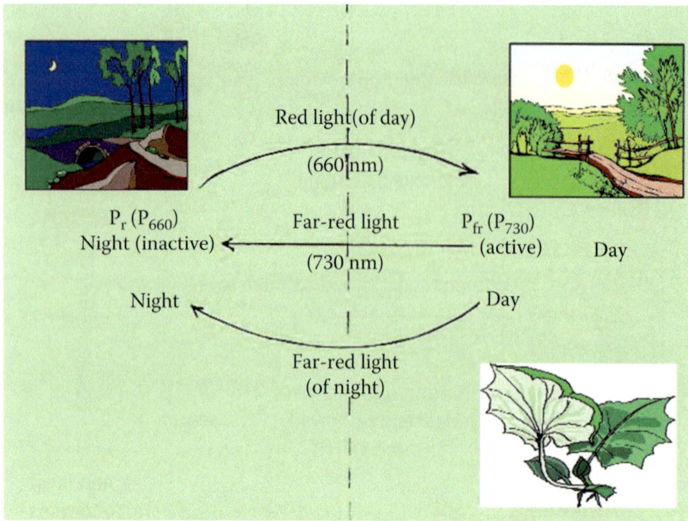

FIGURE 6.20.10
Phytochrome control in plants. P_{730} and P_{660} are alternate forms of phytochrome.

The brains of youngsters are very plastic, with brain neural synapse formation and pruning occurring dynamically. Synapse pruning and formation appear to result from two opposing and competitive processes mediated by brain glial cells and proteins that they produce (Piore, 2016).

6.20.5 Dead Zone

> We tend to overestimate the effect of a technology in the short run and underestimate the effect in the long run.
>
> **—Roy Amara**

Extremely precise control using antagonistic action is not energy efficient, and often precise control is not warranted as long as the operating point of the control system is in proximity to the desired position. The Mechanostat to control bone strength illustrates this very well. There is a mechanism to increase bone strength, and an opposite mechanism to decrease bone strength. If both of these operated together, bone strength would be very precisely controlled, but for what purpose? Bones need to be able to resist a range of forces usually applied periodically. Hence, there is no real reason for precise control within the somewhat wide range of 500–3000 microstrain.

This region in which the control system is essentially free-floating (or open-loop) is called the dead zone. Within this zone, primary active control does not happen.

A different example is the thermoneutral zone within the thermoregulatory systems of humans and other warm-blooded animals. Sweating to remove heat and thermogenesis to produce extra heat are both energy consuming. There is a temperature range where neither of these mechanisms operates, the thermoneutral zone. Temperature control within the thermoneutral zone is accomplished by vasodilation or vasoconstriction of cutaneous blood vessels. The amount of heat lost by this means is determined by controlling the skin surface temperature (see Section 2.7, Heat Transfer).

On the other hand, if the muscles of the limbs of the body were to have a dead zone, then there would be times when the limbs would hang loose and flop around. A dead zone in this case would be undesirable, and so antagonistic muscle tone is usually maintained at low levels.

6.20.6 Time Delays

> The biggest expenditure of the Alabama legislature the first year after the (civil) war was for wooden legs.
>
> **—Donald McCaig**

A characteristic of a control system where the signal must be sensed remotely and then the control action must occur remotely, perhaps at a third location, is that there are time delays built into the system. This is especially evident where the signal is a chemical produced someplace, and the chemical is transported to the sensor via a flow system (such as the blood). We have seen in Section 4.4 that Cheyne-Stokes breathing results from a long-time delay somewhere in the loop. For the case where carbon dioxide is produced in the muscles and must be transported by the blood to chemoreceptors in the neck, there can be a 30 s or more delay between the onset of exercise and the signal to increase respiration. This may not be too bad in some circumstances, but, apparently, it led to disastrous consequences for our forebears. The respiratory system adjusts to the onset of exercise much faster than would be expected, probably using some combination of neural signals from proprioceptors located in the muscles and feedforward control of CO_2 levels (Johnson, 2007).

6.20.7 Working with Biological Control

> At every crossway on the road that leads to the future, each progressive spirit is opposed by a thousand men appointed to guard the past.
>
> **—Maurice Macterlinck**

The biological engineer who attempts to modify the behavior of a living system must be aware of control actions of the system. To expect the system to respond passively without resisting the changes sought is futile. These systems are highly developed and highly capable. The biological engineer must either work with the control system or expect dire consequences. On the other hand, interfering with proper control may be the way to eradicate a particular BU. Understanding this is a major key to successful manipulation of living systems.

When the control systems fail, the result is disaster. BU may maintain internal stability despite wide fluctuations in environmental variables, but when the control mechanisms are overwhelmed, and stability can no longer be maintained, failure is often cataclysmic. In most instances, the BU dies. If death does not immediately ensue, then there is a severe impairment that pervades the organism. This happens when animals suffer heat stroke in extremely hot circumstances, or when blood pressure falls during hemorrhage, or when plants wilt irreversibly during drought. Loss of control means death unless some external means intervenes. Even then, help may arrive too late to do any good.

PHAGES TO THE RESCUE

Great fleas have little fleas upon their backs to bite 'em,
And little fleas have lesser fleas, and so ad infinitum.
And the great fleas themselves, in turn, have greater fleas to go on,
While these again have greater still, and greater still, and so on.

—Augustus De Morgan

Competitive inhibition may not be the only way to keep pathogenic bacteria in check. *Phages* are viruses that naturally victimize bacteria by entering bacterial cells, using bacterial mechanisms to replicate inside, and then destroy the cell when the phages are released into the environment. Each new phage can repeat the process with a new bacterial cell.

Antibiotic-resistant bacteria can be destroyed with phage therapy that bathes infections with solutions containing phages (Svoboda, 2009). If the phages delivered are tailored to the specific type of bacteria to be controlled, then dramatic results can be obtained.

Some bacteria, such as *E. coli* and *Salmonella*, have tough outer membranes that make them particularly resistant to environmental challenges, including antibiotics. These are called Gram-negative bacteria because they do not become stained in the laboratory with a certain blue dye. Other bacteria, such as *Streptococcus* and *Staphylococcus* do not have this outer coating; they can be stained and are called Gram-positive. Gram-negative bacteria are often more pathogenic, but some Gram-positive bacteria may persist in the environment for many years as endospores.

Phages work with both kinds of bacteria. If bacteria were to develop resistance to particular types of phages, the viruses are capable of rapid mutations to overcome bacterial resistance.

There are also parasitic viruses (called *virophage*) that infect other viruses. The viruses they infect belong to a recently discovered class of giant viruses as large as some prokaryotic cells (Van Etten, 2011). These viruses contain the genes necessary for protein synthesis, including a complete set of genes that code for enzymes to ensure delivery of amino acids to sites of protein synthesis (Leslie, 2017).

FEVER THERAPY

In January 1893, Dr. William B. Coley began the administration of a heat-sterilized combined culture of *Streptococcus pyogenes* (the virulent cause of the disease erysipelas) and *Serratia marcenscens* (a mild pathogen involved in eye and urinary infections), His patient was a 16 year old boy with a large inoperable abdominal sarcoma tumor (a *sarcoma* is a cancerous growth derived from muscle, bone, cartilage, or connective tissue). The boy developed the chills, headache, fever, local redness, and swelling of an erysipelas infection. The tumor shrank by 80% and the patient remained cancer-free for more than 20 years (Hobohm, 2009).

(*Continued*)

FEVER THERAPY (Continued)

A review of Coley's records showed a remission rate of 64% and a five-year survival rate of 44% of 170 patients treated in this way. Treatment success correlated with the length of therapy and induced fevers. The higher the fever the better, with 102–104 degrees Fahrenheit being necessary.

It has been thought that the fever mobilized the immune system to all-out warfare. Cancer cells carry many of the characteristics of the person in which they reside, but they also have differences. A mobilized immune system can recognize these differences and fight the cancer. Epidemiological studies suggest that people with several lifetime infections with high fever have significantly reduced likelihood of developing cancer later in life.

Attempts to reproduce and extend Dr. Coley's results have been mixed. For one thing, his patients required multiple treatments that took time to work. For another, radiation or chemotherapy weakens the immune system so that it doesn't respond the same to fever. However, his tantalizing results hold promise for cancer vaccine development and improved cancer treatment strategies.

Example 6.20.1: Using Antibodies

The unique structure of antibodies, with one end specific to a particular protein and the other end being uniform among antibodies of a particular class ought to be a useful tool for engineering designs. Suggest ways in which antibodies may be useful.

Solution

Antibodies can indeed be harnessed for useful devices and processes. Biosensors can be constructed with the standard end of the antibody fixed to a substrate of some sort, with the antigen-sensing end free to detect specific proteins in solution. When the antigen links to a protein, its conformation changes somewhat, and that change can be detected electronically with light, or other electrical resistance, or other radiation. The result is a biosensor very specific to particular antigen detection.

Antibodies can also be used to neutralize specific proteins in a medium. Once the antibodies link with the antigen, the antigen is rendered ineffective. Antibodies that bind to one specific antigen are called *monoclonal* antibodies; those that bind to a number of antigens are called *polyclonal* antibodies. So, antibodies can be used to make an unsafe medium safe.

There are other uses for antibodies that make use of their unique configurations. Producing useful quantities of antibodies can become the next challenge to be overcome. Recombinant DNA biotechnology is often used with bacteria or other organisms to produce specific antibodies that can be separated and purified for later use.

Example 6.20.2: Malariotherapy to Cure Difficult Diseases

Austrian Julius Wagner–Jauregg received the 1927 Nobel Prize for using malariotherapy to control the previously incurable disease of neurosyphilis. Malariotherapy involves injecting sick patients with a curable form of malaria. Although the exact mechanism is not known, the fever–chill cycles of malaria, and the possible recruitment of the immune system seem to help the body to fight other ailments. This may be an example of *competitive exclusion* (see Section 6.17), although the competition is induced after the target microbe is well established. After about 3 weeks, the malaria is cured with inexpensive drugs.

Dr. Henry Heimlich, promoter of the famous Heimlich maneuver for choking victims, used malariotherapy on Chinese HIV-positive patients between 1993 and 1996 (Herzog, 2004). He reported that all patients were alive 2 years later with normal immune system cell counts. These experiments raised much controversy in the medical community, and malariotherapy has not been accepted as a cure for AIDS (acquired immune deficiency syndrome).

Example 6.20.3: Myoelectric Control of Prostheses

Contraction of muscles, like that of the heart, is accompanied by a movement of ions and an electrical depolarization (see Section 4.4.3, Communications). This electrical event can be detected on the surface of the skin as a myoelectric signal.

Movement of powered prosthetic devices such as arms, hands, and legs can be controlled with the surface myoelectric signal (MES) in a very natural way. The person with the prosthesis only has to attempt to make the intended movement and this produces minute contractions of residual muscles similar to contractions in an intact limb.

Using MES for prosthetic control requires three important aspects of control (Englehart and Hudgins, 2003):

1. Accuracy to give a faithful realization of the user's intent.
2. Natural means to control the system, which requires the system to learn muscle activation patterns normally leading to the intended motion.
3. Immediacy, which means that the system will not perceptibly delay the response once a signal is detected. The threshold for unacceptable delay is taken to be 300 ms.

The challenge for engineers is to develop a pattern recognition system that uses the amplitude, duration, and rates of change of MES to discriminate among the many possible intended functions, and to classify the most likely intended movement. Multiple channels of MES can be used, and a feature set can be formed to extract as much information as possible from the MES. A continuous classifier would then determine which prosthetic movement to make at which time. Artificial neural networks (see Section 4.4.5, Open Loop) can be helpful in this process.

Example 6.20.4: Antigens versus Antibodies

Disease is caused when the number of antigens overcomes the immune response. Marchuk (1983) gave a simple mathematical description of the disease process, calling, for the sake of simplicity, all antigens to be "viruses" and all immune responses to be "antibodies" (Bell, 1971).

The change in the number of viruses (V) in an organism depends both on the foundational number of viruses present and the concentration of antibodies (A) acting to neutralize viruses:

$$dV = \beta V\,dt - \gamma AV\,dt$$

or,

$$\frac{dV}{dt} = (\beta - \gamma A)V$$

where β and γ are proportionality coefficients. This equation represents the rate of change of viruses in time.

Antibodies are produced by circulating plasma cells (c) that have a normal level (c^*). Their growth is stimulated by the concentration of the antibody–viral complex (AV). A second equation describing the rate of growth of plasma cells is

$$\frac{dc}{dt} = \alpha A\,(t-\tau)V\,(t-\tau) - \mu\,(c - c^*)$$

where:

α is a coefficient allowing for the probability of a virus–antibody encounter, and stimulation of cell (c) growth

τ is a delay time during which cells are poised for growth

μ is a coefficient of cell death

A third equation relates antibody reactions to plasma cell numbers:

$$\frac{dA}{dt} = rc - (d + \eta\gamma V)A$$

where:

r is the rate of antibody production by one cell

d is the decay rate of antibody effectiveness

$\eta\gamma$ is the rate at which antibodies are used up when they react with viruses (each virus requires η antibodies for neutralization)

There are many aspects not accounted for in this model. Among these is the weakening of the immune system by the viral infection. As with most models, the essential elements can usually be described by the simplest equations. More realistic results must be obtained with increasing model complexity.

Example 6.20.5: Paleobiologists Using Immunities

There is epidemiological as well as forensic interest in identifying ancient parasites, bacteria, and viruses. Learning about these can shed light on the causes of famed pandemics, reasons that historical figures suffered or died, and the coevolution of microbes and hosts. The detection, identification, and characterization of ancient microbes in the environment or human remains have largely been based on analysis of tiny recovered samples of ancient DNA (aDNA). However, potential contamination by modern DNA and altered aDNA can make these methods unreliable. An alternative is to use nonnucleotidic biomolecules, including mycolic acids and proteins remaining from ancient microbes or other tissues. Assaying techniques include immunohistochemistry, immunochromatography, enzyme-linked immunosorbent assays (ELISA), and mass spectrometry.

Example 6.20.6: Oncolytic Virotherapy

An Italian woman in 1904 who was diagnosed with cervical cancer suffered a bite from a rabid dog. She was given the live virus rabies vaccine for the bite, and subsequently, her large tumor disappeared. She lived cancer-free for eight more years.

The idea that viruses could be useful cancer therapy grew out of this case. Viruses can either be selected or engineered to selectively attack and kill cancer cells while leaving normal healthy cells unaffected. These viruses can do this because there are a number of characteristics of cancer cells that are different from normal cells, among them numerous receptors on cancer cell surfaces, and the large amount of viral-supporting raw materials inside the rapidly growing cancer cells.

The viruses attack cancer cells with multiple mechanisms, and so do not depend on a single means of killing cancer. When the viruses have matured within the cancer cell, they burst forth, killing the cell and releasing compounds that are then recognized by the body's immune system as foreign. The immune system T cells contribute to the extermination of the cancer. Thus, oncolytic virotherapy engages the redundancy so successfully used by living things to fight off infections (Mahoney et al., 2014).

Example 6.20.7: Antibiotic Effects Are More Profound When Given Early in Life

The hygiene hypothesis was supported by research results that showed that two common antibiotics given to neonatal mice significantly affected their internal microbiota diversity and made them more susceptible to severe allergic asthma. The same antibiotics administered to adult mice had almost no effect on either the microbiota or asthma susceptibility (Russell et al., 2012).

APPLICATIONS AND PREDICTIONS

1. Control mechanisms must be overcome to force a living system from its normal operating mode. It will be much easier to work with the control system to accomplish objectives.
2. If a control system is to be disrupted, it can either be at the receptor stage, the communication pathway, or at the actuator.
3. Chemicals can interfere with normal control operation.
4. Larger portions of the brain are required to control finer movements.
5. Overcoming the immune response is the main problem of organ implantation.
6. The process of learning is slow and deliberate; once something is learned, it is automatic.

6.21 Biological Units Go Through Natural Cycles

Beware of the young doctor and the old barber.

—**Benjamin Franklin**

BU are all far from equilibrium with their environments. This nonequilibrium condition requires the constant expenditure of energy to maintain biological order. If BU were in a state of equilibrium, they would be in chaotic disorder, and they would exist at the lowest possible energy state. Unfortunately, they would also have no characteristics of what we call life.

Nothing, not even living things, exist in the same form forever. Death is a part of living, its terminal condition. Regeneration or birth is also part of this process, as well as many stages of life in between (Figure 6.21.1).

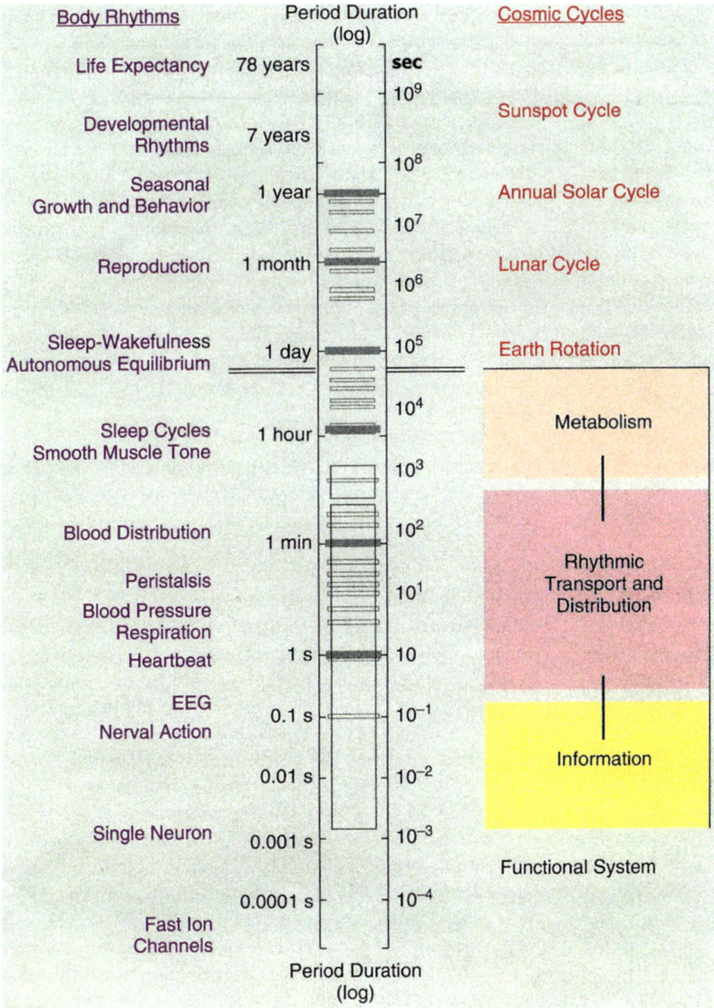

FIGURE 6.21.1
Biological rhythms come in many speeds, from slow to fast. (Redrawn from Moser, M. et al., *IEEE Eng. Med. Biol. Soc. Mag.*, 27, 29–37, 2008.)

6.21.1 Regeneration

> You see things; and you say, "Why?"
> But I dream things that never were; and I say, "Why not?".
>
> —**George Bernard Shaw**

Birth, maturation, senescence, and death are milestones in the BU cycle. Birth is the act of producing a new BU (Figure 6.21.2). We often think of birth as *parturition* in humans and other mammals, whereby the infant is born from a mother who has carried the embryo for the required *gestation* interval (see tables in the Appendix). However, birth can also mean the hatching of birds' eggs, the formation of new daughter prokaryotic cells by binary fission, formation of new daughter eukaryotic cells by mitosis, spore formation and

FIGURE 6.21.2
The cycle of life in simple terms. (Courtesy of Rick Detorie, Creators Syndicate Inc., Los Angeles, CA.)

germination in fungi, seed formation and germination in angiospermic plants, and other mechanisms to form new individuals of the same species. While not behaving completely like living organisms, viruses undergo a somewhat birth-like stage in the *lytic cycle* that results in the release of new *phages* by death or lysis of the host cell. Birth can also mean the formation of a new ecosystem at a restoration site, or the introduction of a new species into a territory where it had never been present before. The formation of a new *pride* of lions, group of monkeys, or family of humans is in a sense a birth process. Birth can be the regeneration of a new organ within an animal capable of this type of formation, or birth can mean the emergence of new plant leaves in the spring. Indeed, birth can also mean the formation of life on Earth, leading to the ultimate ecological system.

6.21.2 Maturation

> The problem with engineering education today is that it is focused almost entirely on science and technology rather than on mankind's needs.
>
> **—Joseph A. Andrade**

While all these acts of generation are different in details, they do result in new BU that must undergo completion in order to become fully mature. Depending on the type of BU we are considering, the process of maturation takes different forms. For humans and other mammals, maturation involves physical changes associated with growth, the readying of defense and reproductive systems, the development of judgment, and the acquisition of knowledge. For long-lived species, the maturation process is also long, and the parental attention given to their offspring is generally also considerable.

For biomes and communities, the maturation process is called *succession*. If succession occurs on newly exposed geological substrates, the process is known as *primary succession*. Primary succession occurs on newly exposed soils when glaciers recede or on new lava flows. *Secondary succession* occurs in areas where a community is destroyed without destroying the soil, as in abandoned agricultural lands, or after a forest fire, or on abandoned urban areas.

Seeds dormant in the soil can sense when a forest fire has burned all the underbrush that would normally successfully compete with the seedlings for light and nutrients. The seeds contain a protein that changes shape in the presence of a chemical compound of smoke and ash. When this component, called *karrikin*, is sensed, the seeds begin to germinate, thus restoring the forest environment (Gammon, 2013).

Once a new area is left undisturbed, a set of pioneer species of plants, animals, and, presumably, microbes, colonize the area. Because the area is not suitable for all local species, only a certain few characteristically colonize the area. Once the pioneer species become established, they modify the environment to make it more suitable for later species. For instance, the first plant species to invade a newly undisturbed area may be a local grass or weed, such as dandelion (Molles, 1999). As they grow and die, these plants leave behind organic matter that encourages insects and additional plant species to colonize the area. Later, tree seedlings may become established, and, as they grow, small animals invade, and birds begin to nest. Eventually, other species characteristic of the area become established. The diversity of plant and animal species increases rapidly at first (Figures 6.21.3 and 6.21.4),

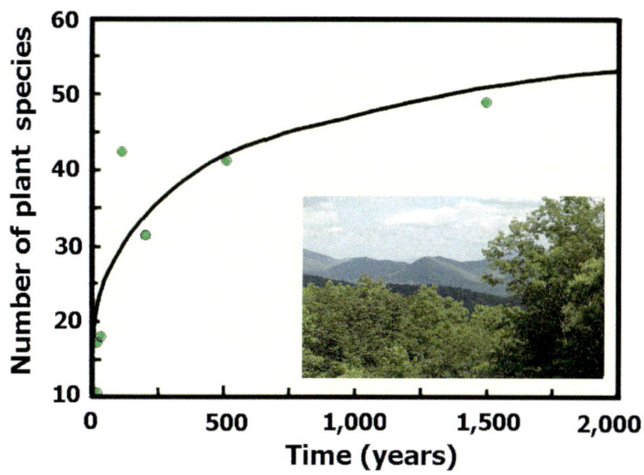

FIGURE 6.21.3
Plant species in a new area uncovered as a glacier receded. The shape of the curve is reminiscent of an exponential shape. (Redrawn from Molles, M.C. Jr., *Ecology: Concepts and Applications*, McGraw-Hill, New York, 1999.)

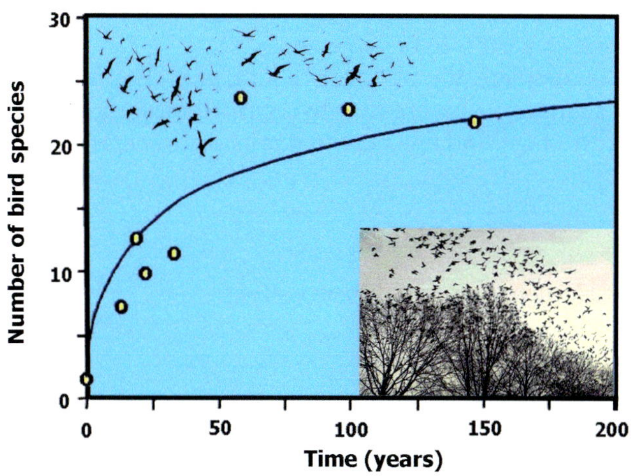

FIGURE 6.21.4
Number of bird species in a new forest area. (Redrawn from Molles, M.C. Jr., *Ecology: Concepts and Applications*, McGraw-Hill, New York, 1999.)

but begins to level off after a time interval that may vary from 1.5 months to 1500 years, depending on climatic conditions (Molles, 1999).

The human microbiome has been found to undergo ecological succession just like with other ecosystems. The distribution of microbes is originally seeded at birth and changes over time with diet, exposure to other bacteria, maturity of the human, medicine use, overall health, and general environment. Eventually, up to 30,000 different microbial strains reside in and on different areas of the body (Dorit, 2014). *Ecobiotics* is the term for small ecologies of bacteria that can change the microbiome or provide missing functions.

The concept of ecological steady-state probably is unrealistic. Long before humans came to become agents of disturbance, fire and wind damage caused forest destruction at frequent intervals (Bormann and Likens, 1979). Although there may have been local areas spared from catastrophic events, most ecological areas have been in various stages of rebuilding and succession throughout their histories. Some forest disturbances may never return to the undisturbed natural state (Shugart and West, 1981).

Understanding growth and maturation is important for skin regeneration, for example. In the adult mammal, an epidermis (the top skin layer) lost through injury regenerates spontaneously as long as a dermal substrate (the lower skin layer) is present underneath. When the dermis is lost, none of the injured mass regenerates; instead, scar tissue, unlike physiologic skin, forms and does not have the same strength of skin. The result is a less-than-satisfactory healed surface.

Especially with the use of stem cells in the alleviation of dysfunctional tissues, the maturation process is important. Stem cells are undifferentiated cells capable of developing differentiated tissue that performs specific functions different from those of other tissues. The ultimate stem cell is the zygote, which can, through successive division, form all the tissues of the body. Adults possess cell lines that have intermediate capacities to form different kinds of tissues: hematopoietic cells (giving rise to various types of blood cells) and intestinal epithelial cells (Jordan and Van Zant, 1995). Both of these must produce large cell populations with short cell lives.

Managing the capacities of these cells to repair damaged tissue involves at least two general stages: (1) maintaining a cultured population outside the body that does not lose the ability to reproduce or to remain undifferentiated, and (2) being able to form the desired type of tissue once injected into the site of damage. In the first stage, maturation is to be arrested; in the second stage, it is to be induced.

Once maturation is complete, the BU enters a long period of stability. Physical and further developmental changes, if any, are slow to occur. There are adequate resources needed to maintain the mature stage, and it is here that organismal reproduction occurs.

6.21.3 Senescence

> Invention is the talent of youth, as judgment is of age.
>
> —**Jonathan Swift**

Senescence (or aging) begins at or shortly after maturity is reached, and its effects accumulate at an ever more noticeable rate (Figure 6.21.5). For animals, this means that the reproductive rate declines, other physical capacities decline; and, for humans, mental capacity declines. Humans have the unusual characteristic that the female loses reproductive capacity roughly in the middle of her potential life span. Other species can (and do) reproduce nearly to the ends of their lives. During senescence, excess capacities of the

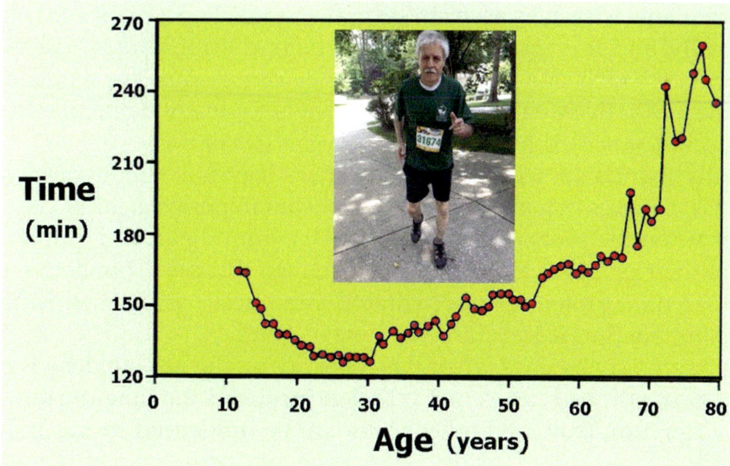

FIGURE 6.21.5
World record marathon times for men ages 10–79. (Redrawn from Fries, J.F. and Crapo, L.M., *Vitality and Aging*, Freeman, San Francisco, CA, 1981.)

organs continually decrease until there is no excess capacity at the end of life. At this point, the smallest external challenge (disease, cold, hot, etc.) can be enough to cause death.

Senescent cells in the human body don't reproduce readily, but do signal neighboring cells to also become senescent. It has been speculated that these cells become senescent when they accumulate enough chromosomal damage to make it likely that they would turn cancerous. To protect against becoming cancer cells, they become passive instead. They age, but do not reproduce, and so avoid disastrous transformations leading to cancer (Davenport, 2009).

Senescence could somehow be linked to the length of the telomeres in the genetic codes of somatal cells (see Section 6.18.4, Telomeres). Or, senescence may be a consequence of natural wear and tear on the body. Nesse and Williams (1994) present arguments pointing to traits that give reproductive advantage to younger individuals that also handicap older individuals. For example, highly acid stomach fluids would possibly protect against ingested infectious microbes, and this would be an advantage during the earlier, most active reproductive years. However, over time, the same stomach acid could lead to ulcers in the aged individual (although many stomach ulcers have been linked to the microbe *Helicobacter pylori*). Menopause, as another example, could have reproductive benefits if it limited the ability of mothers to have additional children in favor of caring for the ones already born. Natural selection may have chosen those genes that favor higher reproductive rates earlier in life that, concomitantly, cause aging effects in the elder years. The higher the reproductive rate of the animal, the shorter the lifespan, in general (i.e., the more severe is senescence). Likewise, the more competitive the animal, and the more its reproductive success depends on its ability to compete in its youth, the more likely it will be that natural selection favors competition over longevity. That is probably the reason women born in the United States live about 7 years longer than males of the same age (Nesse and Williams, 1994).

The oxidation of fatty acids in cell walls is a strong determinant of aging. Vitamin E acts as an antioxidant to help reduce the rate of fatty acid oxidation.

Senescence in plants is seen to occur in several ways. When a plant ages, its tissues cannot heal as rapidly, and it is much less likely to defend itself against disease. *Xylem* vessel elements and cork cells age and die before they can assume their intended functions (Campbell et al., 1999). Eventually, this decline leads to death. Some plants die after one reproductive cycle (*annuals*), others after many cycles (*perennials*).

The genetically mandated senescence that causes leaves to die involves many coordinated steps. Loss of leaves in temperate zones is meant to protect against desiccation of the plant in winter when the roots cannot absorb water from the frozen ground. The separation of leaves at their points of attachment is called *leaf abscission*, but before this happens, many of their essential proteins, carbohydrates, and nucleic acids need to be transported back into the plant via the *phloem* (Taiz and Zeiger, 1998).

Fruits, too, undergo senescence when they overripen. The gas ethylene is important for fruit ripening, and continued exposure to ethylene prolongs the ripening process. Ethylene is produced by ripening fruit, and other fruits can be influenced by gas produced inside each of them. Thus, fruits stored inside bags or other enclosures can be stimulated to ripen as ethylene accumulates. Ethylene is also important in the process of leaf abscission.

COMPREHENSIVE THEORY OF AGING

A theory of aging has been proposed by DePinho and colleagues (Sahin et al., 2011) that incorporates many possible mechanisms having an effect on aging and symptoms of aging. This theory is diagrammed in the figure, and incorporates telomere length, the p53 guardian of the genome gene, two master protein regulators of metabolism, mitochondrial number and function, free radical production and effects, atherosclerosis, and DNA mutation. Although this is still only a theory of aging, it does illustrate the interactions that can occur among the various components associated with aging effects.

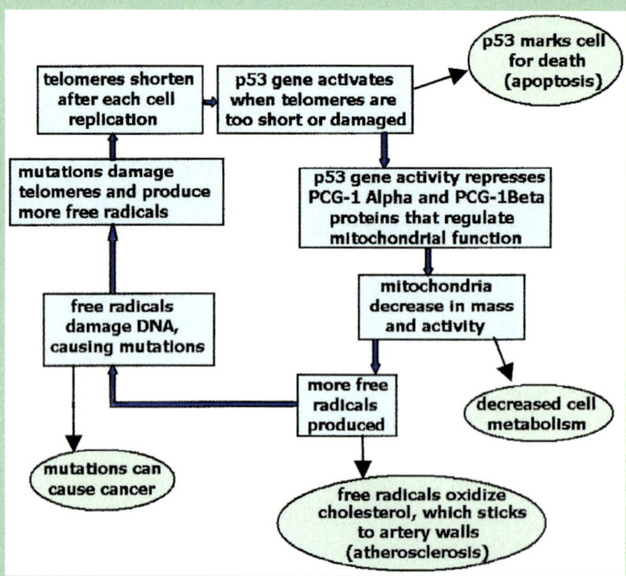

Comprehensive theory of aging as proposed by Sahin et al., *Nature*, 470, 359–365, 2011.

6.21.4 Death

But O heart! heart! heart!
O the bleeding drops of red.
Where on the deck my Captain lies
Fallen cold and dead.

—**Walt Whitman**

Death is the last stage in the life cycle. There is no clear purpose to death except for the fact that it releases bound nutrients. These become available for recycling by other BU.

Death may be genetically programmed or accidental. Accidental deaths are most prevalent just after birth; the period of greatest risk depends on the maturation rate of the specific BU. For humans, there is a great risk of death within the first year of life; for annual plants, their entire lives take less than a year to complete and the greatest risk is within a few weeks of germination.

Accidental deaths are those that include diseases as well as traumatic injuries. As the BU matures, it will usually develop better defenses against diseases and better avoidance of traumatic injuries. So, by the time the BU reaches maturity, the risk of death is usually minimal. Thereafter, the risk increases slowly as the BU ages. Deaths toward the end of the natural lifespan can usually be considered to be both accidental and genetic. Clearly, at least at this point, there are no known means to extend the human lifespan beyond 120 years. In the case of the annual plant, such as corn or marigolds, death begins shortly after the seeds ripen.

One can think of an apple: it is born from a flower, grows to maturity while green and astringent, becomes mature when internal starches turn to sugar and flavor develops fully, exhibits senescence as its texture becomes mushy and soft, and finally it dies as it disintegrates into a brown rotted pile of slime. The apple reveals all the stages of the cycle of life.

6.21.5 Annual Cycles

It will not always be summer; build barns.

—**Hesiod**

Annual cycles are important in many species. There is, of course, the annual cycle for plants that lose their leaves in autumn, remain dormant in the winter, and regain their leaves in the spring and summer. Reproduction usually occurs for these plants during the time of active growth. Tree rings are formed in the trunks of woody plants that correspond to annual growth periods. Climatological influences on growth can be inferred from size and appearance of the rings. Such a field of study is called *dendrochronology*. Annual growth rings also occur in aquatic species, such as corals, clams, and mussels. Study of these is called *sclerochronology* (Jones, 1983). From sclerochronology comes knowledge about the historical aquatic environment.

Insects that depend for their food on angiospermic plants also have an annual cycle. These sometimes become dormant at a very low activity level during the winter, only to regain their vigor in the spring. Honeybees stop rearing young in the late fall, and form a tight cluster around the queen in order to survive. In the spring, at the first signs of nectar and pollen availability, the queen bee again lays eggs and larvae are again raised. The colony increases in numbers and activity, in concert with the availability of food and warmer temperatures.

Some insects do not overwinter at all, but instead lay eggs in the fall in some protected place. In the spring, these eggs hatch and begin the cycle anew.

Small animals that depend upon insects or plants for their food also undergo an annual cycle of feeding, bodily energy storage, reproduction, and lethargy or hibernation. Even larger animals such as bears show this cyclic behavior. Carnivores must adapt to different conditions and different prey throughout the year, and winter becomes a time for survival instead of growth and vigor.

Humans, too, are affected by annual changes in events. Bone growth shows a tendency to annular rings similar to the growth rings in a tree trunk. This would imply that there is an annual cycle in the level of pituitary growth hormone. Other regions of the body are affected, also, by the decreasing amount of light in the autumn that can cause emotional as well as physiological changes. Shortening of the days decreases overall activity level that translates into changes in weight and physical fitness. Changes in temperature also cause acclimation to heat in the summer to change to cold acclimation in the winter. These are associated with different hormonal levels (such as thyroxin and epinephrine) and different levels of thirst and hunger. Although modern living has allowed humans to modify their environments enough that they do not have to be subjected to the most obvious seasonal changes, many subtle environmental clues still drive many of these annual cycles.

Migrating birds, mammals, and insects are affected very much by annual cycles. Migration is genetically controlled in most cases; the most convincing is the migration of Monarch butterflies each fall to their overwintering grounds in Mexico, followed by their return north in the spring. Several generations of butterflies are involved, and the butterflies that fly south are not the same ones that return north (and vice versa). How they know (if, indeed, "knowing" is the word to use) how to navigate to the same locations every year is still a mystery, but it must be genetically determined.

Birds of a feather flock together each year to migrate long distances from their summer breeding grounds to their overwintering sites. The energy use of migration is very high (Tucker, 1975; Johnson, 1991), so birds must store energy as fat in their bodies during the summer in order to prepare for the long journey.

Even birds that do not migrate are affected by annual light cycles. Chickens, for instance, lay eggs much more readily during lengthening days compared to shortening days. Humans keeping chickens for the eggs they lay must compensate by adding artificial light in the autumn.

Reproduction of many species is governed by light levels and changes in light levels. Sheep, deer, and other animals breed in the autumn when light levels are declining. Some plants flower in the spring and not in the fall; for other plants, it is just the opposite. Reproduction requires the coordination of the entire organism, so it can be assumed that the tissues and organs of the organism are themselves subjected to annual rhythms of hormones, nutrients, and physicochemical demands.

Even in the tropics, where there is little change of temperature or light levels throughout the year, annual cycles still exist. There are often changes of moisture levels or nutrient levels that trigger annual cycles. Competition for food in the tropics is more intense when migrating species arrive during the winter. Annual cycles in tropical regions may be more or less intense compared to temperate or polar regions, but they still occur.

WAVES OF WILDEBEESTS

The spectacular migration of 1 million wildebeests and a smaller number of zebras in the Serengeti ecosystem begins each year near the end of the wet season. The journey will take them from their wet-season range in the south to the northwestern dry-season range, then to the southwest, and finally back to the south (Figure). The animals travel a total of 3700 km.

There is a huge energy cost to this migration and the wildebeest mortality cost has been estimated at 3%. Clearly some powerful environmental factors must be driving this migration, but what could they be?

Evidence is accumulating that the quality of water in stagnant pools left after the rains stop is at least a contributing factor. Salinity of the water increases as water evaporates and leaves salt behind. When salinity reaches a level of about 30 parts per thousand, migration begins. Accompanying increasing salinity is increased turbidity, pollution with decaying organic matter, lower oxygenation, and thermal stratification.

Predicting animal movement is important economically and ecologically. Management of this great resource of our planet has become an important goal, if for no other reason than the ecotourist dollars that are brought into this region. As more people are convinced of the superiority of seeing these animals compared to shooting them, managers will be required more and more to protect their numbers and sustain their health (Wolanski et al., 1999).

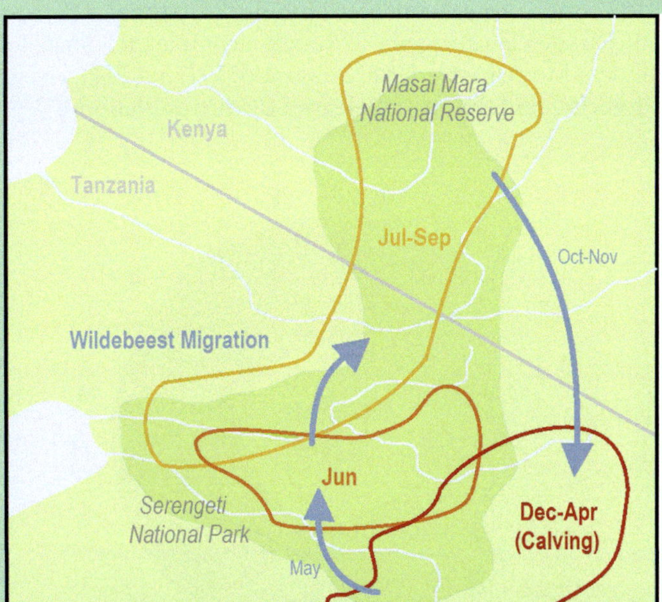

Migration route of wildebeests.

6.21.6 Monthly Cycles

> "I find", said 'e, "things very much as 'ow I've Is found,
> For mostly they goes up and down or else goes round and round."
>
> —**Patrick Reginald Chalmers**

There are, in addition, monthly biological cycles supposedly linked to the rotation of the moon around the Earth (28 days). If, indeed, this is the cause of monthly cycles, the stimulus may be a combination of increased nighttime light and larger tides. The light may stimulate physical activity and contact with cooperators and competitors. Light makes reproduction more likely if a mate can be more readily located, and light (even a little light) stimulates hormonal production and release. Higher tidal activity mixes water from deeper locations and makes nutrients more readily available. Deep churning also may bring up organisms that live on the bed of a sea or lake (*benthic* organisms), and so increase food availability for surface feeders.

When the sun and moon are aligned either on one or both sides of the Earth, as during the full and new moons, tidal fluctuations are greatest (called *spring tides*). When the moon and sun are at right angles to each other, as during the first and third quarters of the moon, tidal fluctuations are at a minimum (called *neap tides*).

Probably the most familiar monthly cycle is the menstrual cycle in human females and several other primates (Campbell et al., 1999). This cycle begins with the thickening of the lining of the uterus, called the *endometrium* Figure 6.21.6, and development of a rich blood supply in preparation for possible implantation of an embryo. In response to gonadotrophin-releasing hormone (GnRH) secreted by the hypothalamus, the pituitary gland secretes small amounts of follicle-stimulating hormone (FSH) and luteinizing hormone (LH). The FSH stimulates several ovarian follicles to grow and the enclosed eggs to enlarge. These follicles secrete estrogens. Through a complex feedback loop, the pituitary gland is then stimulated to release a large amount of LH and FSH, causing final maturation of one of the follicles, and the egg is released during ovulation.

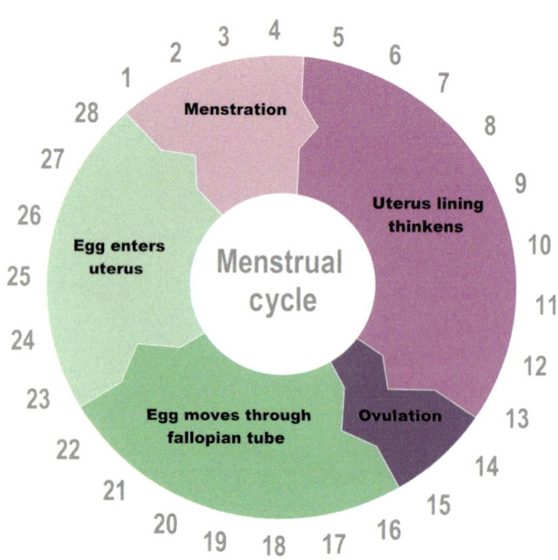

FIGURE 6.21.6
The menstrual cycle in humans.

LH then stimulates the remaining follicular tissue to transform into a glandular structure called the *corpus luteum*. It is the corpus luteum that continues to secrete estrogen and begins to secrete a second steroid hormone called progesterone. The endometrium is stimulated to develop even further. This combination of hormones inhibits the pituitary from releasing LH and FSH, and the corpus luteum eventually disintegrates. Then estrogen and progesterone levels decline sharply, spasms in arteries in the uterine lining deprive the endometrium of blood, and the excess endometrium disintegrates. Menstrual bleeding then removes the excess tissue from the body. This entire process takes about 28 days, and is accompanied by a shallow cycle of body temperature changes.

6.21.7 Diurnal Cycles

> The "Early-to-Bed Mouse," in which a gene was altered to match a human mutation, demonstrates not only how altering a specific gene can change the circadian rhythms of mice but how the process works in humans.
>
> **—Terri Peterson Smith**

Differences in light stimulation, temperature, and food availability over the course of a day can stimulate cycles in activities and responses. Tides are mostly semidiurnal (two high tides and two low tides per day), but in some places, including the Gulf of Mexico and the South China Sea, tides are diurnal. The moon is the most influential heavenly body on the tides.

Diurnal cycles are very prominent. Humans have daily fluctuations in body temperature, adrenocortical function, sodium and potassium excretion, and urine volume (Ganong, 1963). Deep body temperature varies by up to 0.8°C in a repeatable pattern throughout the day. Body temperature decreases at night and remains low during sleep. It rises again in the morning just before the normal time of awakening. These changes probably reflect different circulating hormonal levels and tissue metabolic rates. They are thought to arise in the *limbic system* of the brain, a very ancient structure that includes the allocortex, the amygdala, the hippocampus, and connecting to the thalamus and hypothalamus. The limbic system controls biologic rhythms, sexual behavior, motivation, and the emotions of rage and fear (Ganong, 1963). The exact site of the origin of the diurnal rhythm in humans and mammals has been identified as the suprachiasmatic nucleus (Figure 6.21.7).

There are active circadian clock genes in livers, pancreases, hearts, kidneys, and adipose (fatty) tissue, as well as in the brain. These many clocks assure that body function is coordinated, sufficient glucose is available, necessary enzymes circulate, fat is either deposited or metabolized, and sodium and potassium ionic content of the blood and tissues are adequate for normal periods of activity or rest. When synchrony among these different clocks is lost, metabolic and psychological disorders can arise (Summa and Turek, 2015).

Circadian rhythm is genetically controlled with the production of a protein that accumulates during the night and degrades during the day. Presence of the protein blocks the gene's activity (Greenwood, 2018).

Obesity, diabetes, cardiovascular disease, liver disease, cancer, and depression may all be related to malfunctions of the body's diurnal rhythm through genetic mechanisms related to protein formation. Women engaged in extended night work have a higher risk of breast cancer. Constant light significantly increases tumor development in animals. Immune responses appear to be stronger at certain times of the day and night than at others (Silver et al., 2012). Circadian rhythm controls protein levels by:

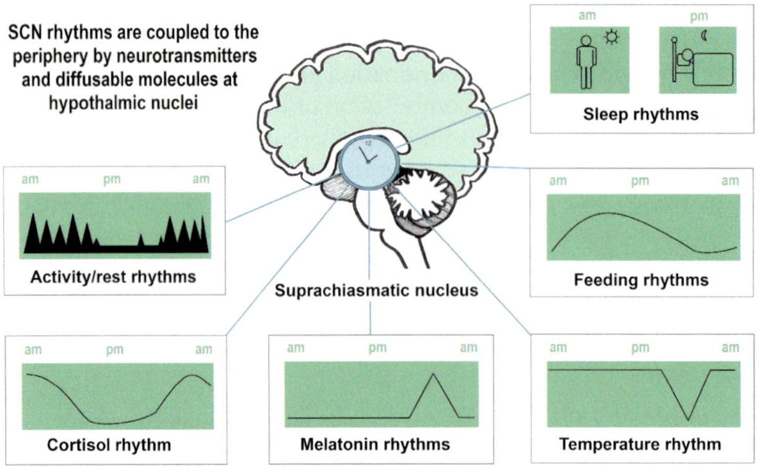

FIGURE 6.21.7
The suprachiasmatic nucleus in the limbic system of the brain is the site of the origin of diurnal rhythms. (Redrawn from Rich, T.A., *IEEE Eng. Med. Biol. Mag.*, 27, 25–28, 2008.)

1. altering when genes are transcribed from DNA into RNA, and
2. influencing when RNA is used to make proteins (Hardin et al., 2013).

Certain genes have DNA-binding sites that allow control by circadian proteins. Genes called *clock*, *period*, and *timeless* help regulate circadian rhythms; *sleepless* and *redeye* control sleep cycles.

Approximately 24 h circadian rhythms have been demonstrated in microorganisms, insects, birds, and other animals, including humans (Lévi, 2008). These rhythms appear to be regulated and involve clock-controlled genes. In humans, there may be some gender differences. There is a reproducible coincidence between times of highest efficacy and least toxicity for most anticancer agents in humans; timed cancer drug therapy can be more effective in men than women.

There is a circadian rhythm coordination among social groups of animals that has been found to be affected by the social environment with others of the same species as well as by cycles of light/dark, cold/warm, and other physical cues (Fuchikawa et al., 2016). These groups of animals eat, sleep, eliminate wastes, and perform group activities together in mostly synchronous fashion.

Circadian rhythms in bacteria have been identified as resulting from the sequential and periodic addition and removal of phosphate molecules from proteins. These proteins can function inside the cell or outside as long as ATP is present. No genes are involved in the regulation of this clock (Chiu, 2008).

Plants, too, have a so-called biological clock controlled by genes and enzymes. Extremely intense light can hinder photosynthesis, so a protein critical to photosynthesis is protectively modified by an enzyme during the morning before the peak of light intensity. This protein can also be damaged by ultraviolet-B radiation, so plants produce a sunscreen made of phenolic compounds. Even this photoprotection can be overwhelmed, so the plant keeps time to anticipate the most likely challenges (Elstein, 2003). Even metabolic and immune responses in plants are regulated to some extent by circadian rhythm (Wang et al., 2011; Teng et al., 2013).

Photosynthetic plants, naturally, have diurnal fluctuations in hormones and activities. *Phytochrome* is a plant pigment that is used to sense light and also to stimulate enzymatic reactions (see Section 6.20.4, Antagonistic Action). Some plants' leaves fold at night and open during the day. This action is stimulated by phytochrome activity. The levels of certain messenger RNA (mRNA) oscillates diurnally in the leaves of pea and wheat (Taiz and Zeiger, 1998). These diurnal activities persist, once established, even in total darkness, they are considered to be *endogenous*. However, without light stimulation, these diurnal cycles tend to last an hour or two longer or shorter than 24 h. The presence of light with a period of 24 h *entrains* the diurnal cycle to be much closer to the duration of a day.

Melatonin, a hormone secreted in the light-sensitive pineal gland in the vertebrate brain, is thought to function in the formation of biological rhythms (Campbell et al., 1999). Melatonin is secreted during the night and seems to decrease neuronal activity in the suprachiasmatic nucleus. It is thought that decreased neural activity in the thalamus allows the brain to be more sensitive to external stimulation, and therefore to be more conscious, as in the daytime.

Melatonin is produced in the pineal gland in the brain. There is a set of photoreceptors in the human eye that is distinct from the rods and cones responsible for vision; this third type of receptor senses light in the blue region and reports to the suprachiasmatic nucleus, which, in turn, prompts the pineal gland to make less melatonin if the level of blue light coming into the eye is higher. There is naturally more blue light in sunshine at noon, less at sunrise and sunset, and none at night. Hence, sleepiness associated with increased melatonin level is naturally keyed to the hours of darkness. Artificial light with significant amounts of blue light can interfere with this natural cycle.

Human blood levels of *adrenocorticotropic hormone* (ACTH), eosinophils, and cortisol vary diurnally and peak during nighttime hours. Urine magnesium, calcium, sodium, potassium, corticosteroids, catecholamines, and metabolites vary diurnally and peak in the daytime. Urine volume is larger during the day. Phosphates in the urine are at their highest levels at night (Wurtman, 1976).

Circadian rhythms are present in organisms at all trophic levels (Johnson and Hastings, 1986). These cycles are internally generated and synchronized (entrained) by sunlight (see Figure 6.21.8). Some organisms have but one internal clock that paces all the periodic physiological variables. Humans, however, appear to have different pacemakers for sleep/wake and body temperature cycles (see Figure 6.21.9).

Circadian rhythms are important to be able to predict biological behavior. Psychological well-being of humans, for instance, may depend on maintaining proper phase relationships in the diurnal cycle. Frequently changing schedules of shift workers tends to disrupt

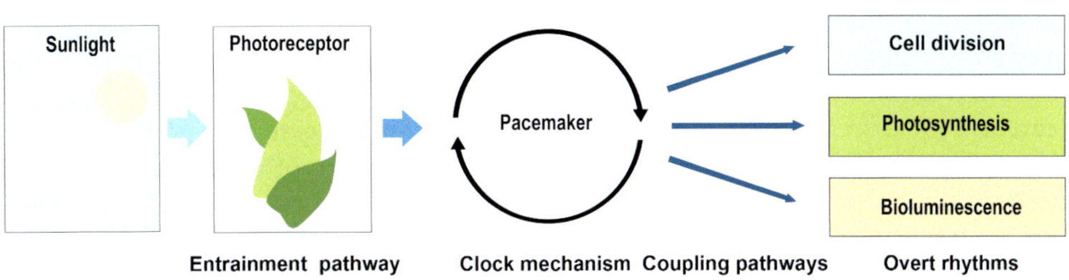

FIGURE 6.21.8
Proposed model of the circadian (diurnal) rhythm of a cell. A biochemical pacemaker maintains the basic rhythm, which is then entrained by exposure to sunlight. (Redrawn from Johnson, C.H. and Hastings, J.W., *Am. Sci.*, 74, 29–36, 1986.)

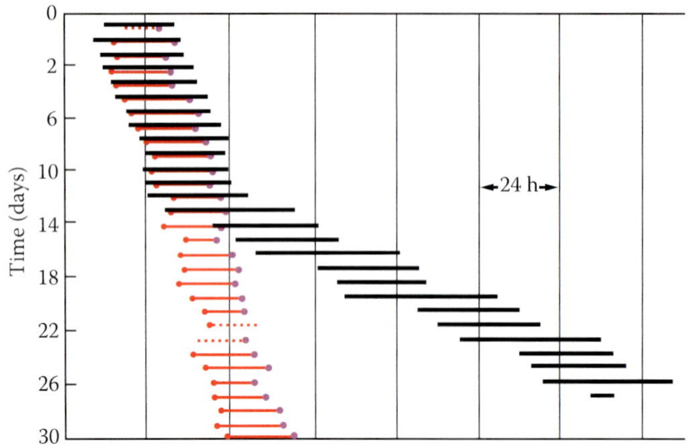

FIGURE 6.21.9
Intrinsic human pacemakers. Shown here are the rhythms of waking and sleeping (black) and body temperature (minimum temperature is given by solid red circles, maximum temperature by pink-filled circles). These data were obtained from a human subject isolated from any time clues. There appear to be two distinct rhythms (25.1 h temperature and 33.4 h waking and sleeping) that uncoupled after 14 days. (Redrawn from Johnson, C.H. and Hastings, J.W., *Am. Sci.*, 74, 29–36, 1986.)

daily cycles and can cause psychological disorder, lowered performance, and increased incidence of accidents. If it is necessary to rotate a schedule, it should be done infrequently and should be a delay rather than an advance (Johnson and Hastings, 1986).

It is no surprise that mammals and birds sleep. But, so do zebra fish and fruit flies (Mason, 2006). Even single cells need to rest. Killer T-cells, part of the immune system, become exhausted if they are overworked, and require a short rest before they are able to function again. Without this rest, they die (Staron et al., 2014). Sleep is so important to life that animals who sleep can die if they don't get it. Energy stores are replenished during sleep, and metabolic toxins are removed. Perhaps most importantly, the nervous system is remodeled during sleep to consolidate memory.

Migrating birds pose some interesting questions regarding sleep. Do birds sleep in flight, or do they forego sleep altogether while flying? Electroencephalogram recordings of the Great Frigatebirds in flight over the ocean for 10 days at a time have shown that they can sleep with either one brain hemisphere at a time or with both hemispheres simultaneously. They also sleep for only a small fraction (7%) of the time that they would have slept on land (Rattenborg et al., 2016).

The efficacy or toxicity of many drugs varies strikingly with the time of day at Ih the drug is administered. The time for maximal toxicity does not necessarily coincide with the time of maximal efficacy. Certain drugs, for instance chemotherapeutic drugs for cancer, can be more effective if they are given at times to minimize toxicity to normal tissues while maximizing destruction of tumors (Levi, 2008). On the other hand, certain drugs interfere with the biological clock or reset it. Such drugs can be used to treat disorders related to abnormal circadian rhythms.

Similarly, chemical lethality on insect pests has been found to vary throughout the day. It can take three times the dose of insecticide to have the same lethal effect on fruit flies if the insects have been exposed to the insecticide in the middle of the day compared to morning, evening, or night. Certain xenobiotic metabolizing (XM) genes are responsible for breaking down and detoxifying insecticide poisons, and these were found to be more active during the times of normal feeding (Hoven et al., 2009).

SLEEP AND HEALTH

Sleep is an important part of health, and not just human health, because other animals, including insects, sleep as well. Plants, too, have resting cycles synchronized to a circadian rhythm. In humans, sleep has profound effects on digestive processing, cancer, heart disease, neurodegenerative diseases, memory formation, and general body repair.

A good night's sleep may do much more than refresh the psyche. *Melatonin* is a hormone that acts as an antioxidant that combats the free radicals (see Section 3.6.7, Organic Reactions) that can cause genetic damage. *Cortisol* is another hormone that helps regulate the immune system. The disruption of normal sleep cycles results in less melatonin production and lower cortisol activity. Perhaps that may be one reason that female shift workers have higher rates of breast cancer than women who sleep normal hours. Regular sleep may help to battle cancer.

Sleep deprivation results in slowed reaction times, reduced vigilance, reduced ability to make decisions, poor judgment, distraction during complex tasks, and reduced or loss of awareness. Lack of sleep impairs performance similarly to alcohol. Being awake for 17 h straight is equivalent to a blood alcohol concentration of 0.05%. Staying awake for 24 h is equivalent to a blood alcohol concentration of 0.10%. The United States defines legal intoxication while driving as 0.08% or greater (Gibson, 2017). Biological engineers may have to deal with sleep deprivation in their designs.

The enumerated benefits of sleep are (Stickgold, 2015):

1. helps the brain remove waste products, including prions, through the *glymphatic system* (Nedergaard and Goldman, 2016)
2. improves immune reactions, antibody production, vaccine effectiveness
3. allows proper functioning of hormones: insulin, ghrelin (stimulates appetite), leptin (inhibits hunger)
4. helps maintain body weight
5. protects against Type 2 diabetes
6. affects negative emotional memories, guards against depression
7. assists memory stabilization, strengthening, integration, and analysis
8. helps improve physical skills
9. allows new connections among pieces of stored information
10. enhances creativity

For sleep apnea, where sleep is often interrupted by periods of obstructed breathing, physicians frequently recommend lifestyle-based changes, such as losing weight and sometimes even playing the didgeridoo, a large Australian wind instrument that strengthens and tones muscles of the tongue. Where these are not effective, Continuous Positive Airway Pressure (CPAP) machines are used to hold the airways open with air pressure, and thus relieve airway obstruction.

6.21.8 Cycles Shorter than a Day

I have made this letter longer than usual because I lacked the time to make it short.

—**Blaise Pascal**

Activities repeated for periods shorter than a day are common in certain places and conditions. The beating of *cilia* (small hair-like fibers) in the respiratory airways to propel mucus and dust particles toward the mouth is in this category. So is the movement of *flagella* (a fine hair-like process of unicellular organism) in microbes (see Section 2.9.2, Fluid Movement).

Neural activity is often periodic with a high frequency. *Thermoreceptors* found in the skin produce a series of pulses, the frequency of which corresponds to a particular temperature (see Section 6.20.1, Senses). Other neural sensors (for taste, smell, and touch, for instance) do the same. It is thought that there are neural circuits in the brain that form timing loops and enable the organism to judge elapsed time.

Certainly, the periodic contraction of the *myocardium* is a cyclic operation of short period. There is a regular series of events beginning with the generation of the cardiac rhythm in the sinoatrial node, a wave of depolarization that spreads across the myocardium, and then contraction of the muscle. Each cycle of this process is typical of the others, and can be studied to learn the details. The respiratory cycle may seem to be similar, but it is initiated in the brain rather than in the respiratory muscles themselves.

6.21.9 Asynchronous Nutrient Cycles

The science of biology is too important to leave to the biologists.

—**Robert M. Nerem**

There is only a limited amount of matter in this world, or in the universe for that matter. If there were no recycling of this matter, we would have used up some elements long ago. Imagine that our bodies did not decay once they died. All that carbon, hydrogen, and oxygen that would be tied up forever in corpses would be unavailable for use by later generations. Fortunately, matter is recycled and reused, again and again. Every atom in our bodies has been in someone else before. Makes you think, doesn't it?

There are cycles for all elements and many compounds. The water cycle is very important to us. We use water to drink, and it is either added to the air by evaporation or it is added to the soil after excretion. From there, it could be used countless times by other organisms, many of which are microscopic in size. Perhaps that water is eventually taken up through plant roots and moves up the stem to the leaves, where it evaporates into the atmosphere. Those same water molecules condense in clouds, and rain upon the land. The water collects in streams and then in rivers. It may be there that the water is pumped into our water systems to be drunk again.

The carbon cycle is another important cycle. Carbon exists in the atmosphere as carbon dioxide, which is used by plant leaves in the presence of sunlight to form simple sugars. These are then further processed to form other carbohydrates, including cellulose for cell walls. Grazing animals devour the plant material, sending it to their stomachs to be digested. There, special bacteria break the cellulose into sugars once more, and these are absorbed into the tissues of the animal. These sugars may either be metabolized into carbon dioxide and water, or they may be used to form muscle and other tissues. Carnivores

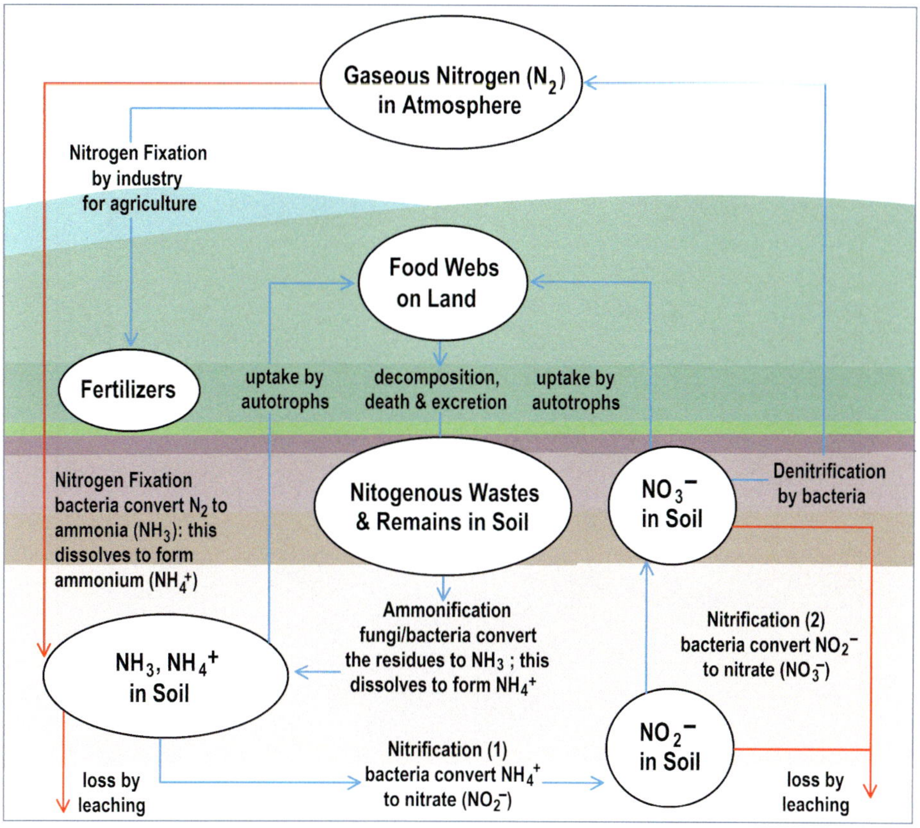

FIGURE 6.21.10
The nitrogen cycle in an ecosystem on land. The action of nitrogen-fixing bacterial species makes some nitrogen available to plants. Other bacteria cycle nitrogen atoms within the ecosystem by converting organic wastes to ammonium and nitrates. The atmosphere is the largest reservoir of nitrogen. (Redrawn from Starr, C., *Biology: Concepts and Applications*, Brooks/Cole, Stamford, CT, 2000.)

prey on these animals and eat their flesh. They digest this material and absorb it into their tissues. Many of these compounds are metabolized into carbon dioxide and returned to the atmosphere through the lungs. The cycle repeats. Notice that, not only does the carbon undergo a cycle, but many organisms usually have positive gains by utilizing the carbon at different stages in the cycle.

There is a nitrogen cycle (Figure 6.21.10). Nitrogen-fixing bacteria in the soil form nitrogen compounds that may become proteins inside the bacteria. When they die, these proteins may decompose into ammonium compounds. Other bacteria change the ammonia into nitrite and then nitrate, which is readily absorbed by plant roots. In the plant now, these compounds form amino acids and proteins. Grazing animals eat the plants, digest the contents into amino acids, which are then absorbed into their systems and form tissue proteins. Some of these proteins are metabolized; others remain in their bodies until they die, and either decompose through the actions of other bacteria, or are incorporated into other living tissues. Eventually, the nitrogen-bearing compounds release their nitrogen into the atmosphere to begin the cycle anew.

There are cycles for all materials important in biological systems. There are iron cycles, sulfur cycles, hydrogen cycles, calcium cycles, potassium cycles, and phosphate

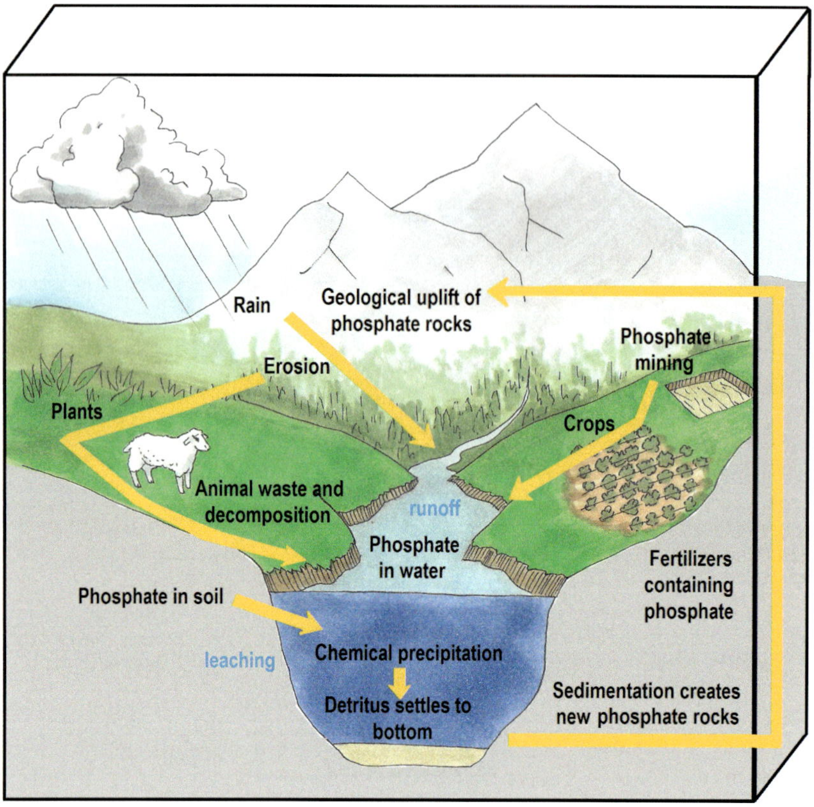

FIGURE 6.21.11
Phosphate cycle including geologic and biotic phases. (Redrawn from Campbell, N.A. et al., *Biology*, 5th ed., Addison Wesley Longman, Menlo Park, CA, 1999.)

cycles (Figure 6.21.11). Iron is particularly scarce in biology. The iron contained in old hemoglobin is recycled within the body and used to form new hemoglobin. Calcium is stored in the bones, and is released during nursing. There is hardly an essential nutrient that isn't recycled, economized, or utilized fully by an organism. The amount of nutrient that does pass from the organism to the environment is usually utilized by other organisms, which treat the nutrient similarly. Not only does this scheme fully utilize limited amounts of compounds present in the environment, but it also tends to check the growth and reproduction of any one species.

The phosphorus cycle is one of the simplest cycles because there is no atmospheric transport of phosphorus (Figure 6.21.11). All organisms require phosphorus as a major constituent of nucleic acids, phospholipids, ATP, creatine phosphate, and (for some organisms) bones and teeth. Plants absorb phosphorus from the soil as inorganic phosphate (PO_4^{-3}), and use it to synthesize organic phosphorus compounds. Animals eat it in this form and use it in their bodies. Decaying organic matter releases phosphorus back to the soil. Some phosphate is leached into water, where it may either be reused or settle out. Eventually it becomes part of sedimentary rocks that, when weathered, release the phosphate back to the soil.

One of the most intriguing concepts associated with a nutrient cycle is the Gaia theory. In this theory, the entire Earth is seen to be a self-correcting, balanced ecosystem on a huge holistic scale (Capra, 1996). There is no assignment of consciousness to the Gaia system, but it does act to maintain conditions suitable for life. Gaia shows that there is a strong interconnection between the Earth's living and nonliving portions.

As an example, we can consider carbon dioxide (CO_2). CO_2 has come from volcanic activity for millions of years. With too much CO_2, the Earth could overheat to the point where life could not be sustained. Plants and animals cycle massive amounts of CO_2 as they photosynthesize, respire, and decay. However, these processes are largely in balance and do not affect the level of CO_2 in the atmosphere.

As rock weathers, carbonates are formed that remove CO_2 from the atmosphere. CO_2 is thus bound in liquid solution by a process that so far has not included living organisms.

Rock weathering is increased by the presence of bacteria, which release carbonates and allow them to be washed to the ocean. There they are incorporated into countless tiny lime shells by small sea creatures. When the creatures die, these shells rain down on the ocean floor, where they form massive sediments of limestone. Some of these shells sink into the mantle of the Earth, melt, and are again spewed out by volcanoes.

The entire cycle of volcanoes to atmosphere to rock weathering to soil bacteria to ocean animals to limestone sediments and back to volcanoes acts as a giant feedback loop to regulate atmospheric CO_2 and the Earth's temperature. As the temperature increases, bacterial action and shell formation increases, thus acting in a self-correcting feedback mode. Wonderful indeed!

Example 6.21.1: Funky Leaf Spot

Most of agriculture is applied biology in a minimally controlled environment. Each time humans attempt to extend their control over growth conditions, they seem to upset some kind of natural balance. Take the case of funky leaf spot in peanuts. There are a number of fungal diseases of this legume, and fungicides are conventionally applied during blooming to control them. Funky leaf spot is a new fungal disease that appears before fungicides are typically applied. This fungus appears to be similar to another disease called early leaf spot. Whether funky leaf spot is a mutation of early leaf spot or not is not yet known, but this could be a biological adjustment to a widespread (at least among peanut farmers) cultural practice that prevents early leaf spot from its full reproductive potential.

Peanut plants seem to be able to recover from funky leaf spot and grow replacement leaves for the ones they lose due to the disease. However, another fungal disease, white mold, is triggered by chemicals released from decaying peanut vegetation. Thus, funky leaf spot, although its effects are not severe, can encourage the appearance of other diseases with more extreme consequences. Such is the way nature works: one organism assisting the other, even inadvertently, to maintain some ecological balance. Cultural practices used by humans to upset this balance can only be thought of as temporary at best.

Example 6.21.2: Modeling the Cycle of 17-Year Cicadas

The emergence of huge numbers of 17-year cicadas in the eastern United States is remarkable. These insects spend 17 years maturing in the soil, emerge by the hundreds or thousands for a few weeks to mate and lay eggs, and the eggs hatch to again grow for another 17 years. The large emergence in the year 2004 will be remembered for many years to come.

What makes these cicadas synchronize their reproductive cycles so precisely at 17 years? Why is there a cycle in the first place? What keeps natural variation from obscuring the population peak? These are just some of the questions that have been asked.

Hayes (2004) has presented results from a simple model of cicada populations. His model included these effects:

1. Natural variation. Each individual cicada can vary in its ability to keep time. Such variation tends to obliterate synchronization.
2. External signal dependence. Cicadas feast on plant roots, and the xylem fluid from plants varies periodically throughout the year. If all cicadas detected the same signals, synchrony would be maintained. However, individual cicadas consume different plants, and may also have individual sensitivity thresholds. Thus, synchrony is weakly maintained.
3. Intergenerational competition. Each cicada generation must compete with previous generations already living in the ground. The total population of cicadas cannot exceed the carrying capacity of the area. This leads to a strong synchronization tendency.
4. Predator satiation. Cicadas are slow, lazy, and nutritious. Predators, such as birds, eat their fills of cicadas, but many more cicadas survive than are eaten if they all emerge together. This strongly leads to synchronization.
5. High infant mortality. If a cicada survives its first 2 years, it will likely survive to maturity (see Section 6.23.1). This leads to synchronization.
6. Avoiding predator cycles. Predator numbers are likely to rise and fall periodically at some few number of years (see Section 6.17.2). The 17-year cycle, being a prime number, has the advantage that it will not likely become resonant with predator population cycles.

Example 6.21.3: The Design Challenge of Fatigue

Fatigue caused by long, tedious work periods or by upsetting of normal circadian rhythms can be a large contributing factor to accidents, misjudgments, and catastrophes. Biological engineering designs that presume a high degree of alertness on the part of the user, whether human or not, can have disastrous consequences if the user is not as alert as required. Technological solutions to user fatigue are often possible as long as the designer considers fatigue and its effects. These solutions are especially important where human life or safety is a risk, as in medicine, transportation, manufacturing, and other applications of human-machine interfacing (Nunes and Cabon, 2015).

APPLICATIONS AND PREDICTIONS

1. Plants and animals will have different requirements at different times of the year and at different times of the day.
2. Ventilation systems will have to be sized according to the time of highest demand.
3. Storing of fruits and vegetables will be more successful if senescence can be controlled.
4. Maintenance of tissue cultures will be more successful if natural cycles can be eliminated.
5. All designs should allow for disposition of dead organisms.

(Continued)

APPLICATIONS AND PREDICTIONS (Continued)

6. Analysis of signals from biological systems will usually involve recognition of cycles.
7. Observations of biological systems taken at one time may not apply to other times. Descriptions of deciduous trees in the winter will not be the same as in the summer.
8. Waste production from a BU will depend on its cycles.
9. Cryogenic preservation suspends all life cycles.

6.22 Biological Units Need Emotional Satisfaction and Intellectual Stimulation

> Scientists announced today that they successfully isolated the gene for loneliness, then felt sorry for it and put it back.
>
> —**M. Nadler**

When biological systems are to be manipulated, we often must consider physical, chemical, nutritive, waste elimination, and other rather tangible needs in order to achieve the desired purpose. That accomplished, we then expect that microbes in a bioreactor will produce the desired biochemical result, animals in a production facility will grow and reproduce, plants in a greenhouse will grow and thrive, organs in a bottle will remain viable, or birds in an enclosure will lay and hatch young. Missing from this is satisfying the emotional and intellectual needs of the biological unit (BU) of interest, and, whereas one might think of emotions and intellect as human attributes, or might extend emotional recognition to some higher-level animals, fulfilling emotional and intellectual needs is important for many BU types.

6.22.1 The Nature of Emotions

> Man is the Animal that Blushes. He is the only one that does it – or has occasion to.
>
> —**Mark Twain**

Emotions, as it turns out, are very ancient (Plutchik, 2001). They are important to single-cell organisms, plants, and higher-level animals. Emotions have been identified in guppies, octopuses, and insects.

When asked to describe an emotion, we all could try and at least cover many aspects of emotion. When asked to define emotion for scientific purposes, we would probably have a much more difficult time. The dictionary defines emotion as "a departure from the normal calm state of an organism of such nature as to include strong feeling…" (Bethel, 1960). Because the dictionary lists the word "feeling" as a synonym for "emotion," this definition is somewhat circular. Even the dictionary guys have a hard time with it.

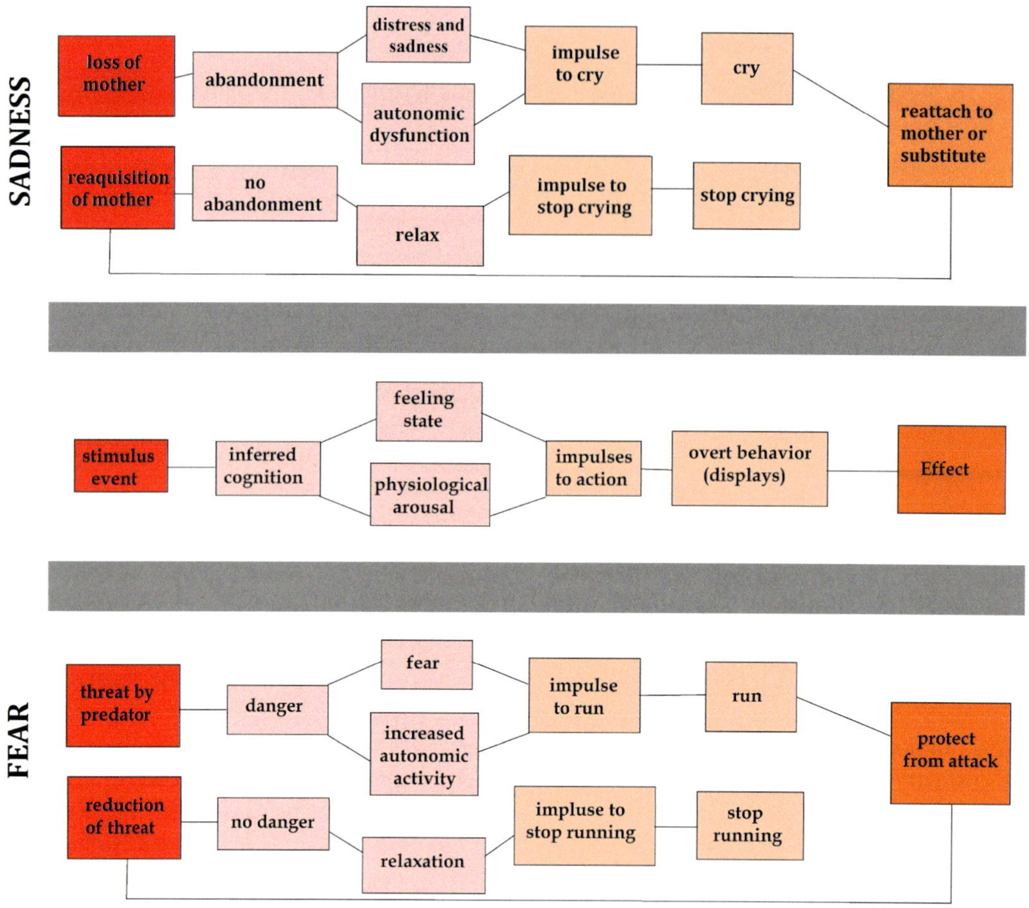

FIGURE 6.22.1
Examples of emotional responses to external stimuli. The result of the emotional response is restoration of equilibrium. (Redrawn from Plutchik, R., *Am. Sci.*, 89, 344–350, 2001.)

One definition is that emotions are a chain of events, made up of feedback loops that are meant to restore an individual to a state of equilibrium (Plutchik, 2001). They are very complex because they are usually chemically mediated and influence many metabolic pathways. They are evolutionarily important because they are responsible for heightened awareness of threats, selection of mates, defense of territory, and maintenance of social order (Figure 6.22.1).

At the evolutionary stage before DNA, RNA had a much broader range of activities in early life-forms than it does today. Survival for these organisms required them to conserve nutrients and shut down unneeded cellular activities (as they still must do). Otherwise, they starve. To do that, RNA would have had to develop the ability to sense its surroundings. It does this by binding to biochemicals (Figure 6.22.2) such as vitamins and amino acids floating in the environment, and subsequently forming different molecules that have different chemical consequences (Chiu, 2007). In this way, RNA illustrates the basis for chemically mediated emotional response.

Fear and anxiety lead to a state of heightened arousal to deal with a predator or a threat to offspring. Aggression can be used to defend territory or achieve dominance over others.

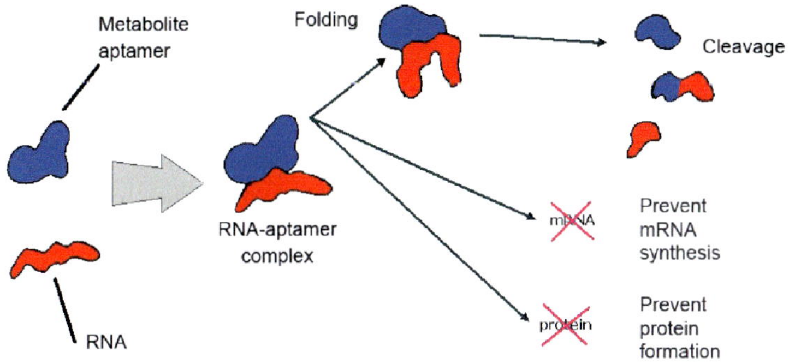

FIGURE 6.22.2
RNA senses metabolites in its environment by binding and preventing protein formation. In this way it controls gene expression.

Love and emotional attachment promote pair bonding, reproduction, and parental care of offspring. Each of these, at least in animals, is mediated by the very ancient limbic system of the brain.

Panic is one emotion that can have serious consequences (Stamm et al., 2004; Stout, 2004). In animals, panic leads to incorrect choices, such as freezing instead of running, or running over cliffs, or drowning in deep bodies of water, or running until physiological limits are exceeded. In humans, the consequences are just as dire, and panic can lead to trampling, drowning, or failure to escape deadly situations. Panic in humans and animals must be considered by the biological engineer whenever the environment places large numbers of individuals together in situations where individuals cannot totally sense the threat or cannot control the outcome. In certain situations, the inability to sense the true nature of the threat can be overcome by information passed through other means of communication. In other situations, designs of structures and procedures can be made to give the individual at least partial control over his actions, regardless of the actions of the crowd.

Cognition is very important to emotion, in that emotions are usually triggered by some stimulus event. This event may be sensed from environmental clues, by surface receptors in single-celled organisms, and by neural receptors in more complex animals.

At the same time, the cognitive ability of a BU is modified by emotional state. Fear focuses attention on only those inputs that are of immediate importance. Other receptor inputs are blocked from being recognized. Contrarily, love, awe, or submissive emotional states may block recognition of inputs related to threat. Thus, there is an emotional coloring to every cognitive act (Capra, 1996). Because of differences in sensing and emotional state, each BU lives in its own distinctive world as a perception of the absolute surrounding environment.

In humans and other higher-level animals, emotions are thought to arise in the brain, but that is not quite true. The nervous system connects the entire body, and the endocrine system of glands secretes hormones that circulate through the entire body. Because of this, tissue and organ BU are washed in biochemicals, and subject to the effects of emotion as is the entire organism.

Peptides (short chains of amino acids, see Section 3.6.3, Amino Acids), comprise the classes of hormones, neurotransmitters, endorphins, and growth factors. They are produced and stored in the nervous system, endocrine system, and immune system (Capra, 1996). They are physiologically active, and increase attention, raise glucose levels, increase heart rate, and deaden pain, among others.

Peptides produced in nerve cells are transported down the long axons to be released at the end, and have a direct effect, not only on other nerve cells, but on the biochemical mix of the blood and interstitial fluid. Transmission of information from nerves is not just by synaptic connection; it is also by direct release of peptides (Capra, 1996).

There are many terms to describe emotions, and to separate them and relate them to one another has been the subject of much psychological research. Plutchik (2001) has presented one model of emotional relationships in the form of a three-dimensional top-like circumplex with eight spokes showing diametrically opposite emotions (Figure 6.22.3). Emotions adjacent to each other are related, and those farther apart are not. In the third dimension are the degrees of emotional intensity. Between the splayed spokes of the flattened circumplex are named emotions that are mixtures of the two primary adjacent emotions.

The importance of emotional responses cannot be minimized, even for engineering purposes. Körner and Matsumoto (2002) have attempted to organize an artificial neural network model to analyze visual signals in a manner similar to the efficient parallel processing of the human brain. Many current approaches to this problem, they claim, have

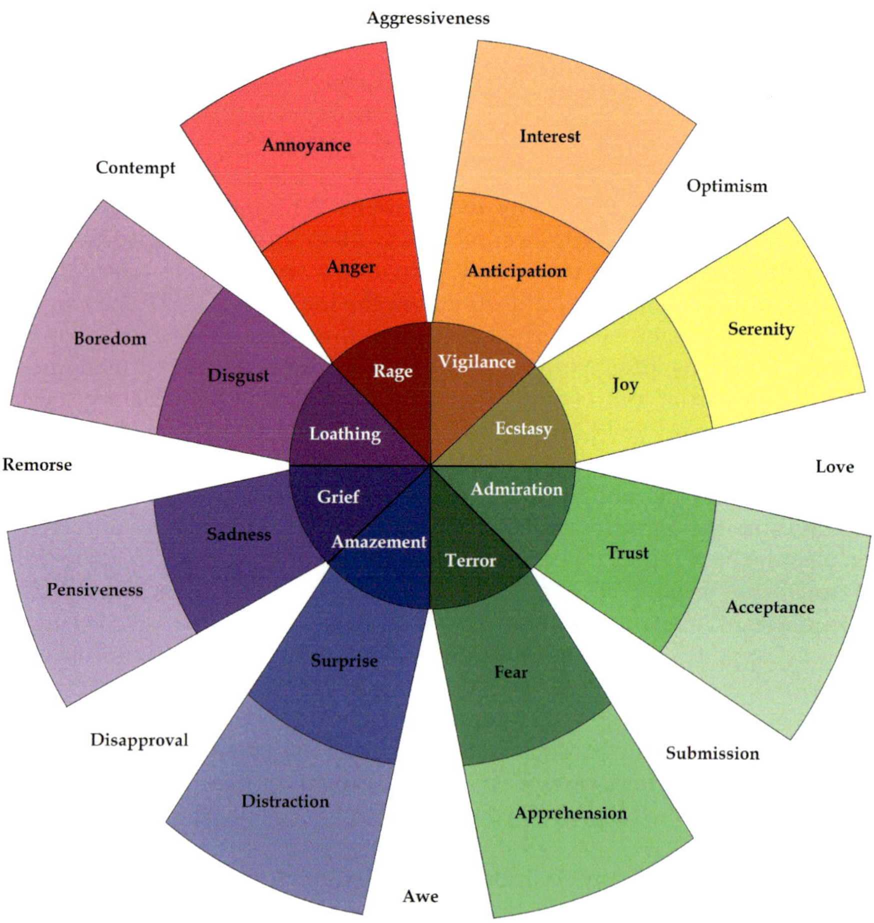

FIGURE 6.22.3
A three-dimensional model of emotions and relationships among them. (Redrawn from Plutchik, R., *Am. Sci.*, 89, 344–350, 2001.)

not attempted to understand the organization of the brain and its functioning. The current thinking about brain functioning is that relatively new brain structures, such as the neocortex is responsible for defining the parameters of brain function. According to this concept, the processing that goes on in the neocortex is responsible for recognition and plotting of responses. Older areas of the brain, the limbic system responsible for basic functions and emotional responses, are confined in their responses by the neocortex. This top-down idea has the psyche of an individual located in the higher centers of the brain.

Körner and Matsumoto (2002) say this top-down concept is wrong. Who a person is, is actually determined by the limbic system. It is the emotional responses of the individual that determine the fears, loves, motivations, joys, and principles at the core of the person. Even the very state of consciousness, according to Antonio Damasio, is derived from feelings and emotions (Pontin, 2014). The important self-image of a person is defined by the older structures of the brain. It is these structures that limit the responses of the newer parts of the brain in bottom-up fashion.

Needless to say, this turns the concept of human brain functioning upside down, and removes an important distinction made by some to divide humans from the rest of the animal kingdom. If true, the innate, hard-wired, patterns in the limbic system that make up the basic emotions have much more importance in human brain functioning than was previously thought. It also might lead to vastly improved engineering means to process information with the efficiency of the human brain (Figure 6.22.4).

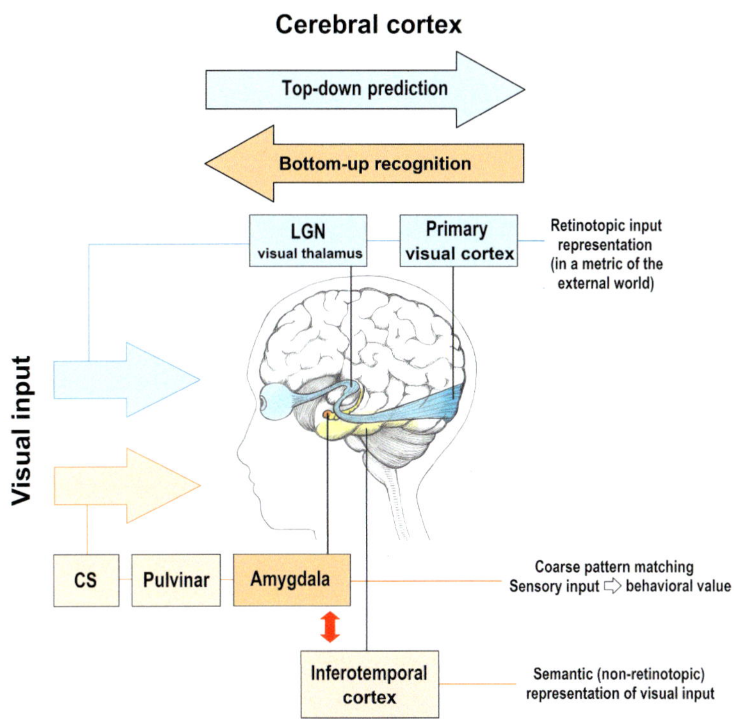

FIGURE 6.22.4
In their concept of brain visual function, Körner and Matsumoto (2002) attribute a visual representation of objects to various older structures of the brain (colliculus superior [CS] and lateral geniculate nucleus [LGN]). Higher-level processing in the inferotemporal cortex is limited by persona attributes defined by the amygdala (part of the limbic system). (Redrawn from Körner, E. and Matsumoto, G., *IEEE Eng. Med. Biol. Mag.*, 21, 121–133, 2002.)

Plants have at times been reported to sense emotions and to be aware of traumatic events going on around them. If true, this would truly be a revolutionary revelation. Although plant cells do, like nearly all other living cells, exhibit a cell potential with the cytoplasm almost 100 mV more negative than the surrounding interstitial fluids, and there have sometimes been found cellular events similar to neural action potentials in animals, a sentient plant would imply some signal integrating mechanism.

Attributing emotions to plants began with results from experiments that were not well controlled. As a result, faulty data were analyzed and found to give statistically significant results. Later attempts to confirm these results by duplication of experimental procedures gave results that contradicted the original conclusions (Galston and Stayman, 1979; Tannous, 1979).

That should have put these ideas to rest, but there exists an antiscientific counterculture that keeps them alive. Scientifically, however, plants sensitive to emotions of other beings cannot be supported. You may talk to your plants, but be aware that it is for your benefit, not for the plants.

THE CONSEQUENCES OF FEAR

Fear is a very powerful emotion useful for survival. Fear alerts humans and animals to potential dangers, and readies them for combat or escape (the so-called *fight or flight reaction*). Fright speeds the heart and constricts some blood vessels to shunt blood to the arms and legs. Some prey animals freeze when frightened, so predators won't see them move. Fear maximizes sensory input by widening the eyes, dilating the pupils, and flaring the nostrils (Lilienfeld, 2010). People recognize facial expressions of fear quicker than those of other emotions, perhaps because fear almost always associates with some kind of danger. Fears distort reality for sufferers. Some of this is due to physiological effects and some is due to psychological changes. People with fear and phobias tend to overestimate real threats (Lilienfeld, 2010).

Women react to fear differently than do men. Whereas men prepare for fight or flight in a fearful situation, women tend to cling to those around them and bond to other females in a collective defensive posture (*tend and bond reaction*).

Some fears appear to be innate, but others are learned. Monkeys with no prior fear of snakes have learned that fear from other monkeys who have experienced the consequences of snake predation (Shumaker, 2010). This is an example of memes in action.

6.22.2 Personality

> There is a part of the male human being that has to be satisfied by the creative and productive part of his life drive. But this is not the whole part. There are lots of parts that have to do with living and people and having fun, and liking yourself and all that stuff.
>
> —**Howard Head**

Human personality and animal temperament are traits related to overall emotional manifestations. We all know that humans have different personalities; we also have been in contact with animals of different temperament. Different breeds of dogs, for instance, often have stereotyped temperaments that lead us to expectations about how they will act.

Law enforcement agencies and societies for the blind have made particular use of these traits when training police dogs or seeing-eye dogs, respectively.

Behavior appears to be at least somewhat genetically determined (Hutchinson, 1981). It is not known to what extent environment influences behavior, but familial relationship does appear to be displayed in behavioral responses of its members.

There is a strong relationship between the state of mind and physiological health. Lonely people don't live as long, on average, as those who have close relationships with others. Those who are married live longer than those who are widowed, and they live longer than those who are divorced, and they live longer than those who were never married. People with positive attitudes are less likely to become sick, and their sicknesses do not last as long as those who are less optimistic. People who imagine themselves as healthy usually are, and are less likely to die in the next year than those who think that they are sick, even with no symptoms apparent. People who exercise and are healthy as a result, have higher levels of circulating prostaglandins and endorphins, substances that have positive psychological effects. They feel better.

DO APES HAVE SOULS?

The question of animal self-awareness is important for more than academic reasons. There are those who argue that nonhuman animals, if they are sentient (that is, capable of feeling and consciousness) should not be treated differently from humans. We cannot keep them in cages, they cannot be exploited for our use, and we certainly cannot use them for biomedical research. So, the question of self-awareness is critical.

How do you test for self-awareness? Linguistic tests given to animals have yielded some interesting results. There are reports of birds able to handle language and put together thoughts with the spoken word. There are also reports of bonobos and chimpanzees able to use sign language to communicate complex thoughts to their handlers. The problem with all of these is that none is definitive and beyond question.

The mirror test has been devised to avoid these difficulties. This test usually begins by placing a mark or dye somewhere on the animal where it cannot ordinarily see. The mark is odorless, tasteless, and without sensory stimulation. The animal is then shown itself in a mirror.

If the animal touches the mark on its own body (usually the face) more often with the mirror than without, then the animal is judged to be self-aware. It is generally accepted that chimpanzees, bonobos, and orangutans can recognize themselves in mirrors (Wynne, 2001). Recent tests on marked dolphins (Ebersole, 2001) have shown that dolphins race to mirrors and twist themselves to see the marks.

There are still those who question self-recognition as equal to self-awareness (Wynne, 2001). Some humans (for instance, blind people) cannot recognize themselves, but are self-aware. Autistic children can recognize themselves, but may not be self-aware.

There is a story that was related to me by a research biologist at Disney World in Orlando, FL. When I asked her if she could comment on reports of animal intelligence, she said she didn't know if all animals were intelligent, but one incident stands out in her mind.

(*Continued*)

DO APES HAVE SOULS? (Continued)

There was, it seemed, a mother chimp and baby in a zoo where she had once worked. The mother had been taught some signs for a number of different words. One day, as this pair was being watched by visitors to the zoo, a human baby dropped its pacifier in a place that was accessible to the baby chimp. The chimp raced over and put the pacifier all the way into its mouth. The mother chimp had seen none of this.

Whenever the mother chimp glanced over at her baby, the baby would hide the pacifier in its mouth; whenever the mother looked away, the baby would open its lips and expose the pacifier to the visitors.

The animal handler saw the pacifier in the baby chimp's mouth, and was concerned that the baby would choke on it. But how to take it from the baby?

The handler signed to the mother the words for "give," "baby," "mouth," and "toy." The mother looked at the baby, but the baby, of course, hid the pacifier. Again, the handler signed the words for "give," "baby," "mouth," and "toy," and again the mother looked at the baby, but the baby was playing it cool. The third time the handler signed to the mother, the chimpanzee looked at the baby, and, seeing nothing, became frustrated and turned her back on the handler. The baby chimp, all this time, was having a great time playing to the audience.

It took a while, but the handler was once again able to catch the attention of the mother. Once again she signed "give," "baby," "mouth," and "toy." This time, however, the mother seemed to have an instant of recognition. She went over to the baby, opened the baby's mouth, retrieved the pacifier, and gave it to the handler.

Would the mother chimp have done the same without the signing? Did she see the baby's antics out of the corner of her eye and realize by herself what the baby was hiding? Or did the sign language tip the mother that her baby was concealing the pacifier? We'll never know, but it makes one wonder.

6.22.3 Neurotransmitters

> To a large degree, dopamine is what makes us human.
>
> **—Li-Huei Tsai**

Much of human and animal behavior is related to biochemicals either naturally present in the brain or administered externally (see Section 4.4.3, Communications).

Dopamine is the brain neurotransmitter associated with pleasure or pain. The joy from a great meal, a job promotion, a winning poker hand, or sexual relations is conveyed partly by dopamine. Dopamine exerts powerful effects on motivation, reward, learning, memory, sexual desire, and pleasure (Gaidos and Keeley, 2006).

Under normal circumstances, dopamine is produced at a relatively constant rate in the brain. Only a portion of dopamine receptors is occupied at any time. Drugs such as cocaine release an avalanche of dopamine, and this activates nearly all of the brain's dopamine receptors. The result is euphoria for a while, but the brain tries to dampen the effect by switching off some of its dopamine receptors. When the drug wears off, the smaller number of functioning receptors means that the person's mood will be depressed. To overcome the depressed feeling, a larger amount of drug must be taken, and thus begins an addiction. An excess of dopamine in other parts of the brain is implicated in schizophrenia and stuttering.

Some people are born with fewer dopamine receptors, and these people are more likely to develop an addiction to cocaine. It sometimes also happens that certain thoughts, actions, or locations become associated with taking drugs, and these are related to dopamine receptor activity. There appears to be little, if anything, that separates physiology from psychology; the mind and body are one.

The neurotransmitter serotonin strengthens signaling between neurons, and is involved in the enhanced neuronal connections associated with learning.

Psychoactive drugs modify the normal functioning of synaptic neurotransmitters in the brain. LSD (D-lysergic acid diethylamide) mimics the neurotransmitter serotonin and can also substitute for dopamine. These drugs excite the same neuronal receptors as serotonin in the primitive part of the brain, the brainstem. Continued use of LSD, however, leads to tolerance and decreased effectiveness (Jacobs, 1987; Jacobs and Trulson, 1979).

Blood glucose levels appear to directly influence brain memory storage (Koehn and Hilbish, 1987). Adrenaline, because it influences blood glucose concentration, also has an effect on memory retention.

GABA

GABA (the shortened name for Gamma-Amino Butyric Acid) is an amino acid extremely important in human physiology despite the fact that is does not form proteins as other primary amino acids do. GABA's major effect is as a neural synapse inhibitory neurotransmitter, and has a general calming effect on the brain. It opens chloride channels in neuron cell membranes, allowing negative ions into the neuron interior and reducing the ability of positive sodium ions to trigger action potentials. Without GABA, nerve cells fire too frequently, and this can cause anxiety episodes, panic attacks, seizures, headaches, addictions, cognitive impairments, and Parkinson's disease tremors. GABA supplementation is said to relieve depression, hyperglycemia, and human growth hormone deficiency. GABA is prescribed to calm Parkinson's disease symptoms. GABA, however, does not in its natural form cross the blood-brain barrier, and so cannot be effective if taken by mouth.

6.22.4 Interpersonal Interactions

> Modern research universities have become segmented. We have scientists over here, humanists and social scientists over there. Knowledge is divided into ever smaller categories, our specialization ever more narrow.
>
> **—Hunter Rawlings**

Communication is another related subject. Plutchik (2001) describes emotion as a social regulation process and that communication plays a big part in assessing and managing that process. Through communication, more risky behavior such as fighting is avoided. He cites the example of a California ground squirrel that provokes a rattlesnake to rattle in order to use the rattling sound to assess the snake's size and body temperature. These two factors determine how dangerous the snake is to squirrel pups.

In another example, animal distress calls communicate to others about a dangerous predator. They not only help to protect a group of animals, but they may startle a predator into releasing its prey or attract a larger predator to compete and possibly allow the prey to escape.

Many animals live in social groups, and communications among individuals are the key to their cohesiveness. They have different vocal calls for things such as discovery of food, sighting predators, types of predators, empathy for others, care of offspring, and social relations with others of the same or opposite sex or status. Examples of these kinds of animals are mammals such as wolves, dolphins, sheep, and prairie dogs, and birds such as crows, chickens, ducks, and parrots. Social groups tend to function better overall than do the same animals living separately (Moehlman, 1987). The information passed from one individual to another in such groups is an example of the memes, cited earlier (see Section 5.4, Competition and Selection).

Love, or something similar, is especially strong among bird and mammal social animals. It appears that love among individuals acts as a cohesive force to keep the group together. When a loved one is lost from one or more of the group, actual grief may ensue. It appears that the capacity for love and mourning are deeply embedded in evolutionary trends (King, 2013).

Unconscious communication in humans is called "body language." While words are often formed by the conscious mind, and convey one message, body language can convey another (see Section 6.11, Crowding). Turning toward a person is usually a sign of attention, turning away conveys disinterest (Scheflen and Ashcraft, 1976). Facial expressions convey deeper thoughts such as disbelief, surprise, and joy. An open posture signals receptiveness, but arms folded and legs crossed denote unreceptiveness. The positions of the head and thorax may be different from that of the pelvis, and this may convey either personal commonality or separation. People sharing a common point of view tend to align their bodies with one another; those on opposing sides tend to face oppositely.

Empathy Quotient (EQ), or *Emotional Intelligence,* is the term used to measure the ability of a person to use emotions of the self or others to achieve certain ends. It is a set of skills with four components (Grewal and Salovey, 2005):

1. Perceiving emotions
2. Facilitating thinking and reasoning with emotions
3. Understanding emotions
4. Managing emotions

The EQ varies across the population as the familiar bell-shaped curve (Figure 6.22.5), or normal distribution. Empathy evolved so that we generally seem to have more for our family and close friends than for others. That stands to reason for early humans who lived in groups in competition with other groups. Competition reduces empathy, or even promotes *schadenfreude* (delight at the bad fortune of others) for competitors (Denworth, 2017).

Females in the general population have slightly higher EQ scores than do males; humanities students score slightly higher than do science students. It appears that empathy arises from the amygdala and at least 10 interconnected regions of the brain (Baron-Cohen, 2011).

The reason that EQ is important to a biological engineer is that, by creating the proper emotional environment, desired ends can be achieved. Creative problems, for instance, are solved better and more easily when the person working on them has positive emotions than when neutral or negative emotions prevail. For environments in schools and

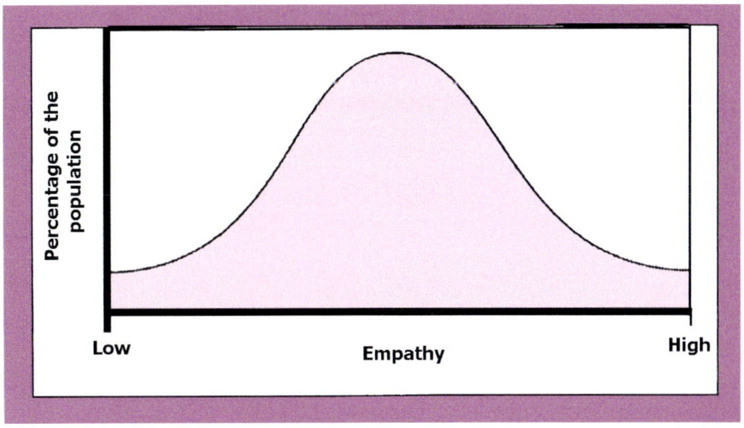

FIGURE 6.22.5
Empathy in the general population is expected to be distributed as a Normal Distribution. Professionals in the humanities, such as musicians and historians, typically score toward the upper side of the distribution, with higher EQs. Professionals who think in logical and systematic ways, such as scientists and mathematicians, typically score lower on EQ testing. Altruism is highly correlated with EQ.

hospitals, for example, this may be critical. The complete biological engineer, one who is fully aware of the likely emotional states of people (or even animals) in an environment that he or she creates, can be more effective with his or her designs or applications.

CHILDREN AT PLAY

Human and animal young need play to develop properly. Curiosity about the environment leads to exploration and familiarization. Play follows.

Play often is characterized by heterogeneity and innovation: "what can I do with this object?" Play is necessary for full development of the individual, because the child learns how to relate to its environment to expand vistas within bounds such as rules for games, to control patterns of thoughts and actions, and to develop physical and psychological complexity. Play is important, and the more social the animal the more necessary play appears to be (Schoggen and Schoggen, 1985).

Play teaches the child how to develop as an individual by mastering the body and the mind. Play develops adult-like behavior and loosens the bond between adult and infant. Exploration and play are thus relevant to the survival of the species.

Environmental stimulation seems to be important for proper development of the brain. The environment might act upon synaptic connections in two ways. First, environmental experience could stimulate or direct the growth of neuronal processes and the formation of new connections between neurons. Second, external stimulation could reinforce connections already intrinsically established. Visual stimulation in the cat, stressful situations in the young rat, and gonadal hormone presence in humans seem to affect brain structure in these animals if they occur during some critical period in brain development (Greenough, 1975).

(Continued)

CHILDREN AT PLAY (Continued)

Facilities accommodating human children or animal young must be able to provide stimulation and allow play. Conditions must allow for both auditory and visual (and perhaps olfactory) stimulation, and the levels of stimulation change with developmental stage. Babies need visual stimulation, but not at such a high density that there is interference. Attention of older infants is centered on their mothers, so the number of other adults in the room is of little consequence. At the toddler stage, the child must be able to return to its mother, either physically or visually, for assurance when exploring new situations. Soon thereafter, the child begins to develop a sense of personal space.

Crowding, stress, or lack of appropriate equipment can interfere with play and learning. Adequate facilities promote cooperative behavior, but lack of resources promotes aggression. Smaller child-adult ratios in the classroom lead to more questioning and object curiosity than higher ratios (Schoggen and Schoggen, 1985).

6.22.5 Brain Development and Learning

Originality is simply a pair of fresh eyes.

—**Thomas Wentworth Higginson**

Learning about the environment and how to respond best to environmental challenges take time. Those animals that learn more and learn better take more time to mature and require longer parental care than do those animals with hard-wired (innate) responses. Humans, in particular, seem to be created to learn; their brains take twenty years or more to mature because immature brains are flexible and can form new neural connections easily. Children explore creatively and easily; adults plan and act effectively. Each of these requires a different brain function and structure (Gopnik, 2010).

The human brain develops unevenly (Wallis, 2004). The maximum brain cell density occurs between the third and sixth month of pregnancy. During the final months before birth, a dramatic pruning of unnecessary brain cells eliminates many neurons. After birth, the number of cerebral neurons continues to decline at a slower pace. New neurons can still be created, but more are lost than gained. Human babies' brains are much more complex than those of adults. Babies have many more neural connections than do adults, but many of these connections are pared away as learning and environmental adaptation are experienced (Tapia et al., 2012).

As an organism first forms its nervous system, it sends out more neurons than necessary to a particular place. Through a series of genetic feedback loops, some of these survive and some do not (Davenport, 2008). There is a balance that occurs between factors causing neural growth and those causing death.

Before the ages of 6 and 12, the neurons grow bushier and create numerous new connections to other neurons. These represent new pathways for signal processing.

Neurons in the brain appear as two different kinds of tissue. Gray matter is composed of neurons and dendrites (see Section 4.4.3, Communications). White matter is made up of neurons and fatty myelin sheaths. The myelin sheaths around neuron axons result in faster and more efficient nerve conduction.

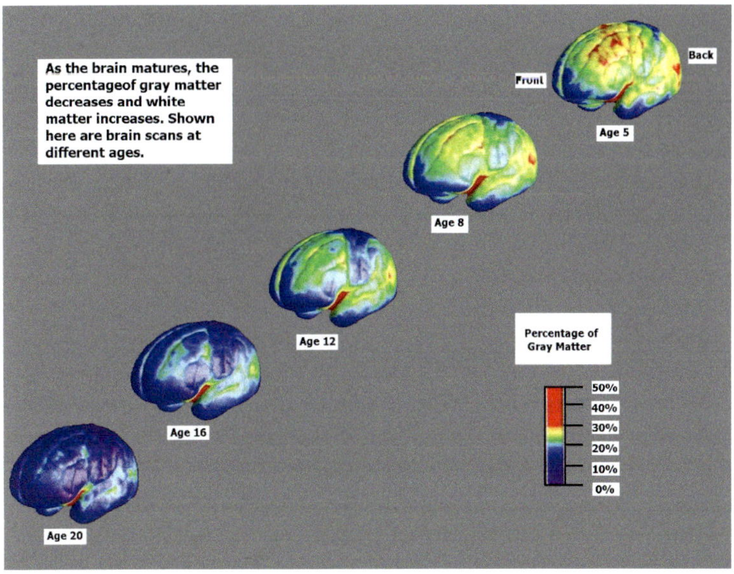

FIGURE 6.22.6
Gray matter wanes as the brain matures. These images over 15 years of time show a shift from red (least white matter) to purple (most white matter) as the brain ages. (Redrawn from Robertson, L.C., *Phi Kappa Phi Forum*, 85, 19–22, 2005.)

Over the years of childhood, white matter replaces gray matter as the brain matures (Figure 6.22.6). This process is most active until about the age of 20, although some myelin formation may occur until the age of 40. As this change takes place, unused or lightly used intraneuronal connections atrophy. Because dendritic connections represent potential processing capability, raw learning power of the brain declines, but processing of already-learned information (as represented by neuronal connections) becomes more efficient as white matter replaces gray matter.

As with other biological characteristics, there is both a genetic and an environmental component to this pruning process. Those connections that are reinforced by repeated use are maintained; those that are not used are lost in a kind of brain adaptation.

Maturation of the brain occurs from the back to the front. In the back are the regions that concern direct contact with the environment through sensory function. Just in front of those regions are areas that coordinate the senses and give a spatial sense. The last portion of the brain to mature in the late teens is the prefrontal cortex, the part that helps to organize thoughts, suppress impulses, plan, set priorities, and weigh the consequences of various actions.

Evidence from rats seems to indicate that new neurons are continually being formed from stem cells in the hippocampus region of the brain (Shor, 2009). This area is particularly involved with learning and memory. These nascent cells respond to learning, especially difficult learning, by forming new neural connections and maturing into fully functioning neurons. If not challenged by new learning, they degenerate and die. It is likely that cancer therapeutic agents, which target rapidly growing cells, can interfere with the formation and maturation of these new neurons and make learning difficult.

The ability to monitor brain activity during language learning in human babies also offers the opportunity to establish biomarkers for language-related disabilities associated with autism, dyslexia, fragile-X syndrome, and others (Kuhl, 2015).

Thus, expectations of the intellectual capabilities of individuals can be rooted in organic organization. Young children have the ability to learn quickly, but forget quickly. Adults are much more capable of acting independently. In between are those years when impulse actions are much more likely. Designing a human–machine interface would be different for each of these age groups.

Individual experience can influence myelin formation (Fields, 2008). Impulses carried by axons can regulate specific neuronal genes so that sticky protein is produced on the surface of the axon. This, then, enables myelin to form. Glial cells monitor pulses moving along an axon, and alter the degree of myelination. White matter in the CNS myelinated neurons is formed by learning as experienced, and the richer the learning, the more white matter is formed. Neglected children, people with mental illnesses, or those exposed to tobacco smoke during late fetal development, all show symptoms of myelination deficiencies.

It has been thought that the right hemisphere of the brain possesses spatial ability and the left hemisphere is involved in verbal skills; the right side is emotional and the left is analytic. Newer information modifies these ideas somewhat.

It seems that each hemisphere is specialized for a certain kind of data processing. The right hemisphere processes data to extract general features, whereas the left hemisphere extracts details (Robertson, 2005). Nearly all sensory inputs have both detail and general features. Somehow the brain pulls these apart to process them separately more efficiently than if all areas produced redundant results (Figure 6.22.7).

Animals and young humans several months old both have the ability to recognize and discriminate among small-valued numbers (Hauser, 2000). Developmental psychologists have constructed nonverbal tests to determine how much subjects understand about numbers. These tests use attention time as a metric of the surprise element of actions taken by researchers regarding the number of objects seen by the subject. For example, a monkey is shown a delectable eggplant, which is then hidden behind a screen, and an additional eggplant is added in full view of the monkey. When the screen is lifted, and

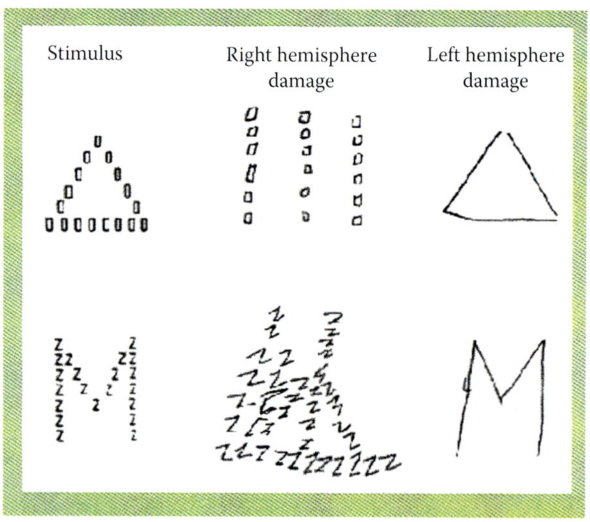

FIGURE 6.22.7
Drawings made by patients after stroke damage to one brain hemisphere or the other. The visual stimulus is on the left. In the middle is what is seen by the left hemisphere, and on the right is what is seen by the right hemisphere. (Redrawn from Robertson, L.C., *Phi Kappa Phi Forum,* 85, 19–22, 2005. With permission.)

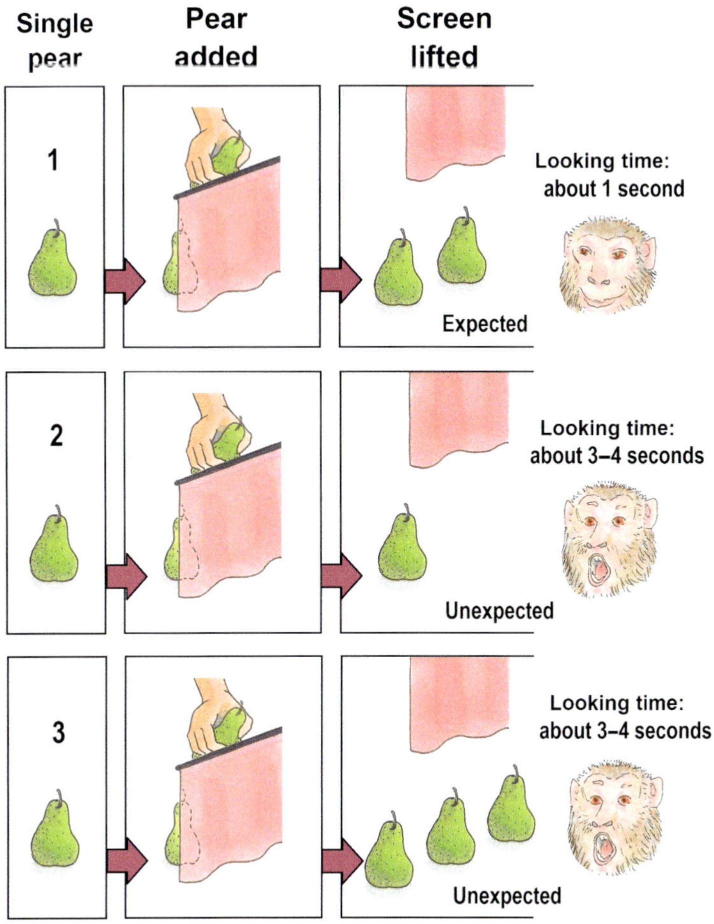

FIGURE 6.22.8
A favorite food item of monkeys is an eggplant, and it is sure to attract the monkey's attention. In an experiment, a screen is lowered in front of an eggplant seen by the monkey. A second eggplant is then placed behind the screen in full view of the monkey. When the screen is raised, the monkey registers surprise (by staring for a longer than normal time) whenever the number of eggplants revealed is different from the expected number of two, as in *b* and *c*. (Redrawn from Hauser, M.D., *Am. Sci.*, 88, 144–151, 2000.)

the monkey sees the expected two eggplants, the time spent by the monkey looking at the eggplants is about a second (Figure 6.22.8). However, if an additional eggplant is added or removed without the monkey's knowledge, then the monkey spends about 3 or 4 s looking at the eggplants after the screen is lifted. The monkey apparently does not expect the results that it sees.

Experimentation, learning, developing perspective, and tool use have all been observed in animals as primitive as fishes. The tuskfish produces water jets to uncover hidden clams and uses rocks to crack them open. The archerfish squirts water into the air to knock down insects and other prey; their behavior is quite flexible, able to track moving prey, adjust for the size of the prey organism, and learn to improve their accuracy by observing other archerfish. Atlantic cods kept in aquaculture tanks have learned by themselves how to use their affixed identification tags to open a feeder door to release food. Even creatures as primitive as bumblebees have been shown to be able to count, understand concepts, use simple tools, and learn in a social context (Loukola, 2017). All of these show that flexible problem-solving abilities are quite pervasive in the animal world (Balcombe, 2016a; 2016b).

ANIMAL INGENUITY

Maya is the name given by researchers to an animal matron of an isolated chimpanzee community in Africa. She has been observed to engage in the most sophisticated form of serial tool use by any nonhuman animal (Foer, 2010):

"Maya arrives at the termite mound, a rock-hard, bulbous structure three times her height, carrying in her mouth several plant stems that she will use to fish out its high-calorie occupants. First she rams a thick twig into a termite hole and widens it by jiggling the stick vigorously. Then she grabs a thin, flexible stem that she plucked off a nearby *Sarcophrynium* plant. Chimps in other parts of Africa are known to fish for termites with implements like this, but Maya goes one step further and modifies the tool. She drags the last six inches of the stem through her teeth to create a wet, frayed end, like a paint brush, and pulls it through her closed fist to straighten out the bristles. With the dexterity of a professional lock picker, she then threads the brush-tipped stem into the same hole, pulls it out, and nibbles off a couple bugs that cling to the wand's frayed edges.

"What's so remarkable about that fishing probe is that it represents a refinement. It's not just that some clever chimp figured out that it could break off a plant stem and use it to fish for termites—an impressive enough discovery in its own right—it's that some other chimp figures out a way to do it even better."

Another incident illustrating animal intelligence involves coyotes and sheep dogs. Shepherds located in regions with many sheep predators often keep sheep guard dogs with their flocks for protection. The dogs are trained to stay with the flock and ward off interlopers. Janet McNally (2010) is one of these shepherds, and described one of the incidents this way:

"I watched and listened as a coyote, just a stone's throw from a young … livestock guard dog, yapped at the dog, enticing her to chase. The [dog] would dash at the coyote, then retreat back toward the sheep. The coyote would come back and yap again, trying to lure the dog out into the woods. The coyote persisted for over 3 h, but the [dog] was only 8 months old. Typical of her breed, she stayed close to the flock and would not be enticed out toward the freeway. The week before, this coyote succeeded in drawing three older, bolder livestock guard dogs out onto the highway where they were promptly struck and killed by cars in three separate incidents over the course of two days. These were older dogs that were familiar with navigating traffic. Is a coyote smart enough to know the cars would kill the dogs?"

Similar results are found when testing young children with colorful dolls. Unexpected results elicit longer attention. What differentiates humans from monkeys, however, is the fact that monkeys can comprehend numbers up to about three or four (Figure 6.22.9), and simple addition/subtraction rules, whereas humans go on to develop a comprehension of larger numbers and more intricate manipulations. Humans, monkeys, and other birds and

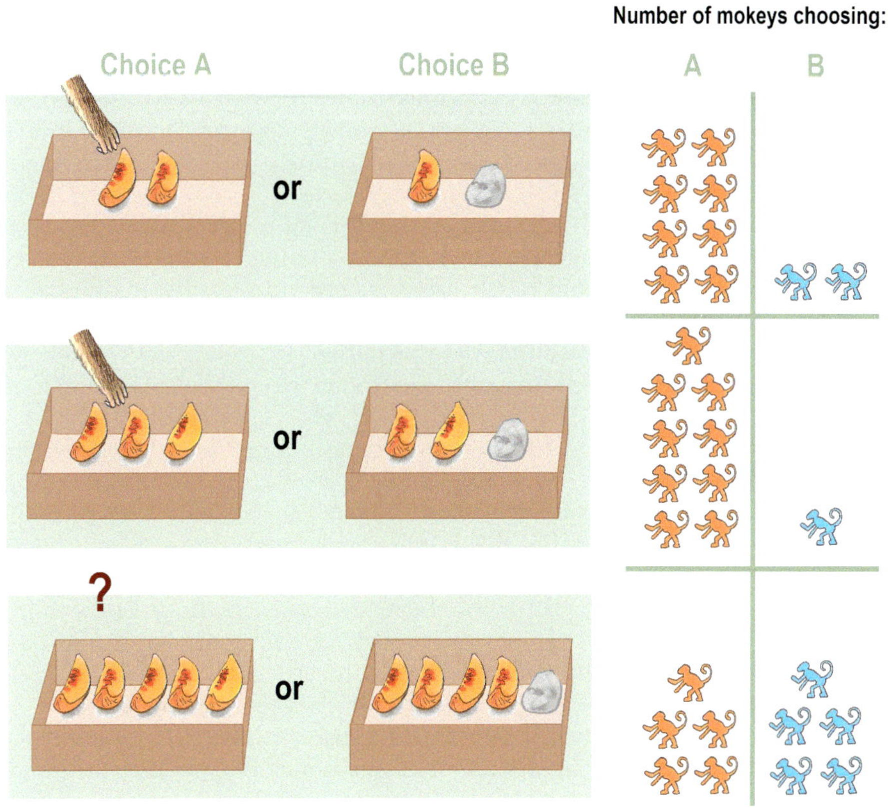

FIGURE 6.22.9
Rhesus monkeys presented with small numbers of apple slices and a rock are overwhelmingly able to choose the larger number of apple slices. When the number of apple slices exceeds 3, the monkeys are no longer able to choose the greater number of food items. (Redrawn from Hauser, M.D., *Am. Sci.*, 88, 144–151, 2000.)

mammals are apparently born with a similar innate (genetically determined) sense, but humans have the ability to develop it further.

Learning in children is influenced strongly by what they already know (Siegler, 1983). If they know about addition but do not completely comprehend temperature, they often predict the temperature of a container with water poured from two other containers each with water at 10°C as 20°C. They likewise tend to add "-ed" to inappropriate words to make past tenses.

In a manner similar to the scientific method (see Section 1.3, Scientific Method), children tend to apply rules based upon what they already know. They then test responses based upon these rules against the real world. Incorrect implications would require modification of their rules. Subsequent rules will presumably be closer to correct.

Therefore, there is a strong link between aptitude for learning and knowledge already possessed. Those with more insightful formulation of hypothetical rules would likely learn quicker than others, but their rates of development would still follow a certain progression. It is possible that animals, too, learn in a similar manner, at least in a limited way.

UNDERSTANDING ANIMALS

Temple Grandin provided a new way to understand the behavior of livestock animals. Born with autism, Grandin used her different perspective to develop insight into animal behavior. She has since become the leading expert on animal handling and production facilities.

Grandin's idea is that people with autism think using imagery and other sensory information instead of language, similar to the way animals think. Autistic people are sensitive to sound, sight, and touch. They are often constantly anxious because of their sensory amplification. Information in the autistic brain is processed more locally than in the normal brain, and more primitive structures, such as the amygdala and hippocampus, are used more and the frontal lobes are used less than in the normal brain.

This has enabled Grandin to identify animal handling problems that have eluded others. For example, cattle were refusing to enter a building despite prodding by their handlers. A visit to the site convinced her that the problem was not with the building but with the American flag that fluttered nearby. Cattle are acutely aware of potential predators, and the flag, with its rapid movement and scary sound, frightened them. She has also designed curved chutes that take advantage of the animals' tendency to move in a circle when they graze. There are no sharp bends that appear to be dead ends to the animals. As a consequence, cattle walk quickly and calmly without the shouting and prodding necessary with conventional facilities.

Her facilities have been judged to be humane as well as effective. By understanding natural tendencies of the animals, she has transformed the way they are handled, and she has illustrated well that it is much better to work with tendencies of biological beings than to fight against them.

6.22.6 Psychological Hierarchy

A thing of beauty is a joy forever.

—**John Keats**

Psychology is the study of the mind and any of its aspects, including the phenomena of consciousness and behavior. While the study of consciousness is not completely relevant to the technologist who wishes to manipulate biological systems, the study of behavior is relevant. We have already considered emotions, personality, and personal spaces. Each of these can influence an engineering or architectural design for structures to enclose humans or animals.

For centuries, humans have tried to interpret their own behaviors in order to understand their purposes and to predict what behaviors would follow (Harbaugh, 1972). There are several general schools of psychological theory that attempt to frame motivation and behavior. One that would seem to be most relevant is the Maslow Hierarchy of Basic Needs (Maslow, 1954):

1. Physiological. Food, drink, sex, requirements to sustain life.
2. Safety. Physical and psychological security, shelter.

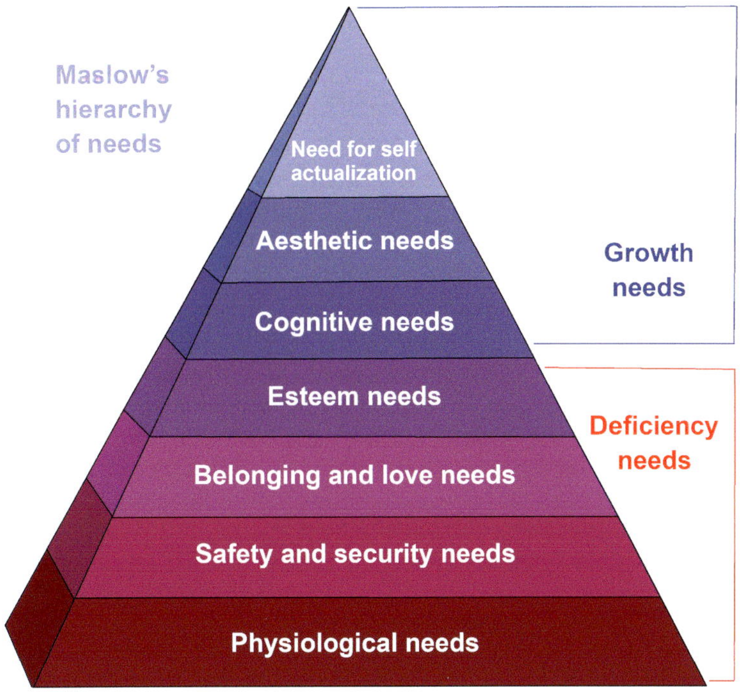

FIGURE 6.22.10
Maslow's hierarchy of human needs and motivation. (Redrawn from Smith, B.D., *Psychology: Science and Understanding*, McGraw-Hill, New York, 1998.)

3. Love. Belongingness, giving and receiving affection, social acceptance.
4. Esteem. Stature, recognition based on achievement.
5. Self-actualization. Development of one's fullest potential, helping others to do the same.

According to Maslow, the items lower on the list must be satisfied before striving for the upper ones (Figure 6.22.10). Those who are deprived of food, for instance, show markedly disturbed behavior. As starvation increases, people become more preoccupied with food and food-related objects; their intellectual functioning is impaired; personality deteriorates. Rehabilitation after starvation reverses all these effects (Harbaugh, 1972). While it is not likely that a biological engineer would design a facility specifically to starve people or animals, safety and security needs must be satisfied before the more social aspects of love, esteem, and self-actualization can be expected to occur. Designs of learning facilities, for instance, must be made recognizing that other more basic needs must also be satisfied.

Maslow's hierarchy appears also to have a developmental component. Lower level needs will be of less concern if they have been habitually satisfied for some time. Thus, the lower levels of physiological and safety and security needs of the average American adult were satisfied in childhood, and have little influence on the adult's motivation (Harbaugh, 1972). Love needs for Americans are often satisfied in adolescence, and esteem needs are thought to be satisfied sometime in early adult life. Self-actualization as a motivational factor may only be important when approaching later life.

It is not clear exactly how much of this is applicable to other animals, but, clearly, some aspects are applicable (Yoerg, 2001). When dealing with biological systems, aspects of consciousness and awareness should probably be assumed to be distributed in a graded continuum rather than as either present or not present.

Maslow's ideas can be considered in evolutionary terms. The most basic levels of Maslow's hierarchy are related to the most fundamental evolutionary need of survival. As the top of the hierarchy is approached, however, the survival needs become less important compared to reproduction. It is here that the center of attention turns from survival of the individual to continued existence of groups of individuals. Cooperation subsumes competition, first within the immediate family, then outward toward the extended family, the community, and eventually to the biosphere (Figure 6.22.11). Commonality extends outward to include more and more environmental surroundings.

Some of these traits can perhaps be seen in the actions of any social animal, from bees to sheep to wolves, but they are certainly more highly developed in human beings, where such global inclusiveness has led to a sense of purpose (Shermer, 2005). Purpose in one's life gives it meaning and transcends the basic biological imperatives of survival and reproduction. Shermer (2005) considers purpose to be personal, but purpose can certainly be shared among individuals. Shermer considers means toward higher goals leading to a sense of purpose:

1. Deep love and family commitment
2. Meaningful work and career
3. Social and political involvement
4. Transcendence and spirituality

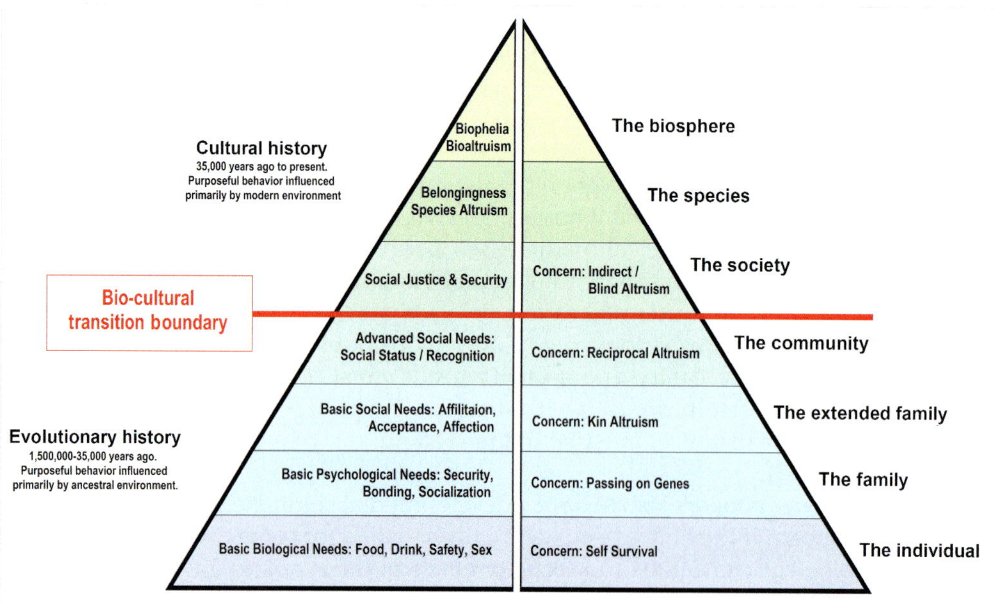

FIGURE 6.22.11
An extension of Maslow's concepts extends to groups and to purpose in life. (Redrawn from Shermer, M., *Am. Sci.*, 93, 101–103, 2005.)

To the biological engineer, involved with the creation of environments for social animals, and especially humans, care must be taken to enable these means toward purpose. In this way, the designs, applications, and operations touched by the engineer will result in the most successful and satisfying conclusion. And the engineer's creations will be most appreciated.

HUMAN FACTORS ENGINEERING

Until the human being is no longer needed in the workplace, in transportation, and in equipment operation, there will continually be an interface where humans and machines come in contact. In many instances, the human supplies the guidance, control, or oversight, and the machine supplements the physical capabilities of the operator, sometimes enabling the operator to perform the work of dozens of men or women.

It is in such situations that the limitations of human understandings and capabilities must be accommodated. These machines have been created to serve humans, instead of the other way around. To design engineers, however, the human is all too often not designed into the system. The result is that the combination of human operator and machine effector does not function as intended. The burden of correct operation is placed on the human operator without recognition that, with some accommodation, the human could perform better, more accurately, or without as much stress.

Expectations of human behavior in response to environmental conditions is called human factors when it relates to technology. Familiarity with human factors helps to design successful systems (Leibowitz, 1985). Human factors engineering is related to ergonomics (the science of physical movements in the workplace), cognitive and behavioral psychology (the study of the mind and how it processes information, learns, and reacts), and anthropometry (the study of sizes of humans and parts of humans). Human factors engineers use information from these related fields and use them in the designs of products and processes that relate to humans.

In many cases, the objective of human factors engineering is to improve human productivity. Also, there are the needs to reduce errors and to improve quality of products produced. Thus, human factors engineers apply their skills to manufacturing processes, to human-computer interfacing, to information transfer systems, to the designs of proper alarms and displays. They also design seating arrangements, thermal environments, functional layouts of controls, and physical accommodations for automobiles, airplanes, buses, and trains.

Human factors are also important in rehabilitation, where various aids for the disabled must be sized correctly, be versatile enough to accommodate special needs, be acceptable to the human user, and be socially acceptable to others. Without proper incorporation of human factors, health problems such as eyestrain, mental stress, and physical injury can result. Cumulative trauma disorder (work-related musculoskeletal disorder) comes about when working repetitively with tools under awkward conditions. Carpel tunnel syndrome is one form of this.

(*Continued*)

HUMAN FACTORS ENGINEERING (Continued)

As workers age, they are less likely to be able to deal with unnatural activities. As workers fatigue, they are more likely to make mistakes. When faced with too many competing stimuli, inaction or incorrect actions become more likely. Thus, human factors must be used to reduce the effects of these shortcomings.

Especially now, when technology has made us so interdependent, the wrong action at the wrong time can have consequences well beyond the immediate operator. Highway traffic accidents kill about 50,000 people per year in the U.S., and tie up traffic for inordinate amounts of time. A misinterpreted railroad signal can cause a derailment that spreads toxic fumes over the countryside. Operator error in Bhopal, India, in 1984 killed over 10,000 people in the surrounding village.

Often looked upon as "soft" engineering not requiring the same level of technical competence as other engineering disciplines, human factors engineering may be just the opposite. It may be more conceptual than mathematical, but that doesn't detract from its importance nor from the elements of design and application that it embraces (Pheasant, 1996; Corlett and Clark, 1995; Wickens et al., 1997).

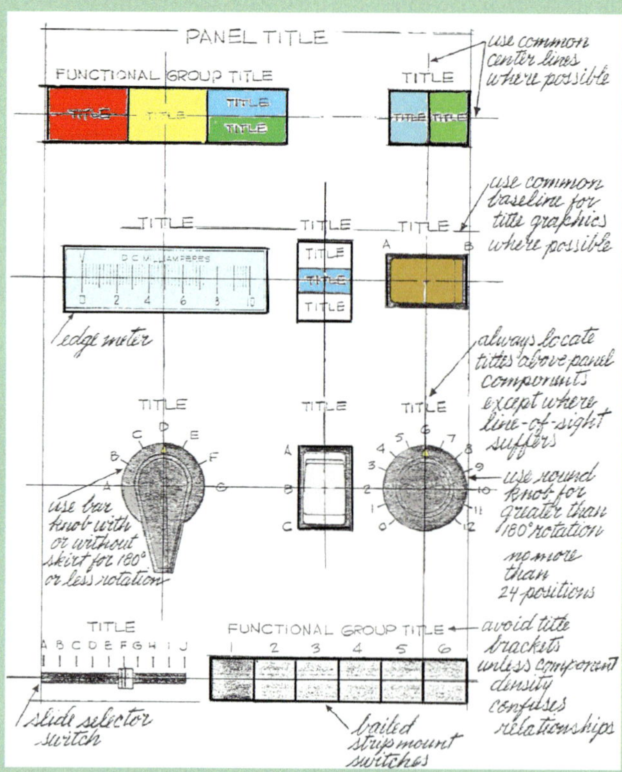

A control panel should be laid out in an orderly fashion, where indicator lights are clearly labeled, colors correspond to the severity of the alarm, and the most important information is presented at the top of the center of view. Attention should conform to the normal left-to-right reading pattern and the configuration should minimize confusion or misinterpretation (Wolfe and Odom, 1976).

6.22.7 Social Infrastructure

> Those who take pains to foster joy accomplish a work as profitable for humanity as those who build bridges, pierce tunnels, or cultivate the ground.
>
> —**Charles Wagner**

"The difference between societies where health expectations are high and those where they are low comes down to the degree to which societies, in their day-to-day living arrangements, honor or violate the basic social character of human existence. No genomic revolution, no matter how profound, will change that basic fact." This statement by Clyde Hertzman sums up the fact that it is socioeconomic infrastructure that plays the most important role in human health (Hertzman, 2001). It is the interconnections of our everyday lives that support us and nurture us that have given us the long lives we live today.

The social structure referred to encompasses many different levels. At the individual level, there is the emotional support given by family and friends. On a political level, there is the security and order provided by governmental structures. On an occupational level, there is the income and the motivation of meaning attendant in our jobs. On a societal level, there are the provisions of clean water, safe and steady food supply, and sanitation services to remove wastes. And, there is the harnessing of technology that continues to improve our lives. It is this overarching infrastructure that has historically allowed humankind to progress in all areas, including health care.

Engineers and technologists might be tempted to point to medical technology as the one major contributor to human health care, but Hertzman (2001) gives evidence to the contrary. For instance, although medicine was credited with defeating tuberculosis, more than 90% of the decline in tuberculosis mortality rates had taken place before a vaccine became available (Figure 6.22.12). Social and economic improvements were likely responsible for the major decline in death rates.

It is known that animals at a lower level on their social hierarchy (Figure 6.22.13) have more stressful and shorter lives (Nesse and Williams, 1994). The socioeconomic trend to egalitarianism in human society has probably produced many positive benefits for human health. There is Maslow's hierarchy for human societies as well as for individuals within society.

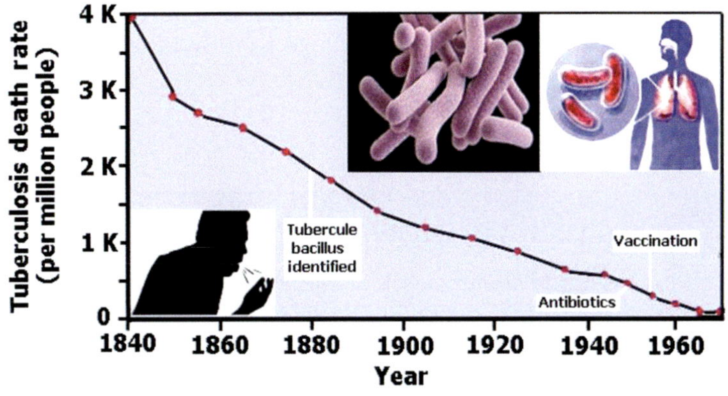

FIGURE 6.22.12
Social infrastructure is extremely important for human health. Here we see that tuberculosis declined markedly before the causative agent was identified, antibiotics were available, or vaccination was developed. (Redrawn from Hertzman, C., *Am. Sci.*, 89, 538–545, 2001.)

FIGURE 6.22.13
The social hierarchy in animals leads to stability and to allocation of resources for maximum reproduction of the most fit individuals, but not to the maximum benefit of all. (Redrawn from Hertzman, C., *Am. Sci.*, 89, 538–545, 2001.)

An interesting link between social behavior and innate tendency (probably genetically determined) was made when it was found that capuchin monkeys possess a strong sense of the difference between fairness and unfairness (Brosnan and DeWaal, 2003). When the monkeys were taught to work for food by giving a rock token to a researcher, they were given a piece of cucumber in exchange. The transaction was accepted satisfactorily by the monkeys until some unfair reward was given to one monkey. If one monkey was given a more desirable grape instead of a cucumber for the work done, or if one monkey was given a grape for no work, the other monkeys reacted strongly. Some slowed their rates of work, some would not take the cucumber, and some would throw the cucumber out of the testing area in a dramatic show of disgust. In dogs, too, unfair rewards lead to uncooperative animals (Range et al., 2009). A similar sense of justice may be present in other species that have a developed social infrastructure, especially when it depends on cooperation. This implies that a sense of fairness may somehow be an artifact of evolution, and that there has been a survival advantage to individuals who are fair in their dealings with others.

MOTHER BEAR MAN

Ben Kilham knows firsthand about the social interactions of black bears and their cubs (Caputo, 2002). He acts as a surrogate mother to cubs orphaned when their mother either dies or abandons them because of an imminent threat to her life. The bear cubs he raises are sensitive, intelligent, and emotional creatures that need more than just food; they also need security, affection, and someone to teach them. They learn about survival and they learn how to interact with each other and the world around them.

(*Continued*)

MOTHER BEAR MAN (Continued)

Kilham drops to all four limbs and chews on an Indian cucumber plant. A cub that he is raising comes over to sniff his mouth to learn that this is the smell of a food plant. Then the cub locates a plant with similar smell and begins eating it. This is the way bear cubs learn from their mothers which of the many choices of food are acceptable and safe. When they first encounter a plant, they hold the plant in their mouths before they eat or discard it. This is probably a survival behavior to check the taste of the plant to determine if it is edible. Bitter plants are usually toxic.

Ben Kilham teaches a young cub that a certain plant is safe to eat by chewing on the plant and allowing the cub to catch the scent of the plant. (Redrawn from Caputo, R., *Natl. Geogr.*, 201(3), 89–101, 2002.)

Bears have special trees that they mark to let other bears know that they have been there. Trees with porous bark, like the red pine, can hold a bear's scent for a long time. If the bears wish to broadcast that they are in the neighborhood, then they bite or rub against one of these trees. Bears have a highly developed sense of smell, and this is one way in which they mark their territories.

Orphaned bear cubs need more than just food.

6.22.8 Mind–Body Interactions

> ...The common cold, if left to itself, ran for a fortnight, but if medically treated, lasted only fourteen days.
>
> **—James Crichton-Browne**

The mind and the physical body are very much interrelated. Psychologically stressful situations have been shown to lead to physical ailments, and physical illness can greatly affect psychological health. It is known that rest and meditation lead to healthier lives; Yoga and other mind-body meditative exercises quiet genes that promote inflation (Buric et al., 2017). People who believe they are healthy usually are healthy, and people who want to recover from an illness or accident usually do better than people who are discouraged.

We must be careful not to attribute cause and effect to the earlier statements, because, although there is a connection, they may only mean that it's only healthy people who think of themselves that way (see Example 1.3.2). Nonetheless, engineering solutions involving people or higher animals need to include the attitudes and emotional needs of their subjects as well as their physical needs.

For instance, older people who suffer from loneliness can be helped by having pets. Engineering and architectural designs could, if possible, include provision for keeping pets. Also, medical devices cannot appear to be threatening or pain-inducing. Packaging can sometimes help here.

Cardiovascular disease in animals has been found to be related to physical, emotional, and behavioral stresses (Nerem et al., 1980). Rabbits that were held, petted, and coddled were found to have a 60% reduction in atherosclerotic lesions compared to their control counterparts, despite comparable serum cholesterol, heart rate, and blood pressure levels. Just the presence of human observers has been shown to change wild animal behavior, allowing them to be less vigilant for the presence of natural predators (Nowak et al., 2014). These results demonstrate that psychosocial environments of animals can have profound effects on health, and that these environments (animal treatments) must be controlled during scientific studies in order to avoid drawing erroneous conclusions.

There has been found a link between a viral infection in the mother during pregnancy and later schizophrenia in her child (Brown et al., 2004). Mothers with high cytokine levels circulating in the blood while pregnant (see Immune System box, Section 6.20.3), indicating the presence of an infectious agent, had a three-fold higher risk of those children developing schizophrenia later in life than did mothers without the marker. A number of viruses, including flu, could be the causative agent. Increased cytokine activity during pregnancy has also been associated with cerebral palsy and autism (see Example 5.3.1 and Fetal Influences box, Section 6.6.5). This is a concrete demonstration of body–mind (as compared to mind–body) interaction.

Humans need to feel in control of their responses and, if possible, of the environmental conditions causing those responses. The feeling of control, however, does not always mean that real control is being exercised. There is evidence that unconscious preparation for physical movement actually precedes the conscious decision to make that movement (Obhi and Haggard, 2004). Thus, it appears as if the movement is somehow preplanned in the brain before the person decides that the action is to happen. Conscious control of the movement then appears to determine not what movement is to take place, but only whether or not the preplanned event will actually happen. Free will, it seems, is actually free won't.

Microbes in the gut may profoundly affect mood and psychological well-being. Gut microbes make neuroactive compounds, including neurotransmitters and metabolites, which also act on the brain (Schmidt, 2015). The gut microbiome thus can help to control mood disorders, depression, stress, anxiety, and autism, among others (Benach et al., 2012; Jones, 2012; Mulle et al., 2013; Reardon, 2013).

Although research along this line has thus far been limited to tests with mice, the evidence is accumulating that personality traits are at least partially determined by the gut microbiome. Anxiety, depression, and even autism may be able to be treated with certain types of microbes, called *psychobiotics* (Schmidt, 2015).

Likewise, results that are coupled in time to stimuli can be considered to be related only if the stimulus appears 1–5 s prior to the response. If the stimulus appears much before this time, or after the response is made, then the two are not perceived to be linked (Obhi and Haggard, 2004).

Consider an instrument that requires an adjustment. Turning a knob changes a reading on the face of the instrument. If there is too much delay between adjustment of the knob (the stimulus) and changes seen in the reading of the meter (the response), then the operator of the instrument will not perceive adjustment of the knob as controlling the meter reading, leading to unsatisfactory performance.

Similarly, *biofeedback* can be used to modify autonomous actions of the body (see Section 4.4.4, Feedback Control). To make biofeedback work, an instrument must make the autonomous action (heart rate, for example) apparent to the subject. The subject then learns how to consciously control the otherwise autonomous action by watching the instrument reading while randomly trying various mental strategies. The key to this procedure is to make the feedback to the subject rapid enough so that the results of different trial strategies can be assessed quickly. The feeling of control must extend to the autonomous action. In this way, heart rate, muscle tension, skin temperature, and a host of other bodily functions can be brought under control.

There is a *blood–brain barrier* that protects the central nervous system (CNS) by excluding large molecules from contacting CNS neurons. It has been reported (Brodie, 2006) that infection or stress can allow certain large molecules access to specific parts of the brain. Bacterial endotoxin or adrenaline have both been shown to breach the blood–brain barrier and permit access to brain nerve cells by autoimmune antibodies. These antibodies normally cause such autoimmune diseases such as lupus, rheumatoid arthritis, or multiple sclerosis (a disease of the peripheral nervous system), but have no effect on brain function. Infections or stress can change that and result in alterations to memory or emotions. This underscores the connections between the body and the brain.

The link between psyche (mind) and soma (body) was confirmed by a commercially available device to control depression. This device consists of an implantable vagus nerve stimulator (Figure 6.22.14), that, when activated, reduced or eliminated depression episodes (Moore, 2005). The same device had been developed to control epileptic seizures, and joins multiple other electrical neural stimulators to control Parkinson's disease, pain, anxiety, chronic headache, bulimia, and others.

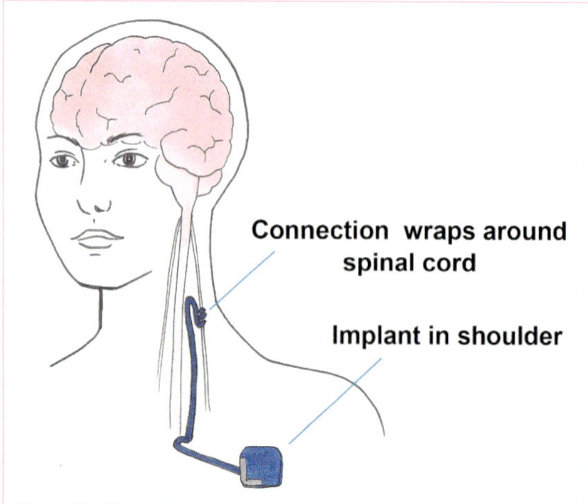

FIGURE 6.22.14
An implanted stimulator presents a train of electrical pulses to the vagus nerve in the chest region. These pulses are then transmitted via the nerve to the brain where they indirectly ameliorate depression. (Redrawn from Moore, S.K., *IEEE Spectr.*, 42, 16–17, 2005.)

ATHLETES AND ILLNESS

Being fit may help to ward off illness, but not when severe exertion or emotion is called for. Whereas most people catch about two upper-respiratory tract infections per year, athletes get twice as many. A contributing factor is probably that breathing through the mouth rather than through the nose increases the likelihood of catching germs (Poultney, 2003). Prolonged strenuous exercise can induce increased levels of stress hormones that inhibit some immune system functions. A study of a major English soccer team showed higher levels of the protective immunoglobulin A (IgA, see Section 6.20, Maintaining Stability) when the team played well. A losing streak was accompanied by a dip in Ig A concentrations. In addition, the numbers of germ-killing T cells decreased as the season wore on.

Nature is good for health. Spending time out in the natural environment has been shown to lower blood pressure (Shanahan et al., 2016; Kawakami et al., 2004), increase the number of cancer-fighting natural killer cells (James et al., 2016; Li et al., 2008; Li et al., 2009), help with attention deficits and cognition (Taylor and Kuo, 2009; 2011; Berman et al., 2008; Piff et al., 2015; Bratman et al., 2015), improve anxiety, depression, and other mood disorders (Dadvand et al., 2015), hasten recovery from surgery (Ulrich, 1984), and promote general human health (Kuo, 2015; Pearson and Craig, 2014). Natural environments are such a major influence on human health that supplying or reproducing green surroundings can become a powerful and inexpensive public health intervention (Larson et al., 2016).

The point of all this is that the biological engineer who attempts to provide optimal conditions to achieve some purpose involving biological systems, be it human health care, nurturing of exotic species, restoration of an ecological biosystem, or at some other trophic level, must consider *all* environmental factors (Figure 6.22.15). The interconnections are many and complex, but together they all form a matrix that can achieve the very best

FIGURE 6.22.15
An environment conducive to human activities must not offend individual sensitivities. (Courtesy of Mother Goose and Grimm, Copyright 2001 Grimmy, Inc., Distributed by King Features Syndicate, New York.)

of expectations (Capra, 1996). So, why must engineers know about psychology, sociology, economics, art, political science, literature, and music? The answer is that each of these is part of the overall structure of the biological system that we are a part of.

PLACEBO EFFECT AND THE THEATER OF MEDICINE

Nothing illustrates the interaction between mind and body better than the placebo effect. A placebo is a dummy medicine used to give the appearance of a real medicine in a drug trial. Placebos frequently contain simple sugar or other substance without known pharmaceutical effect. Giving some of the trial subjects the real medicine and some the placebo is supposed to keep the subjects from knowing which treatment they have received, and the results of the trial should then reflect only the difference between the active drug and no drug.

The problem is that 40%–65% of those people receiving placebos also report improvement of their physical or mental conditions. Just the anticipation of relief, and the care by others, is enough to alleviate suffering. There is evidence that the placebo effect is centered in the brain and changes the activity levels of areas associated with pleasant and unpleasant sensations. Brain imaging has shown that metabolism of these different areas changes when cure is expected. History is full of medicines and medical devices that were claimed to effect cures, but only did so because of the placebo effect. Giving people hope is one of the most powerful medicines available.

Whether it is the placebo effect, conditioned response, or other brain-body interaction, the expectation of healing is very powerful (Vance, 2016). Pain can be eliminated or at least ameliorated, immune responses can be enhanced, and illnesses can be cured when the patient expects those outcomes, even when no other treatment is given. An important element in raising these expectations is the provision of an environment with clues associated with healing. An example is the look and feel of a hospital that can provide comfort for those in distress.

(Continued)

PLACEBO EFFECT AND THE THEATER OF MEDICINE (Continued)

The biomedical engineer who understands this will provide patients with more than just an improved instrument or a new therapy. The trappings of a curative environment should also be provided. Instruments should have a "medical" appearance. Therapies should use methods reminiscent of clinics. Procedures should be performed as the patient expects them to be performed in a medical setting.

It is not enough to provide the best engineering solution for a new treatment or measurement; the expectations of the patient must also be satisfied for the most successful outcome.

Example 6.22.1: Harvesting Broiler Chickens

Broiler chickens are raised in buildings containing 100,000 birds. When they have grown to the proper size in 5–6 weeks, they are caught and placed into wooden crates by crews of workers. From there they will be transported to a processing plant. Suggest a way that the workers can use to catch the birds with minimum disturbance.

Solution

It has been found that catching the chickens at night when they are normally asleep poses the least disturbance when they are caught using conventional techniques. Even so, a lot of feathers still fly.

It could be suggested to condition the birds to congregate in a catching pen when a bell is rung or a whistle is blown. Letting the birds catch themselves would probably pose the least disturbance.

Remark

There have been attempts to build catching machines to harvest the birds, but these did not take advantage of the birds as intelligent living beings. The machines failed to perform as desired, of course.

Example 6.22.2: Wild Animal Display

In a free-range animal display, where African animals are on exhibit for sightseers, animals that are natural enemies are to be displayed closer together than they would be in the wild. Suggest means to keep the animals separated. Prioritize your list, beginning with the most desirable way.

Solution

First, we'll assume that there are elements of Maslow's hierarchy that apply to animals. If that is the case, then supplying attractive food to the meat eaters would attenuate their tendencies to hunt the grazers. Over the course of time, these animals will become less fit to hunt, and the young will not be taught hunting techniques.

Prey animals must be constrained from running, or they would attract predators who would run after them for the sport. The preferred way would be through satiation of prey hunger. Especially attractive food could be given at specific locations several times during the day.

Controlling reproduction is important. Predators with young are more likely to hunt and fight to protect their young than are those without young. Prey cannot overpopulate an area, which leads to overgrazing and searching for more food. Thus, contraception is very necessary, especially because animals in prime condition tend to bear more young.

There must be physical barriers, else one unfortunate incident may counteract all work done to that point. Moats, fences, and barrier strips devoid of tasty vegetation would serve this purpose. Naturally, these must be camouflaged as much as possible.

Tranquilization or surgery could be used on both predators and prey, but this is not very desirable and could be expensive. Electrical constraints, such as radio frequency collars and buried wires could be used, but require that animals be trained to their use, and they will not work in the event of electrical failure.

Remark

Knowledge of natural tendencies can lead to better designs than can brute force impositions.

Example 6.22.3: Toys for Captive Animals

Living free is a stimulating experience, with newness nearly always present. Animals in captivity need stimulation, too, and animal enrichment has been found to keep animals healthy (O'Brien, 2002).

Boredom is why horses chew on wooden doors to their stalls and caged tigers pace from side to side. Bored animals have more illness and are less likely to breed than their stimulated counterparts.

Constructing playthings has become an activity in the more progressive and confined animal facilities. Toys provide an opportunity for animals to adapt to their environments and be creative.

Safety is a major factor in the design and construction of toys. They cannot have sharp edges, and they must be designed with each animal's behavior in mind. The marmoset's rope ladder cannot contain artificial dyes, because the monkeys could chew on the rope and get sick. Marmosets (the world's smallest monkeys, about the size of chipmunks) seem to gnaw on just about everything. Toys built for dolphins must be made of durable material that can endure bites from mouths with 100 teeth each. Atlantic sea turtles have very powerful jaws that can easily break many common materials.

Introducing the toys must be carefully executed, also. They must not upset an animal, which can happen if the toys are large and menacing. They must sometimes be introduced slowly in order for the animals to try to explore them without feeling threatened. And, toys are sometimes removed after a few hours so that the animals won't tire of them too quickly.

Example 6.22.4: An Ergonomic Solution

Once a year in a large vehicle repair shop, mechanics had to prepare used delivery trucks for resale. Part of the process involved scraping off old decals that covered the trucks using a small razor blade tool. It was a slow and tedious process removing about an inch at a time. One or two days were typically required to remove all the decals from a single truck.

One mechanic was assigned to clean about 20 trucks, and this would usually involve a month's worth of scraping. On the second day his arm started aching. Was there anything that could be done to mitigate his pain?

As it turns out, there was. They brainstormed some ideas and finally tried using the nearby power wash to heat the body of the vehicle. When they did, the decal peeled right off. Preheating with the power wash reduced the time to remove the decals to 1 or 2 h per truck and saved three weeks of vehicle preparation time. At least as important, it eliminated what might have been a serious hand and arm injury to the employee. Working smart often beats working hard.

Example 6.22.5: Human Factors Shortcomings at Three Mile Island

The partial meltdown of the reactor core at Three Mile Island nuclear power plant was mainly due to human error (Heppenheimer, 2002). On March 28, 1979, maintenance workers were cleaning sludge from a small pipe when they inadvertently blocked the flow of cooling water. Heat in the core flashed some water into steam and the resulting pressure surge popped a relief valve. Emergency pumps started up to restore water flow, but two valves to the reactor core had been left closed. The water poured, instead, out of the relief valve. Unaware of the open valve, and having a faulty indication that the core was full of water, technicians shut down other emergency pumps. The core ran out of water and began to melt. Workers were unable to obtain correct instrument readings. Finally, cooling water was restored to the core, but the overly hot zirconium tubes that held the nuclear fuel had begun to oxidize and release hydrogen. If oxygen had mixed with hydrogen, the whole reactor could have detonated.

The review in the aftermath of the emergency found that control room panels contained hundreds of indicator lights with no standardized color code. Red, green, amber, and white lights all had no correspondence between color and seriousness. Important gauges were hidden behind consoles, and the most significant displays were scattered rather than grouped where they could all be read at once. The indicator light for the fateful relief valve showed merely that the valve had been commanded to close, not that it actually was closed. There was no means to double-check the valve status. A better designed control panel might have allowed the worst nuclear accident in the United States to be avoided.

Example 6.22.6: Dominance is Reinforced by Body Position

There is a psychological, physiological, and behavioral positive feedback system working in the minds and bodies of human beings that reinforces correlations among these three variables such that they cannot be easily separated. Just the act of smiling has been found to influence disposition toward a more positive outlook; nodding the head "yes" inclines a person to be more open to suggestion.

It has been found that the simple act of posing in a powerful position (exhibiting dominance or expanding occupied space) results in higher levels of the aggression hormone testosterone, lower levels of the stress hormone cortisol, and greater assumption of risk than posing in less powerful positions (closed, receptive, and subordinate). As shown in Figure 6.22.16, these two types of positions by themselves had profound physiological and physical consequences.

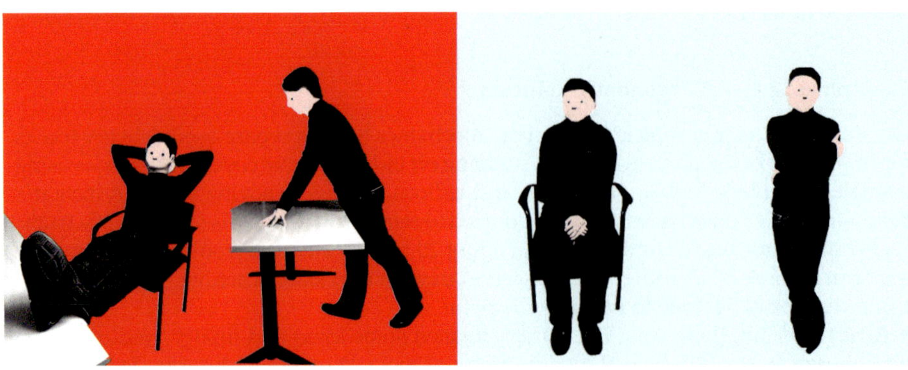

FIGURE 6.22.16
On the left are two poses found to demonstrate power and enhance testosterone and reduce cortisol. On the right are two poses that demonstrate conciliation, withdrawal, or subservience. Testosterone for the right-hand poses was lower and cortisol higher. (Redrawn from Carney, D.R. et al., *Psychol. Sci.*, 21, 1363–1368, 2010.)

It had been known that higher levels of testosterone correlate with human dominance, and that lower levels of cortisol are related to lower stress levels. This research showed that these hormonal adjustments can be induced solely by the act of demonstrating power.

The relationship between social dominance and poses extends to other animals: peacocks spread their tail feathers and strut to attract mates; cats advance sideways to give the appearance of size to ward off intruders; chimpanzees inflate and thrust their chests forward to intimidate others. It is likely that these poses were enhanced by testosterone, and it is likely that testosterone was enhanced by the assumption of these poses.

Example 6.22.7: Sensory Overload in the Driver's Seat

Modern automobiles are designed and built for comfort by insulating the driver from noise, road bumps, and ambient conditions. The driver is isolated from the surrounding environment so well that a myriad of warning systems have had to be installed to indicate when there is trouble. There are dashboard lights for doors ajar, gas filler access flaps open, parking brake engaged, low fuel, engine trouble, air-bag malfunction, anti-lock brake condition, and many others.

When engineers decided to improve safety by warning drivers of nearby vehicles traveling in the blind spot of the driver, that place just to the left and slightly behind the driver (at least for autos driven on the right-hand side of the road), they faced a decision about the best means to warn the driver (Corley, 2010). The visual sense was already overloaded while driving.

Engineers, instead, installed vibrators in the driver's seat. Surrounding vehicles activated vibrators in seat locations corresponding to positions of nearby cars; vehicles to the left activated left vibrators and vehicles to the right activated right-side vibrators. The closer the approaching vehicle, the more intense was the vibration. Engineers hope that using tactile feedback for drivers will help them take more appropriate actions in a timely manner.

Example 6.22.8: Healing Gardens

Gardens on hospital grounds have been shown to facilitate patient healing and recovery (Franklin, 2012). These areas should allow young children to have places for moderate activity, have interesting, contemplative areas for middle-aged adults, and have stimulatory locations for older adults. Research has led to the following checklist for hospital gardens:

1. **Keep It Green.** Lush, layered landscapes with shade trees, flowers, and shrubs at various heights should take up roughly 70% of the space; concrete walkways and plazas about 30%.
2. **Keep It Real.** Abstract sculptures do not soothe people who are sick or worried.
3. **Keep It Interesting.** Mature trees that draw birds and chairs that can be moved to facilitate private conversation foster greater interaction.
4. **Engage Multiple Senses.** Gardens that can be seen, touched, smelled, and listened to soothe best. But avoid strongly fragrant flowers or other odors for patients undergoing chemotherapy.
5. **Mind the Walkways.** Wide, meandering paths that are tinted to reduce glare allow patients with low eyesight, wheelchairs, or walkers to get close to nature. Paving seams must be narrower than one eighth of an inch to prevent trips by patients trailing wheeled IV poles.

6. **Water with Care.** Fountains that sound like dripping faucets, buzzing helicopters, or urinals do not relax anyone, and neither does the strong smell of algae.
7. **Make Entry Easy.** Gardens should not be far away or behind doors that are too heavy for frail or elderly people to open.

Example 6.22.9: Helping Autistics Achieve a Quality Life

Autism of many different degrees of severity affects one in 68 children in the U.S., with four times the incidence in boys compared to girls. Autism appears to be related to new mutations in DNA that affect neurological development and function (Hall, 2015). Autism disorders cover a spectrum of atypical behaviors ranging from somewhat affected to severe intellectual disability.

For those with mild cases, strengths of autistic thinking include:

1. Bottom-up thinking that renders unrelated details unimportant.
2. Associative thinking that places together all similarities.
3. Creative thinking that puts details together differently from others.

Therefore, autism can be considered, not a disease, but rather a neurological difference.

Biological engineers can help provide facilities to accommodate the special needs of autistic people (Robison, 2015):

1. Help them organize their lives, manage their schedules, and deal with sensory overload by supplying quiet spaces and calming lighting.
2. Provide engineering solutions to autistic deficiencies, such as formulating verbal communications for those who cannot speak, and designing machines that can read expressions of those around them for those who cannot interpret body language.

Example 6.22.10: Pecking Hens Attacking

The term "pecking order" originated from observations of chickens raised domestically. More dominant chickens assert themselves by pecking those at lower social levels. As long as the victim of pecking can escape, all this usually does is to establish a social order among the birds.

Chickens raised in confinement can carry the pecking order to an extreme. If the victim of pecking cannot escape, then other chickens sometimes join in the original fray, and mob the unfortunate bird. It starts with pecking of feathers, but may proceed farther. Especially if blood is drawn, other chickens may continue pecking the victim in a frenzy, until the victim is killed. Once a flock of chickens establishes this type of behavior, it will continue with more birds killed.

Farmers have resorted to clipping beaks, dimming lights, or changing foraging material, but these measures are either costly, troublesome for animal rights groups, or ineffective.

Pattern recognition software has been applied with video technology to identify which hens are more likely to cause this trouble (Laursen, 2010). Movement patterns at an early age foretell behavior later on. Recording the movements of flocks of hens over time, and using appropriate pixel-to-pixel algorithms can identify stressed and violent flocks before pecking begins. There are predictable early behavioral differences between those hens that are prone to peck and those that are not, and the video system allows this to be identified without human monitors engaged in a boring job.

APPLICATIONS AND PREDICTIONS

1. Almost all actions will evoke an emotional response. Engineering designs can evoke an emotional response also.
2. The engineered system that evokes a negative emotional response will be unsuccessful.
3. Basic needs of animals and humans will have to be satisfied before higher-level needs. School breakfast and lunch programs have been established to meet this need.
4. Medical instruments that are too complicated to operate will not be accepted.
5. Animal habitats must account for social order.
6. The placebo effect can help cure illness or pain.
7. Human emotions may need to be programmed into a computer before truly lifelike robots can be made.
8. Environmental stress can influence sensitivity to environmental toxins.
9. Contact with another similar BU can enhance emotional stability.
10. A comfortable and emotionally satisfying work environment will encourage the motivation and productivity of employees.

6.23 Biological Units Die

> This obsession with fighting germs has gotten ridiculous. It's gotten so bad that before they give a lethal injection they swab the spot with alcohol.
>
> —**George Carlin**

6.23.1 What Does "Dead" Mean?

> Do not go gentle into that good night.
> Rage, rage against the dying of the light.
>
> —**Dylan Thomas**

Just as it is very difficult to define life and consciousness, so it is with death. In a conversation I had years ago with the physicist Otto Schmitt, he repeatedly referred to a "life force" to distinguish between the living and the dead. It may be clear someday exactly what physical or chemical processes are present in living systems that are not present in dead systems, but to distinguish between the living and the dead in this way is an exercise in complexity (Figure 6.23.1).

All living systems must metabolize to obtain chemical energy in a form suitable to repair damage and maintain control over their inner workings. Any condition that interferes with these biochemical processes, such as oxygen deprivation, hypothermia, hyperthermia, or

FIGURE 6.23.1
Death comes to Mr. Potatohead. (Courtesy of Mother Goose and Grimm, Copyright 2001 Grimmy, Inc., Distributed by King Features Syndicate, New York.)

toxins, can cause death. So, we can usually know what conditions can lead to death even if we don't know the precise mechanisms involved (Kleiber, 1975).

For more complex animals, death may be defined as the lack of a heart beat because blood no longer brings oxygen and glucose to the tissues and waste products from the tissues. However, death of the internal tissues does not occur all at once, and it may take weeks or even months for fingernails and hair to stop growing. The fact that death of parts of the organism is delayed after organismal death allows for harvest of organs and tissues for transplant, but does cloud the meaning of what we mean by "dead."

In humans, especially, death has been defined as no cortical activity. The brain has no oxygen or energy reserves, and it ceases to function adequately a few seconds after blood ceases flowing (Simpson et al., 1957). A nervous system is necessary for proper performance of all advanced animals, and when it no longer functions, death is said to occur.

And then there are the psychosocial definitions of death, meant to satisfy complex legal requirements of human society. *Clinical death* occurs when there is no heartbeat and no spontaneous breathing (Smith, 1998). *Brain death* is characterized by no electroencephalogram (EEG) activity, a complete lack of reflexes, total failure to respond to stimuli and no breathing without a respirator. *Social death* involves terminally ill patients who are hospitalized for long periods before dying and require elaborate life-support equipment to maintain any semblance of life.

Uprooting a plant shows how tenacious is life. The plant will begin drying almost immediately, the leaves wilt, and the nonwoody stems become flaccid. Yet, even several days later for some plants, plunging them back into water can revive them. Clearly, they were not completely dead.

Dead microbes appear under a microscope to be identical to live microbes. The same structures are present, but there is no movement, no cytoplasmic streaming, and no metabolism. Microbes that form spores appear even more dead than alive, yet they can grow and reproduce should suitably favorable environmental conditions be encountered. What it means to be dead here is not easy to explain.

Food sterilization depends on eliminating all live microbes present in the food. Because this is statistically impossible, the number of marker microbes in food is regulated by the U.S. government to be reduced by a factor of 10^{12} from the original numbers. If the original numbers were small, the 10^{12} reduction means that (practically) there are no live microbes in the food. Statistically, however, there is still a very small chance of a viable microbe to be present.

How can we be sure that most of the microbes in food are killed? Experiments are performed on representative microbes to determine conditions that will kill them (Burns, 2007). If no microbes grow in media after treatment, then they are defined as dead. To achieve the 10^{12} reduction, sterilization treatments are made more severe than the experimental environmental conditions, and the experimental results are extrapolated. Death, in this case, is determined by the inability of microbes to grow and reproduce.

Sometimes, it is desired to put the microbes in a state of suspended animation from which they can recover. Common baker's yeast is commercially available in dry form, and will be reactivated when mixed with warm water. Cattle sperm are fast frozen at cryogenic temperatures and thawed for artificial breeding. Human tissues such as corneas are frozen as well, in order to be preserved for later use. There is interest in preserving human stem cells by drying. If successful, the stem cells could be used therapeutically to cure wounds and diseases.

These latter examples illustrate where death is not desired, and specific steps must be required to maintain the ability to return to life after preservation. Special media are often required to preserve cells to be frozen. Drying cells requires special metabolic steps in certain orders in order not to kill the cells. The study of ways in which extremophiles survive harsh environments can help here.

In the preceding examples, death is defined as lack of viability; the inability to grow and reproduce. This is a relatively simple distinction between life and death. However, it does not exactly define whether dried yeast cells are living or dead.

Despite the difficulties that we may have in determining exactly what death is, it is inevitable for complex BU. Prokaryotic bacteria that reproduce by binary fission may be said to be able to live forever, and so can genetic material, but all other cells and tissues inevitably die (Jurassic Park, notwithstanding [Crichton, 1990]). And once death occurs, it cannot be reversed. This is a simple statement, one that may be readily apparent, but one that is profound nevertheless.

In the overall scheme of things, death is not bad. Death is the means by which resources are able to be recycled; death is necessary to supply food for survivors; and death is the means to weed out the weak and unfit so that only the fittest survive. Death, therefore, is an individual price to pay for continuation and improvement of life in general. It is as necessary to life as is life itself.

CRIME SCENE INVESTIGATION: USING BIOLOGY TO SOLVE A MYSTERY

The murder can't be solved unless investigators know the exact time of death. Here's where knowledge of human and insect biology can help to determine the time a human has been dead:

1. Livor mortis. If the body exhibits red, purple, or blue skin discoloration, the person has been dead less than a half-hour.
2. Rigor mortis. The body begins to stiffen within 2 hours of death due to hardening of the muscular tissues. The body becomes completely stiff within 6–12 hours, and then relaxes a bit, depending on ambient temperature, with 26–48 hours. It disappears completely in 5–6 days when decomposition begins.

(*Continued*)

CRIME SCENE INVESTIGATION (Continued)

3. Stomach contents. Food in the stomach can indicate the last meal, and from there an estimate of time of death can be made.
4. Brain hardening. Brain tissue begins to harden after death at a predictable rate.
5. Insects. There is a specific pattern in the decomposition of the body (Figure). One insect feeds on the body shortly after death, then another comes once the body decomposes a bit, and then another somewhat later. The common blowfly is almost always one of the first insects to arrive. Depending on temperature, humidity, and sunlight, the stage one maggot is reached within 30 hours. Stage two and three larvae are reached within 52–85 hours, respectively. The pupa stage is reached in 279 hours, and the adult emerges within 500 hours. Even when gone, the exoskeletons they leave behind can reveal the season of death.

And how old was the victim? Counting osteons, the concentric layers of bony tissue in the skeletal remains, provides the person's age at the time of death.

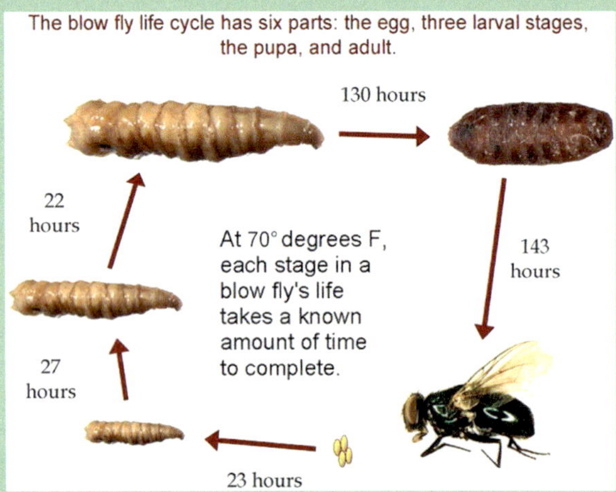

Stages in the life of a fly.

6.23.2 Reliability Theory and Death Rates

> My dad is 102 and he says it's better to wear out than rust out.
>
> —**Chuck Juran**

The pattern of death rate of humans and many other higher order plants and animals is similar to the failure rate of manufactured products (Figure 6.23.2). There is the initial *burn-in period*, due to the use of defective components during construction, when the failure rate is relatively high. With time, the number of failures decreases as products with defective components are repaired or are removed from the product pool. After the burn-in period, there is a more or less lengthy normal *working period* wherein product reliability is highest and failure rate is the lowest that it will be. After this, long time use begets an *aging period*, and failure rates slowly climb again.

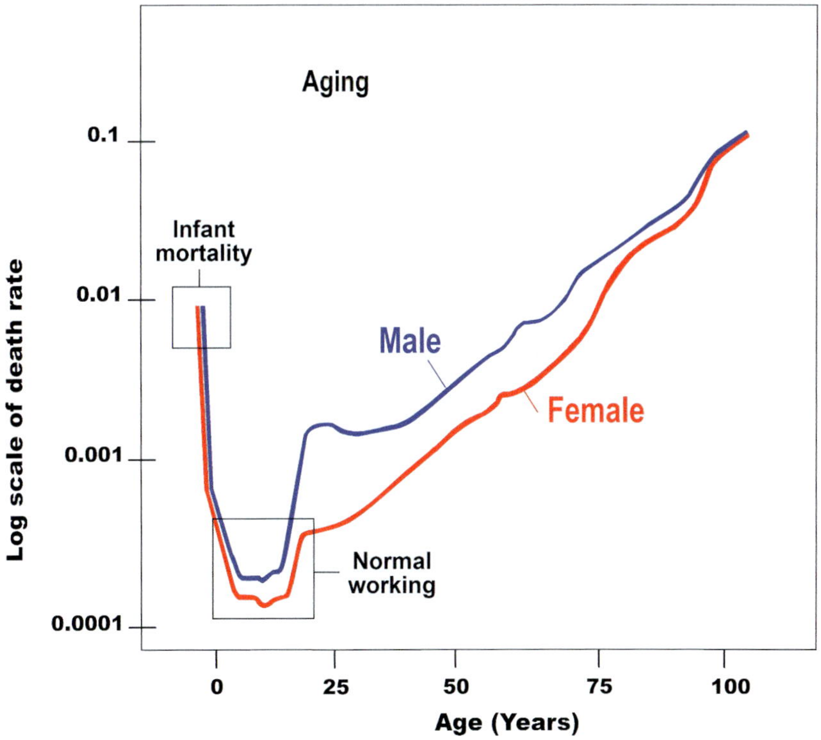

FIGURE 6.23.2
Stages of life illustrating infant mortality, normal working period, and aging. Although this data is for the U.S. population in 1999, the same shape curve is representative of failure of machines. (Redrawn from Gavrilov, L. and Gavrilova, N., *IEEE Spectr.*, 41, 30–35, 2004.)

Humans, plants, and animals undergo the same processes (Gavrilov and Gavrilova, 2004): infant mortality is high and decreases throughout childhood. Young adulthood is relatively healthy and death rates are low. With aging, a lifetime of stresses and challenges takes its toll; death rates rise once more.

Reliability theory has been formulated to deal with product failure, and this has been extended to death rates of living organisms (Gavrilov and Gavrilova, 2004). Some of the basic tenets of this theory are:

1. Weibull power law. It had been observed that the logarithm of materials failure rates increases linearly with the logarithm of age:

$$\text{Log (failure rate)} = k \text{ Log(age)} \tag{6.23.1}$$

   In Figure 6.23.2, it can be seen that the log of the human death rate during the aging period is linearly related to age, not to the log of age. That means that human mortality increases more rapidly than the failure rate of machines.

2. Gompertz law of mortality. Originally given for use in the life insurance business, this law states that the logarithm of death rates increases linearly with age:

$$\text{Log (death rate)} = k\text{(age)} \tag{6.23.2}$$

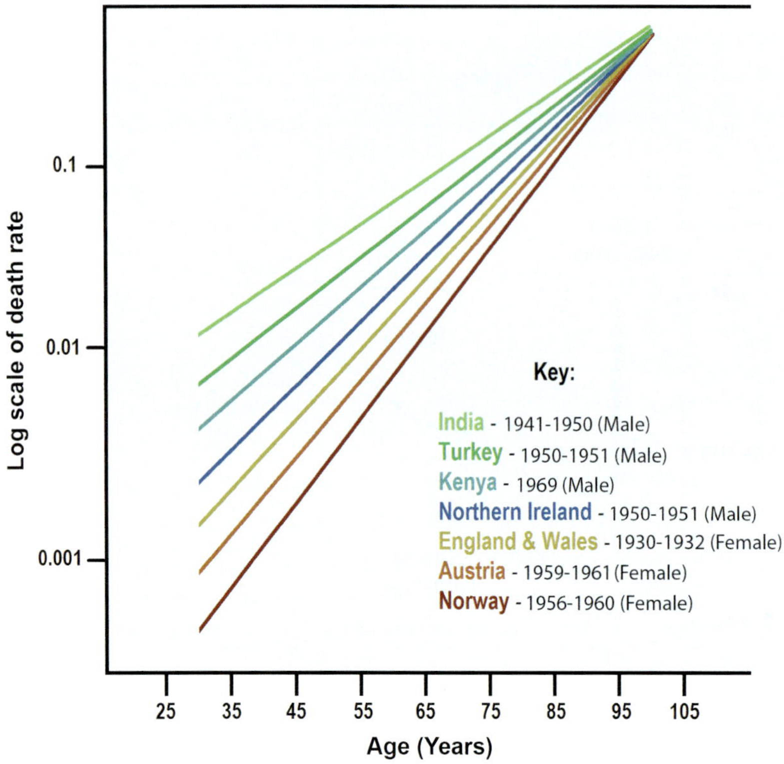

FIGURE 6.23.3
Although early adulthood death rates for different populations differ in magnitude, death rates converge for older people. (Redrawn from Gavrilov, L. and Gavrilova, N., *IEEE Spectr.*, 41, 30–35, 2004.)

In Figure 6.23.2, this can be seen to be true for humans. This law has also been found to be applicable to fruit flies, nematodes, lice, flour beetles, mice, rats, dogs, horses, mountain sheep, baboons, and many other species.

3. Compensation law of mortality. It has been observed that relative differences in mortality rates of different populations decrease with age (Figure 6.23.3). People who live a long life in an undeveloped country, for instance, are hardly at a disadvantage of living longer than a person of the same age in a developed country. Mortality rates converge for different populations.
4. Late-life mortality deceleration law. For both humans and manufactured products, death rates of the very old depart from the logarithmic relationship and instead tend to plateau (Figure 6.23.4). This plateau appears in humans at about age 100, meaning that the chance of dying at age 110 is not much different from the chance of dying at age 105.

Gavrilov and Gavrilova (2004) have explained human failure (death) rates as characteristic of a system composed of redundant faulty parts (Figure 6.23.5). The human body has many such redundancies, from the many cells that comprise a tissue to the presence of two kidneys, two arms, and two lungs. Each of these components, however, is imperfect, so as they fail, which they inevitably must, the degree of redundancy diminishes. At some

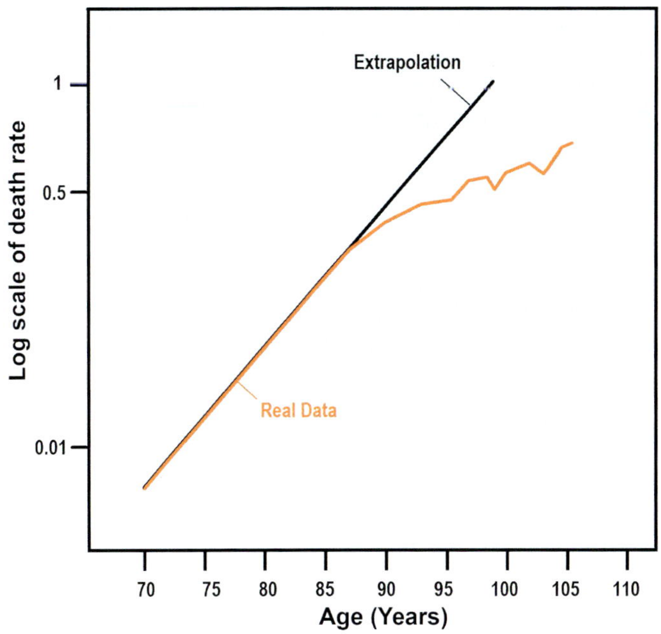

FIGURE 6.23.4
Death rates at an advanced age diverge from the extrapolated rate and plateau. (Redrawn from Gavrilov, L. and Gavrilova, N., *IEEE Spectr.*, 41, 30–35, 2004.)

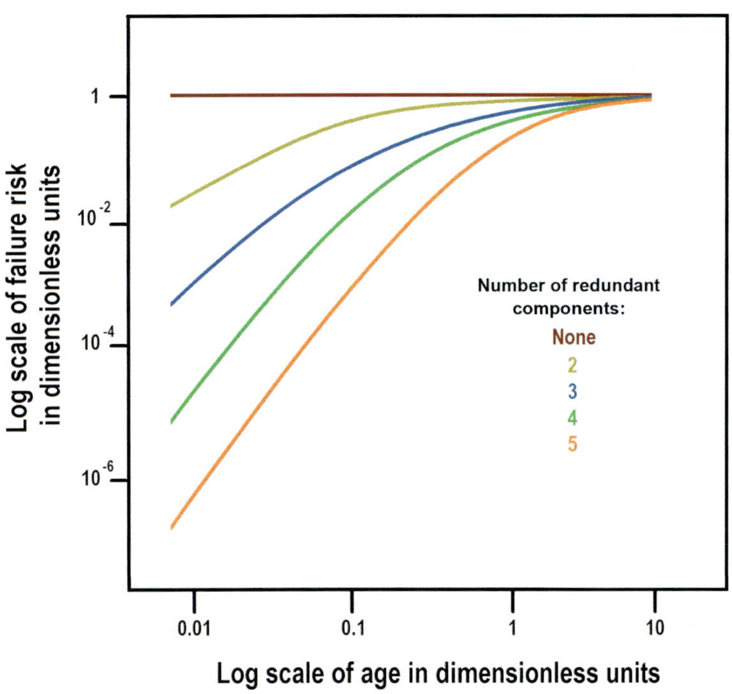

FIGURE 6.23.5
Redundancy of imperfect parts leads to low risk of failure in early life and rising failure rate with age. Near the end, the risk of failure levels off. (Redrawn from Gavrilov, L. and Gavrilova, N., *IEEE Spectr.*, 41, 30–35, 2004.)

point, there is no remaining redundancy, and failure of the last remaining component means death for the entire organism.

Fetal influences (see Section 6.6.5, Adaptations) can be the basis for some of the imperfections of redundant human components. So can faulty genetic codes, mutations (see Section 5.10, Mutations), and other environmental influences (see Sections 6.4, (Wastes), 6.6 (Adaptations), 6.12 (Chemical Stress), and 6.13 (Mechanical Stress)). Reliability theory suggests that aging may be inherently due to the organization of the human body and its dependence on redundancy. Thus, slowing the rate of damage to various components, through proper living with diet, exercise, and stress management, should be able to extend life. Repair or replacement of damaged organs or tissues is also suggested. With proper management, then, there may be no natural limitation to the life span of a human.

6.23.3 Is There a Natural Limit to Life Span?

> At the end of the 19th century, one of the leaders of physics, Lord Kelvin, cautioned his students against going into science because all the important work had already been done and only engineering questions remained to be answered.
>
> —**Bruce J. West**

Despite the trans-species tendency for larger animals to live longer, larger individuals within a species are less likely to live as long as smaller individuals. This has been shown for dogs (Li et al., 1996), mice (Bartke, 2015), and humans (He et al., 2014).

Improvements in medical treatments have clearly extended the average life span of human beings, but the extension of life span has most benefited the young (Fries and Crapo, 1981). Older people have seen hardly any increase in expected years remaining (Kent, 2008). Changes in morbidity have come mostly in life-threatening diseases early in life. As time has passed, and medicine has become more effective, there has been a compression of human mortality rates toward a seemingly natural limit of about a century (Figure 6.23.6). This would indicate that there is some natural limit to life span. Perhaps,

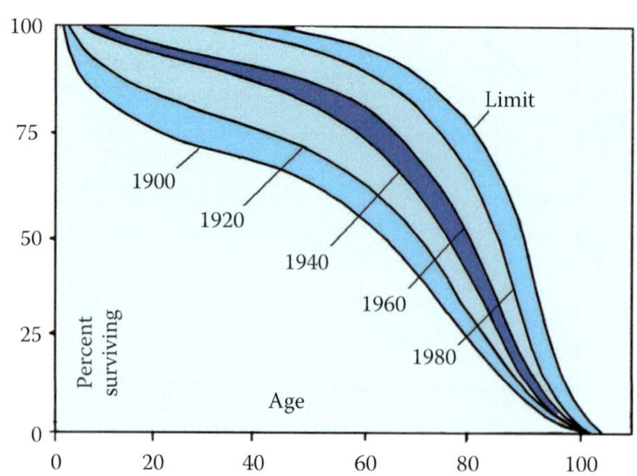

FIGURE 6.23.6
Most of the improvements in medicine have benefited the young, leading to speculation that human life span has a natural limit. This figure shows how human survival rate has changed over the years. (Redrawn from Kent, D.M., *Am. Sci.*, 96, 358–360, 2008.)

the disappearance of the telomeres (Section 6.18.4, Telomeres) or accumulation of free radical damage (Section 3.6.7, Organic Reactions) is the key ingredient of this natural limit.

There is a possibility that, as more fundamental information becomes known about aging processes, means will be found to modify aging and age-related morbidity. Antioxidants such as resveratrol have been observed to extend life span in a variety of laboratory animals (Yuan and Marmorstein, 2013).

There are opportunities here for biological engineers. The first is production of biochemicals, drugs, and medicines, probably in bioreactors filled with recombinant DNA organisms, on a large scale. The second is administration of these products to a large portion of the human population. The third is dealing with geriatric infirmities that will increase as the population ages. The fourth will be managing fertility rates to avoid overcrowding. The fifth will be to develop robotic servants and other assistive technologies to fill the need for personal care of the aged (Micera et al., 2008). There will be environmental, sanitary, nutritional, and habitat opportunities, as well.

EXTENDING NATURAL LIFE

Various studies have shown that natural life span can be extended, or at least not shortened, by following the following list of environmental recommendations:

Take antioxidants such as rapamycin and resveratrol.
Limit exposure to UV light.
Lengthen the telomeres.
Receive blood transfusion from someone much younger.
Exercise moderately.
Refrain from smoking.
Think young thoughts.
Remain optimistic with a positive attitude.
Reduce stress; practice meditation.
Get adequate sleep.
Have some compelling interest in life.
Eat a Mediterranean diet—minimal animal protein; high in plants, nuts, fruit, fat, and some protein.
Practice fasting; restrict caloric intake.
Consume some, but not much, alcohol.
Marry for emotional and physical support.

Example 6.23.1: Canary on a Chip

Canaries were used in mines to indicate the presence of deadly concentrations of carbon monoxide. The birds are more sensitive than humans to carbon monoxide poisoning, so, when the bird collapsed, it was time for mine workers to escape.

Canaries are no longer used in modern mines, but living cells immobilized on a semiconductor chip could serve the same purpose. It has been found that as the cell dies, there is a short increase in electrical resistance of its membrane. Presence of toxic gases can be detected by this type of biosensor. The challenge now is to keep the cell alive for a long time in storage until the biosensor can be used as a detector of toxins.

Example 6.23.2: The Use of Animals for Food

Temple Grandin has made several observations related to animal use and animal welfare (Klein, 2010):

1. the natural cycle of birth and death means that for one living thing to survive, another living thing must die.
2. animals in the wild usually die a violent death from starvation, predators, or exposure.
3. wild animals seldom die from natural causes or old age.
4. killing animals for food at a slaughter house can be much gentler and more humane than death in nature.
5. prey animals hide their pain to keep predators from singling them out; however, they don't hide fear.
6. livestock have been bred to supply food for humans. They exist in their present forms because of human breeding. They deserve a decent life and a quick, painless death.

APPLICATIONS AND PREDICTIONS

1. All living things will die. Designs must account for this fact.
2. Dead material can pollute live material.
3. Death is necessary to recycle nutrients.
4. Death is an element of control.
5. Death is necessary for the processes of natural selection and evolution.
6. Death motivates the transmission of information and resources from one generation to the next.
7. The death of living things follows the same reliability curve of other products.

Questions

6.0.1 Imagine some harsh environments. What adaptations would be required for life to exist in these environments?

6.0.2 Choose a specific BU. List elements of its physical, chemical, and biological environment and the types of interactions that the BU has with each of these elements. Remove one of the listed elements. How is the remainder of interactions affected?

6.1.1 Choose a liquid different from water. How would life be different if it were based on this liquid?

6.1.2 If spoilage microbes produce metabolic water that enhances further spoilage, suggest ways to break or avoid this cycle.

6.1.3 Where is the most likely barrier against the introduction of water-based drugs into a BU from the outside? What ways are there to circumvent this barrier?

6.1.4 Add to the list of Applications and Predictions.

6.2.1 Are there other sources of oxygen, beside atmospheric oxygen, that can be used by BU?

6.2.2 If solubility of oxygen in water decreases as temperature increases, predict the activity level of aquatic species as temperature changes. Draw a graph of your prediction.

6.2.3 Adding oxygen to a fish tank by bubbling air through the water is common. Think of other ways to oxygenate water and list them. Would these methods be useful for bioreactors or soil bioremediation?

6.2.4 How can the actual oxygen needs of BU be determined?

6.2.5 Why is it that people drowning in icy water can survive longer than people drowning in warm water? Give at least two reasons.

6.2.6 If hyperbaric oxygen is dangerous to life, suggest ways to use this to advantage.

6.2.7 Add to the list of Applications and Predictions.

6.3.1 List the essential elements and foods that contain them.

6.3.2 Would it be possible for a space traveler to survive on a totally artificial diet? Why would this be possible or not?

6.3.3 If the population of microbes suddenly expands, what would you suspect would be the cause?

6.3.4 What are vitamins? How are vitamins related to the genetic code of an organism?

6.3.5 Draw a diagram of the nutrient needs and nutrient outputs of an ecosystem.

6.3.6 How would you determine the identity of an essential nutrient?

6.3.7 Food supplies (1) energy needs, and (2) essential nutrients. Which of the two is more important in the short run?

6.3.8 If you were in charge of developing a new food for human consumption, what factors would you need to consider?

6.3.9 What is the heaviest naturally occurring element necessary for life?

6.3.10 Explain how evolution has produced animals that rarely need elements that plants do not use.

6.3.11 How important are microbiomes to humans? Mammals? Invertebrates? Plants?

6.3.12 Speculate about whether bacteria can have microbiomes.

6.3.13 As a biological engineer, how can the microbiome be protected?

6.3.14 Explain how a proper growth medium can be developed to grow a particular type of bacteria in a bioreactor.

6.3.15 Add to the list of Applications and Predictions.

6.4.1 Suggest likely reasons why waste products are toxic to the BU that produced them.

6.4.2 Suggest ways to reduce excess water in the environment.

6.4.3 Why would diseases be more prevalent in the presence of wastes?

6.4.4 Imagine that wastes could be used directly by the BU that produced them. What physical or chemical principles would be violated?

6.4.5 Add to the list of Applications and Predictions.

6.5.1 Animals or humans in locked cars in the summer often die because they cannot lose enough heat. Make a list of the physical and chemical changes that occur in the body when it is overheating. Suggest ways to protect a confined animal or child from overheating.

6.5.2 If life were based upon alcohol rather than water, what would be the implications for heat loss?

6.5.3 Give an example of a BU that has moved its range over time in response to a temperature change.

6.5.4 Imagine that BU did not have an optimum temperature to live and grow. What would be the biological implications? Would there be more or less need for biological engineers as a consequence?

6.5.5 If a BU that normally lives in a temperature range of 30°C–40°C is cooled below this temperature, what can be expected of that BU over the short term? The long term?

6.5.6 What are some of the ways heat maintenance is more efficient for humans adapted to low temperatures compared to humans adapted to high temperatures?

6.5.7 Why are the heat index and wind chill index not as important for the well-being of plants and bacteria?

6.5.8 Add to the list of Applications and Predictions.

6.6.1 Draw two graphs. As the independent variable, use the ratio of Viceroy butterflies to Monarch butterflies. As the dependent variables, plot the expected population numbers of each kind of butterfly. Make the curves relative to the maximum population as the ratio varies. Justify why you drew the curves as you did.

6.6.2 Speculate on possible advantages for high altitude plants to be shorter than lower altitude plants. Of these, which do you think would be the most important, and why?

6.6.3 Is it possible for a blind person to experience Seasonal Affective Disorder? Why do you say this?

6.6.4 List ways in which BU adaptations to light can be used to advantage by engineers and other technologists.

6.6.5 How would you design an environment more protective of fetal development than the womb?

6.6.6 List similar adaptations of plants and animals native to

The tropics

Cold climates

Undersea world

6.6.7 How would you induce a microbe to produce an industrially useful biochemical in large quantities?

6.6.8 Pulmonary hypertension is a disease of the pulmonary blood system that often leads to death. Are there physical means to ameliorate pulmonary hypertension?

6.6.9 Add to the list of Applications and Predictions.

6.7.1 Imagine what the Earth would be like if plants could not modify their environments. How widespread would plant populations be?

6.7.2 Give an example where a nonhuman BU can be used to modify its environment in a way that's useful to humans.

6.7.3 What is a hormone? Where is its effect felt?

6.7.4 Is the gut inside or outside the body?

6.7.5 What kind of materials can come in contact with BU without being modified by them?

6.7.6 Predict what modifications humans will make to Mars.

6.7.7 How does a cancerous tumor modify its environment to enhance its own growth?

6.7.8 Stem cells are modified by their environment to become different kinds of differentiated cells. Do stem cells modify their environment?

6.7.9 Add to the list of Applications and Predictions.

6.8.1 Give examples of adaptations that draw energy away from growth and reproduction.

6.8.2 Global warming is causing the migration of some plant species toward more northern or higher altitude areas. What can you say about the reproduction rate of these plants under such circumstances?

6.8.3 If rapidly growing plants produce less toxin, would you expect Monarch butterflies to feed on young or older plants?

6.8.4 Is the reproduction rate of a species expected to be higher or lower during periods of turmoil? Why?

6.8.5 Imagine a species that is rapidly adapting to a new environment, but that is reproducing at as high a rate as it did when the environment was static. Where would the excess energy come from?

6.8.6 Add to the list of Applications and Predictions.

6.9.1 Describe the movement of slime mold. What adjustments must be made by the slime mold in order to do so?

6.9.2 Animals held captive in zoos are likely to be depressed if they are restrained from moving to conditions more to their liking. How can you overcome this depression?

6.9.3 How do stem cells know the location where they are needed? Speculate on some of the mechanisms for this to happen.

6.9.4 Urban development may act as a barrier to movements of plant and animal species. Predict the consequences.

6.9.5 What are the conditions leading to formation of a new species, and what are the conditions leading to its extinction? Estimate the relative time for each to happen.

6.9.6 List ranges of some BU, and relate these to their sizes.

6.9.7 "Divide and conquer" is a means to eliminate disease-causing microbes or to lead to the extinction of species. Discuss the possibilities of using the divide-and-conquer approach to cure cancer.

6.9.8 Add to the list of Applications and Predictions.

6.10.1 What are the three conditions necessary for evolution of an organism?

6.10.2 Try to think of means to destroy bacteria that will not result in bacterial evolved immunity.

6.10.3 What are some strategies to minimize immunity development? What are some strategies to maximize immunity development? Give examples where each of these would be desirable.

6.10.4 Give examples of beneficial evolution influenced by humans.

6.10.5 What is the effect of evolution on resources available to an organism?

6.10.6 Under what conditions will a breed of fast cats evolve?

6.10.7 Which will evolve faster, a petunia or an oak tree? Why?

6.10.8 Imagine a world without biological evolution. Describe such a world. How would it differ from our present world?

6.10.9 List the steps in the process to improve a product design based upon evolutionary principles.

6.10.10 Add to the list of Applications and Predictions.

6.11.1 Why would it be biologically advantageous that the most aggressive individual in a clan has the greatest opportunity to reproduce?

6.11.2 Do you think that antisocial behavior induced by crowding affects only the higher animals, or do you think it is widespread throughout all of biology? Why do you think so? Where do you think is the line between BU that suffer from crowding and those that don't?

6.11.3 Why do you suppose crowding effects in humans are more severe for males, first-born, and children over ten?

6.11.4 What are some possible biological engineering designs where avoiding crowding effects should become a consideration for design?

6.11.5 Estimate your own personal space. What considerations do you use to make this determination? Do you think your personal space is more, less, or about the same as average?

6.11.6 Look up a tutorial article or book chapter on small group ecology. Write down several interesting details.

6.11.7 What are some technological ways to counteract hypersensitivity to sound? Light? Touch?

6.11.8 Design an experiment that you think would establish psychological stress in a crowded animal. How developmentally advanced do you think the animal would have to be in order to react to the experimental conditions?

6.11.9 Speculate on conditions other than crowding that cause rapid spread of virulent diseases among humans.

6.11.10 What is the carrying capacity of the environment? Is it the same for all BU?

6.11.11 Why do you suppose somatal cells can live at higher densities than can single bacteria?

6.11.12 In what ways do animal welfare considerations influence a biological engineering design? How do biological engineers need to deal with animal welfare advocates when considering design approaches involving animals?

6.11.13 Add to the list of Applications and Predictions.

6.12.1 Look up an example of a cofactor or coenzyme system that is affected by toxic chemicals.

6.12.2 What is the effect on toxicity of a compound of larger molecular size? Polarity of the molecule? Oxidation? Reduction? Degradation? Conjugation?

6.12.3 What is meant by threshold response?

6.12.4 Compare the toxicity of two compounds, the first with an LD_{50} of 9.63 and the second with an LD_{50} of 15.2.

6.12.5 What are the economic and health implications of the linear no-threshold hypothesis for toxins compared to the threshold hypothesis?

6.12.6 If you were given the job to find a new class of bacterial poisons to sterilize medical instruments, where would you look for a source of natural poisons?

6.12.7 At what stage in life is the sensitivity to toxins likely to be greatest? What implications would this have for manufacturers?

6.12.8 What would life be like without natural toxins?

6.12.9 Give examples of measures taken to protect BU against the effects of toxic substances.

6.12.10 Rank the expected relative toxicity of the following plants: cactus, borage, poinsettia, larch, African violet, orchard grass. Take into account the opportunity to be eaten by grazing herbivores. Why did you rank them this way?

6.12.11 If washing hands decreases the protection against absorption of solvents, what procedures would you recommend?

6.12.12 Explain why animals have developed the ability to detoxify small amounts of ingested cyanide. Why is cyanide poisonous?

6.12.13 Speculate on why it is difficult to remove medicines, drugs, and other pharmaceuticals from wastewater.

6.12.14 Add to the list of Applications and Predictions.

6.13.1 Describe what happens to cells as shear rate in surrounding fluid increases.

6.13.2 Are there fluid dynamic means to dissolve atherosclerotic plaque?

6.13.3 Estimate the wall tensile stress for blood vessels of different sizes. Do this by estimating bursting pressures, wall thicknesses, and wall radii.

6.13.4 What can be done to attenuate shear stresses on cells in a bioreactor?

6.13.5 List ways to increase bone strength in patients unable to exercise.

6.13.6 How can plants resist wind loads during hurricanes and other storms?

6.13.7 What would you do to protect a human against severe acceleration?

6.13.8 Add to the list of Applications and Predictions.

6.14.1 What happens to the energy saved by an efficient BU?

6.14.2 Why is walking energetically inefficient? Suggest ways to improve the efficiency of walking.

6.14.3 Explain why bicycles improve efficiency of human locomotion.

6.14.4 Suggest other energy-using physiological processes that could benefit from optimization. If possible, determine if these are optimized.

6.14.5 Discuss the relative advantages and disadvantages of broad versus narrow optima. Give examples where each type of optimum is expected to appear in biological systems.

6.14.6 Diagram simple interactions among species in an ecosystem. Which processes are likely to be optimized? Of what benefit is this optimization to the biological engineer who wishes to use the ecosystem to solve a problem?

6.14.7 Show how the scheme in Figure 6.14.6 can be used to regulate your body weight.

6.14.8 Make a list of technological solutions that were inspired by efficient biological mechanisms.

6.14.9 How much risk is there for a species perfectly adapted to it environment to carry some genes that are not optimum?

6.14.10 Add to the list of Applications and Predictions.

6.15.1 Contrast torpor, hibernation, and estivation. What is similar, and what is different among them? What environmental conditions trigger each?

6.15.2 Warm-blooded animals (homeotherms) maintain body temperatures within close tolerances. Yet, during altered states, body temperature may drop by a significant fraction. Are these animals still classified as warm-blooded when this happens? Why?

6.15.3 List the advantages of torpor, hibernation, and estivation. How would these three states influence the designs of enclosures in which to keep animals that experience altered states?

6.15.4 What challenges to biological engineers are posed by endospores?

6.15.5 Some say that endospores are a means of reproduction. What is the real function of endospores?

6.15.6 What methods could be used to detect and kill anthrax endospores?

6.15.7 Most plants reproduce by seeds or spores. Name other means for plants to reproduce.

6.15.8 What do hibernation, endospore formation, plant seeds, and severe blood hemorrhage have in common?

6.15.9 Speculate on the connections between post-traumatic stress disorder (PTSD) and biological engineering.

6.15.10 Construct a flow diagram of the fate of a mold spore, including

Movement through the air

Settling on a surface

Growth

Ot hers

6.15.11 Add to the list of Applications and Predictions.

6.16.1 Speculate on what would happen if all the cells of the human body had to perform all necessary functions (breathing, biochemical production, communication, etc.).

6.16.2 Express in your own words the meaning of symbiosis. How can symbiotic relationships be exploited to produce solutions to problems involving living systems?

6.16.3 When raising plants in greenhouses, what facts must be known about plant reproduction?

6.16.4 Considering coevolution, why must care be taken when depleting one or two species from a pristine area? What is the ultimate effect of such depletion?

6.16.5 Why is knowledge about biofilms important to engineers who work to improve food safety?

6.16.6 Communal efforts by biological species can be both helpful and detrimental to human exploitation. Make a list of positive and negative examples.

6.16.7 Why must all interactions of a natural organ of the body be known before an artificial organ can be developed?

6.16.8 List some human or animal groups that seem to have been able to survive at least in part due to the cooperation among members. Discuss what advantages were conferred upon the members by cooperating.

6.16.9 Discuss the relationship between probiotics and competitive inhibition.

6.16.10 Add to the list of Applications and Predictions.

6.17.1 Make a list of ways in which some BU exploit other BU and add means of defense against those exploitations. For each of these, consider the ways that the exploitations and defenses affect utilization by biological engineers.

6.17.2 What is the effect of a BU with an unchecked competitive advantage? Give some examples.

6.17.3 Compare the relative advantages of dealing with an unchecked non-native species using biological or nonbiological (perhaps chemical or physical) control.

6.17.4 What are the differences among predators, parasites, and pathogens? What are the similarities? How can these characteristics be used by biological engineers?

6.17.5 Give an example where human technology has not been capable thus far of protecting against another BU. Speculate on technological advances that would need to be made to deal with this problem.

6.17.6 Add to the list of Applications and Predictions.

6.18.1 What are the advantages and disadvantages of asexual reproduction?

6.18.2 Speculate on exploiting the modes of transformation, conjugation, and transduction for biological engineering designs.

6.18.3 What effects do you suppose that growth factors, density-dependent inhibition, and anchorage dependence would have on attempts to grow somatal cells *in vitro*?

6.18.4 Under what circumstances would biological engineers exploit oncogenes?

6.18.5 It is stated in the box "Cancer in Humans" that the number of cellular generations in the average life span of a 75-year-old human is 780. It was also stated in Section 6.18.4 that cells *in vitro* reproduce only 20–50 times before they die. Are these two statements in conflict? Why?

6.18.6 What are the implications for administering telomerase to extend the life of humans or animals?

6.18.7 Provide reasons for sexual dimorphism.

6.18.8 If you were asked to design a facility where rare species were to be bred, what considerations would have to be incorporated to be successful?

6.18.9 Selection of one or the other sex is important for both plants and animals when they are to be reared for food (males usually grow larger) or used for research purposes (some traits are sex-based). What are the difficulties encountered in separating the sexes for sequential hermaphrodites, and how can these be overcome?

6.18.10 Speculate on means that humans could be given the ability to regrow limbs the same way that plants can replace lost tissues.

6.18.11 List characteristics of cancer that can be used against it.

6.18.12 Pick an animal species and discuss special arrangements necessary to transport that specific animal.

6.18.13 Add to the list of Applications and Predictions.

6.19.1 What is the difference between human intelligence and squirrel intelligence or even spider intelligence?

6.19.2 If you were a honey bee, how would your perception of reality differ from what you now perceive? How could this difference influence the biological engineering approach to problems involving honey bees?

6.19.3 How can the means of courtship of fireflies be used to advantage by biological engineers?

6.19.4 Look through the pages of a newspaper that announces engagements and weddings. Note the similarities between faces of the couple. Why do you suppose similar features are attractive? Explain in terms of reproductive advantage to the species.

6.19.5 It has long been dreamed to communicate verbally with animals, or, if that is not possible, at least understand what they are saying to each other. How would you go about trying to do this? Why should you know the frequency content of sounds made by the animals before you go further?

6.19.6 Imagine a pheromone used to trap unwanted insects. What would you think would be the long-term effects of widespread use of the pheromone?

6.19.7 If chemical communication exists as a system rather than as an isolated unit, what are the components of such a system?

6.19.8 Describe the odor of mom's apple pie (if she baked). Are words adequate to describe the odor?

6.19.9 Pick a BU. Describe what modes of communication it uses.

6.19.10 List the human senses. Describe the importance of each in the various stages of eating.

6.19.11 How can the ability of the blind to visualize the locations of objects be used in designs for the blind?

6.19.12 Explain the use of Young's Principle. What are the advantages of broad sensitivity of sensors rather than very selective sensors?

6.19.13 Trace the history of imaging and speculate on the next big steps.

6.19.14 Give examples of the uses of each kind of imaging technique.

6.19.15 What importance does the just noticeable difference have on biological engineering designs?

6.19.16 Give ideas for designs of a rattlesnake trap.

6.19.17 Explain how magnetic resonance imaging can be used to detect active metabolism of tissues.

6.19.18 Explain why it is important for the immune system to develop in the presence of beneficial microbes.

6.19.19 Add to the list of Applications and Predictions.

6.20.1 Draw the block diagram of a means to sense touch through the antennae of insects and steer a robot around a room.

6.20.2 What can be said about those areas of the body with sensor concentrations? Why are the sensors concentrated there?

6.20.3 What would be the advantage of sensor rate of change sensitivity? What would be the advantage of sensor adaptation?

6.20.4 Conditioning produces fast and efficient responses to environmental stimuli. When are these responses not desired? What happens then?

6.20.5 What are the control advantages of redundancy? Which components of a control system in a human are redundant? Which components of a control system in a biome are redundant?

6.20.6 A foreign bacterium enters the body. Describe with a diagram the possible immune system responses.

6.20.7 Give an example where an immune-like response can help solve an unrelated problem involving living things.

6.20.8 Discuss the possible uses of immune system promoters, such as induced malaria. How does the use of helminths contrast with the effects of induced malaria?

6.20.9 What nonstandard uses can you think of using antibodies?

6.20.10 How does stress change immune system response?

6.20.11 How do new vaccine-making techniques differ from older techniques?

6.20.12 Describe progress made in controlling prostheses as naturally as possible.

6.20.13 Of what use can a mathematical model of immune system responses be put?

6.20.14 Give examples of antagonistic control action not mentioned in the text. Your examples may or may not involve living things.

6.20.15 Give an example where antagonistic action can be used to an advantage to control some process.

6.20.16 List advantages and disadvantages of a dead zone in a control system.

6.20.17 A control system including a time delay can be unstable. Take, for instance, the production of a hormone in the liver and the brain as the target organ. What can be done to minimize the time delay in a flow system?

6.20.18 Add to the list of Applications and Predictions.

6.21.1 Speculate about the effects on the natural life cycle of an organism of one stage in the cycle (for example, maturity) is prolonged by human intervention. What are the ethical considerations for such an intervention?

6.21.2 Contrast the stages of ecological succession with the stages of life of a human. What commonalities are there?

6.21.3 Why are stem cells able to maintain the ability to form many kinds of tissues? Suggest ways to use this ability.

6.21.4 What would you expect might happen to male mammals if genetic modifications caused them to live longer?

6.21.5 Describe differences in appearance between old and young plants. How can these differences influence the types of use to which the plants can be put?

6.21.6 How can ethylene gas be used to advantage?

6.21.7 In what areas of our lives is it important to have knowledge of the likelihood of dying?

6.21.8 Name BU with dominant annual cycles. Where can these BU be used?

6.21.9 Pick a BU. Speculate on the effects that annual temperature variations have on various physical and emotional (if appropriate) traits of the BU. Consider such things as growth, hormone production, fitness, reproduction, movement, etc.

6.21.10 What are the connections between natural cycles and

Economics

Reproduction

Drug effectiveness

Disease

Comfort

Nutrition

Legal system

Education

Rate of accidents

6.21.11 If you were to design a facility to produce shellfish, do you think it would or would not be necessary to include simulated tides? Why or why not?

6.21.12 What evidence is there for an intrinsic sense of time in BU?

6.21.13 How is knowledge of nutrient cycles useful for a biological engineer? What kinds of designs are influenced by nutrient cycles?

6.21.14 Add to the list of Applications and Predictions.

6.22.1 Choose the perception of some common physical attribute of the environment (such as temperature, vision of a flower, smelling the scent of an apple, pH of a bathing medium, etc.). Comment on the effect that different emotions would have on that perception.

6.22.2 What impact would different emotions have on designs involving living things?

6.22.3 If human emotions arise partly in response to hormonal levels, what effect would you expect there to be to periodic hormonal supplementation (such as insulin injections, thyroxine pills, or human growth hormone injections) as part of medical intervention?

6.22.4 How would accounting for personality affect designs involving living things?

6.22.5 How would you design a means to supplement neural transmitters in a human being?

6.22.6 How can intraspecies communication be used to affect the state of an individual? How can it be used to signal the state of an individual to a human observer?

6.22.7 Is there an analog to emotion for ecosystems?

6.22.8 What effect will brain maturation have on a biological engineering design?

6.22.9 What effect would a bottom-up theory of brain action have on a biological engineering design?

6.22.10 Of what benefit is the incorporation of learning in a design involving animals? What disadvantages are there?

6.22.11 Describe the hardships encountered in Biosphere II. Do any of these relate to Maslow's hierarchy?

6.22.12 Make a list of products where knowledge of human factors would be of benefit. Indicate on your list the aspects of the products for which human factors engineering can be used.

6.22.13 Can human factors information be generalized to other levels of BU? How? Give examples.

6.22.14 Explain what impact each of these has on designs involving living things:

Psychology

Sociology

Economics

Art

Political science

Literature

Music

History

Philosophy

6.22.15 Explain the ways in which people like Ben Kilham recognize the intelligence and individuality of animals.

6.22.16 Temple Grandin has established a reputation as someone who designs superior cattle handling systems. Find out about her work, and write a short essay on her methods.

6.22.17 Which if the following do you think could be trained to elicit a conditioned response? Why do you say this?

Streptococcus

Snake

Spider

Squirrel

Sheep

Shetland pony

Squid

6.22.18 Give an example of a biological engineering design that could elicit a negative emotional response from the biological object of the design.

6.22.19 The appearance of a prosthetic device may have a large effect on some patients. If it looks good, they will accept it more readily and perhaps have more confidence in its use. Explain how a rehabilitation engineer might improve the appearance of a prosthetic device.

6.22.20 Do we have "will power" or "won't power"?

6.22.21 Describe the necessary ingredients for biofeedback. How can these be used in prosthesis design and use?

6.22.22 Give some means to protect against disease, both internal and external to the body to be protected.

6.22.23 Explain the differences between the perceptions of visual stimuli by the left and right hemispheres of the human brain.

6.22.24 What is the function of the blood–brain barrier?

6.22.25 Add to the list of Applications and Predictions.

6.23.1 How would you define death for the following:

Clostridium botulinum cells

Pancreatic cells

Nerve cells

Human stem cells

6.23.2 How does knowing what represents death influence a design involving living things?

6.23.3 How can we measure life and death?

6.23.4 Describe what are the expected failure rates at different stages of a product. How can high failure rates be overcome?

6.23.5 How are high failure rates related to component redundancies?

6.23.6 Add to the list of Applications and Predictions.

Section IV

Scaling Factors

7

Allometric Relationships

One of the earliest forms of prediction in biology was the use of scaling factors, those proportions, based on morphology, that can be used to extend general knowledge about one creature to another. There are similarities in nature, and these are often caused by physical or chemical requirements that must be met by all critters large and small.

In the paragraphs to follow, there will be some of the scaling factors known at present. Many of these will apply only to restricted classes of biological organisms, but that does not detract from their utility if designing systems related to those types of organisms.

This section is more detailed and quantitative than other sections in this book. The reasons for this are twofold: first, the biological engineer must know how to extend known information from one set of circumstances to another, as given by the second goal in Section 1.6 (Expectations). Second, these scaling relationships do not appear in common places, and so, as a convenience to the reader who may sometime need to apply them, they are given here in some detail.

Scaling relations can be very important in the engineering design process. One difference between an engineering design and a complete guess about what will happen is that the engineering design incorporates the best and most specific prediction based on available facts. This usually involves a calculation step to yield quantitative predictions. It may be that specific quantitative information about a particular characteristic of some species cannot be found easily. Short of performing experiments and making your own measurements, useful information can often be obtained from scaling relationships among similar organisms. Thus, answers to questions about food and oxygen requirements, waste and carbon dioxide production, natural densities, organ masses, locomotion speed, and life-cycle times can all be predicted based on scaling relationships. Engineering related to biology often has a need to know such information. Designs based upon these predictions have a much better chance of success than less enlightened guesses.

7.1 Allometric Relationships from Evolutionary Pressure

Prediction is very difficult, particularly about the future.

—**Niels Bohr**

Allometry is defined as the change of proportions with an increase in the size of a single species or between adults of related groups (Li, 2000). Allometric relations only exist when there is similarity of structure and function between biological units of different size. If completely different mechanisms are involved (e.g., the locomotion of bacteria compared to horses), then no allometric relationship would be expected.

There appear to be universal biological principles at work in scaling relationships, although the natures of these principles have not yet been fully explored. Allometric relationships among very divergent species seem to be scaled with body mass to some simple multiples of one-quarter power ($m^{1/4}$). Thus (Brown et al., 2000),

The leaf area of trees $\propto m^{3/4}$

The radii of mammalian aortas and radii of tree trunks $\propto m^{3/8}$

The circulation time of mammal blood, and of tree sap, and the cycle time of respiratory, cardiac, gestation, postembryonic development, life span $\propto m^{1/4}$

The biological rates, including mammalian heart rate and respiration rate $\propto m^{-1/4}$

Natural selection seems to have led to an economy of the design of structures and functions so that they just meet maximum demands. Any greater capacity would be biologically uneconomical (Brown et al., 2000). If evolution results in allometric relationships among BU, then it is only because the benefit-to-cost ratio of the function in question has been optimized.

Another way to think of allometry is to consider that if organisms do *not* change their form as they change in size, their function is altered, and such functional shifts might be a source of evolutionary innovation (Koehl, 2000).

There are scale-invariant features of biological tissues, such as bone strength, wood strength, or maximum muscle stress, which require size-dependent changes in other features. These changes may be in sizes of limbs or muscle masses, or may be changes in posture and mechanical advantage (Biewener, 2000). A more upright posture puts less bending stress on a limb by reducing the moment arm (Figure 7.1.1). Larger plants and

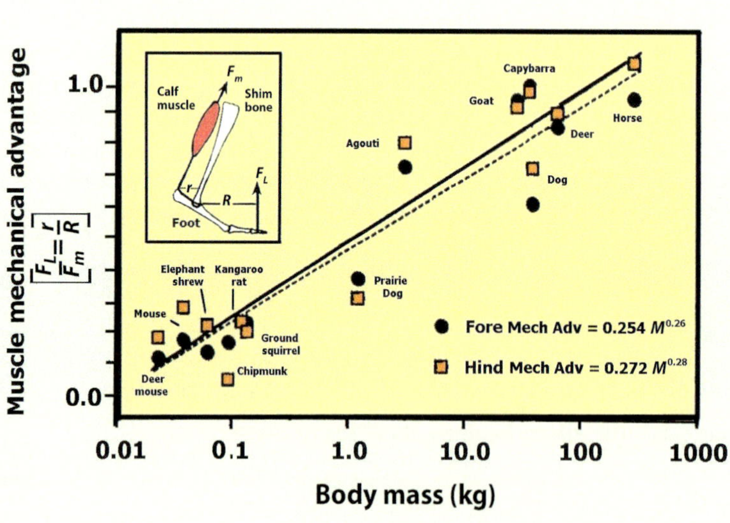

FIGURE 7.1.1
Scaling of posture-related muscle mechanical advantage (r/R) in terrestrial mammals. Both fore- and hind-limb mechanical advantages scale with body mass. Changes in crouched (small mammals) and upright (larger mammals) postures tend to equalize peak bone and muscle stresses of different species (Biewener, 2000). Mechanical advantages of less than 1.0 play to the strengths of muscles: they can produce large forces (F_m), but cannot move fast or far. Thus, fractional mechanical advantages allow higher load forces (F_L) than muscle forces at faster speeds and over greater distances.

animals may scale according to strength, but smaller plants and animals may scale according to stiffness: very slender elements can bend, thus putting associated muscles at a disadvantage (Biewener, 2000).

7.2 Dimensional Analysis

I'll tell thee everything I can:
  There's little to relate.
I saw an aged man,
  A-sitting on a gate.
"Who are you, aged man?" I said.
  "And how is it you live?"
And his answer trickled through my head
  Like water through a sieve.

—Lewis Carroll

Dimensional analysis (see Section 4.1, Equality) is a time-honored technique used in engineering research to group pertinent variables into dimensionless numbers to generalize empirical correlations. Thus, equations for fluid flow often are given in terms of the dimensionless Reynolds number, $dv\rho/\mu$ and forced convection heat transfer equations usually relate the dimensionless Nusselt number, hd/k, to the Reynolds number and dimensionless Prandtl number, $c_p\mu/k$ (Johnson, 1999). In these, d is the object diameter, v is the fluid velocity, ρρ is the fluid density, μ is the fluid viscosity, h is the convection coefficient, k is the fluid thermal conductivity, and c_p is the fluid-specific heat. The beauty of these relationships is that they are valid no matter which of the individual variables in the dimensionless numbers is changing. Hence, the Reynolds number, for instance, is computed at a certain value depending on the magnitudes of d, v, ρ, and μ. If d doubles and ρ halves, there is no difference in the Reynolds number value. Contrarily, if the Reynolds number increases to five times its original value, it makes no difference which of the component variables is the cause of the increase. Dimensionless numbers, then, are extremely useful for empirical relationships and can be used, as well, to understand similarities in physical systems.

The use of dimensionless numbers improves the efficiency of experimental biology. With the use of dimensionless numbers, the number of experiments performed with all the combinations of parameter values may be reduced considerably because only the dimensionless numbers themselves must vary, not all components of the dimensionless numbers. This greatly reduces the effort needed to derive valid experimental data and to relate many experimental results together in the inductive process of forming predictive equations.

For instance, dimensional analysis applied to the heart yields two pi (dimensionless) terms:

The Lamé relation for thick-walled ventricles

$$\pi_1 = \frac{\text{wall tension}}{(\text{ventricular pressure})\,(\text{wall thickness})} \tag{7.2.1}$$

and

$$\pi_2 = \frac{\text{wall thickness}}{\text{wall radius}} \tag{7.2.2}$$

The ratio of the two is the Law of Laplace:

$$\frac{\pi_1}{\pi_2} = \frac{(\text{wall tension})}{(\text{ventricular pressure})\,(\text{wall radius})} \tag{7.2.3}$$

seen in Section 2.9.5 (Static Pressure).

There have been very few attempts to apply dimensional analysis to biological systems. However, as quantitative biological research becomes ever more sophisticated, dimensional analysis will help to discover similarities and produce new scaling factors.

7.3 Golden Ratio

> It has been observed that the height of a man from the crown of the head to the sole of the foot is equal to the distance between the tips of the middle fingers of the two hands when extended in a straight line.
>
> **—Pliny the Elder**

If a line is divided into two segments such that

$$\frac{\text{total length}}{\text{longer segment}} = \frac{\text{longer segment}}{\text{shorter segment}} \tag{7.3.1}$$

then the ratio of the longer segment to the shorter segment can be solved algebraically to be an irrational number $(1+\sqrt{5})/2 = 1.618\ldots$. This number was called the "golden section" by the ancient Greeks (West and Goldberger, 1987). Kepler called it the "divine proportion." We will refer to it as the *golden ratio.*

The golden ratio can also be generated from a set of Fibonacci numbers, which is a series of positive integers beginning with 0 and 1, such that each number is the sum of the two preceding numbers (Figure 7.3.1). This gives a series 0, 1, 1, 2, 3, 5, 8, 13, 21, 34, 55, ... The ratio of each number to its immediate predecessor approaches the golden ratio (13/8 = 1.625, 21/13 = 1.615, 55/34 = 1.618, for example).

A Fibonacci blueprint and the golden ratio appear to be able to describe plant growth and leaf spacing. The reproduction of honeybees, cows, and rabbits appears to be related to the Fibonacci sequence, as is the arrangement of cauliflower and broccoli florets. The ratio of the head-to-toe height in humans to the height from navel to toes also approximates the golden ratio. There is also Fibonacci scaling in the lung. When each of the upper airways divides in two, it does not divide equally. Instead, one division is usually longer than the other (Figure 7.3.2). The ratio of the long tube length to the shorter tube length is very close to the golden ratio (Goldberger et al., 1985).

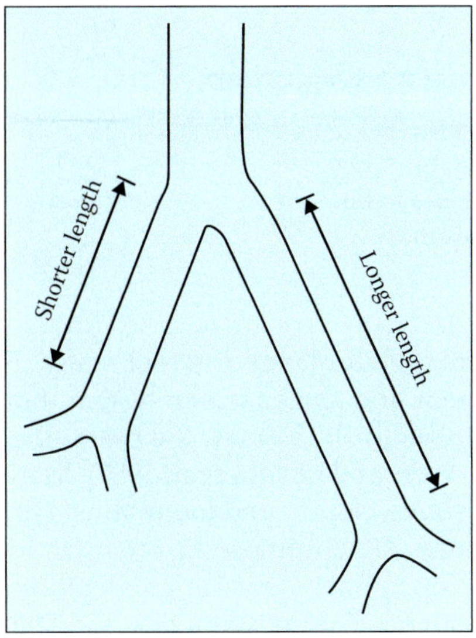

FIGURE 7.3.1
Division of an airway in the lung is asymmetrical, with one daughter tube being longer than the other. The ratio of the longer length to the shorter length approximates the golden ratio.

FIGURE 7.3.2
Having fun with Fibonacci numbers. (Courtesy of Bill Amend, Copyright FOXTROT 2009; Andrews McMeel Syndication.)

7.4 Fractal Scaling within an Organism

> So, Nat'ralists observe, a Flea
> Hath smaller Fleas that on him prey
> And these have smaller Fleas to bite 'em,
> And so proceed ad infinitum.
>
> **—Jonathan Swift**

Many natural phenomena are self-similar at different scales (Figure 7.4.1). This is certainly true for the respiratory system and the vascular system, but also true for the bile duct system, urinary collecting tubes in the kidney, the brain, the lining of the bowel, neural networks, and the placenta (West and Goldberger, 1987). Many plants have the same self-similar structure, as do shorelines, clouds, and mountains. Other fractally organized (scale independent) biological features are chromosome organization (Babbit, 2011) and correlation lengths among birds flying in flocks (swarm intelligence; Hayes, 2011).

Self-similar structures do not have a single length scale. The property of interest (say the diameters of the respiratory airways, for instance) will then be dependent on the scale of interest, usually decreasing in value as the scale becomes smaller (West and Goldberger, 1987). The result is that instead of an exponential function where the ratio of larger scale to smaller scale is a constant, the larger scale is related to the smaller scale through a power law relationship.

To illustrate this point, we consider diameters of the respiratory airways. As the airways divide, the tubes decrease in diameter. If airway diameters were to decrease in the same proportion (e.g., a ratio of ½ would give diameters of 1, ½, ¼, ⅛, … for airway generations 0, 1, 2, 3, …), then there would be a negative exponential relationship between airway diameter and airway generation. An exponential relationship plots as a straight line on a semi-log plot. It can be seen from Figure 7.4.2 that it is only for the lower airway generations that the airway diameters approximate a straight line.

A power law relationship

$$\text{diameter} = (\text{generation})^b \tag{7.4.1}$$

plots as a straight line on a log–log plot. It can be seen from Figure 7.4.2 that a straight line is indeed approximated by the data. This would seem to indicate that self-similarity exists

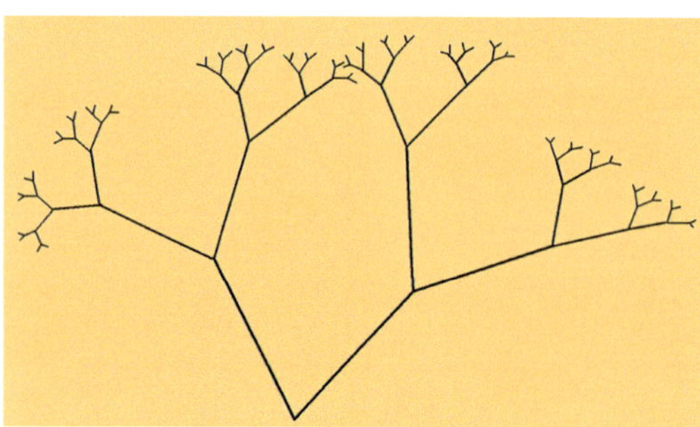

FIGURE 7.4.1
This is an example of a self-similar geometry. Patterns apparent at larger scales are repeated as smaller scales. A pattern such as is illustrated here is typical of the respiratory system and of many plants.

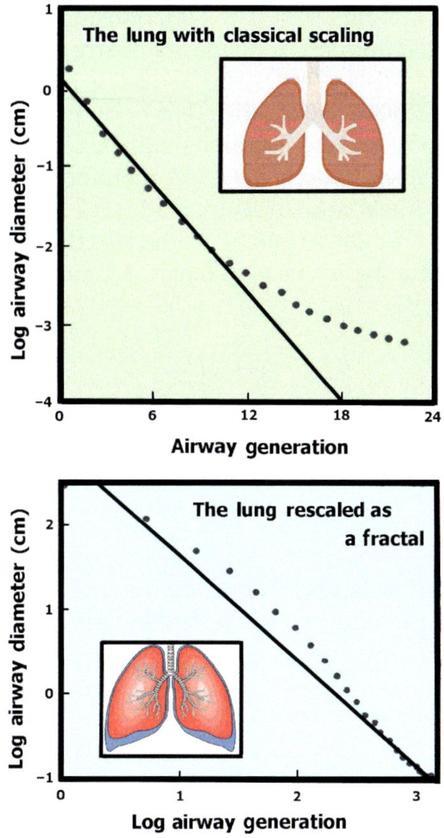

FIGURE 7.4.2
Airway diameter plotted against airway generation does not plot as a straight line on a semi-log plot (upper plot), but does approximate a straight line on a log–log plot (lower plot), therefore confirming the power law relationship characterizing fractal self-similarity. (Redrawn from West, B.J. and Goldberger, A.L., *Am. Sci.*, 75, 354–365, 1987.)

in lung airway diameters. This type of self-similarity has been called a *fractal* because of the fractional dimensions that can be attributed to this geometric relationship (see Section 4.2.2, Self-Similar Data).

Fractals can also be inferred from *morphogenesis* (the development of form or structure), where a single time scale does not adequately address all time-dependent processes. The electrocardiogram seems to have fractal time properties, as well as electrical activity of a single neuron and beat-to-beat variability of the heart rate. There are also fractal (power law) variations in blood neutrophil counts. Further research will probably turn up other cases of self-similarity.

7.4.1 Body Mass

> There is not a fiercer hell than the failure in a great object.
>
> —**John Keats**

Probably the most obvious scaling factor is the relationship between body size and mass. Body densities among similar animals or plants are nearly the same (body density of non-obese human males is about 1070 kg/m^3), so body mass is related directly to body volume. Within some small amount of variation, different body parts usually scale in size with overall body size, so that body volume (V) can usually be considered as related to the cube of some significant body length (L). Body mass (m), therefore, is thus related to L^3. For now, we can consider L to be any linear measurement made on a body part that has some reason to be important in scaling. L might be height, or the length of the leg, or girth circumference, if any of these has meaning for us (see Tables A.1 and A.2). Body mass is related to body length by

$$m = \rho L^3 \tag{7.4.2}$$

where:
- m is the body mass, kg
- ρ is the body density, kg/m^3
- L is the body length, m

Solving for L in this equation is one way to define body length for an oddly shaped organism. Body density varies between 800 kg/m^3 for plants to 1100 kg/m^3 for mammals (Johnson, 1999).

7.4.2 Body Surface Area

> A moment's insight is sometimes worth a life's experience.
>
> **—Oliver Wendell Holmes, Sr.**

Body surface area of animals is difficult to measure. It is much easier to weigh them. Thus, a relationship between body mass and surface area would be useful.

We know that body mass must be proportional to body volume, and that volume has dimensions of length cubed (L^3). Surface area has dimensions of length squared (L^2). Thus, we would expect surface area to be related to body mass to the two-thirds power:

$$A \propto L^2 = (L^3)^{2/3} \propto m^{2/3} \tag{7.4.3}$$

The relationship given by Cena (1974) and Johnson (1999) is

$$A = 0.09\, m^{2/3} \tag{7.4.4}$$

where:
- A is the body surface area, m^2
- m is the body mass, kg

Tabled values may also be used (Table A.3).

Body surface area of humans is given by the DuBois formula (Johnson, 1999):

$$A = 0.2025\, m^{0.425} H_t^{0.725} \tag{7.4.5}$$

where H_t is the height, m.

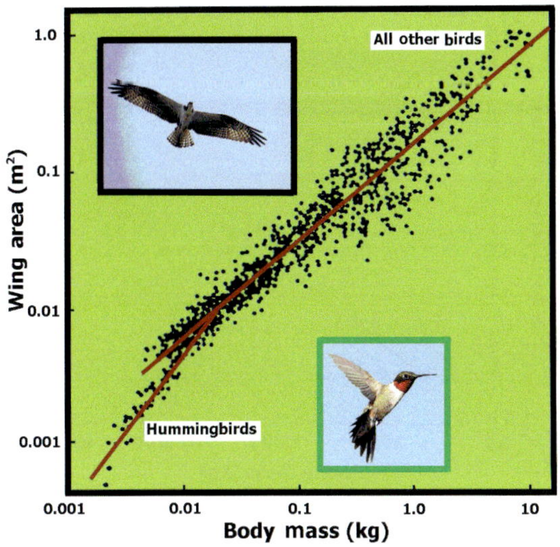

FIGURE 7.4.3
Log–log plot of bird wing area against body mass for birds. (Redrawn from Rayner, J.M.V., *Curr. Ornithol.*, 5, 1–66, 1987.)

Wing surface areas for birds have been measured and plotted on logarithmic coordinates against body mass (Figure 7.4.3) Wing areas should be proportional to the two-thirds power of body mass. Lines fitted to the data are

$$A_w = 0.64\ m^{1.09} \tag{7.4.6}$$

for hummingbirds, and

$$A_w = 0.16\ m^{0.72} \tag{7.4.7}$$

for other birds (Alexander, 2003), where A_w is the wing area, m^2, and m is the body mass, kg.

Plant leaf areas may be found in Table A.5.

7.4.3 Body Dimensions

Life's most urgent question is: what are you doing for others?

—Martin Luther King

Mammalian bone strength is about 1500 kg/cm^2 (Economos, 1983). For larger animals, at least, strength considerations should give a mass–length power relation exponent of about 0.250 (elastic similarity). Geometric similarity would give a mass–length exponent of 0.333 (mass being proportional to length cubed).

The dimensions of various body parts should be proportional for *homologous* (having the same relative structure and proportions) BU of different sizes. Indeed, they should also be proportional to the one-third power of body mass. A log–log graph of total head and body length of whales against body mass appears in Figure 7.4.4 (Alexander, 2003). The graph appears as a straight line, as expected, with a slope of 0.34, also as expected.

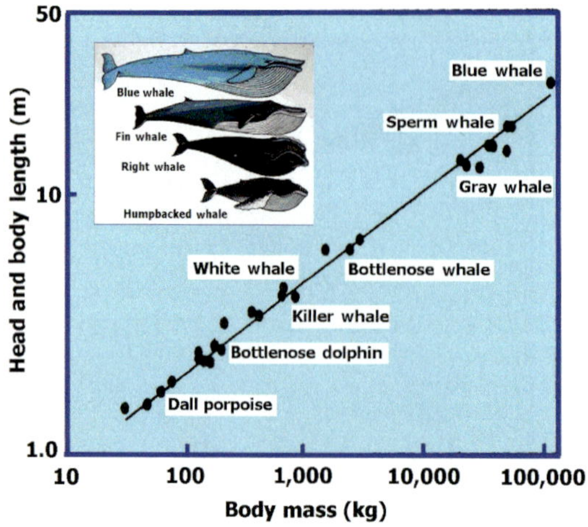

FIGURE 7.4.4
A graph on logarithmic coordinates of total head and body length against body mass for whales. (Redrawn from Economos, A.C., *J. Theor. Biol.*, 103, 167–172, 1983.)

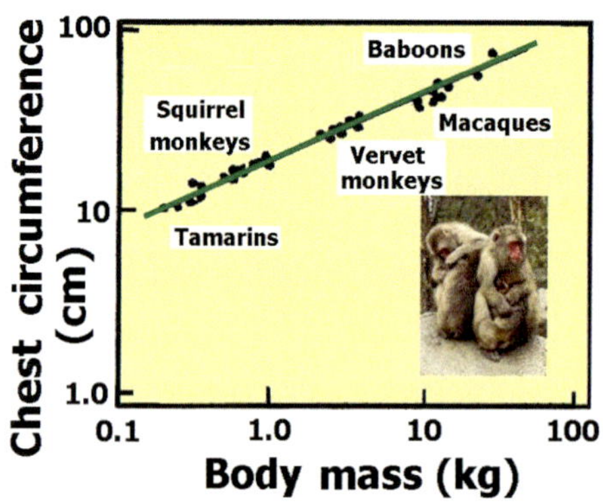

FIGURE 7.4.5
Plot of primate chest circumference against body mass. (Redrawn from Stahl, W.R. and Gummerson, J.Y., *Growth*, 31, 21–34, 1967.)

A log–log plot of chest circumference for primates (Figure 7.4.5) gives a straight line (McMahon, 1984):

$$L_{chest} = 0.171\,m^{0.37} \tag{7.4.8}$$

where L_{chest} is the chest circumference of primates, m. Other primate dimensions have been found to be related to body mass to various fractional powers (Table 7.4.1).

Length of the hind-limb bones (McMahon, 1975a) for adult *ungulates* (hooved mammals) is (Figure 7.4.6)

$$L_{leg} = 0.65\,d^{0.65} \tag{7.4.9}$$

where:

L_{leg} is the sum of femur, tibia, and metatarsal length, m
d is the anterior–posterior midshaft diameter of the femur, m

TABLE 7.4.1

Mass Dependence of Body Measurements in Five Species of Adult Primates

Measurement	Dependence
Trunk height	$m^{0.28}$
Vertex–heel height	$m^{0.30}$
Chest circumference	$m^{0.37}$
Thoracic width	$m^{0.32}$
Maximum thigh girth	$m^{0.38}$
Maximum calf girth	$m^{0.35}$
Knee girth	$m^{0.36}$
Elbow girth	$m^{0.37}$
Neck girth	$m^{0.36}$
Length of femur	$m^{0.34}$
Length of fibula	$m^{0.24}$

Source: McMahon, T.A., *Muscles, Reflexes, and Locomotion*, Princeton University Press, Princeton, NJ, 1984.

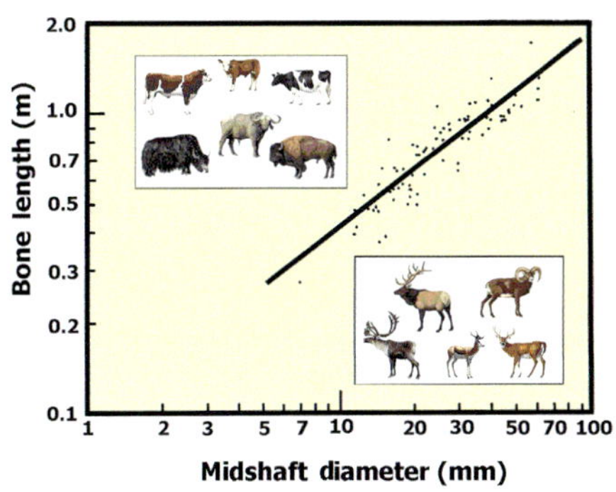

FIGURE 7.4.6
Plot of total leg bone length against femur midshaft thickness. (Redrawn from McMahon, T.A., *Am. Nat.*, 109, 547–563, 1975a.)

Body dimensions for *bovids* (ruminants with paired nondeciduous hollow horns, including antelopes, cattle, sheep, and goats) are also related to body mass (Table 7.4.2).

Low-birth-weight human infants have higher mortality, more hospitalizations in the first 12–20 months of life, and lower intelligence (Morris et al., 1998). The Ponderal Index (100 times birth weight in grams divided by length in centimeters cubed, or 100 g/(cm)^3) is a means to classify infant weights at birth. Morris et al. (1998) used a Ponderal Index value of 2.6 as normal. Values lower than this indicate underweights.

There are two major determinants to plant size (Enquist et al., 2000). The first is the need for mechanical support, and the second is the increased vascular hydrodynamic resistance with longer distances between soil and leaves. To resist buckling from wind or gravity, biomechanical principles lead to this relation between lengths and diameters of trunks and branches (Enquist et al., 2000):

TABLE 7.4.2

Dependences of Body Dimensions on Body Mass of Bovids

Length Measure	Dependence
Body length	$m^{0.27}$
Shoulder height	$m^{0.28}$
Heart girth	$m^{0.37}$
Hock height	$m^{0.22}$
Hind height	$m^{0.24}$

Source: McMahon, T.A., *Muscles, Reflexes, and Locomotion*, Princeton University Press, Princeton, NJ, 1984.

$$l = k\left(\frac{E}{\rho}\right) d^{2/3} \tag{7.4.10}$$

where:

l is the length of the segment, m
E is Young's modulus for the wood, N/m^2
ρ is the density of the wood, kg/m^3
d is the diameter of the segment, m
k is the constant, $s^2/m^{1.67}$

Tree height is proportional to $m^{0.25}$. However, the mass of plants is hard to measure; something like branch diameter is much easier. Thus, many of the allometric relations in plants involve branch diameter:

$$\text{number of leaves} \propto d^2 \tag{7.4.11a}$$

$$\text{number of branches} \propto d^{-2} \tag{7.4.11b}$$

$$\text{reproductive biomass} \propto d^2 \tag{7.4.11c}$$

Additional scaling relations for plants can be found in Enquist et al. (2000).

The diameter of a tree is proportional to the height of the tree raised to the two-thirds power (Bonner and Horn, 2000):

$$d \propto h^{2/3} \tag{7.4.12}$$

Therefore, larger trees are disproportionately thick in order to avoid buckling in the wind. Small plants, for which buckling is not a problem, have stem diameters proportional to their heights to the power of 1.0.

7.4.4 Metabolic Rate and Related Temperatures

> Opportunity is missed by most people because it is dressed in overalls and looks like work.
>
> —**Thomas A. Edison**

Body temperature for nearly all warm-blooded animals is about the same. Mammals have body temperatures regulated to within a range of 36°C–38°C independent of body size (Kleiber, 1975). There is hardly any difference in body temperature between those mammals living in the Arctic or in a tropical jungle. Hibernating mammals, however, are an exception. Birds have a slightly higher body temperature, of about 41°C, compared to mammals.

The maintenance of characteristics such as nearly identical body temperatures in all mammals requires that surface area (proportional to heat loss) and metabolic rate (the amount of heat lost) are linked. Since similar body shapes and nearly identical body densities lead to a relationship between body mass and surface area, metabolic rate is related to body size.

Basal metabolic rate is the summation of heats from all chemical and mechanical processes that must occur to sustain life at a very low level (Johnson, 2007). It is the smallest reproducible amount of heat generated by the body. Other sources of heat are due to the ingestion and digestion of food, and to muscular activity.

Basal metabolic rate (Figure 7.4.7) has been found to vary with body mass (Kleiber, 1975):

$$\mathrm{BMR} = 3.39\, m^{0.75} \tag{7.4.13}$$

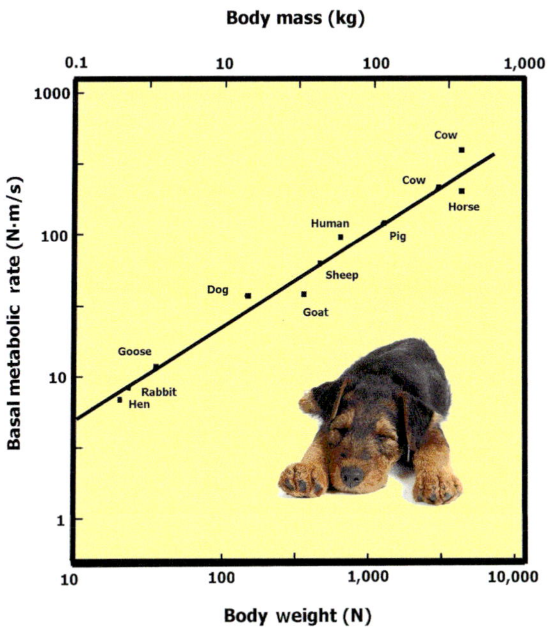

FIGURE 7.4.7
Basal metabolic rate is related to body mass. (Redrawn from Johnson, A.T., *Biomechanics and Exercise Physiology: Quantitative Modeling*, CRC Press/Taylor & Francis, Boca Raton, FL, 2007.)

where:

BMR is the basal metabolic rate, N · m/s

m is the body mass, kg

This similarity gives some guide to heat generated by various species of homeothermic animals (birds and mammals), but does not completely predict heat production. Any muscular activity whatsoever or food eaten within 12–24 h can increase the amount of heat generated. In addition, there is a change of BMR with age, with BMR increasing rapidly in the early stages of life, but declining thereafter for some species and increasing again for others.

The dependence of metabolic rate on mass to the three-quarters power (Figure 7.4.8) extends over almost 27 orders of magnitude from the largest mammal down through mitochondria to the individual molecules catalyzing metabolism (West et al., 2000).

We might suspect that since the rate of heat loss depends on body surface area (see Section 2.7, Heat Transfer), there might be an equivalence of heat production per unit area for all warm-blooded species that maintain their bodies at about the same temperatures. However, this is not exactly true. Taking Equation 7.4.13 for BMR and dividing by Equation 7.4.3 for body surface area shows that BMR per unit area is proportional to $m^{0.75}/m^{0.67} = m^{0.08}$. Thus, a horse is expected to have a slightly higher BMR per unit area than that of a mouse.

The metabolic rates of plants are related directly to their rates of growth and reproduction. These rates of plants scale as $m^{0.75}$, similar to that of animals.

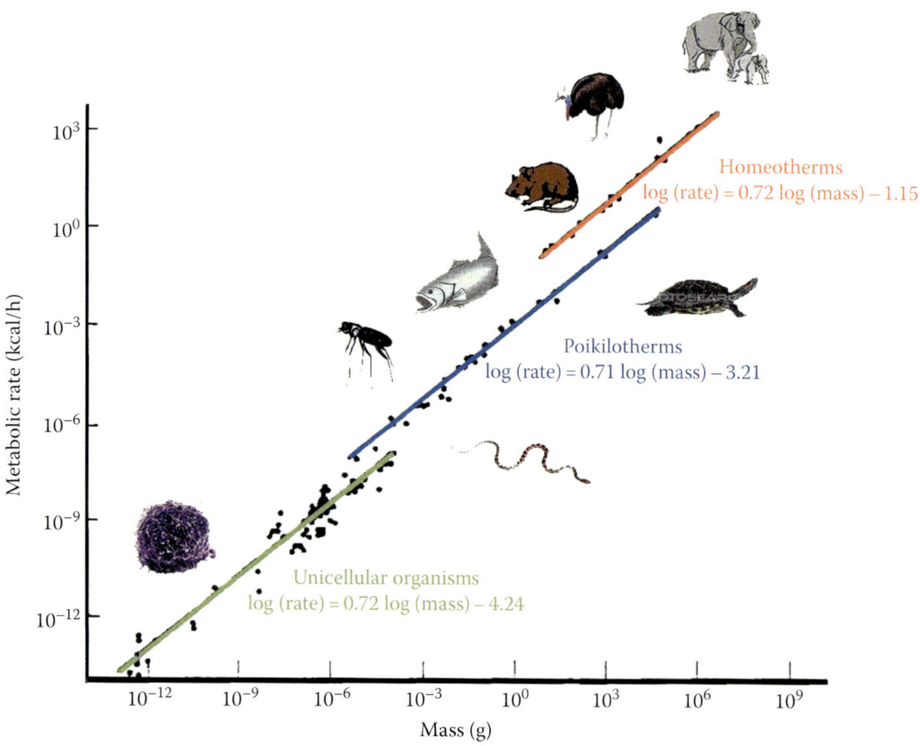

FIGURE 7.4.8

Metabolic rates for a wide range of organisms. Although different classes of organisms have metabolic rates offset from the others, the slopes of the lines ($\propto m^{3/4}$) persist over 20 orders of magnitude.

7.4.5 Oxygen Consumption

Really new ideas are resisted by the experts.

—Charles Townes

Oxygen consumption to support resting metabolic needs is (Dawson, 1991)

$$\dot{V}_{O_2} = 1.87 \times 10^{-7}\, m^{3/4} \tag{7.4.14a}$$

where:
$\dot{V}_{O_2}$ is the basal oxygen consumption, m^3/s
m is the body mass, kg

This relationship was obtained by assuming 4.8 kcal (20,000 N · m) of heat to be equivalent to 1 L of oxygen. Adolph (1949) gives the scaling relation for basal oxygen consumption of mammals as

$$\dot{V}_{O_2} = 1.68 \times 10^{-7}\, m^{0.734} \tag{7.4.14b}$$

with the same units as Equation 7.4.14a.

The oxygen consumption rate for 1 g of average mouse tissue is 9.7×10^{-10} m^3/s, while that for 1 g of average rabbit tissue is 0.12×10^{-10} m^3/s (Dawson, 1991). Thus, 1 g of mouse tissue consumes oxygen (and produces heat) at a rate nearly eight times greater than that of 1 g of average rabbit tissue. More generally, the smaller the animal, the higher its rate of oxygen consumption per unit mass. The fraction of the cardiac output delivered to this tissue must also be in proportion to the oxygen consumption. This leads to the relation, which is nearly invariant with body mass:

$$\left(\frac{\dot{V}_{O_2}}{\text{hr} \cdot m}\right) = 4.9 \times 10^{-8} \tag{7.4.15}$$

where:
$\dot{V}_{O_2}$ is the oxygen consumption of unit mass of tissue, m^3/s
hr is the heart rate, beats/s
m is the body mass, kg

The liver is one of the most metabolically active organs of the body (Table A.9), and, consequently, one of the greatest consumers of oxygen. Adolph (1949) gave the oxygen consumption of the liver of mammals to be

$$\dot{V}_{O_{2L}} = 1.87 \times 10^{-7}\, m^{0.77} \tag{7.4.16}$$

where:
$\dot{V}_{O_{2L}}$ is the liver oxygen consumption, m^3/s
m is the body mass, kg

Maximum oxygen consumption (Figure 7.4.9) of exercising mammals has been plotted and analyzed to give (McMahon, 1984)

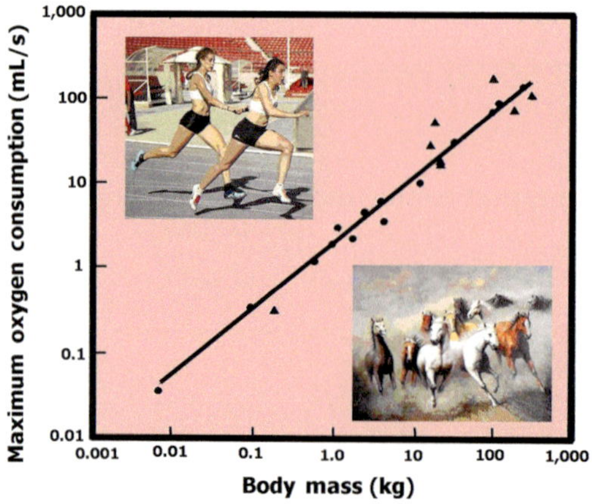

FIGURE 7.4.9
Maximum oxygen consumption of exercising mammals. (Redrawn from Taylor, C.R. et al., *Respir. Physiol.*, 44, 25–37, 1981.)

$$\dot{V}_{O_2\max} = 1.92\times 10^{-6}\, m^{0.809} \tag{7.4.17}$$

where:

$\dot{V}_{O_2\max}$ is the maximum rate of oxygen consumption, m^3/s
m is the body mass, kg

7.4.6 Heat Loss

> When President Kennedy said we would go to the moon, we had the vision but not the technology. Today we have the technology but not the vision.
>
> **—Christopher Reeve**

Heat loss from animals depends not only on the surface area, but also on temperature difference between the body interior and the environment, and on the thickness of the insulating layer including the skin, underlying tissues, and surrounding still air (see Section 2.7, Heat Transfer):

$$\dot{q} = \frac{kA}{L}(T_b - T_{\text{amb}}) \tag{7.4.18}$$

where:

$\dot{q}$ is the heat loss, N · m/s
A is the body surface area, m^2
L is the equivalent thickness of insulating layer, m
k is the thermal conductivity, N/(s · °C)
T_b is the body temperature, °C
T_{amb} is the ambient temperature, °C

The ratio of thermal conductivity to thickness (k/L) is called *conductance*, and is often used in heat transfer calculations on animals where it is difficult to estimate both surface area and insulation thickness.

It might be expected that insulation thickness would vary as L or $m^{1/3}$. Such is not the case, however, because it has been empirically observed that skin thickness is proportional to $m^{1/8}$ (Kleiber, 1975). Conductances have been measured on the carcasses of birds and mammals and found to be (Kleiber, 1975)

$$C = 4.87\ m^{1/2} \quad \text{for birds} \tag{7.4.19a}$$

$$C = 5.91\ m^{1/2} \quad \text{for mammals} \tag{7.4.19b}$$

where:
- C is the thermal conductance, N · m/(s · °C)
- m is the mass, kg

7.4.7 Cardiovascular Factors

> Defects or imperfections as small as 10 microns [in an artificial heart] were sufficient to form platelet clumps leading to thromboembolic complications.
>
> —**Victor L. Poirier**

We start with the heart, which is the organ that propels the blood throughout the body. Larger animals (Figure 7.4.10) have larger hearts (Dawson, 1991):

$$m_H = 0.0043\ m \tag{7.4.20a}$$

Adolph (1949) gives (for mammals)

$$m_H = 0.00575\ m^{0.98} \tag{7.4.20b}$$

where:
- m_H is the mass of the heart, kg
- m is the body mass, kg

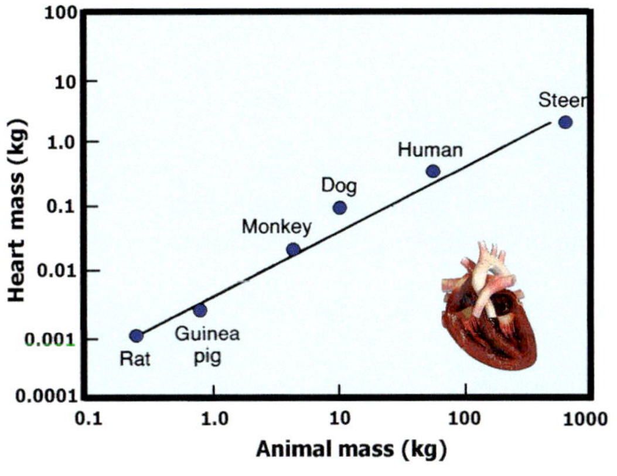

FIGURE 7.4.10
Mass of the heart is related to body mass. (Redrawn from Dawson, T.H., *Engineering Design of the Cardiovascular System of Mammals*, Prentice-Hall, Englewood Cliffs, NJ, 1991.)

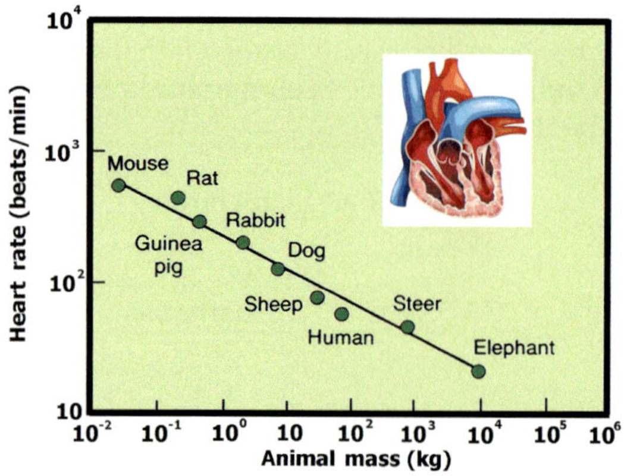

FIGURE 7.4.11
Heart rate is inversely related to body mass. (Redrawn from Dawson, T.H., *Engineering Design of the Cardiovascular System of Mammals*, Prentice-Hall, Englewood Cliffs, NJ, 1991.)

As a rule, smaller animals have faster heart rates than larger animals (Dukes, 1955). The allometric relationship (Figure 7.4.11) for this is (Dawson, 1991)

$$\text{hr} = \frac{3.82}{m^{1/4}} \tag{7.4.21a}$$

or (Adolph, 1949)

$$\text{hr} = \frac{3.62}{m^{0.27}} \tag{7.4.21b}$$

where hr is the resting heart rate, beats/s.

The mass of blood in the body is given by (Dawson, 1991)

$$m_{\text{bl}} = 0.056\ m \tag{7.4.22a}$$

or (Adolph, 1949)

$$m_{\text{bl}} = 0.513\ m^{0.99} \tag{7.4.22b}$$

where:

m_{bl} is the mass of blood, kg
m is the body mass, kg

The difference between these two equations is not large.

For humans, the volume of circulating blood is (Johnson, 1991)

$$V_{\text{bl}} = 80 \times 10^{-6}\ m \tag{7.4.23}$$

where V_{bl} is the blood volume, m^3.

The mass of circulating hemoglobin in the blood is also related to body mass (Adolph, 1949):

$$m_{\text{Hb}} = 0.012\ m^{0.99} \tag{7.4.24}$$

where m_{Hb} is the mass of hemoglobin, kg.

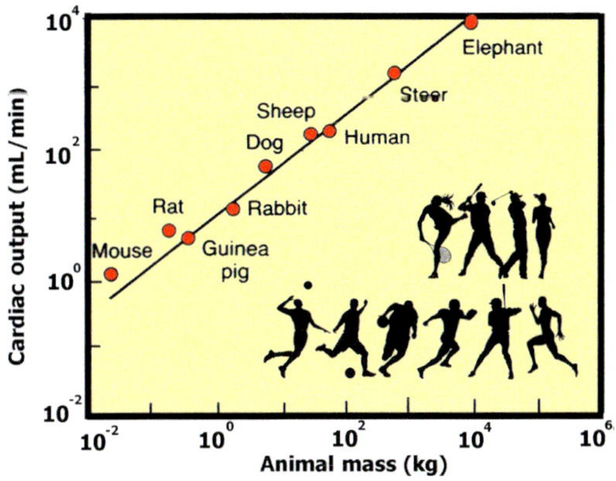

FIGURE 7.4.12
Cardiac output is related to body mass. (Redrawn from Dawson, T.H., *Engineering Design of the Cardiovascular System of Mammals*, Prentice-Hall, Englewood Cliffs, NJ, 1991.)

If it is assumed that about 5 mL of oxygen are delivered by every 100 mL of circulating blood, then the *cardiac output*, or the rate of blood delivered per unit time (Figure 7.4.12), is expected to be about 20 times the average rate of oxygen consumption (Equation 7.4.14a) at rest (Dawson, 1991):

$$\text{CO} = 3.73\times10^{-6}\ m^{3/4} \tag{7.4.25a}$$

where CO is the cardiac output, m^3/s (see also Table A.7).

The linear velocity of blood through the aorta (the arterial vessel leading from the heart) is invariant with body size (Kleiber, 1975). The volume rate of flow through a closed vessel equals the velocity times the cross-sectional area of the vessel. Thus, cardiac output, which is the volume rate of flow in the aorta, is

$$\text{CO} = 3.73\times10^{-6}\ m^{3/4} = v_a A_a \tag{7.4.25b}$$

where:
v_a is the aortic blood velocity, m/s
A_a is the cross-sectional area of aorta, m^2

Hence, with v_a constant, aortic cross-sectional area must be proportional to $m^{3/4}$, and aortic diameter must be proportional to $m^{3/8}$ (Dawson, 1991):

$$d_a = 0.36\ m^{3/8} \tag{7.4.26}$$

where d_a is the aorta diameter, m. Similar mass dependence is true of all the larger blood vessels (arteries and veins) in the body.

Dawson (1991) gave a scaling relationship for the length of the large vessels as $L_a \propto m^{1/4}$, and for the number of large vessels as $n_a \propto m^0$ (or, the number is invariant with body mass). Kleiber (1975) gave scaling relationships for the same three large vessel parameters as

$$d_a \propto m^{1/3} \tag{7.4.27a}$$

$$L_a \propto m^{1/3} \tag{7.4.27b}$$

$$n_a \propto m^0 \tag{7.4.27c}$$

where:

d_a is the large vessel diameter, m
L_a is the large vessel length, m
n_a is the number of large vessels

Diameters of dichotomous branching vessel segments of the vasculature obey Murray's law:

$$d_{\text{parent}}^3 = d_{\text{child}_1}^3 + d_{\text{child}_2}^3 \tag{7.4.28}$$

where:

d_{parent} is the diameter of the vessel before it branches
d_{child} is the diameter of the vessel after the branch

There is another group of vessels in close proximity to the cells of the body called *capillaries*. Because of the diffusional considerations, the capillaries must be nearly the same size in all animals (see Section 2.8, Materials Movement), but the number of capillaries must be proportional to the amount of tissue (Kleiber, 1975). The connecting vessels interposed between larger blood vessels and the capillaries (arterioles and venules) must have intermediate numbers and sizes.

Dawson (1991) gives scaling relationships for the capillaries as

$$d_c \propto m^{1/12} \tag{7.4.29a}$$

$$L_c \propto m^{5/24} \tag{7.4.29b}$$

$$n_c \propto m^{5/8} \tag{7.4.29c}$$

where:

d_c is the capillary diameter, m
L_c is the capillary length, m
n_c is the number of capillaries, unitless

Hemodynamic resistance of the large vessels can be approximated by the Hagen–Poiseuille law as (Kleiber, 1975)

$$R \propto \frac{L}{A^2} \tag{7.4.30}$$

From Kleiber's relationships,

$$R \propto \frac{m^{1/3}}{m^{4/3}} = \frac{1}{m} \tag{7.4.31a}$$

From Dawson's relationships,

$$R \propto \frac{m^{1/4}}{m^{6/4}} = \frac{1}{m^{5/4}} \tag{7.4.31b}$$

The difference between these two is not large.

Arterial blood pressure of animals is independent of body size.

Lastly, the partial pressure of oxygen in the blood is the pressure of oxygen that would be in equilibrium with oxygen dissolved in the blood if the blood were in an open vessel. Oxygen partial pressure is related to the saturation level of hemoglobin in the red blood cells. This is given by Dawson (1991) as

$$pO_2 \propto m^{-1/12} \tag{7.4.32}$$

where pO_2 is the oxygen partial pressure, N/m^2. This relation indicates that there is a slight inverse influence of body mass on oxygen partial pressure at any given hemoglobin saturation level.

Because life span increases as $m^{-1/4}$ and heart rate depends on $m^{1/4}$, the number of heart beats in a lifetime is the same (about 1.5×10^9) for all mammals (West et al., 2000). Likewise, because metabolic rate is also proportional to $m^{1/4}$, the total energy needed to support a given mass of an organism during its lifetime is the same (about 70 MJ/kg) for all mammals.

A summary of cardiovascular allometric relationships is given in Tables 7.4.3 and 7.4.4. For further scaling considerations for the cardiovascular system, consult Dawson (1991).

TABLE 7.4.3

Scaling of Selected Cardiovascular Parameters

Cardiovascular Dimension	Observed Mass Dependence
Aorta, blood velocity	$m^{0.07}$
Aorta, pressure	$m^{0.032}$
Aorta, radius	$m^{0.36}$
Blood, circulation time	$m^{0.25}$
Blood, oxygen affinity	$m^{-0.089}$
Blood, volume	m
Capillaries, density	$m^{-0.095}$
Capillaries, number	$m^{0.75}$
Cardiac output	$m^{0.74}$
Cardiac stroke volume	m
Circulation distance	$m^{0.25}$
Heart rate	$m^{-0.25}$
Peripheral resistance, total	$m^{-0.76}$

Source: West, G.B. et al., The origin of universal scaling laws in biology, In *Scaling in Biology*, J.H. Brown and G.B. West (Eds.), Oxford University Press, New York, pp. 87–112, 2000.

TABLE 7.4.4

Allometric Relations of Some Hemodynamic Parameters

Parameter	Equation
Aorta, diameter (m)	$d_a = 4.10 \times 10^{-3}\ m^{0.36}$
Aorta, length (m)	$L_a = 0.175\ m^{0.31}$
Heart mass (kg)	$m_h = 0.0066\ m^{0.98}$
Heart rate (s^{-1})	$\text{hr} = 3.60\ m^{-0.27}$
Metabolic rate (N·m/s)	$M = 3.41\ m^{0.73}$
Pressure, arterial (N/m^2)	$p_a = 1.17 \times 10^4\ m^{0.033}$
Pulse velocity (m/s)	$v_p = 4.46\ m$
Stroke volume (L)	$V_s = 6.6 \times 10^{-4}\ m^{1.05}$

Source: Li, J.K.-J., Scaling and invariants in cardiovascular biology, In *Scaling in Biology*, J.H. Brown and G.B. West (Eds.), Oxford University Press, New York, pp. 113–128, 2000.

7.4.8 Respiration

Air which has thus served the purpose of animal respiration is no longer common air; it approaches to the nature of fixed air [air containing CO_2 and not O_2] in as much as it is capable of combining with lime-water and precipitating the lime from it, in the form of a calcareous earth; but it differs from fixed air.

—Antoine Lavoisier describing the work of Priestley

Lung capacities, volumes, and function capabilities are all related to body size through metabolic rates. As metabolic rate per unit body mass increases, lung size as a percentage of body volume must usually increase to supply the oxygen needed to support metabolism. Oxygen and carbon dioxide are transferred between the air and the blood in the bottom reaches of the lung called the *alveoli*. Alveolar oxygen must move from the airway to the alveolar wall by the process of diffusion, and this process takes some time to accomplish (see Section 2.8, Convection and Diffusion). Similarly, carbon dioxide must move from the capillary blood to the alveolar wall, and then from the alveolar wall to the airway by diffusion. Higher metabolic rates require that these diffusional processes occur faster, and so one might expect alveolar size to be smaller to sustain higher metabolic rates. Such is the case, as can be seen from Figure 7.4.13 (Johnson, 2007).

It is especially of interest in pulmonary function labs to be able to predict normal ranges of various lung volumes and flow rates in order to diagnose respiratory diseases (see Tables A.8 and 7.4.5). Lung function has been measured and correlated to body size (mass and height) and to age (Bartlett, 1973). Many of these, however, appear as nomograms and graphs rather than equations (Figure 7.4.14).

Total lung mass has been given by McMahon (1984)

$$m_L = 0.0113\ m^{0.99} \tag{7.4.33a}$$

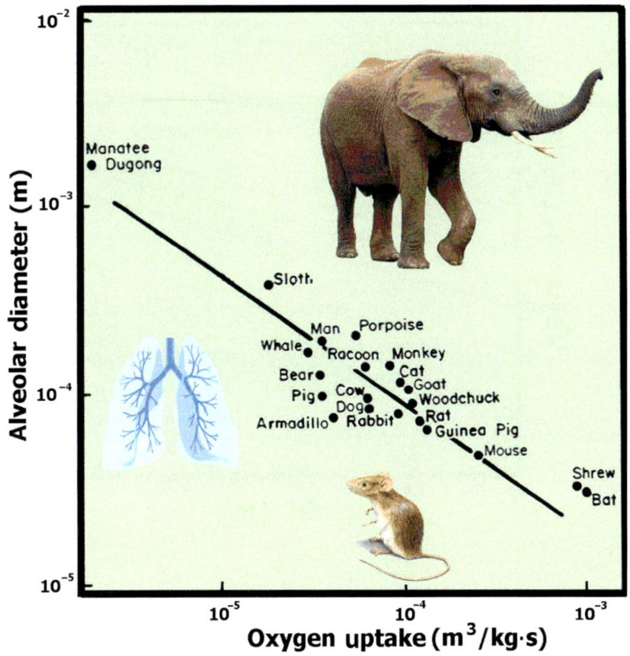

FIGURE 7.4.13
Alveolar diameter is smaller for higher rates of oxygen consumption. (Redrawn from Johnson, A.T., *Biomechanics and Exercise Physiology: Quantitative Modeling*, CRC Press/Taylor & Francis, Boca Raton, FL, 2007.)

TABLE 7.4.5

Prediction Equations for Human Vital Capacity

Subject	Equation
Infants, <1 week, resting	$V = 0.139 + 0.00367\,m$
Males, 10–17 year	$VC = 0.0562\,H + 0.0097\,Am - 6.27$
140–160 cm	$VC = 0.0492\,H - 4.82$
160–190 cm	$VC = 0.0799\,H - 9.75$
Males, standing	$VC = 0.025\,H^2$
Females, standing	$VC = 0.020\,H^2$
Males	$MBC = [86.5 - (0.522 \times A)] \times A^3$
Females	$MBC = [71.3 - (0.474 \times A)] \times A^3$
Both sexes	$MBC = H \times (1.24 - 0.0095\,A)^3$

Source: Spector, W.S., (Ed.), *Handbook of Biological Data*, WADC Technical Report 56-273, Wright-Patterson Air Force Base, OH, 1956.

Notes: Equations are applicable for the following conditions: 37°C, ambient pressure, saturated with water vapor. MBC = maximum breathing capacity, L/min; A = body surface area, sq m; V = pulmonary ventilation, L/min; m = mass, kg; VC = vital capacity, L; H = height, cm; A = age, years. Weight of infants in ounces. Age correction for vital capacity: 46–55 years = −4%; 56–65 years = −8%; 66–75 years = −16%; over 75 years = −30%.

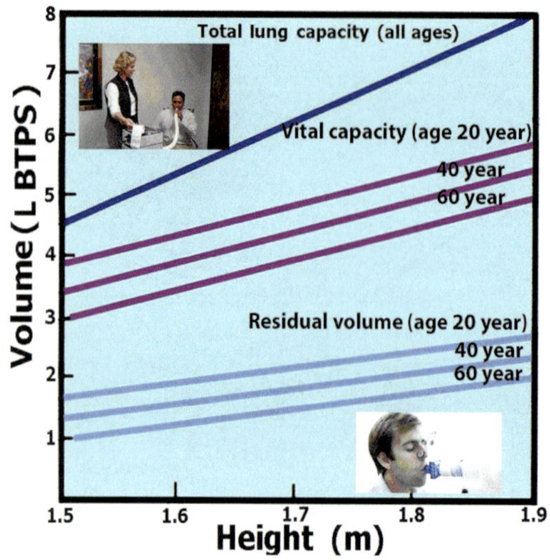

FIGURE 7.4.14
Lung function measurements related to body height and age in males. TLC, total lung capacity; VC, vital capacity; RV, residual volume. (Redrawn from Bartlett, R.G., Respiratory system, In *Bioastronautics Data Book*, J.F. Parker Jr. and V.R. West (Eds.), NASA, Washington, DC, pp. 489–531, 1973.)

Adolph (1949) gave

$$m_L = 0.0116\ m^{0.99} \tag{7.4.33b}$$

where:
 m_L is the lung mass, kg
 m is the body mass, kg

Tidal volume, or the amount of air exhaled during each breath, is, for mammals at rest (McMahon, 1984):

$$V_T = 7.69 \times 10^{-6}\ m^{1.04} \tag{7.4.34a}$$

or (Adolph, 1949)

$$V_T = 6.64 \times 10^{-6}\ m^{1.01} \tag{7.4.34b}$$

where:
 V_T is the tidal volume, m^3
 m is the body mass, kg

And resting respiration rate for mammals is (McMahon, 1984)

$$\text{RR} = 0.892\ m^{-0.26} \tag{7.4.35a}$$

or (Adolph, 1949)

$$\text{RR} = 0.854\ m^{-0.28} \tag{7.4.35b}$$

where:
RR is the respiration rate, breaths/s
m is the body mass, kg

Minute volume, given as the volume of gas exhaled per unit time, is given by

$$\dot{V}_E = (V_T)(\text{RR}) \tag{7.4.36}$$

where $\dot{V}_E$ is the resting minute volume, m³/s
and thus,

$$\dot{V}_E = 6.86 \times 10^{-6}\, m^{0.78} \tag{7.4.37a}$$

or (Adolph, 1949)

$$\dot{V}_E = 5.53 \times 10^{-6}\, m^{0.74} \tag{7.4.37b}$$

A summary of allometric dependencies is given in Table 7.4.6.

Two important parameters of respiratory mechanics are resistances and compliances (compliance is the name given to capacity in fluid flow systems—see Sections 2.1, Effort and Flow, and 2.9, Fluids). Resistances to air flow are distributed throughout the respiratory airways, lung tissue, and chest wall. The upper (larger) airways are usually considered separately from the lower airways because they have different mechanical properties and higher flow rates. The relationship between pulmonary resistance

TABLE 7.4.6

Scaling of Respiratory Parameters

Respiratory Parameter	Observed Mass Dependence
Alveoli, number of	$m^{0.75}$
Alveolus, radius of	$m^{0.13}$
Alveolus, surface area	$m^{0.083}$
Alveolus, volume of	$m^{0.25}$
Intrapleural pressure	$m^{0.00}$
Lung volume	$m^{1.05}$
Lung, surface area	$m^{0.95}$
Oxygen-diffusing capacity	$m^{0.99}$
Power dissipated	$m^{0.78}$
Respiration rate	$m^{-0.26}$
Tidal volume	$m^{1.04}$
Total airway resistance	$m^{-0.70}$
Tracheal air velocity	$m^{0.02}$
Tracheal diameter	$m^{0.39}$

Source: McMahon, 1984.

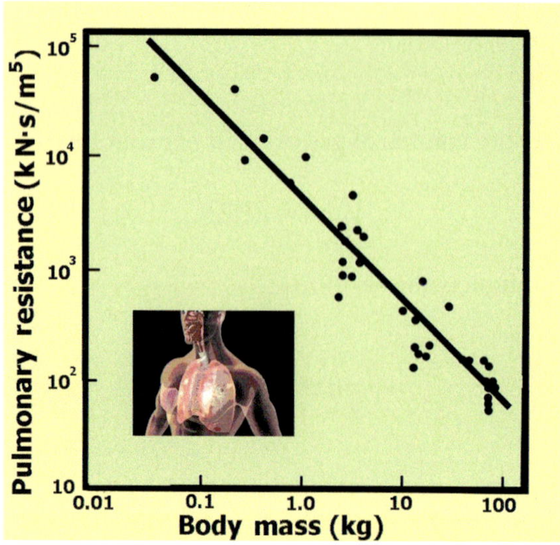

FIGURE 7.4.15
Variation of pulmonary resistance (excluding upper airway resistance) with body mass. (Redrawn from Johnson, A.T., *Biomechanics and Exercise Physiology: Quantitative Modeling*, CRC Press/Taylor & Francis, Boca Raton, FL, 2007.)

(excluding upper airway resistance) and body mass is seen in Figure 7.4.15, and equations are (Johnson, 2007)

$$R_{\mathrm{aw}} - R_{\mathrm{uaw}} = (21.0 \times 10^5)\, m^{-0.862} \tag{7.4.38a}$$

$$R_p - R_{\mathrm{uaw}} = (40.4 \times 10^5)\, m^{-0.903} \tag{7.4.38b}$$

$$R_r - R_{\mathrm{uaw}} = (21.0 \times 10^5)\, m^{-0.393} \tag{7.4.38c}$$

where:

R_{aw} is the airway resistance, N s/m^5
R_{uaw} is the upper airway resistance, N s/m^5
R_p is the pulmonary resistance, N s/m^5
R_r is the respiratory resistance, N s/m^5
m is the body mass, kg

Like resistance terms, compliance terms vary with body mass:

$$C_{\mathrm{tot}} = (1.50 \times 10^{-5})\, m \tag{7.4.39a}$$

$$C_{\mathrm{lt}} = (1.50 \times 10^{-5})\, m^{1.20} \tag{7.4.39b}$$

$$C_{\mathrm{cw}} = (1.50 \times 10^{-5})\, m^{0.898} \tag{7.4.39c}$$

where:

m is the body mass, kg
C_{tot} is the total lung and chest wall compliance, m^5/N
C_{lt} is the lung tissue compliance, m^5/N
C_{cw} is the chest wall compliance, m^5/N

Specific compliance, which is defined as compliance divided by functional residual capacity (FRC), or the resting volume of the lung, appears to assume a nearly constant value of 0.00082 m^2/N (0.08/cm H_2O) for the whole size spectrum of mammals from bats to whales (Mines, 1981).

7.4.9 Walking and Running

> When there is ever a true breakthrough you can find a time period when the consensus was, "Well that's nonsense" ... a true, creative researcher has to have confidence in nonsense.
>
> **—Burt Rutan**

The stride frequency of running vertebrates and swim undulation frequency of fish both scale approximately the same with $m^{-0.17}$. The velocity of running animals and the speeds of flying birds have approximately the same relation to body masses, $m^{-0.17}$. There is also a similarity of muscle forces used for running, flying, and swimming; with very little variation, there is a value of about 60 N/kg developed (Bejan and Marden, 2006). These facts lead to the conclusion that there are some underlying factors regulating aspects of locomotion (Figure 7.4.16).

FIGURE 7.4.16
Flying, running, and swimming each have similar properties based on maximizing distance moved per unit of energy expenditure.

The freely swinging leg during walking can be visualized as a physical pendulum pivoted at the top end. The period of oscillation of this physical pendulum is (Davidovits, 1975)

$$T = 2\pi\sqrt{\frac{2L}{3g}} \tag{7.4.40}$$

where:
T is the period of oscillation, s
L is the leg length, m
g is the acceleration due to gravity, 9.81 m/s^2

In the most effortless walk (called *ballistic walking*), the legs swing at their natural frequency and the time for one step is $T/2$. Walking faster or slower requires more muscular effort and is not likely to happen without some special motivation.

The speed of walking is proportional to the product of the number of steps taken in a given time and the length of the step. The step length is proportional to the length of the leg. Thus, the walking speed is

$$v_{walk} \propto \frac{L}{T} \propto \frac{L}{\sqrt{L}} = \sqrt{L} \tag{7.4.41}$$

where:
v_{walk} is the walking speed, m/s
L is the length of leg, m

Hence, the speed of the natural walk of a person or animal is dependent on the length of the legs. A small animal may move its small legs faster (smaller T), but its walking speed will be slower than that of a tall animal that moves its legs slower.

Running is different. Torque to move the legs during running comes from muscular effort rather than from gravity. It will be assumed that the length of the leg muscles is proportional to the length of the leg (L), and that the cross-sectional area of the muscles is proportional to the length squared (L^2). The muscular mass is thus proportional to length cubed (L^3).

Maximum muscular force is proportional to the area of the muscle ($F_{max} \propto L^2$). This is true because all animal muscles are nearly the same.

Maximum muscular torque depends on the maximum force (F_{max}) and the length of the leg (L). Thus, maximum torque is proportional to $F_{max} \cdot L \propto L^2 \cdot L = L^3$.

The period of oscillation of a physical pendulum (the leg) moving in response to a torque is given by the moment of inertia of the leg and the maximum torque:

$$T \propto \sqrt{\frac{I}{F_{max} \cdot L}} \tag{7.4.42}$$

where:
T is the period of oscillation, s
I is the moment of inertia, kg m^2
F_{max} is the maximum force, N
L is the leg length, m

The moment of inertia of the physical pendulum pivoted at the top end is (Davidovits, 1975)

$$I = \frac{mL^2}{3} \tag{7.4.43}$$

But, because leg mass (m) is proportional to L^3,

$$I \propto L^3 \cdot L^2 = L^5 \tag{7.4.44}$$

Hence,

$$T \propto \sqrt{\frac{L^5}{L^2 \cdot L}} = L \tag{7.4.45}$$

Maximum running speed is given as before

$$v_{\text{run}} \propto \frac{L}{T} \propto \frac{L}{L} = 1 \tag{7.4.46}$$

There is no leg length dependence of the maximum running speed. Any similarly built animal will run at nearly the same maximum speed. There are running speed differences, of course, but the differences are relatively small and do not depend upon size.

Walking in the reduced gravity of the Moon or Mars can be shown to be different from that on Earth with the help of Equation 7.4.40. With reduced gravity, the period of oscillation is longer, and the natural speed of walking is slower. On the Moon, for instance, the force of gravity is only one-sixth that on Earth:

$$\frac{T_{\text{Moon}}}{T_{\text{Earth}}} = \sqrt{6} = 2.45 \tag{7.4.47}$$

Thus, natural walking speed is nearly 2½ times slower (or 40% as fast) on the Moon. If one tries to walk faster than this, one either stubs a toe or must expend muscular effort. *Apollo* astronauts on the Moon preferred, instead, to move about in a series of low jumps a few centimeters high (McMahon, 1984).

Stride frequencies for mammals are given (McMahon, 1984) for walking (Figure 7.4.17):

$$f_{\text{walk}} = 0.9\, L_{\text{sh}}^{-0.571} \tag{7.4.48a}$$

for trotting:

$$f_{\text{trot}} = 1.67\, L_{\text{sh}}^{-0.527} \tag{7.4.48b}$$

and for cantering:

$$f_{\text{cantor}} = 2.0\, L_{\text{sh}}^{-0.493} \tag{7.4.48c}$$

where:

f is the stride frequency, cycles/s

L_{sh} is the shoulder height, m

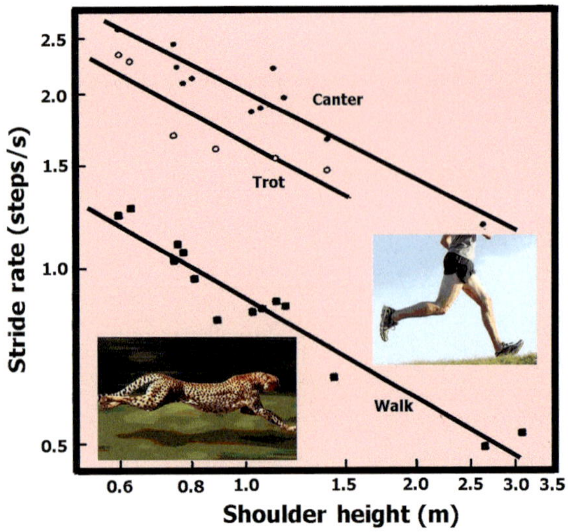

FIGURE 7.4.17
Stride frequencies versus shoulder heights in 14 species of mammals. (Redrawn from Pennycuick, C.J., *J. Exp. Biol.*, 63, 775–799, 1975.)

Locomotion is produced by muscular shortening (see Section 4.4.2, Actuators). The intrinsic speed of muscular sarcomere shortening for mammals has been found to depend on body mass (Figure 7.4.18). Equations relate mass and maximum shortening speed of the extensor *digitorum longus* muscle (McMahon, 1984):

$$v_{\mathrm{EDL}} = 35.4 \times 10^{-6}\, m^{-0.13} \tag{7.4.49a}$$

and soleus muscle:

$$v_{\mathrm{SOL}} = 15.1 \times 10^{-6}\, m^{-0.17} \tag{7.4.49b}$$

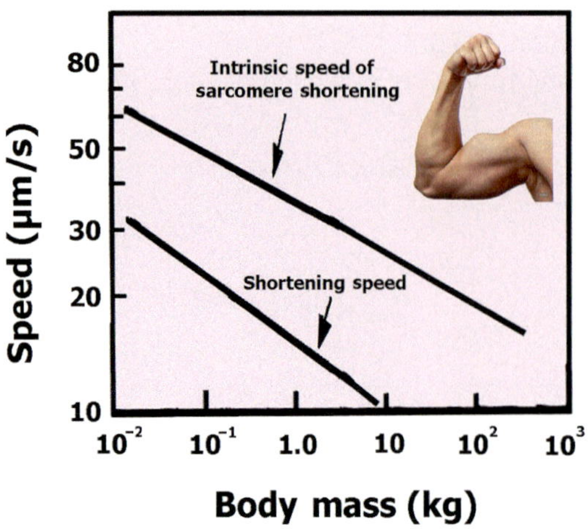

FIGURE 7.4.18
Shortening speeds of unloaded muscles. (Redrawn from McMahon, T.A., *J. Appl. Physiol.*, 39, 619–627, 1975b.)

where:

v is the shortening speed, m/s
m is the body mass, kg

Loaded muscles will shorten at slower rates. Maximum muscle stress is approximately 3.0×10^5 N/m^2 (Alexander, 2000) and does not depend on body mass.

The angular excursions of the rear limbs of quadrupedal mammals have also been measured during running (McMahon, 1975b; McMahon, 1984). These are given as

$$\phi_\pi = 1.30\, m^{-0.10} \tag{7.4.50a}$$

or

$$\phi_o = 74.4\, m^{-0.10} \tag{7.4.50b}$$

where:

ϕ_π is the angular excursion, radians
ϕ_o is the angular excursion, degrees
m is the body mass, kg

The oxygen cost of running appears to be scaled with the size of the animal and does not appear to depend primarily on overall shape of the animal (Figure 7.4.19). From McMahon (1984),

$$\dot{V}_{O_{2R}} = 5.36 \times 10^{-3}\, vm^{0.60} + 2.96 \times 10^{-3}\, m^{0.75} \tag{7.4.51}$$

where:

$\dot{V}_{O_{2R}}$ is the oxygen consumption of running, m^3 O$_2$/s
v is the speed, m/s
m is the body mass, kg

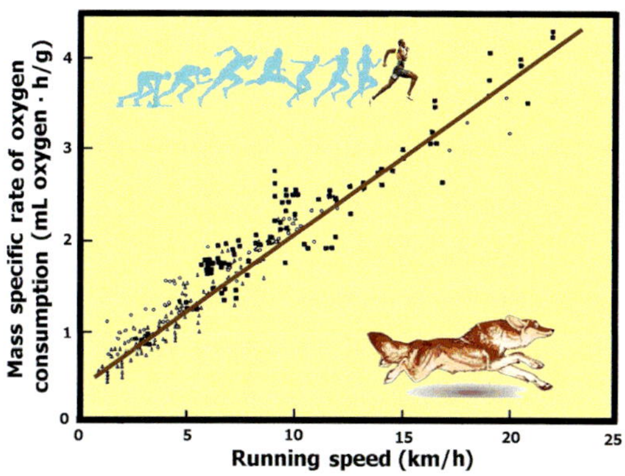

FIGURE 7.4.19
The oxygen cost of running is related to running speed. (Redrawn from Taylor, C.R. et al., *Am. J. Physiol.*, 227, 848–850, 1974.)

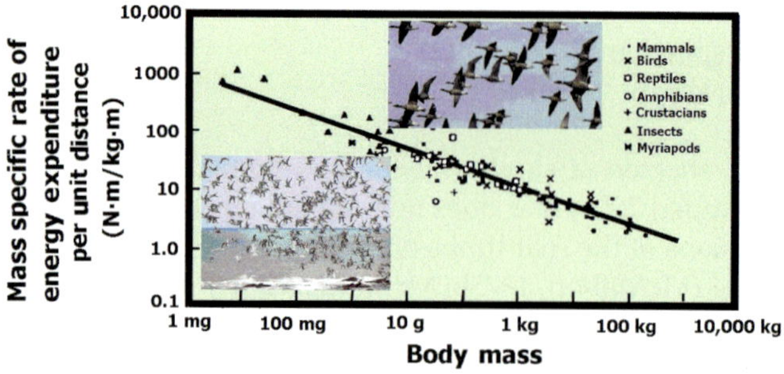

FIGURE 7.4.20
Cost of transport for various species. The line represents the minimum cost of transport. (Redrawn from Full, R.J. and Tu, M.S., *J. Exp. Biol.*, 156, 215–231, 1991.)

Alexander (2003) gave the metabolic cost of transport (power per unit mass per unit speed) for the animal kingdom (see Figure 7.4.20) as

$$M_{\text{transport}} = 10.7\ m^{-0.32} + \frac{6.0\ m^{-0.30}}{v} \tag{7.4.52a}$$

where:

$M_{\text{transport}}$ is the metabolic cost of transport, $\text{N} \cdot \text{m}/(\text{kg} \cdot \text{m})$
m is the body mass, kg
v is the speed, m/s

At high speeds, the metabolic cost of transport approaches its asymptotic minimum value of

$$M_{\text{transport}} = 10.7\ m^{-0.32}, \tag{7.4.52b}$$

which can be obtained from Equation 7.4.52a by letting $v \to \infty$.

Calculated mechanical efficiencies (mechanical output divided by energy cost, see Section 2.4.3, Efficiency) can also be seen to depend on body mass as

$$\eta = 10\ m^{0.307} \tag{7.4.53}$$

where η is the efficiency, percent. Using Equation 7.4.53, and a human body mass of 70 kg, gives a maximum muscle efficiency of 37%.

7.4.10 Relations Involving Time

> Life is very short and very uncertain; let us spend it as well as we can.
>
> —**Samuel Johnson**

Smaller organisms usually take a shorter time to grow, reach sexual maturity, reproduce, and to die. Some of this can be seen in Figure 7.4.21, in which generation time, the time

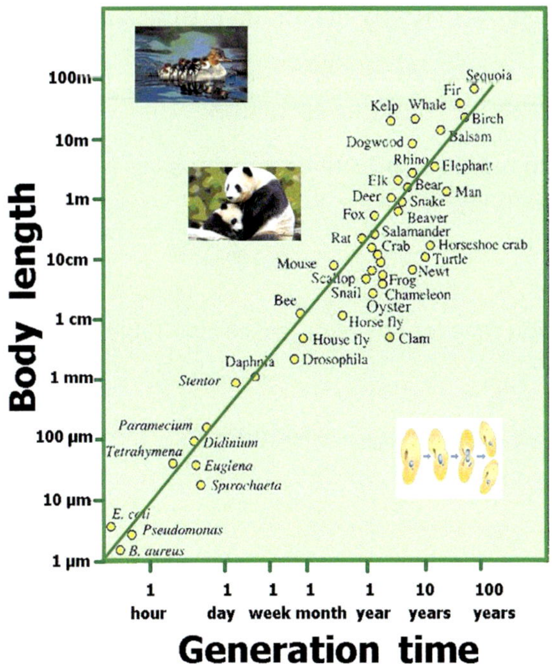

FIGURE 7.4.21

Generation time for smaller organisms is shorter. Notice that data in this graph includes plants, animals, and bacteria, among others. (Redrawn from Molles, M.C. Jr., *Ecology: Concepts and Applications*, McGraw-Hill, New York, 1999.)

between births of successive generations, is plotted against organismal length. A best-fit line through this data is

$$t_{\text{gen}} = 1.70 \times 10^8\, L^{0.935} \tag{7.4.54}$$

where:

t_{gen} is the generation time, s
L is the length of organism, m

The graph and the equation indicate to us that even over a very broad range of sizes, and over many types of organisms, there is a relationship between size and generation time.

McMahon (1984) gives the following for 50% growth time for mammals:

$$t_{50} = 1.11 \times 10^7\, m^{0.25} \tag{7.4.55}$$

where:

t_{50} is the time to grow to one-half maximum mature size, s
m is the body mass, kg

And, for 98% growth time (mammals)

$$t_{98} = 3.81 \times 10^7\, m^{0.26} \tag{7.4.56}$$

where t_{98} is the time to grow to nearly maximum mature size, s. The time to reproductive maturity is

$$t_{mat} = 1.75 \times 10^7 \, m^{0.18} \tag{7.4.57}$$

where t_{mat} is the time to reach reproductive maturity, s.

Gestation time in mammals is

$$t_{ges} = 5.64 \times 10^6 \, m^{0.25} \tag{7.4.58}$$

where t_{ges} is the gestation time for offspring, s (see also Tables A.9 and A.10). And mammal life span in captivity is

$$t_{life} = 3.66 \times 10^8 \, m^{0.20} \tag{7.4.59}$$

where:

t_{life} is the life span, s
m is the body mass, kg

Note that animal life span in the wild can be much shorter than this.

7.4.11 Food and Waste

> ...if by design one engineers a cell's genetic program to produce a certain type of cell behavior or function, this is as much engineering as any of the more traditional activities we readily accept as engineering.
>
> **—Robert M. Nerem**

Food is required to maintain body functions, move about, grow, repair tissue damage, and to bear young. The minimum food requirement is that to support the basal metabolic rate, which we have seen in Equation 7.4.13 to be related to $m^{3/4}$. Thus, between two animals requiring similar food, the one with the larger body mass should receive a larger food portion based on the ratio of body masses to the three-fourths power.

The energy content of food rations can be calculated from the energy equivalent of glucose, 2.816×10^6 N · m/mol glucose (Johnson, 1999). Each mole of glucose has a mass of 180 g, so each gram of glucose yields 15,644 N · m of energy. If we assume a digestion efficiency of 50%, then each gram of glucose equivalent carbohydrate will yield about 7800 N · m of energy:

$$\text{Energy} = \left(\frac{2.816 \times 10^6 \ \text{N} \cdot \text{m/mol glucose}}{180 \ \text{g/mol glucose}} \right)(0.5) = 7822 \ \text{N} \cdot \text{m/g} \tag{7.4.60}$$

More active animals, or those that have had significant tissue damage (due to fighting or accidents), or those that are pregnant, or those actively growing will need more food.

Solid wastes produced will depend on the type of food (highly refined vs. rough plant material) and on the amount of food. Biological engineers designing facilities to hold animals will have to account for wastes produced.

Waste excretion has been given by McMahon (1984) as

$$\dot{V}_{\text{urine}} = 7.04 \times 10^{-10}\, m^{0.75} \tag{7.4.61a}$$

or, from Adolph (1949),

$$\dot{V}_{\text{urine}} = 5.13 \times 10^{-10}\, m^{0.82} \tag{7.4.61b}$$

where:

$\dot{V}_{\text{urine}}$ is the volume rate of urine production, m^3/s
m is the body mass, kg

Production of urine varies throughout the day, so Equations 7.4.61a and 7.4.61b are probably best used by multiplying $\dot{V}_{\text{urine}}$ by 86,400 to obtain the volume of urine produced in 24 h.

Kidney function is usually designated as *clearance*. Clearance is the ratio between the rate of excretion of a substance (kg/s) and its concentration in body fluid (kg/m^3). The units of clearance are thus m^3/s; to determine the rate of excretion (and thus the rate of accumulation outside the body) requires that the bodily concentration be known. Adolph (1949) gives the following maximum clearance rates:

$$\text{CL}_{\text{urea}} = 6.38 \times 10^{-8}\, m^{0.72} \tag{7.4.62}$$

$$\text{CL}_{\text{creatinine}} = 1.37 \times 10^{-7}\, m^{0.69} \tag{7.4.63}$$

$$\text{CL}_{\text{diodrast}} = 2.78 \times 10^{-7}\, m^{0.89} \tag{7.4.64}$$

$$\text{CL}_{\text{hippurate}} = 3.77 \times 10^{-7}\, m^{0.80} \tag{7.4.65}$$

where:

CL is the clearance rate, m^3/s
m is the body mass, kg

Urea is formed in the body from ammonia that comes from deamination of amino acids; it permeates through the entire body except the brain. Excreted urea can be used as a nitrogen source by microbes, plants, and animals. Creatinine is the nitrogenous waste product of muscle creatine; phosphorylated creatine (also called *phosphagen*) is an energy source alternative to ATP (see Section 3.10, Energy-Rich Compounds). Diadrast® is the commercial name for iodopyracet, an iodinated dye used to determine kidney function; it has the property that a very high proportion is filtered and excreted from the kidney in a very short time. Hippurate is a salt of hippuric acid, and is used as a test of liver function.

The average rate of total nitrogen output for mammals is (Adolph, 1949)

$$N_{\text{tot}} = 3.30 \times 10^{-9}\, m^{0.735} \tag{7.4.66}$$

where N_{tot} is the total nitrogen output, kg/s.

Excreted sulfur can form odoriferous wastes and toxic gases. Total sulfur output of mammals is (Adolph, 1949)

$$S = 7.88 \times 10^{-11}\, m^{0.74} \tag{7.4.67}$$

where S is the sulfur output, kg/s.

The water intake needed to support excretion, evaporation, and other bodily functions is (Adolph, 1949)

$$\dot{V}_{H_2O} = 1.21\times 10^{-9}\, m^{0.88} \tag{7.4.68}$$

where $\dot{V}_{H_2O}$ is the water input, m^3/s.

7.4.12 Bird Songs

The nightingale has a lyre of gold,
The lark's is a clarion call,
And the blackbird plays but a boxwood flute.

—William Henley

Bird censuses are often taken by listening for their songs. The sound power output (P, mW/cm^2) at 1 m from the bird has been found to be related to body mass (g):

$$P = 108\, m^{1.14} \tag{7.4.69}$$

The center frequency of bird songs is inversely related to body mass (Calder, 2000):

$$f \propto m^{-0.24} \tag{7.4.70}$$

Larger birds have lower frequency calls.

Defended territorial area (A, m^2) is also allometrically related to mass (Calder, 2000):

$$A = 1068\, m^{1.17} \tag{7.4.71}$$

This means that the power output per unit defended area (P/A) is nearly independent of body mass.

7.4.13 Plant Growth

Plants grow by converting sugars produced by photosynthesis into other carbohydrates that are sturdy and structurally robust. The rate of plant growth therefore depends on the rate at which sugars can be formed through photosynthesis.

The basic metabolic pathway for photosynthetic sugar formation is called the Calvin cycle: carbon dioxide is enzymatically added to an acceptor molecule and then converted into two 3-carbon sugar molecules with the help of photochemically generated ATP and NADPH (Taiz and Zeiger, 1998). Plants that use this mechanism exclusively are called C3 plants.

Some plants possess an additional metabolic pathway to concentrate carbon dioxide before it enters the Calvin cycle. These are called C4 plants because of the 4-carbon molecules that are formed as intermediates. C4 plants comprise about 1% of plant species and 5% of the Earth's plant biomass. They have higher photosynthetic efficiency than do C3 plants. C4 plants are found in habitats with high daytime temperatures and intense sunlight. They also have an advantage in dry climates because their leaf stomata do not have to be as open to admit CO_2 as those for C3 plants. Open stomata allow loss of water to the atmosphere. C4 crops thus use far less water and also less nitrogen than do C3 crops; they typically have 50% higher yields.

The C4 pathway seems to have evolved many times as an example of convergent evolution. Thus, there are few gross physical features distinguishing C4 plants from C3 plants. A few examples of C4 plants are

Crabgrass
Maize
Sugarcane
Sorghum

and

Cacti
Bryophyllum
Epiphytic bromeliads (and pineapple)
Sedums

The top group isolates the C4 pathways in different portions of their leaves; the bottom group separates the C4 pathways in time.

There is an approximately linear relationship between sugar production and light available for absorption. C3 plants produce about 1.9 g of sugar per MJ (one million watt seconds) of light absorbed; C4 plants produce about 2.4 g sugar per MJ of light energy (Sinclair, 2009). A bright day in a temperate climate during the growing season averages about 20 MJ/m^2. The energy available to plants depends on wavelengths (the wavelengths from 400 to 700 nm are utilized in photosynthesis, and are called *photosynthetically active radiation*), the angle of the light striking the plant, the actual plant area intercepting the light, and the degree of cloudiness or haze (Taiz and Zeiger, 1998).

Plants, such as maize (C4), rice (C3), and wheat (C3), produce about 0.75 g of plant mass per gram of photosynthetic sugar. The total plant mass produced is approximately

$$m''_{pl} = 0.75\ \eta_{photo} E_{sol} d \tag{7.4.72a}$$

where:
m''_{pl} is the plant mass, g/m^2
η_{photo} is the photosynthetic efficiency, 1.9 g sugar/MJ for C3 plants and 2.4 g sugar/MJ for C4 plants
E_{sol} is the solar energy, MJ/m^2
d is the number of days

Oil crops, such as sunflower (C3) and peanuts (C3), convert only about 42% of their sugars into plant mass (Sinclair, 2009). Thus,

$$m''_{pl} = 0.42\ \eta_{photo} E_{sol} d \tag{7.4.72b}$$

The earlier calculations assume that sufficient water, light, and nutrients are available for plant growth; they do not limit the plant's potential.

Water necessary for plant growth is mainly determined by the transpiration rate of water lost from the plant to the atmosphere. This rate is

$$H_2O'' = \frac{m''_{pl}\,\Delta p_{vap}}{\eta_{water}} \tag{7.4.73}$$

where:

H_2O'' is the transpiration water need, g/m^2

Δp_{vap} is the vapor pressure deficit = saturated vapor pressure − vapor pressure of the air, N/m^2

η_{water} is the water efficiency coefficient, N/m^2

The water efficiency coefficient value has been determined for each species. Its value is about 9 N/m^2 for a C4 grass, 6 N/m^2 for a C3 grass, or 5 N/m^2 for a C3 oil species. The higher the water efficiency coefficient, the smaller the water requirement.

Typical vapor pressure deficits are 2500 N/m^2 in an arid environment and 1500 N/m^2 for more humid areas (Sinclair, 2009). Vapor pressure deficits can be obtained from information in psychometric charts (Johnson, 1999).

The transpiration amount should be considered the smallest amount of water required to sustain plant growth. It must be supplied at a rate needed by the plant at various stages of growth. The rate of water use at any time can be calculated from Equation 7.4.73 by substituting the time rate of growth for m''_{pl}.

Plants must also have sufficient amounts of the macronutrients (nitrogen, phosphorus, and potassium) and micronutrients (sulfur, calcium, magnesium, iron, manganese, boron, copper, zinc, and molybdenum) in order to thrive. Considering plant leaves only, 4 g of nitrogen per m^2 of ground surface area are needed for C4 plants, and 8.8 g of nitrogen per m^2 of ground surface area (leaves) are needed for C3 plants. Plant stems require 12 mg nitrogen per gram of stem (Sinclair, 2009). Some of this nitrogen may come from previous crops or rainstorms, but most has to be applied in the form of fertilizer.

Example 7.4.1: Frog Jumping Model (Alexander, 2000)

The assumption for this model is that the greater the jumping speed, the more the survival advantage. Frogs with greater jumping speeds can jump faster and longer, and can escape predators more effectively.

Greater muscle masses and stronger legs result in faster jumps, but they also have greater metabolic costs. Thus, the mathematical expression for the difference between benefit (jumping speed) and cost (metabolism) is the term to be maximized in this model (Figure 7.4.22).

Modeling of the metabolic costs include:

Stronger muscles require stronger bones and bigger legs.
Muscle tissue has a metabolic rate several times the resting metabolic rate of the body.
Accumulation of muscle energy is several times the metabolic rate of the body.
There is a typical maximum muscle stress of 300 kN/m^2 that can be exerted.
There is a maximum shortening speed of the muscle.
Resting metabolic rates of muscles are proportional to maximum shortening speed.

Results of this model show that jumping speed generally increases as muscle mass increases up to an unrealistically high muscle mass of 35% or more of body mass. The optimal leg length should be proportional to body mass to the one-third power:

$$L \propto m^{1/3}$$

FIGURE 7.4.22
Diagram of the model of a jumping frog. (Redrawn from Alexander, R.M., Hovering and jumping: Contrasting problems in scaling, In *Scaling in Biology*, J.H. Brown and G.B. West (Eds.), Oxford University Press, New York, pp. 37–50, 2000.)

Example 7.4.2: Monkeys in Space

In the early days of the U.S. space program, monkeys were used to test the first occupied satellites in Earth orbit. If you worked for the National Aeronautics and Space Administration (NASA) at that time, you may have been asked to determine the amount of life support necessary to sustain the monkey named "Ham" while in orbit. Calculate the metabolic rate, heat production, oxygen consumption, food and liquid requirements, and expected rate of waste production.

Solution

From Table A.1, the mass of a rhesus monkey is about 3.5 kg. From Equation 7.4.13, the basal metabolic rate is calculated as

$$\text{BMR} = 3.39\,(3.5)^{0.75} = 8.67\ \text{N}\cdot\text{m/s}$$

Heat production would be expected to be identical to the basal metabolic rate, because, in space, muscle energy required to maintain posture against gravity would be minimal. Thus, heat production is 8.67 N · m/s.

Oxygen consumption to support Ham would be, from Equation 7.4.14a,

$$\dot{V}_{O_2} = 1.87\times10^{-7}(3.5)^{3/4} = 4.79\times10^{-7}\ \text{m}^3\text{/s}$$

From Equation 7.4.14b,

$$\dot{V}_{O_2} = 1.68\times10^{-7}(3.5)^{0.734} = 4.21\times10^{-7}\ \text{m}^3\text{/s}$$

To be conservative, the higher number should be selected unless the weight of oxygen to be orbited is paramount.

Food requirements would be minimal for a short trip into space. For a longer visit, the glucose equivalent of the food would be, from Equation 7.4.60,

$$\frac{\text{BMR}}{7822} = \frac{8.67\,(\text{N}\cdot\text{m/s})}{7822(\text{N}\cdot\text{m/g})} = 1.11\,\text{mg/s}$$

$$= 4.0\,\text{g/h glucose equivalent (pure sugar)}$$

Fruits and vegetables are less energy dense than glucose. Bananas, for instance, contain 85 kcal food energy per 100 g (Watt and Merrill, 1963). The number of N · m per kcal is 4184 (Johnson, 1999). Thus, bananas without skins contain

$$(85\,\text{kcal})\,(4184\,\text{N}\cdot\text{m/kcal})/100\,\text{g} = 3556\,\text{N}\cdot\text{m/g}$$

Bananas would have to be supplied at a rate of

$$\frac{\text{BMR}}{3556} = \frac{(8.67\,\text{N}\cdot\text{m/s})}{(3556\,\text{N}\cdot\text{m/g})}$$

$$= 2.44\,\text{mg/s}$$
$$= 8.8\,\text{g/h}$$

Liquid requirements could probably be supplied by the bananas that contain 75.7% water (Watt and Merrill, 1963). This means that the monkey would ingest

$$(8.8\,\text{g/h})\,(0.757) = 6.64\,\text{g/h of water} = 6.64 \times 10^{-6}\,\text{m}^3\text{/h of water}$$

Additional water would come from metabolism of food.

The estimates of solid waste cannot be determined from the information given, because solid waste production depends on the type of food eaten hours or days before. If we assume that Ham had been placed on a banana diet for a sufficiently long time before his flight, then the solid waste produced would equal the indigestible part of the bananas (fiber and ash). From Watt and Merrill (1963), fiber and ash of bananas without skins totals 1.3 g/100 g. Ingesting 8.8 g bananas per hour would result in an average solid waste production of (8.8 g/h) (0.013), or 0.11 g waste per hour.

Urine production can be estimated from Equation 7.4.59a as

$$\dot{V}_{\text{urine}} = (7.04 \times 10^{-10})\,(3.5)^{0.75} = 1.8 \times 10^{-9}\,\text{m}^3\text{/s}$$
$$= (1.8 \times 10^{-9}\,\text{m}^3\text{/s})\,(10^6\,\text{g/m}^3)(3600\,\text{s/h})$$
$$= 6.4\,\text{g/h}$$

And, from Equation 7.4.59b,

$$\dot{V}_{\text{urine}} = (5.13 \times 10^{-10})\,(3.5)^{0.82} = 1.43 \times 10^{-9}\,\text{m}^3\text{/s}$$
$$= 5.2\,\text{g/h}$$

Both these values are close to the ingested water mass from the bananas.

One additional waste product can be anticipated: carbon dioxide. The metabolism of glucose yields 6 mol of CO_2 for each mole (180 g) of glucose (see Section 3.10 (Energy-Rich Compounds) and Equation 3.8.1):

$$(CH_2O)_6 + 6O_2 \rightarrow 6CO_2 + 6H_2O$$

Thus, moles of CO_2 produced are

$$\frac{(4.0 \text{ g/h})6}{180 \text{ g/mol}} = 0.13 \text{ mol } CO_2/\text{h}$$

The volume of this gas depends on its temperature and pressure, according to the ideal gas law (see Section 2.3, States of Matter).

Example 7.4.3: Life Support in a Large Animal Veterinary Medical Clinic

A large agricultural animal veterinary medical clinic would be expected to perform surgery and provide life support and monitoring for animals ranging in size from baby goats (kids) to adult horses and cows. Determine the range of needs for respiratory, metabolic, and cardiac monitoring and support.

Solution

From Table A.1, the body mass of a baby goat would be about 3 kg and the body mass of an adult horse would be about 770 kg. These masses will establish the range of requirements.

Respiration

From Equation 7.4.37a, minute volume is

Baby Goat	**Adult Horse**
$\dot{V}_E = 6.86 \times 10^{-6}\,(3)^{0.78}$	$\dot{V}_E = 6.68 \times 10^{-6}\,(770)^{0.78}$
$= 1.6 \times 10^{-5}\ \text{m}^3/\text{s}$	$= 1.2 \times 10^{-3}\ \text{m}^3/\text{s}$

From Equation 7.4.37b

$\dot{V}_E = 5.53 \times 10^{-6}\,(3)^{0.74}$	$\dot{V}_E = 5.53 \times 10^{-6}\,(770)^{0.74}$
$= 1.2 \times 10^{-5}\ \text{m}^3/\text{s}$	$= 7.6 \times 10^{-4}\ \text{m}^3/\text{s}$

Respiration rates are obtained from Equation 7.4.35a:

$\text{RR} = 0.892\,(3)^{-0.26}$	$\text{RR} = 0.892\,(770)^{-0.26}$
$= 0.67\ \text{br/s}$	$= 0.16\ \text{br/s}$

or from Equation 7.4.35b:

$\text{RR} = 0.854\,(3)^{-0.28}$	$\text{RR} = 0.854\,(770)^{-0.28}$
$= 0.63\ \text{br/s}$	$= 0.13\ \text{br/s}$

Oxygen consumption for resting animals (BMR) can be obtained from Equation 7.4.14a:

$\dot{V}_{O_2} = 1.87 \times 10^{-7}\,(3)^{0.75}$	$\dot{V}_{O_2} = 1.87 \times 10^{-7}\,(770)^{0.75}$
$= 4.3 \times 10^{-7}\ \text{m}^3/\text{s}$	$= 2.7 \times 10^{-5}\ \text{m}^3/\text{s}$

(Continued)

Baby Goat	Adult Horse
and from Equation 7.4.14b:	
$\dot{V}_{O_2} = 1.68 \times 10^{-7}\ (3)^{0.734}$	$\dot{V}_{O_2} = 1.68 \times 10^{-7}\ (770)^{0.734}$
$= 3.8 \times 10^{-7}\ \text{m}^3/\text{s}$	$= 2.2 \times 10^{-5}\ \text{m}^3/\text{s}$
Metabolism: From Equation 7.4.13	
$\text{BMR} = 3.39\ (3)^{0.75}$	$\text{BMR} = 3.39\ (770)^{0.75}$
$= 7.7\ \text{Nm/s}$	$= 496\ \text{Nm/s}$
Cardiac: Heart rate can be obtained from Equation 7.4.21a:	
$\text{hr} = 3.82/(3)^{0.25}$	$\text{hr} = 3.82/(770)^{0.25}$
$= 2.9\ \text{beats/s}$	$= 0.73\ \text{beats/s}$
or from Equation 7.4.21b:	
$\text{hr} = 3.62/(3)^{0.27}$	$\text{hr} = 3.62/(770)^{0.27}$
$= 2.7\ \text{beats/s}$	$= 0.60\ \text{beats/s}$
Cardiac output is obtained from Equation 7.4.25a as	
$\text{CO} = 3.73 \times 10^{-6}\ (3)^{0.75}$	$\text{CO} = 3.73 \times 10^{-6}\ (770)^{0.75}$
$= 8.5 \times 10^{-6}\ \text{m}^3/\text{s}$	$= 5.5 \times 10^{-4}\ \text{m}^3/\text{s}$

Other parameters can be calculated and used as the basis for design or selection of life support systems. Where there is a disagreement between values calculated from one equation or another, the most conservative choice can be made.

Example 7.4.4: Cats Lapping Water

When cats lap water, they do not cup their tongues, as dogs do. They use their tongues to slap the water, rather than scooping the water into their mouths. When a cat lifts its tongue, a column of water adheres to the tongue, and is drawn into its mouth by inertia (Reis et al., 2010).

Lapping frequency was found to be dependent upon cat body mass:

$$f = 4.6\ m^{-0.181}$$

where:

f is lapping frequency, in laps/sec
m is body mass, in kg.

Larger cats lap water slower than smaller cats. This makes sense, because larger cats have larger tongues that require more power to move as fast as smaller tongues.

Example 7.4.5: Estimate the Amount of Biofuel Biomass That Can Be Produced in a Season

Solution

We assume a C4 species of grass will be produced. An average solar energy flux of 20 MJ/m^2 will be assumed for a growing season of 120 days.

From Equation 7.4.72a,

$$\begin{aligned} m''_{pl} &= 0.75\ \eta_{photo} E_{sol} d \\ &= 0.75 \times 2.4 \times 20 \times 120 \text{ days} \\ &= 4320 \text{ g mass/m}^2\text{, or 43.2 metric tonnes/hectare} \end{aligned}$$

Remark

This yield estimate is about the same as the maximum yield of sugarcane grown in the United States (Sinclair, 2009).

Example 7.4.6: King Kong's Structural Constraints

King Kong was a fictional movie character who appeared as a gorilla much larger than normal. All of his proportions were the same as a normal gorilla.

In real life, structural proportions of bone limit animal sizes and proportions. If King Kong were five times as tall as a normal gorilla, then his mass would scale as his volume, or $5^3 = 125$ times the mass of a normal gorilla. However, the strengths of his bones (limited by critical stresses) would depend on their cross-sectional areas, or $5^2 = 25$ times as strong as normal bones. Thus the forces on his leg bones would be $125/25 = 5$ times larger than normal; his bones would be crushed. Apes can be smaller or larger, but structural considerations impose constraints (Willmore, 2010).

Example 7.4.7: Animals Shaking Off Water

Dogs aren't the only animals that shake when wet. Other animals also use this means to dry their bodies. The smaller the animal, the faster it shakes. Shaking frequency data in Table 7.4.7 appeared in *National Geographic* magazine (Bloch, 2011). Investigate to see if an allometric relationship can be developed between shaking frequency and body mass.

TABLE 7.4.7

Shaking Frequencies of Animals

Animal	Body Mass (kg)	Log (mass)	Shaking Frequency (cps)	Log (frequency)
Mouse	0.023	−1.64	29	1.46
Rat	0.25	−0.602	18	1.26
Guinea pig	0.345	−0.462	14	1.15
Cat	3.3	0.519	9	0.954
Poodle	12	1.08	6	0.778
Lab retriever	14	1.15	5	0.699
Sumatran tiger	160	2.20	4	0.602
Brown bear	550	2.74	4	0.602

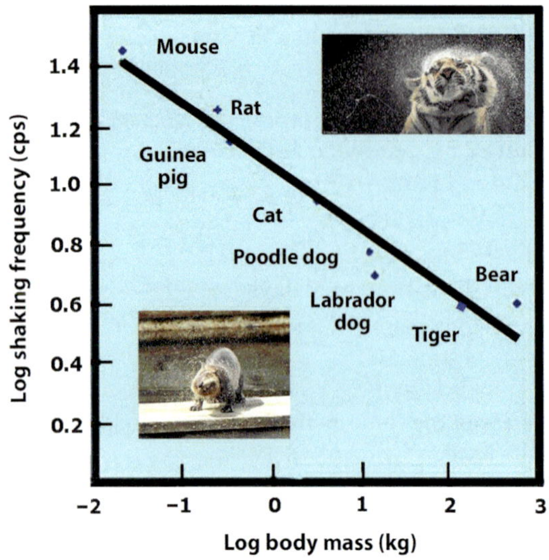

FIGURE 7.4.23
Graph of the logarithmic relationship between shaking frequency (cps) and body mass (kg) for several animals.

Solution

Nominal body mass data were obtained from Table A.1 in the appendix, and are given in Table 7.4.7. Because it is reasonable to suspect a power-law relationship between shaking frequency and body mass, data for each of these variables were converted into logarithms and the values plotted (Figure 7.4.23). The apparent straight-line relationship is consistent with a power-law equation.

Using the least-squares technique explained in Section 4.2.4 (Statistics), the best-fit equation was found to be:

$$\log f = 1.07 \log m^{-0.21} \tag{7.4.74a}$$

Converting this into a power-law equation requires taking the antilog of 1.07 to yield:

$$f = 11.8\, m^{-0.21} \tag{7.4.74b}$$

This equation predicts a shaking frequency of 4.1 cycles per second for the tiger and a frequency of 4.8 cps for a 70 kg human.

7.5 Self-Similarity for Tissues and Organs

7.5.1 Organs

The brain is simply an organ which excretes feeling as the kidneys excrete urine.

—Irving John Good

There are good reasons for being able to predict the masses of organs in the body. Among these are the need to know the dosage of drugs to administer for directed, local delivery;

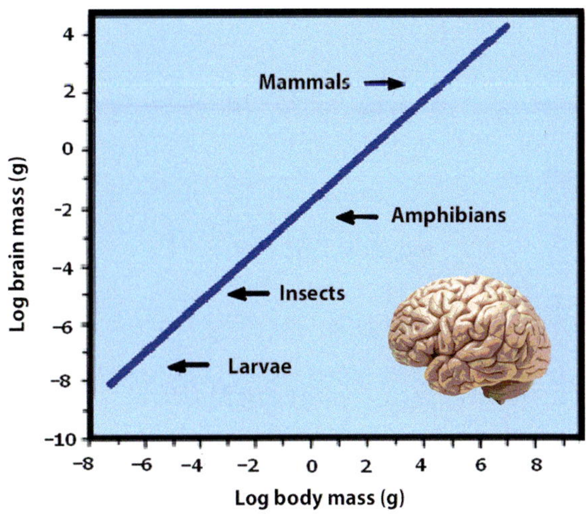

FIGURE 7.5.1
Brain mass related to body mass. (Redrawn from Eberhard, W.G. and Wcsilo, W.T., *Amer. Sci.*, 100, 226–233, 2012.)

the ability to cool isolated organs quickly and effectively; and the need to calculate the amounts of biochemicals found only in those organs. Power law equations for organs were first used by DuBois (1898) and Lapique (1898) for brain weights. Many allometric relations are summarized in Brody (1945) and Huxley (1932). These allometric relations for mammals to follow are each taken from Adolph (1949). All masses are in units of kilograms.

The mass of the brain is (Figure 7.5.1)

$$m_{\text{brain}} = 0.0102\ m^{0.70} \tag{7.5.1}$$

The mass of the liver is

$$m_{\text{liver}} = 0.0334\ m^{0.87} \tag{7.5.2}$$

The mass of the stomach and intestines is

$$m_{\text{gut}} = 0.0740\ m^{0.94} \tag{7.5.3}$$

The peristaltic motion of the gut beats with a duration of

$$t_{\text{gut}} = 2.20 \times 10^{-9}\ m^{0.31} \tag{7.5.4}$$

where t_{gut} is the time for successive waves, s.

The mass of the thyroid gland is

$$m_{\text{thyroid}} = 1.27 \times 10^{-4}\ m^{0.92} \tag{7.5.5}$$

The mass of the adrenal gland is

$$m_{\text{adrenal}} = 2.76 \times 10^{-4}\ m^{0.80} \tag{7.5.6}$$

The mass of the pituitary gland is

$$m_{\text{pituitary}} = 2.48 \times 10^{-5}\, m^{0.76} \tag{7.5.7}$$

The mass of each kidney is

$$m_{\text{kidney}} = 7.52 \times 10^{-3}\, m^{0.85} \tag{7.5.8}$$

The diameter of the renal (kidney) body is

$$d_{\text{kidney}} = 1.41 \times 10^{-4}\, m^{0.08} \tag{7.5.9}$$

where d_{kidney} is the diameter of the renal corps, m.

A nephron is the basic filtration unit of the kidney. The number of nephrons present in both kidneys is

$$n_{\text{nephron}} = 1.88 \times 10^{5}\, m^{0.62} \tag{7.5.10}$$

where n_{nephron} is the number of nephrons in the animal.

Additional allometric relationships for bone lengths, bone weights, and organ weights for primates are given by Stahl and Gummerson (1967).

7.5.2 Tissues

> ...Biological information, or bioinformatics, requires more than knowing the genetic make-up of a biological system. The greater challenge is to relate cell, tissue, and organ function to a cell's genetic program.
>
> **—Robert M. Nerem**

The mass of hemoglobin has already been given by Equation 7.4.24. Myoglobin is the oxygen-binding molecule in the muscles. It has one heme group rather than the four in hemoglobin. Oxygen is transferred from red blood cell hemoglobin to solution in the blood to muscle myoglobin. The mass of myoglobin in the body of mammals is (Adolph, 1949)

$$m_{\text{my}} = 3.32 \times 10^{-4}\, m^{1.31} \tag{7.5.11}$$

where m_{my} is the mass of myoglobin, kg.

Cytochrome is a protein pigment that acts as an electron carrier in the electron transport system of the mitochondria. The amount of cytochrome in the body of mammals is (Adolph, 1949)

$$m_{\text{cyto}} = 3.31 \times 10^{-5}\, m^{0.84} \tag{7.5.12}$$

where m_{cyto} is the cytochrome mass, kg.

Example 7.5.1: Human Kidney Mass

Calculate the expected mass of a human kidney.

Solution

The standard mass for a male adult human is usually taken to be 70 kg. From Equation 7.5.8

$$\begin{aligned} m_{\text{kidney}} &= 7.52 \times 10^{-3}(70)^{0.85} \\ &= 0.28 \text{ kg} \end{aligned}$$

7.6 Self-Similarity in Populations

> Trees, when they are lopped and cut, grow up again in a short time, but men, being once lost, cannot easily be recovered.
>
> **—Plutarch**

There are also scaling relationships among populations of organisms that help to answer questions about the effects of human activities on the environment. These relationships are related to such matters as global warming, biodiversity, and sustainability.

7.6.1 Number of Species

> Important problems are not solely mathematical, but must be viewed with a foundational understanding of the basic principles involved.
>
> **—Albert Einstein**

There is a power-law relationship between the expected number of species in an area and the size of that area (Figure 7.6.1).

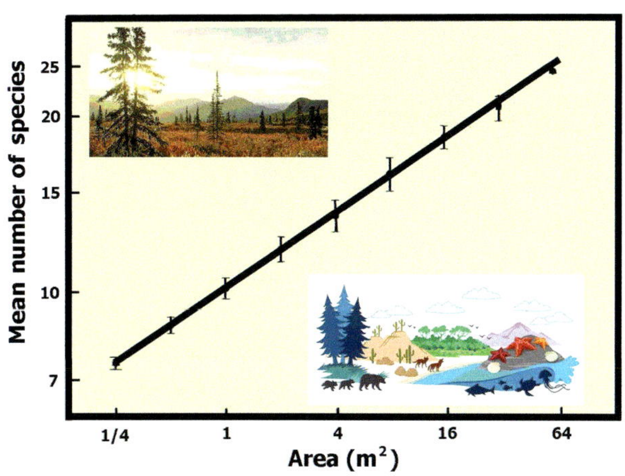

FIGURE 7.6.1
The number of species expected in a grassland ecosystem is related to the area of that system through a power-law relationship. (Redrawn from Harte, J., *Phys. Today*, 55, 29–34, 2002.)

An approximate line for the grassland ecological system in the figure is

$$\mathrm{NS} = 10\,A^{0.215} \tag{7.6.1}$$

where:
NS is the number of species
A is the area, m^2 (Harte, 2002)

Different ecological systems would be expected to yield different exponent values, but the self-similarity, indicated by the power-law relationship, would still be expected to be followed.

7.6.2 Species Range

> Science has a strong theoretical basis, and is expected to provide supporting numbers, in contrast with folklore, emotion, and demagoguery.
>
> —**William A. Calder**

The natural range of an individual within a species is also related fractally to the species population (Harte, 2002). In Figures 7.6.2 and 7.6.3 are seen the range sizes (m^2) and populations (numbers of individuals) for ferns in Quebec and native birds in the United Kingdom. Range is given as

$$R = k(\mathrm{pop})^n \tag{7.6.2}$$

where:
R is the range, m^2
pop is the population

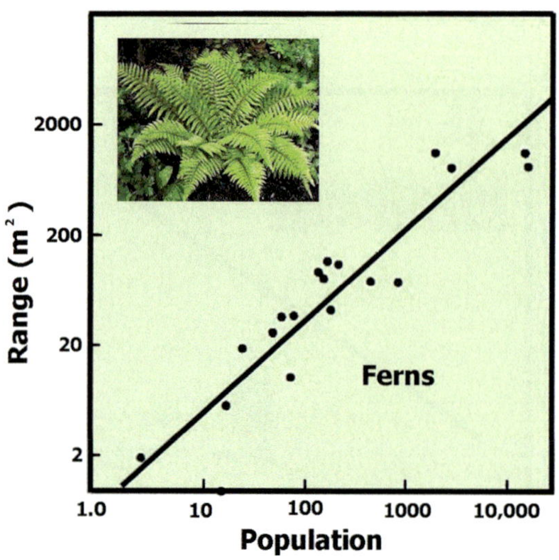

FIGURE 7.6.2
The ranges of different species of ferns in a Quebec forest is related to the total populations of those species. (Redrawn from Harte, J., *Phys. Today*, 55, 29–34, 2002.)

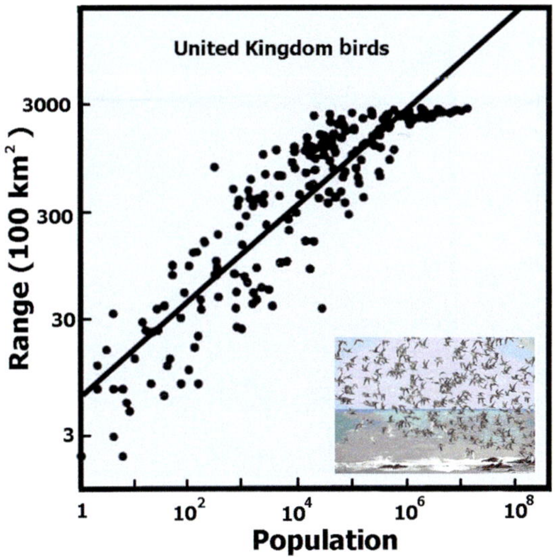

FIGURE 7.6.3
The ranges of different species of birds native to the United Kingdom related to their total populations. (Redrawn from Harte, J., *Phys. Today*, 55, 29–34, 2002.)

Best-fit values for n are 0.85 for the ferns and 0.43 for the birds. Self-similarity theory predicts n values of 0.89 for the ferns and 0.42 for the birds (Harte, 2002).

7.6.3 Population Densities

> It is our intelligence—our interest in figuring things out, our ability to do so, coupled with our manipulative abilities, our engineering talents—that is responsible for our success.
>
> —**Carl Sagan**

There is a relationship between the size and natural density of organisms in the wild (Molles, 1999). As size increases, natural density decreases. In Figure 7.6.4 is shown a logarithmic relationship between the masses of herbivorous mammals and their densities. As expected, the largest herbivores take up the most space, and use the most resources, and so populate any given area with the smallest number of individuals. The shape of the curve relating species type to relative abundance in an ecosystem is lognormal (Dorit, 2014).

In Figure 7.6.5 is shown the density–mass relationship for several animal groups. Again, there is an inverse relationship between population density and body mass. However, there are a number of differences among the groups. Mammals, for instance, tend to live at higher population densities than birds. Also, aquatic invertebrates live at higher population densities than other animals of comparative size.

Plant population density scales as $m^{-3/4}$, the same as for animals (Enquist et al., 2000). Densities of both plants and animals are dictated by the rate of resource supply. When the plant density and resources used per plant are taken into account, the total resource use for an ecosystem is almost constant over a 12 order of magnitude range of plant masses (Enquist et al., 2000).

In Figure 7.6.6, the relationship between population density and body mass for plants is shown. However, unlike the graphs for animal relationships that included data for many

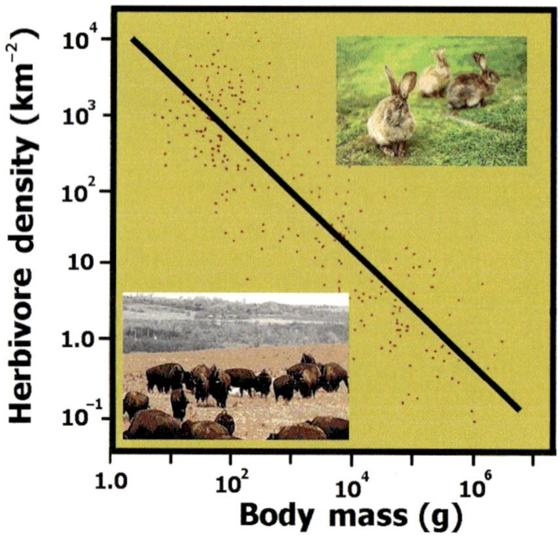

FIGURE 7.6.4
Body size and population density of herbivorous mammals. (Redrawn from Molles, M.C. Jr., *Ecology: Concepts and Applications*, McGraw-Hill, New York, 1999.)

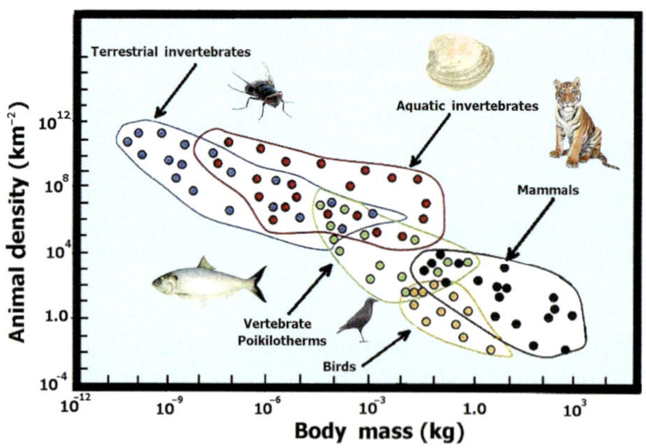

FIGURE 7.6.5
Animal size and population density. (Redrawn from Molles, M.C. Jr., *Ecology: Concepts and Applications*, McGraw-Hill, New York, 1999.)

different species, the plant data in Figure 7.6.6 are mostly from different sized individuals of the same species.

One might expect that population density should decrease as body mass increases because the larger organisms tend to use more of the available resources than do smaller organisms. For plants, one of the most critical resources is space itself, because space represents light availability. For herbivores, the link to space is nearly as strong because space implies a certain amount of plant food availability. For carnivorous animals, the tie to space as a primary resource is not as direct.

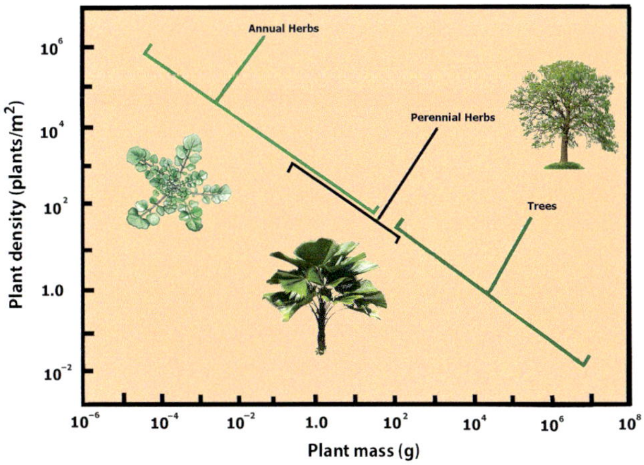

FIGURE 7.6.6
Plant size and population density. (Redrawn from Molles, M.C. Jr., *Ecology: Concepts and Applications*, McGraw-Hill, New York, 1999.)

For those organisms for which space itself is a primary requirement, we might expect that population density would be inversely proportional to the area subsumed by each individual. If body mass (m) is proportional to body volume (V), and body volume is proportional to some characteristic length (L) cubed,

$$m \propto V \propto L^3 \tag{7.6.3}$$

and body area is proportional to L^2, then density would be expected to be inversely proportional to body mass to the 2/3 power (see Section 4.1):

$$\text{density} \propto \frac{1}{\text{Area}} \propto \frac{1}{L^2} \propto \frac{1}{(L^2)^{2/3}} \propto \frac{1}{m^{2/3}} \tag{7.6.4}$$

Actual lines drawn through the data in Figures 7.6.4 through 7.6.6 give

$$\text{density} = 1.00 \times 10^8\, m^{-0.779} (\text{herbivores}) \tag{7.6.5}$$

$$\text{density} = 4.74 \times 10^{11}\, m^{-0.946} (\text{animals}) \tag{7.6.6}$$

$$\text{density} = 6.20\, m^{-0.690} (\text{plants}) \tag{7.6.7}$$

where:
density = organisms per m²
m is the mass, kg

The class of organisms that come closest to expectations are the plants, followed by herbivores. Table 7.6.1 summarizes some of these relationships.

TABLE 7.6.1

Average Population Densities

Class	Equation
All birds	Density = 3.09 $m^{-0.49}$ or Density = 1.39 $m^{-0.60}$
Birds in conifer forests	Density = 0.104 $m^{-0.68}$
All mammals	Density = 52.5 $m^{-0.78}$ or Density = 17.9 $m^{-0.69}$
Herbivorous mammals	Density = 91.2 $m^{-0.73}$

Source: Calder, W.A., Diversity and convergence: Scaling for conservation, In *Scaling in Biology*, J.H. Brown and G.B. West (Eds.), Oxford University Press, New York, pp. 297–323, 2000.

Note: Density is in individuals per square kilometer.

7.6.4 Population Doubling Time

I don't want to grow young again. I just want to keep on growing old.

—Madame de Rothschild

Population doubling time is an important measure of the recovery time for a threatened species. Population doubling time (t_{2n}, in years) has been given by Calder (2000) for mammals as

$$t_{2n} = 0.86\ m^{0.26} \tag{7.4.8a}$$

$$t_{2n} = 1.00\ m^{0.26} \tag{7.4.8b}$$

$$t_{2n} = 0.69\ m^{0.33} \tag{7.4.8c}$$

$$t_{2n} = 0.21\ m^{0.21} \tag{7.4.8d}$$

Each of these equations is based on a different sample of mammals.

Calder (2000) gives doubling times for birds of

$$t_{2n} = 0.62\ m^{0.17} \tag{7.6.9}$$

Therefore, bird populations take roughly three times as long to double as do mammalian populations.

Example 7.6.1: Bison Population Recovery

The goal of an endangered species recovery plan is to increase numbers to sustainable levels (Calder, 2000). This requires an adequate area of suitable habitat and protection from the historic causes of decline.

The American bison (*Bison bison*) was at one time extremely abundant in the midsection of the North American continent. Due mainly to overhunting and the loss of suitable grass rangeland, the population of bison fell precipitously in the mid-1800s. A recovery plan was put into place, and bison numbers have returned to the point where bison are no longer considered close to extinction.

If we imagine a small isolated population of 15 individuals, and determine that at least 100 bison are necessary to sustain that population, how long will we have to have our recovery plan in place?

Solution

The population doubling time has been given in Equation 7.6.8c as

$$t_{2n} = 0.69\ m^{0.33}$$

From Table A.1, the mass of a male bison is found to be 1000 kg. The population doubling time is calculated as

$$t_{2n} = 0.69(1000)^{0.33} = 6.7\,\text{years}$$

Population growth occurs with an exponential relationship:

$$n = n_0 e^{rt}$$

The intrinsic growth rate (r) is the effective compounded growth rate at which the population is capable of growing (Calder, 2000). Growth rate can be obtained from doubling time as

$$r = \ln\left(\frac{n}{n_0}\right)t = \ln\frac{2}{t_{2n}} = \frac{0.69}{6.7}\text{year} = 0.102\,\text{year}^{-1}$$

The growth of bison numbers from 15 to 100 yields a ratio of

$$\left(\frac{n}{n_0}\right) = \frac{100}{15} = 6.67$$

Thus, the number of years to reach the sustainable population of 100 is

$$t_n = \ln\frac{(n/n_0)}{r} = \ln\frac{(6.67)}{(0.102)} = 18.6\,\text{year}$$

We should expect to have to protect the bison herd for at least 19 years.

Example 7.6.2: Poaching of Rhinos

Powdered rhinoceros horn is prized in some parts of the world for its supposed ability to cure fever, arthritis, high blood pressure, and cancer. This demand has led to the unauthorized and illegal killing of rhinos in Africa and Asia. The horns are sawed off the carcass and smuggled to horn dealers.

The numbers of rhinos have plummeted due to this poaching. African black rhinos numbered about 100,000 in the early 1960s, but the population had waned to just 2400 in the early 1990s (Beech and Perry, 2011). At the beginning of the twentieth century, there were only about 50 African white rhinos alive. Now, thanks to conservation efforts, the number is about 20,000. The Sumatran and Javan rhinos, however, are on the verge of extinction.

Saving the African black rhino requires protection for a long enough time for the population to build up to sustainable levels. The mass of a black rhino is about 1200 kg (Clauss et al., 2005). The maximum population doubling time calculated from Equations 7.6.8a–d is 7.2 years. The actual number of black rhinos had doubled to about 4800 by 2011, giving a population doubling time of about 20 years. The difference can probably be attributed to the lasting effect of illegal killing.

APPLICATIONS AND PREDICTIONS

1. Almost all parameters for animals will depend on body mass, and designs involving animals should take these into account.
2. If animals other than the ones for which the equations are specified are to be involved, use the equations as approximations.
3. Biological variation will mean that particular animals' responses could differ considerably from those given by the equations.
4. Fractal (self-similar) scaling will be very common in biology, no matter what the level.

Questions

7.0.1 Why is it important to know about scaling relationships in biology? Give examples.

7.1.1 What exactly are allometric relationships?

7.1.2 Why is it not surprising that allometric relationships exist among organisms?

7.1.3 What is the most common parameter upon which allometric relationships are built? Why? Is there a better parameter to use?

7.1.4 What properties can you think of to limit the sizes and configurations of organisms? What adjustments can be made in other properties to accommodate these limitations?

7.2.1 List dimensionless numbers important in biological engineering. Which of these involve biological parameters not used for nonbiological analysis?

7.2.2 What is the advantage to the use of dimensionless numbers to describe empirical data?

7.3.1 Draw a diagram illustrating the golden ratio.

7.3.2 Is the appearance of the golden ratio due to some underlying principle, or is it just a coincidence?

7.3.3 Where might you expect to find dimensions related by the golden ratio? Speculate on places in addition to the ones mentioned in the text.

7.3.4 Make some measurements of body dimensions. Which appear to form values close to the golden ratio?

7.3.5 Of what importance is knowledge about the golden ratio to the biological engineer?

7.4.1 Why is the principle of self-similarity important to a biological engineer? What type of mathematical expression would be expected for a self-similar characteristic?

7.4.2 How are body mass and physical dimensions related? Why is this truer for living things than for inanimate objects?

7.4.3 Measure the dimensions of several facial features and plot against body mass. Does the plot exhibit fractal scaling? Why or why not?

7.4.4 Why would you expect body density to be invariant with size? Why might it change?

7.4.5 Calculate your body surface area with the Cena relation and with the DuBois formula. Which value do you think is more correct?

7.4.6 If there were an extra-large primate the size of a bottlenose whale, how big would you expect its body length and chest diameter to be? How much confidence should you have in the answer?

7.4.7 What is basal metabolic rate (BMR)? When does it apply to an animal and when does it not? What is the BMR of a domestic cat?

7.4.8 If a horse is expected to have a higher BMR per unit area than a mouse, what implication does this hold for an engineering design?

7.4.9 If a smaller animal consumes more oxygen per unit mass than does a larger animal, how does this influence an engineering design?

7.4.10 Over what range of masses would you expect natural variability to be more important than allometric differences?

7.4.11 For what engineering situations would you need to know the oxygen consumption of the liver?

7.4.12 Why are there two equations for mass of the heart? Of what importance is this?

7.4.13 If the aortic diameter scales with body dimensions, which of these vessel groups would also be expected to scale with body mass?

Arteries

Arterioles

Capillaries

Venules

Veins

7.4.14 Oxygen partial pressure for normal resting human blood is usually taken to be 100 mmHg. What values of oxygen partial pressure would you expect to find in a dog and cat?

7.4.15 If larger lungs are necessary to support the higher metabolic rates of larger animals, what does this mean for a biological engineering design?

7.4.16 Compare expected walking and running speeds of a fox and hare.

7.4.17 Maximum muscle stress is stated as approximately 3.0×10^5 N/m^2. Is this tensile stress, shear stress, bending stress, torsional stress, or compressive stress?

7.4.18 Of what biological engineering importance is the fact that muscular efficiency depends on body mass?

7.4.19 Develop a relationship between generation time and body mass.

7.4.20 Describe various situations where prediction of life span, maturation time, reproductive maturity, and generation time would be important.

7.4.21 Why are food needs and waste needs important in the intensive care unit (ICU) of a hospital?

7.4.22 If kidney function is to be assessed in a large animal clinic, of what value is it to know how kidney clearance rate is body mass dependent?

7.4.23 Consider the advantages and disadvantages of C3 and C4 plants with regard for increased food needs in the future. Also consider global warming effects.

7.5.1 Why would a biological engineer want to know the liver mass of a rat? Of a chimpanzee?

7.5.2 As body mass increases, how would you expect the masses of tissues and organs to change?

7.5.3 Identify where in the body is located each organ or tissue named in Section 7.5.

7.6.1 How many different species would you expect to find in a hectare of grassland? How many more species would there be in 100 ha?

7.6.2 Compare the range of a species of Quebec forest fern with a total population of 1000 individual ferns with that of a species with 100 individuals. Could both coexist comfortably in a 1 km^2 area?

7.6.3 Compare the natural animal densities of birds and mammals each with a body mass of 1 kg. Are these estimates for adults or juvenile animals?

7.6.4 Why do you suppose there is an inverse relationship between herbivore density and body mass?

7.6.5 What is the natural animal density for adult crayfish?

7.6.6 How do you suppose the presence of other species affects population density of a target species?

7.6.7 Why do you think it takes longer for bird populations to double compared to mammals?

Section V

Utilizing Living Systems

8

Biological Engineering Solutions

Herbert Hoover, President of the United States 1929–1933, and himself a civil engineer, said of engineering:

> It is a great profession. There is the fascination of watching a figment of the imagination emerge through the aid of science to a plan on paper. Then it moves to realization in stone or metal or energy. Then it brings jobs and homes to men. Then it elevates the standards of living and adds to the comforts of life. That is the engineer's high privilege.
>
> The great liability of the engineer compared to men of other professions is that his works are out in the open where all can see them. His acts, step by step, are in hard substance. He cannot bury his mistakes in the grave like the doctors. He cannot argue them into thin air or blame the judge like the lawyers. He cannot, like the politicians, screen his shortcomings by blaming his opponents and hope the people will forget. The engineer simply cannot deny he did it. If his works do not work, he is damned.
>
> On the other hand, unlike the doctor, his is not a life among the weak. Unlike the soldier, destruction is not his purpose. Unlike the lawyer, quarrels are not his daily bread. To the engineer falls the job of clothing the bare bones of science with life, comfort, and hope. No doubt as the years go by people forget what engineer did even if they ever knew it. Or some politician puts his name on it. Or they credit it to some promoter who used other people's money. But the engineer himself looks back at the unending stream of goodness, which flows from his successes with satisfaction that few professions may know. And the verdict of his fellow professional is all the accolade he wants.

Biological engineers are afforded the privilege, not only to utilize their engineering skills for the betterment of humankind, but also to be in awe of life and its workings. The ultimate goal of living systems is to use any possible means to dominate their environments. Biological engineers must be aware of this at all times in order to use biological systems effectively.

8.1 Systems Approach

> Computers can do all the left hemisphere processing better and faster than the human brain. So what's left for the human brain is global thinking, creative thinking, intuitive problem solving, seeing the whole picture.
>
> **—Betty Edwards**

If there is any overall guiding principle to a biological engineering creation, be it a design, application, or other utilization of living things, it must be that a whole system must be considered. Living things depend on physical, chemical, and other biological influences. Living things are never self-contained, and they do not affect just one or two qualities of their environments. They are complex, they are redundant, they are exquisite, and they are largely unstoppable at some things that may or may not be the object of the biological engineering creation. Not to take all of these aspects into account can lead to certain failure of the engineering work. The emphasis should be on integrating the many influences characterizing biological problems to form coherent analyses that are worthy of the complex natures of these problems.

Life science research has tended toward the reductionist focus on smaller and more fundamental issues. Biomedical engineering, bioprocess engineering, and others have often followed this trend; and so, we find the subspecialties of tissue engineering and metabolic engineering, among others, describing engineers who have focused on the cellular or subcellular level.

That reductionism is useful, because it feeds the revolution going on today in biological knowledge. We need to know more and more about smaller and smaller issues because many of the foundations of biological essences are found at the lowest possible levels. However, engineering application of this science requires that we work with the entire organism or collection of organisms. Thus, biological engineering designs should be able to maintain a systems perspective when utilizing living systems. The systems perspective provides the unification tying bioenvironmentalists with biomedical specialists.

Part of this philosophy is embodied by the Physiome Project advocated by Bassingthwaighte (1998). The physiome has been proposed as a means to explain or predict phenotypical attributes of an organism through physiological models that begin with genetic sequences (Bassingthwaighte, 2000; Dao et al., 2000). While such a model would need to be systems oriented, a full systems approach would require that environmental effects, usually considered to be as important as genetic factors, be accounted for. The model would likely then begin as rather shallow, but broad. As with many physiological models, 80%–90% of the set of final results can be obtained with the simplest of assumptions; the last 10%–20% requires a great deal of complexity. A systems analyst familiar with biology would realize that effort on more complex versions of a physiome model would be limited by the law of diminishing returns in the context of natural biological variation.

Equally important to the development of this process is the development of a coherent system for describing the connections and interactions between the elements of the system. A biological equivalent to the circuit diagrams used in electrical engineering or the bond graphs used in mechanical engineering must be developed. The development of a system for describing complex biological networks is further complicated by the difficulty in identifying the behavior of the discrete components. Historically, engineering elements have been designed to behave in a consistent fashion. For example, resistors are designed to have nearly the same resistance throughout the entire range of their expected operating conditions. This is not the case with living materials. They are often extremely nonlinear in their behavior. Analysis of biological systems is further complicated by the fact that they not only respond to their environments but also alter their environments. Furthermore, the biotic components of the system respond and adapt to each other.

POPULATION DYNAMICS

(Berryman, 1999)

Studies of populations of organisms have concerned ecologists for many years, and they have developed theories to help with understanding. Because this text takes the viewpoint that the microecology of cells, tissues, organs, and so on, is essentially no different from the macroecology of populations of organisms, these principles apply to all biological levels.

Principle 1: Exponential Growth. If each biological unit begets a similar unit in a certain time, and this process continues unabated by age, resource limits, predation, disease, or death, then the population will continue to grow indefinitely in an exponential fashion (see Section 4.3.3). Such populations are said to be *unstable*.

Principle 2: Cooperation. Groups of organisms often assist members to obtain resources or defend against enemies. It is easier for group members to mate. The chances of individuals in a group to be killed by predators are reduced as the group enlarges. Larger groups can be more productive because certain members of the group can specialize, in a similar manner to economies of scale in organizations. Cooperation can be *intraspecific* or *interspecific*.

Principle 3: Competition. The struggle to obtain limited resources, to avoid enemies, or to find suitable mates can lead to competition, which may be either programmed or inadvertent. Competition limits the population, and can lead to an *extinction threshold*, wherein population numbers can become so low that the population ceases to exist.

Principle 4: Circular Causality. Populations are affected by their environments, but they also directly affect the environment. Because of this two-way action, cycles are formed involving environment and populations. Some environmental factors are constant and nonreactive and some are reactive. Some effects are positive, others negative. For instance, the number of predators usually increases in response to increased numbers of prey (positive response), but, as the number of predators increases, the number of prey decreases (negative response). There is a *carrying capacity* of the environment.

Principle 5: Limiting Factors. Of course, there is no environment that can supply all the needs of an expanding population without limit. Usually there are several resources that are limited, but it is the most limiting resource that acts as the ultimate cap to growth.

These five principles give rise to several positive and negative feedback loops that determine population density. In addition, there is a natural variability that is often considered to be random, although the causes may be deterministic but just presently unknown.

8.2 Relationships between Engineering and Biology

> The fellow that can only see a week ahead is always the popular fellow, for he is looking with the crowd. But the one that can see years ahead, he has a telescope, but he can't make anybody believe he has it.
>
> **—Will Rogers**

There are five general domains in which living things are utilized or affected (Table 8.2.1):

1. Living systems are an integral part of the utilization.
2. Living systems are used as models to solve problems not involving biological components.
3. Biological methods are used to solve biological problems.
4. Processes and devices are applied to living things for their benefit.
5. Living things are affected inadvertently by other projects.

The first of these can be considered a generalization of the field of *bionics*; the second as a generalization of the field of *biomimetics*; the third is generally the domain of *biotechnology*; the fourth has no title, but is found mainly in human and animal health care and may be termed *biomedical engineering*; the fifth is often encountered in environmental matters.

Ecology is the study of plants and animals in relation to their total environment, and *ecological engineering* is the manipulation of a total biological system for a defined goal. We have seen throughout this text that the environment can consist of physical, chemical, and psychosocial factors all relating to an individual organism. It is not such a great extension of these ideas to include, within ecology, the responses of individual cells, tissues, or organs at the smaller scales, or whole biomes at the larger scales. Each of these biological units responds to environmental influences in very complex ways, both influencing the environment and being influenced by the environment. In this respect, there is a certain equivalence between ecological engineering and biological engineering because both are related to the utilization of biological systems in the contexts of their total surroundings. Ecological engineering, however, usually implies interests dealing with whole organisms as found in nature, and where humans are either absent or are no more than just another species in the system. Ecological engineering does not appear in Table 8.2.1 because its basic ideas permeate the entire table.

TABLE 8.2.1

Categories of Biological Utilization

		Object of the Design	
		Biological	**Nonbiological**
Solution to the problem	Biological	Biotechnology	Bionics, hybrid systems Biomimetics Some biotechnology
	Nonbiological	Biomedical	Inadvertent effect

8.2.1 Living Things as the Solution (Bionics, or Hybrid Systems)

> The march of invention has clothed mankind with powers of which a century ago the boldest imagination could not have dreamt.
>
> —**Henry George**

Bionics began as augmentation or replacement of operations and functions of human extremities through machinery controlled by human neural systems (Bionics Symposium, 1960). Bionics has also come to mean the application of knowledge of living organisms to the solution of engineering problems. The essential aspect of bionics was, at least at one time, considered by the military for advanced weaponry that combined the strengths of artificial mechanical systems with those of living human or animal brains.

Hybrid systems is the modern term used to describe a system that incorporates a biological unit as an essential part of a nonbiological design, and thus is essentially equivalent to the older bionics descriptor.

Knowing how living organisms have dealt with common challenges can provide inspiration for biological engineering creations. Using living systems as components of otherwise nonbiological designs is a possibility that should not be overlooked. Thus, microbes have been used to remediate pollution, dolphins have been trained to explore dangerous underwater sites, and honeybees can be used as monitors of dispersed chemicals in the environment.

Although the term bionics has often come to mean control of a nonbiological mechanism by a biological means, in its more general definition, it can be considered to be the utilization of living systems to solve problems of importance to humans. Thus, living things become another set of tools to be used whenever it is advantageous to do so.

Environmental engineering is concerned with such areas as waste water management, air pollution control, recycling, waste disposal, radiation protection, industrial hygiene, environmental sustainability, public health, environmental engineering law, and the environmental impact of proposed construction projects. Environmental engineering may or may not be dependent on a thorough study of biological systems and their implications. *Ecological engineering* integrates ecology and engineering, using engineering tools to design, monitor, and construct whole ecosystems (Mitsch and Jorgensen, 1989; Mitsch, 1996). Ecological engineering definitely needs an appreciation for biological systems and their interactions.

Nowhere has the means to utilize living systems been more enthusiastically developed than in the area of environmental applications. Of these, bioremediation is a major application.

Bioremediation is the use of biological organisms to remove pollutants from the environment (Metcalf and Eddy, 1991). For most natural pollutants, there is generally a microbe that can metabolize each of them. These can be selected and grown specifically to deal with that type of pollutant, for example, crude petroleum, or they can be assumed to be present naturally in small numbers and only require a little time before they increase enough to remove the pollutant. The advantage of selecting certain strains of microbes is that they are often very efficient at remediating the pollution problem. If higher effectiveness is required, it may be possible to bioengineer the microbes by genetic manipulation. The disadvantage of this approach is that it can be expensive and not immediately available.

The advantage of using natural populations is that they are usually adapted to local environmental conditions, and so will be robust in nature. Although it may take some time for the population of these microbes to become large enough to deal with a serious pollution problem, exponential growth of microbial populations can be expected when unlimited by water, oxygen, nutrients, or heat. However, it may be that the pollutant spill is so large that a serious diffusion of the pollutant occurs before the natural microbial population has had a chance to grow to an effective size.

One cannot depend on microbes to deal with contaminants unless the contaminants are nontoxic, and all other needs are met. Fortunately, most common pollutants, such as petroleum products, represent energy sources. In order for microbes to utilize the energy in these carbon compounds, there must be sufficient water available, enough oxygen (for aerobic microbes), and the environment must be warm enough to allow chemical reactions to occur at a sufficiently high rate. In addition, the microbes will need a source of nitrogen to form protein, phosphorus for ATP, and micronutrients for enzymes. All these must be supplied either by the environment (not reliable) or by some system designed by the environmental engineer.

A special case of bioremediation uses plants to mine certain elements from the soil. This is called *phytoremediation*. Some examples of phytoremediation are truly remarkable. When a fertilizer was spilled in a train derailment in North Dakota, a special variety of alfalfa was used (Figure 8.2.1) that had been developed to absorb nitrogen only from the soil, rather than the air (Russelle, 2001). In another instance, plant biologists have developed a genetically engineered plant that thrives in salty irrigation water (Anonymous, 2001). Not only does this plant produce useful tomatoes, but it can actually remove salt from the soil, thus improving the soil for subsequent crops. Sugar beets can be used to remove excess nutrients from soils to reduce leaching of these nutrients into the groundwater (Kaffka et al., 2001). Thus, ground water quality

FIGURE 8.2.1
A special variety of alfalfa was used to phytoremediate a fertilizer spill in North Dakota. (Redrawn from Russelle, M.P., *Am. Sci.*, 89, 252–261, 2001.)

FIGURE 8.2.2
Sugar beets can be used to absorb more nutrients from the soil than is added by fertilizer. They can thus be used to remove excess nutrients left from previous crops. (Redrawn from Kaffka, S.R. et al., *Calif. Agric.*, 55, 42–47, 2001.)

is protected in dry areas where problems sometimes occur (Figure 8.2.2). Duckweed (*Lemna gibba*) can be used to remove pollutants from waste treatment lagoons. They are small, fast growing, efficient, and nutritious, and so have many advantages (Hillman and Culley, 1978).

Water running off asphalt-covered parking lots is contaminated with many pollutants. Particularly foul is the first wave of runoff. An ingenious phytoremediation scheme to cleanse pollutants from this runoff is called bioretention, or rain gardening (Bitter and Bowers, 1994; Davis et al., 2001; Horton, 2002). At the edges of parking lots are located small islands of greenery where water runoff is directed. Strategically placed mulch, soil and sand, and plants filter everything from grease and fertilizer to toxic metals from the water before it can enter streams or other nearby bodies of water. This has advantages over retention ponds because it is a distributed filtering system with less negative impact on the local environment.

Trees have been used for years as windbreaks in open spaces and sound barriers next to major highways. Thus, they form physical obstacles to unpleasant and unwanted environmental effects.

There are other applications utilizing living systems in a central role. One proposed method would use special strains of bacteria pumped into oil wells to decompose heavy petroleum that sticks to underground rocks (Greene, 2001). As much as two-thirds of the petroleum in a well is too thick to recover normally. Actions of these bacteria would allow the loosened oil to be flushed out with water (Figure 8.2.3).

Sheep have been found to improve survivability of newly established reforestation areas (McCreary, 2001). They do this by eating plants that would ordinarily compete successfully with tree seedlings (Figure 8.2.4). In addition, sheep will often graze in low areas where

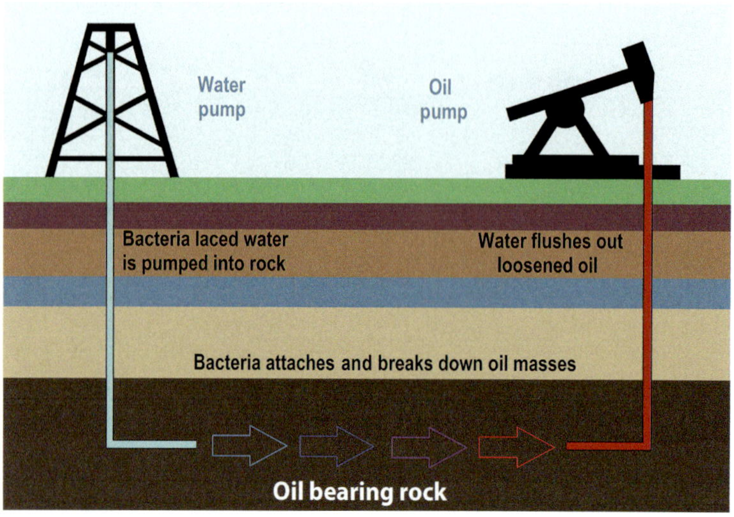

FIGURE 8.2.3
Specially bred bacteria could increase oil production by breaking down thick petroleum into lighter molecules.

FIGURE 8.2.4
Sheep can improve the ecology for growing trees by reducing competition.

plant growth is greatest because of water and nutrient availability. Sheep tend to rest on high spots, where they leave their manure. They thus become agents for transporting nutrients from low areas to high areas against the action of rain and water run-off.

We have already considered one application of living system utilization in medicine, where genetically engineered anaerobic bacterial spores seek out cancerous tissue (see Example 6.15.1). In another application, multiple sclerosis (MS) can be helped by transplanting cells from one part of the body to another (Thieme, 2001a). MS is the result of the destruction of myelin sheaths around nerves in the brain and spinal cord. This slows the conduction of impulses greatly (see Section 4.4.3, Communications). Schwann cells taken from the limbs of patients and injected into MS-affected areas can produce new myelin and reverse the most devastating aspects of the disease.

Nature can also be used to form computers (Hayes, 2001). Messenger RNA (mRNA, see Section 5.3, Genetics) causes proteins to be formed at ribosome sites. Before mRNA can be transcribed from genetic codons, an enzyme, RNA polymerase, must be present. RNA polymerase is blocked from its activity by a repressor protein, which binds to the DNA downstream from the target codon and stands in the way of the polymerase.

Certain metabolites, such as those of lactose, bind to the repressor protein, change its shape, and cause it to loosen its grip on the DNA. mRNA and proteins can then be formed. However, there may be promoter substances also necessary to this process. The result is that proteins are formed only if both inputs allow it. A genetic NAND gate is illustrated in Figure 8.2.5. The NAND gate is a basic unit of computers, so genetic computers can, in theory, be formed from simple bacterial DNA.

Military uses of bionics abound. The military is studying different creatures to learn how they might be useful to solve problems or as models of certain actions (Table 8.2.2). Once the means that these subjects use are determined, the military will attempt to duplicate them.

There is enormous potential to use living organisms to solve problems in environment, medicine, and industry. One of the first questions that a biological engineer should ask is if there is an existing organism that can solve the problem at hand. If so, then use of that organism is often the best and most efficient solution to the problem.

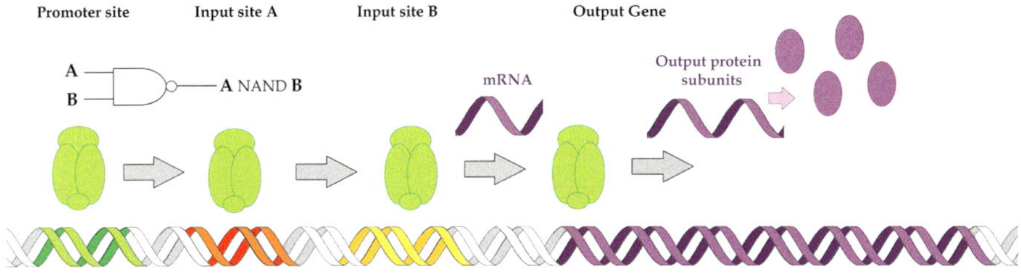

FIGURE 8.2.5
Formation of a NAND gate from bacterial DNA can lead to construction of a genetic computer. (Redrawn from Hayes, B., *Am. Sci.*, 89, 204–208, 2001.)

TABLE 8.2.2
Possible Bionic Models for Military Purposes

Creature	Skill	Mission
Honeybees	Seeking pollen, they attract various airborne chemicals and take them back to the hive	Find land mines. Explosive devices leak small amounts of chemicals into the air. Researchers hope that by analyzing the chemical content in a hive, they can determine if explosives are in the area
Parasitic wasps	They can learn to detect various volatile compounds that they associate with food	Trained to associate food with compounds given off by chemical and biological weapons, wasps may swarm to areas where they're hidden
Giant sphinx moths	Males are extremely sensitive to pheromones given off by females	Provide information on how to detect very low levels of chemicals. Moths could also be trained like wasps
Flies	They're stable and maneuverable. Can land on ceilings, fly backward, and hover with pinpoint accuracy	Help scientists learn more about flight control. The information may improve the aerodynamic capabilities of micro-air vehicles (planes with a 6 in. wingspan)
Beetles	Their infrared detection systems can locate forest fires many miles away	Give scientists a better understanding of infrared detection systems, which could then be used to improve the infrared capabilities of missiles
Lobsters	Their locomotive abilities allow them to live in turbulent waters	Help scientists learn about maneuvering in surf zone, where most mines are. Researchers hope to build robots based on the lobster to locate mines
Bluegill fish	Slight impurities drive big changes in bluegill's vitals. Sensors in the tank monitor the fish and sound an alarm when things get out of whack	The fish stand guard at reservoirs in several locations where they've detected pesticides and a diesel spill
Wasps and bees	When bees detect the target odor, they extend their proboscises. A camera records the positive response, and a computer alerts an operator	Commercial systems are close to reality to detect airborne compounds
Roaches	In an emergency, workers could release roaches in a building, or collect bugs already there, and test them for toxins to determine the next step	A roach-based detection system could be ready within a few years
Butterflies and moths	A chip, implanted during the pupal stage, could control locomotion, monitor the air, and override instincts to feed, mate, and avoid certain environments	Scientists are in the early stages of research; a final product is years away
Elephant seals, bluefin tuna, white sharks	Ocean monitors	Smell satellite tags glued to their heads record ocean depth, temperature, and salinity as they swim

COMPUTING WITH DNA

Computers can be used as tools to manipulate biological data (called bioinformatics) and control devices used with biological systems (bioinstrumentation). These are applications of computers to biologically relevant problems.

Biology itself, however, can be used as a computing tool (Adleman, 1998). One of the first conceptual models of a computer is a "Turing machine" described in 1936 by Alan M. Turing. In this machine, there is an input tape containing data and an output tape containing results. A mechanism (which in a modern digital computer are the instructions written in the program) moves along the input tape, reads the data, and then translates the data into the output results, which are then written on the output tape.

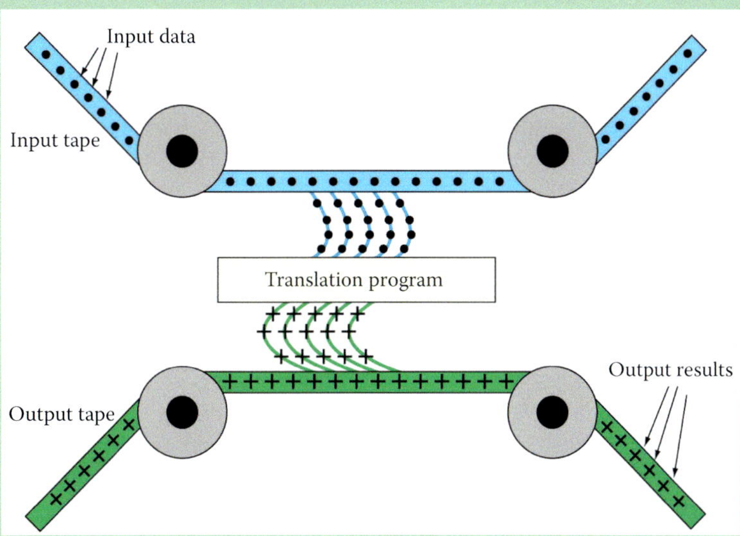

Diagram of a simple Turing machine.

A DNA strand could act as the input tape, containing a sequence of the bases A, T, G, and C. This sequence could be artificially created as a code for any kind of data. Modern digital computers use two states, 0 and 1, as the basis for representation of input data. A string of 0's and 1's, properly weighted, can represent any number or letter of the alphabet, for instance. With four different bases, a DNA computer could be quarternary rather than binary. Of course, the input DNA strand could be restricted to use only two of the four bases, which would make it easier to translate program instructions already developed for a binary computer into the program for DNA computation, but the use of all four bases makes data representation much more compact. For instance, the numbers 0, 1, 2, and 3 can be represented in binary by 00, 01, 10, and 11, or in the DNA four bases as A, T, G, and C. This saves one "space" in the operation of the computer.

(*Continued*)

COMPUTING WITH DNA (Continued)

A DNA strand can act as the input tape and DNA polymerase can be the translation program.

Next, there must be a program of instructions. Presently, we know of one such program, and it is physically represented by the enzyme *DNA polymerase*. This enzyme moves along the input DNA strand and produces a complimentary DNA strand: a C yields a G, a G yields a C, a T yields an A, and an A yields a T. This is not very complicated, but does illustrate the Turing principle. The original DNA strand (input tape) is translated into a complementary DNA strand (output tape) by means of some translation mechanism (DNA polymerase).

While relatively simple, Adleman (1998) has used this mechanism to solve a Hamiltonian Path Problem to determine the sequence of cities to be visited only once each given restrictions on the list of available flights. The practical application of this DNA biocomputer was messy and time-consuming, involving creating an initial DNA strand, creating the complementary strand, removing chance variants, multiplying the number of output strands, and using gel electrophoresis in order to identify the DNA sequence of the output strand. Generation of the output strand was accomplished almost immediately, but extracting the answer so that it could be read took seven days. Future DNA biocomputers could, presumably, be much more sophisticated, and have better presentation capabilities.

8.2.2 Living Things as Models (Biomimetics)

> It's the combined intelligence of dozens or hundreds or thousands of small brains that allow them to function, and to function without any leadership.
>
> **—Michael Crichton**

Biomimetics is the term used for the technique of imitating a biological material or function in the construction of artificial devices or systems. The living system thus becomes the prototype and the artificial device is the copy. Realizing that competition and survival of the most fit response can often lead to the most robust, enduring, and efficient solution to a particular environmental challenge, biological engineers should look to living systems for inspiration, if not outright duplication to solve particular problems.

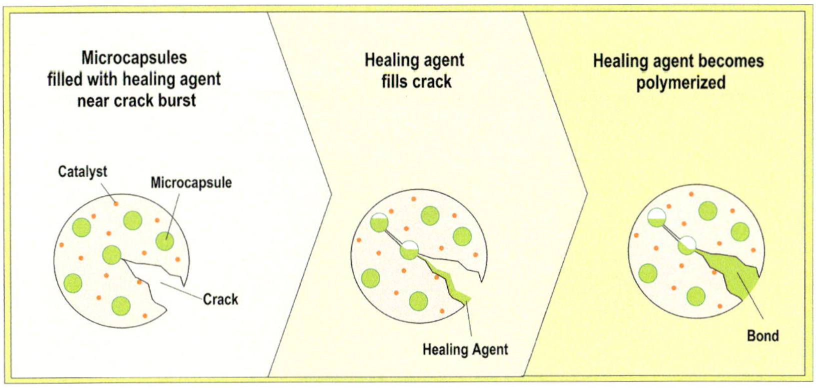

FIGURE 8.2.6
A new material that heals itself was inspired by human wound healing.

Biomimetics has been the inspiration for new materials because materials in living systems are remarkably strong, lightweight, and effective (Andrade, 2000). Proteins in spider silk are being copied for their strength. Composite materials that can serve several functions simultaneously were inspired by materials in living things. A synthetic material that seals tiny cracks that develop within it was inspired by the way wounds heal (Sparks, 2002). The plastic material relies on a healing agent embedded in tiny spheres distributed throughout, and a hardening catalyst present in the plastic. When cracks develop, they release the healing agent, which hardens when it contacts the catalyst (Figure 8.2.6). Besides encapsulating repair materials in microcapsules, vascular capillaries can be incorporated in the material to deliver healing agents to the required location. Other techniques used are integrating the healing agents directly in the object and using microcapsuled bacteria to fabricate repair materials on the spot (Grose, 2014). Concrete may be made to self-heal by including in the concrete encapsulated bacteria and starch. When a crack appears and breaks the capsules, water reaches the bacteria, which eat the starch and excrete calcium carbonate. This seals the crack (Petroski, 2016).

An artificial lung has been developed based upon principles of gas exchange with the blood exhibited in a natural lung (Thieme, 2001b). A balloon wrapped with hollow fibers is implanted in the vena cava by means of a catheter inserted in the leg (Figure 8.2.7). The balloon inflates and deflates as many as 300 times a minute. This causes blood to pump back and forth over the fibers, allowing oxygen to enter the blood and carbon dioxide to be removed.

The field of biotechnology is replete with biomimetic examples. Bioreactors and organs have much in common (Grossmann, 1967). They each serve as a structure in which reactions occur. They can be compared for construction, heat exchange, and mass transfer. Improvements in reactor vessel technology can come from living organs.

There are more exotic examples of biomimetics. For example, birds in flocks and fishes in schools move together very precisely. To explain this coordinated movement, a theory was proposed that assumed each bird or fish was constantly making small errors and corrections (O'Brien, 2001). Over time, individual birds or fish change position relative to those around them. This means that a bird or fish doesn't have to rely on positioning information being relayed across the flock or school. Information instead is passed from different neighbors at different times. Although the mistakes make it difficult for birds

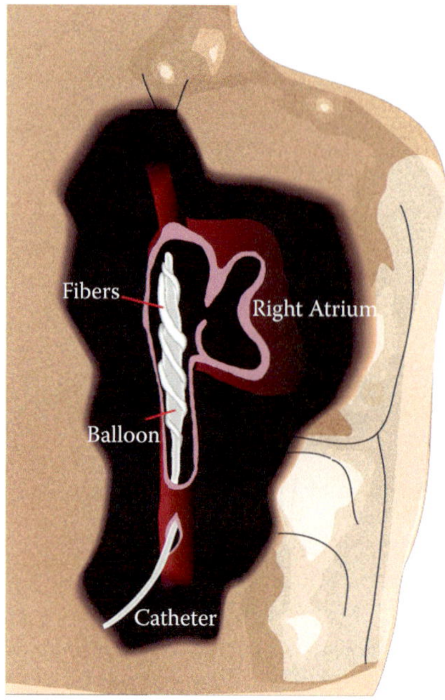

FIGURE 8.2.7
A lung-assist device that performs the functions of the lung while the damaged lung heals is based on biomimetic principles.

to fly together, positioning is transmitted very efficiently. These results can be applied to automobile traffic flow and aircraft flying in formation.

Biomimetic digital ants that patrol computer programs and swarm to security threats can be used to detect possible faults efficiently. The digital ants wander through software and call other ants whenever they find evidence of danger. This swarming can be detected by computer scientists (Grose, 2009).

ANTS, EVOLUTION, ENVIRONMENT, AND EFFICIENCY

(Reprinted from Gordon, 2016)

Collective behavior in ants has evolved in response to how resources such as food are distributed in the environment, as well as to the costs of foraging and the behavior of other species that they meet. Some resources are clustered together in a single patch, whereas others are scattered at random. Ants of many species excel at exploiting patchy resources such as picnics. They use interactions based on pheromones in which one ant follows another, producing recruitment trails. Recruitment makes sense when the resource is patchy—after all, where there are sandwiches, there are likely to be cookies. In contrast, ants that forage for scattered resources, like seeds, do not use recruitment trails, because finding one seed is no guarantee of finding another nearby.

For a robot to be truly useful, it must be able to operate tools used by humans. Thus, the robot must be able to imitate human manipulative ability. For versatility, the robot should demonstrate an intelligent adaptability. Hence, the ideal robot would become a reflection of human abilities, with the exception that it would have fewer limitations and be more expendable.

There is a promise, although presently little more than that, that once brain operation is understood, it can be mimicked. The field of *neural engineering* aims to determine the operation of the brain and its neural connections, and how we interact with our physical environment and with each other. Development of this field would have profound effects on life and living.

NEURAL ENGINEERING

(Foster et al., 2003)

The last unconquered frontier of understanding of human activities is the brain and the mind. Yet, even that is beginning to yield to persistent study. Functional Magnetic Resonance Imaging (fMRI) can form images of brain metabolic functions. MRI is used to detect radio frequency (RF) signals of nuclear protons when in a strong magnetic field and perturbed by pulses of RF energy. Blood hemoglobin that has lost its oxygen affects the emitted signal, so fMRI can sense blood oxygen levels.

Not only can fMRI detect brain tumors, strokes, and other seriously damaged brain tissue, but it is also beginning to be used to diagnose psychiatric disorders such as schizophrenia by detecting spatial differences in brain metabolic activity. fMRI has also been found to be able to be used to distinguish between lying or telling the truth by imaging metabolism of those regions of the brain important to paying attention, and controlling errors.

Positron Emission Tomography (PET) is another imaging technique that employs radioactive tracers to image brain activity. PET can detect and map the presence of glucose, neurotransmitters, and a dozen other chemicals critical to brain function. Subtle changes in brain structure or function that correlate to diseases have been used to distinguish brain chemistry changes associated with Alzheimer's disease, schizophrenia, alcoholism, anxiety disorders, and post-traumatic stress disorder. PET can also be used to detect emotional responses and perceptions of emotion.

A technique called Brain Fingerprinting can be used to detect information stored in memory, and perhaps be used similarly to the polygraph. In Brain Fingerprinting, a stimulus such as a photograph or audio recording is presented to the person under test. Microvolt electrical signals, known as event-related potentials (ERPs), are measured on the scalp, and are distinctively different for knowledge known to the subject. Thus, the investigator can tell the difference between information recognized by the subject and that which is irrelevant.

(*Continued*)

NEURAL ENGINEERING (Continued)

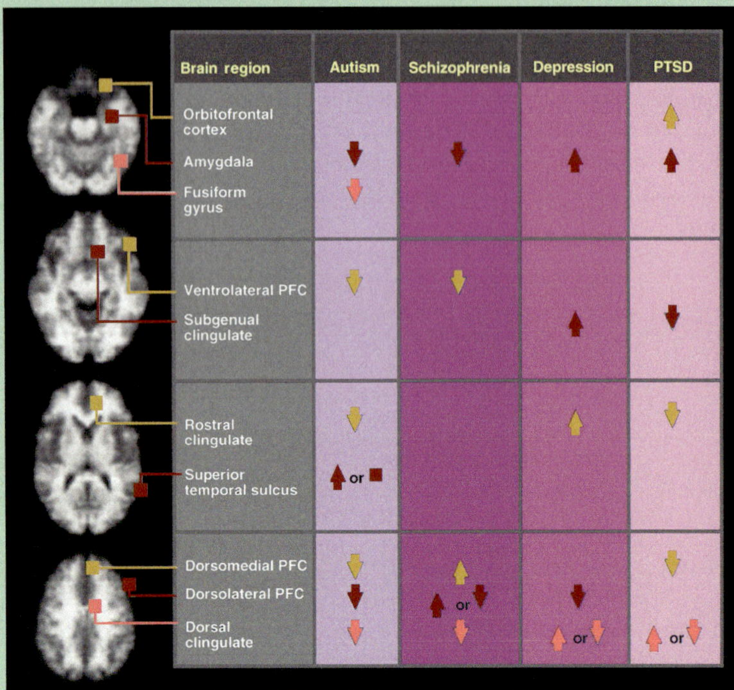

Brain images contain information about function. Ten regions of the brain have been identified as important to deal with complex social interactions, and people with some common psychiatric disorders show levels of brain activities in these areas different from those of people with normal, healthy brains. Medical imaging can detect these deviations. Up arrows indicate increased activity, down arrows indicate depressed activity, and a box indicates no difference. PFC = prefrontal cortex; PTSD = post-traumatic stress disorder. (Redrawn from Foster, K. R. et al., IEEE Spectr, 40(6), 34–39, 2003.)

Computer-brain interfaces can work two ways. Cochlear implants have been developed to detect sound with an external microphone and relay the electrical signal to electrode arrays that directly stimulate inner ear nerve fibers. A visual prosthesis promises to similarly help the blind by applying electrical signals from a camera to an array of microelectrodes implanted into the visual cortex of the brain. Electrical signals from the brain can be used to control prosthetic limbs, computer software, or robots. Electrodes implanted into the pleasure centers of the brains of rats have been used to train rats to respond to investigators' commands.

These technologies certainly promise tremendous benefits as well as ethical challenges. Whether they will result in good or ill will depend largely on the motivation of the scientists and engineers and by public acceptance. Nonetheless, the technology is developing.

(*Continued*)

NEURAL ENGINEERING (Continued)

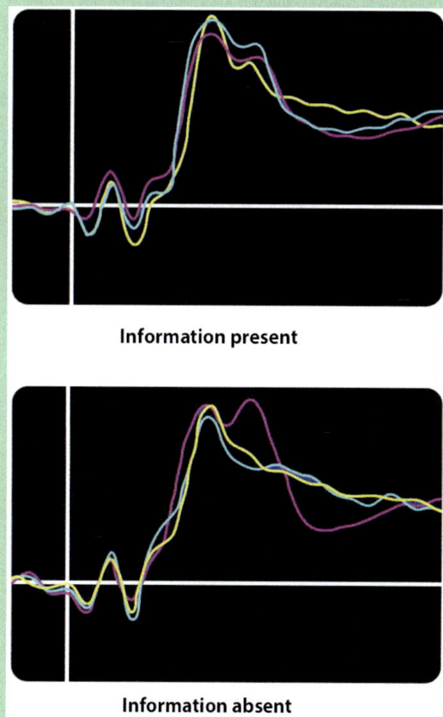

Brain fingerprinting detects the presence of specific information in the brain by measuring microvolt signals at the scalp. By analyzing the differences among signals prompted by three different types of informational cues, the system determines whether the information is present (left) or absent (right) in the brain (Redrawn from Foster, K. R. et al., *IEEE Spectr*, 40(6), 34–39, 2003.)

Complex, self-organizing systems continuously adapt to and change with their environments but do so in ways that are impossible to predict. Because of this, people have begun to realize that industrial firms operate very much like ecosystems (Roston, 2001). The result of this has been *complexity theory*, explaining how a complex adaptive systems approach can help make policy, business, education, and research decisions. Studying how ants find the closest food source, for instance, has helped industries to find efficient solutions to marketing soap, schedule movement of casks of whiskey, and control crowds at amusement parks (Figure 8.2.8).

Ecological theory applied to urban areas uses metrics energy input–output ratios and the area of surrounding land necessary to produce the energy and nutrients required to sustain a city (Collins et al., 2000). The results can be useful for public policy related to the development of different regions and support of city inhabitants.

Imitation of biological systems is a viable method to solve many nonbiological problems. The biological engineer should ask if there is an analogous problem faced by some living system. If the answer is "yes," then it is likely that the living system has already solved the problem (as long as it yet survives), and copying relevant parts of this solution may appear to be an act of genius.

FIGURE 8.2.8
Studying the way ants organize themselves to obtain food leads to more efficient managerial algorithms.

FORAGING THEORY

Foraging for food is not such a simple endeavor. *Foraging theory* was developed in the 1970's to explain foraging patterns and strategies based on a cost-benefit analysis. The animal examines the available food and weighs the amount of energy required to obtain the food. When the cost (or energy) becomes too high compared to the benefit (or food), foraging animals move to new territory. Computation of the cost-to-benefit has been selected over generations until the techniques are fixed in neural pathways of the foragers.

The theory of *information foraging* has been inspired by food foraging theory. People search for information on the internet similarly to how they search for food, and the same strategies are used. Information is gleaned from a web site or database until the cost of obtaining further information becomes too high. When it does, people move on to another site. They rely on certain cues to tell them whether a particular site contains the data they seek. These are certain words or pieces of information left by previous users (such as reviews or ratings). The relationship and interaction between people and the online environment is called the *Internet Ecology* (McFedries, 2003).

Search patterns are similar across the entire animal kingdom, including dinoflagellate microbes, foraging honeybees, wandering albatrosses, reindeer, and early human hunter-gatherers, and these patterns appear to be optimized for efficient use of energy resources, in order to search for food, search for a mate, avoid predators, visit a watering hole, look for a den site, maintain and mark territorial claims, migrate, or search for a site (or host) on which to lay eggs. These patterns, called *Lévy search patterns*, alternate between clusters of short steps interspersed with longer steps to allow the searcher to cover large [target] areas (Viswanathan et al., 1999). The search procedure exhibited by honey bees has led to the *Honey Bee Algorithm* used by major web hosting organizations to streamline internet services (Wolf et al., 2016).

8.2.3 Biological Solutions to Biological Problems (Biotechnology)

> The cloning technique is remarkable, without a doubt, but I believe it is wrong to classify a cloned creature as an invention. Doing so somehow implies that a clone is different and inferior to other living creatures merely because the method of creation has changed. A clone is just another member of its species.
>
> —**Laura White**

Solving a biological problem with a biological solution is one of the most obvious means to utilize biological systems. This is the area of *biotechnology*, and often involves the use of *genetic manipulation*, and *recombinant DNA*. This is also the one area where engineering methods, those that largely involve constant characteristics and linearity, cannot easily be applied. Most techniques in biotechnology have been developed by biological scientists using methods from their laboratory investigations. Engineers have contributed mostly in the scale-up phase, where production was increased from laboratory scale to industrial scale, or in the development of measurement methods.

Several applications where biotechnology can be utilized have already been given in Section 8.2.1, Bionics. The distinctions between this section and Section 8.2.1 are not very clear cut, because the use of microbes, for instance, in bioremediation is often considered to be biotechnology. The term biotechnology is more commonly applied when microbes that have been genetically modified are used (Figure 8.2.9).

The steps involved for recombinant DNA are often as follows:

1. Cellular DNA is isolated.
2. The DNA strand is cut by special enzymes (*restriction enzymes*).
3. New genes are inserted into the gap in DNA.
4. The DNA strand is reconnected.
5. Identification methods, such as electrophoresis, are used to confirm that the new gene was successfully placed in the correct location.

The process of splicing genes can be highly inefficient and the yield of cells containing the desired additional genetic material has historically been very small; small, that is, until the advent of CRISPR technology. To select the successful from the unsuccessful cells with previous methods, a second gene was added to the one being spliced. This second gene was chosen to confer resistance to an antibiotic no longer widely used (such as kanamycin). When the group of cells was added to a medium containing the antibiotic, those that had not acquired the new characteristics were killed, and only the successful cells survived.

Genetically edited organisms (GEOs) can be made with a system called *CRISPR* (for clustered regularly interspaced short palindromic repeats, a name that describes the distribution of target viral RNA or DNA segments throughout bacteria threatened by viruses from which the segments come). CRISPR is not only efficient but also applicable to virtually any living species. The *gene editing* using CRISPR allows the desired insertion of genes from different strains of the same species that convey beneficial characteristics to the target organism rather than the insertion of genes from other species, as in genetically modified organisms. This method can be used for adding, disrupting, changing the sequence of specific genes, or gene regulation in many species at all hierarchical levels. The CRISPR system uses certain genetic patterns to identify the specific site on the chromosome where the desired gene is to be inserted. Using this marker sequence, a *restriction enzyme* is used to cleave the double-stranded DNA

Harnessing Microbes

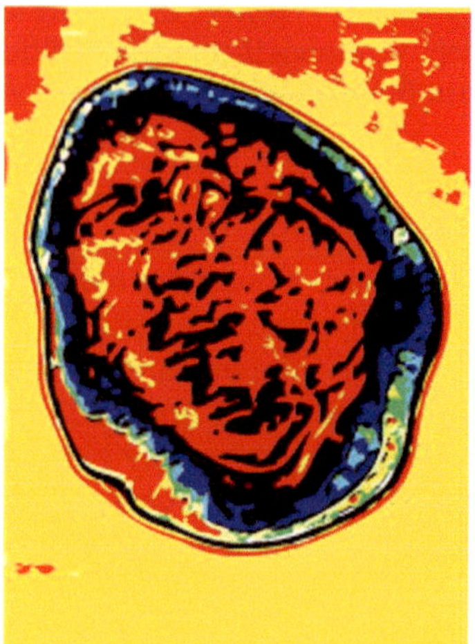

Some of the most infectious disease-causing microbes, like this measles virus, are being retrained to attack tumor cells instead of healthy ones.

C. botulinum
The bacterium that causes food poisoning also thrives in the oxygen-deprived depths of tumors, where it delivers a cancer-killing toxin.

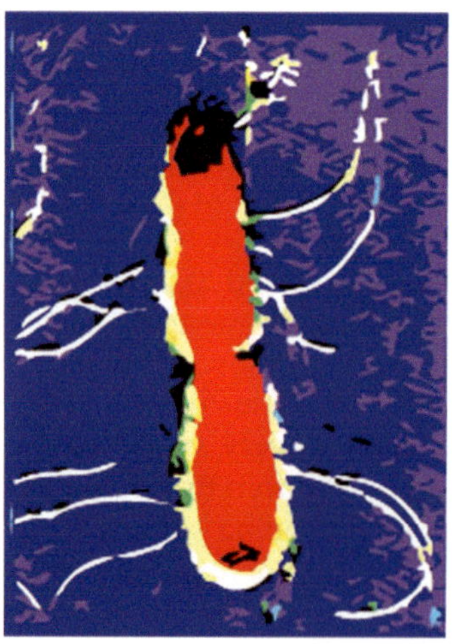

Salmonella
Another bacterial resident of unclean food, it has a natural affinity for tumor cells and can be genetically altered to produce a powerful anticancer agent.

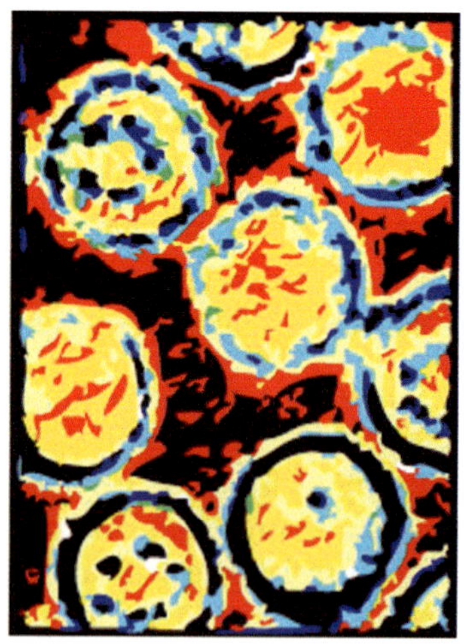

Epstein–Barr
This virus triggers the production of killer immune cells that can be directed to destroy any number of different cancer cells.

FIGURE 8.2.9
These types of bacteria and viruses are being harnessed for useful ends.

at the sequence-specific designated site, and the gene of interest is inserted where it is supposed to go. *Guide RNA* (gRNA), or *single guide RNA* (sgRNA), and a special protein (Cas9) that functions as a restriction enzyme to cleave the DNA molecule at a specific target site (Figure 8.2.10). gRNA consists of a string of nucleotides that are complementary to the target cleavage sequence. Cellular mechanisms are enacted to repair the breaks including the newly

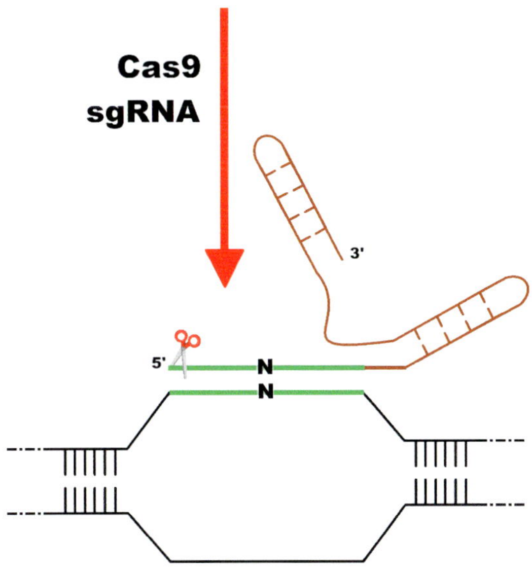

FIGURE 8.2.10
The restriction enzyme, Cas9, cuts a DNA strand in a place indicated by a guide RNA segment that matches up with the complementary DNA strand in a particular place. With the DNA gap established, genetic material can be inserted at that location, or a portion of the DNA strand can be eliminated, if desired.

inserted gene. Gene editing is possible in specific local tissues. The CRISPR/Cas9 gene editing technique is generally functional in all tissues, but, using a tissue-specific promoter, the guide RNA (gRNA) can be made to function in only certain target tissues (Ablain et al., 2015). However, the CRISPR method has the potential to cause additional accidental and unwanted gene edits to be made. The problems with CRISPR are:

1. Genetic repair mechanisms are uncontrolled, and completely up to the cell. When CRISPR induces a double-stranded genetic break, the break is repaired via random end joining processes.
2. Many off-target mutations (edits) are possible because DNA molecules are negatively charged and the Cas9 protein is positively charged, so the Cas9 molecule is attracted to off-target sites.

With *light activated CRISPR* (LACE), the Cas9 effector system uses blue light to control transcription of target DNA sequences. Antibiotic selection is not necessary with this gene editing technique (Knox, 2014).

A somewhat older gene editing method is called TALEN (Transcription Activator-Like Effector Nucleases). Both TALEN and CRISPR use endonucleases that initiate double-strand breaks (DSBs) at virtually any genomic target sequence, and are used for many applications, including gene knockout, transgene knock-in, gene tagging, and correction of genetic defects (Davis, 2014). Another gene editing technique, ZFN (Zinc-Finger Nucleases), is even older, but much more expensive and difficult to use (Perkel, 2013).

Using techniques such as CRISPR, synthetic biology can now be used to modify existing organisms or create new organisms to produce therapeutic medicines, biofuels, and environmental remediators (Leake, 2015; Waldholtz, 2017). CRISPR can be used to advance the growth of human replacement parts from animals. CRISPR is used to knock out the genes controlling the formation of particular organs in fertilized eggs of animals (usually pigs), and then induced pluripotent stem cells from humans are injected into the young animal embryo. The embryo forms a chimera with both animal and human cells. The human genes are therefore the ones controlling the formation of the target organ(s). The organs can be harvested at the particular level of maturity that is needed (Belmonte, 2016).

Gene therapy can be used to correct imperfect organ operation, albeit at the basic level of cell function (Le Doux et al., 1995). Gene therapy is the introduction of nucleic acids into cells for a therapeutic effect. Portions of the DNA strand are transferred into the host cell to produce proteins that the unaided cell cannot produce. There are over 4000 known human genetic diseases, and it is likely several hundred of these will be able to be successfully treated with gene therapy (Le Doux et al., 1995). Gene therapy is still in its infancy, and its success is still largely dependent upon chance. Steps in this process involve

1. Identifying the gene controlling the function of interest
2. Isolating and reproducing the gene
3. Selecting a *vector* to introduce the gene to the recipient
4. Introducing the gene into the vector
5. Causing the vector to contact the recipient's cells
6. The vector penetrating the cell
7. The genetic material from the vector incorporating into the genetic material of the recipient

With so many complex steps involved, it is no wonder that gene therapy must still undergo much more development.

Sometimes, the disease is caused by expression of a gene that should be turned off. A biotechnological technique that is used in this case is to take the gene, scramble its genetic code so that, when reinserted into the genome, it can no longer code for a normal product. This is called *antisense DNA* (Vogel, 2003). This technique is used to interfere with the cellular formation of undesired substances.

Recombinant DNA techniques target the germ cell lines of plants and animals in order to modify these for all future generations. Human gene therapy, however, has been ethically constrained to correct only somatic cells. If possible, source cells, such as bone marrow cells that produce all blood cells, are targeted. Once the therapeutic gene has been introduced into these cells, all future daughter cells will themselves carry the gene. If it is not possible to introduce the gene into source cells, then the gene therapy procedure will have to be repeated periodically, with unequal results likely.

Gene therapy is often conducted *ex vivo* (or, outside the body). The therapeutic gene is introduced to the recipient cells, and then they are returned to the body. There are generally four choices used to deliver nucleic acid (DNA or RNA) into a target cell:

1. Transfection reagents and endocytosis
2. Electroporation
3. Viruses
4. Biolistics (shooting DNA- or RNA-coated pellets into the cell)

Of these, the first three are usually used with animal cells and the fourth can be used for agricultural work. *Lentiviruses* are sometimes used to transfect genetic material into non-dividing cells. Carbon nanotubes shaped like tiny needles have been found to transect the cell membrane, both by endocytosis and passive diffusion. Drugs and genetic material can be incorporated within the nanotubes to be delivered inside the cell (Al-Jamal, 2015).

Human gene therapy often uses retrovirus (a virus that infects animal cells and copies its genetic code from RNA to DNA of the host cell using the enzyme *reverse transcriptase*) and adenovirus (a virus of the respiratory system) vectors. These have been modified to be nonreproductive and to elicit as little of an immune response as possible. Other *ex vivo* human vectors are possible, and are being developed to reduce adverse reactions possible with these viruses (Le Doux et al., 1995). An advantage of using viral vectors, especially adenoviruses, for cell transfection is that whole animal transfections are easily accomplished. An advantage of using viral vectors, especially adenoviruses, for cell transfection is that whole animal transfections are easily accomplished.

A host of other recombinant DNA introduction systems are possible with plants and animals. These include *Agrobacterium* microbes, which invade the plants and naturally cause root disease, and the gene gun (or ballistic, also called biolistic) methods that shoot tiny heavy metal pellets (e.g., tungsten) into the cell. These pellets are coated with DNA material that combines with the host cell DNA, if the pellet reaches the cell nucleus, does not cause too much damage, and all other conditions are correct. Electroporation (inducing pore formation in the cell membrane through an electrical current) may also be used to transfer DNA material (see Section 2.11, Electricity).

Possibly, the largest triumph of biotechnology is in the introduction of transgenic crops, an advance that has revolutionized crop production. The first commercial introduction of *genetically modified organisms* (GMO) occurred in 1996. By 2000, transgenic soybeans, corn, and cotton accounted for more than half the area planted to these crops (Bruening, 2000).

The first GMO crops were formulated for disease and pest resistance. The microbe *Bacillus thuringiensis* (Bt) is a soil-borne organism that causes death in the caterpillar stage of *Lepidoptera* (butterflies and moths) after ingestion. Many economically important insect pests of field crops are *Lepidoptera*, and the isolated Bt gene has been inserted into these crops to protect against these pests without spraying with pesticides (Dandekar and Gutterson, 2000).

Glyphosate is a popular herbicide commercially called "Round-Up." It kills plants by translocation to roots, where it has lethal effects. Glyphosate is then harmlessly decomposed in the soil. GMO crops incorporate a gene that allows them to produce a modified enzyme that makes them tolerant of glyphosate. Thus, fields can be sprayed with this herbicide, and the only plants supposed to die are the weeds. Glyphosate-resistant weeds have developed, however, and cause problems for crop producers (See Example 1.6.4). New transgenic crops based on other herbicides have been introduced.

Other specific disease resistance has been incorporated in economically important crops. The papaya industry in Hawaii was headed for annihilation because of the deadly papaya ringspot virus. A GMO papaya variety gave resistance to this virus (Figure 8.2.11).

Other transgenic crops have been produced for better nutrition. Corn containing higher than normal levels of the amino acid lysine, rice with a gene to produce beta carotene, nuts with healthy oils, tomatoes containing vaccines against disease, and other foods with high levels of antioxidants will soon be available (Dandekar and Gutterson, 2000).

There are transgenic crops to resist frost damage, have longer shelf life (or, keeping ability), are seedless, and do not age in the field. Plants and animals are being designed to produce higher yields, produce biochemicals important for industry, and are drought- and salt-tolerant.

Biotechnologists have inserted genes from humans or other organisms into common plants in order to produce medicines for human or animal uses. Some early examples of these are a human gene to produce an antibody against the Herpes virus inserted into

FIGURE 8.2.11
Transgenic cultivars of papaya resistant to the papaya ringspot virus have saved the Hawaiian papaya industry.

corn, genes for hepatitis vaccines into tomatoes and potatoes, genes for non-Hodgkin's lymphoma medicine and anti-Ebola antibodies (Branswell, 2015) into tobacco (Figure 8.2.12), and genes for lysozyme production from rice.

Genetically modified switchgrass can grow polyhydroxybutyrate (PHB) that can be used to fashion plastic bags, thermal foam, food packaging, and other objects (Cooper, 2008). PHB grown in this way is easy to harvest and does not require further chemical alteration. The switchgrass by product can be used as a biofuel. These crops are mass produced in the ordinary way, but instead of being used for food, they are sent to processing plants for biochemical extraction. These biochemicals are called *biopharmaceuticals*, and their production is called *biopharming*.

Gene drive is the phrase used to describe the increased prevalence of one or more genes due to natural or human genetic modification causes. They are genetic systems that circumvent traditional rules for genetic propagation; they greatly increase the odds that the genes will be passed on to offspring. This can allow them to spread to all members of a population even if they reduce the chance that each individual organism will reproduce. Gene drive has been mentioned as the means to eliminate disease-carrying insect vectors, such as the *Aedes aegypti* mosquito, which infects humans with the Zika virus.

All this is not without concern, however. Is it ethical to extirpate an entire species, as with the *Aedes aegypti* mosquito? Also, faced with the vast challenge of Bt crops, insects are almost sure to develop resistance, and then Bt will no longer be able to control pests organically. The use of

FIGURE 8.2.12
Tobacco can be used as a preferred crop to modify to produce pharmaceutical products. It is not likely that tobacco products would find their way into the human food supply.

GMO crops is sure to change the economic structure of agriculture, and the safety of some of these crops has been questioned. Not everyone agrees that cloning ought to be done, especially because there is always the threat of human duplication. Worse, perhaps, is the prospect of human–other animal hybrids developed for growth in the production of human replacement organs or for other purposes. The ethics of these situations have not been settled.

There are many economic advantages to growing transgenic crops that produce pharmaceutical products, but there are problems as well. Food can be contaminated when transgenes are not isolated, plants intended to produce modified products can accidentally be commingled with food plants, or pollen and seeds can be consumed by wild insects, birds, and animals. Failure to properly control the appearance of volunteer transgenic plants (those that grow wild from previous year's seed) has resulted in contaminated feedstocks (Marvier, 2007). Regulatory oversight of this industry may not be adequate to protect people and the environment from unfortunate accidents. Human error appears to be unavoidable.

Artificial endosymbiosis, where one type of microbial cell is manipulated to become an intrinsic part of another type of host cells, has the potential to dramatically alter host cell physiology, organismal biology, and ecosystem functioning without directly manipulating the genome. Hence, endosymbiosis could be an alternative to genetically manipulated

CRISPR HISTORY

(Adapted from Lane, 2016)

In 2007, some dairy scientists working for the Danish food industry company, Danisco, had a problem with their yogurt. It seems that some of their yogurt batches had an off-taste. After studying the bacteria that ferments the milk, they found that when those bacteria were infected by viruses, the yogurt had weird flavors.

(*Continued*)

CRISPR HISTORY (Continued)

When they mapped the bacterial genome, they saw regions that contained short DNA sequences that were repeated quite frequently. Interspersed between these sequences were spacer segments of DNA and each spacer was different from the others. This, actually, was not new. Scientists had known about repeating segments in bacteria for a couple of years and had named them "Clustered Regularly Interspaced Short Palindromic Repeats"—CRISPR. Researchers originally thought that the duplicate DNA sections were like backup sequences in case something went wrong.

But, in 2007, those Danisco scientists found something very strange. They discovered that the spacer segments were actually DNA sequences of viruses, not the host bacteria. And, through some excellent deductive logic, they concluded that the bacteria used this chromosome region as a library of viral DNA, and also that these sequences were part of some type of bacterial immune defense system against those viruses. This really made sense when they discovered a bacterial enzyme that killed viruses. The enzyme would use the viral DNA as a search sequence inside the cell—which was the invading virus—and the enzyme would cut that DNA and thus destroy the virus.

This was fascinating information, but the only people interested in it were scientists who studied viral infections of bacteria, not a very large group of investigators.

Jump to the summer of 2012. Scientists from two widely separated labs, the University of California at Berkley and Umea University in Sweden, worked together on the properties of a very special enzyme called Cas9. This is one of the powerful cutting enzymes in the CRISPR system. Even though Cas9 is a bacterial enzyme, these researchers found that they could insert this enzyme into the nucleus of mammalian cells so that it could work on mammalian DNA. These cells also have sophisticated repair mechanisms.

These researchers found that if they preloaded the Cas9 enzyme with a guide RNA, the enzyme would search for the corresponding sequence in the chromosomal DNA and cut The DNA at that point. The cell would try to repair the DNA, but it may

(Continued)

CRISPR HISTORY (Continued)

not repair it properly. Thus, cutting the DNA would effectively silence that gene. And by strategically choosing an appropriate guide RNA sequence, they could program the Cas9 enzyme to cut DNA at nearly any specific point in the chromosome. They had essentially discovered a chromosomal scissors that could be used in all kinds of living cells—mammals, birds, insects, worms, and even plants.

organisms (GMOs) without the close regulatory scrutiny attracted by GMO organisms (Kearney et al., 2017).

Cloning of genetically identical animals is a technique that uses somatic cell DNA transferred to germ cells (Figure 8.2.13). By this technique, a single animal with desired genetic makeup can be duplicated as often as one would like to improve livestock traits (Murray and Anderson, 2000).

Although DNA in the nuclei of cloned cells is identical with that of the donor somatic cells, mitochondrial DNA remains identical with the germ cell into which the nuclear DNA was placed. Thus, a clone will never be exactly identical with its donor (unless the donor was the germ cell). Mitochondria control cellular metabolic functions, which can influence phenotype.

Because mitochondria are passed from one generation to the next only through the maternal line, and it sometimes happens that there are genetic disorders of the mitochondria, these disorders are transmitted to subsequent generations no matter how healthy the paternal mitochondria. One solution is to use in vitro fertilization (IVF) with three

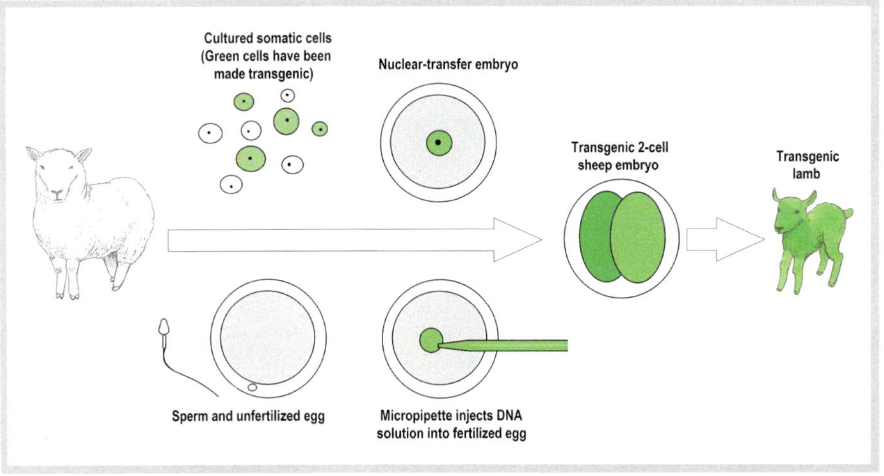

FIGURE 8.2.13
Transgenic farm animals are produced by direct microinjection of DNA into young embryos and by cloning from transgenic somatic cells. Microinjection procedures use recently fertilized eggs, which for some species can be obtained from *in vitro* fertilization procedures, before the first cell division. If the foreign DNA becomes integrated into the embryonic genome at the one-cell stage, as the embryo develops, all of its cells will contain the transgene. The offspring that is born after transfer of the embryo to the reproductive tract of a recipient female will be transgenic. Alternatively, somatic cells can be collected from an animal, cultured in the laboratory, and exposed to foreign DNA. Some cells will become transgenic, and these cells can be selected for use as nuclear donors in nuclear-transfer procedures. The resulting nuclear-transfer embryo will be transgenic, as will the offspring born after embryo transfer and term development in the reproductive tract of a recipient female. (Redrawn from Murray, J.D. and Anderson, G.B., *Calif. Agric.*, 54, 57–65, 2000.)

gamete cells. The nucleus of the intended maternal cell with mitochondrial disease can be inserted to replace the nucleus of a second maternal cell with healthy mitochondria and then the resulting hybrid cell can be fertilized with the male gamete. The result is an offspring with intended nuclear genetics and healthy mitochondria that are passed down to subsequent generations. Louise Brown in the U.K. was the first IVF baby born in 1978. Zain Rajani

COPY CAT

Animal clones are genetically identical to their predecessors (here is where the word "parent" really loses its meaning). Clones have been made for quite a few years by separating the cells of immature embryos and implanting each cell in the womb of a separate mother. All such animals are identical, making this process attractive for reproduction of livestock animals with special traits (such as the ability of dairy cows to produce a very large amount of milk or to produce exotic biochemicals in their milk).

The first animal cloned from a somatic cell of an adult animal was the sheep named Dolly, created in Scotland. A calico cat called "CC" for Copy Cat was created in a similar way in 2001 at Texas A&M University (Figure). The donor of eggs was an anonymous female cat, and the donor of somatic cells was a cat named Rainbow, a calico female (Figure). CC was genetically identical to Rainbow. Interestingly enough, although Rainbow and CC have the same nuclear genes, their colorings are much different. This illustrates that environment plays a role in development.

CC herself. (Redrawn from Kluger, J., *Time* 159(8), 58–59, 2002.)

(*Continued*)

COPY CAT (Continued)

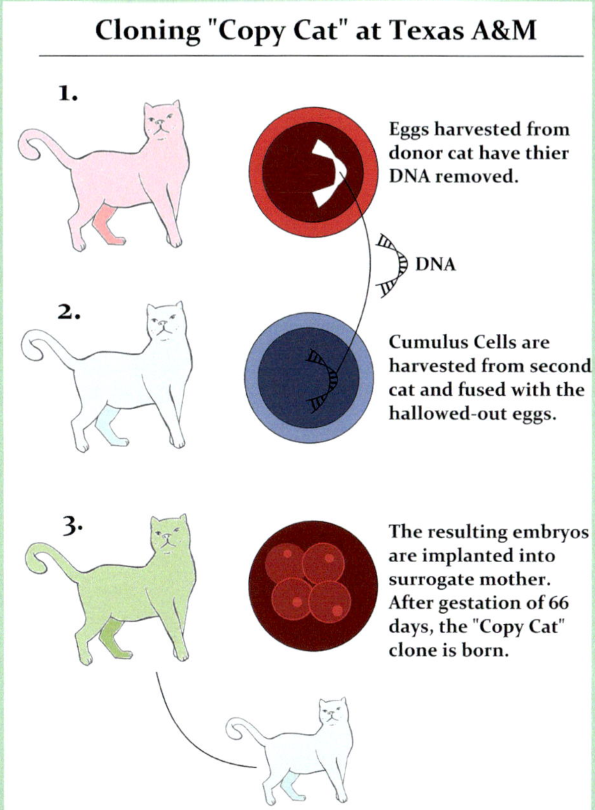

The method used to create CC. (Redrawn from Kluger, J., *Time* 159(8), 58–59, 2002.)

So far, animal clones developed from somatic cells have shown several weaknesses that clones from separated embryonic cells have not shown. One of these is premature aging, possibly caused by the shortened telomeres on the ends of the DNA strands taken from the original somatic cells. There have been other physical weaknesses noted, as well. Thus, animal cloning is still in its infancy.

GREEN FLUORESCENT PROTEIN

Imagine being able to see cellular proteins as they move around the cell. Or imagine a visual marker for transgenic plants. Or, even more, imagine an animal pet that glows in black light. All these are possible because of Green Fluorescent Protein (GFP).

(*Continued*)

GREEN FLUORESCENT PROTEIN (Continued)

GFP was isolated from a jellyfish *Aequorea victoria*, where it caused the jellyfish to glow a bright green color when targeted by ultraviolet light (Lippincott-Schwartz and Patterson, 2003). When a high-energy photon hits the protein, several chemical bonds are changed, and, when they return to their unexcited form, lower energy light is liberated. GFP requires no coenzymes or coreactants and so is self-contained in its ability to fluoresce (Goodsell, 2004).

Because of this, GFP can be fused to almost any protein of interest to analyze protein location, movement, and chemistry in living cells. GFP has been used to probe protein function, monitor protein interactions, and act as biosensors for protein presence. GFP can be attached to specific antibodies to detect the presence of foreign antigens (Ryan, 2003), or attached to viruses to track their movement throughout the body.

The gene that produces GFP has been isolated, and can be coupled to other genes inserted into genetically modified organisms (see Section 8.2.3, Biotechnology). Thus, transgenic organisms can be monitored under field conditions by illumination with ultraviolet light to see if they emit green fluorescent light. This has been used, for example, as a biomarker for tracking the Bt gene for insect resistance in field-grown crops. As long as the GFP gene and Bt gene remain linked, one can be used as an indicator of the presence of the other (Harper et al., 1999).

Various mutations of the GFP gene have resulted in fluorescent proteins colored blue, cyan, yellow, and red, in addition to green (Lippincott-Schwartz and Patterson, 2003). Other fluorescent proteins from reef coral have been discovered and have been optimized for expression in mammalian cells (BD Biosciences, 2004)

GFP is distinct from the glow-in-the-dark chemiluminescent protein-enzyme system luciferin-luciferase that produces the glow in fireflies and glow worms (Krones et al., 1990). This system does not require ultraviolet or blue light for excitation, but does require adenosine triphosphate (ATP) (see Section 3.10, Energy-Rich Compounds). It has thus been used as an indicator of metabolic activity.

Among the more exotic, if controversial, uses for GFP are the creation of fluorescent plants, fishes, rats, mice, frogs, flies, worms, and many other living things.

in Canada in 2015 was the first baby born with Augment, a system that implants mitochondria into a mother's egg cells.

8.2.4 Living Things as Recipients (Biomedical Engineering)

> …there are two possible ways in which the young graduate can acquire both specialized bioengineering expertise and adequate exposure to the life sciences. One can either take a master's degree in bioengineering directly, or a "sandwich course," which intersperses hospital experience with academic training. In either case one must learn to THINK LIKE A BIOLOGIST.
>
> **—David J. Dewhurst**

The fourth general class of biological engineering solutions applies devices or processes to living things. The devices themselves may be entirely chemical or physical and not biological, but the challenge is to use them to correct the problem while not interfering with other functions performed by the biological object of the design.

Most biomedical devices fall into this category. Specifications for these devices are often numerous and changing. They sometimes must serve multiple uses and satisfy multiple users (Shade and Johnson, 2003). In addition, certain of these devices must be fit for individuals with disabilities; the devices may need to be modified substantially to perform the intended function on the specific person.

Future advances in biological engineering will involve robots to complete numerous tasks both repetitive and not (The word *robot* was coined in 1920 by Czech writer Karel Capek in his dramatic play R.U.R., which stands for Rossum's Universal Robots). There is a growing involvement of robotics in medicine and engineers have a large role to play in their development. Robots will be used in situations hazardous to human health, such as natural disaster searches and recovery, dangerous chemical spill cleanups, as means to collect data from otherwise inaccessible environmental locations, or as means to complete unsafe or highly repetitive tasks. They can be used for individual health care delivery. Robots are also used for companionship and emotional support of lonely or infirm individuals. Robots have been used in the operating room for surgery or for medical surveillance or for ailment diagnosis. Robots can be used for training of animals or pruning of plants. Exoskeletons, similar to robots, are being developed to assist in walking, weight-lifting, and manipulation. The range of applications is endless. Robots are even being trained to spot changes in patients that relate to medical conditions such as pain or depression. This is called *Behaviomedics*. Biological engineers can be involved in robot design and training, in order to make them operate in ways compatible with their human companions.

Artificial Intelligence (AI) will be used with robotic systems to assist with diagnosis and correction of medical afflictions and disorders. Engineers are attempting to develop AI systems with learning capabilities (*Deep Learning*) similar to those of human brains. This has led to development of design principles for such systems. Three principles of neocortex learning (Hawkins, 2017) are:

1. Learning by rewiring. New memories are formed by the formation of new synapses between neurons.
2. Sparsely distributed representations. Having different neuronal representations of elemental attributes in different places in the brain that together can determine similarities or differences.
3. Sensor-motor integration. Keeping track of many sensory inputs simultaneously to form a complete concept of an image, or object.

Deep Learning systems use *Convolutional Neural Networks* (CNNs), similar to Artificial Neural Networks (see Section 4.4.5, Open-Loop Systems), but CNNs develop rough classifications of information and refine these classifications more and more deeper in the network.

Engineering solutions to correct deficiencies found in living systems take on several forms, depending on the maturity of the technology (Figure 8.2.14). At first, these designs are based on first principles from physics and chemistry, and they must then be modified to meet the special requirements of the living things to which they are to be applied. They may be mechanisms that already exist, but are being applied to living things for the first

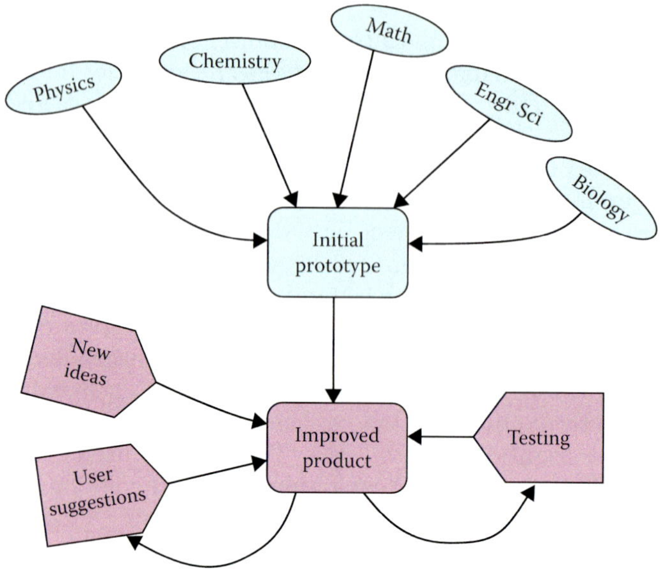

FIGURE 8.2.14
Initial attempts at an engineering design usually involve principles from the basic sciences and mathematics. Thereafter, improvements come about by application of empirical knowledge obtained from user experiences and structured testing.

time, or they may begin as entirely new. They must be made to fit constraints of size, power consumption, configuration, materials biocompatibility, effectiveness, reliability, and other issues important to the living system that is the object of the attention (King and Fries, 2003). Biological engineering designs at this stage are likely to be relatively crude and unreliable, with marginal advantages over other means to solve the same problem.

These designs soon enter an empirical phase, wherein further improvements are incremental and based upon experiences with these first models. Modifications in existing models may have no better justification than the fact that they were tried and seem to work better than the original. Improvements in materials, improvements in reliability, and improvements in efficacy come about this way. This stage in the design is worked on by specialists such as biomedical engineers or perhaps by nonengineering technologists rather than by biological engineering generalists. Companies producing these products or processes amass large amounts of much applied proprietary data coming from sources such as laboratories and customers. The person working at this design stage is often working on a small part of the overall product and utilizes specialized knowledge of the particular product and its application.

Even with a lot of improvement, replacement organs or tissues are imperfect compared to the original. Biological Units (BUs) are robust and redundant. Organs usually have more capacity than is needed to perform their functions; instead of one gene to repair and rebuild errors, there are often several; organisms can adjust to vastly different environments. BUs are wonderful. Therefore, it is unusual for BU to function so poorly that they need replacement.

Much of the effort to develop replacement BU has come about to serve human needs, and, to a lesser extent, the needs of animals that humans care for. Attempts have been made to replace nearly all tissues and organs of the human body, and these attempts will continue for the foreseeable future.

There have been two competing approaches to replacement BU. The first is transplantation of a similar BU. These can be *autographs* (tissue from the same person), *allographs* (tissue from the same species), or *xenographs* (tissue from a different species). A blood transfusion from one person to another is an example of an allograph. Banking one's own blood for a later surgical operation is an autograph. Using blood from swine is a xenograph.

As you might suspect, the farther removed the source species is from the species receiving the transplantation, the more difficulty there is in having the BU accepted. Successful cross-species transplantation is extremely rare; transplantations from individual to individual of the same species must be matched for several tissue protein markers [the major histocompatibility complex, or MHC (see Section 6.20, Controls)], and even then the recipient must usually take cyclosporin or other suppressors of the immune system in order to avoid rejection. General immune system suppression leaves the recipient susceptible to infection and cancer, but drugs such as cyclosporin A or FK506 interfere only with helper T cell activation and do not cripple nonspecific immune responses (Campbell et al., 1999).

One successful cross-species transplant, in the loose sense of the word, is the use of insulin for the treatment of diabetes. Insulin has been obtained from the pancreases of slaughtered cows, sheep, and pigs, purified, and injected into humans (Fenster, 1999). The only differences between insulin from these species are a few small chemical groups, called *residues*, in three positions on one of the amino acid chains of the molecule (White et al., 1959). These differences are so small that they have little effect on the efficacy of this hormone.

The *stratum corneum*, or tough outside layer of skin, is nearly impenetrable to chemicals of any kind, but especially to large molecules. The stratum corneum is composed of flat, dead, epithelial cells arranged as bricks with fatty lipid mortar. A brief burst of ultrasonic energy can liquefy the lipids, thus making the skin permeable to complex molecules such as drugs, or allowing blood glucose to rise to the skin surface for monitoring (Figure 8.2.15).

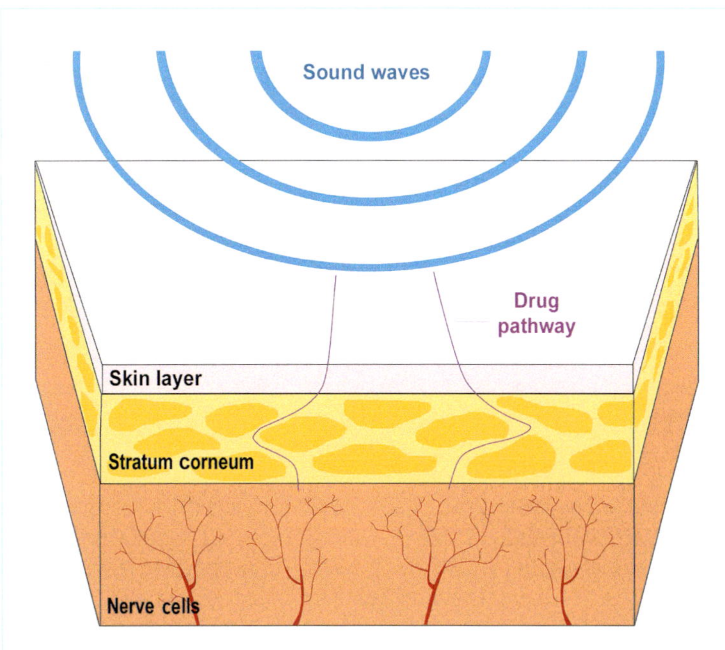

FIGURE 8.2.15
Ultrasound energy can make the skin temporarily permeable to drugs such as insulin.

THE DISCOVERY OF INSULIN

Call it luck, ignorance, or naiveté, but whatever it was, it led Dr. Frederick Banting to discover insulin as a treatment for diabetes. He had been able to think of a solution to diabetes because he had been unknowing enough to view it as a simple problem (Fenster, 1999).

Before Banting, it was known that pancreas dysfunction had something to do with diabetes, but trying to isolate the culprit posed a real problem. The digestive secretions of the pancreas were so powerfully destructive that they would destroy any other agents in the liquefied pancreas. Diabetes, meanwhile, was moving from the twenty-eighth leading cause of death in 1900, to twelfth in 1920, to seventh in 1940. The most effective treatment at the time was slow starvation of 500–1000 calories food intake per day.

What Banting did was to tie off the secretory ducts of the pancreases of dogs until the digestive secretory tissues atrophied, and then what was left was the insulin-producing part. Injecting pancreas extract into a dog with diabetes induced by removing its pancreas cured the disease. These experiments were performed despite obstacles and active opposition by one of the foremost experts in diabetes at the University of Toronto. Banting received a Nobel Prize in 1923 for his work.

As a farmer's son, Banting knew that slaughterhouses could be an abundant source for pancreatic material. Banting's assistant, Charles Best, developed a means to extract the insulin from full-grown cows.

Almost all the insulin available now comes from recombinant DNA modified *E. coli* bacteria. These bacteria that include genes for human insulin production yield insulin that, when purified, is less likely than animal insulin to cause allergic reactions. Human insulin was the first major product produced by the biotechnology industry.

There has been, and continues to be, progress toward performing organ function through artificial means. Actually, we should say *bioartificial* means, because there usually is some combination of human-made materials and biologically based materials incorporated into a biologically inspired design scheme.

Replacement organs can either be located external to the body or inside it. An internal placement usually has the advantage that it allows normal individual movement; the disadvantages are that there is little room to implant a replacement device and there is a serious design constraint on energy delivery and material compatibility with the human or animal body.

THE MOST MUNDANE REPLACEMENT ORGAN

Charles Lindbergh has been credited with being the first person to think seriously about ways to replace the functions of the heart and lung in order to allow surgery on hearts temporarily relieved of their life-sustaining responsibilities (Fenster, 1998). His sister-in-law had a serious cardiac condition that could not be corrected without the heart surgery that was impossible in the years following his solo flight across the Atlantic in 1927. He joined Dr. Alexis Carrel at the Rockefeller Institute and became a very visible spokesman while also bringing a practical engineering view to the project.

(*Continued*)

THE MOST MUNDANE REPLACEMENT ORGAN (Continued)

He showed that if ever the challenge of artificial replication of an organ were to be realized, medicine and engineering would have to do it as a team.

It was not as a team that comfortable and affordable dentures were brought to the public. Early dentures were made of ivory or hippopotamus tooth and required springs to hold them in place against the gums (Prioli, 1991). These dentures were ill-fitting, and uncomfortable, as best exemplified by Stuart's 1796 portrait of George Washington that adorns each U.S. one-dollar bill.

When Charles Goodyear accidentally invented vulcanized rubber between 1839 and 1851, he was given the idea of using that rubber (called Vulcanite) to make bases for artificial teeth. Within a few years, Vulcanite replaced the gold, silver, ivory, and animal bone that had been used for five centuries. However, in 1868 Vulcanite that cost $4 per pound could be used to make $150 worth of dentures. This ratio is not unusual for materials used in medical devices: cost of the materials is often a very small part of the final cost.

An unscrupulous man named Josiah Bacon, who took control of the Goodyear Company, began to enforce a licensing agreement for dentists based upon Goodyear's patents. When dentists across the nation expressed outrage and refused to pay, Bacon launched a ruthless campaign in the courts. The courts repeatedly backed Bacon's position.

One dentist, Dr. Samuel P. Chalfant, was hounded so much that he had abandoned lucrative practices in Delaware and Missouri. He had finally established himself in San Francisco, but Bacon again tracked him down. On Easter Sunday morning in April 1879, Chalfant walked into Bacon's hotel room and shot him dead. The aggressive legal campaign against dentists ended with Bacon's death, and dentists were then free to bring this important advance in dental prosthetics to their patients.

Since then, dental prosthetics, or false teeth, have been made of other materials, including celluloid and acrylic that can be molded to fit their wearers in comfort.

THE ARTIFICIAL KIDNEY

Dr. William Kolff was a hospital intern in Groningen, Holland when World War II began in Poland. When the German Army conquered his country a year later, Kolff transferred to the small Municipal Hospital in quiet Kampen to escape the close scrutiny of the Nazis. It was there, during the desperation days of Nazi occupation, that Kolff developed the first artificial kidney (Fenster, 1998).

Kidneys remove waste urea and excess water from the blood stream. They do this by filtering molecules and water through glomeruli membranes that act as sieves to retain cells and large proteins in the blood. When the kidneys fail, coma and death follow.

Dr. Kolff had learned that cellophane has some of the same sieve-like properties of the walls of the glomerulus, but he had no access to cellophane in his Nazi-controlled country. Where could he get some? The only person who had any was the local sausage maker who used cellophane for sausage casings. So, Kolff talked him into supplying some casings.

(Continued)

THE ARTIFICIAL KIDNEY (Continued)

The blood of kidney failure patients was to flow through the casings, but it needed to be pressurized. To exploit centrifugal force on the blood, he used a rotating drum that was donated to him by the director of an enamel pot factory.

The tubing between the stationary patient and the rotating drum had to include a sealed joint, but the joint between the two tended to leak. Kolff did not know how to deal with this problem for a while until he realized that automobiles were faced with the same problem. Upon his request, the Kampen Ford repair shop provided a water pump seal that worked perfectly.

So that all the water and solutes in the blood were not removed as they passed through the sausage casing, the casing had to be bathed by a special solution that controls the final concentration in the blood (Galletti et al., 1995c; Johnson, 1999). Clandestine work was conducted in seclusion during the day and secretly at night by Kolff and his small number of associates to perfect a dependable artificial kidney.

Kolff tried his machine on the first humans in 1943. The first sixteen cases were helped but not saved by the device. The seventeenth patient, however, was a 67 year old woman dying from kidney failure who was in prison when she became ill. She had been accused by her husband of being a Nazi collaborator. She was already comatose and on the verge of death when Dr. Kolff bypassed her kidneys with his artificial device. After twelve hours he leaned close to her face and asked if she was able to hear him. The woman opened her eyes, looked at Kolff, and said, "I am going to divorce my husband." She lived, and she did (Fenster, 1998).

Take the implantable artificial heart, for instance. It should be a pump small enough to fit within the body and powerful enough to maintain the correct flow of blood. The power source must be reliable and unobtrusive. The material it is made from must be flexible, durable, and compatible with living tissue (see Section 3.12, Free Energy). The total device must be rugged and reliable (Figure 8.2.16).

The first artificial heart implanted in a human being was the Jarvik-7 heart invented by Dr. Willem Kolff (yes, the same Dr. Kolff) and Dr. Robert Jarvik in Salt Lake, Utah. The recipient was Dr. Barney Clark, who lived for 112 days. The Jarvik-7 did not satisfy all the design objectives. It was pneumatically driven, and the power source and compressor occupied a very large cart. This effectively tethered the patient to a nearly fixed location. The tubes supplying air transpierced the chest, thus giving opportunity for infection. Blood clotting was a problem, as well as the ability of the patient to tolerate the blood pressure increased over that to which his body had adjusted. The visible prominence of the shortcomings of the device led to the suspension of artificial heart human implantation for nearly 20 years. In the interim, human implantation of devices to assist the left ventricle only or devices intended as bridge-to-transplant (temporary devices meant to sustain the patient until a replacement heart was available for transplant) provided the opportunity to improve device quality and reliability. Present devices are much closer to the ideal heart replacement than earlier devices, and improvements are constantly being made (Rosenberg, 1995).

Functional artificial organs have been in existence since about 1940 or 1950. As such, they are rather primitive devices with a different set of characteristics from the natural organs

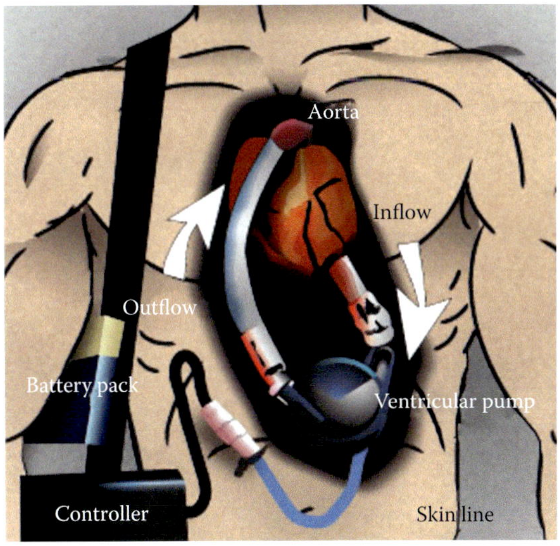

FIGURE 8.2.16
An implantable artificial heart must meet many requirements. Pictured here is a left ventricle assist device (LVAD) that does not completely replace the entire natural heart.

they are to replace (Galletti, 1995). Natural organs have been refined over thousands of generations. They usually have extra capacity to perform their intended function and so can adjust well to bodily stresses. Artificial organs are designed to perform specific functions within certain design constraints. They may or may not be anatomically similar to the organs they are meant to replace, and they often use different mechanical, electrical, or chemical processes to achieve the same functional objectives as natural organs. They use materials that are usually inferior to the materials that constitute natural organs, and artificial organs do not repair themselves. Nonetheless, people are benefiting from improved artificial organs and prostheses.

Until 1950, wooden legs, corrective glasses, and dental prostheses represented most of the artificial organs in use. Within a decade, the artificial kidney (Galletti et al., 1995b; Lysaght and Moran, 1995), the heart–lung machine (Galletti and Colton, 1995), the cardiac pacemaker (Forde and Ridgely, 1995; Jeffrey, 1997; Greatbatch, 2000), the arterial graft (Ku and Allen, 1995), the prosthetic cardiac valve (Yoganathan, 1995), and the artificial hip joint (Lord and Turner-Smith, 1995) were all being used. Now, we have the membrane lung (Voss and Butterfield, 1995), the implantable lens, finger and tendon prostheses, total knee replacements, and soft-tissue implants for reconstructive surgical use (Yannas, 1995a; Nakamura and Shimizu, 1995). Artificial pancreases (Galletti et al., 1995a), artificial blood (Intaglietta and Winslow, 1995), liver support systems (Galletti and Jauregui, 1995), nerve guidance channels (Valentini, 1995), transcutaneous electrical nerve stimulation (TENS) to quell pain, electrical neural stimulation to counter the effects of epilepsy (Fuller, 2001) or Parkinson's disease (Peckham and Smith, 1995), implantable cardiac defibrillators (Duffin, 1995), and mechanical ventilators (Behbehani, 1995) are all becoming more common or are showing promise. Electrical neural stimulation techniques can be used to treat somatic pain, post-traumatic stress disorder (PTSD), effects of stroke, tinnitus, immune responses, inflammatory bowel disease, hormonal release, and other conditions. Most of these are treated with vagal stimulation (Grifantini, 2016).

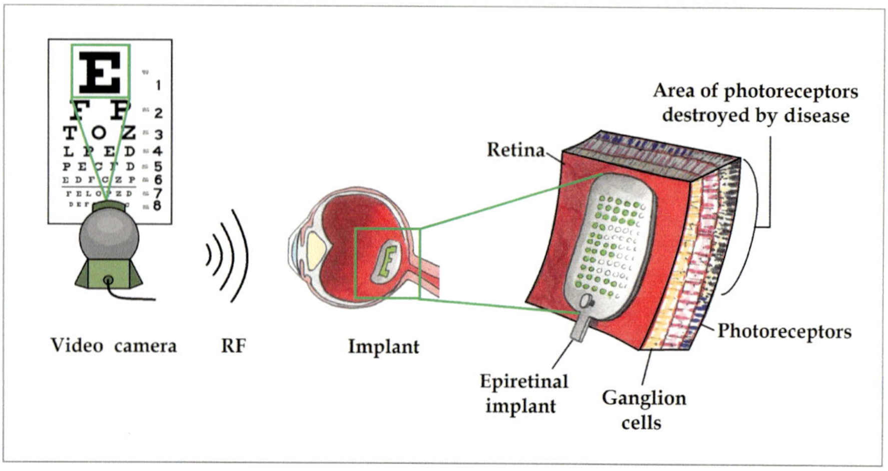

FIGURE 8.2.17
An artificial retina can convert light images into electrical signals that can be recognized in the brain. (Redrawn from Weiland, J.D. and Humayun, M.S., *IEEE Eng. Med. Biol. Mag.*, 24, 14–21, 2005.)

Cochlear implants are becoming more sophisticated and allow people to hear who cannot hear in any other way, including hearing aids (Heppenheimer, 2001).

Biomimetic prostheses can take many forms. There are continuing developments of artificial retinas (Figure 8.2.17) (Rodger and Tai, 2005; Weiland and Humayun, 2005), artificial limbs, reanimation of paralyzed limbs (Loeb and Davoodi, 2005), and sound processing to improve hearing. Ambitious projects are underway to interface directly with the central nervous system to ameliorate deafness, blindness, paralysis, epilepsy, and the tremors of Parkinson's disease (Stieglitz et al., 2005; Wise, 2005). Even the confusion and lost cognitive function of Alzheimer's disease may be helped by bypassing functionally defective brain regions with microelectronic signal processing units (Figure 8.2.18) (Berger et al., 2005).

Tissue engineering is the manipulation of living cells to produce tissues and organs to replace or augment those naturally present (Bellamkonda and Aebischer, 1995; Berthiaume and Yarmush, 1995; Brooks et al., 1995; Freed and Vunjak-Novakovic, 1995; Humes, 1995;

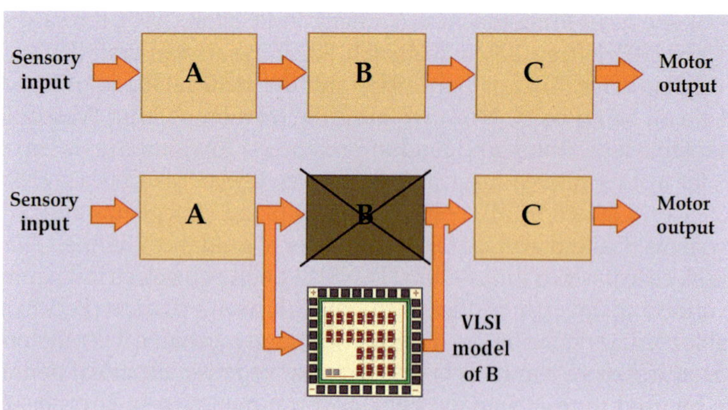

FIGURE 8.2.18
Microelectronic signal processing units may one day be able to replace functions of damaged or diseased neuronal areas of the brain or spinal cord. (Redrawn from Berger, T.W. et al., *IEEE Eng. Med. Biol. Mag.*, 24, 30–44, 2005.)

Kim and Vacanti, 1995). Cells harvested from donor tumors can be propagated for many generations in cell culture media without degradation, because their telomeres do not shorten when the cells reproduce (see Section 6.18.4, Telomeres). These cells can be used to fashion bioartificial tissues such as artificial skin or vascular grafts grown *in vitro* (literally translated as "in glass," or in laboratory containers).

Another category of tissue engineering is to provide an artificial extracellular matrix, or *scaffold*, into which natural tissues grow. The matrix may be either permanent or absorbable. Bone and cartilage implants use this method, wherein the matrix may be seeded with *fibroblasts*, undifferentiated connective tissue cells that may differentiate into *chondroblasts* (giving rise to cartilage), *collagenoblasts* (giving rise to collagen), or *osteoblasts* (produces the calcified intercellular material of bone). Artificial hearts and other organs can now be grown as cells supported on scaffolds; the scaffolds can be made of a material that is absorbed over time.

Biomaterials replace parts or functions of the body in safe, reliable, economic, and physiologically acceptable manners (Park, 1995). The challenge for biomaterials is to be strong enough, pliable enough, chemically inert, biologically compatible, nontoxic, and failure-free. It was not recognized at first that materials in contact with the blood often caused thrombus (blood clot) formation, or that some materials can irritate the surrounding tissue and cause inflammation. The field of biomaterials is a specialty by itself, and there are many challenges left to overcome (see Section 3.11, Temperature and Pressure).

There is a natural progression of improvements in artificial organs or medical interventions as time goes on. The first artificial kidneys, heart lung machines, and hip replacements were crude, but, as work on specific problems proceeded, they improved with each new design. The future looks bright for further advances combining engineering and biology working as a team.

There is a burgeoning activity in electrophysiology and applications of electronics to diagnostics and health maintenance systems. Modern biomedical instrumentation has found applications as wide-ranging as deep brain stimulation, artificial retina, cochlear implant, artificial heart and left ventricular-assist devices (LVADs), neurostimulators for pain relief, prosthetics and sensors with neural attachments to give feeling to amputees, incontinence-control implants, and chips for drug delivery. Brain implants can allow quadriplegic patients to control robotic arms to move as the patients wish (Strickland, 2015). Intracranial implanted electrical stimulators have been used to treat epileptic seizures. Other electrical systems stimulate the Vagus nerve for the same purpose (Denison et al., 2015).

Electrical stimulation of specific neural circuits has been found to alleviate the body's inflammatory responses by blocking release of excess secretions from various organs, such as the spleen, pancreas, thyroid, and others (Tracy, 2015). Bioelectric medicine is likely to be a safer and more effective alternative to drug therapies for such diseases as arthritis, diabetes, auto-immune diseases, hepatitis, cancer, and others.

Many mental illnesses appear to be related to the synchrony of neural processes in the brain (Perdue, 2015). When all neurons are firing at a rate of about 10 Hz, the brain is most energy efficient and functions optimally. At frequencies either higher or lower than this, mental processes are disturbed, leading to depression, anxiety, bi-polar disorder, autism, and others. Magnetic resonance therapy (MRT) produces a series of repetitive magnetic pulses that induce small electrical currents in the brain to restore neural synchrony at the optimal 10 Hz rate. MRT shows promise to alleviate mental disorders without medicines.

Wearable sensors are being developed at a fast pace to measure physiological parameters, such as heart rate, blood pressure, sweat composition, blood glucose level, and many others, during normal living activities. These sensors make it possible to assess health in real time and under routine conditions for animals, plants, and people of all ages (Lee, 2015; Perry, 2015).

When these wearable sensors are networked with smartphone applications (apps), health monitoring can become sophisticated and widespread (Ackerman, 2015; Champlin et al., 2017). There are apps for such things as body vital signs, eye exams, eardrum inspection, electrocardiogram (ECG), and others. Constant monitoring of human medical physiology is possible and leads to huge amounts of historical data (big data). Measurement aberrations hint to disease states (Campbell, 2017).

All these sensors deliver vast amounts of information in real time. That makes it likely that humans receiving the information may suffer from sensory overload, and not be capable of acting upon this information in a timely way. Driving modern automobiles of piloting present-day airplanes, for instance, can be overwhelming for some people. Even robotics in the form of self-driving cars may not help in the event of an emergency when the complacent and distracted passenger in the car may suddenly have to assume control. The biological engineer must learn how to design systems that account for all contingencies, including information overload.

Computer-aided diagnostic tools, called Clinical Decision Support Systems (CDSSs), assist clinicians in providing the best medical care in the most efficient manner (Mertz, 2015). Operating rooms can be simulated for training and management purposes (Phitayakom et al., 2015). Virtual reality is used in medicine to simulate surgery and disease symptoms (Anscombe, 2015).

For the environment, new remote sensors are also being used for precise environmental monitoring of forest and animal conditions, especially in view of water shortages and global warming (Miller, 2015). The same types of sensors monitoring human health and fitness are also being used to track parameters from plants, animals, and the general environment for increases in efficiency, sustainability, and agricultural production. Sensors improve breeding efficiency of rare species, growth rates of livestock, longevity of valuable animals and plants, and better understanding of microbial interactions with the environment.

8.2.5 Living Things Inadvertently Affected

> What's wrong with technology is that it's not connected in any real way with matters of the spirit and of the heart. And so it does blind, ugly things quite by accident and gets hated for that.
>
> **—Robert Pirsig**

There is a fifth class of interactions involving engineers with living systems. This is the case where some human action inadvertently involves living things, and the biological engineer may be called upon to minimize any harm done. The design is not intended to use living things to solve a problem, to use living things as models of the means to solve a problem, or to correct deficiencies or enhance the capabilities of existing living things, but the design is intended to maintain or restore living things caught up in a chain of events of human origin. This category is most often of an ecological nature.

Works of civil construction often produce severe disturbances in ecological balance. Some of these have double the potential for damage. An example of this is the dredging of harbors or shipping channels. Not only does the original dredging site severely disturb benthic organisms, but then the spoil is often deposited in another body of water. This disturbs a second site (Rhoads et al., 1978). Knowing the needs and interactions of native species in the area can help to identify those sites where spoil disposal is not as critical as it would be in other places. Freshwater ecosystems, which provide vital goods and services such as clean water, biodiversity, flood protection, and recreation, are as endangered as any other type of ecosystem on the planet, given all the damming, dumping, draining, and diverting to which they have been subjected.

Global warming has been a continuing concern. Human activities, especially burning fossil fuels to produce carbon dioxide, and keeping methane-releasing cattle, release gases into the atmosphere that blanket the Earth and act as barriers to the escape of long-wave heat radiation. The result is that more heat is trapped in the ecosystem, and the Earth warms. The degree of this warming is still in question.

Engineers assist in the solution to this problem by formulating predictive models, designing energy-efficient devices, and scrubbing CO_2 from power-plant exhausts. They can help the switch to alternative energy sources and reduce the need for travel by improving web communications.

Not all human activities harm the environment. Numerous species, such as deer, coyotes, woodchucks, and pigeons, thrive in urban environments. The need in these places is to limit their fecundity, and biological engineers can assist in the effective delivery of contraceptives where they are needed.

Humans strive for *sustainability*—to have a minimum impact on the outside world. Under this concept, food is to be locally grown, waste is to be used locally, and energy sources are to be renewable. Sustainable communities are most likely an unattainable ideal, but biological engineers can help to reach toward that goal.

In the end, humans are part of the ecology of this planet. Human activities have caused adjustments in the types and number of biological responses, but the same principles are as true under new conditions as they were without humans. It may be impossible to maintain the natural environment as it was before human intervention, but it is certainly possible to manage the environment as a compromise between the economic, health, and social needs of humans and the rest of the biological world.

Example 8.2.1: Biotechnology Brings Good Things to Life

The plastics you use may now be based on polylactic acid (PLA) from renewable corn rather than from nonrenewable petroleum sources (ASEE, 2002). PLA-based plastics will be used to manufacture everything from food wrap to car parts to clothing. Because the products are sugar-based, they are biodegradable.

Manufacture of PLA starts with the corn plant and enzymes from bioengineered microbes. When it was found that standard enzymes did not act fast enough at the relatively low temperatures required by the microbes, essential genes were inserted into thermophilic bacteria.

Example 8.2.2: Hearing Aids

Producing a more satisfactory hearing aid is a goal of biomedical engineers. The first attempts at hearing aids amplified all sounds, which made everything louder, including acoustic noise. It was still difficult to distinguish meaning from vocal sounds, because the amplified noise sometimes obscured vocal information. This was especially true if the hearing impairment was not uniform across the frequency spectrum.

Acoustic sounds are described by formants, which are clusters of frequencies that together determine the type of sounds we hear. Fundamental formant frequencies are usually not related to each other, but harmonic frequencies within each formant are multiples of the fundamental frequencies of each formant. The vowels are represented by frequencies lower than consonants, and contain relatively few formants. Measurements of auditory nerve responses to spoken vowels indicate the important formants for each sound.

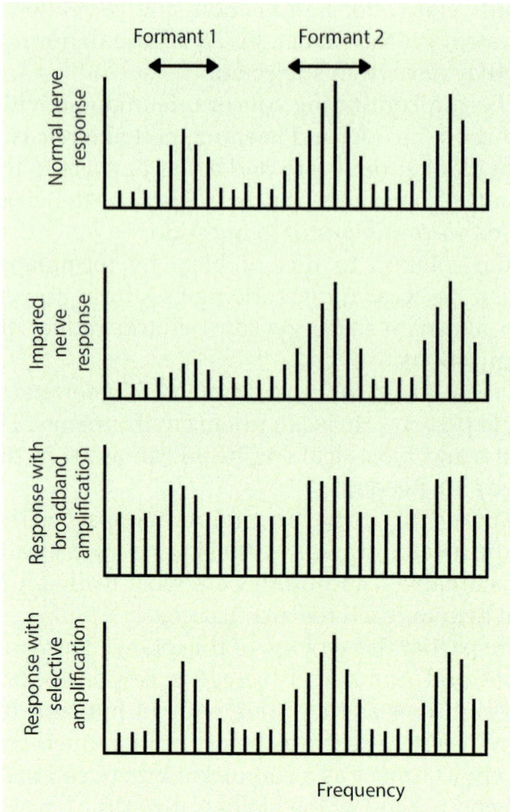

FIGURE 8.2.19
Hearing aid response based on formants.

An adaptive hearing aid can be designed that predicts the normal auditory nerve response to spoken language, defines the important formants, and amplifies only the frequencies contained in the important formants (Figure 8.2.19). Thus, the amplified sound will not contain extraneous amplified noise. It will only contain those sounds that make it easier to perceive what is being said. In this way, imitating normal auditory nerve output can be used to overcome selective frequency hearing impairment (Sachs et al., 2002).

Example 8.2.3: Cottonwood Trees to the Rescue

Danbury, Connecticut has a mercury problem. At one time, there were many factories there that made felt hats. Mercury was used to soften animal pelts to make the felt, and, when the mercury bath was spent, factories just dumped it and started again. Over time, the level of methyl mercury in the soil rose, and is now commonly in the range of 1–315 ppm. The city wants to clean up this property and bring the mercury level to no more than 20 ppm for residential sites.

Instead of digging up the soil and treating it, the city will use genetically modified cottonwood trees to phytoremediate the site. Cottonwood trees quickly develop extensive root systems and grow quickly. It is intended that the trees will absorb the mercury, and be cut down in 4–5 years before they reach reproductive age. Cutting them before they can produce seeds assures that the modified genetic material will not be released into the environment.

Phytoremediation can occur in one of three ways:

1. *Phytoextraction*, or *phytoaccumulation*. The plants absorb the contaminant with their roots and store the contaminant in their wood.
2. *Phytovolatilization*. The plants absorb the contaminant and pass it into the atmosphere.
3. *Phytostabilization*. The plant roots immobilize the contaminant and change it into a nontoxic form in the soil.

Cottonwood trees phytoremediate through the process of phytoextraction. The mercury has long since been transformed from metallic mercury into methyl mercury by microbial activity in the soil. The trees store it in the same form. Wood that is harvested from the trees must therefore be treated as toxic waste.

Example 8.2.4: Louisiana Crawfish Ponds

As hunting and the loss of natural wetlands have contributed to the decline of egrets, ibises, and herons in southern Louisiana, crawfish farm ponds have reversed the downward trend (Huner, 2000). Crawfish ponds supply food and beneficial environments for many different types of water birds (Figure 8.2.20). Clearly, crawfish aquaculture is one human activity able to compensate for others.

Heavy predation from birds can be tolerated as long as crawfish prices are high enough. Increased competition from abroad, however, has resulted in lower prices. Environmentalists contend that culling by birds reduces crawfish competition and should allow the remaining crawfish to grow larger and command better prices.

The problem is that experimental evidence shows that the remaining crawfish will not grow bigger unless moved to underpopulated ponds. There is apparently some factor in their original habitat that limits their growth. Unless this combination of economic and environmental problems can be solved, the farmers will turn to growing other crops that don't require flooding. What would you do?

FIGURE 8.2.20
White-faced ibis fly over snowy egrets wading in a Louisiana crawfish pond. These ponds can promote breeding of native birds, but are maintained only as long as they produce economic benefits to their owners.

Example 8.2.5: Restoring Balance

Mechanoreceptors, those neurons that translate mechanical forces or position into electrical impulses, can become less sensitive with age. The result is faulty balance in older people. Sensitivity can be restored by applying small, random electrical or mechanical stimulation to skin surfaces. This is called *stochastic resonance* (Harry et al., 2005).

Many sensors inside and outside the body are threshold devices that do not respond unless a signal reaches a certain minimum threshold level. For instance, instruments in a smoothly flying glider do not immediately indicate changes in altitude or speed until the change is large enough to overcome mechanical stiction of the dials. In order to make these instruments respond more rapidly, a mechanical vibrator is installed in the instrument panel. The vibrator keeps the dials from sticking.

Likewise, mechanoreceptors in the skin can be made more sensitive by applying a small-amplitude random signal (Figure 8.2.21). By itself, this electrical or mechanical noise is less than the sensor threshold, and not detectable. When added to a small primary signal, however, some of the noise pulses exceed the threshold, and the sensor fires. Thus, the magnitude of the adequate stimulus has been made smaller.

Applying a steady background signal does not have the same effect. Over time, the sensor would adjust to a constant signal and lose sensitivity to it. With random noise, no such adjustment is observed.

The result of this is improved balance for elderly people, greater foot sensitivity for diabetics, improved recovery from strokes, and better ability to walk and climb stairs for those with impaired function (Harry et al., 2005).

Example 8.2.6: Bioleaching of Metal Ores

Bacteria are increasingly being used to extract metals from low-grade ores or from ores difficult to mine. Strains of the bacterium *Thiobacillus ferrooxidans* thrive on mineral-rich rock. Many of these rocks contain heavy metal sulfides, and form acidic solutions when

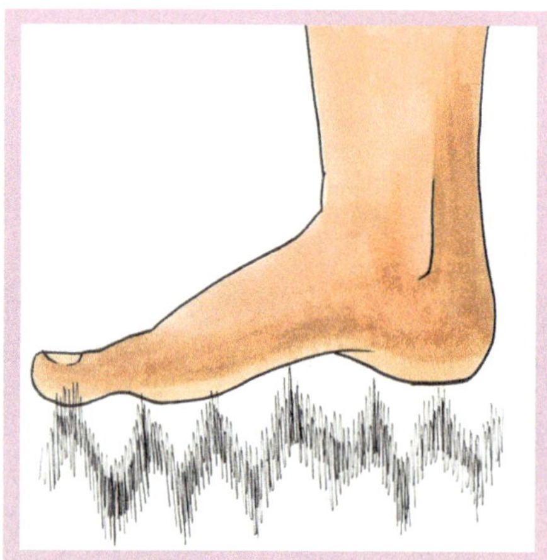

FIGURE 8.2.21
Adding a small amount of noise (spikes) to a primary signal (here given by a sinusoidal variation just for illustration) allows a subthreshold signal to exceed the sensor threshold. Just the right amount of noise makes the signal able to be detected most of the time. Too little noise doesn't help; too much noise washes out the signal. (Redrawn from Harry, J.D. et al., *IEEE Spectr.*, 42, 37–41, 2005.)

rainwater leaches through them. *T. ferrooxidans* facilitates this leaching when it uses the compounds for its metabolism. This process is called *bioleaching.*

Low-grade ores that contain concentrations of metals uneconomical to extract through conventional smelting can be ground and piled on a water-impermeable surface, and treated with *T. ferrooxidans.* The metals are recovered economically from the solutions of acidic water runoff coming from the pile (Rawlings and Silver, 1995).

More than 25% of the world's copper is extracted using bioleaching. Bioleaching also shows promise for recovering gold, cobalt, and uranium. Because *T. ferrooxidans* survives in the harsh environments of acidic rock, this bacterium has been nominated as a means to extract valuable minerals from the surfaces of other planets (Olsson-Francis and Cockell, 2010).

Example 8.2.7: Cyborg Beetles

Tiny robots that could fly inside caves, tunnels, and barricaded rooms could provide needed reconnaissance for the military, law enforcement personnel, or emergency first responders. Fully synthetic micromechanical robotic fliers require too much energy to be powered by present battery technology. Beetles have very efficient flying mechanisms, but they need to be able to carry miniature cameras and to be controlled to fly where they are needed. Beetles large enough to carry miniature radios, batteries, and controllers have been used for this purpose (Maharbiz and Sato, 2010). Several very thin wires implanted in the insect's brain and flight muscles permit humans to control beetles to start or stop flying, and to turn left or right in flight (Figure 8.2.22).

Example 8.2.8: Scientists Create a Bionic Leaf

Using electricity from solar panels, *Ralstonia eutropha* microbes, and a cobalt and phosphorus catalyst, a bionic leaf has been constructed to scrub carbon dioxide from the air and form isopropanol fuel at a 10% efficient rate of energy conversion into chemical product. This is roughly 10 times the efficiency of natural photosynthesis occurring in a leaf (Liu et al., 2016).

FIGURE 8.2.22
This beetle carries a miniature radio, controller, and battery and can be controlled in flight. (Redrawn from Maharbiz, M.M. and Sato, H., *Sci. Amer.*, 303, 94–99, 2010.)

APPLICATIONS AND PREDICTIONS

1. Artificial organs will become more like their natural counterparts as time goes on. This includes size, function, and energy sources, among others.
2. Medical interventions will someday eliminate the forms of disease most common today, but then other diseases will become important in the future.
3. There will still be natural limitations on the span of a quality life, despite all the medical interventions possible.
4. Societal hazards will become increasingly important as medical interventions become better at repair and elimination of disease and damage. These hazards will include automobile accidents, crimes, euthanasia, and self-inflicted trauma. Similarly, nonmedical aspects of our environments will assume a greater importance.
5. There will continue to be a tension between medical technology and affordability.

8.3 The Completed Design

> Nothing in the world can take the place of persistence. Talent will not; nothing is more common than unsuccessful men with talent. Genius will not; unrewarded genius is almost a proverb. Education will not; the world is full of educated derelicts. Persistence and determination alone are omnipotent. The slogan "Press on" has solved and always will solve the problems of the human race.
>
> **—Calvin Coolidge**

First off, there is no such thing as a completed design (Figure 8.2.11). New research results, new materials, new technologies, and new methods often make designs obsolete as soon as they are completed. Concurrent engineering, wherein engineers work in groups together with marketing, manufacturing, packaging, testing, and other experts, can reduce development time of new products. However, many biological engineering creations utilizing living things are unique, so do not offer advantages of scale. They do, however, provide sufficient challenge to avoid failures that should have been anticipated.

SUBMARINE BUILT FOR TWO

The two-person submersible, called DualWorker, built by Nuytco Research Limited is an example of a biological engineering design of several kinds. DualWorker is equipped with four or six thrusters of 750 W each that can propel the 3 m long craft at up to 1.5 m/s. It features hydraulic manipulator arms with a reach of 1.5 m and it can remain submerged for up to 3 days. It is used for exploring underwater shipwrecks and exploring the ocean floor.

(*Continued*)

SUBMARINE BUILT FOR TWO (Continued)

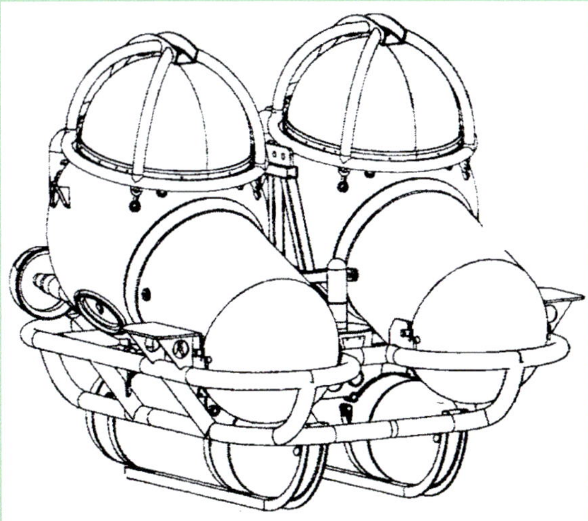

Drawing of the submarine. (Redrawn from ASEE, *Prism*, 12, 12, 2002. With permission.)

This represents four levels of biological engineering. First, in order to support the two passengers in the vessel, life support systems had to be installed to supply their physiological needs. This includes the supply of adequate amounts of oxygen and removal of carbon dioxide. Food and water must be available, as well as waste storage and removal. Emergency life support had to be included. Next, human factors had to be considered. Adequate room had to be incorporated in order to accommodate various sizes of people. There must be the ability to see their surroundings without undue distortion. Controls had to be placed conveniently and their functions clearly delineated. The possibility of error, such as opening the craft underwater, had to be eliminated. Third, the hydraulic manipulator arms are an example of biomimetics, where the arm movement mimics the natural movement of human arms and hands. Last, the submersible can be used as a tool for biologists to explore the deep and learn more about undersea life. In this respect, its use will benefit human knowledge of underwater biology, and (inadvertently) affect other biological species.

(*Continued*)

SUBMARINE BUILT FOR TWO (Continued)

The two-person submarine is an example of biological engineering design on several levels. (Courtesy of Shelley Acuff, Nuytco Research Ltd., North Vancouver, Canada.)

There is a strong parallel between the evolution of a product and biological evolution. In both cases, the original version is relatively crude compared to later models (although the original version may be very sophisticated in design). Improvements occur incrementally and often modularly. Improvements are usually not made unless there is a clear advantage in doing so. Biological evolution is driven by a survival and reproductive advantage in a competitive environment. Product improvement occurs with an economic advantage in a competitive environment. Biological evolution takes place by some rearrangement that has a large, although not wholly, random cause. Product evolution takes place by some rearrangement that is largely driven by definite design goals with some, although small, element of randomness. In both cases, the perfect organism or the perfect product can never be recognized as existing. There are always improvements to be made. Evolution of products, like evolution of living things, can be punctuated by sudden paradigm shifts that result in either completely new technologies or new classes of organisms. The biological engineer who appreciates the parallelism between these two processes has a valuable perspective that can serve him or her well.

It is the biological engineer's high privilege to work with such fascinating and resilient entities as are living things. The engineer or technologist working with biology must know what to expect from living things, and that may be to expect the unexpected.

BIONANOTECHNOLOGY

Nanotechnology involves making and modifying extremely small sized tools and products, sized of the order of biological molecules such as proteins and DNA (a nanometer is roughly equivalent to five atoms in a row). Biological applications of nanotechnology include microfluidics (controlling fluid motion), drug delivery, integration of semiconductors and biology for sensing and control, bioprocessing on a nanoscale (production of biochemicals on site), biosensors, drug delivery at the micro-location where they are most needed, genetic assays, aids to healing, selective filters, and nanorobotics. Indeed, almost all functions that can be performed on a larger scale can be performed faster and cheaper with bionanotechnology. Smaller means more selective, smaller sample sizes, and quicker reactions. Cells themselves can be considered to be a self-replicating collection of nanomachines, and cells can be harnessed to perform desired functions.

One interesting example of bionanotechnology is a diminutive motor made from a bacterium's flagellum. Remember that bacteria are extremely small and have very low Reynolds numbers when they move (see Section 2.9.2, Fluid Movement). They must have effective means to propel themselves through fluid. Some bacteria are propelled by means of a corkscrew-shaped flagellum (or tail-like organelle) that rotates continuously to move the microbe. This flagellum can be used to rotate a small mass to form a motor to perform rotational mechanical work. The body of the bacterium must be immobilized and ATP must be supplied as an energy source (see Section 3.10, Energy-Rich Compounds). The motor may then be used to power a nanopump, coil a nanowire, or turn a nanopropeller. There are a myriad of potential nano-uses for such a motor.

As another example, the function of low-density lipoprotein (LDL) is to carry hydrophobic substances, such as cholesterol, through the body. Cancer cells, as it turns out, have high cholesterol requirements. There are many LDL receptors on the surfaces of cancer cells, but few on normal cells. If a means can be found to link anti-cancer drugs with LDL, the drugs could be targeted to the cells that they are intended for. The LDL could then be used as a nano-delivery truck for chemotherapy. Venoms from bees, snakes, scorpions, or others can become effective anticancer agents when encased in nanoparticles that target cancer cells.

In an environmental application, nanoparticles can be used to scavenge bacteria, heavy metals, and other impurities from polluted water. The result is pure water. When their work is completed, the particles can be collected and reused. Nanoparticles can also be used to help grow food (Raliya et al., 2016). Scientists have shown that adding zinc oxide nanoparticles to the leaves of mung bean plants can enhance phosphorous uptake from the soil by 11%. Phosphorous is a necessary plant nutrient, but its supply is limited. Thus, adding these nanoparticles to plants has the potential to save phosphorous and at the same time reduce phosphorous runoff (Ramesh et al., 2016).

(*Continued*)

BIONANOTECHNOLOGY (Continued)

Bionanotechnology can also be used to form semiconductor circuits that assemble themselves (Fairley, 2003). One means to do this is to use a virus as a carrier of specific materials. The virus must first be able to stick to a substrate (for example, zinc sulfide). Various proteins on the surfaces of viruses have a range of affinities for the substrate material. A selection process is used whereby viruses are exposed to the substrate material and then washed with a dilute acid. Those viruses whose protein coats have a natural affinity for the substrate remain and the others are gone. The viruses that remain are isolated and allowed to multiply by infecting bacteria. The process is repeated; however, a stronger acid solution is used for each cycle. In this way, only the viruses that stick most strongly to the substrate remain after several cycles.

Once these viruses are selected, they can be used to precipitate other materials on the substrate in patterns determined by the patterns of viral binding. This can be controlled by other semiconductor processes. After these materials are precipitated, the viruses can be vaporized with short bursts of heat.

As another example, thin sheets of nanoscale materials, such as graphene, can be cut (called *kirigami*) and folded (called *origami*) to form microscopic structures that can be integrated with biological systems. Potential uses include graphene springs to measure electrical signals from neurons and microscopic containers to release medicines on cue (Saunders, 2015).

Example 8.3.1: Feeding Cattle

Providing adequate feeding facilities for dairy cattle highlights some of the intricacies involved when dealing with biological individuals. Some of the factors that should be kept in mind when designing such a system include (Albright and Arave, 1997; Graves, 1998)

- Feeding and watering systems must be placed where young or inexperienced animals can find them.
- When cows eat in groups, they each eat more than when they are fed separately (an example of social facilitation).
- Cows fed in groups are less fearful than when fed individually.
- Cows eating with their heads in the downward position produce 17% more saliva, which affects rumen function, than cows eating with heads horizontally oriented.
- Sufficient room must be available between feeding stations. Required space changes as animals mature. Pregnant animals require 10% more width.
- Leftover feed that has been wetted by saliva can spoil or provide a haven for the growth of unwanted microbes. Unused feed must be easily scraped away after cattle eat.
- Cattle feeding on both sides of a bunk feeder are positioned similar to a natural head-down, aggressive posture, encroaching on each other's personal space. This should be avoided.
- Cows are mobile and can assume different positions when feeding.
- Painful or unpleasant experiences in prior feeding are remembered and affect present behavior.

Example 8.3.2: Green Roofs Mitigate Adverse Effects of Urban Buildings

Green roofs are designed to absorb rainfall, delay water runoff, save on summertime cooling, reduce rooftop maintenance, and improve the aesthetic appearances of urban building rooftops (Figure 8.2.23). They consist of water-impermeable rooftop layers covered with a thin layer of plant growth medium (soil), and covered with plants grown for their hardiness under difficult conditions. The plants must be able to absorb water very quickly during storms either frequently or infrequently timed, but survive during drought conditions between rainfall events. Thus, many of the plants used for green roofs are succulents that can internally store large amounts of water. Irrigation may be used to help the plants survive long periods between rains, but the water use for irrigation actually helps to cool the building underneath as it is evaporated through transpiration. Roofs are usually the sight of the greatest heat loss in the winter and the hottest temperatures in the summer. Green roofs reduce the amount of stormwater runoff and also delay the time at which runoff occurs, resulting in decreased stress on sewer systems at peak flow periods. The plants used on green roofs improve urban air quality and turn otherwise unusable roof space into pleasant outdoor areas for social and private occasions (GRHC, 2016).

Whereas there are similarities among green roofs, each installation is unique. Hence, all technical performance details provided will vary by region, climate, building, and green roof type. Biological engineers can specialize in green roof design and construction. Certification of green roof design and construction is performed in the U.S. by the Leadership in Energy and Environmental Design (LEED) organization.

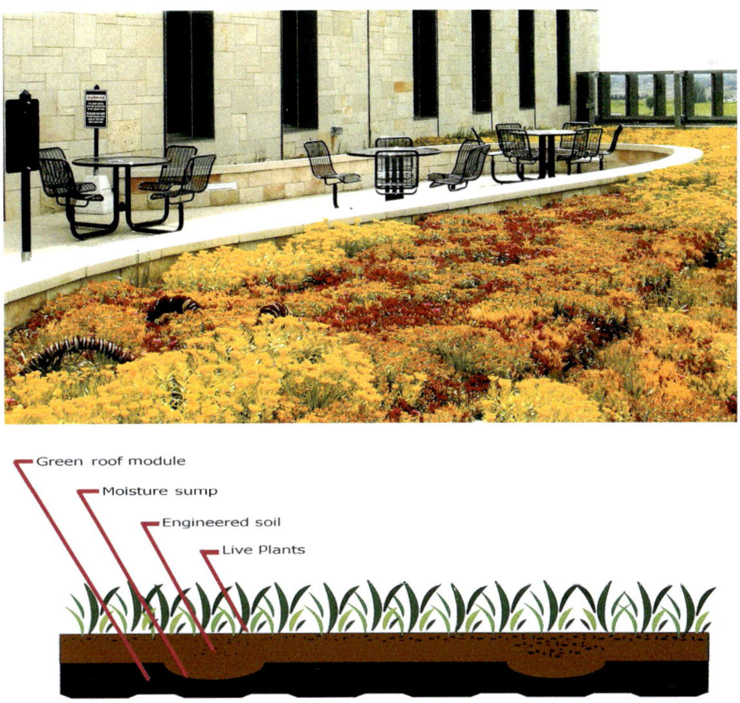

FIGURE 8.2.23
Plants comprising a green roof provide many advantages, including reducing immediate runoff of rain, building cooling, and forming an aesthetically pleasing space. (Redrawn from GRHC, Green Roof Benefits, Green Roofs for Healthy Cities, http://www.greenroofs.org/index.php/about/greenroofbenefits, 2016.)

8.4 Dénouement

> I have not failed – I have just found 10,000 ways that won't work.
>
> **—Thomas Edison**

> Science is more than a body of knowledge, it's a way of thinking.
>
> **—Carl Sagan**

> The job of the artist is to fight against entropy. Entropy in its simplest form refers to the tendency of everything to collapse. If you smash a vase and throw the pieces in the air, they won't re-land in the shape of a perfect vase. But the job of the artist is to smash the vase and then fashion something coherent out of those shards.
>
> **—William Kentridge**

> He has achieved success who has lived well, laughed often and loved much; who has enjoyed the trust of pure women, the respect of intelligent men and the love of little children; who has filled his niche and accomplished his task; who has left the world better than he found it, whether by an improved poppy, a perfect poem or a rescued soul; who has never lacked appreciation of Earth's beauty or failed to express it; who has always looked for the best in others and given them the best he had; whose life was an inspiration; whose memory a benediction.
>
> **—Bessie Anderson Stanley**

No consideration of biology for engineers, where the objective is to manipulate and use what is known about living things to produce useful new products and processes, can be complete without some consideration about whether the manipulation should be done. Certainly, with the tools that are now known and about to be known, a good number of capabilities can lead to unacceptable results, not unacceptable from a biological viewpoint, because biology is effectively neutral regarding new applications, but they can be unacceptable from a *bioethics* point of view. As Randy Oliver has said:

> *"Nature does not favor specific traits nor reward success. Nature only penalizes failure. Those left after penalization are called survivors, or 'more fit'."*

What makes these new applications acceptable or not is up to the general societal consensus of acceptability. It is up to the biological engineer to limit her or his activities to the confines of ethical acceptability.

There are so many applications possible these days that test the bounds of acceptability (King et al., 2015):

1. New assisted reproductive technologies (Fischer, 2015)
2. Human brain manipulation
3. Prosthetic devices with independent control
4. What to do about degrees of sentience in other non-human species
5. Environmental control of ecosystems
6. Introduction of new life forms

7. Possibility of human immortality
8. Legal ramifications of free will
9. Introduction of human emotions into machines
10. Conflict between food production and environment and many others. The burden is then on the shoulders of those who understand the implications of the technology to help identify those areas of questionable ethical concern and help to guide the conversation towards rational decisions of acceptability. That means you.

Questions

8.0.1 Read the statement by Herbert Hoover, and interpret it in terms of biological engineering.

8.1.1 What do we mean by a "systems approach?" Why is the systems approach appropriate for biological engineering?

8.1.2 Speculate on the types of tools to classify biological information that could be of most usefulness in biological engineering design.

8.1.3 Relate the five principles of population dynamics to other sections in the book.

8.2.1 Relate an example from each of the different classifications in Table 8.2.1. In each case, what are/were particular biological engineering challenges?

8.2.2 Give a specific hybrid systems example.

8.2.3 What would be the differences between a human who employs robotic parts as prosthetic organs and a robot who employs human parts as prosthetic organs?

8.2.4 What are the considerations necessary for a successful phytoremediation solution?

8.2.5 Would you consider sheep to be a biological engineering design solution? Why or why not?

8.2.6 What would you say is the future of DNA computers?

8.2.7 Give a specific example of a biomimetic design. What considerations went into the design?

8.2.8 Is there a present-day problem that can be solved biomimetically? What is it, and what is the biological model to use as a guide?

8.2.9 Discuss social engineering based on biological systems. Are biological systems good models for such things as economics? Remember that humans are biological creatures too.

8.2.10 How do you search the Internet? Do you see a parallel between your methods and food foraging?

8.2.11 What kind of things can eventually be accomplished with further understanding of the human mind? How will engineers use this understanding?

8.2.12 What do you think about the fact that genetically modified organisms are immune to certain antibiotics?

8.2.13 Propose an efficient means to transfer genetic material into a cell nucleus.

8.2.14 Where do engineers fit into biotechnology?

8.2.15 List advantages and disadvantages of biopharming.

8.2.16 How would you protect against biopharmed crops being mixed with the food supply? Why is this necessary?

8.2.17 What uses do you suggest for the plant residues left after the extraction of biopharmaceuticals?

8.2.18 Cloning farm animals means that, theoretically at least, if you liked eating that steak last night, you can eat it again tonight. Why won't this be true?

8.2.19 Describe the process to design a biomedical instrument. At what stage should potential customers be consulted?

8.2.20 Why are human-designed replacement organs unlikely to function as well as the originals?

8.2.21 What are the design goals for an artificial organ?

8.2.22 What is tissue engineering, and where is the engineering?

8.2.23 How can a systems approach minimize inadvertent ecological damage by human activities?

8.2.24 What are some available means to minimizing environmental effects of human activities?

8.2.25 Describe what you think it would be like to have eyes located on both sides of your head, like a bird or horse, instead of in the front of your head. How would your perspective of reality change? How does this different perspective influence a biological engineering design?

8.2.26 Imagine being a slithering animal, such as a snake or a worm. What environmental characteristics would be most important to you? How would these influence a biological engineering design involving this type of animal?

8.2.27 If you were cold-blooded, rather than warm-blooded, how would environmental controls for your residence and place of work differ from what they are now?

8.2.28 Give examples where the following could be incorporated into an engineering design:

Slithering animal

Plant

Animal that communicates verbally

Bacteria

Virus

Intestinal wall cell

Muscle cell

Flying insect

Finfish

8.2.29 Suggest ways to distinguish between GMO grain and non-GMO grain.

8.2.30 What are some biological ways that GMO grain can be made to be relatively easily distinguished from non-GMO grain?

8.2.31 Robots have been used successfully in medical procedures, especially in surgery. Investigate current use of robots and speculate on some of their advantges over traditional surgical techniques.

8.2.32 Robots can be used in agriculture. How?

8.2.33 Wearable sensors can give up-to-the-minute physiological and health information. Is this good or bad or both?

8.2.34 Discuss the value of big data in predicting plant and animal disease epidemics. What is the role of the biological engineer is this process?

8.2.35 Why is it important to know how deep learning occurs? Hint: How does deep learning depend on the accuracy of its algorithms and trial data?

8.2.36 Suggest where biological engineers might apply robotics and deep learning to new situations of importance.

8.2.37 Add to the list of Applications and Predictions.

8.3.1 How do you think bionanotechnology will change the way engineers solve biologically related problems?

8.3.2 The cases of Karen Ann Quinlan in 1975 and Terri Schiavo in 2003 illustrate the conundrum posed by modern medical technology. In both of these cases, technology supported their physiological needs, but left them with no consciousness. Long court battles were fought to allow them to die by withholding life support. Technology created these problems; what can technology do to rectify the situation?

8.3.3 In 2003 it was announced that electrodes implanted in a monkey's brain were able to be used to control a disconnected robotic arm without any nervous connection between the brain and the arm. The monkey only needed to think about moving the arm. There will probably come a time when humans using robotic prostheses and robots using human prosthetic devices will be hard to distinguish from each other. What role does technology play in protecting the essence of humanity? How can it do this?

8.3.4 Animatronics is the technology for making sophisticated human or animal substitutes. They may include robotic pets or animated displays. Choose an example of animatronics, write a set of design goals for the example, and relate these design goals to sections of this book.

8.3.5 The following is an actual e-mail request made in 2004:

> I am an Industrial Hygienist and received a call this morning to test for mold. The customer is a Periodontal Doctor with ten employees. The building is two floors with a crawl space below them. The professional office building itself is visibly clean, including signs of water intrusion and visible mold. The Crawl space is dry, the wood is clean with some suspect insulation, which I took a bulk sample of. I collected several Air-O-Cell samples and expect to have data within 48 h. I have advised my client that these are not symptoms that have been associated with mold contamination.
>
> Three of the occupants working for my client have been to the ER, or personal Physician. All of these individuals are complaining of tightening of chest, chest pains and shortening of breath. The Periodontal Doctor visited the ER over a year ago with the same symptoms. The recent occurrences are all within the last few weeks. Further,

there are two additional occupants in the office space on the second floor that have now had the same occurrence.

One hint, the office manager stated that she had recently been to Dr. Andrew Weil, (he has a web page if you would like to see what I am dealing with). Dr. Andrew told her this was an environmental issue and she has "Custro Contronitis." I cannot find Custro Contronitis in any of my books.

I originally thought Flu. Why would one occur last year? And, why would flu symptoms of five individuals all complain of Heart Pain?

CO readings are "0 ppm."

CO2 readings are under "900 ppm."

RH ranged between 30 and 35.

Temperature inside was 68 F.

The heating system is forced air and looked clean. It does have some type of electrostatic attachment; no ozone is being produced.

Any Advice?

8.3.6 The following is excerpted from the *Baltimore Sun*, February 17, 2004. As a biological engineer, what would you recommend?

...So far, nothing but poison gas has dissuaded the monk parakeet, [Florida's] most prolific and rapidly growing exotic species, from building its giant condo-style nests on power lines and high-voltage substations—a habit that occasionally triggers outages.

After spending nearly $300,000 to encourage the lime-green urbanites to take up residence elsewhere, Florida's largest utility is scratching its collective corporate head.

...While the majority [of monk parakeets] in most locales still prefer living in palms or other trees, that's not the case in South Florida. Here, monks favor a more structured jungle. [Florida Power and Light] counts about 1,500 high-rise parakeet condos along the South Florida power grid, some weighing hundreds of pounds and housing dozens of raucous pairs, each with their own entryway and three-cavity chamber.

...The only parrot that builds nests from scratch, monks love anchoring their elaborate twig piles in angles, and transmission lines and substations offer plenty of them.

They also like staying out of reach of predators, including humans. In 1999, a Pinellas County man was severely burned while trying to capture hatchlings in a Clearwater substation.

Then there's the bane of all city dwellers, urban crowding. As monks grow in number—one study says the wild population doubles every five years—their nesting stock diminishes. As palms and other trees fill up, monks look for alternatives.

8.4.1 After studying the material in this book, have you gotten a better idea about the biological engineering role in biology?

8.4.2 Revisit Question 1.8.2. Compare your answer before reading further with your present response.

8.4.3 What do you hope to accomplish in your professional career?

8.4.4 Do you see the many promises and opportunities available in engineering applied to biology and medicine?

Appendix

TABLE A.1A

Body and Organ Masses for Vertebrates

Species	Sex and Number	Body Mass (kg)	Adrenals (g/100 g)	Brain (g/100 g)	Eyes (g/100 g)	Heart (g/100 g)
Man						
Australian aborigine	m 1	76	0.005	1.77		
Chinese	m 1	84	0.009	1.76		0.66
Filipino	m 1	43		2.57		0.46
Indian, Maya Quiche	m 1	42	0.03	3.02		0.52
Indian, Maya Quiche	f 1	46	0.02	2.18		0.49
Negro	m 7	47	0.03	2.73		0.81
White, American	m 7	67	0.02	1.96		0.42
White, European	m 4	49	0.03	2.53		0.64
Agouti, brown (*Dasyprocta punctata*)	m, f 5	2.6	0.03	0.58	0.50	0.51
Antbear (*Cyclops didactylus*)	? 1	0.09	0.84	4.77		
Anteater (*Tamanduas tetradactyla*)	m, f 4	2.2		1.09		0.03
Armadillo (*Dasypus novemcinctus*)	m, f 12	3.3	0.03	0.25		0.28
Ass (*Equus asinus*)	f 1	150	0.01	0.27		0.55
Bat, vampire (*Desmodus rotundus*)	m, f 5	0.028	0.04	3.34		
Bear, brown (*Ursus americanus*)	m 1	550	0.001			
Bear, grizzly (*U. horribilis*)	f 1	140	0.04	0.16		0.79
Beaver (*Castor canadensis*)	m 1, f 1	5	0.009	0.45		0.43
Bison, American (*Bison bison*)	f 1	55	0.01	0.61	0.08	0.66
Bison, American (*Bison bison*)	m	1000				
Bison, American (*Bison bison*)	f	600				
Buffalo, African (*Syncerus caffer*)	m 3, f 1	700	0.005	0.09	0.008	0.47

(Continued)

TABLE A.1A (*Continued*)

Body and Organ Masses for Vertebrates

Species	Sex and Number	Body Mass (kg)	Adrenals (g/100 g)	Brain (g/100 g)	Eyes (g/100 g)	Heart (g/100 g)
Bushbok (*Tragelaphus scriptus*)	m 1, f 1	44	0.01	0.37	0.11	0.76
Camel, bactrian (*Camelus bactrianus*)	m 1	450	0.005	0.12	0.02	
Caribou, ground (*Rangifer arcticus*)	m 3, f 1	98	0.004	0.30	0.04	0.90
Cat, domestic (*Felis catus*)	m 7, f 3	3.3	0.02	0.77	0.32	0.45
Cattle, Holstein (*Bos taurus*)	m 5	900	0.005	0.05		0.37
Cattle, Holstein (*B. taurus*)	f 198	600	0.006	0.07		0.37
Cheetah (*Acinonyx jubatus*)	m 2	21	0.009	0.39	0.16	0.51
Chimpanzee (*Pan troglodytes*)	m 1	52	0.02	0.84		0.48
Chimpanzee (*P. troglodytes*)	f 1	44	0.02	0.74		0.50
Chipmunk (*Tamias striatus*)	m 2	0.07	0.04	2.96	0.74	7.96
Coati (*Nasua nasua*)	m 2	5.1	0.009	0.66	0.04	0.38
Coyote (*Canis latrans*)	f 2	8.5				0.85
Deer, white-tailed (*Odocoileus virginianus*)	m 1	65		0.32		0.97
Dog (*Canis familiaris*)	m 2, f 2	13	0.01	0.59	0.10	0.85
Elephant (*Loxodonta africana*)	m 1	6600	0.01	0.08	0.001	0.39
Fox, gray (*Urocyon cineroargenteus*)	m 1	3.8	0.01	0.99	0.11	0.58
Fox, red (*Vulpes fulva*)	f 1	4.6	0.01	1.15	0.09	0.90
Gazelle (*Gazella thomsoni*)	m 2	24	0.008	0.38	0.11	1.00
Giraffe (*Giraffa camelopardalis*)	m 1	1200	0.006	0.06	0.01	0.41
Goat (*Capra hircus*)	m 1	28		0.41	0.11	
Gorilla (*Gorilla gorilla*)	m 1	180	0.02			
Guinea pig (*Cavia porcellus*)	m 58	0.26	0.07	1.33		0.53
Guinea pig (*C. porcellus*)	f 10	0.43	0.08	0.92	0.24	0.39
Hamster, golden (*Mesocricetus auratus*)	m 2, f 2	0.12	0.02	0.88	0.18	0.47

(*Continued*)

TABLE A.1A (*Continued*)
Body and Organ Masses for Vertebrates

Species	Sex and Number	Body Mass (kg)	Adrenals (g/100 g)	Brain (g/100 g)	Eyes (g/100 g)	Heart (g/100 g)
Hare, African (*Lepus capensis*)	f 1	2.9	0.02	0.35	0.25	1.02
Hippopotamus (*Hippopotamus amphibious*)	f 1	1350	0.004	0.05	0.003	0.34
Horse, Percheron (*Equus caballus*)	m 1	635	0.006	0.10	0.02	0.88
Horse, Percheron (*E. caballus*)	f 1	770	0.004	0.08	0.02	0.61
Hyena, spotted (*Crocuta crocuta*)	m 2	62	0.02	0.28	0.06	0.72
Hyrax (*Heterohyrax brucei*)	m 1	0.75	0.02	1.64		0.48
Jackal (*Canis mesomelas*)	m 2	2.8	0.02	1.61	0.24	0.75
Jaguar (*Felis onca*)	f 1	34	0.02	0.43	0.05	0.54
Kinkajou (*Potos flavus*)	f 1	2.6	0.007	1.18	0.07	0.54
Lemming, rock (*Dicrostonyx rubricatus*)	m 4	0.05	0.03	0.17	0.28	0.59
Leopard (*Panthera pardus*)	m 1	48	0.01	0.28	0.06	0.42
Lion (*P. leo*)	m 4	125	0.01	0.19		0.85
Lion (*P. leo*)	f 3	97	0.01	0.20	0.04	0.54
Lynx (*Lynx baileyi*)	m 1	7.4	0.08			0.33
Manatee (*Trichechus manatus*)	m 1	425		0.08	0.001	0.29
Manatee (*T. manatus*)	f 1	560	0.002		0.01	0.22
Mole (*Scalopus aquaticus*)	m 1	0.04	0.04	2.93		0.69
Mongoose (*Ichneumia albicauda*)	m 1	4.4	0.01	0.64	0.09	0.64
Monkey, black howler (*Alouatta palliata*)	m, f 28	6.2	0.01	0.81		0.33
Monkey, rhesus (*Macaca mulatta*)	m 4	3.3	0.02	2.78	1.06	0.38
Monkey, rhesus (*M. mulatta*)	f 7	3.6	0.03	2.57		0.34
Mouse, jumping (*Zapus hudsonicus*)	m 1, f 3	0.018	0.04	3.57	0.14	1.03
Mouse, meadow (*Microtus drummondi*)	m, f 67	0.023	0.03	0.29	0.10	0.68
Muskrat (*Ondatra zibethica*)	m 1	0.90	0.01	0.59	0.21	0.36

(*Continued*)

TABLE A.1A (*Continued*)
Body and Organ Masses for Vertebrates

Species	Sex and Number	Body Mass (kg)	Adrenals (g/100 g)	Brain (g/100 g)	Eyes (g/100 g)	Heart (g/100 g)
Opossum, woolly (*Philander laniger*)	m 1, f 1	190	0.53			1.58
Porcupine (*Erethizon dorsatum*)	m 1, f 3	2.9	0.01	0.78	0.09	0.55
Porpoise (*Phocaena phocaena*)	m 1	140	0.007	1.22	0.04	0.52
Rabbit, giant Flemish (*Lepus* spp.)	m 2	3.7	0.01	0.29		0.29
Rabbit, giant Flemish (*Lepus* spp.)	f 22	2.5	0.02	0.40		0.35
Raccoon (*Procyon lotor*)	m 1	5.2	0.02	0.82		0.81
Raccoon (*P. lotor*)	f 1	2.2	0.07	1.51	0.15	0.89
Rat, Norway (*Rattus norvegicus*)	m 2, f 1	0.25	0.05	1.22	0.10	0.52
Reedbuck (*Redunca redunca*)	m 2	31	0.006	0.34	0.10	0.76
Seal, ringed (*Phoca hispida*)	m 3, f 2	39	0.007	0.63	0.18	0.73
Shrew (*Blarina brevicauda*)	m 29	0.02	0.02	1.87	0.009	1.02
Shrew (*B. brevicauda*)	f 39	0.017	0.02	2.11	0.006	1.05
Skunk (*Mephitis mephitis*)	m 1, f 2	2.1	0.01	0.33		0.58
Sloth, three-toed (*Bradypus tridactylus*)	m, f 6	1.8	0.01	0.75		
Squirrel, red (*Sciurus hudsonicus*)	m 4	0.18	0.03	2.57	0.27	0.86
Squirrel, red (*S. hudsonicus*)	f 4	0.25	0.03	2.02	0.21	0.73
Steinbok (*Raphicerus campestris*)	m 2	8.6	0.01	0.57	0.17	0.84
Swine (*Sus scrofa*)	f 36	102	0.004			0.32
Tapir (*Tapirella bairdii*)	m 1, f 1	11.4	0.02			0.85
Tiger (*Panthera tigris*)	f 1	160	0.01	0.14		0.27
Walrus (*Odobenus rosmarus*)	m 1, f 3	600	0.002	0.17	0.003	0.68
Warthog (*Phacochoerus aethiopicus*)	m 1	65	0.01	0.19	0.03	0.50
Weasel, arctic (*Mustela arctica*)	m 3, f 1	0.18	0.01	2.80	0.08	1.71
Whale, white (*Delphinapteus leucas*)	m 4	447	0.006	0.52	0.007	0.55
Whate, white (*D. leucas*)	f 2	300	0.009	0.78	0.007	0.57

(*Continued*)

TABLE A.1A (*Continued*)

Body and Organ Masses for Vertebrates

Species	Sex and Number	Body Mass (kg)	Adrenals (g/100 g)	Brain (g/100 g)	Eyes (g/100 g)	Heart (g/100 g)
Wildebeest (*Connochaetes taurinus*)	m 2	210	0.003	0.21		0.62
Wolf (*Canis lupus*)	m 1	22	0.01	0.52	0.08	1.08
Zebra (*Equus quagga*)	m 3, f 1	280	0.008	0.20	0.03	1.42
Blackbird (*Quiscalus quiscala*)	f 1	0.08	0.02	3.56	0.23	0.14
Bluebird (*Sialia sialis*)	m 1, f 1	0.03	0.55	4.24		1.39
Buzzard, steppe (*Buteo vulpinus*)	m 1	0.56	0.05	1.41		0.82
Catbird (*Dumatella carolinensis*)	f 1	0.03	0.01	0.43		0.99
Canary (*Serinus canaries*)	m 1, f 1	0.016	0.04	4.72	1.75	1.29
Cowbird (*Molothrus ater*)	f 1	0.07	0.02	4.08		1.61
Crane, gray (*Grus canadensis*)	m 1	106	0.01	0.52	0.66	1.15
Crow (*Corvus brachyrhyncos*)	m 1	0.33	0.02	2.76		0.95
Duck, pintail (*Anas acuta*)	f 1	0.67	0.01	0.74	0.25	1.24
Eagle, tawny (*Aguila rapax*)	m 2, f 3	2.4	0.45	0.59	1.34	0.63
Egret, great white (*Casmerodius albus*)	f 1	10	0.02	0.59		0.90
Flamingo (*Phoeniconaias minor*)	m 3, f 2	15	0.02	0.49	0.22	0.94
Fowl, domestic (*Gallus domesticus*)	m 8	0.73	0.009	0.40	0.58	0.57
Fowl, domestic (*G. domesticus*)	f 16	0.61	0.01	0.44	0.58	0.63
Fowl, white leghorn, "germ-free"	?	0.9–1.2	0.009			0.35
Goose, Egyptian (*Alopochen aegypticus*)	f 1	1.9	0.02	0.39		0.96
Guineafowl (*Numida meleagris*)	m 1	1.6	0.02	0.26		0.88
Gull, herring (*Larus argentatus*)	f 2	0.53	0.02	0.95	1.45	0.98
Hawk, red-tailed (*Buteo borealis*)	f 3	1.0	0.01	0.97	2.06	0.67
Hummingbird (*Amazilia tzacati*)	f 1	0.005	0.007	4.16	2.50	2.37

(Continued)

TABLE A.1A (*Continued*)

Body and Organ Masses for Vertebrates

Species	Sex and Number	Body Mass (kg)	Adrenals (g/100 g)	Brain (g/100 g)	Eyes (g/100 g)	Heart (g/100 g)
Ostrich, masai (*Struthio camelus*)	m 1	125	0.02	0.03	0.08	0.98
Owl, horned (*Buteo virginianus*)	m 1	1.2	0.01	1.16		0.73
Partridge (*Francolinus sephaena*)	m 1	0.21	0.02	0.72		0.70
Pelican (*Pelecanus occidentalis*)	f 2	3.3	0.03	0.54	0.38	0.67
Pheasant (*Phasianus coichicus*)	m 1	0.62	0.02	0.53	0.85	0.90
Pigeon (*Columba livia*)	m 3, f 1	0.27	0.16	0.95		1.75
Raven (*Corvus coraz*)	f 1	1.25		2.81		0.85
Robin (*Turdus migratorius*)	m 2	0.07	0.03	3.01		1.46
Sparrow (*Passer domesticus*)	m 75	0.024	0.03	4.36	1.95	1.73
Sparrow (*P. domesticus*)	f 11	0.023	0.03	4.38	2.23	1.69
Starling (*Sturnus vulgaris*)	m 15	0.06	0.02	3.26	1.46	1.62
Starling (S. *vulgaris*)	f 10	0.06	0.02	3.13	1.80	1.49
Stork, European (*Ciconia ciconia*)	m 2, f 1	3.3	0.01	0.47	0.51	0.92
Alligator (*Alligator mississippiensis*)	m 2	190	0.004	0.007	0.01	0.15
Crocodile (*Crocodylus acutus*)	m 1, f 1	110	0.004	0.01		0.12
Iguana lizard (*Iguana iguana*)	f 1	1.3	0.02			0.19
Lizard (*Lacerta viridis*)	m, f 15	0.05	0.04	0.24		0.12
Snake, black (*Coluber constrictor*)	m 1, f 2	0.43	0.03	0.07	0.05	0.22
Snake, boa (*Boa imperator*)	f 1	1.8	0.008	0.02	0.03	0.31
Snake, green (*Zamenis viridis*)	m 3, f 3	0.022	0.36	0.95		
Snake, python (*Python molurus*)	m 1	6.1	0.04	0.02	0.02	0.30
Snake, water moccasin (*Ancistrodon pisci*)	f 1	0.73	0.14	0.09	0.08	0.65
Toad, horned (*Phrynosoma cornutum*)	m 2, f 3	0.025	0.03	0.52	1.28	0.44

(*Continued*)

TABLE A.1A (*Continued*)

Body and Organ Masses for Vertebrates

Species	Sex and Number	Body Mass (kg)	Adrenals (g/100 g)	Brain (g/100 g)	Eyes (g/100 g)	Heart (g/100 g)
Turtle (*Aromochelys tristycha*)	m 1	0.12				0.43
Turtle (*A. tristycha*)	f 2	0.09				0.48
Turtle (*Testudo graeca*)	m, f 30	0.32	0.009	0.09		
Turtle, cumberland (*Chrysemys elegans*)	m 21	0.84				0.32
Turtle, cumberland (*C. elegans*)	f 1	0.86				0.31
Frog, bull (*Rana catesbiana*)	m 7	0.49	0.02	0.93	0.48	0.32
Frog, leopard (*R. pipiens*)	m 10	0.036				0.43
Frog, leopard (*R. pipiens*)	f 19	0.038				0.48
Barracuda (*Sphyraena barracuda*)	m 3, f 3	8.8		0.04	0.44	0.24
Carp (*Cyprinus carpio*)	m 2, f 4	1.05		0.12	0.28	0.15
Codfish (*Gadus morrhua*)	f 1	10.6		0.05	0.6	0.15
Haddock (*G. aeglefinus*)	f 6	3.3		0.06	0.2	0.17
Mackerel (*Scomber vernalis*)	m 1	0.76		0.08		
Mackerel (*S. vernalis*)	f 2	1.5		0.11	0.54	0.20
Perch (*Perca flavescens*)	m 6	0.17		0.15	0.55	0.23
Perch (*P. flavescens*)	f 1	0.19		0.17	0.73	0.77
Pike (*Esox lucius*)	m 4, f 3	0.42		0.12	1.15	0.15
Salmon (*Salmo salar*)	m 3	3.4		0.03	0.23	0.36
Salmon (*S. salar*)	f 5	5.4		0.02	0.15	0.19
Trout, rainbow (*Salmo irideus*)	m 2	0.26		0.17	0.69	0.17
Trout, rainbow (*S. irideus*)	f 4	0.22		0.19	0.70	0.13

Source: Spector, W.S., ed., *Handbook of Biological Data*, WADC Technical Report 56-273, Wright-Patterson Air Force Base, OH, 1956.

Note: Organ masses are relational, grams per 100 grams body mass.

TABLE A.1B

Body and Organ Masses for Vertebrates (cont.)

Species	Kidneys (g/100g)	Liver (g/100g)	Lungs (g/100g)	Spleen (g/100g)	Stomach and Intestines (g/100g)	Thyroid (g/100g)
Man						
Australian aborigine						0.005
Chinese	0.37	2.43		0.21		0.02
Filipino						0.06
Indian, Maya Quiche	0.44	2.48	3.13	0.14		0.08
Indian, Maya Quiche						0.02
Negro	0.51	2.81				0.06
White, American	0.41	2.30	0.73	0.25		0.04
White, European						0.03
Agouti, brown (*Dasyprocta punctata*)	0.59	2.84	0.22			
Antbear (*Cyclops didactylus*)						
Anteater (*Tamanduas tetradactyla*)	0.82	2.64	1.05	0.14		
Armadillo (*Dasypus novemcinctus*)	0.48		0.70	0.22		0.009
Ass (*Equus asinus*)	1.04	0.84	0.83	0.39		0.006
Bat, vampire (*Desmodus rotundus*)						0.04
Bear, brown (*Ursus americanus*)	0.03					0.004
Bear, grizzly (*U. horribilis*)	0.38			0.21		0.04
Beaver (*Castor canadensis*)	1.08	3.03	0.97			
Bison, American (*Bison bison*)	0.47	1.27	2.17	0.27		0.09
Buffalo, African (*Syncerus caffer*)	0.24	0.98	0.94	0.33	20.90	0.005
Bushbok (*Tragelaphus scriptus*)	0.40	1.95	1.64	0.26	12.62	0.01
Camel, bactrian (*Camelus bactrianus*)				0.08		
Caribou, ground (*Rangifer arcticus*)	0.13	1.83	2.10	0.26		
Cat, domestic (*Felis catus*)	1.07	3.59	1.04	0.29	10.37	0.01
Cattle, Holstein (*Bos taurus*)	0.20	0.92	0.69	0.15	3.14	0.010
Cattle, Holstein (*B. Taurus*)	0.24	1.20	0.72	0.16	4.83	0.006

(Continued)

TABLE A.1B (*Continued*)

Body and Organ Masses for Vertebrates (cont.)

Species	Kidneys (g/100g)	Liver (g/100g)	Lungs (g/100g)	Spleen (g/100g)	Stomach and Intestines (g/100g)	Thyroid (g/100g)
Cheetah (*Acinonyx jubatus*)	0.47	3.22	1.16	0.15		0.02
Chimpanzee (*Pan troglodytes*)						0.009
Chimpanzee (*P. troglodytes*)	0.48	2.75	1.36		14.09	0.01
Chipmunk (*Tamias striatus*)	0.10	7.4	0.96	0.29	6.2	0.04
Coati (*Nasua nasua*)	0.08	1.63	0.47	0.07		0.03
Coyote (*Canis latrans*)	0.94	3.44	0.72	0.17		
Deer, white-tailed (*Odocoileus virginianus*)		1.57				
Dog (*Canis familiaris*)	0.30	2.94	0.94		4.83	0.02
Elephant (*Loxodonta africana*)	0.27	1.62	2.08		13.88	0.01
Fox, gray (*Urocyon cineroargenteus*)	0.46	1.35	0.51		11.31	
Fox, red (*Vulpes fulva*)						0.003
Gazelle (*Gazella thomsoni*)	0.43	2.15	1.15			0.007
Giraffe (*Giraffa camelopardalis*)	0.18	1.56	0.99	0.18	10.65	0.005
Goat (*Capra hircus*)		1.90				
Gorilla (*Gorilla gorilla*)						0.003
Guinea pig (*Cavia porcellus*)	1.17	5.14	1.18			
Guinea pig (*C. porcellus*)	0.86	3.86	1.07	0.21		0.02
Hamster, golden (*Mesocricetus auratus*)	0.53	5.16	0.46			0.006
Hare, African (*Lepus capensis*)	0.42	1.77	0.61		10.07	0.006
Hippopotamus (Hippopotamus amphibious)	0.23	1.75	0.84	0.23	27.70	0.008
Horse, Percheron (*Equus caballus*)	0.27	1.34	0.90	0.54		0.006
Horse, Percheron (*E. caballus*)	0.23	0.87	0.70	0.20		0.007
Hyena, spotted (*Crocuta crocuta*)	0.64	5.12	10.92		10.91	0.01

(Continued)

TABLE A.1B (*Continued*)

Body and Organ Masses for Vertebrates (cont.)

Species	Kidneys (g/100g)	Liver (g/100g)	Lungs (g/100g)	Spleen (g/100g)	Stomach and Intestines (g/100g)	Thyroid (g/100g)
Hyrax (*Heterohyrax brucei*)	0.86	4.20	0.74		34	0.01
Jackal (*Canis mesomelas*)	0.81	4.30	1.05		10.81	0.03
Jaguar (*Felis onca*)	0.48	2.59	1.67	0.18		0.004
Kinkajou (*Potos flavus*)		3.76	2.99			0.02
Lemming, rock (*Dicrostonyx rubricatus*)	1.48	5.05	1.59	0.40		0.008
Leopard (*Panthera pardus*)		1.87	1.04	0.22		0.10
Lion (*P. leo*)			2.12			
Lion (*P. leo*)	0.53	3.24	2.06	0.22		
Lynx (*Lynx baileyi*)	0.33					0.004
Manatee (*Trichechus manatus*)		1.30	0.72			0.01
Manatee (*T. manatus*)	0.24	1.12	0.67			0.01
Mole (*Scalopus aquaticus*)	1.59	3.91	1.86			0.02
Mongoose (*Ichneumia albicauda*)	0.79	1.39	1.32		6.25	0.004
Monkey, black howl (*Alouatta palliata*)	0.58	3.25	0.63	0.74		
Monkey, rhesus (*Macaca mulatta*)		2.09				0.02
Monkey, rhesus (*M. mulatta*)			1.89			0.01
Mouse, jumping (*Zapus hudsonicus*)	1.26	5.63	1.34			0.01
Mouse, meadow (*Microtus drummondi*)	1.53	4.56	1.70			0.01
Muskrat (*Ondatra zibethica*)	0.83	2.44	0.48		1.95	0.001
Opossum, woolly (*Philander laniger*)	2.10	4.74	1.58	0.79		
Porcupine (*Erethizon dorsatum*)	0.96	4	0.98			0.02
Porpoise (*Phocaena phocaena*)		2.08	3.69	0.04	9.28	0.01
Rabbit, giant Flemish (*Lepus* spp.)	0.61	2.66				0.02
Rabbit, giant Flemish (*Lepus* spp.)	0.70	3.19	0.53			

(*Continued*)

TABLE A.1B (*Continued*)

Body and Organ Masses for Vertebrates (cont.)

Species	Kidneys (g/100g)	Liver (g/100g)	Lungs (g/100g)	Spleen (g/100g)	Stomach and Intestines (g/100g)	Thyroid (g/100g)
Raccoon (*Procyon lotor*)	0.68	3.58	0.28			
Raccoon (*P. lotor*)	1.61	6.29	0.87		10.11	0.008
Rat, Norway (*Rattus norvegicus*)	1.09	3.35	0.79	0.29	2.52	0.001
Reedbuck (*Redunca redunca*)	0.32	1.65	1.34			0.004
Seal, ringed (*Phoca hispida*)	0.70	2.81	1.85	0.32		0.008
Shrew (*Blarina brevicauda*)	1.08	5.81	2.24			0.01
Shrew (*B. brevicauda*)	1.25	5.45	2.19			0.01
Skunk (*Mephitis mephitis*)	0.28	2.69	1.59			0.004
Sloth, three-toed (*Bradypus tridactylus*)						
Squirrel, red (*Sciurus hudsonicus*)	0.62	2.18	1.45			0.01
Squirrel, red (*S. hudsonicus*)	0.53	2.68	1.28			0.01
Steinbok (*Raphicerus campestris*)	0.45	2.03	1.74		6.38	0.01
Swine (*Sus scrofa*)	0.26	1.51		0.13	1.95	0.007
Tapir (*Tapirella bairdii*)	1.30	3.07	2.10	1.12		
Tiger (*Panthera tigris*)		1.14	0.64	0.57		
Walrus (*Odobenus rosmarus*)	0.68	2.92	1.36		4.42	0.01
Warthog (*Phacochoerus aethiopicus*)	0.46	2.30	0.84		15.23	0.005
Weasel, arctic (*Mustela arctica*)	0.99	4.74	2.08			0.05
Whale, white (*Delphinapteus leucas*)	0.49	1.52	2.70	0.04	2.70	0.02
Whale, white (*D. leucas*)	0.61	1.59	2.62	0.05	3.06	0.02
Wildebeest (*Connochaetes taurinus*)	0.23	1.07	1.34		17.54	0.005
Wolf (*Canis lupus*)	0.82	2.76	3.56			
Zebra (*Equus quagga*)	0.35	1.67	0.80	0.41		0.007
Blackbird (*Quiscalus quiscala*)	0.16	3.21	0.21	0.06	7.78	0.01
Bluebird (*Sialia sialis*)						0.02

(*Continued*)

TABLE A.1B (*Continued*)
Body and Organ Masses for Vertebrates (cont.)

Species	Kidneys (g/100g)	Liver (g/100g)	Lungs (g/100g)	Spleen (g/100g)	Stomach and Intestines (g/100g)	Thyroid (g/100g)
Buzzard, steppe (*Buteo vulpinus*)	0.60	1.94	0.83			0.03
Catbird (*Dumatella carolinensis*)			1.84			0.01
Canary (*Serinus canaries*)	0.16	5.39	0.15	0.11	14.17	0.008
Cowbird (*Molothrus ater*)						0.02
Crane, gray (*Grus canadensis*)	0.71	1.78	0.93	0.04	4.76	0.008
Crow (*Corvus brachyrhyncos*)			2.96			0.01
Duck, pintail (*Anas acuta*)	1.21	4.53	2.56	0.13	14.84	0.008
Eagle, tawny (*Aguila rapax*)	0.50	1.82	1.04		6.36	0.01
Egret, great white (*Casmerodius albus*)	0.79	3.20	3.21		13.11	0.01
Flamingo (*Phoeniconaias minor*)	1.18	2.68	1.47			0.03
Fowl, domestic (*Gallus domesticus*)	0.62	2.21	0.60	0.13		0.01
Fowl, domestic (*G. domesticus*)	0.68	2.36	0.61	0.15		0.01
Fowl, white leghorn, "germ-free"		1.53	0.51			
Goose, Egyptian (*Alopochen aegypticus*)	0.50	1.77	1.80			0.02
Guineafowl (*Numida meleagris*)	0.45	1.76	1.79			0.02
Gull, herring (*Larus argentatus*)		5.12				0.007
Hawk, red-tailed (*Buteo borealis*)	0.30	1.37	0.9		1.79	0.09
Hummingbird (*Amazilia tzacati*)	0.81	5.23	0.20			0.009
Ostrich, masai (*Struthio camelus*)		1.66	2.36			
Owl, horned (*Buteo virginianus*)			0.91			0.007
Partridge (*Francolinus sephaena*)	1.30	4.16				0.009

(Continued)

TABLE A.1B (*Continued*)

Body and Organ Masses for Vertebrates (cont.)

Species	Kidneys (g/100g)	Liver (g/100g)	Lungs (g/100g)	Spleen (g/100g)	Stomach and Intestines (g/100g)	Thyroid (g/100g)
Pelican (*Pelecanus occidentalis*)		2.22	0.91		7.75	0.005
Pheasant (*Phasianus coichicus*)	0.77	1.46			9.04	0.008
Pigeon (*Columba livia*)		1.76				0.11
Raven (*Corvus coraz*)	0.71					0.009
Robin (*Turdus migratorius*)			2.42			0.01
Sparrow (*Passer domesticus*)	1.46	5.12	1.56	0.18	11.45	0.02
Sparrow (*P. domesticus*)	1.53	4.67	1.72	0.18	11.53	0.02
Starling (*Sturnus vulgaris*)	1.71	3.46	1.87	0.11	9.15	0.01
Starling (*S. vulgaris*)	1.85	3.76	1.87	0.07	9.69	0.01
Stork, European (*Ciconia ciconia*)	0.65	1.92	1.11			0.01
Alligator (*Alligator mississippiensis*)		0.38	0.54	0.07	2.95	0.006
Crocodile (*Crocodylus acutus*)		1.02	1.00			0.004
Iguana lizard (*Iguana iguana*)		2.49	0.28		3.18	0.009
Lizard (*Lacerta viridis*)	0.12	5		0.16		0.02
Snake, black (*Coluber constrictor*)	0.60	0.60	0.80	0.18	2.79	0.02
Snake, boa (*Boa imperator*)	0.52	1.66	0.76			0.008
Snake, green (*Zamenis viridis*)	8.77	2.19		0.57		0.20
Snake, python (*Python molurus*)						0.02
Snake, water moccasin (*Ancistrodon pisci*)	1.85	8.85	3.12	0.76	41.24	0.07
Toad, horned (*Phrynosoma cornutum*)						
Turtle (*Aromochelys tristycha*)	0.43	2.8	0.85	0.18	2.729	
Turtle (*A. tristycha*)	0.47	2.9	0.76	0.17		
Turtle (*Testudo graeca*)	0.48	2.66		0.06		0.01

(*Continued*)

TABLE A.1B (*Continued*)

Body and Organ Masses for Vertebrates (cont.)

Species	Kidneys (g/100g)	Liver (g/100g)	Lungs (g/100g)	Spleen (g/100g)	Stomach and Intestines (g/100g)	Thyroid (g/100g)
Turtle, cumberland (*Chrysemys elegans*)	0.32	5.43	1.07	0.22	6.23	
Turtle, cumberland (*C. elegans*)	0.36	5.92	0.84	0.47	7.32	
Frog, bull (*Rana catesbiana*)	0.29	2.75	0.53	0.07	4.72	0.007
Frog, leopard (*R. pipiens*)	0.43	2.81	0.85	0.18	3.50	
Frog, leopard (*R. pipiens*)	0.47	2.88	0.76	0.17	3.77	
Barracuda (*Sphyraena barracuda*)	1.82	0.69		0.15	4.17	0.002
Carp (*Cyprinus carpio*)	0.55			0.23	7.92	0.0008
Codfish (*Gadus morrhua*)	0.19	1.52		0.95	5.47	0.006
Haddock (*G. aeglefinus*)	0.34	4.05		0.08	8.90	0.002
Mackerel (*Scomber vernalis*)						
Mackerel (*S. vernalis*)		0.43		0.12	0.20	0.002
Perch (*Perca flavescens*)	0.27	0.88		0.09	2.90	0.002
Perch (*P. flavescens*)		1.54			4.12	0.001
Pike (*Esox lucius*)	0.42	0.86		0.09	5.10	0.002
Salmon (*Salmo salar*)	1.05	2.02		0.32	6.29	0.005
Salmon (*S. salar*)	0.74	1.73		0.22	4.30	0.0007
Trout, rainbow (*Salmo irideus*)	0.55	0.99		0.21	9.58	0.003
Trout, rainbow (*S. irideus*)	0.50	0.99		0.27	8.89	0.002

Source: Spector, W.S., ed., *Handbook of Biological Data*, WADC Technical Report 56-273, Wright-Patterson Air Force Base, OH, 1956.

Note: Organ masses are relational, grams per 100 grams body mass.

TABLE A.2
Body Masses of Insects

Species	Fresh Weight		
	Larva (mg)	**Pupa (mg)**	**Adult (mg)**
Orthoptera			
Cockroach, American (*Periplaneta americana*)			m 890 (550–1470); f 1050 (650–1720)
Cockroach, giant (*Micropanesthia rhinoceros*)			19.3 g (18.4–19.5 g)
Cockroach, German (*Blattelia gemanica*)			m (39–45); f (70–73)
Cockroach, Hawaiian (*Nyctobbora noctivaga*)			(1290–2190)
Cockroach, Oriental (*Blatta orientalis*)			m 400 (323–515); f 750 (540–870)
Grasshopper, differential (*Melanoplus differentialis*)			m 854 (436–1232); f 1428 (812–2607)
Grasshopper, Rocky Mt. (*M. mexicanus mexicanus*)			m (153–161); f (156–165)
Locust, migratory (*Locusta migratoria*)	(372–578)		m 1350 (1200–1400); f 2500 (2250–2900)
Walking stick (*Dixippus morosus*)			1050 (900–1100)
Walking stick (*Sphodromantis bioculata*)			2079
Odonata			
Fly, big green dragon (*Anax junius*)	(1200–1500)		(500–900)
Anoplura			
Louse, bloodsucking (*Enderleinellus zonatus*)			0.005
Hemiptera			
Bug, boxelder (*Leptocoris trivittatus*)			(32–39)
Bug, milkweed, large (*Oncopeltus fasciatus*)			m 47 (31–75); f 66 (40–95)
Lepidoptera			
Cabbageworm, imported (*Pieris rapae*)	156 (110–165)		
Hornworm, tomato (*Protoparce quinquemaculata*)	8.3 g (6.2–10.5 g)		
Moth, bee (*Galleria mellonella*)	175 (85–310)		
Moth, codling (*Carpocapsa pomonella*)	47 (42–63)		
Moth, raisin (*Ephestia figulilella*)	19.4		
Moth, striped sphinx (*Deilephila euphorbiae*)	4038	2609	1263
Moth, webbing clothes (*Tineola biselliella*)	(4.9–9.2)	(3.0–5.5)	m (2.1–4.3); f (4.1–9.4)
Silkworm (*Bombyx mori*)	1770	1170	

(Continued)

TABLE A.2 (*Continued*)
Body Masses of Insects

Species	Fresh Weight		
	Larva (mg)	Pupa (mg)	Adult (mg)
Diptera			
Fly, black blow (*Phormia regina*)	44 (22–63)		m (38–40); f (42–50)
Fly, black cherry fruit (*Rhagoletis fausta*)			m 3.0 (1.8–3.9); f 4.1 (1.5–6.8)
Fly, bluebottle (*Calliphora erythrocephala*)			m (50–69); f (60–77)
Fly, giant crane (*Tipula abdominalis*)	1200 (800–1600)		
Fly, greenbottle (*Lucilia sericata*)	0.05	28 (24–35)	m 27; f 40
Fly, house		(22–27)	m (12–17); f (16–21)
Maggot, apple (*Rhagoletis pomonella*)			m 4.6 (2.6–7.5); f 6.9 (1.7–11)
Maggot, cherry (*R. cingulata*)			m 3.8 (2.4–5.2); f 5.1 (3.2–7.0)
Mosquito (*Culex tarsalis*)		2.2	
Mosquito, malaria (*Anopheles quadrimaculatus*)	1.91	m 2.99; f 3.45	m 1.37; f 1.66
Mosquito, yellow-fever (*Aedes aegypti*)			(3.6–3.9)
Coleoptera			
Beetle, Colorado potato (*Leptinotarsa decemlineata*)	98 (83–110)		m 160 (146–176); f 173
Beetle, confused–flour (*Tribolium confusem*)	2.0 (1.5–2.4)		2.1 (1.4–3.5)
Beetle, convergent lady (*Hippodamia convergens*)			21.3
Beetle, diving (*Dysticus marginalis*)	1366 (1305–1498)	1950	1356 (1186–1466)
Beetle, goliath (*Goliathus goliathus*)			(40–100 g)
Beetle, Japanese (*Popillia japonica*)	(196–276)	(176–222)	m 114; f 146
Beetle, lady (*Ceratomegilla fuscilabrus*)			13.1
Mealworm, yellow (*Tenebrio molitor*)	(83–180)		
Weevil, bean (*Acanthoscelides obtectus*)			m (2.9–6.8); f (2.8–8.3)
Hymenoptera			
Ant, giant (*Camponotus gigas*)			(75–347)
Bee, honey, worker (*Apis mellifera*)	137 (158–171)	105 (147–176)	120 (87–134)
Wasp, paper-nest (*Polistes variatus*)			149
Queen Wasp, parasitic (*Caraphractus cinctus*)			0.005

Source: Spector, W.S., ed., *Handbook of Biological Data*, WADC Technical Report 56-273, Wright-Patterson Air Force Base, OH, 1956.

TABLE A.3

Estimating Surface Area of Mammals

Animal	Body Mass (g)	*K*-Value (Constant)
Antelope	6,300	14.1
Bat	64.6 (12.7–36.4)	57.5 (54.0–59.8)
Bat	83 (5–116)	44.5 (44.0–45.0)
Cat	1,550 (1,500–1,600)	8.7 (8.6–8.9)
Cat	100 (84–116)	10.0 (9.9–10.0)
Cat	708 (219–1,389)	10.7 (9.5–11.9)
Cattle	375,000 (163,000–641,000)	11.0 (9.0–13.8)
Cattle	476,000 (208,000–762,000)	9.3 (8.1–10.8)
Hereford, thin	241,000 (89,000–407,000)	9.9 (9.3–10.5)
Hereford, med	315,000 (78,000–493,000)	9.4 (8.8–10.0)
Hereford, fat	314,000 (171,000–549,000)	8.6 (8.3–9.0)
Cattle	695,000 (476,000–815,000)	7.6 (7.3–7.9)
Dog	1,070 (130–3,650)	10.1 (9.3–11.0)
Dog	1.80	11.0
Dog	9,500 (8,900–10,000)	9.9 (9.85–9.9)
Dog	12,700 (3,200–29,800)	11.6 (10.2–12.5)
Dog	14,310 (3,390–32,640)	11.2 (10.3–12.1)
Dog	27,000	12.3
Fox	6,200 (6,100–6,300)	13.0 (12.9–13.2)
Goat	15,100	10.5
Guinea pig	157 (123–191)	10.4 (10.1–10.8)
Guinea pig	206 (123–269)	9.5 (8.4–10.8)
Guinea pig	256 (235–269)	8.6 (8.4–8.9)
Guinea pig	323 (160–810)	8.9 (7.9–9.6)
Guinea pig	373 (148–650)	9.6 (9.0–9.9)
Guinea pig	400 (380–420)	7.1
Hedgehog	200	7.5
Horse		9.0
Horse	(47,000–555,000)	10.5
Horse	(70,000–750,000)	(8.2–10.3)
Lion	64,200	12.3
Marten, pine	1,400	8.8
Monkey	2,670 (800–6,600)	11.8 (10.8–13.2)
Mouse	12.9	6.9
Mouse	14.7 (6.0–26.5)	7.9
Mouse	15.9 (10.7–19.7)	10.5 (10.4–10.5)
Mouse	16.4 (10.4–22.0)	11.4 (9.7–13.3)
Mouse	20.2	6.3
Mouse	(16.0–24.8)	9.0 (8.4–9.4)
Mouse, deer	22.0	8.5
Mouse, field	29.0 (26.0–31.0)	6.9 (6.5–7.2)

(Continued)

TABLE A.3 (*Continued*)
Estimating Surface Area of Mammals

Animal	Body Mass (g)	*K*-Value (Constant)
Mouse, red back	22.0	7.1
Opossum	1,200 (1,000–1,300)	11.3 (10.5–11.8)
Rabbit	32 (26–40)	8.5
Rabbit	560 (70–925)	9.7
Rabbit	1,130 (1,120–1,140)	10.0 (9.0–11.0)
Rabbit	2,600	5.7
Rat, white	25 (23–28)	9.5 (9.4–9.6)
Rat	42 (35–53)	10.5 (10.1–10.8)
Rat	80 (50–129)	9.9 (9.6–10.4)
Rat	95 (22–164)	7.6 (7.3–8.8)
Rat	125 (24–366)	7.5 (6.6–8.3)
Rat	133 (70–310)	11.6 (10.9–12.1)
Rat	137 (47–295)	9.0
Rat	170 (164–177)	7.15
Rat	176 (25–461)	11.4 (9.6–13.0)
Rat	(19–418)	9.0
Rat	(65–335)	10.5
Sheep	17,680	11.0
Sheep	(21,800–29,100)	10.7
Sheep	(2,200–68,000)	8.3
Sheep	(23,600–37,700)	8.5
Sheep	(3,780–50,400)	9.1
Shrew, long-tailed	3.5	8.0
Shrew, short-tailed	20	7.0
Swine		8.8
Swine	40,110	15.3
Swine	(25,000–330,000)	9.0
Swine	48,300 (1,100–123,000)	9.9 (8.6–12.4)
Whale, fin	160,000 (115,000–220,000)	8.3 (7.5–8.9)
Whale	43,000,000	11.1
Woodchuck	1,236	9.3

Source: Spector, W.S., ed., *Handbook of Biological Data*, WADC Technical Report 56-273, Wright-Patterson Air Force Base, OH, 1956.

Notes: Area (in cm^2) is determined from the formula $A = Km^{2/3}$, where m is body mass in g. Ranges in parentheses include 95% of the sample population.

TABLE A.4
Surface Area per Leaf for Various Plants

Species	Leaf Surface (sq cm/Leaf)
Alfalfa (*Medicago sativa*), early spring leaf	2
Alfalfa (*M. sativa*), late spring leaf	2
Alfalfa (*M. sativa*), secondary leaf	1.2
Alfalfa (*M. sativa*), tertiary leaf	0.6
Alfalfa (*M. sativa*), quaternary leaf	0.4
Apple, laxton (*Pyrus malus*)	18
Atalantia (*Atalantia citroides*)	9
Atalantia (*A. disticha*)	3
Balsamocitrus (*Balsamocitrus gabonensis*)	12
Banana (*Musa paradisiacal sapientum*)	2,700–5,200
Basswood (*Tilia americana*)	73
Bean (*Phaseolus vulgaris*), 15 day old	49
Beech (*Fagus* sp.)	22
Box-orange (*Severinia buxifolia*)	3.5
Catalpa (*Catalpa speciosa*), small leaf	29–71
Catalpa (*C. speciosa*), medium leaf	135
Catalpa (*C. speciosa*), large leaf	240–380
Cherry-orange (*Citropsis schweinfurthi*)	10
Citron (*Citrus medica*)	20–33
Clappia (*Clappia suaedifolia*)	0.7
Coldenia (*Coldenia hispidissima*)	0.2
Corn (*Zea mays*)	600–1,320
Cottonwood (*Populus fremontii wislizenli*)	50
Crownbeard (*Verbesina encelioides*)	12
Cucumber (*Cucumis sativus*), cotyledons	15
Cucumber (*C. sativus*), first leaf	18
Cucumber (*C. sativus*), second leaf	29
Cucumber (*C. sativus*), third leaf	33
Date (*Phoenix dactylifera*), 12 year old	43,700
Desert-lime (*Eremocitrus glauca*)	60 sq mm
Elm (*Ulmus americana*)	54
Fingerlime (*Microcitrus australasica*)	40 sq mm*
Fir (*Abies lasiocarpa*), moist habitat	80 sq mm*
Fir (*A. lasiocarpa*), dry habitat	55 sq mm*
Frankenia (*Frankenia jamesii*)	0.2
Gooseberry (*Ribes rotundifolium*)	2–16.5
Gourd (*Cucurbita foetidissima*)	560
Grape, malaga (*Vitis vinifera*)	125–150
Grape, Muscat (*V. vinifera*)	88
Grapefruit (*Citrus paradisi*)	40–45
Greggia (*Greggia camporum*)	2
Groundsel (*Senecio spartioides*)	5
Lemon (*Citrus limonia*)	37–40
Lime (*C. aurantifolia*)	13–14.5

(Continued)

TABLE A.4 (*Continued*)

Surface Area per Leaf for Various Plants

Species	Leaf Surface (sq cm/Leaf)
Kumquat (*Fortunella margarita*)	9
Maple (*Acer saccharum*)	73
Milkweed (*Aschlepias arenaria*)	48
Musa (*Musa acuminata*)	0.7–2 sq m
Morning glory (*Ipomoea purpurea*), first leaf	35
Morning glory (*I. purpurea*), third leaf	50
Morning glory (*I. purpurea*), fifth leaf	80
Morning glory (*I. purpurea*), seventh leaf	100
Oleander (*Nerium oleander*), 300 ft-c	10.5
Oleander (*N. oleander*), 86 ft-c	11
Orange (*Citrus sinensis*), 3 year old	3–130
Orange (*C. sinensis*), 29 year old	2–48
Pepperweed (*Lepidium alyssoides*)	5
Periwinkle (*Vinca rosea*), 300 ft-c	4.5
Periwinkle (*V. rosea*), 86 ft-c	4
Pine (*Pinus banksiana*), 36 year old	
Unthinned forest	52.4 sq mm*
Heavily thinned	51.6 sq mm*
Moderately thinned	49.7 sq mm*
Pine (*P. contorta*), dry habitat	128 sq mm*
Pine (*P. contorta*), moist habitat	232 sq mm*
Pine (*P. resinosa*), 27 year old	401 sq mm*
Pine (*P. strobes*), 27 year old	122 sq mm*
Pummelo (*Citrus grandis*)	40
Ragweed (*Ambrosia trifida*)	100
Redbud (*Cercis canadensis*)	65
Saltbush (*Atriplex canescens*)	1
Seepweed (*Suaeda suffrutescens*)	59 sq mm
Spruce (*Picea engelmannii*), dry habitat	44 sq mm*
Spruce (*P. engelmannii*), moist habitat	69 sq mm*
Sumac (*Rhus trilobata*)	21
Sunflower (*Helianthus annuus*)	38
Tabog (*Chaetospermym glutinosum*)	8.6
Taro (*Colocasia antiquorum*)	9,100
Trifoliate-orange (*Poncirus trifoliata*)	5
Wheat (*Triticum aestivum*), first leaf	5
Wheat (*T. aestivum*), third leaf	13
Wheat (*T. aestivum*), fifth leaf	15
Yellow-poplar (*Liriodendron tulipifera*)	130
Zinnia (*Zinnia grandiflora*)	0.8

Source: Spector, W.S., ed., *Handbook of Biological Data*, WADC Technical Report 56-273, Wright-Patterson Air Force Base, OH, 1956.

Notes: Values for needle leaves, indicated by an asterisk (*), represent the entire surface; values for flat leaves must be multiplied by 2 to obtain the entire externally exposed leaf surface (both upper and lower). Values are sq cm unless otherwise specified.

TABLE A.5
Surface Areas and Number of Leaves per Plant

Species	Leaves	
	No./Plant	**sq cm/Plant**
Alfalfa (*Medicago sativa*) (per stem)	88	16,000
Apple (*Pyrus malus*), 2 year old		11,000
Apple (*P. malus*), 5 year old		294,000
Apple (*P. malus*), 9 year old	20,000	318,000
Barley (*Hordeum vulgare*), 49 day old		900
Bean (*Phaseolus vulgaris*), 15 day old	2	98
Beech (*Fagus* sp.), in forest	35,000	780,500
Beech (*Fagus* sp.), in open field	200,000	4,460,000
Beet, mangold (*Beta vulgaris*)		3,050
Beet, sugar (*B. vulgaris*)		4,080
Catalpa (*Catalpa speciosa*)	26,024	1,952,000
Coldenia (*Coldenia hispidissima*)	11,560	2,300
Corn (*Zea mays*)		7,900
Cotton (*Gossypium* sp.)		535–1,200
Crowbeard (*Verbesina encelioides*)		560
Cucumber (*Cucumis sativus*), 135 ft-c		16
Cucumber (C. sativus), 270 ft-c		18
Cucumber (*C. sativus*), full sun		180–1,100
Date (*Phoenix dactylifera*), 12 year old	120	525 sq m
Fir (*Abies alba*), trunk diameter 40 cm	15,000,000	
Fir (*A. pectinata*)		12,000
Frankenia (*Frankenia jamesii*)	9,487	2,182
Gourd (*Cucurbita foetidissima*)	1,617	911,976
Grape, malaga (*Vitis vinifera*)	32	4,000–4,900
Grape, muscat (*V. vinifera*)	26	2,300–2,400
Milkweed (*Asclepias arenaria*)	36	1,700
Morning glory (*Ipomoea purpurea*)	9	750
Oleander (*Nerium oleander*), 86 ft-c	24	268
Oleander (*N. oleander*), 300 ft-c	23	241
Orange (*Citrus sinensis*), 3 year old	16,419	344,000
Orange (*C. sinensis*), 6 year old	37,257	590,000
Orange (*C. sinensis*), 29 year old	172,613	2,000,000
Peach (*Prunus persica*), 5 year old		922,000
Peach (*P. persica*), 5 year old		750,000
Pepperweed (*Lepidium alyssoides*)	1,318	6,700
Periwinkle (*Vinca rosea*), 86 ft-c	163	630
Periwinkle (*V. rosea*), 300 ft-c	257	1,150
Pine (*Pinus banksiana*), 36 year old		
Unthinned forest	744,924	390,000*
Heavily thinned	1,628,022	840,000*
Moderately thinned	891,211	443,000*

(Continued)

TABLE A.5 (*Continued*)

Surface Areas and Number of Leaves per Plant

Species	Leaves No./Plant	Leaves sq cm/Plant
Pine (*P. ponderosa*), 2 year old		72*
Pine (*P. resinosa*), 27 year old, 2 × 2	19,439	77,900*
Pine (*P. resinosa*), 27 year old, 4 × 4	51,060	205,000*
Pine (*P. strobus*), 27 year old, 2 × 2	121,805	149,000*
Pine (*P. strobus*), 27 year old, 4 × 4	186,644	228,000*
Potato (*Solanum tuberosum*)		17,800
Raspberry (*Rubus occidentalis*), shoot		13,100
Raspberry (*R. occidentalis*), fruiting cane		16,300
Saltbush (*Atriplex canescens*)		47,000
Seepweed (Suaeda suffrutescens)		3,210
Sorghum, red amber (*Sorghum vulgare*)		4,840
Spruce (*Picea* sp.), 4 year old	6,577	350*
Strawberry (*Fragaria chiloensis*)		
Dunlap, with runners		1,440
Dunlap, without runners		896
Sumac (*Rhus trilobata*)	3,000	63,240
Sunflower (*Helianthus annuus*)	59	2,260
Taro (*Colocasia antiquorum*)	10	90,730
Wheat (*Triticum aestivum*), 600 ft-c	5	46–65

Source: Spector, W.S., ed., *Handbook of Biological Data*, WADC Technical Report 56-273, Wright-Patterson Air Force Base, OH, 1956.

Notes: Values for needle leaves, indicated by an asterisk (*), represent the entire surface; values for flat leaves must be multiplied by 2 to obtain the entire externally exposed leaf surface (both upper and lower). Values are sq cm unless otherwise specified.

TABLE A.6

Blood Flow and O_2 Consumption of Various Organs in a 63 kg Adult Human

Region	Mass (kg)	Blood Flow (mL/min)	Arteriovenous O_2 Difference (mL O_2/Liter blood)	O_2 Consumption (mL/min)	Percentage of Total: Cardiac Output	Percentage of Total: Oxygen Consumption
Whole body	63.0	5400	46	250	100.0	100.0
Brain	1.4	750	62	46	13.9	18.4
Heart muscle	0.3	250	114	29	4.7	11.6
Kidneys	0.3	1260	14	18	23.3	7.2
Liver	2.6	1500	34	51	27.8	20.4
Skeletal muscle	31.0	840	60	50	15.6	20.0
Skin	3.6	462	25	12	8.6	4.8
Rest of body	23.6	336	129	44	6.2	17.6

TABLE A.7
Basal Metabolism of Humans

	Metabolic Rate (kcal/sq m/h)	
Age (Year)	**Males**	**Females**
3	60.1 (51.8–63.3)	54.5 (47.0–62.0)
4	57.9 (49.9–65.9)	53.9 (46.5–61.3)
5	56.3 (48.5–64.1)	53.0 (45.7–60.3)
6	54.0 (46.5–61.5)	51.2 (44.1–58.3)
7	52.3 (45.1–59.5)	49.7 (42.8–56.6)
8	50.8 (43.8–57.8)	48.0 (41.4–54.6)
9	49.5 (42.7–56.3)	46.2 (39.8–52.6)
10	47.7 (41.1–54.3)	44.9 (38.7–51.1)
11	46.5 (40.1–52.9)	44.1 (38.0–50.2)
12	45.3 (39.0–51.6)	42.0 (36.2–47.8)
13	44.5 (38.4–50.6)	40.5 (34.9–46.1)
14	43.8 (37.8–49.8)	39.2 (33.8–44.6)
15	43.7 (37.7–49.7)	38.3 (33.0–43.6)
16	42.9 (37.0–48.8)	37.7 (32.5–42.9)
17	41.9 (36.1–47.7)	36.2 (31.2–41.2)
18	40.5 (34.9–46.1)	35.7 (30.8–40.6)
19	40.1 (34.6–45.6)	35.4 (30.5–40.3)
20	39.8 (34.3–45.3)	35.3 (30.4–40.2)
21	39.4 (34.0–44.8)	35.2 (30.3–40.1)
22	39.2 (33.8–44.6)	35.2 (30.3–40.1)
23	39.0 (33.6–44.4)	35.2 (30.3–40.1)
24	38.7 (33.4–44.0)	35.1 (30.3–39.9)
25	38.4 (33.1–43.7)	35.1 (30.3–39.9)
26	38.2 (32.9–43.5)	35.0 (30.2–39.8)
27	38.0 (32.8–43.2)	35.0 (30.2–39.8)
28	37.8 (32.6–43.0)	35.0 (30.2–39.8)
29	37.7 (32.5–42.9)	35.0 (30.2–39.8)
30	37.6 (32.4–42.8)	35.0 (30.2–39.8)
31	37.4 (32.2–42.6)	35.0 (30.2–39.8)
32	37.2 (32.1–42.3)	34.9 (30.1–39.7)
33	37.1 (32.0–42.2)	34.9 (30.1–39.7)
34	37.0 (31.9–42.1)	34.9 (30.1–39.7)
35	36.9 (31.8–42.0)	34.8 (30.0–39.0)
36	36.8 (31.7–41.9)	34.7 (29.9–39.5)
37	36.7 (31.6–41.8)	34.6 (29.8–39.4)
38	36.7 (31.6–41.8)	34.5 (29.7–39.3)
39	36.6 (31.5–41.7)	34.4 (29.7–39.1)
40	36.5 (31.5–41.5)	34.3 (29.6–39.0)
50	36.0 (31.0–40.0)	33.4 (28.8–38.0)
55	35.4 (30.5–40.3)	32.9 (28.4–37.4)
60	34.8 (30.0–39.6)	32.4 (27.9–36.9)

(Continued)

TABLE A.7 (*Continued*)
Basal Metabolism of Humans

Age (Year)	Metabolic Rate (kcal/sq m/h)	
	Males	Females
65	34.0 (29.3–38.7)	31.8 (27.4–36.2)
70	33.1 (28.5–37.7)	31.3 (27.0–35.6)
75 and above	31.8 (27.4–36.2)	31.1 (26.8–35.4)

Source: Spector, W.S., ed., *Handbook of Biological Data*, WADC Technical Report 56-273, Wright-Patterson Air Force Base, OH, 1956.
Notes: Values are means of basal kilocalories per sq m surface area per hour. Standard ranges are enclosed in parentheses.

TABLE A.8A
Basal and Resting Energy Metabolism for Vertebrates

Animal	Stage	Approx. Age		Body Mass Kg		Body Surface Area Sq m		Resting Metabolism Cal/Kg/Day		Resting Metabolism Cal/sq m/Day	
		M	F	M	F	M	F	M	F	M	F
Man	Adult			65	56	1.83	1.65				
Baboon	Adult			6.2		0.40					
Beef cattle	Young	1.8 months		70	60	1.4	1.3	43.8	51.1	2190	2385
	Half-grown	11 months		300	250	3.2	2.9	25.6	26.3	2420	2295
	Adult	1.7 years		500	400	4.2	3.7	21.2	21.1	2515	2270
Cat	Adult			3.0		2.0					
Chimpanzee	Adult			38		1.1					
Dairy cattle	Young	6 months			150		2.4		34.4		2100
	Half-grown	1.2 years			300		3.6		26.0		2170
	Adult	2 years			500		5.0		21.1		2180
Dog	Adult			15.5	11.7	0.65	0.58				
Elephant	Adult			3670		23.8					
Elephant, sm	Adult			1360		13.7					
Goat	Young			2				130	125		
	Half-grown			20				63	54		
	Adult			70				42	34		

(*Continued*)

TABLE A.8A (*Continued*)
Basal and Resting Energy Metabolism for Vertebrates

Animal	Stage	Approx. Age M	Approx. Age F	Body Mass Kg M	Body Mass Kg F	Body Surface Area Sq m M	Body Surface Area Sq m F	Resting Metabolism Cal/Kg/Day M	Resting Metabolism Cal/Kg/Day F	Resting Metabolism Cal/sq m/Day M	Resting Metabolism Cal/sq m/Day F
Guinea pig	Young	30 days		0.2		0.029		115	120	780	805
	Half-grown	82 days	87 days	0.4		0.046		90	95	800	825
	Adult	290 days	270 days	0.8		0.071		63	68	710	765
Horse	Young	2.8 months	2.9 months	200		2.8		32.2	32.7	2280	2320
	Half-grown	9.5 months	8.5 months	350		4.0		24.6	25.3	2150	2210
	Adult	4.2 years	2.8 years	650		5.9		25.2	24.7	2770	2710
Macaque	Adult			4.2		0.31					
Marmot	Adult			2.6		0.18					
Monkey, rhesus	Adult			3.2		0.26					
Mouse, albino	Adult			0.02		0.005					
Mouse, dwarf	Adult			0.008		0.004					
Mouse, obese	Adult			0.06		0.01					
Mule	Young	4 months		200		2.9		39.2		2700	
	Half-grown	13 months		400		4.5		30.6		2705	
	Adult	38 months		600		5.8		26.4		2710	
Pony, Shetland	Adult			280		4.4					
Rabbit	Adult			3.5		0.2					
Rat	Young	29 days		0.05		0.013		280		1085	
	Half-grown	50 days	60 days	0.15		0.026		195	165	1120	970
	Adult	60 days	120 days	0.2		0.031		155	135	1000	870
Sheep	Adult			49.5	42.7	1.10	0.95				
Swine	Young	9.4 months	8.3 months	75		1.5		37.3	40.5	1880	2040
	Half-grown	1.3 years	1.1 years	150		2.3		28.9	25.1	1880	1625
	Adult	2.1 years		250		3.2		23.7	17.8	1860	1390
Birds, wild	Adult			3.0		0.2					
Canary	Adult			0.016		0.006					
Chicken	Young	4 weeks		0.25	0.2	0.04	0.03	195	210	1220	1230
	Half-grown	13 weeks		1.1	0.9	0.11	0.1	105	100	1020	900
	Adult	25 weeks		2.6	2.0	0.21	0.17	95	75	1160	880

(*Continued*)

TABLE A.8A (*Continued*)

Basal and Resting Energy Metabolism for Vertebrates

		Approx. Age		Body Mass Kg		Body Surface Area Sq m		Resting Metabolism Cal/Kg/Day		Resting Metabolism Cal/sq m/Day	
Animal	**Stage**	**M**	**F**	**M**	**F**	**M**	**F**	**M**	**F**	**M**	**F**
Dove, ring	Adult			0.15		0.03					
Duck	Adult			0.93		0.1					
Goose	Adult			5.0	3.3	0.29	0.23				
Parakeet	Adult			0.03		0.009					
Pigeon	Adult			0.28		0.04					
Lizard				1.2		0.11					
Turtle				0.14		0.02					
Frog				To 0.05							
Toad				To 0.05							
Fish				To 0.25							
Fish, sturgeon				1400		11.8					

Source: Spector, W.S., ed., *Handbook of Biological Data*, WADC Technical Report 56-273, Wright-Patterson Air Force Base, OH, 1956.

TABLE A.8B

Basal and Resting Energy Metabolism for Vertebrates

		Oxygen Consumption (Resting Metabolism) L/kg/Day		Basal Metabolism Cal/kg/Day		Basal Metabolism Cal/sq m/Day		Oxygen Consumption (Basal Metabolism) L/kg/Day	
Animal	**Stage**	**M**	**F**	**M**	**F**	**M**	**F**	**M**	**F**
Man	Adult			25.5	23.2	910	790		
Baboon	Adult			48		760			
Beef cattle	Young	9.1	10.6						
	Half-grown	5.3	5.5		18.2		1595		3.8
	Adult	4.4	4.4		15.2		1635		3.2
Cat	Adult			50		750			
Chimpanzee	Adult			29.2		980			
Dairy cattle	Young		7.1						
	Half-grown		5.4						
	Adult		4.4						

(*Continued*)

TABLE A.8B (*Continued*)

Basal and Resting Energy Metabolism for Vertebrates

		Oxygen Consumption (Resting Metabolism)		Basal Metabolism				Oxygen Consumption (Basal Metabolism)	
		L/kg/Day		Cal/kg/ Day		Cal/sq m/ Day		L/kg/Day	
Animal	**Stage**	**M**	**F**	**M**	**F**	**M**	**F**	**M**	**F**
Dog	Adult			33.5	38.5	800	770		
Elephant	Adult			13.3		2060			
Elephant, sm	Adult			11.8		1170			
Goat	Young								
	Half-grown								
	Adult								
Guinea pig	Young	23	24	110	110	735	765	23	24
	Half-grown	29	19	85	90	755	790	18	19
	Adult	13	14	60	63	675	710	13	13
Horse	Young	6.7	6.8						
	Half-grown	5.0	5.2						
	Adult	502	5.1						
Macaque	Adult			49.3		675			
Marmot	Adult			28.3		420			
Monkey, rhesus	Adult			48.4		610			
Mouse, albino	Adult			170		525			
Mouse, dwarf	Adult			125		280			
Mouse, obese	Adult			130		550			
Mule	Young	8.1							
	Half-grown	6.3							
	Adult	5.5							
Pony, Shetland	Adult			16.7		1060			
Rabbit	Adult			47		810			
Rat	Young		60	240		930		51	
	Half-grown	39	34	160	155	930	890	34	33
	Adult	32	28	140	120	905	760	30	25
Sheep	Adult			26.3	25.7	1180	1160		
Swine	Young	7.7	8.4	30.9	30.2	1550	1520	6.4	6.3
	Half-grown	6.0	5.2	21.9	19.7	1420	1210	4.5	3.9
	Adult	4.9	3.7	17.4	14.1	1360	1100	3.6	2.9
Birds, wild	Adult			57		830			
Canary	Adult			310		760			

(*Continued*)

TABLE A.8B (*Continued*)

Basal and Resting Energy Metabolism for Vertebrates

		Oxygen Consumption (Resting Metabolism)		Basal Metabolism				Oxygen Consumption (Basal Metabolism)	
		L/kg/Day		Cal/kg/Day		Cal/sq m/Day		L/kg/Day	
Animal	**Stage**	**M**	**F**	**M**	**F**	**M**	**F**	**M**	**F**
Chicken	Young	40	43						
	Half-grown	21	20	90	90	870	830	19	19
	Adult	19	16	85	70	1075	800	18	15
Dove, ring	Adult			130		700			
Duck	Adult			90		855			
Goose	Adult			54	61	940	880		
Parakeet	Adult			225		690			
Pigeon	Adult			100		670			
Turtle				11.4		64			
Frog						130			
Toad						130			
Fish						33			
Fish, sturgeon				0.3		31			

TABLE A.9

Tissue Oxygen Consumption for Animals

Tissue	**Oxygen Consumption Rate mm (O_2/mg Dry Tissue/h)**
Man	
Cerebral cortex	6.0–10.3
Decidua	2.5
Lung, embryo	3.7
Lymph nodes	3.8–5.9
Mucosa, gastric	9.6
Muscle, smooth, gastric	1.3
Uterine	0.6
Salivary gland	6.3
Skin, adult	2.1
Fetus	1.8
Sperm	0.54
Tonsil	5.1

(*Continued*)

TABLE A.9 (*Continued*)
Tissue Oxygen Consumption for Animals

Tissue	Oxygen Consumption Rate mm (O_2/mg Dry Tissue/h)
Rat	
Adrenal	10.0
Brain cortex	26.3
Cerebral cortex	2.9
Cerebral cortex	10.8
Cerebral cortex	8.0
Cerebral cortex	13.6
Cerebral cortex	9.5
5 day old	6.2
50 day old	14.7
Adult	8.5–17.1
Chorion	13.5
Diaphragm	6.3
Diaphragm	5.4
Diaphragm	9.4
Diaphragm	6.3
Embryo, 1–3 mg	10.5–14.6
13–14 day old	7.2–11.0
Ganglion, dorsal root	8.0
Hypothalamus	10.4
Kidney	15.8
Kidney	38.0
Kidney	23.2
Kidney	23.1
Kidney	34.0
Kidney	26.0
Kidney cortex	38.2
Liver	7.2
Liver	8.1
Liver	9.0
Liver	10.7
Liver	26.0
Liver	17.2
Liver, fetus	7.1
3–21 days old	13.2
Adult	9.8–10.2
Adult	6.5–11.6
Lung	8.6
Adult	7.9
Adult	4.4–7.8
Embryo	10.0
Lymph nodes	4.4

(*Continued*)

TABLE A.9 (*Continued*)
Tissue Oxygen Consumption for Animals

Tissue	Oxygen Consumption Rate mm (O_2/mg Dry Tissue/h)
Mammary, term. of preg.	1.3
Mammary, 15–22 days lactation	10.0
2 days after weaning	5.5
Medulla, 5 days old	3.4
50 days old	9.0
Adult	2.5–4.9
Mucosa, colon	3.4–14.6
Duodenal	8.8
Gastric	7.2
Ileum	3.7
Jejunal	12.4
Muscle, diaphragm	4.1–5.9
Diaphragm	5.9
Heart	3.8
Skeletal	2.3–3.1
Smooth, gastric	3.5
Smooth, intestinal	6.3
Smooth, intestinal	7.1
Ovary	5.7
Pancreas	5.2
Pituitary, anterior	5.9
Posterior	6.6
Young	12.0
Placenta	3.9
20 days old	7.3
Prostate	7.6
Retina	22.0–32
Salivary gland	11.6–16.6
Skin, newborn	3.5
10–36 days old	4.9–3.6
79 days, adult	1.8–1.2
Sperm	7.7
Spleen	12.7
Spleen	7.2–12.9
Spleen	13.0
Testis	7.2–14.3
Testis	11.0
Thymus	5.5–5.8
Thyroid	12.5–13
Uterus	7.6
Uterus	3.7
Castrate	5.2
Castrate, plus estrogen E	7.9

(*Continued*)

TABLE A.9 (*Continued*)
Tissue Oxygen Consumption for Animals

Tissue	Oxygen Consumption Rate mm (O_2/mg Dry Tissue/h)
Mouse	
Adrenal	6.0
Brain cortex	32.9
Cerebral cortex	11.0
Embryo	10.4
Kidney cortex	46.1
Liver	8.8–13.8
Liver	18.7
Liver	23.1
Lung	7.3–8.0
Lung	12.0
Ovary	9.0
Placenta, 0.4 mg	7.5
10.9–13.7 mg	6.4
Pituitary	8.0
Skin, newborn	6.1
Spleen	16.9
Guinea pig	
Adrenal	6.0
Brain cortex	27.3
Cerebral cortex	6.9
Cerebral cortex	11.7
Epithelium	6.1
of castrate	2.8
Kidney cortex	31.8
Liver	13.0
Liver	8.1
Liver	5.0
Fatty	7.4
Lung	6.1
Lung	7.4
Lung	8.5
Muscle, smooth	1.7
of castrate	1.4
Pancreas	2.7
Salivary gland	5.0
Skin	3.0
Sperm	8.0
Sperm	18.4
Spleen	8.13

(*Continued*)

TABLE A.9 (*Continued*)

Tissue Oxygen Consumption for Animals

Tissue	Oxygen Consumption Rate mm (O_2/mg Dry Tissue/h)
Rabbit	
Brain cortex	28.2
Cerebral cortex	7.3–10.4
Embryo	8.5
Ganglion, celiac	4.0
Kidney cortex	34.5
Liver	11.6
Liver	4.2–7.7
Lung	8.0
Lung	6.7
Marrow, erythroid cells	9.0
Myeloid cells	6.0
Mucosa, colon	11.1
Uterine	6.1
Muscle, diaphragm	2.4
Smooth, intestinal	2.6
Pancreas	4.6
Placenta, fetal side	5.3
Uterine side	3.4
Sperm, ejaculated	4.4
Spleen	14.2
Testis	7.7
Thyroid	11.7
Cat	
Brain cortex	26.9
Cerebral cortex	8.5–12.2
Kidney cortex	22.7
Liver	13.2
Lung	3.9
Lung	3.9
Medulla	3.5
Muscle, heart	2.5
Smooth intestinal	1.4
Pancreas	5.8
Salivary, acetylcholine stimulation	13.6
Eserine + acetylcholine	22.7
Resting	10.3
Spinal cord	1.3
Spleen	8.4

(*Continued*)

TABLE A.9 (*Continued*)

Tissue Oxygen Consumption for Animals

Tissue	Oxygen Consumption Rate mm (O_2/mg Dry Tissue/h)
Dog	
Brain cortex	21.2
Caudate nucleus	1.36
Cerebellum	1.07
Cerebral cortex	1.16
Cerebral cortex	6.7
Heart	2.6
Heart	2.7
Heart	4.6
Heart	6.3
Kidney cortex	27.0
Liver	6.0
Liver	11.7
Lung	4.9
Medulla	0.69
Muscle, skeletal	1.2
Muscle, skeletal	1.3
Muscle, skeletal	1.7
Diaphragm, young	1.9
Diaphragm, juvenile	4.2
Heart	2.6
Pancreas	3.2
Retina	20.8
Salivary	10.6
Spinal cord	0.5
Spleen	6.6
Thalamus	1.01
Thyroid	2.0
Midbrain	0.92
Horse	
Brain cortex	15.7
Kidney cortex	21.5
Liver	5.4
Liver	2.1
Lung	4.4
Spleen	4.2
Cattle	
Brain cortex	17.2
Kidney cortex	23.5
Liver	8.2
Cow	2.6

(*Continued*)

TABLE A.9 (*Continued*)

Tissue Oxygen Consumption for Animals

Tissue	Oxygen Consumption Rate mm (O_2/mg Dry Tissue/h)
Lung	4.3
Retina, ox	10.7
Sperm	6.6
Sperm	11.2
Sperm	12.8
Sperm, epididymal	2.6
Spleen	4.4
Thyroid, calf	2.6
Swine	
Cerebral cortex	
29–60 day old fetus	5.5
99 day old fetus	6.5
Birth to adult	8.5
Retina	17.7
Thyroid, hog	2.1
Sheep	
Brain cortex	19.7
Kidney cortex	27.5
Liver	2.5
Liver	8.5
Lung	5.4
Sperm, ejaculated	9.0
Spleen	6.9
Trigeminal, nerve	0.5
Ganglion	0.3
Chick	
Allantois	22.3
Embryo, 0.1–1.2 g	15.9–21.4
4.7 g	8.1
5–6 days old	10–12
19 days old	7.7
Brain	25.0
Heart, 4 days old	30
Heart, 6–7 days old	14.9
Liver, 6 days old	7.5
Liver, 12 days old	4.5
Liver, 20 days old	1.5
Fowl	
Liver, hen	14.5
Sperm, ejaculated	2.8

(*Continued*)

TABLE A.9 (*Continued*)

Tissue Oxygen Consumption for Animals

Tissue	Oxygen Consumption Rate mm (O_2/mg Dry Tissue/h)
Pigeon	
Cerebral cortex	14.6
Lung	3.6
Muscle, skeletal	2.1
Pancreas	8.7
Frog	
Hippocampus	2.4
Muscle, skeletal, resting	0.18–0.24
Electrically stimulated	0.79–4.24
Smooth, intestinal	0.28
Nerve, sciatic	0.3
Retina	3.5
Spinal Cord	2.3

Source: Spector, W.S., ed., *Handbook of Biological Data*, WADC Technical Report 56-273, Wright-Patterson Air Force Base, OH, 1956.

Note: Tissue water contents are approximately 70%.

TABLE A.10

Respiration Rates for Adult Animals in a Resting or Basal State

Organism	cc O_2/kg Body Mass/h
Mammalia	
Man, maximum work	4,000
Man, resting	200
Anteater, spiny	1,100
Armadillo	201
Bat, brown, big	800
Bat, brown, little	1,500
Bear, polar, cub	700
Cat, Australian, native	560
Cow	390
Dog	580
Dormouse, awake	852
Dormouse, hibernating	15
Elephant, Indian, 37 year (female)	155
Fox, arctic, white	505
Guinea pig	1,250
Hamster	1,050
Hamster, golden, awake	2,900
Hamster, golden, hibernating	70

(*Continued*)

TABLE A.10 (*Continued*)
Respiration Rates for Adult Animals in a Resting or Basal State

Organism	cc O_2/kg Body Mass/h
Horse	130
Lemming	1,700
Manatee, Florida sea cow	120
Marmoset	1,040
Monkey, night	510
Mouse, house, basal	1,530
Mouse, house, resting	3,500
Opossum, Australian	700
Platypus, duckbilled	460
Porpoise	360
Rabbit	460–850
Raccoon	395
Rat	2,000
Rat, kangaroo	950
Seal	540
Sheep	340
Shrew, Monterey	7,200
Shrew, long-tailed	13,700
Shrew, short-tailed	5,200
Shrew, Sonoma (male)	6,100
Shrew, Sonoma (female)	5,500
Sloth, two-toed	216
Sloth, three-toed	168
Squirrel, Arctic, ground	600
Squirrel, flying	2,000
Swine	220
Weasel	5,000
Woodchuck, awake	262
Woodchuck, hibernating	14
Aves	
Bunting, snow	3,350
Canary	2,900
Dove	950
Duck	800
Fowl, hen	630
Goose	547–592
Gull, arctic	1,640
Hawk, night	1,750
Hummingbird, day	13
Manakin	4,620
Pigeon	710
Sparrow	2,100

(*Continued*)

TABLE A.10 (*Continued*)

Respiration Rates for Adult Animals in a Resting or Basal State

Organism	cc O_2/kg Body Mass/h
Reptilia	
Alligator lucius, 25°C	64
A. mississippiensis, 22°C	8.9
Coluber matrix, 20°C	92–150
Constrictor constrictor, 16°C	4.9
C. constrictor, 22°C	10
C. constrictor, 30°C	24
Crotalus atrox, 16°C	6.8
C. atrox, 22°C	16.4
C. atrox, 30°C	35.5
Drymarchon corais couperi, 16°C	10.1
D. corais couperi, 22°C	20
D. corais couperi, 30°C	47
Iguana tuberculata, 22°C	22.2
I. tuberculata, 30°C	52
Lacerta agilis, 20°C	1,980
Malaclemys centrata, 24°C	35
Python reticulatum, 22°C	12.2
Storeria dekayi (male), 20°C	266
S. dekayi (female), 20°C	183
Testudo vicina, 22°C	22
Amphibia	
Molge vulgaris, 20°C	123
Rana esculenta, winter, 20°C	85
R. esculenta, summer, 20°C	437
R. fusca, winter, 20°C	100
R. fusca, summer, 20°C	210
R. temporaria, winter, 19°C	85
R. temporaria, summer, 19°C	554
Pisces	
Anguilla vulgaris, 25°C	128
Arapaima gigas, 25°C	9
Cobitis fossilis, 20°C	51
Crenichthys baileyi, 37°C	546
Cyprinus tinca, 20°C	104
Esox lucius, 18°C	102
Heliasis chromis, 20°C	162
Lepidosiren paradoza, 20°C	42
Salmo trutta, 12°C	226
Scomber scrombrus, 20°C	726
Serranus scriba, 20°C	151

(*Continued*)

TABLE A.10 (*Continued*)

Respiration Rates for Adult Animals in a Resting or Basal State

Organism	cc O_2/kg Body Mass/h
Sphoeroides maculatus, 20°C	62
Stenotomus chrysops, 20°C	174
Tautoga onitis, 20°C	62
Tautogolabus adspersus, 21°C	120
Cephalochordata and Tunicata	
Amphioxus lanceolatus, 16°C	35
A. lanceolatus, 20°C	45
Ascidia mentula, 25°C	4.8
Salpa pinnata, 16°C	8
S. pinnata, 20°C	12
S. tilesii, 16°C	2.0
S. tilesii, 20°C	2.8
Arthropoda	
Apis mellifera, resting, 20°C	17,466
A. mellifera, true flight, 20°C	87,000
Asellus aquaticus, 17°C	348
Astacus leptodactylus, 20°C	70
Callianax subterranea, 15°C	930
Carcinus maenus, 15°C	625
Cryptocercus punctulatus, 5°C	28.5
Culex sp., 20°C	575
Dronia vulgaris, 15°C	3,000
Drosophila americana, resting, 20°C	1,560
D. americana, true flight, 20°C	21,800
Emerita talpodia, 20°C	112
Eriphia spinifrons, 15°C	1,828
Formica sp., 20°C	532
Geotrupes sp., 21°C	447
Homarus americanus, 15°C	507
Ilia nucleus, 15°C	253
Limnophilus vittatus, 10°C	500
Lucilia sericata, 20°C	95,600
Maja verrucosa, 15°C	1,460
Melanotus communis, 27°C	2,400
Melolontha sp., 20°C	724–960
Musca sp., 20°C	3,200–5,112
Ocypode albicans, 26°C	139
Paguristis maculata, 15°C	1,600
Palaemon squilla, 19°C	128
Palinurus vulgaris, 15°C	12,874
Pandalina brevirostrus, 15°C	20
Passalus cornutus, 17°C	30

(*Continued*)

TABLE A.10 (*Continued*)

Respiration Rates for Adult Animals in a Resting or Basal State

Organism	cc O_2/kg Body Mass/h
Periplaneta orientalia, 20°C	277
Pilumnus hirtellus, 15°C	160
Arthropoda	
Pugettia producta, 15°C	100
Sicyonia sculpa, 15°C	443
Spirontocaris cranchi, 15°C	6
Talorchestis megalopthalma, 17°C	180
Venessa sp., resting, 20°C	600
Venessa sp., true flight, 20°C	100,000
Zootermopsis angusticollis, 20°C	400
Annelida	
Arenicola sp., 12°C	30
Chaetopterus pergamentaceus, 15°C	8
Glycera siphonostoma, 25°C	15
Lumbricus communis, 21.5°C	206
L. herculeus, 10°C	45
L. terrestris, 20.5°C	138
Nereis virens, 15°C	26
Sipunculus nudus, 16°C	50
Tubifex sp., 15°C	200
Mollusca	
Aplysia limacina, 16°C	30
Eledone moschata, 16°C	181
Helix pomatia, 20°C	94
Limax agrestis, 20°C	350
Mytilus edulis, 14°C	13
M. galloprovincialis, 25°C	18
Octopus vulgaris, 16°C	47–87
Pleurobranchea meckeli, 25°C	36
Pterotrachea coronata, 16°C	7.8
Sepia officinalis, 15°C	320
Tethys leporina, 16°C	12
Echinodermata	
Asterias rubens, 15°C	30
Holothuria impatiens, 25°C	17
Ophioderma longicauda, 25°C	8–32
Strongylocentrotus lividus, 25°C	15
Nemathelminthes	
Ascaridia galli, 37°C	525
Ascaris lumbricoides, large, 37°C	72

(*Continued*)

TABLE A.10 (*Continued*)
Respiration Rates for Adult Animals in a Resting or Basal State

Organism	cc O_2/kg Body Mass/h
A. lumbricoides, small, 37°C	156
A. lumbricoides (male), 37°C	112
A. lumbricoides (female), 37°C	61
Heterakis spumosa, 38°C	880
Litomosoides carinii, 37.5°C	800
Nematodirus sp., 37°C	1,070
Neoaplectana glaseri, 30°C	2,600
Nippostrongylus muris, 37°C	1,430
Ostertagia circumcincta, 38°C	1,480
Setaria equinum, 38°C	250
Strongylus equines, 38°C	511
Syphacia obvelata, 38°C	1,010
Platyhelminthes	
Dendrocoelum lacteum, 2.5°C	4.4
D. lacteum, 25°C	26.3
Diphyllobothrium latum, 37°C	243
Fasciola hepatica, 37.5°C	330
Paramphistomum cervi, 38°C	3
Planaria torva, 2.5°C	18.9
P. torva, 25°C	75.8
Triaenophorus nodulosus, 22°C	418
Ctenophora	
Beroe ovata, 16°C	5
Cestus veneris, 16°C	2.6
C. veneris, 25°C	25
Coelenterata	
Anemonia sulcata, 18°C	13.4
Aurelia surita, 13°C	3.4
Carmarina hastata, 16°C	6
Rhizostoma pulmo, 16°C	7.2
Porifera	
Suberites massa, 22.4°C	0.0241
Protozoa	mm^3 O_2/Million Cells/h
Amoeba chaos chaos, 20°C	7,050
A. chaos chaos, 25°C	9,010
A. chaos chaos, 30°C	13,244
Astasia klebsii, young, 25.2°C	3.8
Chilomonas paramecium, 25°C	16.4
Leishmania braziliensis, 32°C	0.32
Leptomonas ctenocephali, 28°C	0.27

(*Continued*)

TABLE A.10 (*Continued*)

Respiration Rates for Adult Animals in a Resting or Basal State

Organism	cc O_2/kg Body Mass/h
Paramecium aurelia, 20°C	354
P. aurelia, 25°C	616
P. aurelia, 35°C	1,512
P. caudatum, 20°C	2,110
P. caudatum, 25°C	3,860
P. multimicronucleatum, 25°C	1,021
Plasmodium cathemerium, 38°C	0.25
P. knowlesi, 38°C	0.34
Strigomonas fasciculata, 28°C	0.37
Tetrahymena geleii, 26.8°C	632.5
Trichomonas foetus, 28°C	2.15
Trypanosoma congolense, 37°C	1.53
T. cruzi, 37°C	1.24
T. gambiense, 37°C	1.70
T. lewisi, old, 37°C	0.51
T. rhodesiense, 37°C	1.94

Source: Spector, W.S., ed., *Handbook of Biological Data*, WADC Technical Report 56-273, Wright-Patterson Air Force Base, OH, 1956.

TABLE A.11

Respiration Rates for Fungi

Species	Temperature (°C)	Specifications	O_2 Consumption Rate (μL/mg dry mass/h)	CO_2 Produced/ O_2 Consumed
Myxomycetes *Physarum polycephalum*	22	Endogenous	2.4[a]	0.75–0.85
Phycomycetes				
Allomyces arbuscula	20	Starved	0.8	
A. moniliformis		Starved	1.0	
Cystopus candidus		Host, host + fungus		0.93, 0.95
Leptomitus lacteus	20	Endogenous, at 0, 4, 8 days	20, 15, 10	0.98
Mucor guilliermondi	25	Endogenous, mycelial phase	5.7–10	
	25	+ Glucose, mycelial phase	5.6–21.4	
	25	Endogenous, yeast phase	7.1–8.6	
	25	+ Glucose, yeast phase	7.8–39	
M. stolonifer	20, 30			1.53, 1.72

(*Continued*)

TABLE A.11 (*Continued*)
Respiration Rates for Fungi

Species	Temperature (°C)	Specifications	O_2 Consumption Rate (μL/mg dry mass/h)	CO_2 Produced/ O_2 Consumed
Ascomycetes				
Ashbya gossypii	30	Endogenous, at 1, 2, 3 days	19, 11, 8	
	30	+ Glucose, at 1, 2, 3 days	32, 20, 12	
	30	+ Sucrose, at 1, 2, 3 days	30, 17, 13	
	30	+ Lactose, at 1, 2, 3 days	0, 0, 0	
	30	+ Pyruvate, at 1, 2, 3 days	8, 3, 0	
	30	+ Ethanol, at 1, 2, 3 days	12, 8, 3	
Erysiphe graminis tritici	22	Host, host + fungus	1.7, 6.0[c]	
Melanospora destruens	30	+ Glucose	6	
Neurospora crassa	30	Endogenous	11–38	
N. tetrasperma	25	Endogenous, dormant	0.2–0.6	
	25	Endogenous, germinating	9–22	
Saccharomyces cerevisiae R		No stored reserves	83–109	
		Fat as reserves	76	
		Glycogen as reserves	0	
S. cerevisiae U		No stored reserves	10–137	
		Fat as reserves	125	
		Glycogen as reserves	47	
Sclerotinia sp.	23–25			1.15
Zygosaccharomyces acidifaciens	28	Endogenous, at 24, 48, 72 h	16, 7, 7	
	28	+ Glucose, at 24, 48, 72 h	60, 35, 35	
Basidiomycetes				
Exidia glandulosa				0.7
Polystictus versicolor	17.5	At 2, 21, 100% O_2	3, 9.5, 10.4[b]	
	29.5	At 2, 100% O_2	7.4, 17.2[b]	
Psalliota campestris	25	1.9–2.9		
Puccinia pruni		Host, host + fungus		1.06, 0.82
Ustilago sphaerogena		Endogenous	75	
		+ Sugars	150–375	
Fungi imperfecti				
Aspergillus clavatus	15–25		12.4	
			44	
A. flavus	30	At 4–6 days	6–7	
A. niger	19, 35	+ Glucose		0.98, 1.30
	18, 35	+ Sucrose		0.91, 1.22
	36	+ Glycerol		0.82–0.86
	35	+ Mannitol		1.20

(*Continued*)

TABLE A.11 (*Continued*)

Respiration Rates for Fungi

Species	Temperature (°C)	Specifications	O_2 Consumption Rate (μL/mg dry mass/h)	CO_2 Produced/ O_2 Consumed
Blastomyces dermatitidis	37	Endogenous, at pH 2, 6, 8	0.5, 12, 11[c]	
	37	+ Glucose, at pH 2, 6, 8	1, 12, 12[c]	
Candida albicans	30	Endogenous	5	
	30	+ Glucose	40	
Fusarium avenaceum	23–25			5.46
F. dianthi	23–25			1.85
F. trichotheciodes	30	Endogenous, at 1–4 h	31–11	
	30	+ Glucose, at 1–4 h	64–56	
Helminthosporium gramineum	23–25	Endogenous		1.31
H. inaequalis	23–25	Endogenous		1.16
Memnoniella echinata	30	Endogenous, at pH 4, 6, 8	1.0, 1.5, 1.3	
		+ Glucose, at pH 4, 6, 8	2.6, 3.1, 2.7	
		+ Lactose, at pH 4, 6, 8	0.7, 1.3, 1.2	
Myrothecium verrucaria	30	+ Glucose, at pH 4, 6, 8	19, 25, 30	
		Endogenous, at pH 4, 6, 8	2.9, 2.6, 5.6	
		Starved, sucrose grown	42	
Penicillium chrysogenum	23–25	3 strains		1.10–1.27
P. digitatum	23–25	4 strains		1.39–1.63
P. notatum	23–25	At 4, 8, 11 days	46, 198 152[c]	
Torulopsis utilis	30	Glycine	3.7[a]	0.86
	30	Urea	3.5[a]	1.15
	30	α-Alanine	5.2[a]	0.89
	30	β-Alanine	4.2[a]	1.16

Source: Spector, W.S., ed., *Handbook of Biological Data*, WADC Technical Report 56-273, Wright-Patterson Air Force Base, OH, 1956.

O_2 consumption rate differs from units indicated:

[a] μL/mg wet mass/h.

[b] μL/sq cm area/h.

[c] μL/10 μL tissue volume/h.

TABLE A.12
Respiration Rates for Bacteria Suspended in the Presence of Glucose

Species	Temperature (°C)	Culture Age (h)	O_2 Consumption Rate (mL/mg Dry Mass/h)
Azotobacter chroococcum	22	36	2,000–10,000
Aerobacter aerogenes	36, 30	17, 48	47, 50
Bacillus cereus (short)	30?	18	42–86
B. cereus (filamentous)	30?	18	3–49
B. subtilis	37	6–8	170
Escherichia coli	40, 32	20	200, 272
Lactobacillus bulgaricus	37, 45	8	34, 55
Leuconostoc citrovorum	38	16	8
Micrococcus luteus	35	30–34	15
M. flavus	35	30–34	8
M. auranticus	35	30–34	14
M. cinnebareus	35	30–34	32
M. freundenreichii	35	30–34	20
Mycobacterium phlei	38	84	28
M. smegmatis	38	84	23
M. stercoris	38	84	15
M. sp. Karlinski	38	84	22
M. ranae	38	84	32
M. leprous kedrowsky	38	84	8
M. butyricum	38	84	13
M. tuberculosis hominis	38	252	4
M. tuberculosis avian	37	84	1
Pneumococcus, Type I	37	18	27
Pseudomonas fluorescens	26	20	58
Streptococcus faecalis, B33A	38	18	106
S. faecalis, 10C1	37	15	57–80
S. faecalis, Lancefield D	37	12–15	7
S. thermophilus, C3	37, 50	8	4, 5
S. thermophilus, MC	37, 50	8	9, 10

Source: Spector, W.S., ed., *Handbook of Biological Data*, WADC Technical Report 56-273, Wright-Patterson Air Force Base, OH, 1956.

TABLE A.13

Respiration Rates for Algae

Species	Temperature (°C)	CO_2 Production Rate (μL/100 mg Dry Mass/h)	O_2 Consumption Rate (μL/100 mg Dry Mass/h)	CO_2 Production/ O_2 Consumption
Cyanophyta (blue green)				
Anabaena sp.	25	460		0.90
Nostoc commune	19	0.40		0.40
Chlorophyta (green)				
Chara vulgaris	18	1.5[a]		
Chlorella ellipsoidea	25		165	
C. pyrenoidea	20	89		0.89
C. vulgaris	20		475–192[b]	
Cladophora rupestris	20		33	
Coelastrum proboscideum	20		170	
Enteromorpha compressa	20	27		3.6
E. linza	19	66		0.62
Haematococcus pluvialis	20		180	
Nitella flexilis	18	1.6		
Scenedesmus obliquus	25		50	
Spirogyra majuscula	10.4		0.5[a]	
S. varians	10.4		0.6[a]	
Ulva lactuca	20	13–16		2.4–6.1
Valonia utricularis	20	8.4		1.5–5.7
Phaeophyta (brown)				
Ascophyllum nodosum	20	1.6[a]		0.80
Chorda tomentosa	9		74	
Cutleria multifida	20	7.2–17		0.5–2.1
Cystoseira abrotanifolia	20	4.5–10		1.2–3.7
C. amentacea	20	17		3.9
C. barbata	20	13–17		2.1–4.0
Desmarestia aculeata	14		24	
D. viridis	14		14[a]	
Dictyota dichotoma	20	9.4–9.2		0.98–1.04

(Continued)

TABLE A.13 (*Continued*)

Respiration Rates for Algae

Species	Temperature (°C)	CO_2 Production Rate (μL/100 mg Dry Mass/h)	O_2 Consumption Rate (μL/100 mg Dry Mass/h)	CO_2 Production/ O_2 Consumption
Ectocarpus siliculosus	12		41[a]	
Fucus serratus	18	18[a]		0.54
F. vesiculosus	17	11		0.60
Laminaria digitata	17		2[a]	
Taonia atomaria	20	6.7–20		0.9–3.1
Rhodophyta (red)				
Ceramium rubrum	17	45		0.89
Chondrus crispus	14		18	
C. crispus	20		28	
Cladostephus spongiosus	20		39	
Cryptonemia lomation	20	7.5–9.9		2.4–3.8
Delesseria alata	20		41	
Furcellaria fastigiata	14		7	
Gelidium corneum	20	13		3.26
Gracilaria compressa	20	9		1.4
Laurencia papillosa	20	18		4.88
Phyllophora nervosa	20	4.6		1.56
Plocamium coccineum	14		21	
Polyides lumbricoides	14		5	
Polysiphonia urceolata	12		10[a]	
P. violacea	11	107		1.02
Porphyra laciniata	17		39	

Source: Spector, W.S., ed., *Handbook of Biological Data*, WADC Technical Report 56-273, Wright-Patterson Air Force Base, OH, 1956.

O_2 consumption rate or CO_2 production rate differs from units indicated:

[a] μL/100 mg fresh mass/h.

[b] μL/10^9 cells/h.

TABLE A.14A

Respiration Rates for Lichens

Species	O_2 consumption Rate (μL/100 mg Dry Mass/h)		
	30°C	10°C	0°C
Alectoria nigricans	33	14	8
Cetraria chrysantha	19	9	3.9
C. glauca	61	31	10
C. islandica	48	19	8
Cladonia scholanderi	13	7.5	3.1
C. sylvatica	24	6.8	2.9
Cornicularia divergens	40	11	5
Lobaria linita	72	22	10
L. scrobiculata	50	29	12
Parmelia nigrociliata	25	13	4
Peltigera aphthosa	90	33	17
Ramalina alludens	13	3.3	2.2
Solorina crocea	43	24	10
Sticta laciniata	28	11	7
S. weigelii	40	14	6.7
Thamnolia vermicularis	28	14	4.2
Umbilicaria cinereorufescens	30	9.8	4.1
U. proboscidea	18	6.5	3.5

Source: Spector, W.S., ed., *Handbook of Biological Data*, WADC Technical Report 56-273, Wright-Patterson Air Force Base, OH, 1956.

TABLE A.14B

Respiration Rates for Lichens (cont.)

Species	Temperature (°C)	O_2 Consumption Rate (μL/100 mg Dry Mass/h)	CO_2 Production/ O_2 Consumption
Cladonia rangiferina	50	10	0.80
Evernia prunastri	50	40	0.78
	60	30	0.88
Orthotrichum affine	55	17	0.70
Pertusaria communis			0.84
Physica aipolia			0.73
P. ciliaris	45	18	0.80
Ramalina farinacea	50	25	0.77

Source: Spector, W.S., ed., *Handbook of Biological Data*, WADC Technical Report 56-273, Wright-Patterson Air Force Base, OH, 1956.

TABLE.A.15

Respiration Rates of Liverworts and Mosses

Species	Temperature (°C)	CO_2 Production Rate (μL/100 mg Dry Mass/h)	O_2 consumption Rate (μL/100 mg Dry Mass/h)
Liverworts (Hepaticae)			
Chiloscyphus fragilis	25		60–100
Marchantia polymorpha	20	0.6[a]	
Riccia fluitans	25		250–300
Mosses (Musci)			
Fontinalis antipyretica	25		70–140
Hylocomium parietinum	30	92	
	20	46	
	0	15	
H. proliferum	30	92	
	20	46	
	0	15	
H. squarrosum	30	100	
	20	61	
	0	15	
Hypnum cupressiforme	18.5		2–30
H. fluitans	18	0.83[b]	
Polytrichum juniperinum, shoot	18	1.2–0.7[b]	
Sphagnum girgensohnii	30	130	
	20	71	
	5	20	

Source: Spector, W.S., ed., *Handbook of Biological Data*, WADC Technical Report 56-273, Wright-Patterson Air Force Base, OH, 1956.

CO_2 production rate different from units indicated:

[a] μL/sq cm/h.

[b] μL/100 mg fresh mass/h.

TABLE A.16

Respiration Rates for Horsetails and Ferns

Species	Material	Temperature (°C)	CO_2 Production Rate (µL/100 mg Wet Mass/h)	O_2 Consumption Rate (µL/100 mg Wet Mass/h)	CO_2 Production/ O_2 Consumption
Horsetail (Equisetineae)					
Equisetum maximum	Shoot	20	6		0.78
	Fruiting shoot	20	100		0.83
	Stem	Room temperature	9.6		0.80
	Branchlet	Room temperature	19		0.69
Ferns (Filicineae)					
Asplenium adiantum-nigrum	Leaf	20	13		0.86
	Leaf with sori	20	17		1.01
	Leaf blade	Room temperature	13.4		0.80
	Petiole	Room temperature	8.3		0.80
Dryopteris austriaca	Leaf	48	122		
	Leaf	30	36		
	Leaf	10	25		
Eupteris aquiline	Leaf	48	168		
	Leaf	30	46		
	Leaf	10	15		
Polypodium virginiana	Leaf	20	10		0.92
	Leaf with sori	20	19		1.06
Pteris aquiline	Leaf	22	19		0.84
	Leaf with sori	22	35		1.01
Scolopendrium scolopendrium	Leaf	30		31	
	Leaf	22		17.5	
	Leaf	13		9.9	
	Leaf	3		2.2	

Source: Spector, W.S., ed., *Handbook of Biological Data*, WADC Technical Report 56-273, Wright-Patterson Air Force Base, OH, 1956.

TABLE A.17

Respiration Rates for Plant Storage Organs in Storage

Species	Material	Temperature (°C)	CO_2 Production Rate (μL/100mg Wet Mass/h)	O_2 Consumption Rate (μL/100 mg Wet Mass/h)	CO_2 Production Rate/ O_2 Consumption Rate
Arrowhead (*Sagittaria latifolia*)	Rhizome	25	4.1		
Artichoke (*Cynara scolymus*)	Tuber	25	1.4–0.8		
Beet (*Beta vulgaris*)	Root	15.5	0.8		
Bur-reed (*Sparganium eurycarpum*)	Rhizome	25		2.3	
Carrot (*Daucus carota*)	Root	24	3.3–1.5		1.10–1.18
		10	1.5–0.5		1.08–1.01
		0.5	0.4–0.2		0.92–1.16
Cattail (*Typha latifolia*)	Rhizome	25	2.4		
Dahlia (*Dahlia variabilis*)	Root	25			0.99
Gladiolus (*Gladiolus* sp.)	Corm	23	8.5[a]		
Milkweed (*Asclepias incarnata*)	Rhizome	25	3.7		
Onion (*Allium cepa*)	Bulb	21	0.7–1		
		10	0.4–0.5		
		0	0.1–0.2		
Oxalis (*Oxalis cenma cernua*)	Rhizome	Room temperature	5		1.18
Potato (*Solanum Solatium tuberosus*)	Tuber	24	0.6–0.3		1.02–0.75
		10	0.2–0.15		0.86–0.99
		0.5	0.07–0.15		0.45–0.66
Sedge (*Scirpus validus*)	Rhizome	25	2.8		

(Continued)

TABLE A.17 *(Continued)*

Respiration Rates for Plant Storage Organs in Storage

Species	Material	Temperature (°C)	CO_2 Production Rate (μL/100mg Wet Mass/h)	O_2 Consumption Rate (μL/100 mg Wet Mass/h)	CO_2 Production Rate/ O_2 Consumption Rate
Sweet flag (*Acorus calamus*)	Rhizome	25	3.7		
Sweet potato (*Ipomoea batatus*) Puerto Rico	Root	35	6.2		
		25	4.0		
		15	1.9		
Triumph	Root	35	5.6		
		25	3.2		
		15	1.4		
Turnip (*Brassica rapa*)	Root	15.5	1.6		
		4.5	0.1		
		0	0.02		
Waterlily (*Nuphar advenum*)	Rhizome	25	3.0		

Source: Spector, W.S., ed., *Handbook of Biological Data*, WADC Technical Report 56-273, Wright-Patterson Air Force Base, OH, 1956.
CO_2 production rate different from units indicated:
[a] μL/100 mg dry mass/h.

TABLE A.18
Respiration Rates for Fruits

Species	Temperature (°C)	CO_2 Production Rate (μL/100 mg Wet Mass/h)	CO_2 Production/ O_2 Consumption
Apple (*Pyrus malus*)			
Delicious	20	1.7–0.4–0.4	
Jonathan	27	7.2–4.5	0.45–0.92
Maiden blush	25	4.2–20	
	0	1.4–2.4	
Winesap	25	1.9–0.9	
	0	1.8–0.6	
Apricot (*Prunus armeniaca*)	18	2.8–4.1	
	4	1.1–1.0	
Avocado (*Persea gratissima*)	5	1	
	25	7–15	
Banana (*Musa paradisca sapientum*)	31	3.1	
	20	1.8	
	12.5	0.9	
	0	0.4	
Barberry (*Berberis vulgaris*)	25		1.20
Bean (*Phaseolus vulgaris*)	24	16.4–6.6	1.1–1.0
	0.5	0.95–0.6	0.94–0.96
Bryony (*Bryonia dioica*)	25	64–8.5	
Cherry (*Prunus avium*)	20	68–2	
Corn (*Zea mays*)	30	21–18	
Cranberry (*Vaccinium*)	24	32–14	
Cucumber (*Cucumis sativus*)	24	2.3–0.8	1.01–0.91
	10	1.0–0.4	1.01–1.10
	0.5	0.2–0.7	0.97–0.88
Elder (*Sambucus nigra*)	18	12	
Grape (*Vitis vinifera*)	28		1.6
Grapefruit (*Citrus grandis*)	38	2.5	2.1
	21	1.0	1.1
	10	0.4	1.4
	0	0.1	1.2
Guava (*Psidium guajava*)	30	20–3.6	
Hawthorn (*Crataegus punctata*)	28		1.26
Ivy (*Hedera helix*)	20	13–50–19	
Lemon (*Citrus limonia*)	38	4.1	1.4
	10	0.5	1.1
	0	0.15	1.2
Lilac (*Syringa vulgaris*)	25	42–8.5	
Oak (*Quercus alba*)	30	14.9[a]	0.71
	10	4.8[a]	0.30
	2.5	2.7[a]	0.16

(Continued)

TABLE A.18 (*Continued*)
Respiration Rates for Fruits

Species	Temperature (°C)	CO_2 Production Rate (μL/100 mg Wet Mass/h)	CO_2 Production/ O_2 Consumption
Oak (*Q. rubra*)	30	6.4[a]	0.46
	10	3[a]	0.13
	2.5	1.6[a]	0.08
Okra (*Hibiscus esculentus*)	30	306–104	
Orange (*Citrus nobilis*)	28		1.07
Orange (*C. sinensis*)			
Washington navel	21	2.0	1.1
	10	0.8	1.1
	0	0.2	1.2
Valencia	38		1.7
	21	1.8	1.0
	0	0.2	1.1
Papaya (*Carica papaya*)	15.6	0.83	
	10	0.46	
	4.4	0.24	
Pea (*Pisum sativum*)	24	20–12	1.32–1.06
	10	7.9–3.1	1.13–1.00
	0.5	2.2–1.4	1.00–0.96
Peach (*Prunus persica*)	18	1.4–2.0	
	4	0.4–0.3	
	25	7–5	
Peach (*Prunus persica*)	25	8–2	
Pear (*Pyrus communis*)	18	6.3–1–1.2	
	18	1–0.9–2.2	
Pepper (*Capsicum frutescens*)	24	4.0–1.4	1.12–0.88
	10	1.2–0.6	1.27–0.88
	0.5	0.4–0.3	0.96–0.96
Persimmon, Hachiya (*Diospyros kaki*)	20–27	1.8	1.2
Persimmon, Fuyu (*D. kaki*)	20–27	1.4	1.1
Pigeon pea (*Kajanus indicus*)	21	30–0	
Pimento (*Pimenta officinalis*)	30	40[a]	
Plum (*Prunus domestica*)	18	8.7	
Poppy (*Papaver somniferum*)	20	38	1.5
Rye (*Secale cereale*)	28	240–12[a]	
Smartweed (*Polygonum scandens*)	30		0.87
Snowberry (*Symphoricarpos ramosus*)	17	33	
Strawberry, missionary (*Fragaria* sp.)	20	3.3–5.1	0.84–0.91
Strawberry (*F. vesca*)	28		1.27
Tobacco (*Nicotiana tabacum*)	28		0.94

(*Continued*)

TABLE A.18 (*Continued*)

Respiration Rates for Fruits

Species	Temperature (°C)	CO_2 Production Rate (μL/100 mg Wet Mass/h)	CO_2 Production/ O_2 Consumption
Tomato (*Lycopersicon esculentum*)	24	2.5–1.6	1.11–1.13
	10	0.8–0.6	1.39–1.06
	0.5	0.4–0.2	1.11–1.02
	28	2.6–2.5	1.8–1.4
Wheat (*Triticum aestivum*)	28	340–8[a]	

Source: Spector, W.S., ed., *Handbook of Biological Data*, WADC Technical Report 56-273, Wright-Patterson Air Force Base, OH, 1956.

CO_2 production rate different from units indicated:

[a] μL/100 mg dry mass/h.

TABLE.A.19

Lung Ventilation for Resting Vertebrates

Species	Respiration Frequency (Breath/Min)	Tidal Volume (mL)	Minute Volume (L)
Man (Homo sapiens)			
Premature	33	12.4 (8.4–17.3)	0.41 (0.28–0.58)
Newborn, asleep	43 (24–116)	16.7 (10.0–27)	0.72 (0.43–1.41)
Adult male	11.7 (10.1–13.1)	750 (757–895)	7.4 (5.8–10.3)
Adult female	11.7 (10.4–13.0)	339 (285–393)	4.5 (4.0–7.0)
Cat (*Felis catus*)	26	12.4	0.32
Cow (*Bos taurus*)	31 (27–40)	2,850 (2,200–3,800)	86 (59–104)
Dog	18 (11–38)	320 (251–432)	5.2 (3.3–7.4)
Goat	19.0	310	5.7
Guinea pig (*Cavia cobaya*)	90 (69–104)	1.8 (1.0–3.9)	0.16 (0.09–0.38)
Hamster (*Mesocricetus auratus*)	74 (33–127)	0.83 (0.42–1.2)	0.054 (0.025–0.083)
Horse	11.9 (10.6–13.6)	9,000 (8,520–9,680)	107
Manatee, Florida (*Trichechus latirostris*)	7.0 (6.0–8.0)	(5,000–9,000)	45 (35–60)
Marmot (*Marmota marmota*)	8.0	22.0	0.17
Monkey (*Macaca mulatta*)	40 (31–52)	21 (9.8–21)	0.86 (0.31–1.41)
Mouse (*Mus musculus*)	163 (84–230)	0.15 (0.09–0.23)	0.023 (0.011–0.036)
Porpoise (*Tursiops truncatus*)	1.1 (0.9–1.3)	9,000 (8,000–10,000)	9.7 (9.0–10.4)
Rabbit (*Lepus cuniculus*)	51 (38–60)	21 (19.3–24.6)	1.07 (0.80–1.14)
Rat (*Rattus norvegicus*)		1.5 (1.4–1.6)	0.100 (0.075–0.130)
Rat, cotton (*Sigmodon hispidus*)	94 (75–115)	0.35 (0.24–0.70)	0.04 (0.023–0.071)
Sloth (*Choleopus hoffmani*)	13.0		0.84 (0.80–1.0)
Sloth (*Bradypus griseus*)	(4.5–8.0)		0.49 (0.33–0.73)
Turtle (*Maleclemys centrata*)	3.7	14.0	0.05

Source: Spector, W.S., ed., *Handbook of Biological Data*, WADC Technical Report 56-273, Wright-Patterson Air Force Base, OH, 1956.

TABLE A.20

Apparent Maximum Rates of Photosynthesis under Natural Conditions

Species	CO_2 Fixation (mg CO_2/100 sq cm/h)
Arctic zone	
Chamaenerium (*Chamaenerium latifolium*); 0°C	10
Cloudberry (*Rubus chamaemorus*)	8
Dock (*Rumex acetosella*); 10°C	12
Willow (*Salix glauca*); 10°C	4
Willow (*S. glauca*); 20°C	6
Temperate zone	
Anemone (*Anemone nemorosa*); in shade	6–11
Apple (*Pyrus malus*)	20
In shade	20
Corn (*Zea mays*)	10
Dock (*Rumex acetosella*)	9
Elder (*Sambucus nigra*)	4, 6
In shade	1, 7
Fern, polypody (*Polypodium virginiana*); in shade	5, 6
Horsebean (*Vicia faba*)	17
Mustard (*Sinapis alba*)	20–26
Oat (*Avena sativa*)	13
Pine (*Pinus taeda*)	14, 3
In shade	2, 5
Potato (*Solanum tuberosum*)	20
Rhododendron (*Rhododendron brachycarpum*); in shade	2, 8
Sunflower (*Helianthus annuus*)	5, 5–24
Tomato (*Lycopersicon esculentum*)	16, 8
Desert zone	
Atriplex (*Atriplex vesicarium*)	10
Date (*Phoenix dactylifera*)	3, 4
Grape (*Vitis vinifera*)	16, 1
Heliotrope (*Heliotropium arguzioides*)	27
Limoniastrum (*Limoniastrum feei*)	1, 3
Oleander (*Nerium oleander*)	10, 3
Tropical zone	
Calophyllum (*Calophyllum inophyllum*)	7, 3
Cassia (*Cassia fistula*)	10, 9
In shade	8, 6
Coconut (*Cocos nucifera*)	0, 9
Mango (*Mangifera indica*)	14, 8
Sugarcane (*Saccharum officinarum*)	5

(*Continued*)

TABLE A.20 (*Continued*)

Apparent Maximum Rates of Photosynthesis under Natural Conditions

Species	CO_2 Fixation (mg CO_2/100 sq cm/h)
Mountain zone	
Barley (*Hordeum sativum*)	≥30
Eurotia (*Eurotia ceratoides*)	≥44
Gentiana (*Gentiana algicola*)	≥100
Geum (*Geum montanum*)	48
Homogyne (*Homogyne alpina*)	43
In shade	18
Soldanella (*Soldanella alpina*)	39
In shade	19
Veronica (*Veronica bellidioides*)	65

Source: Spector, W.S., ed., *Handbook of Biological Data*, WADC Technical Report 56-273, Wright-Patterson Air Force Base, OH, 1956.

TABLE A.21

Apparent Maximum Rates of Photosynthesis for Near-Optimum Conditions

			CO_2 Fixation (g CO_2/h)		
Species	CO_2 in Air (%)	Temperature (°C)	Per 100 g Wet Mass	Per 100 g Dry Mass	Per sq dm × 1000
Basswood (*Tilia cordata*)	5	25	1.88	5.8	28
Elder (*Sambucus nigra*)					
Green leaves	5	25	1.96	5.3	34
Yellow leaves	5	25	0.88	4.7	18
Maple (*Acer pseudoplatanus*)					
Young leaves	5	25	0.98	3.0	16
Old leaves	5	25	2.07	5.8	26
Poplar (*Populus pyramidalis*)	5	25	1.90	6.0	40
Sunflower (*Helianthus annuus*)	5	25	2.30	13.4	80
Chlorella (*Chlorella pyrenoidosa*)					
In shade				11.5	
In light				13.4	
Sea lettuce (*Ulva lactuca*)		25			11.8
Sea weed (*Gigartina harveyana*)		16			14
Water net (*Hormidium flaccidum*)		20			

Source: Spector, W.S., ed., *Handbook of Biological Data*, WADC Technical Report 56-273, Wright-Patterson Air Force Base, OH, 1956.

TABLE A.22
Apparent Rates of Photosynthesis for Specified Conditions

Species	Temperature (°C)	Light (ft-c)	CO_2 in Air	Photosynthesis Rate[a]
Barley (*Hordeum vulgare*)	24	500	Natural conditions	9–16 mg CO_2/sq dm/h
Bean (*Phaseolus vulgaris*)	25	1,400	Natural conditions	5, 8–16.6 mg CO_2/sq dm/h
Dogwood (*Cornus florida*)	30	2,000	Natural conditions	2(3.1) mg CO_2/sq dm/h
Laurel (*Prunus laurocerasus*)	29.5	Noon sun	Natural conditions	23.2 mg CO_2/sq dm/h
Lemon (*Citrus limonia*)		1,300	1.5%	3–5 mL O_2/sq dm/h
Oak (*Quercus rubra*)	30	2,000	Natural conditions	5(6) mg CO_2/sq dm/h
Orange (*Citrus sinensis*)		1,300	1.5%	4–6 mL O_2/sq dm/h
Pelargonium (*Pelargonium zonale*)	24	500	Natural conditions	5.3 mg CO_2/sq dm/h
Pine (*Pinus taeda*)	30	2,000	Natural conditions	2(3.9) mg CO2/sq dm/h
Potato (*Solanum tuberosum*)	24	>5,000	Natural conditions	16–20 mg CO_2/sq dm/h
Spruce (*Picea pungens*)	24	2,200	Natural conditions	0.03 mg CO_2/100 leaves/h
Sugar cane (*Saccharum officinarum*)	36	Natural	Natural conditions	3–6 mg CO_2/sq dm/h
Sunflower (*Helianthus annuus*)		4,460	5%	(80) mg CO2/sq dm/h
Sphagnum (*Sphagnum girgensohnii*)		110–260		2.8 mg CO_2/g dry mass/h
Chlorella (*Chlorella vulgaris viridis*)	22.4	26,700 lux	Carbonate buffer 9	195 cu mm O_2/100 mil cells/h
Chlorella (*C. saccharophila*)	22.4	26,700 lux	Carbonate buffer 9	452 cu mm O_2/100 mil cells/h
Kelp (*Macrocystis pyrifera*)		Low	Sea water	17.5 mL O_2/sq dm/h

Source: Spector, W.S., ed., *Handbook of Biological Data*, WADC Technical Report 56-273, Wright-Patterson Air Force Base, OH, 1956.

[a] Values in parentheses are maximum rates; all others, average rates.

TABLE A.23

Estimated Annual Carbon Production

Region	Area (sq km)	Carbon Fixed (Ton/Year)	
		Per sq km	Total
Forest	44×10^6	250	11×10^9
Cultivated land	27×10^6	149	4.3×10^9
Grassland	31×10^6	43	1.1×10^9
Desert	47×10^6	7	0.2×10^9
Total land	149×10^6		16.6×10^9
Ocean	361×10^6	340	16.6×10^9

Source: Spector, W.S., ed., *Handbook of Biological Data*, WADC Technical Report 56-273, Wright-Patterson Air Force Base, OH, 1956.

TABLE A.24

Efficiency of Photosynthesis

Specification	Value
Energy utilized in photosynthesis by one acre of corn plants in synthesis of 8732 kg glucose (total sugar, as glucose, manufactured by one acre of corn plants)	3.3×10^7 kcal
Total solar energy available on the acre during growing season	2.043×10^9 kcal
Photosynthetic efficiency of corn plants, i.e., percent of available energy used in photosynthesis	1.6%
Energy equivalent of earth's carbon production	$(13.6 \pm 8.1) \times 10^{17}$ kcal
Mean solar radiation	7.4×10^{20} kcal
Photosynthetic efficiency of the world	0.18 (±0.12)%

Source: Spector, W.S., ed., *Handbook of Biological Data*, WADC Technical Report 56-273, Wright-Patterson Air Force Base, OH, 1956.

TABLE A.25

Typical Unstressed Heart Rates

Animal	Beats/min
Vertebrates	
Mammalia	
Man (*Homo sapiens*)	
Embryo, 5 months	156 (150–160)
6 months	154 (141–155)
7 months	149 (118–156)
8 months	142 (129–152)
9 months	146 (131–173)
Newborn, premature	(110–185)
Newborn	134 (101–160)
2 years	108 (84–134)
4 years	103 (80–133)
6 years	93 (72–128)
8 years	89 (72–114)
10 years	87 (56–106)
15 years	83 (66–112)
20 years	71 (59–99)
20–24 years	74 (41–100)
25–30 years	72 (52–102)
30–35 years	70 (58–104)
35–40 years	72 (56–100)
40–45 years	72 (50–104)
45–50 years	72 (49–100)
50–55 years	72 (52–94)
55–60 years	75 (48–108)
60–65 years	73 (54–100)
65–70 years	75 (52–96)
70–75 years	75 (54–104)
75–80 years	72 (50–94)
80 and over	78 (63–98)
Recumbent	66 (40–100)
Sitting	73 (31–110)
Standing	82 (54–124)
Sleeping	Male 59, female 65
Waking, male	78 (61–119)
Waking, female	84 (67–121)
Ass (*Equus asinus*)	50 (40–56)
Badger (*Taxidea taxus*)	138 (128–144)
Bat (*Plecotus auritus*)	750 (100–970)
Beaver (*Castor canadensis*)	140
Camel (*Camelus bactrianus*)	30 (25–32)

(*Continued*)

TABLE A.25 (*Continued*)
Typical Unstressed Heart Rates

Animal	Beats/min
Cat (*Felis catus*)	120 (110–140)
Newborn	(168–300)
Cattle (*Bos taurus*)	70 (40–100)
Chipmunk (*Eutamias minimas*)	684 (660–702)
Dog (*Canis familiaris*)	(100–130)
Newborn	(168–180)
Dolphin (*Delphinus* spp.)	150
Dormouse (*Muscardinus* sp.)	646 (580–780)
Elephant (*Elephas maximus*)	35 (22–53)
Ferret (*Mustela vison*)	231 (216–242)
Giraffe (*Giraffa camelopardalis*)	66
Goat (*Capra hircus*)	90 (70–135)
Newborn	(145–240)
Guinea pig (*Cavia porcellus*)	280 (260–400)
Hamster (*Cricetus cricetus*)	450 (300–600)
Hedgehog (*Erinaceus europaeus*)	300 (189–320)
Hibernating	(3–15)
Horse (*Equus caballus*)	44 (23–70)
Newborn	(100–120)
Hyena (*Hyaena hyaena*)	(55–58)
Lemming (*Dicrostonyx rubricatus*)	416 (348–465)
Lion (*Panthera leo*)	40
Manatee (*Trichechus* spp.)	(50–60)
Marmot (*Marmota marmota*)	180 (120–206)
Monkey (*Macaca mulatta*)	192 (165–240)
Mouse (*Mus musculus*)	600 (328–780)
Mouse, deer (*Peromyscus* sp.)	534 (324–858)
Opossum (*Didelphis virginiana*)	(120–240)
Panther (*Felis concolor*)	60
Porcupine (*Erethizon dorsatus*)	(280–320)
Porpoise (*Tursiops truncates*)	150
Rabbit (*Oryctolagus cuniculus*)	205 (123–304)
Rat (*Rattus* spp.)	328 (261–600)
Newborn	161 (81–241)
Seal (*Phoca vitulina*)	100 (10–140)
Sheep (*Ovis aries*)	75 (60–120)
Shrew (*Blarina brevicauda*)	699 (618–780)
Shrew (*Sorex cinereus*)	782 (588–1320)
Skunk (*Mephitis mephitis*)	166 (144–192)
Squirrel (*Sciurus vulgaris*)	354 (320–372)
Squirrel (*Citellus* spp.)	249 (96–378)
Swine (*Sus scrofa*)	(55–86)
Newborn	227

(*Continued*)

TABLE A.25 (*Continued*)

Typical Unstressed Heart Rates

Animal	Beats/min
Tapir (*Tapirus indicus*)	44
Tiger (*Panthera tigris*)	64
Wallaby (*Macropus* spp.)	125
Weasel (*Mustela frenata*)	182 (172–192)
Weasel, shorttail (*M. erminea*)	357 (300–420)
Whale (*Beluga* spp.)	16 (12–23)
Aves	
Blackbird (*Turdus merula*)	(390–590)
Bramblefinch (*Fringilla* sp.)	(900–920)
Buzzard (*Buteo* spp.)	300 (206–351)
Canary (*Serinus canarius*)	(514–1000)
Cassowary (*Casuarius galeatus*)	70
Catbird (*Dumetella* sp.)	330 (318–354)
Cardinal (*Richmondena* sp.)	(375–800)
Chaffinch (*Fringilla coelebs*)	700
Chickadee (*Parus atricapillus*)	(480–1000)
Cowbird (*Molothrus* sp.)	(315–779)
Crane (*Arthropoides paradisea*)	120
Crow (*Corvus cornix*)	378 (312–492)
Dove (*Columba* spp.)	282 (185–300)
Duck (*Anas* spp.)	268 (212–317)
Falcon (*Falco senchris*)	367
Finch (*Carduelis elegans*)	920 (914–925)
Fowl (*Gallus* spp.)	312 (178–458)
Goose (*Anser a. domesticus*)	(80–144)
Goshawk (*Accipiter gentiles*)	347
Greenfinch (*Chloris hortensis*)	740 (703–848)
Gull (*Larus canus*)	401 (360–483)
Hawk (*Astur palumbarius*)	347
Hummingbird (*Archilochus colubris*)	615
Jackdaw (*Corvus monedula*)	342 (326–358)
Kestrel (*Tinnunculus alaudarius*)	367
Kingfisher (*Alcedo ispida*)	440
Kite (*Milvus* spp.)	258
Ostrich (*Struthio camelus*)	(60–70)
Parrot (*Psittacus erithacus*)	320
Penguin (*Aptenodytes* spp.)	240
Pigeon (*Columba* spp.)	170 (141–244)
Redstart (*Ruticilla phoenicurus*)	890
Robin (*Turdus migratorius*)	570 (520–620)
Rook (*Corvus frugilegus*)	380 (352–440)
Sparrow (*Passer domesticus*)	804 (745–850)

(*Continued*)

TABLE A.25 (*Continued*)

Typical Unstressed Heart Rates

Animal	Beats/min
Sparrow, song (*Melospiza* sp.)	(450–1020)
Starling (*Sturnus vulgaris*)	388 (375–400)
Stork (*Ciconia* sp.)	161
Swan (*Sthenelides olor*)	257
Thrasher, brown (*Toxostoma* sp.)	278 (270–294)
Titmouse (*Parus major*)	870
Towhee (*Pipilo erythrophthalmus*)	(445–810)
Turkey (*Meleagris gallopavo*)	211 (93–330)
Vulture (*Gyps fulvus*)	199
Wren (*Troglodytes aedon*)	(450–950)
Reptilia	
Adder (*Bitis* sp.)	40
Adder (*Tropidonotus natrix*)	(23–68)
Blindworm (*Anguis fragilis*)	64
Crocodile (*Crocodilus acutus*)	(10–70)
Lizard (*Lacerta viridis*)	64 (60–66)
Ringsnake (*Coluber natrix*)	(23–41)
Tortoise (*Testudo* spp.)	(11–60)
Turtle (*Emys orbicularis*)	(6–9)
Turtle (*Terrapene* spp.)	(6–70)
Amphibia	
Frog (*Rana pipiens*)	(4–50)
Salamander (*Salamandra* spp.)	(30–40)
Toad (*Bufo* spp.)	(40–50)
Pisces	
Barbel (*Barbus fluviatilis*)	(35–90)
Bass (*Micropterus salmoides*)	20 (5–50)
Blenny (*Zoarces viviparous*)	(71–86)
Bullhead (*Ameiurus* sp.)	22 (5–50)
Carp (*Cyprinus* spp.)	(40–78)
Chub (*Leuciscus dobula*)	18
Cod (*Gadus morrhua*)	(26–40)
Dragonet (*Callionymus lyra*)	(60–84)
Eel (*Anguilla* spp.)	(39–68)
Eel, marine (*Conger conger*)	(33–50)
Goldfish (*Carassius auratus*)	(36–40)
Gurnard (*Trigla hirundo*)	(62–86)
Haddock (*Melanogrammus* sp.)	(30–40)
Perch (*Perca fluviatilis*)	(52–66)
Pike (*Esox lucius*)	(30–54)

(*Continued*)

TABLE A.25 (*Continued*)
Typical Unstressed Heart Rates

Animal	Beats/min
Plaice (*Pleuronectes platessa*)	(54–76)
Pogge (*Agonus cataphractus*)	(81–90)
Ray (*Tetranarce* spp.)	(16–50)
Rockling (*Montella mustela*)	(64–82)
Scorpion fish (*Scorpaena scrofa*)	(11–24)
Sculpia (*Cottus scorpius*)	(55–74)
Shad, gizzard (*Dorosoma* sp.)	20 (5–50)
Shark (*Carcharodon carcharias*)	(18–30)
Shark (*Squalus acanthias*)	(16–50)
Skate (*Raja* spp.)	(16–50)
Stickleback (*Gasterosteus* sp.)	(60–100)
Tench (*Tinca tinca*)	(31–42)
Trout (*Salmo trutta*)	(30–46)
Wrasse (*Labrus mixtus*)	(40–81)
Tunicates	
Sea squirt (*Molgula* sp.)	(43–80)
(*Appendicularis* sp.)	250
(*Ascidia depressa*)	(31–33)
(*A. mentula*)	(16–20)
(*Ciona intestinalis*)	(17–32)
(*Clavellina lepardiformis*)	(23–50)
(*Cyclosalpa pinnata*)	(26–30)
(*Perophora annectens*)	43
(*Phallusia mammillata*)	(9–12)
(*Pyrosoma giganteum*)	(38–58)
(*Salpa bicaudata*)	(13–40)
(*S. fusiformis*)	(41–69)
Invertebrates	
Arthropoda	
Aphid (*Aphis* sp.)	74 (66–80)
Beetle (*Chrysopa* sp.)	(53–63)
Botfly, larva (*Gastrophilus equi*)	(40–44)
Cockroach (*Periplaneta* sp.)	(60–90)
Crab (*Limulus polyphemus*)	20 (8–28)
Arthropoda	
Crab (*Maia* sp.)	(25–46)
Crab, cocoa-nut, larva (*Porcellana* sp.)	170
Crayfish (*Astacus marinus*)	50 (30–87)
Crayfish (*Cambarus clarkii*)	116 (75–136)
Daphnia (*Daphnia pulex*)	(140–166)

(*Continued*)

TABLE A.25 (*Continued*)

Typical Unstressed Heart Rates

Animal	Beats/min
Lobster (*Homarus gammarus*)	60 (50–100)
Lobster, rock (*Palinurus* sp.)	35 (30–50)
Moth (*Cossus cossus*)	15
Moth, larva (*Sphinx ligustri*)	61 (39–82)
Shrimp (*Mysis* sp.)	260 (140–320)
Shrimp (*Lysmata seticaudata*)	175 (50–200)
Silkworm, larva (*Bombyx mori*)	(30–40)
Spider (*Eperia diadema*)	132 (130–134)
Mollusca	
Chiton (*Cryptochiton* sp.)	(5–7)
Chiton (*Ischnochiton* sp.)	(12–25)
Clam (*Dreissena polymorpha*)	(30–60)
Clam (*Pisidium* sp.)	(60–75)
Clam (*Mya arenaria*)	(5–14)
Cuttlefish (*Sepia officinalis*)	(18–40)
Cuttlefish (*Loligo* spp.)	(60–80)
Mussel (*Anodonta cygnea*)	(2–29)
Mussel (*Mytilus edulis*)	(15–25)
Octopus (*Octopus vulgaris*)	(12–59)
Oyster (*Ostrea edulis*)	(25–30)
Sea hare (*Aplysia* sp.)	(8–35)
Sea hare (*Pterotrachea* sp.)	67 (50–80)
Scallop (*Pecten jacobeus*)	(22–50)
Slug (*Ariolimax agrestris*)	(20–40)
Snail (*Helix pomatia*)	(10–60)
Snail (*Haliotis tuberculata*)	(40–45)
Snail (*Valvata piscinalis*)	100
Snail (*Natica* sp.)	(5–7)
Snail (*Bulla* sp.)	(10–18)
Snail (*Limnaea auricularis*)	(42–100)
Snail, land (*Helix hortensis*)	40
Whelk (*Sycotypus* sp.)	(5–8)
Miscellaneous	
Clamworm (*Nereis virens*)	8
Earthworm (*Lumbricus terrestris*)	17 (15–20)
Leech (*Hirundo medicinalis*)	6
Lugworm (*Arenicola* sp.)	7 (6–8)

Source: Spector, W.S., ed., *Handbook of Biological Data*, WADC Technical Report 56-273, Wright-Patterson Air Force Base, OH, 1956.

Note: Values in parentheses are estimates of the range of 95% included heart rates.

TABLE A.26

Cardiac Output for Vertebrates under Various Conditions

Species	Body Mass (kg)	Surface Area (sq m)	Physiological State	Heart Rate (Beats/s)	Stroke Volume (mL/Beat)	O_2 Consumption (L/min)	Cardiac Output (L/min)	Cardiac Index (L/sq m/min)
Man	18.0	0.82	Basal, age 5	96	23.8		2.28	2.78
	31.2	1.19	Basal, age 10½	90	36.3		3.27	2.75
	66.0	1.79	Basal, age 16	60	86.2		5.18	2.90
	68.5	1.85	Basal, age 25	65	78.4		5.10	2.76
	70.0	1.83	Basal, age 33	68	62.8		4.27	2.33
	72.4	1.85	Basal, age 47	72	55.8		4.02	2.17
	69.8	1.79	Basal, age 60	80	57.0		4.55	2.54
	65	1.78	Sitting			0.285 (0.256–0.309)	6.6 (4.7–8.0)	3.7
			Standing		29.2 (21.9–35.2)		2.21 (1.61–2.85)	
		1.84	Tilting to 70°			0.248 (0.220–0.272)	4.86 (3.94–6.20)	2.7 (2.0–3.9)
			Acute hypoxia, 760 mmHg	59.8	72.7		4.35	
			Acute hypoxia, 600 mmHg	66.8	76.0		5.08	
			Acute hypoxia, 376 mmHg	75.9	81.3		6.17	
			Acute hypoxia, 260 mmHg	85.7	105.9		9.08	
			Hemorrhagic shock	113	31	0.154 (0.119–0.223)	3.5	2.20 (1.38–2.99)
			Vagotonia	40	42	0.204	1.7	
			Sympatheticotonia	106	110	0.327	9.1	
	65	1.78	Bicycle ergometer 102 kg m/min			0.630 (0.621–0.638)	8.8	4.91
	68.1	1.81	Basal, age 23	63.3 (49–80)	54.9 (29–72)		3.43 (2.0–5.3)	1.90

(*Continued*)

TABLE A.26 (*Continued*)
Cardiac Output for Vertebrates under Various Conditions

Species	Body Mass (kg)	Surface Area (sq m)	Physiological State	Heart Rate (Beats/s)	Stroke Volume (mL/Beat)	O_2 Consumption (L/min)	Cardiac Output (L/min)	Cardiac Index (L/sq m/min)
			Basal, 150 kg m/min	77.5 (51–99)	78.0 (54–101)		6.08 (3.3–8.6)	3.36
			Basal, 300 kg m/min	86.3 (59–113)	88.4 (68–119)		7.62 (4.6–10.2)	4.21
			Basal, 450 kg m/min	97.8 (69–127)	108.1 (77–170)		10.62 (5.9–17.9)	5.87
	64	1.75	Basal, 840 kg m/min			2.08 (2.05–2.10)	19.2 (19.0–19.4)	11.0
	75	1.98	Basal, 1200 kg m/min			2.92 (2.83–2.98)	23.9 (23.0–25.1)	12.1
	75	1.98	Basal, 1680 kg m/min			3.84 (3.79–3.94)	33.8 (28.9–37.3)	17.1
			Treadmill, 6 MPH, 0°			2.41 (2.120–2.900)	21.6 (18.3–27.5)	
			Treadmill, 7 MPH, 0°			2.657 (2.530–3.530)	27.0 (22.0–33.0)	
Bass				20		(0.05–0.09)		
Cat	2.5		Anesthetized		2.0–2.4		0.271	1.55 (0.36–2.74)
	4.1		Basal	178.8			(0.168–0.352)	
Catfish				19	3.15	0.56		
Chick embryo					0.12–0.2			
Cow				60			(0.3–1.0)	
Dog	6.4	0.39	Basal		244		14.6	
	14.4	0.66	Anesthetized			0.057 (0.042–0.068)	1.12 (0.65–1.57)	2.9
	23.9	0.93	Anesthetized				1.82 (1.14–2.50)	2.8 (1.97–4.18)
Dog	11.8		Basal	170.5	6.91 (4.42–9.74)		2.66 (2.00–3.32)	2.9
				(123.3–226.0)			1.18 (0.80–1.59)	
	16.1	0.71	Basal			0.106 (0.063–0.184)	2.21 (1.20–3.84)	3.1
	15	0.68	Shivering			0.234 (0.232–0.236)	4.49 (3.66–5.31)	6.6
	21.6	0.87	Treadmill, 3 MPH, 0°			0.459 (0.393–0.616)	5.30 (4.2–7.8)	6.1

(*Continued*)

TABLE A.26 (*Continued*)

Cardiac Output for Vertebrates under Various Conditions

Species	Body Mass (kg)	Surface Area (sq m)	Physiological State	Heart Rate (Beats/s)	Stroke Volume (mL/Beat)	O_2 Consumption (L/min)	Cardiac Output (L/min)	Cardiac Index (L/sq m/min)
	21.5	0.87	Standing			0.210 (0.193–0.238)	3.9 (3.6–4.1)	4.5
	21.5	0.87	Treadmill, 3 MPH, 5°			0.609 (0.598–0.620)	5.8 (5.7–5.9)	6.7
	21.5	0.87	Treadmill, 5 MPH, 10°			1.402 (1.380–1.420)	12.15 (12.1–12.2)	14.0
Dogfish				36	(0.4–1.5)		(0.014–0.054)	
Ferret				227			0.139 (0.082–0.20)	
Goat, kid			Basal				(0.13–0.20)	
Goat, mature	23.7	0.91	Basal			0.176 (0.078–0.329)	3.1 (1.37–5.60)	3.4
Horse	283	4.30	Standing			1.364	18.8	4.4
	283	4.30	Treadmill, 47.6 m/min, 0°			2.965	31.4	7.3
	342	4.90	Treadmill, 56.9 m/min, 6.5°			4.414 (4.315–4.513)	53.1 (46.6–59.5)	10.8
	342	4.90	Standing			2.480	24.0	4.9
Monkey, rhesus				120	8.8		1.06	
Rabbit, white	3.2 4.13	0.21	Anesthetized Basal	214.5 (169.5–288.5)	214.5 (169.5–288.5)	0.021 (0.014–0.028)	0.35 (0.26–0.48) 0.53 (0.25–0.75)	1.7
Rat	0.18	0.03	Anesthetized		1.3–2.0		0.047 (0.015–0.079)	1.6

Source: Spector, W.S., ed., *Handbook of Biological Data*, WADC Technical Report 56-273, Wright-Patterson Air Force Base, OH, 1956.

TABLE A.27
Gestation Periods of Mammals

Species	Breeding Season	Gestation Period (Days)	Litter Size	Litter No./Year
Man (*Homo sapiens*)	All year	267	1, 2, 3	
Antelope (*Hippotragus niger*)	All year	(270–281)		
Ape (*Macaca sylvanus*)		210		
Armadillo (*Dasypus novemcinctus*)	Summer	(150–240)	4	1
Ass (*Equus asinus*)	March–August	365	1	
Baboon (*Papio porcarius*)	All year	210	1	
Badger (*Taxidea taxus*)	August–September	(183–265)	3 (1–7)	
Bat (*Eptesicus fuscus*)	Fall	35	2 (1–4)	1–3
Bat, vampire (*Desmodus rotundus*)	All year		1	1+
Bear (*Ursus horribilis*)	June–July	208	2 (1–4)	
Beaver (*Castor canadensis*)	January–February	120	4 (1–6)	
Bobcat (*Lynx rufus*)	Late February	50	3 (1–4)	
Buffalo (*Bison bison*)	All year	275	1 (1–2)	1
Camel (*Camelus* spp.)	All year	(315–410)	1	
Cat (*Felis catus*)	February–July	63	4	
Cattle (*Bos taurus*)	All year	281	1	1
Chinchilla (*Chinchilla laniger*)	All year	(105–115)		(1–8)
Chimpanzee (*Pan troglodytes*)	All year	237	1	
Chipmunk (*Tamias striatus*)	March–July	31	(3–6)	2
Deer (*Cervus elaphus*)	September–October	234	1	
Dog (*Canis familiaris*)	Spring–Fall	63	7 (1–22)	
Eland (*Taurotragus oryx*)	All year	260	1	
Elephant (*Elephas maximus*)		624	1	
Elk (*Alces alces*)	September–October	(240–250)	1	
Ferret (*Mustela putorius*)	March–August	(42–45)	8.5	(5–13)
Fox (*Vulpes fulva*)	December–March	52	(4–9)	
Gibbon (*Hylobates lar*)	All year	210	1	
Giraffe (*Giraffa camelopardalis*)	July–September	(400–480)	1	
Gnu (*Connochaetes taurinus*)	June	(240–270)		
Goat (*Capra hircus*)	September–Winter	148	(1–5)	
Gopher (*Geomys breviceps*)	February–August	(40–50)	3 (1–9)	1–3
Guinea pig (*Cavia porcellus*)	All year	68	3 (1–8)	
Hamster (*Mesocricetus auratus*)	All year	16	(1–12)	3
Hare (*Lepus americanus*)	March–August	30	3 (1–7)	1
Hedgehog (*Erinaceus europaeus*)	March–September	(35–49)	5 (3–7)	1
Hippopotamus (*Hippopotamus amphibius*)	All year	231	1	
Horse (*Equus caballus*)	All year	336	1	
Hyena (*Crocuta crocuta*)	All year	(91–110)	1	
Jaguar (*Panthera onca*)	September–October	(93–110)	(2–4)	

(*Continued*)

TABLE A.27 (*Continued*)

Gestation Periods of Mammals

Species	Breeding Season	Gestation Period (Days)	Litter	
			Size	No./Year
Kangaroo (*Macropus rufus*)	Once a year	(38–40)	1	
Leopard (*Panthera pardus*)	All year	(92–105)	3 (1–4)	
Lion (*P. leo*)	All year	(105–113)	(1–6)	
Lynx (*Lynx canadensis*)	Early March	60	(1–5)	
Marmoset (*Hapale jacchus*)		146	2 (1–3)	
Marten (*Martes foina*)	July–August	(255–285)	(3–5)	
Monkey (*Macaca mulatta*)	All year	164	1	
Mouse (*Mus musculus*)	All year	19	6 (1–12)	(4–6)
Muskrat (*Ondatra zibethica*)	April–October	30	7 (1–11)	2
Nutria (*Myocaster coypus*)	All year	(120–150)	10	
Otter (*Lutra canadensis*)	February	60	(1–4)	
Panther (*Felis concolor*)	All year	(90–93)	3	(1–2)
Swine (*Sus scrofa*)	All year	114	9 (6–15)	
Platypus (*Ornithorhynchus paradoxus*)	July–October		2	
Porcupine (*Erethizon dorsatus*)	November–December	112	1 (1–4)	
Rabbit (*Oryctolagus cuniculus*)	All year	31	8 (1–13)	
Raccoon (*Procyon lotor*)	January–June	63	4 (1–6)	
Rat (*Rattus norvegicus*)	All year	21	12 (4–20)	2
Reindeer (*Rangifer tarandus*)	September–October	230	1	
Rhinoceros (*Didermocerus sumatrensis*)	July–October	210	1	
Sable (*Martes zibellina*)	June–August	(270–285)	(1–4)	
Sea lion (*Zalophus californianus*)	June–July	(348–365)	1	
Seal (*Arctocephalus pusillus*)	October–December	(330–360)	2 (1–2)	
Sheep (*Ovis aries*)	Fall	151	(1–4)	
Shrew (*Sorex araneus*)	March–September	(35–49)	5 (3–7)	1
Skunk (*Mephitis mephitis*)	March	62	(3–8)	
Squirrel (*Sciurus carolinensis*)	December–August	44	4 (1–6)	(1–2)
Tapir (*Tapirus terrestris*)	Seasonal	397		
Tiger (*Panthera tigris*)	All year	(105–112)	3 (1–6)	
Vole (*Microtus agrestis*)	February–October	21	4	
Walrus (*Odobenus rosmarus*)	April–July	365		
Wapiti (*Cervus canadensis*)	All year	255	1	1
Waterbuck (*Kobus ellipsiprymnus*)	May–July	240	1	
Whale (*Balaenoptera borealis*)	May–August	360	1	
Wolf (*Canis lupus*)	December–April	63	(3–9)	2
Woodchuck (*Marmota monax*)	March–April	28	4	
Woodrat (*Neotoma micropus*)	April–December	33	2 (2–3)	
Zebra (*Equus quagga*)	March–November	(340–365)	1	

Source: Spector, W.S., ed., *Handbook of Biological Data*, WADC Technical Report 56-273, Wright-Patterson Air Force Base, OH, 1956.

TABLE A.28
Breeding Habits of Aquatic Invertebrates

Species and Location	Sexual Maturity Age	Sexual Maturity Size (mm)	Eggs or Young per Brood	Breeding Season
Xiphosura				
Crab, horseshoe (*Limulus polyphemus*), Delaware Bay area	9–11 year	m 178–258	Few to 1000	April–July
Crustacea				
Crab, blue (*Callinectes sapidus*), Chesapeake Bay area	m, f 13 month	m 135–215	1,750,000	July–August
Crayfish (*Orconectes immunis*), New York State area	m, f 15 month	m 40–60	102 (84–195)	June–October
Cyclops (*Cyclops viridis*), Germany	m 41–132 day	f 1.5–5	75 (20–160)	All year
Lobster (*Homarus americanus*), Delaware–Newfoundland	4–5 year	m 17–60	8500	July–September
Waterflea (*Daphnia longispina*), Northeastern United States	f 75–86 h	m 1.2, f 1.9	28 (4–35)	All year
Mollusca				
Chilton, gray (*Ischnochiton magdalenensis*), California–Mexico	2 year	35–36	57,970	
Clam, hard shell (*Venus mercenaria*), Baja California	1–2 year	5–7		July–August
Clam, pismo (*Tivela stultorum*), Pacific area	5 year	10–12	750,000	
Clam, razor (*Siliqua patula*), eastern Pacific area	2–4.2 year	10–14		
Clam, soft shell (*Mya arenaria*),	1–2 year			May–August
Arctic–N. Carolina; Pacific Drill, oyster (*Urosalpinx cinerea*), Delaware Bay area	15–25 month	15–24	300–960	April–November
Mussel (*Mytilus edulis*), cosmopolitan	1–2 year			May–September
Nudibranch, red (*Rostanga pulchra*), California			9–156	December–February
Ormer (*Haliotis tuberculata*), Channel Islands–Europe	3 year	5	10,000	July–September
Oyster (*Crassostrea virginica*), Delaware Bay area	1 year	25–50	½–1 million	June–August
Periwinkle (*Litorina littorea*), North Atlantic–Florida			1–3	
Scallop, bay (*Aequipecten irradians*), western Atlantic	12 month	78		
Scallop, giant (*Placopecten magellanicus*), North Atlantic	3–4 year	50–70		June–October

(*Continued*)

TABLE A.28 (*Continued*)

Breeding Habits of Aquatic Invertebrates

Species and Location	Sexual Maturity Age	Sexual Maturity Size (mm)	Eggs or Young per Brood	Breeding Season
Snail, burrowing (*Polynices duplicatus*), northern United States–Mexico	2 year	>12		June–August
Snail, edible (*Helix pomatia*), Europe	33–39 month		40–200	May–July
Snail, pond (*Lymnaea stagnalis*), Wisconsin	4–14 month	50–60	6,000	July–October
Top–shell, great (*Trochus niloticus*), Indo–Pacific	2 year	6–7		
Whelk, channeled (*Busycon canaliculatum*), Cape Cod–Mexico			360–6240	
Echinodermata	1–2 year	60–210	Thousands	June–August
Starfish (*Asterias forbesi*), Long Island Sound				

TABLE A.29

Incubation and Care of Young in Birds

Order, Family	Common Name	Incubation Duration (Days)	Time from Hatching until Birds Leave Nest (Days)
Struthioniformes Struthionidae	Ostriches	42	0
Rheiformes Rheidae	Rheas	35–42	0
Casuariiformes Dromicelidae	Emus, cassowaries	58–61	0
Apterygiformes Apterygidae	Kiwis	75	6
Tinamiformes Tinamidae	Tinamous	21	0
Sphenisciformes Spheniscidae	Penguins	38–56	56–112
Gaviiformes Gaviidae	Loons	28–30	0
Colymbiformes Colymbidae	Grebes	21–27	0
Procellariiformes			
Diomedeidae	Albatrosses	63–80	150–251
Procellariidae	Fulmars, shearwaters	51–58	49–95
Hydrobatidae	Storm petrels	38–50	56
Pelecanoididae	Diving petrels	56	54

(*Continued*)

TABLE A.29 (*Continued*)
Incubation and Care of Young in Birds

Order, Family	Common Name	Incubation Duration (Days)	Time from Hatching until Birds Leave Nest (Days)
Pelecaniformes			
Phaethontidae	Tropic–birds	28	62
Pelecanidae	Pelicans	28–42	14–35
Sulidae	Boobies	42–45	45+
Phalacrocoracidae	Cormorants	24–25	35–42
Ciconiiformes			
Ardeidae	Herons, bitterns	18–28	10–52
Scopidae	Hammerheads	21	42
Ciconiidae	Storks	30–38	63
Threskiornithidae	Ibises	21–24	42+
Phoenicopteridae	Flamingos	30–32	3–4
Anseriformes			
Anhimidae	Screamers	42	0
Anatidae	Swans, geese, ducks	21–35	0
Falconiformes			
Cathartidae	Vultures	39–56	56–70
Accipitridae	Hawks	28–56	28–133
Falconidae	Falcons	28–29	25–35
Galliformes			
Megapodiidae	Megapodes	57–70	0
Cracidae	Chachalacas	22–24	0
Tetraonidae	Grouse, ptarmigans	21–27	0
Phasianidae	Quail, pheasants	21–28	0
Meleagrididae	Turkeys	28	0
Gruiformes			
Mesoenatidae	Monias		
Turnicidae	Bustard–quail	12–13	0
Gruidae	Cranes	29–32	0
Rallidae	Rails	19–24	0–2
Rhynochetidae	Kagus	36	
Eurypygidae	Sun-bitterns	27	21
Otididae	Bustards	20–25	0
Charadriiformes			
Jacanidae	Jacanas	23	
Haematopodidae	Oyster-catchers	24–27	0
Charadriidae	Plovers	23–28	0
Scolopacidae	Sandpipers	18–29	0
Recurvirostridae	Avocets	23	0
Phalaropodidae	Phalaropes	20–21	0

(*Continued*)

TABLE A.29 (*Continued*)
Incubation and Care of Young in Birds

Order, Family	Common Name	Incubation Duration (Days)	Time from Hatching until Birds Leave Nest (Days)
Burhinidae	Thick-knees	26–27	0
Stercorariidae	Skuas, jaegers	23–26	0
Laridae	Gulls, terns	20–34	0 to several
Rhynchopidae	Skimmers		0
Alcidae	Murres, puffins, guillemots	24–42	2–49
Columbiformes			
Pteroclididae	Sand grouse	22–28	
Raphidae	Solitaires	49	
Columbidae	Pigeons, doves	12–19	10–35
Psittaciformes			
Psittacidae	Parrots	17–31	28–36
Cuculiformes			
Cuculidae	Cuckoos	11–18	6–22
Strigiformes			
Tytonidae	Barn owls	30–34	56–64+
Strigidae	Owls	27–35	21–35
Caprimulgiformes			
Caprimulgidae	Goatsuckers	16–20	0
Apodiformes			
Apodidae	Swifts	17–21	20–42
Trochilidae	Hummingbirds	16–17	19–25
Trogoniformes			
Trogonidae	Trogons	18–19	16–30
Coraciiformes			
Alcedinidae	Kingfishers	21–23	24–35
Momotidae	Motmots	17–21	28–30
Meropidae	Bee-eaters	22	30
Coraciidae	Rollers	18–19	26–28
Upupidae	Hoopoes	16	29
Bucerotidae	Hornbills	28–40	75
Piciformes			
Galbulidae	Jacamars	19–22	20–26
Capitonidae	Barbets	13–15	
Ramphastidae	Toucans	16	43–45
Picidae	Woodpeckers	11–18	19–35
Passeriformes			
Dendrocolaptidae	Wood hewers	15±	19
Furnariidae	Oven birds	15–20	13–29

(Continued)

TABLE A.29 (*Continued*)
Incubation and Care of Young in Birds

Order, Family	Common Name	Incubation Duration (Days)	Time from Hatching until Birds Leave Nest (Days)
Formicariidae	Antbirds	14–17	9–13
Cotingidae	Cotingas	18–19	19–25
Pipridae	Manakins	19–21	13–15
Tyrannidae	Tyrant flycatchers	14–21	12–24
Menuridae	Lyrebirds	28	42
Alaudidae	Larks	11–12	9–12
Hirundinidae	Swallows	14–18	18–28
Oriolidae	Old World orioles	14–15	14–15
Corvidae	Crows	16–20	15–38
Paradiseidae	Birds of paradise	14–18	18–31
Paridae	Titmice	13–15	14–21
Sittidae	Nuthatches	15	22–24
Certhiidae	Creepers	15	15–20
Chamaeidae	Wren-tits	15–16	15–16
Timaliidae	Babbling thrushes	21±	
Cinclidae	Dippers	16	19–24
Troglodytidae	Wrens	14–19	13–22
Mimidae	Thrashers, catbirds	12–13	11–14
Turdidae	Thrushes	12–14	12–18
Sylviidae	Old World warblers	12–15	9–14
Regulidae	Kinglets	14–17	18–20
Muscicapidae	Old World flycatchers	12–20	10–20
Prunellidae	Accentors, hedge sparrows	12–15	13
Motacillidae	Pipits	13–14	10–15
Bombycillidae	Waxwings	12	16
Ptilogonatidae	Silky flycatchers	15	18–19
Laniidae	Shrikes	16	15–21
Sturnidae	Starlings	12	20–21
Meliphagidae	Honey-eaters	13–18	14–18
Nectariniidae	Sun birds	12–13	
Zosteropidae	White-eyes	11–12	9–11
Vireonidae	Vireos	13–14	11–13
Coerebidae	Honey creepers	12–14	14–19
Parulidae	Wood warblers	11–17	8–14
Ploceidae	Weaver finches	12–16	13–19
Icteridae	Blackbirds	11–14	9–34
Thraupidae	Tanagers	12–16	10–24
Fringillidae	Finches, sparrows	11–14	8–17

Source: Spector, W.S., ed., *Handbook of Biological Data*, WADC Technical Report 56-273, Wright-Patterson Air Force Base, OH, 1956.

TABLE A.30
Breeding Habits of Reptiles

Species	Sexual Maturity[a] (Year)	Breeding Season[b]	Gestation or Incubation Time[c]	Broods Size	Broods No./Year
Crocodila					
Alligator, American (*Alligator mississippiensis*)	5–10	January–September	56–66	29–88	1
Sauria					
Chameleon, false (*Anolis carolinensis*)	Male 2, female 1	April–August	6–7 weeks	1	8–10
Gila monster (*Heloderma suspectum*)			30	5–13	1
Lizard, alligator (*Gerrhonotus. multicarinatus*)		May	51–60	8–20	1
Lizard, collared-, western (*Crotaphytus collaris*)	<1–<3	May–June	8–13 weeks	4–24	1
Lizard, night desert (*Xantusia vigilis*)	Female 3	May–June	3 months	1–3	
Lizard, Pacific fence (*Sceloporus occidentalis*)	2	March–April	2 months	6–13	1
Lizard, sagebrush (*Sceloporus graciosus*)		April–May	62	2–7	1
Lizard, Texas horned (*Phrynosoma cornutum*)		April–May	39–47	23–37	1
Racerunner, common tessellated (*Cnemidophorus tigris*)		May–June	80	2–4	1–2
Skink, common five-lined (*Eumeces fasciatus*)	<2	May–June	4–9 weeks	2–18	1
Slowworm (*Anguis fragilis*)	3–4	May–June	3 months	7–19	1
Uta, northern ground (*Uta stansburiana*)		April–May	61–67	2–4	1
Serpentes					
Bullsnake (*Pituophis catenifer*)	<3	April–May	64–71	3–19	1
Copperhead (*Ancistrodon contortrix*)	3–4	April–May		142	2–10
Cottonmouth (*A. piscivorus*)		March–April, fall	5 months	3–15	1
Racer, western blue (*Coluber constrictor*)		May–June	1–2 months	15–25	1
Rattlesnake, Gt. Basin (*Crotalus viridis*)	3–4	April–June	4–5 months	3–13	0.5
Snake, brown (*Storeria dekayi*)		Mar–Apr	4 months	13–24	1
Snake, common garter (*Thamnophis sirtalis*)	2–3	March–May, fall	87–116	6–51	1–2
Snake, hog-nosed (*Heterodon platyrhinos*)		April–May		8–40	1
Snake, mud (*Farancia abacura*)		July	110	22–104	1
Snake, Pacific rubber (*Charina bottae*)	2–3	June		3–5	1

(Continued)

TABLE A.30 (*Continued*)

Breeding Habits of Reptiles

Species	Sexual Maturity[a] (Year)	Breeding Season[b]	Gestation or Incubation Time[c]	Broods	
				Size	No./Year
Snake, water (*Natrix erythrogaster*)		April–May	120–150	8–27	1
Testudinata					
Terrapin, diamondback (*Malaclemys terrapin*)	5–6	Spring	3 months	4–9	1–3
Tortoise, desert (*Gopherus agassizii*)	15–20	May	80–120	2–13	1
Turtle, Atlantic loggerhead (*Caretta caretta*)		March–July	31–65	120–130	2–3
Turtle, common box (*Terrapene carolina*)	3–5	April–May	70–114	2–7	1
Turtle, common musk (*Sternotherus odoratus*)	Male 2–3, female 9–11	April–October	60–75	1–6	1–2
Turtle, common snapping (*Chelydra serpentina*)		April–November	81–90	8–80	1–2

Source: Spector, W.S., ed., *Handbook of Biological Data*, WADC Technical Report 56-273, Wright-Patterson Air Force Base, OH, 1956.

[a] Males in some species mature before females.

[b] Varies with geographical distribution.

[c] Accepted average expressed in days; other values in weeks or months are approximations.

TABLE A.31
Breeding Habits of Amphibians

Species	Breeding Season	Eggs or Young per Clutch	Period of Growth to Sexual Maturity		
			Egg	Larva	Adult
			Days		
Urodela					
Eel, congo (*Amphiuma tridactylum*)	May–November	42–150			2–3 years
Eel, mud (*Siren intermedia*)		224–706			2 years
Hellbender (*Cryptobranchus alleganiensis*)	August–September	300–450	64–84	540	5–6 years
Mudpuppy (*Necturus maculosus*)	September–June	18–180	38–63		5 years
Newt, eastern (*Diemictylus viridescens*)	April–June	200–375	20–35	80	2 years
Salamander, Eschscholtz (*Ensatina eschscholtzii*)	October–April	12–14			
Salamander, European (*Salamandra salamandra*)	July	12–72	90–150		4–5 years
Salamander, four-toed (*Hemidactylium scutatum*)	April–October	22–64	38–60	48	2 ½ years
Salamander, green (*Aneides aeneus*)	May–June	10–26	84–91		
Salamander, marbled (*Ambystoma opacum*)	September–January	50–200	30–180	180–240	410–510
Salamander, slimy (*Plethodon cinereus*)	October–December	3–13			2 years
Salamander, spotted (*Ambystoma maculatum*)	March–April	To 250	31–54	61–110	2 years
Salamander, tiger (*A. tigrinum*)	January–March	23–110	24–30	75–118	360
Salamander, two–toed (*Eurycea bislineata*)	January–April	12–41	60–70	2–3 years	
(*Hynobius chinensis*)	May	35–70		60	
Amura					
Bullfrog (*Rana catesbeiana*)	February–August	10,000–25,000	4–5	2 years	2–3 years
Frog (*Zachaenus parvulus*)	July	30		17	
Frog, chirping (*Arthroleptella hewitti*)	October	36	10		
Frog, common (*Rana temporaria*)	February–April	1,500–4,000	14–21	90–180	4–5 years
Frog, cricket (*Acris gryllus*)	All year	250	4	50–90	2 years
Frog, leopard (*Rana pipiens*)	February–December	1,200–1,500		60–90	1–2 years
Frog, New Zealand (*Leiopelma hochstetteri*)	December	6–18	30		

(Continued)

TABLE A.31 (*Continued*)
Breeding Habits of Amphibians

Species	Breeding Season	Eggs or Young per Clutch	Period of Growth to Sexual Maturity		
			Egg	Larva	Adult
			Days		
Toad, bell (*Ascaphus truei*)	May–September	28–50	30	365	
Toad, common tree (*Hyla versicolor*)	April–July	1,000–2,000	4–5	45–65	1–3 years
Toad, Fowler's (*Bufo woodhousei*)	April–August	8,000	2–4	40–60	
Toad, narrow mouth (*Microhyla carolinensis*)	May–September	850	2	20–70	2 years
Toad, spadefoot (*Scaphiopus holbrookii*)	January–December	100+	0.5+	18–28	
(*Discoglossus pictus*)	January–October	300–1,000	2–4	30–60	
(*Xenopus laevis*)	September–October	<100–1,000	3	35–300	Male ½; female 2 years

Source: Spector, W.S., ed., *Handbook of Biological Data*, WADC Technical Report 56-273, Wright-Patterson Air Force Base, OH, 1956.

TABLE A.32
Breeding Habits of Fish

Species	Spawning Season	Number of Eggs
Bass, largemouth (*Micropterus salmoides*)	Spring, summer	2,000–26,000
Bowfin (*Amia calva*)	Spring	23,000–64,000
Bullhead, brown (*Ameiurus nebulosa*)	Spring	2,000–10,000
Carp (*Cyprinus carpio*)	Spring, summer	500,000–2,000,000
Cod, Atlantic (*Gadus callarias*)	Winter, spring	3,000,000–9,000,000
Dogfish, Atlantic spiny (*Squalus acanthias*)	All year	2–11
Eel, American (*Anguilla rostrata*)	Winter	5,000,000–20,000,000
Eel, Conger (*Conger oceanica*)	Summer	3,000,000–7,900,000
Flounder, winter (*Pseudopleuronectes americanus*)	Winter, spring	500,000–1,500,000
Gar, longnose (*Lepisosteus osseus*)	Spring	36,500
Haddock (*Melanogrammus aeglefinus*)	Winter, spring	169,000–1,840,000
Hagfish, Atlantic (*Myxine glutinosa*)	All year	19–30
Halibut, Atlantic (*Hippoglossus hippoglossus*)	Spring, summer	2,183,000
Herring, Atlantic (*Clupea harengus*)	Spring, summer, autumn	20,000–40,000
Lamprey, sea (*Petromyzon marinus*)	Spring	236,000
Mackerel, Spanish (*Scomberomorus maculatus*)	Spring, summer	20,000
Mummichog (*Fundulus heteroclitus*)	Spring, summer	460
Pike, northern (*Esox lucius*)	Spring	10,000–100.000

(*Continued*)

TABLE A.32 (*Continued*)

Breeding habits of Fish

Species	Spawning Season	Number of Eggs
Pumpkinseed (*Lepomis gibbosus*)	Spring, summer	Several thousand
Ray, southern sting (*Dasyatis americana*)		3–5
Salmon, Atlantic (*Salmo salar*)	Spring	7,000
Sea horse, northern (*Hippocampus hudsonius*)	Spring	150
Shad (*Alosa sapidissima*)	Spring	100,000–156,000
Shark, hammerhead (*Sphyrna zygaena*)	Summer	29–37
Shark, man-eater (*Carcharodon carcharias*)	Summer	9
Skate, little (*Raja erinacea*)	All year	6
Stickleback, three spine (*Gasterosteus aculeatus*)	Spring	100–150
Sturgeon, Atlantic (*Acipenser sturio*)	Spring, summer	1,000,000–2,500,000
Trout, brown (*Salmo trutta*)	Autumn, winter	200–6,000
Trout, rainbow (*S. gairdnerii*)	Spring, summer, autumn	400–3,000

Source: Spector, W.S., ed., *Handbook of Biological Data*, WADC Technical Report 56-273, Wright-Patterson Air Force Base, OH, 1956.

TABLE A.33

Reproduction of Insects

		Duration (Days)				
Species	Eggs per Female	Egg	Larva or Nymph	Pupa	Adult	Generations per Season
Anoplura						
Louse, cattle (four species)	30–50	10–30	9–15		10–30	7–10
Louse, human (three species)	50–300	5–21	7–10		10–30	10–12
Coleoptera						
Beetle						
Black carpet (*Attagenus piceus*)	42–114	6–11	238–638	6–24	32–72	<1
Carpet (*Anthrenus scrophulariae*)	32	10–18	66	14		1–3
Cigarette (*Lasioderma serricorne*)	30	>6	>30	14–21	21	5–6
Colorado potato (*Leptinotarsa decemlineata*)	>500	4–9	10–21	5–10		1–3
Confused flour (*Tribolium confusum*)	300–400	4–14	>22	5–18	1000	5–6
Dried-fruit (*Carpophilus hemipterus*)	>80	>3	28–120	>14	15	>8
Drug-store (*Stegobium paniceum*)			20–150	12–18		3–6
Flea (*Systena* spp.)	500–900	5–8	14–21	10–14		2–3
May (*Phyllophaga fiscal*)		21	>730	30	21–28	⅓
Mexican bean (*Epilachna varivestis*)	250–1200	5–14	20–35	10		3–4

(*Continued*)

TABLE A.33 (*Continued*)
Reproduction of Insects

Species	Eggs per Female	Duration (Days) Egg	Larva or Nymph	Pupa	Adult	Generations per Season
Red-legged ham (*Necrobia rufipes*)	400–1000	>3	>17	>13	420	6–10
Saw-toothed grain (*Oryzaephilus surinamensis*)	45–285	>8	>30	>6	180–1100	6–7
Striped cucumber (*Acalymma vittata*)			14–40	7		1–4
Borer, lesser grain (*Rhyzopertha dominica*)	300–500					8–12
Cadelle (*Tenebroides mauritanicus*)	436–1000	7–10	39–414	8–25	365	3
Curculio, plum (*Conotrachelus nenuphar*)	41–175	4–10	17–48	8–30	±365	1 to >1
Mealworm, yellow (*Tenebrio molitor*)	276	12–16	>600	18–20	60–90	>1
Weevil						
Alfalfa (*Hypera postica*)	200–800	13–17	17–21	7–14	14–21	1–2
Bean (*Acanthoscelides obtectus*)	200	5–20	11–40	5–18	14–63	2–5
Cotton boll (*Anthonomus grandis*)	80–200	3–5	7–12	3–5	30–300	4–10
Cowpea (*Callosobruchus maculates*)	82–196	4–6	9–240	5–18	15	8–10
Granary (*Sitophilus granarius*)	50–250	4–8	19–34	5–16	210–250	8–12
Sweetpotato (*Cylas formicarius elegantulus*)		7	14–21	7		6–8
Diptera						
Fly						
Horn (*Siphona irritans*)	50–400	1–4	4–8	4–8	5–20	4–10
Horse (*Tabanus atratus*)	100–400	2–5	100–600	5–20	5–20	1–2
House (*Musca domesticus*)	75–200	1–3	4–10	4–18	10–50	4–18
Stable (*Stomoxys calcitrans*)	20–100	2–5	11–30	5–20	5–30	4–10
Vinegar (*Drosophila* spp.)		<1	3–11	2–8	14	5–6
Grub, cattle (*Hypoderma* spp.)	100–500	3–10	250–280	18–70	1–25	1
Maggot, seed-corn (*Hylemya cilicrura*)	100	1–8	10–16	10–20	30–35	2–5
Mosquito (hundreds of species)	100–1036	2–1800	5–15	2–5	5–300	1–17
Yellow fever (*Aedes aegypti*)		2–365	6	2–3	15–60	
Screwworm (*Callitroga americana*)	100–300	1–2	4–5	5–40	5–30	2–12
Hemiptera						
Bug, harlequin (*Murgantia histrionica*)	75–100	4–15	40–60			3–4
Bug, squash (*Anasa tristis*)	200–300	7–14	28–42		15–110	1

(*Continued*)

TABLE A.33 (*Continued*)
Reproduction of Insects

Species	Eggs per Female	Duration (Days) Egg	Larva or Nymph	Pupa	Adult	Generations per Season
Homoptera						
Aphid, melon (*Aphis gossypii*)		90–120	3–7		7–28	20
Aphid, pea (*Macrosiphum pisi*)		90–120	10			7–20
Cicada, periodical (*Magicicada septendecim*)		42–49	13–17 yr		30–40	
Greenbug (*Toxoptera graminum*)	3–7	90–120	6–30		26–60	5–20
Leafhopper, beet (*Circulifer tenellus*)	300–400	5–40	25–52		120–150	3–5
Leafhopper, potato (*Empoasca fabae*)	60–90	10	14		30	2–4
Hymenoptera						
Bee, honey (*Apis mellifera*)		3	8	9	35–40	
Sawfly, wheat stem (*Cephus cinctus*)	50	7–10	300	7–10	7	1
Wasp, digger (*Tiphia vernalis*)	50–75	8–9	120–180	180–240	30–42	1
Lepidoptera						
Bollworm, pink (*Pectinophora gossypiella*)	200	4–10	14–21	12–18	14	3–6
Borer						
European corn (*Pyrausta nubilalis*)	400	4–9	30–40	10–14	10–24	1–3
Peach tree (*Sanninoidea exitiosa*)	200–600	7–48	270–380	16–25	4–16	1
Squash vine (*Melittia cucurbitae*)	150–200	7–14	30	Winter		1–2
Sugarcane (*Diatraea saccharalis*)	200	4–9	20–30	6–7	7–14	4–5
Cabbageworm, imported (*Pieris rapae*)	200–500	7	14	7–14		3–6
Earworm, corn (*Heliothis armigera*)	1000	2–8	13–28	14	12	1–7
Hornworm, tobacco (*Protoparce sexta*)	200–300	7	21–28	14–28		1–3
Leafworm, cotton (*Alabama argillacea*)	400–600	3–20	7–21	7–21	10–24	3–8
Moth						
Casemaking clothes (*Tinea pellionella*)	>40	>6	30–100	10–90	7–28	1–2
Codling (*Carpocapsa pomonella*)	6–100	4–14	15–72	7–40	3–20	>1 to >3
Indian-meal (*Plodia interpunctella*)	200–400	1–2	13–288	>8	18	5

(*Continued*)

TABLE A.33 (*Continued*)

Reproduction of Insects

Species	Eggs per Female	Duration (Days)				Generations per Season
		Egg	Larva or Nymph	Pupa	Adult	
Mediterranean flour (*Ephestia kuehniella*)	116–700	>3	40	5–7	3–4	>6
Oriental fruit (*Grapholitha molesta*)	100–200	3–28	10–26	5–35	2–34	4–7
Mallophaga						
Louse, cattle biting (*Bovicola bovis*)	20–50	7–30	15–25		10–30	7–10
Orthoptera						
Cockroach, American (*Periplaneta americana*)	200–1000	35–100	200–400		212–303	
Grasshopper (many species)	300–400	90–120	40–60		>30	1
Siphonaptera						
Flea (many species)	50–400	2–13	7–30	7–35	8–150	3–5
Thysanoptera						
Thrips, gladiolus (*Taeniothrips simplex*)	150	4–12	4–12	3–8	26–32	6

Source: Spector, W.S., ed., *Handbook of Biological Data*, WADC Technical Report 56-273, Wright-Patterson Air Force Base, OH, 1956.

TABLE A.34

Life Spans of Animals

Species	Recorded Life Span	
	Average (Year)	**Maximum (Year)**
Vertebrata		
Mammalia		
Agouti (*Dasyprocta aguti*)	6	>10
Alpaca (*Lama pacos*)	12	>17
Anteater, spiny (*Tachyglossus aculeatus*)		50
Antelope, pronghorned (*Antilocapra americana*)	8	15
Ape, black (*Cynopithecus niger*)	2	18
Ass, African wild (*Equus asinus taeniopus*)	14.6	19.3
Baboon, sacred (*Papio hamadryas*)	15	Female-24
Badger, American (*Taxidea taxus*)	11	>13
Bat, American brown (*Eptesicus fuscus*)		2
Bear, brown (*Ursus arctos*)		34
Bear, grizzly (*U. horribilis*)	20	>31
Bear, polar (*Thalarctos maritimus*)	16	33
Beaver, American (*Castor canadensis*)		Female-19
Blackbuck (*Antilope cervicapra*)	7	Male-15
Buffalo, American (*Bison bison*)	10	Female >22
Buffalo, African (*Syncerus caffer*)	10	>15
Camel, dromedary (*Camelus dromedarius*)		>25
Cat, domestic (*Felis catus*)	13–17	21
Cheetah (*Acinonyx jubatus*)	6	>15
Chimpanzee (*Pan troglodytes*)	15–20	37
Chinchilla (*Chinchilla laniger*)	4	7
Chipmunk, eastern (*Tamias striatus*)	2.5	>7
Civet (*Viverra* spp.)		>15
Coyote (*Canis latrans*)	9	14
Deer, fallow (*Dama dama*)	10	Female-15
Dingo (*Canis dingo*)	3	Female >12
Dog, domestic (*C. familiaris*)	13–17	34
Dolphin (*Delphinus delphis*)	25–30	
Dormouse, garen (*Eliomys quercinus*)	2–3	>5
Elephant, African (*Loxodonta africana*)	24	36
Elephant, Indian (*Elephas maximus*)		57
Elk, European (*Alces alces*)	15–20	25
Fox, arctic (*Alopex lagopus*)	8	14
Fox, red (*Vulpes fulva*)		12
Gazelle, Korin (*Gazella rufifrons*)		Male-11
Genet (*Genetta pardina*)	>7	12.5
Gibbon (*Hylobates* spp.)		>23
Giraffe (*Giraffa camelopardalis*)	14	>28
Gnu, brindled (*Connochaetes taurinus*)		16

(*Continued*)

TABLE A.34 (*Continued*)
Life Spans of Animals

Species	Recorded Life Span	
	Average (Year)	**Maximum (Year)**
Goat (*Capra hircus*)	8–10	18
Gorilla (*Gorilla gorilla*)		>7
Guinea pig (*Cavia porcellus*)	>2	>6
Hamster, common (*Cricetus cricetus*)	2	2.5
Hamster, golden (*Mesocricetus auratus*)	1	1.8
Hedgehog, European (*Erinaceus europaeus*)		2
Hippopotamus (*Hippopotamus amphibius*)	40	49
Horse, domestic (*Equus caballus*)	20–30	62
Hyena, spotted (*Crocuta crocuta*)	12	Female-25
Ibex, Nubian (*Capra nubiana*)	8.5	>10
Jackal, black-backed (*Canis mesomelas*)	8	13
Jaguar (*Panthera onca*)	14	>22
Kangaroo, red (*Macropus rufus*)		16.3
Lemur, black (*Lemur macaco*)	10	Male-21
Leopard (*Panthera pardus*)	14	23
Lion (*P. leo*)	20–25	29
Llama (*Lama glama*)	15	20
Loris, slow (*Nycticebus coucang*)		10
Lynx, Canadian (*Lynx canadensis*)	6	>11
Marmoset (*Hapale jacchus*)	11	16
Marmot, alpine (*Marmota marmota*)	7	>13
Marten, pine (*Martes martes*)	10	>13.5
Mongoose, zebra (*Mungos mungo*)	5	Female >8
Monkey, bonnet (*Macaca radiata*)	12	13
Monkey, rhesus (*M. mulatta*)	15	29
Mouse, harvest (*Micromys minutus*)	2	2.5
Mouse, house (*Mus musculus*)	1–2	>3
Opossum (*Didelphis* spp.)		>7
Orangutan (*Pongo pygmaceus*)	8	Male-26
Otter (*Lutra* spp.)	20–25	>15
Ox (*Bos taurus*)		30
Peccary, collared (*Pecari tajucu*)	5	>15
Platypus (*Ornithorhynchus paradoxus*)		
Porcupine, African (*Hystrix cristata*)	8–12	>20
Prairie dog (*Cynomys ludovicianus*)	4	>8
Puma (*Felis concolor*)	9	16
Rabbit, European (*Oryctolagus cuniculus*)	5–6	>13
Raccoon (*Procyon lotor*)	4	>13
Rat, house (*Rattus rattus*)	2–3	4
Reindeer (*Rangifer tarandus*)		12
Rhinoceros, Indian great (*Rhinoceros unicornis*)	40–45	47

(*Continued*)

TABLE A.34 (*Continued*)

Life Spans of Animals

Species	Recorded Life Span	
	Average (Year)	Maximum (Year)
Sea lion, California (*Zalophus californianus*)	13	19
Seal, cape fur (*Arctocephalus pusillus*)	13	Female >20
Seal, common (*Phoca vitulina*)	10–15	>14
Sheep, domestic (*Ovis aries*)		20
Shrew, jumping (*Elephantulus rozeti*)		3.4
Skunk, Canadian (*Mephitis mephitis*)	9	6
Slith, two-toed (*Choloepus didactylus*)		>11
Squirrel, gray (*Sciurus carolinensis*)		14–15
Swine (*Sus scrofa*)	16	27
Tapir, Brazilian (*Tapirus terrestris*)	<6	9
Tiger (*Panthera tigris*)	11	19
Weasel (*Mustela nivalis*)	24–37	>7
Whale, arctic (*Balaena mysticetus*)		
Wolf, European (*Canis lupus*)	12	14
Woodchuck (*Marmota monax*)	22	>9
Yak (*Peophagus gruuniensis*)		22
Zebra, mountain (*Equus zebra*)		>25
Aves		
Bird of paradise (*Paradisea apoda*)		>12
Blackbird, European (*Turdus merula*)		18
Bluebird (*Sialia sialis*)		4.5
Bunting, red-headed (*Emberiza luteola*)		>13
Buzzard, African (*Buteo desertorum*)		>18
Canary, house (*Serinus canarius*)		24
Cardinal (*Richmondena cardinalis*)		22
Chickadee (*Parus atricapillus*)		>7
Cockatoo, slender-billed (*Kakatoe tenuirostris*)		85
Condor (*Vultur gryphus*)		52
Coot, slaty (*Fulica ardesiaca*)		3
Cormorant (*Phalacrocorax carbo*)		23
Cowbird, bay-winged (*Molothrus badius*)		>12
Crane, common (*Grus communis*)		>42
Dove, collared (*Turtur risorius*)		30–40
Duck, domestic (*Anas platyrhynchus domesticus*)		19
Eagle, Chilean (*Geranoaetus melanoleucus*)		>42
Egret, American snowy (*Leucophoyx thula*)		>16
Emu (*Dromiceius novaehollandiae*)		40
Finch, chestnut-eared (*Amadina castanotis*)		>8
Flamingo, European (*Phoenicopterus roseus*)		>22
Fowl, domestic (*Gallus domesticus*)		30

(*Continued*)

TABLE A.34 (*Continued*)
Life Spans of Animals

Species	Recorded Life Span	
	Average (Year)	Maximum (Year)
Goldfinch, European (*Carduelis carduelis*)		27
Goose, Canadian (*Branta canadensis*)		32
Gull, herring (*Larus argentatus*)		44
Heron (*Ardea cinerea*)		>24
Hornbill, Indian great (*Buceros bicornis*)		33
Jay, blue (*Cyanocitta cristata*)		>4
Kingfisher, laughing (*Halcyon* sp.)		11
Kiwi (*Apteryx australis*)		20
Lovebird, gray-headed (*Agapornis cana*)		>8
Lyrebird (*Menura superba*)		>8
Macaw, blue-and-yellow (*Ara ararauna*)		43
Macaw, red-and-blue (*A. macao*)		64
Magpie (*Pica pica*)		12
Mockingbird (*Mimus polyglottos*)		>6
Nightingale (*Luscinia luscinia*)		3.8
Nuthatch, white-breasted (*Sitta carolinensis*)		>8
Ostrich, African (*Struthio camelus*)		50
Owl, barn (*Tyto alba*)		>13
Owl, snowy (*Nyctea nyctea*)		24.5
Parrot, Maximilian's (*Pionus maximiliani*)		9
Parakeet, ring-necked (*Psittacula torquatus*)		>20
Partridge, European (*Perdix perdix*)		>5
Pelican, Australian (*Pelecanus conspicillatus*)		52
Penguin, king (*Aptenodytes patagonica*)		26
Pheasant, ring-necked (*Phasianus colchicus*)		>27
Pigeon, domestic (*Columba livia domestica*)		35
Plover, Old World golden (*Pluvialis apricaria*)		>1
Puffin, Atlantic (*Fratercula arctica*)		8
Quail, European (*Coturnix coturnix*)		10
Raven (*Corvus corax*)		69
Robin, American (*Turdus migratorius*)		>12
Rook (*Corvus frugilegus*)		>14
Skylark (*Alauda arvensis*)		24
Sparrow, Italian (*Passer italiae*)		20
Starling (*Sturnus vulgaris*)		>15
Stork, black (*Ciconia nigra*)		30
Swallow (*Hirundo rustica*)		<1
Swan, trumpeter (*Cygnus buccinator*)		>29
Thrush, song (*Turdus musicus*)		>11
Titmouse, great (*Parus major*)		9
Turkey (*Meleagris gallopavo*)		>12

(*Continued*)

TABLE A.34 (*Continued*)
Life Spans of Animals

Species	Recorded Life Span	
	Average (Year)	Maximum (Year)
Vulture, griffon (*Gyps fulvus*)		41
Waxbill, Amandava (*Estrilda amandava*)		>10
Reptilia		
Alligator, American (*Alligator mississippiensis*)		56
Anaconda (*Eunectes murinus*)		28
Black snake (*Coluber constrictor*)		5.3
Boa constrictor (*Constrictor constrictor*)		>23
Caiman, black (*Caiman niger*)		28
Chameleon (*Chameleon* sp.)		>3.5
Cobra, African black (*Naja melanoleuca*)		>26
Cooter (*Pseudemys scripta*)		7
Copperhead, North American (*Ancistrodon contortrix*)		18.5
Cottonmouth, North American (*A. piscivorus*)		21
Crocodile, American (*Crocodylus acutus*)		13.5
Garter snake (*Thamnophis sirtalis*)		6
Gecko, Moorish wall (*Tarentola mauritanica*)		7.4
Gila monster (*Heloderma suspectum*)		20
Iguana, Galapagos (*Conolophus subcristatus*)		15.1
King.snake (*Lampropeltis getulus californiae*)		14.8
Lizard, European glass (*Ophisaurus apus*)		24
Lizard, long-tailed (*Latastia longicaudata*)		2.3
Lizard, monitor (*Varanus salvator*)		10.8
Matamata (*Chelys fimbriata*)		10.3
Puff adder (*Bitis arietans*)		13.9
Python, Frican rock (*Python sebae*)		15.5
Rattlesnake, North American (*Crotalus atrox*)		18.6
Skink, sand (*Chalcides ocellatus*)		>9.5
Terrapin, Reeve's (*Geoclemys reevesii*)		24.3
Tortoise, European pond (*Emys orbicularis*)		66
Tortoise, Galapagos (*Testudo elephantopus*)		177
Tuatara (*Sphenodon punctatus*)		>28
Turtle, common box (*Terrapene carolina*)		123
Turtle, green (*Chelonia mydas*)		21.0
Turtle, musk (*Sternotherus odoratus*)		53.1
Turtle, snapping (*Chelydra serpentina*)		20
Water mocassin (*Ancistrodon piscivorus*)		21
Water snake (*Natrix sipedon*)		7
Whip snake (*Masticophis flagellum*)		13.4
Viper, cape (*Causus rhombeatus*)		>6.5

(*Continued*)

TABLE A.34 (*Continued*)
Life Spans of Animals

Species	Recorded Life Span	
	Average (Year)	Maximum (Year)
Amphibia		
Coecilian, South American (*Siphonops annulatus*)		9.5
Congo "eel" (*Amphiuma means*)		26.8
Frog, African speckled (*Rana adspersa*)		5.4
Frog, bull (*R. catesbeiana*)		>15.6
Frog, clawed (*Xenopus laevis*)		15
Frog, green (*Rana clamitans*)		>10.1
Frog, leopard (*R. pipiens*)		>5.9
Frog, palaearctic grass (*R. temporaria*)		>4.3
Frog, palaearctic water (*R. esculenta*)		>5.1
Frog, red-spotted (*Leptodactylus pentadactylus*)		15.7
Frog, South African (*Phrynomerus bifasciata*)		>0.4
Hellbender (*Cryptobranchus alleganiensis*)		>28.5
Mudpuppy, North American (*Necturus maculosus*)		>8.8
Newt, California (*Taricha torosus*)		21
Newt, common (*Triturus viridescens*)		2.9
Newt, European crested (*T. cristatus*)		17
Newt, Pyrenean (*Euproctus asper*)		>7
Proteus, European (*Proteus anguinus*)		15
Salamander, Asiatic (*Megalobatrachus maximus*)		55
Salamander, European (*Salamandra atra*)		3
Salamander, long-tailed (*Eurycea lucifuga*)		>1
Salamander, spotted (*Ambystoma maculatum*)		25
Salamander, tiger (*A. tigrinum*)		11
Siren, North American (*Siren lacertian*)		25.5
Toad, American (*Bufo americanus*)		12–23
Toad, Cuban (*B. peltocephalus*)		13
Toad, Degen's (*B. vittatus*)		2.4
Toad, giant (*B. alvarius*)		2
Toad, northwestern (*B. boreas*)		6
Toad, Surinam (*Pipa pipa*)		>7.8
Tree-frog, Florida (*Hyla gratiosa*)		5.9
Tree-frog, giant (*H. septentrionalis*)		6.75
Tree-frog, palaearctic (*H. arborea*)		14
Tree-frog, rain (*H. versicolor*)		6.7
Tree-frog, South American (*H. raddiana*)		2.3
Pisces		
Anchovy, northern (*Engraulis mordax*)		7
Bass, large-mouth (*Micropterus salmoides*)		11
Bass, small-mouth (*M. dolomieu*)		11

(*Continued*)

TABLE A.34 (*Continued*)
Life Spans of Animals

Species	Recorded Life Span	
	Average (Year)	Maximum (Year)
Bass, striped (*Roccus saxatilis*)		24
Bullhead, black (*Ameiurus melas*)		9
Carp, mirror (*Cyprinus carpio* var.)		47
Carp, Prussian (*C. carpis*)		<6.5
Catfish, flathead (*Pilodictis olivaris*)		15
Cod, Atlantic (*Gadus morhua*)		13
Dogfish (*Scyllidae* sp.)		>2
Eel, electric (*Electrophorus electricus*)		11.5
Eel, North American (*Anguilla rostrata*)		6
Flounder (*Pleuronectes flesus*)		10
Flounder, starry (*Platichthys stellatus*)		>8
Flounder, winter (*Pleuronectes americanus*)		1
Gar, longnose (*Lepisosteus osseus*)		30
Goldfish (*Cyprinus carassius auratus*)		25
Grayling, American (*Thymallus signifer*)	4	11
Grunion (*Leuresthes tenuis*)		4
Haddock (*Melanogrammus aeglefinus*)	1.9	15
Halibut, Atlantic (*Hippoglossus hippoglossus*)		40
Herring, Pacific (*Clupea pallasi*)	5.9	19
Lamprey, American brook (*Lampetra lamottei*)		5
Lungfish, African (*Protopterus annectens*)		>17
Lungfish, South American (*Lepidosiren paradoxa*)		8.3
Mackerel, jack (*Trachurus symmetricus*)		>20
Mackerel, Pacific (*Pneumatophorus diego*)		11
Minnow, European (*Phoxinus phoxinus*)		13
Paddlefish (*Polyodon spathula*)		14
Perch (*Perca fluviatilis*)		>10.8
Perch, yellow (*P. flavescens*)	3.4	13
Pickerel, chain (*Esox niger*)	3.3	8
Pike (*E. lucius*)	4.6	24
Roach (*Rutilus rutilus*)		12
Salmon, Atlantic (*Salmo salar*)		13
Salmon, pink (*Oncorhynchus gorbuscha*)		1.8
Sardine, California (*Sardinops caerulea*)	3.3	13
Seahorse (*Hippocampus guttulatus*)		6
Shark (*Scyliorhinus stellaris*)		18
Shiner, common (*Notropis cornutus*)		6
Smelt, American (*Osmerus mordax*)	2.3	5
Smelt, jack (*Atherinopsis californiensis*)		9–10
Sole, Dover (*Microstomus pacificus*)		15
Squawfish, Sacramento (*Ptychocheilus grandis*)		9
Sturgeon, white (*Acipenser transmontanus*)		50

(*Continued*)

TABLE A.34 (*Continued*)
Life Spans of Animals

Species	Recorded Life Span	
	Average (Year)	Maximum (Year)
Sunfish, green (*Lepomis cyanellus*)	2.2	9
Tautog (*Tautoga onitis*)		8
Trout, brown (*Salmo trutta*)	3.2	18
Trout, lake (*Salvelinus namaycush*)	7.6	41
Trout, rainbow (*Salmo irideus*)		>3.9
Tuna, bluefin (*Thunnus thynnus*)		7
Walleye, yellow (*Stizostedion vitreum*)	4.8	18
Whitefish, lake (*Coregonus clupeaformis*)	5.3	26
Invertebrata		
Echinodermata		
Asteroidea		
Starfish (*Asterias rubens*)		>5
Holothuroidea		
Sea cucumber (*Cucumaria planci*)		>10
Mollusca		
Pelecypoda		
Clam, giant (*Tridacna gigas*)		60–100
Mussel, edible (*Mytilus edulis*)		<1
Mussel, fresh-water (*Margaritana margaritifera*)		100
Mussel, fresh-water (*Pisidium* spp.)		2–4
Mussel, fresh-water (*Unio crassus*)		12
Mussel, pearl (*Pinctada* spp.)		8
Mussel, pond (*Anodonta fluviatilis*)		10
Oyster (*Ostrea edulis*)		10
Scallop (*Pecten yessoensis*)		7
Shell, fingernail (*Musculium* spp.)		2–4
Shell, fingernail (*Sphaerium* spp.)		2–4
Shipworm (*Teredo* spp.)		<1
Gastropoda		
Opisthobranch (*Gasteropteron meckelii*)		1
Periwinkle (*Littorina littorea*)		20
Prosobranch, freshwater (*Neritina* spp.)		5
Sea hare (*Aplysia* spp.)		1
Shell, moon (*Lunatia heros*)		30
Slug (*Milax marginatus*)		2–3
Slug (*Limax* spp.)		1–3
Slug, marine (*Doris* spp.)		1
Slug, marine nudibranch (*Aeolidia* spp.)		1
Slug, marine nudibranch (*Janolus instatus*)		1
Snail, edible (*Helix pomatia*)		18

(*Continued*)

TABLE A.34 (*Continued*)

Life Spans of Animals

Species	Recorded Life Span	
	Average (Year)	Maximum (Year)
Snail, fresh-water (*Ancylus* spp.)		4–5
Snail, fresh-water (*Lymnaea* spp.)		4–5
Snail, fresh-water (*Planorbis* spp.)		4–5
Snail, land (*Campylaea cingulata*)		4–5

Source: Spector, W.S., ed., *Handbook of Biological Data*, WADC Technical Report 56-273, Wright-Patterson Air Force Base, OH, 1956.

TABLE A.35

Cell Division Frequency of Bacteria and Viruses

Organism	Culture Medium	Temperature (°C)	Generation Time (min)
Aerobacter aerogenes	Broth or milk	37	16–18
	Glucose + peptone	37	17.3
	Peptone	37	22–30
	Synthetic	37	29–44
Azotobacter spp.	Mineral salts + sugar	25–30	240–348
A. chroococcum	Glucose broth	27–39	
	Sugar + urea	28	74
Bacillus cereus	Broth	37	18.8
	Glucose broth	37	17–24.5
B. megatherium	Broth	30	31
B. mycoides	Broth	37	28
B. subtilis	Glucose broth		26–32
B. thermophilus	Broth	55	18.3
	Tryptophan + broth	54.5	16
Clostridium amylobacter	Corn mash	37	51
C. botulinum	Glucose broth	37	35
C. welchii	Milk	37	35
Corynebacterium diphtheriae	Serum + glucose broth	37	34
C. pseudodiphtheriae	Broth	37	37
Diplococcus mucosus	Milk	37	32
D. pneumoniae I	Broth	37	24.5
	Serum	37	29
	Serum + broth	37	20.5
D. pneumoniae II	Broth	37	33
	Glucose broth	37	30
	Serum + broth	37	23
Erwinia carotovora	Broth	37	57

(*Continued*)

TABLE A.35 (*Continued*)
Cell Division Frequency of Bacteria and Viruses

Organism	Culture Medium	Temperature (°C)	Generation Time (min)
	Glucose broth	37	42
E. amylovora	Broth	30	71–94
Escherichia coli	Broth	37	17
	Lactose broth	37	16
	Milk	37	12.5
E. coli communior	Broth	37	16
Lactobacillus acidophilus	Milk	37	66–87
L. bulgaricus	Glucose milk	37	41–75
	Milk	37	39–74
	Peptone milk	37	50
	Tomato juice + milk	37	38–40
	Yeast extract	37	37
L. casei	Milk	25	38
L. delbruckii	Wort	45	83
L. pentoaceticus	Yeast extract	28	67
Mycobacterium tuberculosis	Synthetic	37	792–932
Pasteurella lepiseptica	Broth + blood		24.4
Phytomonas campestris	Broth	23–25	165
P. campsetre	Broth	25	98
P. glycineum	Broth	23–25	95
P. phaseoli	Bean broth	25	160
	Broth	25	150
	Glucose broth	25	138
P. sojae	Broth	23–25	82
Phytomonas tabacum	Broth	25	81
Proteus vulgaris	Broth	37	21.5
	Peptone + phosphate	37	40
Pseudomonas fluorescens	Broth	30	40
	Glucose broth	37	34–34.5
P. pyocyanea	Broth	37	34
	Glucose broth	37	31
	Lactose broth	37	34
Rhizobium japonicum	Mineral salts + yeast + mannitol	25	344–461
R. leguminosarum	Mineral salts + mannitol	25	79–187
R. meliloti	Mineral salts + yeast + mannitol	25	107
R. trifolii	Mineral salts + yeast + mannitol	25	101
Salmonella enteritidis	Broth	42	21.5
S. paratyphi	Broth	37	23
	Peptone	37	28
S. suipestifer	Broth	37	26

(*Continued*)

TABLE A.35 (*Continued*)
Cell Division Frequency of Bacteria and Viruses

Organism	Culture Medium	Temperature (°C)	Generation Time (min)
S. typhi	Bile + pus	37	24.5
	Broth	37	23.5
	Glucose broth	37	29
	Glucose + peptone	37	33
Serratia marcescens	Milk	37	37
Shigella dysenteriae	Milk	37	23
	Peptone + phosphate	37	37
Spirochaeta spp.	Modified thioglycolate	37	528
Staphylococcus albus	Glucose broth	37	24–25
S. aureus	Broth	37	27–30
	Glucose broth	37	32
Streptococcus fecalis	Glucose-citrate broth	37	27
	Milk	37	26.5
S. hemolyticus	Beef heart broth	37	32
	Glucose broth	37	26
	Glucose serum broth	37	34
S. lactis	Glucose milk	37	26
	Lactose broth	30	48
	Milk	37	26
	Peptone milk	37	37
S. liquefaciens	Milk	37	27
S. mastitidis	Glucose milk	37	35–37
S. thermophilus	Milk	37	27
Treponema pallidum	Rabbit skin	37	1800
	Rabbit testes	37	1980
Vibrio comma	Broth	37	21.2–38
Vibrio costatus	Broth	27	42
Viruses			
Influenza A (PR-8)	Allantoic membrane	37	330–510
A (5 strains)	Allantoic membrane	37	300–360
B (3 strains)	Allantoic membrane	37	480–600
Swine	Allantoic membrane	37	360

Source: Spector, W.S., ed., *Handbook of Biological Data*, WADC Technical Report 56-273, Wright-Patterson Air Force Base, OH, 1956.

TABLE A.36

Cell Division Frequency of Protozoa

Organism	Substrate	Temperature (°C)	Cell Division per Day
Astasia	Tryptophan + acetate	25	3.1
Chilomonas paramecium	NaAc + mineral salts	24	3.5
Didinium nasutum	Hopkin's + paramecium	21	3.6
Euglena gracilis	In dark, no acetate	10	0.03
	In dark + acetate	23	0.47
	Wheat infusion	25	3.5
Glaucoma pyriformis	Yeast extract	25	6.1
	Yeast extract	24.2	6.9
	Yeast + yeast extract	25.2	7.6–8
Leucophrys patula	Glaucoma	25	3.7
Paramecium aurelia	Lettuce + bacteria	20±	0.72
P. caudatum	Mineral salts + bacteria	25–28	1.8
	Oat medium + bacteria	26	2.3
Polytomella uvella	Aerated peptone	22	4.4
	Non-aerated peptone	22	1.8
Stentor coeruleus	Peter's + ciliates	19	0.6–0.9
	Modified Peter's (+ citrate)	18–20	0.7–2.1
	Hetherington's + citrate	22	0.65
Stylonychia pustulata		25?	4.5–5
Tetrahymena geleii		24	5.7–10.9

Source: Spector, W.S., ed., *Handbook of Biological Data*, WADC Technical Report 56-273, Wright-Patterson Air Force Base, OH, 1956.

TABLE A.37

Water Balance for Resting Terrestrial Animals

Animal	Body Mass (g)	Water Intake (g/100 g Body Mass per Day)	Metabolic Water (g/100 g Body Mass per Day)	Output: Urine (g/100 g Body Mass per Day)	Output: All Other (g/100 g Body Mass per Day)
Man	65,000	3.5	0.5	1.9	2.1
Cat	2,900	7.2	1.2	4.1	4.3
Cattle					
Brahman, dry	409,000	5.5	0.6		
Holstein, dry	745,000	6.0	0.7		
Holstein, milking	529,000	14.8	1.1		
Jersey, milking	403,000	11.8	1.0		
Steer	584,000	4.0	0.8	0.9	3.9
Dog	18,600	4.6	1.4	1.9	4.1
Elephant	3,670,000	4.2	0.4	1.3	3.3
Guinea pig	450	14.5	2.5		
Hamster, golden	70	18.4	3.2		
Horse	420,000	5.5	0.7	1.2	4.3
Monkey, rhesus	4,900	7.0	1.2	5.3	2.9
Mouse, albino	21	10.1	10.3	4.3	16.1
Mouse, deer	20	9.0	6.4		
Rabbit	3,670	11.3	1.7	7.4	5.6
Rat, albino	225	13.9	2.4	5.8	10.5
Rat, cotton	130	17.7			
Rat, kangaroo	106	5.5	2.7		
Vole	29	21.1	6.2		
Chicken	1,550	13.0	3.1		

Source: Spector, W.S., ed., *Handbook of Biological Data*, WADC Technical Report 56-273, Wright-Patterson Air Force Base, OH, 1956.

TABLE A.38
Water Balance for Resting Aquatic Animals

Animal	Temperature (°C)	Body Volume (mL)	Water Intake plus Metabolic Water (% Body Volume/Day)
In fresh water			
Amphibia			
Bufo vulgaris		22	117
Rana esculenta		65	22
Rana pipiens	20	32	40
Rana temporaria		9	100
Salamandra		20	53
Triton marmoratus		5	43
Insecta			
Chironomus larva		0.1	22
Corethra larva	20	6.2	19
Crustacea			
Cambarus		13	5.3
Eriocheir	13	60	3.6
Potamobius		46	4.1
Annelida			
Lumbricus	19	4	60
Protozoa			
Amoeba proteus	23	3×10^{-6}	360
Cyclidium		2×10^{-9}	2.2×10^{4}
Euplotes	25	3×10^{-7}	10^{4}
Lembus	26	2×10^{-9}	6×10^{4}
Leucophrys	21	4.7×10^{-7}	3,300
Paramecium	22	1.9×10^{-7}	6,200
Rhabdostyla	15	8×10^{-9}	1.1×10^{4}
Spirostomum		2.2×10^{-6}	550
Zoöthamnium	15	1.4×10^{-8}	5,500
In sea water			
Pisces			
Anguilla		250	6.5
Myoxocephalus		180	11.5

(Continued)

TABLE A.38 (*Continued*)

Water Balance for Resting Aquatic Animals

Animal	Temperature (°C)	Body Volume (mL)	Water Intake plus Metabolic Water (% Body Volume/Day)
Crustacea			
Cancer		300	6.5
Carcinus		40	10.0
Maia		2,200	2.7
Protozoa			
Amoeba mira		6×10^{-9}	4,300
Cothurnia	15	1.2×10^{-8}	700
Zoöthamnium marinum	15	1.2×10^{-7}	750

Source: Spector, W.S., ed., *Handbook of Biological Data*, WADC Technical Report 56-273, Wright-Patterson Air Force Base, OH, 1956.

References

The mind is bigger than the sky.

—Emily Dickinson

Ablain, J., E. M. Durand, S. Yang, Y. Zhou, and L. I. Zon, 2015, A CRISPR/Cas9 vector system for tissue-specific gene disruption in zebrafish, *Dev. Cell* 32(6): 756–764.

Ackerman, D., 2001, The mystery of pollen, *Parade*, September 9, p. 12.

Ackerman, E., 2015, The race to build a real-life tricorder, *IEEE Spectr.* 6(15): 46–49, 85.

Ackerman, J., 2006, Dragonflies strange love, *Nat. Geogr.* 209(4): 105–119.

Ackerman, J. M., C. C. Nocera, and J. A. Bargh, 2010, Incidental haptic sensations influence social judgments and decisions, *Science* 328: 1712–1714.

Adams, B., 2007, From uncertainty to practice: The latest developments in nanotechnology, *Synergist* 18(8): 40–42.

Adleman, L. M., 1998, Computing with DNA, *Sci. Am.* 279(2): 54–61.

Adler, P. N. and J. Nathans, 2016, The cellular compass, *Sci. Am.* 314(3): 66–71.

Adolph, E. F., 1949, Quantitative relations in the physiological constitutions of mammals, *Science* 109: 579–585.

Aiello, J. R. and A. Baum, 1979, *Residential Crowding and Design*, Plenum Press, New York.

Albright, J. L. and C. W. Arave, 1997, *The Behavior of Cattle*, CAB International, Oxon, UK.

Alder, F., 2002, Bacteria-resistant parts, *Resource* 9(1): 13–14.

Alegre, M.-L. and T. F. Gajewski, 2016, Germ warfare, *Sci. Am.* 314(4): 50.

Alexander, C., 2015, The invisible war on the brain, *Nat. Geogr.* 227(2): 30–51.

Alexander, R. M., 2000, Hovering and jumping: Contrasting problems in scaling, in *Scaling in Biology*, J. H. Brown and G. B. West, (Eds.), Oxford University Press, New York, pp. 37–50.

Alexander, R. M., 2003, *Principles of Animal Locomotion*, Princeton University Press, Princeton, NJ.

Al-Jamal, K. T., 2015, Carbon nanotubes deliver in medicine, *Am. Sci.* 103(2): 122–127 .

Allen, L., 2004, Fightin' squirrels, *Pop. Sci.* 265(3): 54.

Allen, S. E., 2016, Healing the burn, *IEEE Pulse* 7(4): 24–29.

Alocilja, E. C., 2002, *Principles of Biosystems Engineering*, Eaudition Books, North Chelmsford, MA.

Amábile-Cuevas, C. F., 2003, New antibiotics and new resistance, *Am. Sci.* 91: 138–149.

Amábile-Cuevas, C. F., 2010, Zeroing in on cancer, *Am. Sci.* 98: 366–368.

Anbar, M., 2001, What is life? *IEEE Eng. Med. Biol.* 4(3): 15–19.

Anderson, A. J. and E. J. Underwood, 1976, Trace-element deserts, in *Readings From Scientific American: Conditions for Life*, W. H. Freeman, San Francisco, CA, pp. 39–45.

Anderson, P. W., 1972, More is different, *Science* 177(4047): 393–396.

Andrade, J., 2000, Summary of biomaterials subjects, talk given at *Biomedical Engineering Educational Summit*, December 7–10, Whitaker Foundation, Arlington, VA.

Anonymous, 2001, Genetically engineered tomato grows in salty water, *Calif. Agric.* 55(5): 4.

Anonymous, 2007, Real-time, small-scale measurements in biological systems, *Biosci. Technol.* 32(4): 20, 22.

Anscombe, N., 2015, A different view, *IEEE-EMBS Pulse* 6(4): 14–19.

Antonacci, G. and S. Braakman, 2016, Biomechanics of subcellular structures by non-invasive brillouin microscopy, *Sci. Rpt.* 6, doi:10.1038/srep37217.

Appel, R. and B. Cocroft, 2014, Plants respond to leaf vibrations caused by insect herbivore chewing, *Oecologia* 175(4): 1257–1266.

Arai, J. A., S. Li, D. M. Hartley, and L. A. Feig, 2009, Transgenerational rescue of a genetic defect in long-term potentiation and memory formation by juvenile enrichment, *J. Neurosci.* 29(5): 1496–1502.

Arjamaa, O. and T. Vuorisalo, 2010, Gene-culture coevolution and human diet, *Am. Sci.* 98: 140–147.

ASABE, 2010, New high-quality maize genome sequence will advance basic and applied research, *Resource* 17(3): 24–25.

ASAE, 2001, Touching plants may alter their future, *Resource* 8(4): 4–6.

ASAE, 2002, Regulator of heat sensitivity discovered in tomatoes, *Resource* 9(8): 6.

Aschehoug, E. T., F. S. Sivakoff, H. L. Cayton, W. F. Morris, and N. M. Haddad, 2015, Habitat restoration affects immature stages of a wetland butterfly through indirect effects on predation, *Ecology* 96: doi:10.1890/14–2403.1.

ASEE, 2002, Submarine built for two, *Prism* 12(2): 12.

Ash, S., C. Olm, C. T. McMillan, A. Boller, D. J. Irwin, L. McCluskey, L. Elman, and M. Grossman, 2015. Deficits in sentence expression in amyotrophic lateral sclerosis. *Amyotroph. Lateral Scler. Frontotemporal Degener.* 16: 31–39.

Ashe, A., A. Sapetschnig, E.-M. Weick, J. Mitchell, M. P. Bagijn, A. C. Cording, A.-L. Doebley et al., 2012, piRNAs can trigger a multigenerational epigenetic memory in the germline of C. elegans, *Cell* 150(1): 88–99.

ASPCA, 2002, Pet tips, www.shadetreemechanic.com/aspca%20news.htm.

Autumn, K., M. Sitti, Y. A. Liang, A. M. Peattie, W. R. Hansen, S. Sponberg, T. W. Kenny et al., 2002, Evidence for van der Waals adhesion in gecko setae, *Proc. Natl. Acad. Sci.* 99: 12252–12256.

Azari, N. P. and R. J. Seitz, 2000, Brain plasticity and recovery from stroke, *Am. Sci.* 88: 426–431.

Babbit, G. A., 2011, Chromatin evolving, *Am. Sci.* 99: 48–55.

Baeuerle, P. A., 2008, Can antibodies help cure cancer? *Drug Discov. Dev.* 11(5): 32–35.

Baier, R., 2002, Thirty years of thromboresistance: How it has been obtained, founders award address, *Proceedings of the Annual Meeting*, Society for Biomaterials, Minneapolis, MN.

Bailey, A. A. and P. L. Hurd, 2005b, Finger length ratio (2D:4D) correlates with physical aggression in men but not in women, *Biol. Psychol.* 68(3): 215–222.

Bailey, A. A. and P. L. Hurd, 2005a, Depression in men is associated with more feminine finger length ratios, *Personal. Individ. Differ.* 39(4): 829–836.

Bak, P. and M. Paczuski, 1997, Mass extinctions vs. uniformitarianism in biological evolution, in *Physics of Biological Systems: From Molecules to Species*, H. Flyvbjerg, J. Hertz, M. H. Jensen, O. G. Mouritsen, and K. Sneppen, eds., Springer-Verlag, Berlin, Germany, pp. 341–356.

Baker, M., 2011, When worlds collide, *HHMI Bull.* 24(1): 8–9.

Balcombe, J., 2016a, Einstein of the sea, *Sci. Am.* 314(6): 66–69.

Balcombe, J., 2016b, *What a Fish Knows: The Inner Lives of Our Underwater Cousins*, McMillan and Co., New York.

Barnes, J. W., 1994, *Statistical Analysis for Engineers and Scientists: A Computer-Based Approach*, McGraw-Hill, New York.

Baron-Cohen, S., 2005, The essential difference: The male and female brain, *Phi Kappa Phi Forum* 85(1): 23–26.

Baron-Cohen, S., 2011, The empathy bell curve, *Phi Kappa Phi Forum* 91(1): 10–12.

Barrie, M. D., J. Baker, M. D. Hoover, and C. L. Geraci, 2017, Nanobiomonitoring and surveillance, *Synergist* 28(2): 28–31.

Bartke, A., 2015, What are we learning from mutant mice that live very long and healthy lives? *Phi Kappa Phi Forum* 95(2): 9–12.

Bartlett, K. H., M. Martinez, and J. Bert, 2004, Modeling of occupant-generated CO_2 dynamics in naturally ventilated classrooms, *J. Occup. Environ. Hyg.* 1: 139–148.

Bartlett, R. G., 1973, Respiratory system, in *Bioastronautics Data Book*, J. F. Parker Jr. and V. R. West, eds., NASA, Washington, DC, pp. 489–531.

Bassingthwaighte, J. B., 1998, Physiome Project, www.physiome.org (accessed July, 2000).

Bassingthwaighte, J. B., 2000, Strategies for the physiome project, *Ann. Biomed. Eng.* 28: 1043–1058.

BD Biosciences, 2004, BD living colors reef coral fluorescent proteins, www.bdbiosciences.com (accessed April 13, 2004).

Beckwith, J., 2001, Triumphalism in science: A review of the triumph of sociobiology, *Am. Sci.* 89: 471–472.

Bedi, G., F. Carrillo, G. A. Cecchi, D. F. Slezak, M. Sigman, N. B. Mota, S. Ribeiro, D. C. Javitt, M. Copelli, and C. M. Corcoran, 2015, Automated analysis of free speech predicts psychosis onset in high-risk youths, *npj Schizophrenia* 1: Article number: 15030, doi:10.1038/npjschz.2015.30.

Beech, H. and A. Perry, 2011, Killing fields, *Time* 177(24): 40–47.

Beech, J. R. and I. C. Mackintosh, 2005, Do differences in sex hormones affect handwriting style? evidence from digit ratio and sex role identity as determinants of the sex of handwriting, *Personal. Individ. Differ.* 39(2): 459–468.

Behbehani, K., 1995, Mechanical ventilation, in *The Biomedical Engineering Handbook*, J. Bronzino, ed., CRC Press, Boca Raton, FL, pp. 1301–1310.

Bejan, A. and J. H. Marden, 2006, Constructing animal locomotion from new thermodynamics theory, *Am. Sci.* 94: 342–349.

Bell, G., 1971, Mathematical model of clonal selection and antibody production II, *J. Theor. Biol.* 33: 339–378.

Bell, T., 2010, Designing against disease, *The Bent of Tau Beta Pi* 101(1): 13–18.

Bellamkonda, R. and P. Aebischer, 1995, Tissue engineering in the nervous system, in *The Biomedical Engineering Handbook*, J. Bronzino, ed., CRC Press, Boca Raton, FL, pp. 1754–1773.

Belmonte, J. C. I., 2016, Human organs from animal bodies, *Sci. Am.* 315(5): 32–37.

Benach, J. L., E. Li, and M. M. McGovern, 2012, A microbial association with autism, *mBIO* 3(1): e00019–12. doi:10.1128/mBio.00019-12.

Ben-Jacob, E., I. Cohen, and A. Czirok, 1997, Smart bacterial colonies, in *Physics of Biological Systems: From Molecules to Species*, H. Flyvbjerg, J. Hertz, M. H. Jensen, O. G. Mouritsen, and K. Sneppen, eds., Springer-Verlag, Berlin, Germany, pp. 307–324.

Bennetto, H. P., 1990, Electricity generation by microorganisms, *Biotech. Educ.* 1(4): 163–168.

Berger, D., 2002, Does biochemistry have to be organic? *Phi Kappa Phi Forum* 82(1): 6–7.

Berger, T. W., A. Ahuja, S. H. Courellis, S. A. Deadwyler, G. Erinjippurath, G. A. Gerhardt, G. Gholmieh et al., 2005, Restoring lost cognitive function, *IEEE Eng. Med. Biol. Mag.* 24(5): 30–44.

Berman, M. G., J. Jonides, and S. Kaplan, 2008, The cognitive benefits of interacting with nature, *Psychol. Sci.* 19(12): 1207–1212.

Berryman, A. A., 1999, *Principles of Population Dynamics and Their Application*, Stanley Thornes, Ltd., Cheltenham, UK.

Berthiaume, F. and M. L. Yarmush, 1995, Tissue engineering, in *The Biomedical Engineering Handbook*, J. Bronzino, ed., CRC Press, Boca Raton, FL, pp. 1556–1566.

Bethel, J. P., ed., 1960, *Webster's New Collegiate Dictionary*, G. & C. Merriam, Springfield, MA.

Biergans, S. D., C. Claudianos, J. Reinhard, and C. G. Galizia, 2016, DNA methylation adjusts the specificity of memories depending on the learning context and promotes relearning in honeybees, *Front. Mol. Neurosci.* doi:10.3389/fnmol.2016.00082.

Biewener, A. A., 2000, Scaling of terrestrial support: Differing solutions to mechanical constraints of size, in *Scaling in Biology*, J. H. Brown and G. B. West, eds., Oxford University Press, New York, pp. 51–66.

Billings, C. E., 1973a, Barometric pressure, in *Bioastronautics Data Book*, J. F. Parker Jr. and V. R. West, eds., NASA, SP-3006 U.S. Government Printing Office, Washington, DC, pp. 1–34.

Billings, C. E., 1973b, Atmosphere, in *Bioastronautics Data Book*, J. F. Parker Jr. and V. R. West, eds., NASA, SP-3006 U.S. Government Printing Office, Washington, DC, pp. 35–63.

Bingham, M. D., C. W. Suggs, and C. F. Abrams, 1992, Vibration attenuation of cushioned gloves, *Appl. Engr. Agr.* 8(1): 4–8.

Bionics Symposium, 1960, Living prototypes—The key to new technology, September 13–15, WADD Technical Report 60-600, Wright-Patterson Air Force Base, OH.

Birch, M. C., 1978, Chemical communication in pine bark beetles, *Am. Sci.* 66: 409–419.

Bitter, S. D. and J. K. Bowers, 1994, Bioretention as a water quality best management practice, *Watershed Prot. Tech.* 1(3): 114–116.

Bloch, H., 2011, Shake it off, *Nat. Geogr.* 220(3): 32.

Blood, R. P., M. G. Yost, J. E. Camp, and R. P. Ching, 2015, Whole-body vibration exposure intervention among professional bus and truck drivers: A laboratory evaluation of seat-suspension designs, *J. Occup. Environ. Hyg.* 12(6): 351–362.

Bocklandt, S., W. Lin, M. E. Sehl, F. J. Sánchez, J. S. Sinsheimer, S. Horvath, and E. Vilain, 2011, Epigenetic predictor of age, *PLoS One* 6(6): e14821.

Boehm, C., 2012, Ancestral hierarchy and conflict, *Science* 336(6083): 844–847.

Boeing, 2004, Desired attributes of an engineer, www.boeing.com/companyoffices/pwu/attributes/attributes.html (accessed September, 2006).

Bogaert, A. F., C. C. Fawcett, and L. K. Jamieson, 2009, Attractiveness, body size, masculine sex roles and 2D:4D ratios in men, *Personal. Individ. Differ.* 47(4): 273–278.

Bogas, D., L. Nyberg, R. Pacheco, N. F. Azevedo, J. P. Beech, M. Gomila, J. Lalucat et al., 2017, Applications of optical DNA mapping in microbiology, *BioTechiques* 62(6): 255–267.

Bonner, J. T. and H. S. Horn, 2000, Allometry and natural selection, in *Scaling in Biology*, J. H. Brown and G. B. West, eds., Oxford University Press, New York, pp. 25–35.

Booth, A., 1976, *Urban Crowding and Its Consequences*, Praeger, New York.

Booth, F. W. and P. D. Neufer, 2005, Exercise controls gene expression, *Am. Sci.* 93: 28–35.

Bormann, F. H. and G. E. Likens, 1979, Catastrophic disturbance and the steady state in northern hardwood forests, *Am. Sci.* 67: 660–669.

Borrell, B., 2012, Dust up, *Sci. Am.* 306(1): 80–83.

Bradley, R., 2014, Reprogrammable cells, *Sci. Am.* 311(6): 47.

Brandtzaeg, P., 2007, Why we develop food allergies, *Am. Sci.* 95: 28–35.

Branswell, H., 2015, Ebola war, *Sci. Am.* 312(3): 48–55.

Bratman, G. N., J. P. Hamilton, K. S. Hahn, G. C. Daily, and J. J. Gross, 2015, Nature experience reduces rumination and subgenual prefrontal cortex activation, *Proc. Nat. Acad. Sci.* 112(28): 8567–8572.

Brill, W. J., 1979, Nitrogen fixation: Basic to applied, *Am. Sci.* 67: 458–466.

Briscoe, N. J., K. A. Handasyde, S. R. Griffiths, W. P. Porter, A. Krockenberger, and M. R. Kearney, 2014, Tree hugging koalas demonstrate a novel thermoregulatory mechanism for arboreal mammals, *Biol. Lett.* 10(6). doi:10.1098/rsbl.2014.0235.

Brodie, C., 2004a, Head in hand, *Am. Sci.* 92: 27–28.

Brodie, C., 2004b, TNT and talking cells, *Am. Sci.* 92: 415–416.

Brodie, C., 2006, Immune to fear, *Am. Sci.* 94: 121–122.

Brodie, C., 2007a, Persistently clean? *Am. Sci.* 95: 25–26.

Brodie, C., 2007b, Sensitive cells, *Am. Sci.* 95: 123.

Brody, S., 1945, *Bioenergetics and Growth*, Reinhold, New York.

Brooks, G. A., T. D. Fahey, and T. P. White, 1996, *Exercise Physiology: Human Bioenergetics and Its Applications*, Mayfield, Mountain View, CA.

Brooks, S. V., N. M. Cole, and J. A. Faulkner, 1995, Tissue engineering of skeletal muscle, in *The Biomedical Engineering Handbook*, J. Bronzino, ed., CRC Press, Boca Raton, FL, pp. 1774–1787.

Brosnan, S. F. and F. B. M. DeWaal, 2003, Monkeys reject unequal pay, *Nature* 425: 297–299.

Brown, A. S., J. Hooton, C. A. Schaefer, H. Zhang, E. Petkova, V. Babulas, M. Perrin, J. M. Gorman, and E. S. Susser, 2004, Elevated maternal interleukin-8 levels and risk of schizophrenia in adult offspring, *Am. J. Psychiatry* 161: 889–895.

Brown, A. S., 2012, Searching for that killer app! *Bent* 103(2): 12–16.

Brown, C. R., 2010, How cliff swallows choose where to live, *Phi Kappa Phi Forum* 90(1): 4–7.

Brown, J. H., G. B. West, and B. J. Enquist, 2000, Scaling in biology: Patterns and processes, causes and consequences, in *Scaling in Biology*, J. H. Brown and G. B. West, eds., Oxford University Press, New York, pp. 1–24.

Brown, K., 2006, Plants are us, *HHMI Bull.* 19(4): 16–19.

Brown-Brandl, T. M., 2008, Heat stress in feedlot cattle, CAB reviews: Perspectives in agriculture, veterinary science, nutrition and natural resources, Vol. 3, Cabi, Boston, MA, p. 16. doi:10.1079/PAVSNNR20083016.

Bruening, G., 2000, Transgenes are revolutionizing crop production, *Calif. Agric.* 54(4): 36–46.

Bruni, C., G. Koch, and R. Strom, 1975, A dynamical model of humoral immune response, *Math. Biosci.* 27: 191–212.

Buckman, R., 2003, *Human Wildlife: The Life That Lives on Us*, Johns Hopkins University Press, Baltimore, MD.

Buerk, D. G., 1993, *Biosensors: Theory and Applications*, Technomic Publishing, Lancaster, PA.
Buric, I., M. Farias, J. Jong, C. Mee, and I. A. Brazil, 2017, What is the molecular signature of mind–body interventions? A systematic review of gene expression changes induced by meditation and related practices. *Front. Immunol.* 8: 670, doi:10.3389/fimmu.2017.00670.
Burke, K. L., 2014, First person: Joan Strassman, *Am. Sci.* 102(5): 327–328.
Burke, K. L., 2015, The challenge of restoring biodiversity's outer edge, *Am. Sci.* 103(3): 166–167.
Burns, S., 2007, Sterilization: New gamma methods increase options for allograft materials, *Med. Dev. Diagn. Ind.* 29(3): 82–92.
Burt, A. and R. Trivers, 2006, *Genes in Conflict: The Biology of Selfish Elements*, Belknap Press of Harvard University Press, Cambridge, MA.
Bushdid, C., M. O. Magnasco, L. B. Vosshall, and A. Keller, 2014, Humans can discriminate more than 1 trillion olfactory stimuli, *Science* 343(6177): 1370–1372.
Buston, P. M. and M. Y. L. Wong, 2014, Why some animals forgo reproduction in complex societies, *Am. Sci.* 102: 290–297.
Buswell, A. M. and W. H. Rodebush, 1976, Water, in *Readings From Scientific American: Conditions for Life*, W. H. Freeman, San Francisco, CA, pp. 75–82.
Calder, W. A., 2000, Diversity and convergence: Scaling for conservation, in *Scaling in Biology*, J. H. Brown and G. B. West, eds., Oxford University Press, New York, pp. 297–323.
Callier, V., 2016, Cancerous co-conspirators, *Sci. Am.* 314(4): 22.
Campbell, N. A., J. B. Reece, and L. G. Mitchell, 1999, *Biology*, 5th ed., Addison-Wesley Longman, Menlo Park, CA.
Campbell, S., 2017, The quantified patient checks in, *IEEE Pulse* 8(4): 4–10.
Caporale, L. H., 2003, Foresight in genome evolution, *Am. Sci.* 91: 234–241.
Capra, F., 1996, *The Web of Life: A New Scientific Understanding of Living Systems*, Anchor Books, New York.
Caputo, R., 2002, Mother bear man, *Nat. Geogr.* 201(3): 89–101.
Carney, D. R., A. J. C. Cuddy, and A. J. Yap, 2010, Power posing: Brief nonverbal displays affect neuroendocrine levels and risk tolerance, *Psychol. Sci.* 21: 1363–1368.
Carslaw, H. S. and J. C. Jaeger, 1959, *Conduction of Heat in Solids*, Oxford University Press, London, UK.
Cartwright, F. F., 1972, *Disease and History*, The New American Library, New York.
Cena, E.-R., A. B. Joy, K. Heneman, and S. Zidenberg-Cherr, 2007, Low-income women in California may be at risk of inadequate foliate intake, *Calif. Agric.* 61(2): 85–89.
Cena, K., 1974, Radiative heat loss from animals and man, in *Heat loss from Animals and Man: Assessment and Control*, J. L. Monteith and L. E. Mount, eds., Butterworths, London, UK, pp. 33–58.
Chadwick, D., 2001, Evolution of whales, *Nat. Geogr.* 200(5): 65–77.
Chamary, J. V. and L. D. Hurst, 2009, The price of silent mutations, *Sci. Am.* 300(6): 46–53.
Chamberlain, W. B., 2003, Bionutrient modeling for design of in situ bioremediation, *Pollut. Eng.* 35: 28–33.
Champlin, C., D. Bell, and C. Shocken, 2017, AI medicine comes to Africa's rural clinics, *IEEE Spectr.* 54(5NA): 42–48.
Chandra, R., H. Zhou, I. Balasingham, and R. M. Narayanan, 2015, On the opportunities and challenges in microwave medical sensing and imaging, *IEEE Trans. Biomed. Engr.* 62(7): 1667–1682.
Chau, T. and K. Parker, 2004, On the robustness of stride frequency estimation, *IEEE Trans. Biomed. Eng.* 51: 294–303.
Chen, I., 2008, Know when to fold 'Em, *HHMI Bull.* 21(2): 8–9.
Chen, J. Y. and A. Y. Sivachenko, 2005, Data mining in protein interactomics, *IEEE Eng. Med. Biol. Mag.* 24(3): 95–102.
Chica, R. A., N. Doucet, and J. N. Pelletier, 2005, Semi-rational approaches to engineering enzyme activity: Combining the benefits of directed evolution and rational design, *Curr. Opin. Biotechnol.* 16: 378–384.
Chiu, L. S., 2007, Modern relics, *HHMI Bull.* 20(2): 32–37.
Chiu, L. S., 2008, Tick-tock goes a bacterial clock, *HHMI Bull.* 21(1): 29.

Chomsky, N., 1965, *Syntactic Structures*, Mouton, The Hague, the Netherlands.

Clabby, C., 2014, Laser-sharp vision, *Am. Sci.* 102(1): 18–19.

Clabby, C., 2011, Cracking cellular motion, *Am. Sci.* 99: 28–29.

Clauss, M., T. Froeschle, J. Castell, J.-M. Hatt, S. Ortmann, W. J. Streich, and J. Hummel, 2005, Fluid and Particle Retention Times in the Black Rhinoceros Diceros bicornis, a Large Hindgut-Fermenting Browser. *Acta Theriol.* 50: 367–376.

Cloud, J., 2010, Why genes aren't destiny, *Time* 175(2): 48–53.

Cohen, S., D. S. Krantz, G. W. Evans, and D. Stokols, 1981, Cardiovascular and behavioral effects of community noise, *Am. Sci.* 69: 528–535.

Collins, J. P., A. Kinzig, N. B. Grimm, W. F. Fagan, D. Hope, J. Wu, and E. T. Borer, 2000, A new urban ecology, *Am. Sci.* 88: 416–425.

Conn, P. M. and J. A. Janovick, 2005, A new understanding of protein mutation unfolds, *Am. Sci.* 93: 314–321.

Conniff, R. and D. A. Murawski, 2001, Deadly silk, *Nat. Geogr.* 200(2): 30–45.

Cooper, A., 2008, Growing plastic, *Pop. Sci.* 279(12): 28.

Corlett, E. N. and T. S. Clark, 1995, *The Ergonomics of Workspaces and Machines: A Design Manual*, Taylor & Francis Group, Bristol, PA.

Corley, A.-M., 2010, The danger-sensing driver's seat, *IEEE Spectr.* 47(7): 12–13.

Corno, C., G. B. Fiore, and M. L. Costantino, 2004, A mathematical model of neonatal tidal liquid ventilation integrating airway mechanics and gas transfer phenomena, *IEEE Trans. Biomed. Eng.* 51: 604–611.

Coulon, M., C. L. Wellman, I. S. Marjara, A. M. Janczak, and A. J. Zanella, 2013, Early adverse experience alters dendritic spine density and gene expression in prefrontal cortex and hippocampus in lambs, *Psychoneuroendocrinology* 38(7): 1112–1121.

Courage, K. H., 2015, Our personal vaccine helpers, *Sci. Am.* 312(6): 22.

Covert, M. W., 2014, Simulating a living cell, *Sci. Am.* 310(1): 44–51.

Cowen, R., 2015, Sound bytes, *Sci. Am.* 312(3): 44–47.

Creighton, L., 2005, The science of fun, *ASEE Prism*: 44–49.

Crichton, M., 1990, *Jurassic Park*, Knopf, New York.

Crockett, J. U., O. Tanner, and R. Christ, 1977, *Herbs*, Time-Life Books, Alexandria, VA.

Cuervo, L. A., 1976, Neuronal and synaptic function, in *Biological Foundations of Biomedical Engineering*, J. Kline, ed., Little, Brown and Company, Boston, MA, pp. 439–478.

Cummins, K., 2004, Biological monitoring and skin exposure assessment, *Synergist* 15(5/6): 35–36.

Cundiff, J. S. and K. R. Mankin, 2003, *Dynamics of Biological Systems*, ASAE, St. Joseph, MI, pp. 4–1–4–64.

Cundiff, J. S. and C. W. Suggs, 1976, Analytical model for rigid body vibration of a hand-held chain saw part I: The gasoline engine, *Trans. ASAE*: 647–653.

Curry, A., 2017, A 9,000-year love affair, *Nat. Geogr.* 231(2): 30–53.

Dadvand, P., M. J. Nieuwenhuijsen, M. Esnaola, J. Forms, X. Basagaña, M. Alvarez-Pedrerol, I. Rivas et al., 2015, Green spaces and cognitive development in primary schoolchildren, *Proc. Nat. Acad. Sci.* 112(26): 7937–7942.

Dalziel, C. F., 1972, Electric shock hazard, *IEEE Spectr.* 9: 41–50.

Dandekar, A. M. and N. Gutterson, 2000, Genetic engineering to improve quality, productivity, and value of crops, *Calif. Agric.* 54(4): 49–56.

Dao, N., P. J. McCormick, and C. F. Dewey Jr., 2000, The human physiome as an information environment, *Ann. Biomed. Eng.* 28: 1032–1042.

Davenport, J. R., 2008, Nerve cell navigation, *HHMI Bull.* 21(3): 32–37.

Davenport, J. R., 2009, When cells grow old, *HHMI Bull.* 22(1): 33–37.

Davidovits, P., 1975, *Physics in Biology and Medicine*, Prentice Hall, Englewood Cliffs, NJ.

Davidson, T. L. and A. L. Riley, 2015, Taste, sickness, and learning, *Am. Sci.* 103(3): 204–211.

Davies, A. G., J. T. Pierce-Shimomura, H. Kim, M. K. VanHoven, T. R. Thiele, A. Bonci, C. I. Bargmann, and S. L. McIntire, 2003, A central role of the BK potassium channel in behavioral responses to ethanol in *C. elegans*, *Cell* 115: 655–666.

Davis, A. P., M. Shokouhian, H. Sharma, and C. Minami, 2001, Laboratory study of biological retention (bioretention) for urban storm water management, *Water Environ. Res.* 73: 5–14.

Davis, C. C. and Z. Xi, 2015, Horizontal gene transfer in parasitic plants, *Curr. Opinion Plant Biol.* 26: 14–19.

Davis, E., 2014, *Genome Editing: Which Should I Choose, TALEN or CRISPR?* GeneCopoeia Online Technical Note, http://www.genecopoeia.com/resource/genome-editing-talen-or-crispr/ (accessed November 10, 2015).

Davis, L. E., 2004, Pass the piranhas, *Sky* (Delta Air Lines) May: 96–99.

Dawkins, R., 1976, *The Selfish Gene*, Oxford University Press, New York.

Dawkins, R., 1989, *The Selfish Gene*, 2nd ed., Oxford University Press, Cambridge, UK.

Dawson, T. H., 1991, *Engineering Design of the Cardiovascular System of Mammals*, Prentice Hall, Englewood Cliffs, NJ.

Dayan, F. E. and E. A. Dayan, 2011, Porphyrins: One ring in the colors of life, *Am. Sci.* 99: 236–243.

de Duve, C., 1995, *Vital Dust: Life as a Cosmic Imperative*, Basic Books, New York.

de Kruif, P., 1926, *Microbe Hunters*, Harcourt Brace and Company, San Diego, CA.

Dean, G. A., 1965, An analysis of the energy expenditure in level and grade walking, *Ergonomics* 8: 31–47.

Deleon, S., R. Halitschke, R. S. Hames, A. Kessler, T. J. DeVoogd, and A. A. Dhondt, 2013, The effect of polychlorinated biphenyls on the song of two passerine species, *PLoS One* 8(9): doi:10.1371/journal.pone.0073471.

Delsemme, A. H., 2001, An argument for the cometary origin of the biosphere, *Am. Sci.* 89: 432–442.

deMenocal, P. B., 2014, Climate shocks, *Sci. Am.* 311(3): 48–53.

Denison, T., M. Morris, and F. Sun, 2015, Building a bionic nervous system, *IEEE Spectr.* (15.2): 33–39.

Denworth, L., 2017, I feel your pain, *Sci. Am.* 317(6): 58–63.

Dewar, H., 2001, Dangerous liaisons in Hawaii, *The Baltimore Sun*, July 22, p. 2A.

DiMilla, P. A., K. Barbee, and D. A. Lauffenburger, 1991, Mathematical model for the effects of adhesion on mechanics and cell migration speed, *Biophys. J.* 60: 15–37.

Ding, Y., M. Fromm, and Z. Avramova, 2012, Multiple exposures to drought 'train' transcriptional responses in arabidopsis. *Nat. Commun.* 3: 740. doi:10.1038/ncomms1732.

Domingue, B. W., J. Fletcher, D. Conley, and J. D. Boardman, 2014, Genetic and educational assortive mating among U.S. adults, *PNAS* 111(2): 7996–8000.

Dorit, R., 2008, All things small and great, *Am. Sci.* 96: 284–286.

Dorit, R., 2014, The superorganism revolution, *Am. Sci.* 102(5): 330–333.

Dove, A., 2010, The search for animal alternatives, *Drug Discov. Dev.* May: 10–13.

Doyle, K., 2015, Cancer in cell therapy's crosshairs, *Genet. Engr. Biotech. News*, 35(7): 1, 30–32.

Drake, J. W., B. W. Glickman, and L. S. Ripley, 1983, Updating the theory of mutation, *Am. Sci.* 71: 621–630.

Duan, F., K. L. Curtis, and J. C. March, 2008, Secretion of insulinotropic proteins by commensal bacteria: Rewiring the gut to treat diabetes, *Appl. Environ. Microbiol.* 74(23): 7437–7438.

DuBois, E., 1898, Ueber die Abhängigkeit des Hirngewichtes von der Köpergrösse bei den Säugethieren, *Arch. Anthrop.* 25: 1–28.

Dudley, S. A. and A. L. File, 2008, Yes, kin recognition in plants! *Biol. Lett.* 4: 69–70.

Duffin, E. G., 1995, Implantable defibrillators, in *The Biomedical Engineering Handbook*, J. Bronzino, ed., CRC Press, Boca Raton, FL, pp. 1284–1300.

Dugatin, L. A. and M. Hasenjager, 2015, The networked animal, *Sci. Am.* 312(6): 50–55.

Dukes, H. H., 1955, *The Physiology of Domestic Animals*, Comstock, Ithaca, NY.

Dunker, A. K. and R. H. Kriwacki, 2011, The orderly chaos of proteins, *Sci. Am.* 304(4): 68–73.

Dunlap, K. A., M. Palmarini, M. Varela, R. C. Burghardt, K. Hayashi, J. L. Farmer, and T. E. Spencer, 2006, Endogenous retroviruses regulate periimplantation placental growth and differentiation, *Proc. Natl. Acad. Sci.* 103: 14390–14395.

Dunn, R., 2014, The evolution of architecture, *Sci. Am.* 311(5): 72–77.

Dunsworth, H. M., A. G. Warrener, T. Deacon, P. T. Ellison, and H. Pontzer, 2012, Metabolic hypothesis for human altriciality, *PNAS* 109(38): 15212–15216.

Durand, P., A. Rashidi, and R. Michod, 2011, How an organism dies affects the fitness of its neighbors, *Am. Nat.* 177: 224–232.

Dutton, G., 2006, Oligos enable new classes of therapeutics, *Genet. Eng. Biotechnol. News (GEN) Suppl.* 26(8): 4–6.

Dyer, F. C. and J. L. Gould, 1983, Honey bee navigation, *Am. Sci.* 71: 587–597.

Ebbinghaus, S., A. Dhar, J. D. McDonald, and M. Gruebele, 2010, Protein folding stability and dynamics imaged in living cell, *Nat. Methods* 7: 319–323.

Eberhard, W. G. and W. T. Wcislo, 2012, Plenty of room at the bottom? *Am. Sci.* 100(3): 226–233.

Ebersole, R. S., 2001, What's on my flipper? *Pop. Sci.* 259(3): 26.

Economos, A. C., 1983, Elastic and/or geometric similarity in mammalian design? (letter to the editor), *J. Theor. Biol.* 103: 167–172.

Elson, M. L., 2009, Nutrition, in the palm of your hand, *Healthy Aging* 4(5): 39–42.

Elstein, D., 2003, Plants' biological clocks help them prepare for the day, *Agric. Res.* 51: 11.

Engen, T., 1987, Remembering odors and their names, *Am. Sci.* 75: 497–503.

Englehart, K. and B. Hudgins, 2003, A robust, real-time control scheme for multifunction myoelectric control, *IEEE Trans. Biomed. Eng.* 50(7): 848–854.

Enquist, B. J., G. B. West, and J. H. Brown, 2000, Quarter-power allometric scaling in vascular plants: Functional basis and ecological consequences, in *Scaling in Biology*, J. H. Brown and G. B. West, eds., Oxford University Press, New York, pp. 167–198.

Erickson, R. P., 1984, On the neural bases of behavior, *Am. Sci.* 72: 233–241.

Eshbach, O. W. and M. Souders, 1975, *Handbook of Engineering Fundamentals*, John Wiley & Sons, New York.

Eslami-Mossallam, B., R. D. Schram, M. Tompitak, J. van Noort, and H. Schiessel, 2016, Multiplexing genetic and nucleosome codes: A computational approach, *PLoS One*. doi:10.1371/journal.pone.0156905.

Everts, S., 2013, Mirror molecules, *Sci. Am.* 308(5): 79–81.

Fairley, P., 2003, Germs that build circuits, *IEEE Spectr.* 40(11): 37–41.

Farrar, J. L., J. J. Nunez, and R. M. Davis, 2002, Repeated soil applications of fungicide reduce activity against cavity spot in carrots, *Calif. Agric.* 56(2): 76–79.

Feeney, R. E., 1974, A biological antifreeze, *Am. Sci.* 62: 712–719.

Feldman, R., Z. Rosenthal, and A. I. Eidelman, 2014, Maternal-preterm skin-to-skin contact enhances child physiologic organization and cognitive control across the first 10 years of life, *Biol. Psych.* 75: 56–64.

Fenn, E. A., 2001, *Pox Americana: The Great Smallpox Epidemic of 1775–82*, Hill & Wang, New York.

Fenster, J. M., 1998, The inner limits, *Am. Herit. Invent. Technol.* 13(3): 10–18.

Fenster, J. M., 1999, The conquest of diabetes, *Am. Herit. Invent. Technol.* 14(3): 48–55.

Fenton, M. B. and J. H. Fullard, 1981, Moth hearing and the feeding strategies of bats, *Am. Sci.* 69: 266–275.

Ferrari, M. and M. T. H. Chi, 1998, The nature of naïve explanations of natural selection, *Int. J. Sci. Educ.* 20(10): 1231–1256.

Fessenden, M., 2010, The fantastic world of plants, *CALS News* Spring: 15–18.

Fick, A., 1881, The development of heat by muscular activity, *Science* 2(79): 618–621.

Fields, R. D., 2008, White matter matters, *Sci. Am.* 298(3): 54–61.

Fink, B., J. T. Manning, and N. Neave, 2004, Second to fourth digit ratio and the 'big five' personality factors, *Personal. Individ. Differ.* 37(3): 495–503.

Fink, B., N. Neave, K. Laughton, and J. T. Manning, 2006, Second to fourth digit ratio and sensation seeking, *Personal. Individ. Differ.* 41(7): 1253–1262.

Fischer, S., 2015, State of the ART, *IEEE-EMBS Pulse* 6(1): 8–13.

Fischer, S., 2016, What lies within, *IEEE Pulse* 7(5): 16–19.

Foer, J., 2010, The truth about chimps, *Nat. Geogr.* 217(2): 130–145.

Fogiel, M., 1999, *The Biology Problem Solver*, Research & Education Association, Piscataway, NJ.

Fonseca, J. M. and S. Ravishakar, 2007, Safer salads, *Am. Sci.* 95(6): 494–501.

Forde, M. and P. Ridgely, 1995, Implantable cardiac pacemakers, in *The Biomedical Engineering Handbook*, J. Bronzino, ed., CRC Press, Boca Raton, FL, pp. 1258–1269.

Foster, K. R., P. R. Wolpe, and A. L. Caplan, 2003, Bioethics and the brain, *IEEE Spectr.* 40(6): 34–39.

Foutz, T., 1996, Adaptation of the skeleton to mechanical usage: The control mechanism, *IBE Newsl.* 1(3): 2–3.

Franklin, D., 2012, Nature that nurtures, *Sci. Am.* 306(3): 24–25.

Freed, L. E. and G. Vunjak-Novakovic, 1995, Tissue engineering of cartilage, in *The Biomedical Engineering Handbook*, J. Bronzino, ed., CRC Press, Boca Raton, FL, pp. 1788–1806.

Fried, G. H. and G. J. Hademenos, 1999, *Theory and Problems of Biology, Schaum's Outlines*, McGraw-Hill, New York.

Friend, T., 2004, *Animal Talk: Breaking the Codes of Animal Language*, Free Press, New York.

Fries, J. F. and L. M. Crapo, 1981, *Vitality and Aging*, W. H. Freeman, San Francisco, CA.

Fromme, P. and J. C. H. Spence, 2017, Split-second reactions, *Sci. Am.* 316(5): 62–67.

Frost, H. M., 1987, Bone mass and the mechanostat, *Anat. Rec.* 219(1): 1–9.

Fuchikawa, T., A. Eban-Rothschild, M. Nagari, Y. Shemesh, and G. Bloch, 2016, Potent social synchronization can override photic entrainment of circadian rhythms, *Nat. Commun.* 7, doi:10.1038/ncomms11662.

Full, R. J. and M. S. Tu, 1991, Mechanics of a rapid running insect: Two-, four-, and six-legged locomotion, *J. Exp. Biol.* 156: 215–231.

Fuller, A., 2001, An electronic victory against epilepsy, *Am. Herit. Invent. Technol.* 17(2): 40.

Gage, F. H. and A. R. Muotri, 2012, What makes each brain unique, *Sci. Am.* 306(3): 26–31.

Gagliano, M., M. Grimonprez, M. Depczynski, and M. Renton, 2017, Tuned in: Plant roots use sound to locate water, *Oecologia* 184(1): 151–160.

Gaidos, S. and J. Keeley, 2006, A life-altering chemical, *HHMI Bull.* 19(1): 10–11.

Gallego-Perez, D., D. Pal, S. Ghatak, V. Malkoc, N. Higuita-Castro, S. Gnyawali, L. Chang et al., 2017, Topical tissue nano-transfection mediates non-viral stroma reprogramming and rescue, *Nat. Nanotechnol.* 12(10): 974–979.

Galletti, P. M. and C. K. Colton, 1995, Artificial lungs and blood-gas exchange devices, in *The Biomedical Engineering Handbook*, J. Bronzino, ed., CRC Press, Boca Raton, FL, pp. 1879–1897.

Galletti, P. M. and H. O. Jauregui, 1995, Liver support systems, in *The Biomedical Engineering Handbook*, J. Bronzino, ed., CRC Press, Boca Raton, FL, pp. 1952–1966.

Galletti, P. M., 1995, Prostheses and artificial organs, in *The Biomedical Engineering Handbook*, J. Bronzino, ed., CRC Press, Boca Raton, FL, pp. 1827–1837.

Galletti, P. M., C. K. Colton, and M. J. Lysaght, 1995b, Artificial kidney, in *The Biomedical Engineering Handbook*, J. Bronzino, ed., CRC Press, Boca Raton, FL, pp. 1898–1922.

Galletti, P. M., C. K. Colton, and M. J. Lysaght, 1995c, Artificial kidney, in *The Biomedical Engineering Handbook*, J. Bronzino, ed., CRC Press, Boca Raton, FL, pp 1898–1922.

Galletti, P. M., C. K. Colton, M. Jaffrin, and G. Reach, 1995a, Artificial pancreas, in *The Biomedical Engineering Handbook*, J. Bronzino, ed., CRC Press, Boca Raton, FL, pp. 1967–1984.

Gallup, G. G., 1979, Self-awareness in primates, *Am. Sci.* 67: 417–421.

Galston, A. W. and C. L. Slayman, 1979, The not-so-secret life of plants, *Am. Sci.* 67: 337–344.

Gammon, K., 2013, Time to sprout, *HHMI Bull.* 26(3): 10–11.

Ganetsky, B., 2000, Tracking down a cheating gene, *Am. Sci.* 88: 128–135.

Ganong, W. F., 1963, *Review of Medical Physiology*, Lange Medical Publications, Los Altos, CA.

Garrett, R. H. and C. M. Grisham, 1999, *Biochemistry*, Saunders, Fort Worth, TX.

Gatlin, L. L., 1972, *Information Theory and the Living System*, Columbia University Press, New York.

Gavrilov, L. and N. Gavrilova, 2004, Why we fall apart, *IEEE Spectr.* 41(9): 30–35.

Gervasi, D. D. and F. P. Schiestl, 2017, Real-time divergent evolution in plants driven by pollinators, *Nat. Comm.* 8, doi:10.1038/ncomms14691.

Gessert, M., 2003, Lamb chats, *Shepherd* 48(6): 20.

Gessner, R. C., C. B. Frederick, E. S. Foster, and P. A. Dayton, 2013, Acoustic angiography: A new imaging modularity for assessing microvasculature architecture, *Int. J. Biomed. Imag.* doi:10.1155/2013/9365.

Gibson, S., 2017, A good night's sleep, *Synergist* 28(1): 30–33.

Gilad, Y., A. Oshlack, G. K. Smyth, T. P. Speed, and K. P. White, 2006, Expression profiling in primates reveals a rapid evolution of human transcription factors, *Nature* 440: 242–245.

Gillespie, R. G., 2005, The ecology and evolution of Hawaiian spider communities, *Am. Sci.* 93: 122–131.

Glover, J., 2009, Crops, roots, and soil biological processes: Synergistic interactions, *AAA Annual Meeting*, Chicago, IL.

Goforth, S., 2009, Piecing together rotavirus's unique approach, *HHMI Bull.* 22(4): 8–9.

Goforth, S., 2012, An intentional life, *HHMI Bull.* 25(1): 18–23.

Goldberger, A. L., B. J. West, T. Dresselhaus, and V. Bhargava, 1985, Bronchial asymmetry and Fibonacci scaling, *Experimentia* 41: 1537–1538.

Goldhamer, D. A., E. Fereres, and M. Salinas, 2003, Can almond trees directly dictate their irrigation needs? *Calif. Agric.* 57(4): 138–144.

Goldman, D. I. and D. L. Hu, 2010, Wiggling through the world, *Am. Sci.* 98: 314–323.

Goodsell, D. S., 2000, Biomolecules and nanotechnology, *Am. Sci.* 88: 230–237.

Goodsell, D. S., 2004, Molecule of the month: Green fluorescent protein, www.resb.org/pdb/molecules/pdb42_1.html (accessed April 13, 2004).

Gopnik, A., 2010, How babies think, *Sci. Am.* 303(1): 76–81.

Gordon, D. M., 2016, Collective wisdom of ants, *Am. Sci.* 314(2): 44–47.

Gorman, C., 2001, New hope for failing hearts, *Time* 158(23): 81.

Gotelli, N. J., 1998, *A Primer of Ecology*, Sinauer Associates, Sunderland, MA.

Gould, F., K. Magori, and Y. Huang, 2006, Genetic strategies for controlling mosquito-borne diseases, *Am. Sci.* 94: 238–246.

Gould, J. L., 1980, The case for magnetic sensitivity in birds and bees (such as it is), *Am. Sci.* 68: 256–267.

Gould, S. J., 1986, Evolution and the triumph of homology, or why history matters, *Am. Sci.* 74: 60–69.

Gould, S. E., 2012, Underground network, *Sci. Am.* 307(4): 28.

Gould, S. J., 1986, Evolution and the triumph of homology, or why history matters, *Am. Sci.* 74(1): 60–69.

Grant, P. R. and B. R. Grant, 2002, Adaptive radiation of Darwin's finches, *Am. Sci.* 90: 130–139.

Graves, R. E., 1998, Design configurations for feeding spaces, in *Dairy Feeding Systems: Management, Components, and Nutrients*, NRAES, Ithaca, NY, pp. 153–166.

Greatbatch, W., 2000, *The Making of the Pacemaker*, Prometheus Books, Amherst, NY.

Green, M. P., L. D. Spate, T. E. Parks, K. Kimura, C. N. Murphy, J. E. Williams, M. S. Kerley, J. A. Green, D. H. Keisler, and R. M. Roberts, 2008, Nutritional skewing of conceptus sex in sheep: Effects of a maternal diet enriched in rumen-protected polyunsaturated fatty acids (PUFA), *Reprod. Biol. Endocrinol.* 6: 21. doi:10.1186/1477-7827-6-21.

Greenberg, R. and B. R. Tufts, 2001, Infecting other worlds, *Am. Sci.* 89: 296–299.

Greene, K., 2001, Unleash the bugs, *Pop. Sci.* 259(5): 34.

Greenough, W. T., 1975, Experiential modification of the developing brain, *Am. Sci.* 63: 37–46.

Greenwood, V., 2016, Beyond reasonable doubt, *Nat. Geogr.* 230(1): 30–55.

Greenwood, V., 2018, The clocks within our cells, *Sci. Amer.* 319(1): 50–57.

Grene, M., 1987, Hierarchies in biology, *Am. Sci.* 75: 504–510.

Grewal, D. and P. Salovey, 2005, Feeling smart: The science of emotional intelligence, *Am. Sci.* 93: 330–339.

GRHC, 2016, Green roof benefits, green roofs for healthy cities, http://www.greenroofs.org/index.php/about/greenroofbenefits (accessed January 17, 2017).

Gribble, G. W., 2004, Amazing organohalogens, *Am. Sci.* 92: 342–349.

Grifantini, K., 2016, Electrical stimulation: A panacea for disease? *IEEE Pulse* 7(4): 30–35.

Grodins, F. A., 1981, Models, in *Regulation of Breathing*, T. F. Hornbein, ed., Marcel Dekker, New York, pp. 1313–1351.
Grose, T. K., 2009, Cyber swarm, *ASEE Prism* 19(4): 19.
Grose, T. K., 2004, Really cricket, *ASEE Prism* 12(4): 38–42.
Grose, T. K., 2014, Damage control, *ASEE Prism* 24(3): 30–33.
Grossmann, P., 1967, Chemical engineering and biology, *Chem. Ing. Tech.* 39(21): 1217–1225.
Gulbins, E. and F. Lang, 2001, Pathogens, host-cell invasion and disease, *Am. Sci.* 89: 406–413.
Guo, Y., Z. Zheng, J. J. LaClair, J. Chory, and J. P. Noel, 2013, Smoke-derived karrikin perception by the α/β-hydrolase KAI2 from *Arabidopsis*, *Proc. Natl. Acad. Sci. U S A* 110(20): 8284–8289.
Gupta, R. A., N. Shah, K. C. Wang, J. Kim, H. M. Horlings, D. J. Wong, M.-C. Tsai et al., 2010, Long non-coding RNA HOTAIR reprograms chromatin state to promote cancer metastasis, *Nature* 464(7291): 957–1094.
Hackel, L., B. B. Doll, and D. M. Amodio, 2015, Instrumental learning of traits versus rewards: Dissociable neural correlates and effects on choice, *Nat. Neurosci.* 18(9): 1233–1235.
Hadley, N. F., 1980, Surface waxes and integumentary permeability, *Am. Sci.* 68: 546–553.
Haefner, J. W., 1996, *Modeling Biological Systems: Principles and Applications*, Chapman & Hall, New York.
Hale, W. G., J. P. Margham, and V. A. Saunders, 1995, *Collins Dictionary of Biology*, Harper Collins, Glasgow, UK.
Hall, S. S., 2015, Solving the autism puzzle, *MIT Technol. Rev.* 118(1): 36–43.
Hamadeh, H. and C. A. Afshari, 2000, Gene chips and functional genomics, *Am. Sci.* 88: 508–515.
Hamady, M., T. H. T. Cheung, K. Resing, K. J. Cios, and R. Knight, 2005, Key challenges in proteomics and proteoinformatics, *IEEE Eng. Med. Biol. Mag.* 24(3): 34–40.
Hamilton, W. D., 1964a, The genetical evolution of social behaviour I, *J. Theor. Biol.* 7: 1–16.
Hamilton, W. D., 1964b, The genetical evolution of social behaviour II, *J. Theor. Biol.* 7: 17–52.
Harbaugh, T. E., 1972, A validation study of the Maslow hierarchy of needs, PhD thesis, University of Maryland, College Park, MD.
Hardin, P. E., J. C. Hall, and M. Rosbash, 2013, Circadian oscillations in period gene mRNA levels are transcriptionally regulated, *PNAS* 89(24): 11711–11715.
Hardison, R., 1999, The evolution of hemoglobin, *Am. Sci.* 87: 126–137.
Harlow, H., 1958, The nature of love, *Am. Psychol.* 13: 673–685.
Harper, B. K., S. A. Mabon, S. M. Leffel, M. D. Halfhill, H. A. Richards, K. A. Moyer, and C. N. Stewart Jr., 1999, Green fluorescent protein as a marker for expression of a second gene in transgenic plants, *Nat. Biotechnol.* 17: 1125–1129.
Harper, H. A., 1963, *Review of Physiological Chemistry*, Lange Medical Publications, Los Altos, CA.
Harris, S., 2001, Entropy cartoon, *Am. Sci.* 89: 431.
Harrison, F., A. Roberts, K. Rumbaugh, C. Lee, and S. P. Diggle, 2015, A 1000-year-old antimicrobial remedy with antisthaphylococcal activity, paper S19Ewe1006, *Society for General Microbiology Annual Conference*, Birmingham, UK.
Harry, J. D., J. B. Niemi, A. A. Priplata, and J. J. Collins, 2005, Balancing act, *IEEE Spectr.* 42(4): 37–41.
Harte, J., 2002, Toward a synthesis of the Newtonian and Darwinian worldviews, *Phys. Today* 55(10): 29–34.
Haurwitz, R. E., M. Jinek, B. Wiedenheft, K. Zhou, and J. A. Doudna, 2010, Sequence- and structure-specific RNA processing by a CRISPR endonuclease, *Science* 329(5997): 1355–1358.
Hauser, M. D., 2000, What do animals think about numbers? *Am. Sci.* 88: 144–151.
Hawkins, J., 2017, What intelligent machines need to learn from the neocortex, *IEEE Spectr.* 54(6NA): 34–39, 70–71.
Hawks, J., 2014, Still evolving (after all these years), *Sci. Am.* 311(3): 86–91.
Hayes, B., 2001, Computing comes to life, *Am. Sci.* 89: 204–208.
Hayes, B., 2004, Bugs that count, *Am. Sci.* 92: 401–405.
Hayes, B., 2011, Flights of fancy, *Am. Sci.* 99: 10–14.
Haynie, D. T., 2001, *Biological Thermodynamics*, Cambridge University Press, Cambridge, UK.

Hazen, R. M., 1999, *The Great Principles of Science*, Part IV, The Great Courses on Tape, The Teaching Company, Springfield, VA.

He, Y. H., X. Lu, M. X. Bi, L. Q. Yang, L. Y. Xu, and Q. P. Kong, 2014, The reduction of vascular disease risk mutations contributes to longevity in the Chinese population, *Meta Gene* 17(2): 761–768.

Heatwole, H., 1978, Adaptations of marine snakes, *Am. Sci.* 78: 594–604.

Hege, J. S. and G. Cole, 1966, A mathematical model relating circulating antibody and antibody forming cells, *J. Immunol.* 97: 34–40.

Heinrich, B., 1976, Bumblebee foraging and the economics of sociability, *Am. Sci.* 64: 384–395.

Hell, B. and K. Päßler, 2011, Are occupational interests hormonally influenced? The 2D:4D-interest nexus, *Personal. Individ. Differ.* 51(4): 376–380.

Heller, S., 2002, *Too Loud, Too Bright, Too Fast, Too Tight*, Harper Collins, New York.

Hellman, H., 1998, *Great Feuds in Science: Ten of the Liveliest Disputes Ever*, John Wiley & Sons, New York.

Hellman, H., 2001, *Great Feuds in Medicine: Ten of the Liveliest Disputes Ever*, John Wiley & Sons, New York.

Heppenheimer, T. A., 2001, Beyond the hearing aid, *Am. Herit. Invent. Technol.* 17(2): 36–42.

Heppenheimer, T. A., 2002, Nuclear power: What went wrong? *Invent. Technol.* 18(2): 46–56.

Herb, B., F. Wolschin, M. Aryee, B. Langmead, G. Amdam, A. P. Feinberg, 2012, Reversible switching between epigenetic states in honeybee behavioral subcasts. *Nat. Neurosci.* 15: 1371–1373.

Hertzman, C., 2001, Health and human society, *Am. Sci.* 89: 538–545.

Herzog, B., 2004, Dr. Eponymous, *Cornell Alumni Mag.* 106(5): 39–43.

Hildebrand, M., 1987, The mechanics of horse legs, *Am. Sci.* 75: 594–601.

Hillman, W. S. and D. D. Culley Jr., 1978, The uses of duckweed, *Am. Sci.* 66: 442–451.

Hinterthuer, A., 2008, Safety dance over plastic, *Sci. Am.* 299(3): 108–111.

Hobohm, U., 2009, Healing heat: Harnessing infection to fight cancer, *Am. Sci.* 97(1): 34–41.

Hodgson, L. and J. M. Tarbell, 2002, Solute transport to the endothelial intercellular cleft: The effect of wall shear stress, *Ann. Biomed. Eng.* 30: 936–945.

Hoekzema, E., E. Barba-Müller, C. Pozzobon, M. Picado, F. Lucco, D. García-García, J. C. Soliva et al., 2016, Pregnancy leads to long-lasting changes in human brain structure, *Nat. Neurosci.* doi:10.1038/nn.4458.

Hoffmann, R., 2001, Thermophiles in Kamchatka, *Am. Sci.* 89: 20–24.

Hölldobler, B. and E. O. Wilson, 1983, The evolution of communal nest-weaving in ants, *Am. Sci.* 71: 490–499.

Hooper, J., 2011, The man who would stop time, *Pop. Sci.* 279(2): 51–57, 79–81.

Hornbein, T. F., 1981, *Regulation of Breathing*, Marcel Dekker, New York.

Horton, T., 2002, 'Rain gardens' filter storm water naturally, *The Baltimore Sun*, February 1, p. 2B.

Houpt, K. A. and T. R. Wolski, 1982, *Domestic Animal Behavior for Veterinarians and Animal Scientists*, Iowa State University Press, Ames, IA.

Hoven, L. A., K. A. Sherman, S. Butler, and J. M. Giebultowicz, 2009, Does the clock make the poison? Circadian variation in response to pesticides, *PLoS One* 4(7): e6469. doi:10.1371/journal.pone.0006469.

Howery, L. D. and T. J. DeLiberto, 2005, Indirect effects of carnivores on livestock foraging behavior and production, *Shepherd* 50(5): 8–13.

Huda, N. M., Z. Lewis, K. M. Kalanetra, M. Rashid, S. M. Ahmad, R. Raqib, F. Qadri, M. A. Underwood, D. A. Mills, and C. B. Stephensen, 2014, Stool microbiota and vaccine responses of infants, *Pediatrics* 134(2): e362–e372.

Hughes, M. P., 2003, Micro- and nano-electrokinetics in medicine, *IEEE Eng. Med. Biol. Mag.* 22(6): 32.

Humes, H. D., 1995, Tissue engineering of the kidney, in *The Biomedical Engineering Handbook*, J. Bronzino, ed., CRC Press, Boca Raton, FL, pp. 1807–1824.

Huner, J. V., 2000, Crawfish and water birds, *Am. Sci.* 88: 301–303.

Hutchinson, G. E., 1981, Random adaptation and imitation in human evolution, *Am. Sci.* 69: 161–165.

Huxley, J. S., 1932, *Problems of Relative Growth*, Methuen Publishing, London, UK.

Ibáñez, A., J. F. Cardona, Y. V. Dos Santos, A. Blenkmann, P. Aravena, M. Roca, E. Hurtado et al., 2012, Motor-language coupling: Direct evidence from early Parkinson's disease and intracranial cortical recordings, *Cortex* doi:10.1016/j.cortex.2012.0.

Insel, P. M. and H. C. Lindgren, 1978, *Too Close for Comfort: The Psychology of Crowding*, Prentice Hall, Englewood Cliffs, NJ.

Intaglietta, M. and R. M. Winslow, 1995, Artificial blood, in *The Biomedical Engineering Handbook*, J. Bronzino, ed., CRC Press, Boca Raton, FL, pp. 2011–2024.

Jablonka, E. and G. Raz, 2009, Transgenerational epigenetic inheritance: Prevalence, mechanisms, and implications for the study of heredity and evolution, *Q. Rev. Biol.* 84(2): 131–176.

Jacobs, B. L. and M. E. Trulson, 1979, Mechanisms of action of LSD, *Am. Sci.* 67: 396–404.

Jacobs, B. L., 1987, How hallucinogenic drugs work, *Am. Sci.* 75: 386–392.

James, P., J. E. Hart, R. F. Barnay, and F. Laden, 2016, Exposure to greenness and mortality in nationwide prospective cohort study of women, *Environ. Health Perspec.* 124(4): doi:10.1289/ehp.1510363

Jeffrey, K., 1997, Many paths to the pacemaker, *Am. Herit. Invent. Technol.* 12(4): 28–39.

Jilek, M., 1971, The number of immunologically activated cells after repeated immunization (a mathematical model), *Folia Microbiol.* 16: 12–23.

Jirtle, R. L. and M. K. Skinner, 2007, Environmental epigenomics and disease susceptibility, *Nat. Rev. Genet.* 8: 253–262.

Johnson, A. T. and C. R. Dooly, 1995, Design of respiratory protective masks to improve human performance, in *The Biomedical Engineers Handbook*, J. D. Bronzino, ed., CRC Press, Boca Raton, FL, pp. 2321–2334.

Johnson, A. T. and D. C. Davis, 1990, Biological engineering: A discipline whose time has come, *Eng. Educ.* 80(1): 15–18.

Johnson, A. T. and W. M. Phillips, 1995, Philosophical foundations of biological engineering, *J. Eng. Educ.* 84: 311–318.

Johnson, A. T., 1991, Curvefitting, in *Digital Biosignal Processing*, R. Weitkunat, ed., Elsevier, New York, pp. 309–336.

Johnson, A. T., 1993, How much work is expended for respiration? *Front. Med. Biol. Eng.* 5: 265–287.

Johnson, A. T., 1995, The change in initial lung volume during exercise, *IEEE Trans. Biomed. Eng.* 42: 278–281.

Johnson, A. T., 1999, *Biological Process Engineering: An Analogical Approach to Fluid Flow, Heat Transfer, and Mass Transfer Applied to Biological Systems*, John Wiley & Sons, New York.

Johnson, A. T., 2007, *Biomechanics and Exercise Physiology: Quantitative Modeling*, CRC Press/Taylor & Francis Group, Boca Raton, FL.

Johnson, A. T., 1991, *Biological Process Engineering; An Analogical Approach to Fluid Flow, Heat Transfer, and Mass Transfer Applied to Biological Systems*, John Wiley & Sons, New York.

Johnson, C. H. and J. W. Hastings, 1986, The elusive mechanism of the circadian clock, *Am. Sci.* 74: 29–36.

Jones, D. S., 1983, Sclerochronology: Reading the record of the molluscan shell, *Am. Sci.* 71: 384–391.

Jones, D., 2012, Autism symptoms reversed in mice, *Nature News*, http://www.nature.com/autism-symptoms-reversed-in-mice-1.11869 (accessed June 30, 2015).

Jones, W. L. and A. L. Ingelfinger, 1973, Atmosphere control, in *Bioastronautics Data Book*, J. F. Parker Jr. and V. R. West, eds., NASA, SP-3006 U.S. Government Printing Office, Washington, DC, pp. 807–845.

Jordan, C. T. and G. Van Zant, 1995, The biology of stem cells, in *The Biomedical Engineering Handbook*, J. D. Bronzino, ed., CRC Press, Boca Raton, FL, pp. 1671–1678.

Juniper, B. E., 2007, The mysterious origin of the sweet apple, *Am. Sci.* 95: 44–51.

Kaffka, S. R., D. Kirby, and G. R. Peterson, 2001, Reducing fertilizer in sugarbeets can protect water quality, *Calif. Agric.* 55(3): 42–47.

Kantrowitz, P., 1972, Electrical safety in hospitals, *Instrum. Technol.* 6: 35–39.

Karam, P. A., 2003, Health effects of worker and fetal ionizing radiation exposure, *Synergist* 14(2): 31–34.

Karban, R. and K. Shiojiri, 2009 Self-recognition affects plant communication and defense, *Ecol. Lett.* 12(6): 502–506.

Karlin, S., 2010, Sanna Gaspard: High tech healer, *IEEE The Institute* 34(3): 16.

Karn, B. and H. S. Matthews, 2007, Nanoparticles without macroproblems, *IEEE Spectr.* 44(9): 55–58.

Kawakami, K., M. Kawamoto, M. Nomura, H. Otani, T. Nabika, and T. Gonda, 2004, Effects of phytoncides on blood pressure under restraint stress in SHRSP, *Clin. Exp. Pharmacol. Physiol.* 31(Suppl. 2): S27–S28.

Kearney, R., Z. Whatley, S. Rivera, and D. Hewitt, 2017, The prospects of artificial endosymbioses, *Am. Sci.* 105(1): 36–45.

Keats, J., 2006, John Koza has built an invention machine, *Pop. Sci.* 268(5): 66–72, 92.

Kemeny, D. M., 1991, *A Practical Guide to ELISA*, Pergamon Press, Oxford, UK.

Kennedy, J. M., 1983, What can we learn about pictures from the blind? *Am. Sci.* 71: 19–26.

Kent, D. and R. Haywood, 2007, When averages hide individual differences in clinical trials, *Am. Sci.* 95: 60–68.

Kent, D. M., 2008, Listening to resveratrol, *Am. Sci.* 96: 358–360.

Kessler, J. O. and N. A. Hill, 1997, Complementarity of physics, biology, and geometry in the dynamics of swimming micro-organisms, in *Physics of Biological Systems: From Molecules to Species*, H. Blyvbjerg, J. Hertz, M. H. Jensen, O. G. Mouritsen, and K. Sneppen, eds., Springer-Verlag, Berlin, Germany, pp. 325–340.

Khakh, B. S. and G. Burnstock, 2009, The double life of ATP, *Sci. Am.* 30(6): 84–92.

Khan, Z., G. Guelich, H. Phan, R. Redman, and S. Doty, 2012, Bacterial and yeast endophytes from poplar and willow promote growth in crop plants and grasses, *ISRN Agronomy* 2012. doi:10.5402/2012/890280.

Kicinger, R. and T. Arciszewski, 2007, Breeding better buildings, *Am. Sci.* 95: 502–508.

Kim, H.-Y., C. R. Estes, A. G. Duncan, B. D. Wade, F. C. Cleary, C. R. Lloyd, W. R. Ellis Jr., and L. S. Powers, 2004, Real-time detection of microbial contaminants, *IEEE EMBS Mag.* 23(1): 122–129.

Kim, T. H. and J. P. Vacanti, 1995, Tissue engineering of the liver, in *The Biomedical Handbook*, J. Bronzino, ed., CRC Press, Boca Raton, FL, pp. 1745–1753.

King, B. J., 2013, When animals mourn, *Sci. Am.* 309(1): 63–67.

King, J., C. Haase-Pettingell, and D. Gossard, 2002, Protein folding and misfolding, *Am. Sci.* 90: 445–453.

King, P. H. and R. C. Fries, 2003, *Design of Biomedical Devices and Systems*, Marcel Dekker, New York.

King, P. H., R. C. Fries, and A. T. Johnson, 2015, *Design of Biomedical Devices and Systems*, 3rd ed, Taylor & Francis Group, Boca Raton, FL.

Kleiber, M., 1975, *The Fire of Life: An Introduction to Animal Energetics*, Robert Krieger Publishing, Huntington, NY.

Klein, L., 2010, Michigan Sheep Breeders Association/Michigan Shepherds Weekend, *Shepherd* 55(2): 27.

Kluger, J., 2002, Here kitty, kitty, *Time* 159(8): 58–59.

Kluger, M., 2007, The power of birth order, *Time* 170(18): 42–48.

Knight, W., 2017, Reinforcement learning, *MIT Technol. Rev.* 120(2): 32–35.

Knox, M., 2014, The gene genie, *Sci. Am.* 311(6): 42–46.

Koch, C. and G. Marcus, 2014, Cracking the brain's codes, *MIT Technol. Rev.* 117(4): 43–46.

Koehl, M. A. R., 2000, Consequences of size change during ontogeny and evolution, in *Scaling in Biology*, J. H. Brown and G. B. West, eds., Oxford University Press, New York, pp. 67–86.

Koehn, R. K. and T. J. Hilbish, 1987, The adaptive importance of genetic variation, *Am. Sci.* 75: 134–141.

Koenig, W. D. and J. M. H. Knops, 2005, The mystery of masting in trees, *Am. Sci.* 93: 340–347.

Körner, E. and G. Matsumoto, 2002, Cortical architecture and self-referential control for brain-like computation, *IEEE Eng. Med. Biol. Mag.* 21(5): 121–133.

Kotz, J. C. and P. Treichel, 1999, *Chemistry and Chemical Reactivity*, Saunders, Ft. Worth, TX.

Kozubek, J., 2016, An on off switch for genes, *Sci. Am.* 314(1): 52–57.

Kresge, N., 2013, Intoxicating research, *HHMI Bull.* 26(3): 6–7.

Kresge, N., 2014, Fly brain filters, *HHMI Bull.* 27(1): 42.

Krones, M. J., A. T. Johnson, and O. J. Hao, 1990, A numerical technique for interpreting bioluminescence ATP assay, *Environ. Technol.* 11: 1107–1111.

Ku, D. N. and R. C. Allen, 1995, Vascular grafts, in *The Biomedical Engineering Handbook*, J. Bronzino, ed., CRC Press, Boca Raton, FL, pp. 1871–1878.

Kuhl, P. K., 2015, Baby talk, *Sci. Am.* 313(5): 64–69.

Kuo, M., 2015, How might contact with nature promote human health? promising mechanisms and a possible central pathway, *Front. Psychol.* 6: 1093. doi.10.3389/fpsyg.2015.01093.

LaBarbera, M. and S. Vogel, 1982, The design of fluid transport systems in organisms, *Am. Sci.* 70: 54–60.

Laird, D. W., P. D. Lampe, and R. G. Johnson, 2015, Cellular small talk, *Sci. Am.* 312(5): 70–77.

Lamar, E., 2012, Cells on the move, *HHMI Bull.* 25(2): 12–17.

Lancaster Farming, 2009, Penn State study of modified crop reveals hidden cost of resistance, *Lancaster Farming* 55(6): A10.

Lancaster Farming, 2010, Weed expert: Roundup-resistant problem must be dealt with, *Lancaster Farming*, 56(1): A50.

Landau, M., 1984, Human evolution as narrative, *Am. Sci.* 72: 262–268.

Lane, W., 2005, Evolutionary ruminations, *Shepherd* 50(10): 22–23.

Lane, W., 2006, Friendly fungi, *Shepherd* 51(4): 32–33.

Lane, W., 2007, Cellulose threads, *Shepherd* 52(1): 24–25.

Lane, W., 2008, Chewing the fat, *Shepherd* 53(11): 20.

Lane, W., 2009, Like a dance for nitrogen, *Shepherd* 54(4): 30–31.

Lane, W., 2010a, Dinosaur dreams, *Shepherd* 55(2): 18–19.

Lane, W., 2010b, Efficient nitrogen, *Shepherd*, 55(5): 10–11.

Lane, W., 2016, Mobs & other types of grazing, *Shepherd* 61(11): 34–35.

Langwig, K. E., J. R. Hoyt, K. L. Parise, W. F. Frick, J. T. Foster, and A. M. Kilpatrick, 2016, Resistance in persisting bat populations after white-nose syndrome invasion, *Philosoph. Trans. Roy. Soc. B: Biol. Sci.* 372(1712). doi:10.1098/rstb.2016.0044.

Lapique, L., 1898, Sur la relation de poids de l'Encéphale au poids du corps, *C. R. Biol.* 10 ser 5: 62–63.

Larson, D. W., 2002, Engineering fish, *Phi Kappa Phi Forum* 82(2): 6–7, 9.

Larson, L. R., V. Jennings, and S. A. Cloutier, 2016, Public parks and wellbeing in urban areas of the Unites States, *PLoS One* 11(4): e0153211. doi:10.1371/journal.pone 0153211.

Lashinsky, A., 2007, Fighting bad taste, *Pop. Sci.* 270(5): 41–42.

Laursen, L., 2010, When chickens attack, *IEEE Spectr.* 47(9): 11–13.

Lave, L. B. and E. P. Seskin, 1979, Epidemiology, causality, and public policy, *Am. Sci.* 67: 178–186.

Le Doux, J. M., J. R. Morgan, and M. L. Yarmush, 1995, Gene therapy in *The Biomedical Engineering Handbook*, J. Bronzino, ed., CRC Press, Boca Raton, FL, pp. 1518–1555.

Leake, D., 2015, Next-gen gene synthesis enables large-scale engineering in biological systems, *IEEE-EMBS Pulse* 6(5): 26–28.

LeCouteur, P. and J. Burreson, 2003, *Napoleon's Buttons: How 17 Molecules Changed History*, Putnam/Tarcher, New York.

Lee, E., 2015, Baby by the numbers, *IEEE Spectr.* 6(15): 24.

Lee, H. H., M. N. Molla, C. R. Cantor, and J. J.Collins, 2010, Bacterial charity work leads to population-wide resistance, *Nature* 467: 82–85.

Lee, R. J. and N. A. Cohen, 2016, Bitter taste bodyguards, *Sci. Am.* 314(2): 38–43.

Lee, S. S., A. M. Paspalof, D. D. Snow, E. K. Richmond, E. J. Rosi-Marshall, and J. J. Kelly, 2016, Occurrence and potential biological effects of amphetamine on stream communities, *Environ. Sci. Technol. 50*(17): 9727–9735.

Lehnert, R., 2010, Use glyphosate with caution, *Good Fruit Grower* 61(14): 18–20.

Leibowitz, H. W., 1985, Grade crossing accidents and human factors engineering, *Am. Sci.* 73: 558–562.

Lemonick, M. D. and A. Dorfman, 2002, Microbes at the extremes may tell us how life began, *Time* 160(5): 42–45.

Lemonick, M. D., 2007, The flavor of memories, *Time* 169(5): 102–104.

Lemonick, M., 2001, Deadly delivery, *Time* 158(18): 32–36, 38.

Lerer, S., 1998, *The History of the English Language*, The great courses on tape, The Teaching Company, Springfield, VA.

Leslie, M., 2008, Genetic balancing act, *HHMI Bull.* 21(2): 26–31.
Leslie, M., 2017, Cell-like giant viruses found, *Science* 356(6333): 15–16.
Lévi, F. A., 2008, The circadian timing system: A coordinator of life processes, *IEEE Eng. Med. Biol. Mag.* 27(1): 17–19.
Levin, L. A., 2002, Deep-ocean life where oxygen is scarce, *Am. Sci.* 90: 436–444.
Levin, P. S. and D. A. Levin, 2002, The real biodiversity crisis, *Am. Sci.* 90: 6–8.
Levin, P. S. and M. H. Schiewe, 2001, Preserving salmon biodiversity, *Am. Sci.* 89: 220–227.
Levine, R. V., 2003, The kindness of strangers, *Am. Sci.* 91: 226–233.
Lewis, C. A., J. Crayle, S. Zhou, R. Swanstrom, and R. Wolfenden, 2016, Cytosine deamination and the precipitous decline of spontaneous mutation during earth's history, *Proc. Nat. Acad. Sci.* 113(29): 8194–8199.
Lewis, D. B. and D. M. Gower, 1980, *Biology of Communication*, John Wiley & Sons, New York.
Li, Y., B. Deeb, W. Pendergrass, and N. Wolf, 1996, Cellular proliferative capacity and life span in small and large dogs, *J. Gerontol. A Biol. Sci. Med. Sci.* 51(6): B403–B408.
Li, J. K.-J., 2000, Scaling and invariants in cardiovascular biology, in *Scaling in Biology*, J. H. Brown and G. B. West, eds., Oxford University Press, New York, pp. 113–128.
Li, J. L., R. S. Cornman, J. D. Evans, J. F. Pettis, Y. Zhao, C. Murphy, W. J. Peng, J. Wu et al., 2014, Systemic spread and propagation of a plant-pathogenic virus in European honeybees, *Apis mellifera*, *mBio* 5(1): 1–11.
Li, Q., M. Kabayashi, Y. Wakayama, H. Inagaki, M. Katsumata, Y. Hirata, T. Shimizu et al., 2009, Effect of phytoncide from trees on human natural killer cell function, *Internat. J. Immopathol. Pharmacol.* 22(4): 951–959.
Li, Q., K. Morimoto, M. Kobayashi, H. Inagaki, M. Katsumata, Y. Hirata, K. Hirata et al., 2008, A forest bathing trip increases human natural killer cell activity and expression of anti-cancer proteins in female subjects, *J. Biol. Regulat. Homeostatic Agents* 22(1): 45–55.
Lieberman-Aiden, E., N. L. van Berkum, L. Williams, M. Imakaev, T. Ragoczy, A. Telling, I. Amit et al., 2009, Comprehensive mapping of long-range interactions reveals folding principles of the human genome, *Science* 326: 289–293.
Liebman, M. N., 2005, An engineering approach to translational medicine, *Am. Sci.* 93: 296–298.
Liebovitch, L. S., 1998, *Fractals and Chaos Simplified for the Life Sciences*, Oxford University Press, New York.
Lilienfeld, S. O., 2010, Fear: Can't live with it, can't live without it, *Phi Kappa Phi Forum* 90(3): 16–18.
Linden, D. J., 2015, *Touch: The Science of Hand, Heart, and Mind*, Viking Press, New York.
Lippincott-Schwartz, J. and G. H. Patterson, 2003, Development and use of fluorescent protein markers in living cells, *Science* 300: 87–90.
Liu, C., B. C. Colón, M. Ziesack, P. A. Silver, and D. G. Nocera, 2016, Water splitting–biosynthetic system with CO_2 reduction efficiencies exceeding photosynthesis, *Science* 352(6290): 1210–1213.
Liubicich, D., 2010, How can jellyfish of the same kind gather in a group if they don't have brains? *HHMI Bull.* 23(3): 45.
Loeb, G. E. and R. Davoodi, 2005, The functional reanimation of paralyzed limbs, *IEEE Eng. Med. Biol. Mag.* 24(5): 45–51.
Loewenstein, W. R., 1999, *The Touchstone of Life: Molecular Information, Cell Communication, and the Foundations of Life*, Oxford University Press, Oxford, UK.
Lord, M. and A. Turner-Smith, 1995, Orthopedic prosthetics and orthotics in rehabilitation, in *The Biomedical Engineering Handbook*, J. Bronzino, ed., CRC Press, Boca Raton, FL, pp. 2055–2070.
Loukola, O. J., C. J. Perry, L. Coscos, and L. Chittka, 2017, Bumblebees show cognitive flexibility by improving on an observed complex behavior, *Science* 355(6327): 833–836.
Lovelock, J., 2006, *The Revenge of Gaia: Earth's Climate Crisis and the Fate of Humanity*, Basic Books, New York.
Ludlow, A. T., J. B. Zimmerman, S. Witkowski, J. W. Hearn, B. D. Hatfield, and S. M. Roth, 2008, Relationship between physical activity level, telomere length, and telomerase activity, *Med. Sci. Sports Exerc.* 40(10): 1764–1771.
Lumma, V., P. E. Pettay, and A. F. Russell, 2007, Male twins reduce fitness of female co-twins in humans, *PNAS* 104: 10915–10920.

Lutz, R. A., T. M. Shank, and R. Evans, 2001, Life after death in the deep sea, *Am. Sci.* 89: 422–431.
Lysaght, M. J. and J. Moran, 1995, Peritoneal dialysis equipment, in *The Biomedical Engineering Handbook*, J. Bronzino, ed., CRC Press, Boca Raton, FL, pp. 1923–1935.
MacDermott, A. J. and G. E. Tranter, 1990, Biomolecular handedness and the weak interaction, in *Symmetries in Science IV: Biological and Biophysical Systems*, Plenum Press, New York, pp. 67–124.
Madore, B. F. and W. L. Freedman, 1987, Self-organizing structures, *Am. Sci.* 75: 252–259.
Maharbiz, M. M. and H. Sato, 2010, Cyborg beetles, *Sci. Am.* 303(6): 94–99.
Maisuria, V. B., Z. Hosseinidoust, and N. Tufenkji, 2015, Polyphenolic extract from maple syrup potentiates antibiotic susceptibility and reduces biofilm formation of pathogenic bacteria, *Appl. Environ. Microbiol.* 81(11): 3782–3792.
Mahoney, D. J., D. F. Stojdl, and G. Laird, 2014, Virus therapy for cancer, *Sci. Am.* 311(5): 54–59.
Malyshev, D. A., D. Kirandeep, T. Chen, N. Dai, J. M. Foster, I. R. Corrêa, and F. E. Romesberg, 2014, A Semi-synthetic organism with an expanded genetic alphabet, *Nature* 509: 385–388.
Manning, J. T., 2002, The ratio of 2nd to 4th digit length and performance in skiing, *J Sports Med. Phys. Fitness* 42(4): 446–450.
Manning, J. T., D. Scutt, J. Wilson, and D. I. Lewis-Jones, 1998, The ratio of 2nd to 4th digit length: a predictor of sperm numbers and concentrations of testosterone, luteinizing hormone and oestrogen, *Hum. Reprod.* 13(11): 3000–3004.
Marchuk, G. I., 1983, *Mathematical Models in Immunology*, Optimization Software, Inc., New York.
Martin. L. B. and C. A. C. Coon, 2010, Infection protection and natural selection, *Science* 330(6004): 602–603.
Marvier, M., 2007, Pharmaceutical crops have a mixed outlook in California, *Calif. Agric.* 61(2): 59–66.
Marx, R. B., R. L. Irvine, and D. Irvine, 2003, Bioreactor troubleshooting, *Pollut. Eng.* 35(2): 24–29.
Maslow, A. H., 1954, *Motivation and Personality*, Harper, New York.
Mason, M., 2006, Man's best model, *HHMI Bull.* 19(4): 26–31, 60.
Matchock, R. L., 2008, Low digit ratio (2D:4D) is associated with delayed menarche, *Am. J. Hum. Biol.* 20(4): 487–489.
Mattson, M. P., 2015, What doesn't kill you, *Sci. Am.* 313(1): 40–45.
Matushansky, I. and R. G. Maki, 2005, Fighting cancer through the study of sarcomas, *Am. Sci.* 93: 414–421.
Maure, F., J. Brodeur, N. Ponlet, J. Doyon, A. Firlej, É. Elguero, and F. Thomas, 2011, The cost of a bodyguard, *Biol. Lett.* 7(11). doi:10.1098/rsbl.2011.0415.
Maurer, A. C., 1979, The therapy of diabetes, *Am. Sci.* 67: 422–431.
May, E. E., 2006, Communication theory and molecular biology at the crossroads, *IEEE Eng. Med. Biol. Mag.* 25(1): 28–29.
May, M., 2010, The synthetic side of protein expression, *Drug Discov. Dev.* May: 14–15.
May, R. M. and J. Seger, 1986, Ideas in ecology, *Am. Sci.* 74: 256–267.
Maynard Smith, J., 1982, *Evolution and the Theory of Games*, Cambridge University Press, Cambridge, UK.
Mazumdar, J., 1989, *An Introduction to Mathematical Physiology and Biology*, Cambridge University Press, New York.
McClain, C., 2010, An empire lacking food, *Amer. Sci.* 98: 470–477.
McCreary, D., 2001, Agroforestry is promising for previously cleared hardwood rangelands, *Calif. Agric.* 55(6): 37–41.
McFedries, P., 2003, The internet ecology, *IEEE Spectr.* 40(4): 68.
McFrederick, Q. S., J. C. Kathilankal, and J. D. Fuentes, 2008, Air pollution modifies floral scent trials, *Atmos. Environ.* 42: 2336–2348.
McGinnis, L., 2010, Helping cattle keep their cool, *Agric. Res.* 58(3): 20–21.
McGregor, A. P., V. Orgogozo, I. Delon, J. Zanet, D. G. Srinivasan, F. Payre, and D. L. Stern, 2010, Morphological evolution through multiple cis-regulatory mutations at a single gene, *Nature* 448: 587–590.
McIntyre, M. H., E. S. Barrett, R. McDermott, D. D. P. Johnson, J. Cowden, S. P. Rosen, 2007, Finger length ratio (2D:4D) and sex differences in aggression during a simulated war game, *Personal. Individ. Differ.* 42(4): 755–764.
McMahon, T. A., 1975a, Allometry and biomechanics: Limb bones in adult ungulates, *Am. Nat.* 109: 547–563.

McMahon, T. A., 1975b, Using body size to understand the structural design of animals: Quadrupedal locomotion, *J. Appl. Physiol.* 39: 619–627.

McMahon, T. A., 1984, *Muscles, Reflexes, and Locomotion*, Princeton University Press, Princeton, NJ.

McMasters, J. H. and R. M. Cummings, 2004, Challenges for education in aerospace & engineering for the 21st century, *The Bent of Tau Beta Pi* 95(3): 12–19.

McNally, J., 2010, The changing predator landscape, *Shepherd* 55(12): 6–7.

Mehal, W. Z., 2015, Cells on fire, *Am. Sci.* 312(6): 44–49.

Mench, J. A. and A. van Tienhoven, 1986, Farm animal welfare, *Am. Sci.* 74: 598–603.

Mengüc, Y., 2017, Smart and squishy robots, *Am. Sci.* 105(3): 143–147.

Meredith, D., 2006, Solving the puzzle of the resilient embryo, *HHMI Bull.* 19(1): 8–9.

Mertz, L., 2015, From annoying to appreciated, *IEEE-EMBS Pulse* 6(5): 4–9.

Metcalf and Eddy, Inc., 1991, *Wastewater Engineering*, John Wiley & Sons, New York.

Meyer, A., 2015, Extreme evolution, *Sci. Am.* 312(4): 70–75.

Micera, S., P. Bonato, and T. Tamura, 2008, Gerontechnology, *IEEE Eng. Med. Biol. Mag.* 27(4): 10–14.

Michalowski, J., 2010, Reality check, *HHMI Bull.* 23(4): 10–11.

Miller, D., 2003, Leveling the fight with beavers, *Prog. Farmer* 118(11): 70–71.

Miller, P., 2015, Pulse of the planet, *Nat. Geogr.* 228(5): 146–159.

Miller, W. R., 2011, Tardigrades, *Am. Sci.* 99: 384–391.

Milsum, J. H., 1966, *Biological Control Systems Analysis*, McGraw-Hill, New York.

Mines, A. H., 1981, *Respiratory Physiology*, Raven Press, New York.

Misiak, M., L. D. Kaczmarek, T. B. Kashdan, M. Zimny, H. Urbański, A. Zasiński, and D. Disabato, 2015, Prenatal exposure to sex hormones predicts gratitude intervention use. examination of digit ratio, motivational beliefs, and online activities, *Personal. Individ. Differ.* 77: 68–73.

Mitsch, W. J., 1996, Ecological engineering: A new paradigm for engineers and ecologists, in *Engineering Within Ecological Constraints*, P. C. Schulze, ed., National Academy Press, Washington, DC, pp. 114–132.

Mitsch, W. J. and S. E. Jorgensen, 1989, Introduction to ecological engineering, in *Ecological Engineering: An Introduction to Ecotechnology*, W. J. Mitsch and S. E. Jorgensen, eds., John Wiley & Sons, New York, pp. 3–12.

Mizrach, A., A. Hetzroni, T. Ignat, J. Grinshpun, M. Mazor, D. Shuman, R. Mankin, N. Epsky, and R. Heath, 2004, Trapping flies with acoustics, *Resource* 11(1): 11–12.

Moehlman, P. D., 1987, Social organization in jackals, *Am. Sci.* 75: 366–375.

Molles, M. C. Jr., 1999, *Ecology: Concepts and Applications*, McGraw-Hill, New York.

Moore, S. K., 2005, Zapping the blues away, *IEEE Spectr.* 42(5): 16–17.

Moran, D. S. and K. B. Pandolf, 1999, *Physiological Strain Index (PSI) to Evaluate Heat Stress*, Technical Report T99-XX, US Army Research Institute of Environmental Medicine, Natick, MA.

Moriarty, L., 2012, SPR solutions for interaction analysis, *GEN: Genetic Engineering and Biotechnology News* 32(12): 24–26, 28.

Morowitz, H. J., 2011, When PPLO became mycoplasma, *Am. Sci.* 99(2): 102–105.

Morrell, V., 2015, Feeding frenzy, *Nat. Geogr.* 228(1): 76–87.

Morris, S. S., C. G. Victora, F. C. Barros, R. Halpern, A. M. B. Menezes, J. A. César, B. L. Horta, and E. Tomasi, 1998, Length and ponderal index at birth: Associations with mortality, hospitalizations, development, and post-natal growth in Brazilian infants, *Int. J. Epidemiol.* 27: 242–247.

Morrow, M. P. and D. B. Weiner, 2010, DNA drugs come of age, *Sci. Am.* 303(1): 49–53.

Moser, M., M. Frühwirth, and T. Kenner, 2008, The symphony of life, *IEEE Eng. Med. Biol. Soc. Mag.* 27(1): 29–37.

Moskowitz, D. S., R. Sutton, D. C. Zuroff, and S. N. Young, 2015, Fetal exposure to androgens, as indicated by digit ratios (2D:4D), increases men's agreeableness with women, *Personal. Individ. Differ.* 75: 97–101.

Motokawa, T., 1989, Sushi science and hamburger science, *Perspect. Biol. Med.* 32: 489–504.

Moyer, M. W., 2013, The myth of antioxidants, *Sci. Am.* 308(2): 62–67.

Mukherjee, B. and M. Sivaprakasam, 2015, An eye toward improving, *IEEE-EMBS Pulse* 6(4): 20–25.

Mulcahy, D. L. and G. B. Mulcahy, 1987, The effects of pollen competition, *Am. Sci.* 75: 44–50.

Mulle, J. G., W. G. Sharp, and J. F. Cubells, 2013, The gut microbiome: A new frontier in autism research, *Curr. Psychiatry Rep.* 15(2): 337. doi:10.1007/s11920–012–0337–0.

Müller, T., A. Pfennig, P. Klein, G. Gradl, M. Jäger, and T. Schnelle, 2003, The potential of dielectrophoresis for single-cell experiments, *IEEE Eng. Med. Biol. Mag.* 22(6): 51–61.

Mumford, J. A. and T. Nichols, 2006, Modeling and inference of multisubject fMRI data, *IEEE Eng. Med. Biol. Mag.* 25(2): 42–51.

Murashov, V. and J. Howard, 2015, Risks to health care workers from n-enabled medical products, *J. Occup. Environ. Hyg.* 12(6): D75–D85.

Murphy, G., 1950, *Similitude in Engineering*, Ronald Press, New York.

Murray, J. D. and G. B. Anderson, 2000, Genetic engineering and cloning may improve milk, livestock production, *Calif. Agric.* 54(4): 57–65.

Myers, V. G., 2017, Where past and present meet, *The Progressive Farmer*, p. B–12.

Nakamura, T. and Y. Shimizu, 1995, Tracheal, laryngeal, and esophageal replacement devices, in *The Biomedical Engineering Handbook*, J. Bronzino, ed., CRC Press, Boca Raton, FL, pp. 1997–2010.

Nedergaard, M. and S. A. Goldman, 2016, Brain drain, *Sci. Am.* 314(3): 44–49.

Nelson, G. E., G. G. Robinson, and R. A. Boolootian, 1970, *Fundamental Concepts of Biology*, John Wiley & Sons, New York.

Nelson, X., 2014, The spider's charade, *Am. Sci.* 311(6): 86–89.

Nerem, R. M., M. J. Levesque, and J. F. Cornhill, 1980, Social environment as a factor in diet-induced atherosclerosis, *Science* 208: 1475–1476.

Nesse, R. M. and G. C. Williams, 1994, *Why We Get Sick: The New Science of Darwinian Medicine*, Vintage Books, New York.

NGS, 2002, Outfoxed by aliens, *Nat. Geogr.* 202(2), Geographica.

NIAA, 2007, Why poisonous plants are sometimes lethal, sometimes not, *Shepherd* 52(9): 22–23.

Nielsen, J. and J. Villadsen, 1994, *Bioreaction Engineering Principles*, Plenum Press, New York.

Nijland, R. and J. G. Burgess, 2010, Bacterial olfaction, *Biotech. J.* 5(9): 974–977.

Noel, C. and R. Dando, 2015, The effect of emotional state on taste perception, *Appetite* (95): 89–95.

Nowak, K., A. leRoux, S. A. Richards, C. P. J. Scheijen, and R. A. Hill, 2014, Human observers impart habituated samango monkeys perceived landscape of fear, *Behav. Ecol.* 25(5): 1199–1204.

Nowak, M. A., 2012, Why we help, *Sci. Am.* 307(1): 34–39.

Nunes, A. and P. Cabon, 2015, The fatigue conundrum, *Am. Sci.* 103(3): 218–223.

Nye, J. V. C. and E. Orel, 2015, The influence of prenatal hormones on occupational choice: 2D:4D evidence from Moscow, *Personal. Individ. Differ.* 78: 39–42.

O'Brien, D., 2002, At the aquarium, the toys aren't just for kids, *The Baltimore Sun*, November 15, p. 2B.

O'Brien, J. D., 2001, How do birds in flocks and fish in schools turn and move together so precisely? *Pop. Sci.* 259(5): 84.

Obhi, S. S. and P. Haggard, 2004, Free will and free won't, *Am. Sci.* 92: 358–365.

Ogle, B. M. and J. L. Platt, 2004, The biology of cell fusion, *Am. Sci.* 92: 420–427.

Oka, Y., M. Ye, and Zucker, C. S., 2015, Thirst driving and suppressing signals encoded by distinct neural populations in the brain, *Nature* 520: 349–352.

Okamura, A. M., C. Simone, and M. D. O'Leary, 2004, Force modeling for needle insertion into soft tissue, *IEEE Trans. Biomed. Eng.* 51: 1707–1716.

Oliver, R., 2007, Fighting varroa: The silver bullet or brass knuckles? *Am. Bee J.* 147(6): 526–530.

Olsen, R. A., R. B. Clark, and J. H. Bennett, 1981, The enhancement of soil fertility by plant roots, *Am. Sci.* 69: 378–384.

Olson, S., 2007, The changing face of DNA, *HHMI Bull.* 20(4): 26–31.

Olsson-Francis, K. and C. S. Cockell, 2010, The use of cyanobacteria for in-situ resource use in space applications, *Planet. Space Sci.* 58(10): 1279–1285.

Owen, D., 2015, The science of delicious, *Nat. Geogr.* 228(6): 61–81.

Palmer, C., 2014, An off-switch for neurons, *HHMI Bull.* 27(3): 26–37.

Palumbi, S. R., 2001, *The Evolution Explosion: How Humans Cause Rapid Evolutionary Change*, W. W. Norton, New York.

Panescu, D., 2004, Conductive keratoplasty, *IEEE EMBS Mag.* 23(4): 16–18.

Park, A., 2001, Telling disease from terror, *Time* 158(17): 26.
Park, A., 2010, The vitamin-D debate, *Time*, August 30, p. 66.
Park, J. B., 1995, Biomaterials, in *The Biomedical Engineering Handbook*, J. Bronzino, ed., CRC Press, Boca Raton, FL, pp. 530–536.
Park, S.-W., E. Kaimoyo, D. Kumar, S. Mosher, and D. F. Klessig, 2007, Methyl salicylate is a critical mobile signal for plant systemic acquired resistance, *Science* 318: 113–116.
Patel, P., 2009, The brain–machine interface, unplugged, *IEEE Spectr.* 46(10): 13–14.
Pauly, D., V. Christensen, R. Froese, and M. L. Palomares, 2000, Fishing down aquatic food webs, *Am. Sci.* 88: 46–51.
Payne, J. and B. Pugh, 2010, On-farm mortality composting of livestock carcasses, *Shepherd*, pp. 26–29.
Pearson, D. G. and T. Craig, 2014, The great outdoors? exploring the mental health benefits of natural environments, *Front. Psychol.* 5: 1178. doi:103389/fpsyg.2014.01178.
Peckham, P. H. and B. Smith, 1995, Implantable stimulators for neuromuscular control, in *The Biomedical Engineering Handbook*, J. Bronzino, ed., CRC Press, Boca Raton, FL, pp. 1270–1283.
Pekar, J. J., 2006, A brief introduction to functional MRI, *IEEE Eng. Med. Biol. Mag.* 25(2): 24–26.
Pell, R. W. and L. B. Allen, 2015, Bringing postnatural history into view, *Am. Sci.* 103(3): 224–227.
Pennisi, E., 2005, Talking about a revolution: Hidden RNA may fix mutant genes, *Science* 307: 1852.
Pennycuick, C. J., 1975, On the running of the Gnu and other mammals, *J. Exp. Biol.* 63: 775–799.
Perdue, M., 2015, Tackling mental illnesses with a novel approach, *GEN* 35(11): 8–9.
Perkel, J. M., 2013, *Genome Editing with CRISPRs, TALENs and ZFNs*, Biocompare Technical Note, http://www.biocompare.com/Editorial-Articles/144186-Genome-Editing-with-CRISPRs-TALENs-and-ZFNs/ (accessed November 10, 2015).
Perry, T. S., 2015, Giving your body a "check engine" light, *IEEE Spectr.* 6(15): 34–39, 84.
Peters, R. P. and L. D. Mech, 1975, Scent-marking in wolves, *Am. Sci.* 63: 628–637.
Petroski, H., 2001, Success and failure in engineering, *Natl. Forum* 81(1): 10–17.
Petroski, H., 2016, The road ahead, *Am. Sci.* 104(4): 236–241.
Pheasant, S., 1996, *Bodyspace: Anthropometry, Ergonomics, and the Design of Work*, Taylor & Francis Group, Bristol, PA.
Phillips, A. L., 2009, Savory individuals, *Am. Sci.* 97: 380–381.
Phitayakom, R., W. Levine, E. Petrusa, B. Daily, E. Eromo, D. Gee, M. Hemingway, R. Minehart, M. Pian-Smith, and J. A. Gordon, 2015, Making it real, *IEEE-EMBS Pulse* 6(4): 10–13.
Piccolo, S., 2014, Twist of fate, *Sci. Am.* 311(4): 74–81.
Piff, P., P. Dietze, M. Feinberg, D. M. Stancato, and D. Keltner, 2015, Awe, the small self, and prosocial behavior, *J. Personal. Social Psychol.* 108(6): 883–899.
Piore, A., 2016, The rogue immune cells that wreck the brain, *MIT Technol. Rev.* 119(3): 30–36.
Piore, A., 2017, The body electrician, *Pop. Sci.*: 64–68, 92.
Piskounova, E., M. Agathocleous, M. M. Murphy, Z. Hu, S. E. Huddlestun, Z. Zhao, A. M. Leitch, T. M. Johnson, R. J. DeBerardinis, and S. J. Morrison, 2015, Oxidative stress inhibits distant metastasis by human melanoma cells, *Nature* doi:10.1038/nature15726.
Plotkin, M. J., 2000, *Medicine Quest: In Search of Nature's Healing Secrets*, Viking Press, New York.
Plutchik, R., 2001, The nature of emotions, *Am. Sci.* 89: 344–350.
Pollack, G. H., 2001, *Cells, Gels, and the Engines of Life: A New, Unifying Approach to Cell Function*, Ebner and Sons, Seattle, WA.
Pontin, J., 2014, Q+A Antonio Damasio, *MIT Technol. Rev.* 117(4): 49–51.
Poultney, R., 2003, Fit as a fiddle, sick as a dog, *Pop. Sci.* 262(1): 38.
Powell, C. S., 2015, Have we found alien life? *Pop. Sci.* 286(2): 32–39, 70.
Powell, J. R. and T. Dobzhansky, 1976, How far do flies fly? *Am. Sci.* 64: 179–185.
Prindle, A., J. Liu, M. Asally, S. Ly, J. Garcia-Ojalvo, and G. M. Süel, 2015, Ion channels enable electrical communication in bacterial communities, *Nature* 527(7576): 59–63.
Prioli, C., 1991, Rubber dentures for the masses, *Am. Herit. Invent. Technol.* 7(2): 28–31, 34–37.
Purdy, J., 2005, Love is the answer, *UGA Res. Mag.* 34(1): 32–43.

Putnam, R. W., 1995, Polarity of cell membranes, in *Cell Physiology Source Book*, N. Speielakis, ed., Academic Press, San Diego, CA, pp. 230–242.

Raeburn, P., 2007, Small talk, *HHMI Bull.* 20(1): 20–25, 56.

Raliya, R., J. C. Tarafdar, and P. Biswas, 2016, Enhancing the mobilization of native phosphorus in the mung bean rhizosphere using ZnO nanoparticles synthesized by soil fungi, *J. Agric. Food Chem.* 64(16): 3111–3118.

Ramesh, R., J. C. Tarafdar, and P. Biswas, 2016, Enhancing the mobilization of native phosphorus in the mung bean rhizosphere using ZnO nanoparticles synthesized by soil fungi, *J. Agric. Food Chem.* 64(16): 3111–3118.

Ramesh, S. A., S. D. Tyerman, B. Xu, J. Bose, S. Kaur, V. Conn, P. Domingos et al., 2015, GABA signalling modulates plant growth by directly regulating the activity of plant-specific anion transporters, *Nat. Comm.* 6: 7879.

Range, F., L. Horn, Z. Viranyi, and L. Huber, 2009, The absence of reward induces inequity aversion in dogs, *Proc. Natl. Acad. Sci.* 106: 340–345.

Rasmussen, S., M. A. Bedau, L. Chen, D. W. Deaner, D. C. Krakauer, N. H. Packard, and P. F. Stadler, 2004, Transitions from nonliving to living matter, *Science* 305: 963.

Rattenborg, N. C., B. Voirin, S. M. Cruz, R. Tisdale, G. Dell'Omo, H.-P. Lipp, M. Wikelski, and A. L. Vyssotski, 2016, Evidence that birds sleep in mid-flight, *Nat. Commun.* 7, Article Number 12468, doi:10.1038/ncomms12468.

Rawlings, D. E. and S. Silver, 1995, Mining with microbes, *Nat. Biotechnol.* 13: 773–778.

Raymann, K., Z. Shaffer, and N. A. Moran, 2017, Antibiotic exposure perturbs the gut microbiota and elevates mortality in honeybees, *PLOS Biol.* 15(3): e2001861. doi:10.1371/journal.pbio.2001861.

Rayner, J. M. V., 1987, Form and function in avian flight, *Curr. Ornithol.* 5: 1–66.

Reardon, S., 2013, Bacterium can reverse autism-like behaviour in mice, *Nat. News*, http://www.nature.com/news/bacterium-can-reverse-autism-like-behaviour-in-mice (accessed June 30, 2015).

Reeves, A. L., 1981, The metabolism of foreign compounds, in *Toxicology: Principles and Practice*, Vol. I, A. L. Reeves, ed., John Wiley & Sons, New York, pp. 1–28.

Regalado, A., 2016a, The extinction invention, *MIT Technol. Rev.* 119(3): 54–59.

Regalado, A., 2016b, Gene-editing swap, *MIT Technol. Rev.* 119(4): 24, 26.

Reis, P. M., S. Jung, J. M. Aristoff, and R. Stocker, 2010, How cats lap: Water uptake by *Felis catus*, www.sciencexpress.org, doi:10.1126/science.1195421.

Reisch, R., 2016, Species in the making, *Sci. Am.* 315(5): 54–61.

Rhoads, D. C., P. L. McCall, and J. Y. Yingst, 1978, Disturbance and production on the estuarine sea floor, *Am. Sci.* 66: 577–586.

Rich, T. A., 2008, Analyzing the symptoms in cancer patients, *IEEE Eng. Med. Biol. Mag.* 27(1): 25–28.

Riegler, B. and K. Rhodes, 2006, Meeting silicone adhesion challenges with primers, *Med. Dev. Diagn. Ind.* 24(9): 80–90.

Roberts, B. A. and M. M. Martel, 2013, Prenatal testosterone and preschool disruptive behavior disorders, *Personal. Individ. Differ.* 55(8): 962–966.

Robertson, L. C., 2005, The bilateral brain: Are two better than one? *Phi Kappa Phi Forum* 85(1): 19–22.

Robison, J. E., 2015. Fixing autism research, *MIT Technol. Rev.* 118(1): 10–11.

Rodger, D. C. and Y.-C. Tai, 2005, Microelectronic packaging for retinal prostheses, *IEEE Eng. Med. Biol. Mag.* 24(5): 52–57.

Rodger, J. C. and B. L. Drake, 1987, The enigma of the fetal graft, *Am. Sci.* 75: 51–57.

Rosen, H., 2008, The gatherers, *Pop. Sci.* 272(1): 60–66, 91.

Rosenberg, G., 1995, Artificial heart and circulatory assist devices, in *The Biomedical Engineering Handbook*, J. Bronzino, ed., CRC Press, Boca Raton, FL, pp. 1839–1846.

Rosenwald, M., 2008, Compassion cure, *Pop. Sci.* 272(3): 27–29.

Ross, G., 2004, Woofers and tweeters, *Am. Sci.* 92: 28–29.

Ross, P. E., 2015, Diabetes has a new enemy: RoboPancreases, *IEEE Spectr.* 6(15): 40–43, 88.

Roston, E., 2001, Nature's bottom line, *Time* 158: Y9.

Röthlisberger, D., O. Khersonsky, A. M. Wollacott, L. Jiang, J. DeChancie, J. Betker, J. L. Gallaher et al., 2008, Kemp elimination catalysts by computational design, *Nature* 453: 190–195.

Roylance, F. D., 2002, Links between El Niño, disease, *The Baltimore Sun* 166(45): 2.

Ruch, T. C. and H. D. Patton, 1966, *Physiology and Biophysics*, W. B. Saunders, Philadelphia, PA.

Russell, E. W., 1961, *Soil Conditions and Plant Growth*, John Wiley & Sons, New York.

Russell, S. L., M. J. Gold, M. Hartmann, B. P. Willing, L. Thorson, M. Wlordarska, N. Gill et al., 2012, Early life antibioticdriven changes in microbiota enhance susceptibility to allergic asthma, *EMBO Rep.* 13: 440–447.

Russelle, M. P., 2001, Alfalfa, *Am. Sci.* 89: 252–261.

Ryan, T. A., 2003, Fluorescent proteins with ties that bind, *Nat. Biotechnol.* 21: 1447–1448.

Sachs, A., 2010, Kitty prozac, *Time* 176(20): 57–58.

Sachs, M. B., I. C. Bruce, R. L. Miller, and E. D. Young, 2002, Biological basis of hearing-aid design, *Ann. Biomed. Eng.* 30: 157–168.

Sadik, O., 2017, The two faces of nanotechnology, *Am. Sci.* 105(4): 208–211.

Sahin, E., S. Colla, M. Liesa, J. Moslehi, F. L. Müller, M. Guo, M. Cooper et al., 2011, Telomere dysfunction induces metabolic and mitochondrial compromise, *Nature* 470: 359–365.

Saltus, R., 2006, A cell's second act, *HHMI Bull.* 19(1): 14–19.

Saltus, R., 2010A, A silver lining, *HHMI Bull.* 23(2): 22–27.

Saltus, R., 2010B, Into the third dimension, *HHMI Bull.* 23(3): 24–27.

Sarpeshkar, R., 2006, Brain power, *IEEE Spectr.* 43(5): 24–29.

Saulnier, B., 2009, Shot in the arm, *Cornell Alumni Mag.* 111(4): 36–37.

Saunders, F., 2015, Making the cut, *Am. Sci.* 103(6): 375.

Scarcelli, G., W. J. Polachek, H. T. Nia, K. Patel, A. J. Grodzinsky, R. D. Kamm. and S. H. Yun, 2015, Noncontact three-dimensional mapping of intracellular hydromechanical properties by brillouin microscopy, *Nature Meth.* 12: 1132–1134. doi:10.1038/nmeth.3616.

Scerri, E. R., 2014, Master of missing elements, *Am. Sci.* 102(5): 358–365.

Scheflen, A. E. and N. Ashcroft, 1976, *Human Territories: How We Behave in Space-Time*, Prentice Hall, Englewood Cliffs, NJ.

Schiefer, H. B., D. G. Irvine, and S. C. Buzik, 1997, *Understanding Toxicology*, CRC Press, Boca Raton, FL.

Schmidt, C., 2015, Thinking from the gut, *Sci. Am.* 312(3): S13–S15.

Schmidt, C., 2017, Don't drink the water, *Sci. Am.* 316(4): 64–69.

Schnabel, J., 2010, Nano-motion pictures, *HHMI Bull.* 23(1): 45.

Schneck, D. J., 1990, *Engineering Principles of Physiologic Function*, New York University Press, New York.

Schneider, T. D., 2006, Claude Shannon: Biologist, *IEEE Eng. Med. Biol. Mag.* 25(1): 30–33.

Schoggen, P. and M. Schoggen, 1985, Play, exploration, and density, in *Habitats for Children: The Impacts of Density*, J. F. Wohlwill and W. van Vliet, eds., Lawrence Erlbaum Associates, Hillside, NJ, pp. 77–95.

Schuster, P., J. Weber, W. Grüner, and C. Reidys, 1997, Molecular evolutionary biology from concepts to technology, in *Physics of Biological Systems from Molecules to Species*, H. Flyvberg, J. Hertz, M. H. Jensen, O. G. Mouritsen, and K. Sneppen, eds., Springer-Verlag, Berlin, Germany, pp. 284–306.

Schwerdtfeger, A. and J. Heer, 2008, Second to fourth digit ratio (2D:4D) of the right hand is associated with nociception and augmenting-reducing, *Personal. Individ. Differ.* 45(6): 493–497.

Scudellari, M., 2014, Errors in division, *HHMI Bull.* 27(3): 12–19.

Secomb, T. W., R. Hsu, E. Y. H. Park, and M. W. Dewhirst, 2004, Green's function methods for analysis of oxygen delivery to tissue by microvascular networks, *Ann. Biomed. Eng.* 32: 1519–1529.

Seeley, T. D., 1983, The ecology of temperate and tropical honeybee societies, *Am. Sci.* 71: 264–272.

Seelig, J. D. and V. Jayaraman, 2013, Feature detection and orientation tuning in the drosophila central complex, *Nature* 203(7475): 262–266.

Serbest, G., J. Horwitz, and K. A. Barbee, 2002, Characterization of neuronal cell injury and neuroprotective effect of poloxamer, in *Proceedings of the 28th Annual Northeast Bioengineering Conference*, Philadelphia, PA, K. Moxon, D. El-Sherif, and S. Kanakasabai, eds., IEEE, Piscataway, NJ, pp. 155–156.

Shade, D. and A. T. Johnson, 2003, Design of respiratory devices, in *Standard Handbook of Biomedical Engineering & Design*, M. Kutz, ed., McGraw-Hill, New York.

Shadwick, R. E., 2005, How tunas and lamnid sharks swim: An evolutionary convergence, *Am. Sci.* 93: 524–531.

Shanahan, D. F., R. Bush, K. J. Gaston, B. B. Lin, J. Dean, E. Barber, and R. A. Fuller, 2016, Health benefits from nature experiences depend on dose, *Sci. Rpt 6*, doi:10.1038/srep28551.

Shannon, C. and W. Weaver, 1949, *The Mathematical Theory of Communication*, University of Illinois Press, Urbana, IL.

Shapin, S., 2004, The accidental scientist, *Am. Sci.* 92: 374–376.

Shepherd, 2010, Scientists look for a link between toxo infections and brain disorders such as schizophrenia, *Shepherd* 55(9): 12–13.

Sherman, P. W. and S. M. Flaxman, 2001, Protecting ourselves from food, *Am. Sci.* 89: 142–151.

Shermer, M., 2005, The soul of science, *Am. Sci.* 93: 101–103.

Shermer, M., 2010, Our neandertal brethren, *Sci. Am.* 303(2): 34.

Shirayama, M., M. Seth, H.-C. Lee, W. Gu, T. Ishidate, D. Conte, and C. C. Mello, 2012, piRNAs initiate an epigenetic memory of nonself RNA in the C. elegans germ line, *Cell* 150(1): 65–77.

Shneiderman, B. and J. Hendler, 2017, It's the partnership, stupid, *Issues Sci. Technol.* 33(4): 37–40.

Shor, T. J., 2009, Saving new brain cells, *Sci. Am.* 300(3): 46–54.

Shugart, H. H. Jr. and D. C. West, 1981, Long-term dynamics of forest ecosystems, *Am. Sci.* 69: 647–652.

Shuler, M. L. and F. Kargi, 1992, *Bioprocess Engineering: Basic Concepts*, Prentice Hall, Englewood Cliffs, NJ.

Shumaker, R. W., 2010, The missing link? Maybe fear, *Phi Kappa Phi Forum* 90(3): 14–15.

Siegler, R. S., 1983, How knowledge influences learning, *Am. Sci.* 71: 631–638.

Sigurdsson, S., P. A. Philipsen, L. K. Hansen, J. Larsen, M. Gniadecka, and H. C. Wulf, 2004, Detection of skin cancer by classification of Raman spectra, *IEEE Trans. Biomed. Eng.* 51: 1784–1793.

Silver, A. C., A. Arjona, W. E. Walker, and E. Fikrig, 2012, The circadian clock controls toll-like receptor 9-mediated innate and adaptive immunity, *Immunity* doi:10.1016/j.immuni.2011.12.017.

Simmons, K., 2006, 05.1116/3 Evolution ecology and biodiversity, Lab Manual Online, www.kentsimmons.uwinnipeg.ca/16com05/16labman05/1b6pg2_files/earthworm2.bmp (accessed November 28, 2007).

Simpson, G. G., C. S. Pittendrigh, and L. H. Tiffany, 1957, *Life: An Introduction to Biology*, Harcourt, Brace and Company, New York.

Sinclair, T. R., 2009, Taking measure of biofuel limits, *Am. Sci.* 97: 400–407.

Sleight, R. G. and M. A. Lieberman, 1995, Signal transduction, in *Cell Physiology Source Book*, N. Sperelakis, ed., Academic Press, San Diego, CA, pp. 117–127.

Smedley, K. D., K. J. McKain, and D. N. McKain, 2014, 2D:4D Digit ratio predicts depression severity for females but not for males, *Personal. Individ. Differ.* 70: 136–139.

Smith, B. D., 1998, *Psychology: Science and Understanding*, McGraw-Hill, New York.

Sneed, A., 2014, How big animals deter cancer, *Sci. Am.* 311(4): 28.

Snyder, A., E. Pamer, and J. Wolchok, 2015, Could microbial therapy boost cancer immunotherapy? *Science* 350(6264): 1031–1032.

Sokolnikoff, I. S. and R. M. Redheffer, 1958, *Mathematics of Physics and Modern Engineering*, McGraw-Hill, New York.

Sommer, R., 1969, *Personal Space: The Behavioral Basis of Design*, Prentice Hall, Englewood Cliffs, NJ.

Sosis, R., 2004, The adaptive value of religious ritual, *Am. Sci.* 92: 166–172.

Southwick, C. H., 1970, *Animal Aggression: Selected Readings*, Van Nostrand Reinhold, New York.

Spanu, P. D., J. C. Abbott, J. Amselem, T. A. Burgis, D. M. Soanes, K. Stüber, E. V. L. van Themaat et al., 2010, Genome expansion and gene loss in powdery mildew fungi reveal tradeoffs in extreme parasitism, *Science*, 330(6010): 1543–1546.

Sparks, H., 2002, When cracks form, they repair like wounds, *Pop. Sci.* 259(7): 54–55.

Spector, W. S., ed., 1956, *Handbook of Biological Data*, WADC technical report 56–273, Wright-Patterson Air Force Base, OH.

Spence, C., 2017, *Gastrophysics: The New Science of Eating*, Penguin Random House, New York.

SRC, 2013, Offspring of stressed mothers develop differently, *Shepherd* 58(7): 24.

Srivastava, P. K., 2008, New jobs for ancient chaperones, *Sci. Am.* 299(1): 50–55.

Stahl, W. R. and J. Y. Gummerson, 1967, Systematic allometry in five species of adult primates, *Growth* 31: 21–34.

Stamm, B. H., F. Tuma, F. H. Norris, N. F. Piland, O. VanDerHart, J. A. Fairbank, H. E. Stamm IV, C. Higson-Smith, L. Barbanel, and R. F. Levant, 2004, The terror part of terrorism, *IEEE EMBS Mag.* 23(1): 149–161.

Starfield, A. M., K. A. Smith, and A. L. Bleloch, 1990, *How to Model It: Problem Solving for the Computer Age*, McGraw-Hill, New York.

Staron, M. M., S. M. Gray, H. D. Marshall, I. A. Parish, J. H. Chen, C. J. Perry, G. Cui, M. O. Li, and S. M. Kaech, 2014, The transcription factor foxO1 sustains expression of the inhibitory receptor PD-1 and survival of antiviral CD8+ T cells during chronic infection, *Immunity* 41(5): 802–814.

Starr, C., 2000, *Biology: Concepts and Applications*, Brooks/Cole, Stamford, CT.

Stec, B., 2004, Living and nonliving matter, *Science* 305: 41.

Steele, R. G. D. and J. H. Torrie, 1960, *Principles and Procedures of Statistics*, McGraw-Hill, New York.

Stein, R. A., 2012, RNA analysis unearths invaluable insights, *Genet. Eng. Biotechnol. News* 32(8): 22, 24, 26.

Stein, R. A., 2014, Reshaping the cancer transcriptome, *Genet. Eng. Biotechnol. News* 34(7): 1, 22, 24–25.

Stein, R. A., 2016, MicroRNAs rise from trash to treasure, *Genet. Eng. Biotechnol. News* 36(4): 14–16.

Stewart, W., 2011, Colostrum: Immunity, warmth, and nutrition, *Shepherd* 56(5): 6–7.

Stickgold, R., 2015, Sleep on it! *Sci. Am.* 313(4): 52–57.

Stieglitz, T., M. Schuettler, and K. P. Koch, 2005, Implantable biomedical microsystems for neural prostheses, *IEEE Eng. Med. Biol. Mag.* 24(5): 58–65.

Stix, G., 2010, Social analgesics, *Sci. Am.* 303: 22–23.

Stix, G., 2014, The "it" factor, *Sci. Am.* 311(3): 72–79.

Stone, N. D., P. B. Siegel, C. S. Adkinson, and W. B. Gross, 1984, Vocalizations and behavior of two commercial stocks of chickens, *Poult. Sci.* 63: 616–619.

Stout, C. E., 2004, The psychology of countering terrorism with technology, *IEEE EMBS Mag.* 23(1): 142–148.

Strickland, E., 2015, Telekinesis made simple, *IEEE Spectr.* 07.15: 11, 14.

Strickland, E., 2016a, Cyborgs go for the gold, *IEEE Spectr.* 01.16: 30–33.

Strickland, E., 2016b, Tech companies mull archiving data in DNA, *IEEE Spectr.* 53(7): 9–11.

Stroh, M., 2005, Life built to order, *Pop. Sci.* 266: 64–70, 98–99.

Strumm, W. and J. J. Morgan, 1981, *Aquatic Chemistry*, John Wiley & Sons, New York.

Summa, K. C. and F. W. Turek, 2015, The clocks within Us, *Sci. Am.* 312(2): 50–55.

Svoboda, E., 2009, The next phage, *Pop. Sci.* 274(4): 60–65, 88.

Taiz, L. and E. Zeiger, 1998, *Plant Physiology*, Sinauer Associates, Sunderland, MA.

Talmage, D. W., 1979, Recognition and memory in the cells of the immune system, *Am. Sci.* 67: 173–177.

Tannous, A. I., 1979, The secrets of nature, *Am. Sci.* 67: 392–393.

Tapia, J. C., J. D. Wylie, N. Kasthuri, K. J. Hayworth, R. Schalek, D. R. Berger, C. Guatimosim, H. S. Seung, and J. W. Lichtman, 2012, Pervasive synaptic branch removal in the mammalian neuromuscular system at birth, *Neuron.* 74(5): 816–829.

Taylor, A. F. and F. E. Kuo, 2009, Children with attention deficits concentrate better after walk in the park, *J. Attention Disord.* 12(5): 402–409.

Taylor, A. F. and F. E. Kuo, 2011, Could exposure to everyday green spaces help treat ADHD? Evidence from children's play settings, *Appl. Psychol.: Health Well-Being* 3(3): 281–303.

Taylor, C. R., A. Shkolnik, R. Dmi'el, D. Baharav, and A. Borut, 1974, Running in cheetahs, gazelles, and goats: Energy cost and limb configuration, *Am. J. Physiol.* 227: 848–850.
Taylor, C. R., G. M. O. Maloiy, E. R. Weibel, V. A. Langman, J. M. Z. Kamau, H. J. Seeherman, and N. C. Heglund, 1981, Design of the mammalian respiratory system. III. Scaling maximum aerobic capacity to body mass: Wild and domestic mammals, *Respir. Physiol.* 44: 25–37.
Teixeira, A., 1992, Thermal process calculations, in *Handbook of Food Engineering*, D. R. Heldman and D. B. Lund, eds., Marcel Dekker, New York, pp. 563–619.
Teng, S. W., S. Mukherji, J. R. Moffitt, S. deBuyl, and E. K. O'Shea, 2013, Robust circadian oscillations in growing cyanobacteria require transcriptional feedback, *Science* 340(6133): 737–740.
Tenner, E., 1996, *Why Things Bite Back*, Vintage Books, New York.
Thayer, R. and J. Ellefson, 2014, Synthetic biology: New letters for life's alphabet, *Nature* 509: 291–292.
Thieme, T., 2001a, New hope for paralysis victims, *Pop. Sci.* 254(1): 37.
Thieme, T., 2001b, A breath of life, *Pop. Sci.* 259(2): 33.
Thornhill, R. and D. T. Gwynne, 1986, The evolution of sexual differences in insects, *Am. Sci.* 74: 382–389.
Thornton, A. and K. McAuliffe, 2006, Teaching in wild meerkats, *Science* 313: 227–229.
Thundimadathil, J., 2013, Drug delivery using cell-penetrating peptides, *Drug Discov. Dev.* 16(1): 22–23.
Tickle, C., 1981, Limb regeneration, *Am. Sci.* 69: 639–646.
Tiffany-Castiglioni, E., 2003, 98.5 percent pure chimp, *Phi Kappa Phi Forum* 83(3): 6–7.
Tiffany-Castiglioni, E., 2004, Chicken liver and the soul, *Phi Kappa Phi Forum* 84(1): 6–7.
Timbrell, J. A., 1995, *Introduction to Toxicology*, Taylor & Francis Group, London, UK.
Tittmann, P., 2015, The wood in the forest: Why California needs to reexamine the role of biomass in climate policy, *California Agriculture* 69(3): 133–137.
Tortora, G. J., B. R. Funke, and C. L. Case, 2001, *Microbiology: An Introduction*, Addison-Wesley Longman, San Francisco, CA.
Tracy, K. J., 2015, Shock medicine, *Sci. Am.* 312(3): 28–35.
Tran, T.-N.-N., G. Aboudharam, D. Raolt, and M. Drancourt, 2011, Beyond ancient microbial DNA: Nonnuceotidic biomolecules for paleomicrobiology, *BioTechniques* 50(6): 370–380.
Traynor, K. S., J. S. Pettis, D. R. Tarpy, C. A. Mullin, J. L. Frazier, M. Frazier, and D. vanEnglesdorf, 2016, In-hive pesticide exposome: Assessing risks to migratory honey bees from inhHive pesticide contamination in the eastern United States, *Scientific Reports* 6, Article Number: 33207. doi:10.1038/srep33207.
Trefil, J. S. and R. M. Hazen, 1998, *The Sciences: An Integrated Approach*, Wiley, New York.
Trefil, J., H. J. Morowitz, and E. Smith, 2009, The origin of life, *Am. Sci.* 907: 206–213.
Tregaskis, S., 2002, The prenatal principle, *Cornell Alumni Mag.* 104(4): 40–45.
Trut, L. N., 1999, Early canid domestication: The farm-fox experiment, *Am. Sci.* 87: 160–169.
Tucker, V. A., 1975, The energetic cost of moving about, *Am. Sci.* 63: 413–419.
Turner, P. E., 2005, Cheating viruses and game theory, *Am. Sci.* 93: 428–435.
Ulrich, R. S., 1984, View through a window may influence recovery from surgery, *Science* 224(4647): 420–422.
Valentini, R. F., 1995, Nerve guidance channels, in *The Biomedical Engineering Handbook*, J. Bronzino, ed., CRC Press, Boca Raton, FL, pp. 1985–1996.
Van Etten, J. L., 2011, Giant Viruses, *Am. Sci.* 99: 304–311.
Vance, E., 2016, Mind over matter, *Nat. Geogr.* 230(6): 30–55.
Vandenbergh, J. G., 2003, Prenatal hormone exposure and sexual variation, *Am. Sci.* 91: 218–225.
Vastag, B., 2010, When passion and skill converge, *HHMI Bull.* 23(3): 8–9.
Veale, J. F., D. E. Clarke, T. C. Lomax, 2010, Biological and psychosocial correlates of adult gender-variant identities: A Review, *Personal. Individ. Differ.* 48(4): 357–366.
Verhoeven, J. E., D. Révész, E. S. Epel, J. Lin, O. M. Wolkowitz, and B. W. J. H. Pennix, 2014, Major depressive disorder and accelerated cellular aging: Results from a large psychiatric cohort study, *Mol. Psychiatry* 19: 895–901.

Vermillion, R. J., 2008, Considering static control for electrostatic discharge-sensitive medical device packaging, *Pharm. Med. Packaging News* 16(2): 54–62.

Viswanathan, G. M., S. V. Buldyrev, S. Havlin, M. G. E. da Luz, E. P. Raposo, and H. E. Stanley, 1999. Optimizing the success of random searches, *Nature* 401(6756): 911–914. doi:10.1038/44831.

Vizard, F., 2001, Inside CSI, *Pop. Sci.* 259(3): 49–52.

Vogel, G. M., 2003, Combating cancer-drug resistance: The ceramide connection, *Phi Kappa Phi Forum* 83(1): 20–24.

Vogel, S., 1988, How organisms use flow-induced pressures, *Am. Sci.* 76: 28–34.

Vogel, S., 2010, The deceptional life, *Am. Sci.* 98: 436–437.

Voss, G. I. and R. D. Butterfield, 1995, Parenteral infusion devices, in *The Biomedical Engineering Handbook*, J. Bonzino, ed., CRC Press, Boca Raton, FL, pp. 1311–1321.

Wagner, M. R., D. S. Lundberg, D. Coleman-Derr, S. G. Tringe, J. L. Dangl and T. Mitchell-Olds, 2014, Natural soil microbes alter flowering phenology and the intensity of selection on flowering time in a wild *Arabidopsis* Relative, *Ecol. Lett.* 17(6): 717–726.

Wainer, H., 2007, The most dangerous equation, *Am. Sci.* 95: 249–256.

Wakeam, E., R. A. Hernandez, D. Rivera Morales, S. R. Finlayson, M. Klompas, and M. J. Zinner, 2014, Bacterial ecology of hospital workers' facial hair: A cross-sectional study. *J. Hosp. Infect.* 87(1): 63–67.

Waldholtz, M., 2017, Transformers, *Sci. Am.* 316(4): 46–53.

Waldman, M., S. Nicholson, N. Adilov, and J. Williams, 2008, Autism prevalence and precipitation rates in California, Oregon, and Washington counties, *Arch. Pediatr. Adolesc. Med.* 162(11): 1026–1034.

Waldrop, M. M., 2017, Inside the moonshot effort to finally figure out the brain, *MIT Technol. Rev.* 120(6): 54–60.

Wallis, C., 2004, What makes teens tick, *Time* 163(19): 56–65.

Wang, K., D. E. Milkie, A. Saxena, P. Engerer, T. Misgeld, M. E. Bronner, J. Mumm, and E. Betzig, 2014, Rapid adaptive optical recovery of optimal resolution over large, *Nat. Methods* 11(6): 625–628.

Wang, R., 2010, Toxic gas, lifesaver, *Sci. Am.* 302(3): 66–71.

Wang, W., J. Y. Barnaby, Y. Tada, H. Li, M. Tör, D. Caldelari, D. Lee, X.-D. Fu, and X. Dong, 2011, Timing of plant immune responses by a central circadian regulator, *Nature* 470(7332): 110–114.

Wang, X., J. Ma, X. Li, X. Zhao, Z. Lin, J. Chen, and Z. Shao, 2015, Optimization of chemical fungicide combinations targeting the maize fungal pathogen, *Bipolaris maydis*: A systematic quantitative approach, *IEEE Trans. Biomed. Engr.* 62(1): 80–87.

Ward, C., 2010, Saving wolves saved yellowstone, *The Baltimore Sun*, Oct 1, p. 17.

Ward, J., 2011, Bodies in motion, *Pop. Sci.* 278(5): 63–69, 93–94, 97.

Warner, R. R., 1984, Mating behavior and hermaphroditism in coral reef fishes, *Am. Sci.* 72: 128–136.

Waterland, R. A. and R. L. Jirtle, 2003, Transposable elements: Targets for early nutritional effects on epigenetic gene regulation, *Mol. Cell Biol.* 23(15): 5293–5300.

Watt, B. K. and A. L. Merrill, 1963, Composition of foods, in *Agriculture Handbook No. 8*, US. Department of Agriculture, U.S. Printing Office, Washington, DC.

Webster, R. G. and E. J. Walker, 2003, Influenza, *Am. Sci.* 91: 122–129.

Wegner, N. C., O. E. Snodgrass, H. Dewar, and J. R. Hyde, 2015, Whole-body endothermy in a mesopelagic fish, the Opah, *Lampris guttatus*, *Science* 348(6236): 786–789.

Weiland, J. D. and M. S. Humayun, 2005, A biomimetic retinal stimulating array, *IEEE Eng. Med. Biol. Mag.* 24(5): 14–21.

Weis, S. E., A. Firker, and J. Hennig, 2007, Associations between the second to fourth digit ratio and career interests, *Personal. Individ. Differ.* 43(3): 485–493.

Weiss, K. M., 2017, Is precision medicine possible? *NAS Issues Sci. Technol.* 34(1): 37–42.

Wenner, A. M., 1998, Odors, wind, and colony foraging: Part III of three parts: Insights from bee hunting, *Am. Bee J.* 138(12): 897–899.

Wepking, C., B. Avera, B. Badgley, J. E. Barrett, J. Franklin, K. F. Knowlton, P. P. Ray, C. Smitherman, and M. S. Strickland, 2017, Exposure to dairy manure leads to greater antibiotic resistance and increased mass-specific respiration in soil microbial communities, *Proc. R. Soc. B* 284: 20162233. doi:10.1098/rspb.2016.2233.

Werner, I., S. L. Clark, and D. E. Hinton, 2003, Biomarkers aid understanding of aquatic organism responses to environmental stresses, *Calif. Agric.* 57(4): 110–115.

West, B. J. and A. L. Goldberger, 1987, Physiology in fractal dimensions, *Am. Sci.* 75: 354–365.

West, B. J., 2006, *Where Medicine Went Wrong: Rediscovering the Path to Complexity*, World Scientific Publishers, New York.

West, G. B., J. H. Brown, and B. J. Enquist, 2000, The origin of universal scaling laws in biology, in *Scaling in Biology*, J. H. Brown and G. B. West, eds., Oxford University Press, New York, pp. 87–112.

West-Eberhard, M. J., 2003, *Developmental Plasticity and Evolution*, Oxford University Press, Oxford, UK.

Whipp, B. J., 1981, The control of exercise hyperpnea, in *Regulation of Breathing*, Part II, T. F. Hornbein, ed., Marcel Dekker, New York, pp. 361–457.

White, A., P. Handler, E. L. Smith, and D. Stetten, 1959, *Principles of Biochemistry*, McGraw-Hill, New York.

White, R. G., J. A. DeShazer, C. J. Tressler, G. M. Borcher, S. Davey, A. Waninge, A. M. Parkhurst, M. J. Milanuk, and E. T. Clemens, 1995, Vocalization and physiological response of pigs during castration with or without a local anesthetic, *J. Anim. Sci.* 73: 381–386.

White-Grindley, E., L. Li, R. M. Khan, F. Ren, A. Saraf, L. Florens, and K. Si, 2014, Contribution of Orb2A stability in regulated amyloid-like oligomerization of *Drosophila* Orb2, *PLOS Biology* doi:10.1371/journal.pbio.1001786.

Wickelgren, I., 2004, Can worms tame the immune system? *Science* 305: 170–171.

Wickens, C. D., S. E. Gordon, and Y. Liu, 1997, *An Introduction to Human Factors Engineering*, Addison-Wesley Longman, New York.

Wilham, J. M., C. D. Orrú, R. A. Bessen, R. Atarashi, K. Sano, B. Race, K. D. Meade-White, L. M. Taubner, A. Timmes, and B. Caughey, 2010, Rapid end-point quantitation of prion seeding activity with sensitivity comparable to bioassays, *PLoS Pathog* 6(12): e1001217. doi:10.1371/journal.ppat.1001217.

Williams, L. E. and J. A. Bargh, 2008, Experiencing physical warmth promotes interpersonal warmth, *Science* 322: 606–607.

Williams, S. C. P., 2009, Membrane awakening, *HHMI Bull.* 22(4): 24–29, 48.

Williams, S. C. P., 2010, Bacteria helping bacteria, *HHMI Bull.* 23(4): 42.

Williams, S. C. P., 2010, Gut reaction, *HHMI Bull.* 23(3): 12–17.

Williams, S. C. P., 2011, Skin sense, *HHMI Bull.* 24(1): 10–11.

Williams, S. C. P., 2013A, Cellular search engine, *HHMI Bull.* 26(1): 10–11.

Williams, S. C. P., 2013B, One foot in front of the other, *HHMI Bull.* 26(1): 21–25.

Williams, T. J., M. E. Pepitone, S. E. Christensen, B. M. Cooke, A. D. Huberman, N. J. Breedlove, T. J. Breedlove, C. L. Jordan, and S. M. Breedlove, 2000, Finger-length ratios and sexual orientation, *Nature* 404: 455–456.

Williamson, D. I. and S. E. Vickers, 2007, The origins of larvae, *Am. Sci.* 95: 509–517.

Willmore, K. E., 2010, Development influences evolution, *Am. Sci.* 98: 220–227.

Wilson, D. S. and E. O. Wilson, 2008, Evolution "for the good of the group", *Am. Sci.* 96: 380–389.

Winford, E. M., J. T. Stevens, and H. D. Safford, 2015, Effects of fuel treatments on California mixed-conifer forests, *Calif. Agric.* 69(3): 150–155.

Winter, G., 2006, Gas plasma: A green alternative for surface engineering, *Med. Technol.* 10(7): 22–24.

Wise, K. D., 2005, Silicon microsystems for neuroscience and neural prostheses, *IEEE Eng. Med. Biol. Mag.* 24(5): 22–29.

Wolanski, E., E. Gereta, M. Borner, and S. Nduma, 1999, Water, migration, and the serengeti ecosystem, *Am. Sci.* 87: 526–533.

Wolf, S., E. Nicholls, A. M. Reynolds, P. Wells, K. S. Lim, R. J. Paxton, and J. L. Osborne, 2016, Optimal search patterns in honeybee orientation flights are robust against emerging infectious diseases, *Scientific Reports* 6, Article number: 32612. doi:10.1038/srep32612.

Wolfe, N. T. and J. A. Odom, 1976, *Control Panel Layout Design Guide*, Honeywell Microswitch, Freeport, IL.

Wolfe-Simon, F., J. S. Blum, T. R. Kulp, G. W. Gordon, S. E. Hoeft, J. Pett-Ridge, J. F. Stolz et al., 2010, A bacterium that can grow by using arsenic instead of phosphorous, *Science* 332(6034): 1163–1166. doi:10.1126/science.1197258.

Wolfshohl, K., 2003, New ways to whip aflatoxin, *Prog. Farmer* 118(10): 42, 44.

Wong, K., 2012, Helpless by design? *Sci. Am.* 307(5): 28.

Woodruff, E. A., 1995, Clinical care of patients with closed loop drug delivery systems, in *The Biomedical Engineering Handbook*, J. D. Bronzino, ed., CRC Press, Boca Raton, FL, pp. 2447–2458.

Wu, C., 2010, Peril in small places, *ASEE Prism* 20(3): 26–31.

Wurtman, R. J., 1976, The effects of light on the human body, in *Readings from Scientific American: Conditions for Life*, W. H. Freeman, San Francisco, CA, pp. 125–134.

Wynne, C. O. L., 2001, The soul of the ape, *Am. Sci.* 89: 120–122.

Xin, H., J. A. DeShazer, and D. W. Leger, 1989, Pig vocalizations under selected husbandry practices, *Trans. ASAE* 32: 2181–2184.

Yang, C., M. W. Tibbit, L. Basta, and K. S. Anseth, 2014, Mechanical memory and dosing influence stem cell fate, *Nat. Mater.* doi:10.1038/INMAT3889.

Yang, W. and S. Hekimi 2010, A mitochondrial superoxide signal triggers increased logevity in *Caenorhabditis elegans, PLOS Biol.* doi:10.1371/journal.pbio.1000556.

Yannas, I. V., 1995a, Artificial skin and dermal equivalents, in *The Biomedical Engineering Handbook*, J. Bronzino, ed., CRC Press, Boca Raton, FL, pp. 2025–2038.

Yannas, I. V., 1995b, Regeneration templates, in *The Biomedical Engineering Handbook*, J. D. Bronzino, ed., CRC Press, Boca Raton, FL, pp. 1619–1635.

Yates, G. T., 1986, How microorganisms move through water, *Am. Sci.* 74: 358–365.

Yelton, D. E. and M. D. Scharff, 1980, Monoclonal antibodies, *Am. Sci.* 68: 510–516.

Yoerg, S. I., 2001, *Clever as a Fox: Animal Intelligence and What It Can Teach Us about Ourselves*, Bloomsbury, New York.

Yoganathan, A., 1995, Cardiac valve prostheses, in *The Biomedical Engineering Handbook*, J. Bronzino, ed., CRC Press, Boca Raton, FL, pp. 1847–1870.

Yoon, S.-H. and S. Park, 2011, A mechanical analysis of woodpecker drumming and Its application to shock-absorbing systems, *Bioinspiration Biomim* 6(1): 016003.

Yosim, A., K. Bailey, and R. C. Fry, 2015, Arsenic, the "king of poisons", in food and water, *Am. Sci.* 103(1): 34–41.

Youngsteadt, E., 2011, Dying generously, *Am. Sci.* 99(3): 208–209.

Yuan, H. and R. Marmorstein, 2013, Red wine, toast of the town (Again), *Science* 339(6124): 1156–1157.

Zak, P. J., 2008, The neurobiology of trust, *Sci. Am.* 298(6): 88–95.

Zheng, Z. and M. J. Cohn, 2011, Developmental basis of sexually dimorphic digit ratio, *PNAS* 108(39): 16289–16294.

Zimmer, C., 2014, Meet nature's nightmare: Mindsuckers, *Nat. Geogr.* 226(5): 36–55.

Zimmerman, A. L., 1995, Visual transduction, in *Cell Physiology Source Book*, N. Sperelakis, ed., Academic Press, San Diego, CA, pp. 506–513.

Zoecklin, B., 2000, Seasonal variation influencing protein stability, Enology Notes #11, Virginia Tech, www.fst.vt.edu/extension/enology/contentextenologynotes11.htm (accessed April 15, 2004).

Zuckerman, C., 2015, Nesting instincts, *Nat. Geogr.* 228(1): 10–11.

Index

Note: Page numbers in italic and bold refer to figures and tables respectively.